Lecture Notes in Networks and Systems

Volume 454

The series "Lecture Notes in Networks and Systems" publishes the latest developments in Networks and Systems—quickly, informally and with high quality. Original research reported in proceedings and post-proceedings represents the core of LNNS.

Volumes published in LNNS embrace all aspects and subfields of, as well as new challenges in, Networks and Systems.

The series contains proceedings and edited volumes in systems and networks, spanning the areas of Cyber-Physical Systems, Autonomous Systems, Sensor Networks, Control Systems, Energy Systems, Automotive Systems, Biological Systems, Vehicular Networking and Connected Vehicles, Aerospace Systems, Automation, Manufacturing, Smart Grids, Nonlinear Systems, Power Systems, Robotics, Social Systems, Economic Systems and other. Of particular value to both the contributors and the readership are the short publication timeframe and the world-wide distribution and exposure which enable both a wide and rapid dissemination of research output.

The series covers the theory, applications, and perspectives on the state of the art and future developments relevant to systems and networks, decision making, control, complex processes and related areas, as embedded in the fields of interdisciplinary and applied sciences, engineering, computer science, physics, economics, social, and life sciences, as well as the paradigms and methodologies behind them.

Indexed by SCOPUS, INSPEC, WTI Frankfurt eG, zbMATH, SCImago.

All books published in the series are submitted for consideration in Web of Science.

For proposals from Asia please contact Aninda Bose (aninda.bose@springer.com).

More information about this series at https://link.springer.com/bookseries/15179

Saad Motahhir · Badre Bossoufi
Editors

Digital Technologies and Applications

Proceedings of ICDTA'22, Fez, Morocco, Volume 1

Springer

Editors
Saad Motahhir
Ecole Nationale des Sciences Appliquées
Fez, Morocco

Badre Bossoufi
Faculty of Sciences
Sidi Mohamed Ben Abdellah University
Fez, Morocco

ISSN 2367-3370 ISSN 2367-3389 (electronic)
Lecture Notes in Networks and Systems
ISBN 978-3-031-01941-8 ISBN 978-3-031-01942-5 (eBook)
https://doi.org/10.1007/978-3-031-01942-5

This Springer imprint is published by the registered company Springer Nature Switzerland AG
The registered company address is: Gewerbestrasse 11, 6330 Cham, Switzerland

We are honored to dedicate the proceedings of ICDTA'22 to all the participants and committees of ICDTA'22.

Foreword

It is with deep satisfaction that we write this Foreword to the proceedings of the ICDTA'22 organized by École Nationale des Sciences Appliquées and Faculté des sciences which belong to SMBA University, Fez, Morocco, on January 28 and 29, 2022.

This conference was bringing together researchers, academics, and professionals from all over the world, experts in digital technologies and their applications. This conference particularly encouraged the interaction of research students and developing academics with the more established academic community in an informal setting to present and discuss new and current work.

The papers contributed the most recent scientific knowledge known in the field of digital technologies and their applications. Their contributions helped to make the conference as outstanding as it has been. The organizing and technical program committees put much effort into ensuring the success of the day-to-day operation of the meeting.

We hope that this book proceedings will further stimulate research in digital technologies such as artificial intelligence, Internet of things, embedded systems, network technology, information processing, and their applications, in several areas including hybrid vehicles, renewable energy, mechatronics, medicine, etc.

We feel honored and privileged to serve the best recent developments to you through this exciting book proceedings.

We thank all authors and participants for their contributions.

S. Motahhir
B. Bossoufi

Preface

This volume contains the first section of the written versions of most of the contributions presented during the conference of ICDTA'22. The conference provided a setting for discussing recent developments in a wide variety of topics including artificial intelligence, Internet of things, embedded systems, network technology, information processing, and their applications, in several areas such as hybrid vehicles, renewable energy, mechatronics, and medicine. The conference has been a good opportunity for participants from various destinations to present and discuss topics in their respective research areas.

ICDTA'22 conference tends to collect the latest research results and applications on digital technologies and their applications. It includes a selection of 169 papers from 350 papers submitted to the conference from universities and industries all over the world. This volume includes half of the accepted papers. All of the accepted papers were subjected to strict peer-reviewing by two–four expert referees. The papers have been selected for this volume because of their quality and their relevance to the conference.

We would like to express our sincere appreciation to all authors for their contributions to this book. We would like to extend our thanks to all the referees for their constructive comments on all papers; especially, we would like to thank organizing committee for their hardworking. Finally, we would like to thank the Springer publications for producing this volume.

S. Motahhir
B. Bossoufi

Acknowledgments

We request the pleasure of thanking you for taking part in the second edition of the International Conference on Digital Technologies and Applications ICDTA'22. We are very grateful for your support, so thank you everyone for bringing your expertise and experience around the conference and engaging in such fruitful, constructive, and open exchanges throughout the two days of the ICDTA'22 conference.

We would like to extend our deepest thanks and gratitude to all the speakers for accepting to join us from different countries. Thank you for being such wonderful persons and speakers. Again, thanks for sharing your insight, knowledge, and experience.

Of course, this event could not be that successful without the effort of the organizing and technical program committees. Therefore, Pr. Badre and I would like to express our sincere appreciation to all of you who generously helped us.

We would like to especially thank all the participants for the confidence and trust you have placed in our conference. We hope we lived up to your highest expectations.

Our humble acknowledgment would be incomplete without thanking our biggest source of support; therefore, our deepest gratitude goes to Prof. Redouane Mrabet the president of Sidi Mohammed Ben Abdellah University, and the same goes for Prof. Lahrach Abderrahim, the director of the National School of Applied Sciences; Prof. Belmlih Mohammed, the dean of the faculty of science; Prof. Mohammed Ouazzani Jamil, the vice president of UPF University; Prof. Abdelmajid Saka, the deputy director of the National School of Applied Sciences; and Professor Elhassouni Mohammed, the vice dean of faculty of sciences. Thank you all for your support and for being all-time all set up to host such scientific events.

S. Motahhir
B. Bossoufi

Contents

Artificial Intelligence, Machine Learning and Data Analysis

The Integration of Industry 4.0 in Operations Management: Toward Smart Lean Six Sigma

Dounia Skalli[1(✉)], Abdelkabir Charkaoui[1], and Anass Cherrafi[2]

[1] Laboratory of Engineering, Industrial Management and Innovation, Faculty of Sciences and Technology, Hassan 1st University, PO Box 577, Settat, Morocco
{d.skalli,abdelkabir.charkaoui}@uhp.ac.ma

[2] Laboratory of Mathematical Modeling, Simulation and Smart Systems (L2M3S), ENSAM Meknes, MoulayIsmailUniversity, Meknes, Morocco
a.cherrafi@ensam.umi.ac.ma

Abstract. Recently, Industry 4.0 "I4.0" has been identified as a priority research area and has attracted the interest of many researchers, universities and companies worldwide. I4.0 is now revolutionizing manufacturing operations management among other areas through the increased availability and accessibility of data.

The capability of digital technologies for speed data collection and analysis promises a new age of continuous improvement and process performance. Lean six sigma "LSS", the widely best continuous improvement concept used in manufacturing will be affected by I4.0 technologies. However, the relationship between the two paradigms is not well developed in the literature and remains to be studied. To fill this gap, this paper aims to address what has already been described in the literature and analyze the scientific research trends related to the combination of I4 and LSS in manufacturing context through a structured literature review. The findings confirm that Lean Six Sigma and Industry 4.0 is a new emerging research field that has attracted a great scientific interest over the last decade.

Our study will relevant to both academics and practitioners, as it describes the main lines of research in the field. Further studies are needed to validate these results.

Keywords: Lean six sigma · Industry 4.0 · Lean · Six Sigma · Digitalization · Literature review

1 Introduction

Since its appearance at the Hannover fair in 2011, the concept of Industry 4.0 as well as digitalization and Smart Manufacturing are widely discussed. Currently, I4.0 is rapidly growing as a new emerging manufacturing paradigm [1].

With I4.0 digital technologies, the way companies exchange information and interact has changed profoundly [2].

The Industry 4.0 advanced technologies will have impact on the production value chain, quality system, logistics, customer satisfaction and industries in general. As a

S. Motahhir and B. Bossoufi (Eds.): ICDTA 2022, LNNS 454, pp. 3–11, 2022.
https://doi.org/10.1007/978-3-031-01942-5_1

result, the competitiveness of company will depend on the level of innovation and continuous improvement but also on the new technologies implemented. The study aims to examine how I4.0 digital technologies combined to LSS improvement methodology can enhance companies' performance leading to operational excellence.

The goal of companies is to continuously increase productivity, quality and service levels [3]. To achieve this and to improve the effectiveness, efficiency and consistency of operations, companies have to invest in continuous process improvement [4]. A widely applied and popular approach to process improvement is Lean Six Sigma methodology [5]. The combination of digitalization and continuous improvement practices can result in a big change [6, 7]. More Industry 4.0 concept gains momentum, manufacturers are more confident in the new benefits of automation and systems integration [3].

Since the field of Industry 4.0 has only emerged in the last few years, the published literature on the subject is still scattered. However, few studies have been conducted on the relationship between Industry 4.0 and LSS [8–14]. Authors discuss the integration and combination of one or two paradigms at the same time. The combined effect of the two paradigms lean and industry 4.0 are the most discussed [3, 15–27].

Despite the great scientific interest in this field, the subject is not yet well explored either in the literature or in practice. It is therefore a new topic that needs to be investigated. To cover this research gap, the current study is conducted. The aim of this research study is to analyze the scientific research trends related to the connection between I4 and LSS by using a structured literature review. The aim is to investigate the existing literature on the integration of the two concepts LSS and I4.0.

The paper is organized as follows: Sect. 2 describes the research methodology used for the study and a descriptive analysis is explored to illustrate the trends of research in Sect. 3. Then, Sect. 4 presents the results and discussions. A conclusion is presented including future perspectives and limitations.

2 Research Method

The literature review is used to examine and evaluate the current literature to find relevant articles and explore research gaps. [28] States that the literature review guides authors in evaluating and analyzing the literature of interest, identifying the conceptual content of the field, and contributing to theory development. A literature search was conducted through the most used database Scopus. This is because Scopus is one of the largest databases of peer-reviewed literature. The database covers research from top publishers (Elsevier, Emerald, Springer, and Wiley) among others. The search terms applied to identify the closest matching publication are listed in Table 1. The review process is illustrated in Fig. 1.

The literatures about LSS and I4.0 published from 2011 to September 2021 were scanned in the Scopus database. This is a very important part of our analysis, as it allows us to extract the most relevant articles referring to the concepts lean six sigma and industry 4.0 and their combination using the selected keywords. After removing all duplicates papers, we screen the titles and abstracts of articles to filter which are relevant to our inclusion criteria and exclude the papers that meet with the exclusion criteria (Table 1). We review all full-text articles for eligibility to be included in the descriptive analysis.

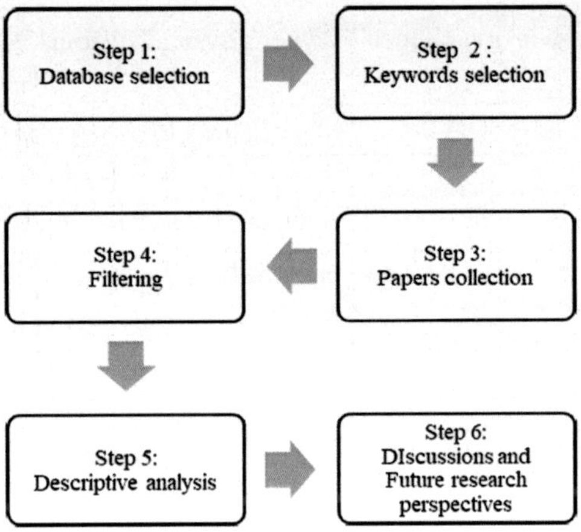

Fig. 1. Review methodology

Table 1. The research process

Research period	2011 To September 2021	
Keywords	Lean six sigma Six sigma DMAIC Lean manufacturing Lean Continuous improvement LSS	Industry 4.0 Digitization Smart factory The 4th industrial revolution Digital technologies Big data/IoT/CPS The fourth industrial revolution
Inclusion criteria	• Journal article, conference paper, book chapter • English language	
Exclusion criteria	• Non-English language publication, No full text • Text–books, unpublished working papers, and conference presentations • Not peer-reviewed	

3 Descriptive Analysis

Through the research process, we identified 72 papers to be analyzed. This section presents a descriptive analysis of the selected publications and highlights the literature findings in terms of keywords occurrence, country analysis, distribution of publications over time, distribution by major journal in the field and title co-occurrence terms.

3.1 Keywords Occurrence in Title and Abstract

"Lean" and "Industry 4.0" represent the majority of keywords identified in publications titles (60%), followed by "lean 4.0" (13%) and "lean six sigma and industry 4.0" (12%). The remaining words used in the titles of publications with low frequency are as follows: "smart" "digital" "lean automation" "DMAIC" "continuous improvement" "operations management" "operational excellence" "cyber physical systems," "IoT," and "big data" (see Fig. 2).

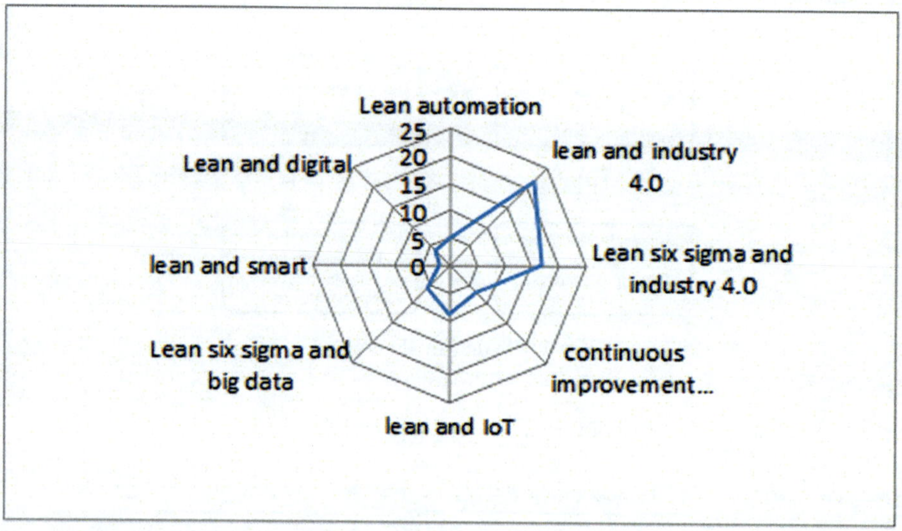

Fig. 2. Most used keywords

3.2 Distribution of Publications Per year

Figure 3 shows the distribution of publications by year. We notice an increase in the number of publications since 2017. The number of papers published from 2019 has remarkably increased. 80% were published from 2019 to September 2021, which shows the increasing interest of researchers for this field of research especially since I4.0 is an emerging concept that has appeared recently.

3.3 Country Analysis

Figure 4 shows the geographic affiliation of the authors most involved in the field. Of the 36 countries that have contributed to reviewed articles, with at least five publications, Brazil leads, followed by Italy, Germany, and Norway. European countries represent 57% of our reviewed articles.

3.4 Distribution Per Research Paper Type

Figure 5 shows the distribution of publications by type of article. (Journal article, conference paper or chapter). Journal articles represent the majority of the publications selected for the literature review (60%). Most of the conference papers are published in IFAC online and IEOM international conference while the six chapters identified are publisher by springer publisher.

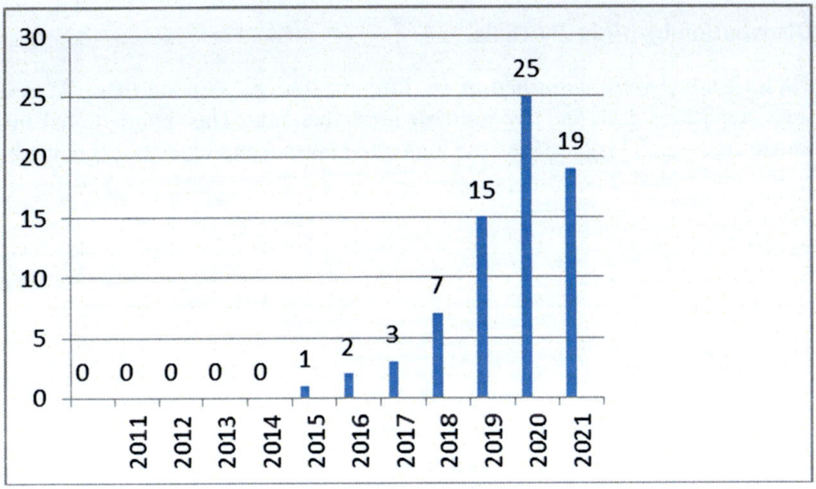

Fig. 3. Year publications

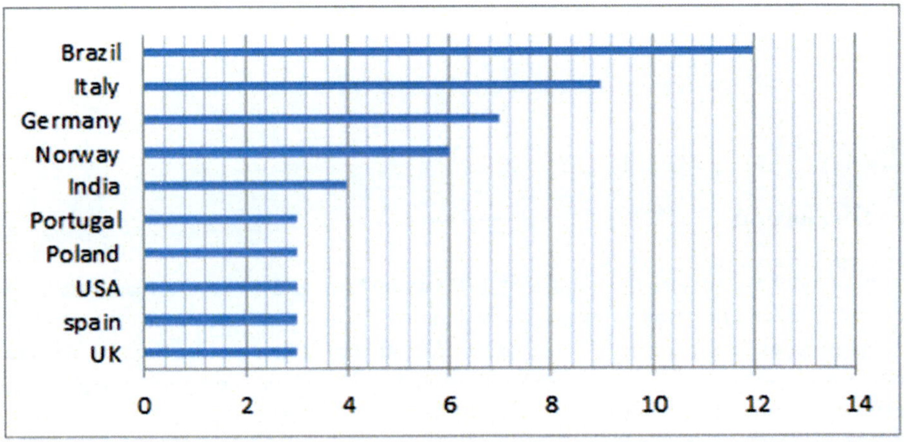

Fig. 4. The most active countries in the field

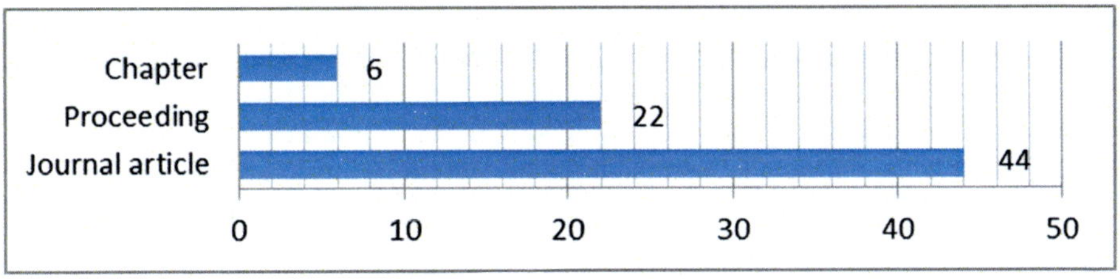

Fig. 5. Distribution of papers by type

3.5 Distribution by Main Journals

The articles studied were published in 46 different sources. Among these, 12 sources published one article each and five journals more than one. The "international Journal of production research" is classified the high cited journal and have the high number of publication in the field (Fig. 6).

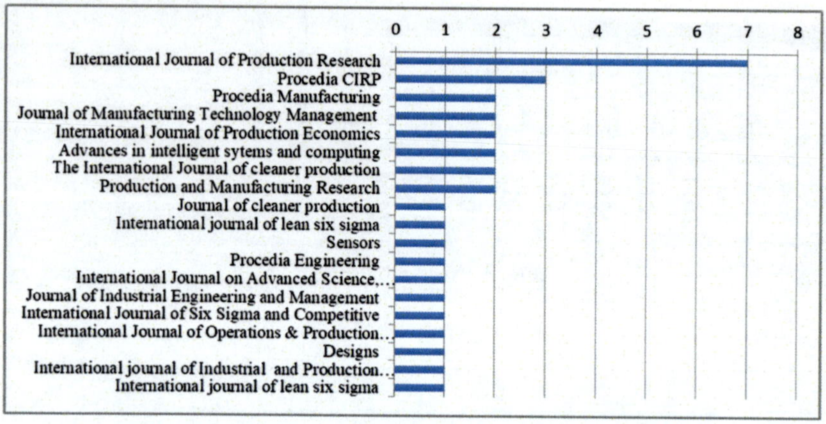

Fig. 6. The top journals with most published articles in the field

4 Discussion

In this study, we review the relationship between LSS and I4 in the manufacturing context. The combination of Lean Six Sigma and industry 4.0 has a positive impact on operational performance. According to [8], authors states that LSS and I 4.0 are considered synergistic concepts because they have a similar focus on eliminating waste, customer satisfaction and process efficiency. The majority of publications is theoretical and concern only basic results on the potential combination of LSS and I4. Out of papers included in the systematic review, the majority (68%) focused on lean and Industry 4.0, while 29% discussed LSS and Industry 4.0 and 3% addressed six sigma with I4.0.

From the main findings of the literature review, it can be concluded that:

- Lean and I4.0 integration and correlation model has been discussed and there are available empirical studies in the field.
- Lean six sigma and I4.0 is an emerging field with few literature review studies.
- Few studies have empirically examined the mutual impact of I4 and lean.
- There are no qualitative, nor quantitative studies available in literature, on the dri vers, barriers and CSFs for the integration of the two paradigms LSS and I4.0.
- Lack of frameworks providing recommendations for LSS and I4.0 integration with well-defined instructions and roadmaps.

5 Conclusion

The purpose of this paper is to provide an overview of the current state of the art in research on the relationship between LSS and I4.0 by conducting a structured literature review on Scopus scientific database. A total of 72 articles were reviewed. Statistical analysis indicates that the integration of "I4.0 and LSS" is an emerging field with an increasing number of publications in recent years. We conclude that the combination of the two paradigms have interest both researchers and professionals over the world. This article may have a limitation resulting from the use of the Scopus database, as it is possible that there are other relevant articles in other scientific databases that have not been considered. Future studies might include other popular databases, hence the results obtained have to be discussed and confirmed by more studies. This study is of mutual benefit to both academics and professionals by providing a focused perspective on the issue.

References

1. Doh, S.W., Deschamps, F., Pinheiro De Lima, E.: Systems integration in the lean manufacturing systems value chain to meet industry 4.0 requirements. In: Borsato, M., Wognum, N., Peruzzini, M., Stjepandic, J., Verhagen, W.J.C. (eds.) Transdisciplinary Engineering: Crossing Boundaries, vol. 4, pp. 642–650. IOS Press, Amsterdam (2016). https://doi.org/10.3233/978-1-61499-703-0-642
2. Nasiri, M., Ukko, J., Saunila, M., Rantala, T.: Managing the digital supply chain: the role of smart technologies. Technovation **96–97**, 102121 (2020). https://doi.org/10.1016/j.technovation.2020.102121
3. Ejsmont, K., Gladysz, B., Corti, D., Castano, F., Mohammed, W.M., Martinez Lastra, J.L.: Towards lean industry 4.0–current trends and future perspectives. Cogent Bus. Manag. **7**(1), 1781995 (2020). https://doi.org/10.1080/23311975.2020.1781995
4. Lameijer, B.A., Pereira, W., Antony, J.: The implementation of Lean Six Sigma for operational excellence in digital emerging technology companies. JMTM **32**(9), 260–284 (2021). https://doi.org/10.1108/JMTM-09-2020-0373
5. Shah, R., Chandrasekaran, A., Linderman, K.: In pursuit of implementation patterns: the context of Lean and Six Sigma. Int. J. Prod. Res. **46**(23), 6679–6699 (2008). https://doi.org/10.1080/00207540802230504
6. Bordeleau, F.-E., Mosconi, E., de Santa-Eulalia, L.A.: Business intelligence and analytics value creation in Industry 4.0: a multiple case study in manufacturing medium enterprises. Prod. Plan. Control **31**(2–3), 173–185 (2020). https://doi.org/10.1080/09537287.2019.1631458
7. Kaivo-oja, J., Virtanen, P., Jalonen, H., Stenvall, J.: The effects of the internet of things and big data to organizations and their knowledge management practices. In: Uden, L., Heričko, M., Ting, I.H. (eds.) Knowledge Management in Organizations. KMO 2015. LNBIP, vol. 224, pp. 495–513. Springer, Cham (2015). https://doi.org/10.1007/978-3-319-21009-4_38
8. Anass, C., Amine, B., Ibtissam, E.H., Bouhaddou, I., Elfezazi, S.: Industry 4.0 and lean six sigma: results from a pilot study. In: Saka, A., et al. (eds.) CPI 2019. LNME, pp. 613–619. Springer, Cham (2021). https://doi.org/10.1007/978-3-030-62199-5_54
9. Bhat, V.S., Bhat, S., Gijo, E.V.: Simulation-based lean six sigma for Industry 4.0: an action research in the process industry. Int. J. Qual. Reliab. Manag. (2020). https://doi.org/10.1108/IJQRM-05-2020-0167

10. Arcidiacono, G., Pieroni, A.: The revolution lean six sigma 4.0. Int. J. Adv. Sci. Eng. Inf. Technol. **8**(1), 141–149 (2018). https://doi.org/10.18517/ijaseit.8.1.4593
11. Belhadi, A., Kamble, S.S., Zkik, K., Cherrafi, A., Touriki, F.E.: The integrated effect of big data analytics, lean six sigma and green manufacturing on the environmental performance of manufacturing companies: the case of North Africa. J. Clean. Prod. **252**, 119903 (2020). https://doi.org/10.1016/j.jclepro.2019.119903
12. Sodhi, H.: When industry 4.0 meets lean six sigma: a review. IEJ **13**(1) (2020). https://doi.org/10.26488/IEJ.13.1.1214
13. Chiarini, A., Kumar, M.: Lean six sigma and industry 4.0 integration for operational excellence: evidence from Italian manufacturing companies. Prod. Plan. Control (2020). https://doi.org/10.1080/09537287.2020.1784485
14. Jayaram, A.: Lean six sigma approach for global supply chain management using industry 4.0 and IIoT, pp. 89–94 (2016). https://doi.org/10.1109/IC3I.2016.7917940
15. Ciano, M.P., Dallasega, P., Orzes, G., Rossi, T.: One-to-one relationships between industry 4.0 technologies and lean production techniques: a multiple case study. Int. J. Prod. Res. (2021). https://www.scopus.com/inward/record.uri?eid=2-s2.0-85091943701&doi=10.1080%2f00207543.2020.1821119&partnerID=40&md5=76344842915a593a1d614c7c7ad30405
16. Deuse, J., Dombrowski, U., Nöhring, F., Mazarov, J., Dix, Y.: Systematic combination of Lean Management with digitalization to improve production systems on the example of Jidoka 4.0. Int. J. Eng. Bus. Manag. **12**, 184797902095135 (2020). https://doi.org/10.1177/1847979020951351
17. Dombrowski, U.: The lean production system 4.0 framework–enhancing lean methods by industrie 4.0. In: IFIP International Federation for Information Processing 2018, pp. 410–416. Springer, Cham (2018)
18. Mrugalska, B., Wyrwicka, M.K.: Towards lean production in industry 4.0, vol. 182, pp. 466–473 (2017). https://doi.org/10.1016/j.proeng.2017.03.135
19. Pereira, A.C., Dinis-Carvalho, J., Alves, A.C., Arezes, P.: How industry 4.0 can enhance lean practices. FME Trans. **47**(4), 810–822 (2019). https://doi.org/10.5937/fmet1904810P
20. Sanders, A., K. Subramanian, K.R., Redlich, T., Wulfsberg, J.P.: Industry 4.0 and lean management – synergy or contradiction? In: Lödding, H., Riedel, R., Thoben, K.-D., von Cieminski, G., Kiritsis, D. (eds.) APMS 2017. IAICT, vol. 514, pp. 341–349. Springer, Cham (2017). https://doi.org/10.1007/978-3-319-66926-7_39
21. Sony, M.: Industry 4.0 and lean management: a proposed integration model and research propositions. Prod. Manuf. Res. **6**(1), 416–432 (2018). https://doi.org/10.1080/21693277.2018.1540949
22. Tortorella, G., Sawhney, R., Jurburg, D., De Paula, I.C., Tlapa, D., Thurer, M.: Towards the proposition of a lean automation framework integrating industry 4.0 into lean production. J. Manuf. Technol. Manag. https://doi.org/10.1108/JMTM-01-2019-0032
23. Tortorella, G.L., Saurin, T.A., Filho, M.G., Samson, D., Kumar, M.: Bundles of lean automation practices and principles and their impact on operational performance. Int. J. Prod. Econ. (2021). https://www.scopus.com/inward/record.uri?eid=2-s2.0-85102765149&doi=10.1016%2fj.ijpe.2021.108106&partnerID=40&md5=9670478b03537bdfa96ec4f780fdd8ab
24. Tortorella, G.L., Pradhan, N., Macias de Anda, E., Trevino Martinez, S., Sawhney, R., Kumar, M.: Designing lean value streams in the fourth industrial revolution era: proposition of technology-integrated guidelines. Int. J. Prod. Res. **58**(16), 5020–5033 (2020). https://doi.org/10.1080/00207543.2020.1743893
25. Tortorella, G.L., Giglio, R., van Dun, D.H.: Industry 4.0 adoption as a moderator of the impact of lean production practices on operational performance improvement. Int. J. Oper. Prod. Manag. **39**(6/7/8), 860–886 (2019). https://doi.org/10.1108/IJOPM-01-2019-0005

26. Tortorella, G.L., Fettermann, D.: Implementation of industry 4.0 and lean production in Brazilian manufacturing companies. Int. J. Prod. Res. (2018). https://www.scopus.com/inward/record.uri?eid=2-s2.0-85031807940&doi=10.1080%2f00207543.2017.1391420&partnerID=40&md5=04c44c67d94dab099764cd52813ca8b0
27. Wagner, T., Herrmann, C., Thiede, S.: Industry 4.0 impacts on lean production systems. In: Tseng, M.M., Tsai, H.Y., Wang, Y. (eds.) Manufacturing Systems 4.0, vol. 63, pp. 125–131. Elsevier Science Bv, Amsterdam (2017). https://doi.org/10.1016/j.procir.2017.02.041
28. Tranfield, D., Denyer, D., Smart, P.: Towards a methodology for developing evidence informed management knowledge by means of systematic review. Br. J. Manag. **14**(3), 207–222 (2003). https://doi.org/10.1111/1467-8551.00375

Adoption of AI in Talent Acquisition: A Conceptual Framework

Sattu Ramesh and Simanchala Das(✉)

Koneru Lakshmaiah Education Foundation, Guntur, AP, India
rameshstpi@gmail.com, drsdas2001@gmail.com

Abstract. Artificial intelligence has limited applicability in human resource management processes despite technological advancements. Talent acquisition is one of the crucial activities where the impact of AI is significantly conspicuous as compared to other HR processes. But implementing AI in this process requires effective adoption of the technology for ensuring operational efficiency. Moreover, there is a lack of conceptual clarity due to the limited scope of the study in this area. Hence, this study proposed a conceptual framework to help academicians and HR professionals adopt AI in talent acquisition and other allied areas of people management. This study reviewed three theories, the Value-based adoption model (V.A.M.), Technology-Organization-Environment (TOE), and the Task-technology fit (TTF) model, to identify the gap and propose a new framework. The traditional models were used in AI adoption, lacking a rational approach. Organizations should adopt Technology that provides maximum value to the organization compared with the costs and benefits. The study also proposed that the traditional TA methods organization and environmental-related factors and stickiness moderated between perceived value and adoption of AI in talent acquisition.

Keywords: Artificial intelligence · Talent acquisition · Value-based adoption model (VAM) · Technology-Organization-Environment (TOE) · And the Task-technology fit (TTF) model

1 Introduction

In an increasingly competitive global talent market, organizations are undergoing unprecedented changes due to technological advancement. Digitalization has made talent acquisition function more sophisticated as recruiting and selecting new employees with the right skills and knowledge help in accomplishing business goals. In the last decade, rapid advances in Artificial Intelligence (AI) have opened up the unprecedented potential for the digital transformation of organizations [9] Human Resource Management is changed by using Artificial Intelligence, which adds new capabilities to human resource management. From 2016 to 2017, investment in cognitive and artificial intelligence grew at a CAGR of 50.1 percent. By 2021, the total investment in AI technology solutions is expected to touch $57.6 billion (Aayog NITI, 2018). Only 22% of Indian

S. Motahhir and B. Bossoufi (Eds.): ICDTA 2022, LNNS 454, pp. 12–20, 2022.
https://doi.org/10.1007/978-3-031-01942-5_2

organizations use AI (Aayog NITI, 2018) for any business processes demonstrating the slow adoption of AI in India.

AI technologies are mainly used in recruiting, training, and employee engagement to reduce costs, save time, and enhance the accuracy and efficiency of HR tasks. AI can also improve talent acquisition, according to a study by Alexander Mann Solutions. Artificial Intelligence is used to manage TA functions particularly in multinational companies barring a few exceptions that fail to adopt the technology. It is important to examine how HR Managers are implementing AI technology in TA as it provides them with more insight into the use of AI for TA.

AI has been adopted widely in various organizations, but the pace of adoption seems to be slow. Thus, slow adoption on the one hand and inherent limitations in resource mobilization on the other have paved the way for researching this area. Through this research, a conceptual framework is proposed which would help the organizations to understand the various critical aspects that should be considered while adopting the new technology. This research is adopted the value-based adoption model, Technology-Organization-Environment (TOE), and the Task-technology fit (TTF) [26] models to propose a conceptual framework.

A typical example of techno-centric adoption models [11], each focuses on individual adoptions of technology and on technologists. The TOE model explains technology adoption as it pertains to the organization from a cognitive perspective [15], and the DTOE model [7] involves technology adoption from an organizational perspective [16]. [17] shows that enterprise adoption of technology is dominated by the TOE framework. [4] also claimed that these traditional models were not sufficient. Moreover, the new ICT users should be recognized as technology users and as consumers as well. Therefore, the study combined the value-based adoption (VAM), TOE, and TTF models to propose a new theoretical framework for AI adoption in organizational settings. Previous studies on AI adoption were limited, lacked the conceptual framework, and failed to provide a holistic model for understanding AI adoption in talent acquisition function.

The paper is organized as follows. Firstly, the previous literature on the adoption of AI in talent acquisition is reviewed, then the theoretical background was explained and a proposal for a conceptual model was made. Next, a conceptual model of AI adoption was formulated with the help of existing models, and theories are presented. An overview of the proposed methodology and expected results are presented in the study's succeeding section. The paper ends with a discussion and conclusion.

2 Theoretical Background

2.1 Related Literature on AI Adoption

In recent years, AI capabilities have become evident in technological advancements in e-recruitment, computer learning, NLP, and deep learning. Automation can be applied to many routine recruitment and selection processes to increase efficiency. It has been noted by [6] that organizations are beginning to recognize the value of investing more in digital tools. Moreover, electronic recruitment has evolved with AI use due to the possibility of substantial process automation such in selection tasks and freeing up manual recruiting [12].

According to [18], organizations have begun incorporating AI in the recruitment process to reduce the need for manual activities like evaluating resumes communicating, scheduling interviews, and making reminders [13]. Recruitment and selection will be automated by artificial intelligence (AI); however motional and empathy qualities will still be needed to create a cultural fit assessment [19]. According to [10], the major concern is a lack of trust in AI-related bias. Additionally, cultural and personal biases are intentionally introduced by coders in algorithms, he opined.

It appears that recruitment agencies place varying levels of importance on the benefits of automation, with 40% emphasizing enhanced efficiencies and 38% emphasizing increased engagement of employees, according to the UK Recruitment Trends Report. A high-touch or high-volume recruitment strategy is available in the recruiting market. According to a 2016 UK Recruitment Trends Report, automation is expected to damage existing jobs. 41% of respondents expect job losses, 30% expect more jobs, and 29% are undecided about their future employment. AI/ML solutions also enable organizations to process the large volume of applicant data speedily and accurately. AI/ML algorithms could similarly contribute to faster and more efficient decision-making in the talent management process.

To investigate IS adoption from an individual perspective, technology acceptance research has used various technology acceptance models [20]. [4] asserted that these models were limited in explaining new ICT acceptance. Therefore, the study combined the value-based adoption and technology acceptance model (T.A.M.) and proposed a new theoretical framework for AI adoption.

2.2 Value-Based Adoption Model (VAM.)

[12] proposed a value-based adoption model. According to this model, adoption is based on maximizing value. Originally, the model was developed to investigate the factors responsible for the adoption of mobile internet. VAM measures how to value perception can influence value adoption from a value maximization angle. The model is widely used in estimating consumer acceptance of mobile internet and smart home services. Since value maximization is a key component of the adoption due to its rationality. By examining the mediating role of perceived value between a customer's beliefs about the benefits and sacrifices of M-Internet adoption and their intention to adopt Mobile-Internet, several studies have demonstrated that value perception determines M-Internet adoption. Organizations evaluate the technology from the cost benefits perspective. Cost is considered the organization's sacrifices while adopting the technology, such as privacy risk, innovation resistance, and trust in the technology. Perceived benefits comprise the usefulness of the technology, relative advantages, and cost-effectiveness in the system. These perceived sacrifices and benefits determine the perceived value derived from the technology and the perceived value determining the adoption intention for the technology.

2.3 Technology-Organization-Environment (TOE) Framework

A multi-perspective model of TOE, an organization-level model, was developed by [15]. With the TOE model, three groups of perspectives are examined including the adoption of technology. Technology's aspect deals with both external and internal technology that organizations need to adopt [21, 22]. In addition to reliability, security, capability, and quality, technology factors are also discussed such as compatibility of IT solutions and the benefits of the technology [22]. An organization's perspective focuses on several items such as the size of the firm and centralization of the organization, the complexity of its organizational structure, the quality of its human resources, and above all, the organization's readiness.

The TOE framework has a key flaw as several of the characteristics in the adoption apply to large firms, where clients can expect continuity and fewer complaints than too small businesses. Nonetheless, the TOE paradigm has gained empirical validity and has served as the foundation for numerous ICT adoption studies, particularly those involving EDI and inter-organizational information systems (IOIS). Scholars who used the TOE framework in their research found that the qualities of the innovation, the technology of the organization, and the external environment are all beneficial in explaining and forecasting adoption rates. Within the T-O-E paradigm, [23] discovered that technological competency, company size, financial commitment, competitive pressure, and regulatory backing are important adoption determinants. [24] confirmed the TOE framework's use in small businesses by proposing a perception-based EDI adoption model with six determinants including perceived usefulness.

2.4 Task-Technology Fit (TTF) Model

According to the TTF model, the extent to which technology assists an individual in performing a portfolio of tasks is the basis of an assessment model for IS users in organizations. [3] discuss how a better-fitting technology results in better performance (TTF). The adoption of technology through TTF is discussed in studies on RFID, big data in a mobile cloud environment for healthcare, mobile banking gamification for training, and virtual learning systems. TTF is used in this study to understand better how teachers use AI Technology (Fig. 1).

3 Hypothesis Formulation and Conceptual Model

SNP (Security and Privacy Issues), IR (Innovation Resistance), TR (Trust), COS (Cost-Effectiveness), USE (Usefulness), REA (Relative Advantage).

3.1 Perceived Benefits

Organizations have the potential benefit of implementing AI in HR processes, specifically in talent acquisition. A global candidate pool can be provided by AI, helping organizations have the best talent across the globe with various skills. Using AI in recruitment is also more cost-effective and efficient in the long run [25]; it reduces paperwork, makes the hiring process faster, and increases the likelihood of finding and hiring the right candidate [14].

With the application of AI, recruiters can focus exclusively on attracting talent. Using strategies like candidate sourcing, organizations can gain greater insight into talent requirements through technologies like resume scanning and removing unconscious

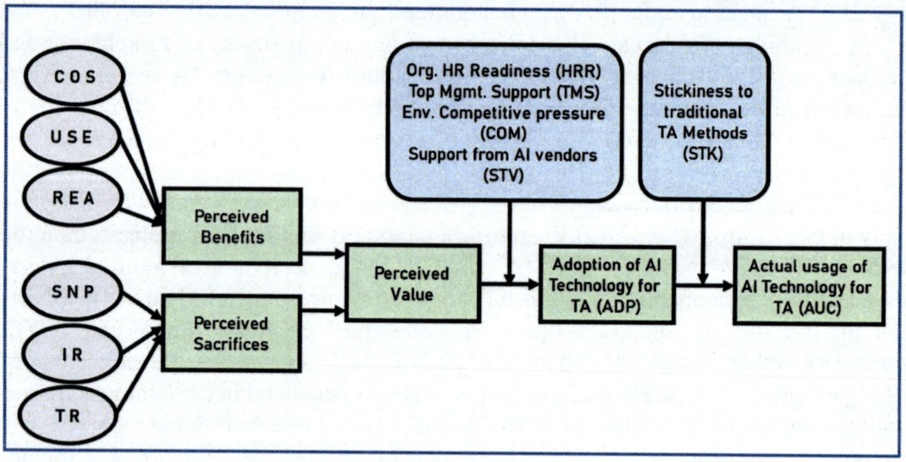

Fig. 1. Conceptual model

bias from screening applications. The AI framework can also be used for applicant selection, staff participation re-engagement, career development, etc. [26]. Therefore, Artificial Intelligence-based tools enable the human resources department to maximize performance.

AI also adds value to organizations by providing high technical support to facilitate talent acquisition. For example, candidates can be engaged by AI assistants. Additionally, they can route suitable applicants directly to recruiters. Subsequently, AI and NLP (natural language processing) and AI assistants can provide a front end for candidate communications. AI enables the recruiters to focus more on important functions of talent acquisition, such as negotiation and other processes. At the same time, AI assistants can focus on candidate capture, screening, scheduling coordination, communication, and engagement. Moreover, AI has altered the entire business landscape as new HRM practices, applications, and methods tend to emerge. It has also changed the administrative functions of HRM, which have become more dynamic, evolving, and modified.

Thus, the role of AI, robotics, advanced technologies in human resource practices can never be underestimated. In the previous studies, AI technology was found to have a positive impact on recruitment, training, and performance evaluation. Based on the above literature, the following hypotheses were formulated:

H1: Usefulness of AI in talent acquisition positively affects perceived value.
H2: Cost-effectiveness of AI in talent acquisition positively affects perceived value.
H3: Relative advantages of AI in talent acquisition positively affect perceived value.

3.2 Perceived Sacrifices

The use of AI addresses many ethical, legal, privacy, and moral objections and the handling of potential candidate ramifications [18]. Because of the big worries about AI-related bias, [10] lack of trust is a matter of concern. The impact of technology and talent analytics on talent management highlights TA's security and privacy concerns while implementing AI.

SNP (Security and Privacy issue) is another concern while exchanging data. As security is one of the determinants of HRIS adoption, studies indicate that adopting technology is hampered due to security concerns. Negatively organizations place a greater emphasis on security, which later affects the adoption of AI [5]. According to [27], AI-based recruitment could alienate candidates due to ethical concerns.

The lack of trust in the technology seems to be a major factor for limited AI applications in recruitment. A further concern is whether AI can produce bias-free decisions with accuracy and reliability. There may be an increased fear that unethical recruitment and selection practices creep into the recruiting and selection process because of this, as discussed previously. [10] warning that AI cannot be trusted to know that algorithms are free of bias" biases? that poses a challenge for the field. [28] warned that it is unclear whether AI recruitment would improve efficiency and fairness.

Resistance is an opposing or retarding force. "Why is resistance to technology in our technological time still present?" one might ask. AI technology is likely to subsume many of the daily tasks performed by knowledge workers, forcing a rethink of the individual's proper role in the organization. Meanwhile, top executives worry about losing control of AI and doing real damage to their organization's core processes and their careers. Given the above literature, the organizations' employees resist innovations such as AI implementation and hence don't consider them more valuable.

The following hypotheses were developed based on the above literature:

H4: Security and privacy concerns reduce the perceived value of AI.
H5: Lack of trust decreases the perceived value of AI.
H6: Employee resistance to innovation decreases the perceived value of AI.

3.3 Moderating Effect of HRR, TMS, COM, STV Between Perceived Value and Adoption of AI for TA

To adopt HR technology, HRR (HR Readiness) is essential. HRR refers to the accessibility of budget with the HR department and the skills and resources required for adopting AIT. Earlier IS studies [19] state that organizational readiness affects IS adoption. However, it has been found that readiness is not responsible for technology adoption. As top managers actively participate in introducing any new technology, they play a vital role in its adoption. TMS affects technology adoption [19, 2], as confirmed in a study [1]. An important and unique technology used for recruitment is AIT for TA. Thus, the top executive may review the details of the AIT for TA before adopting it. As a part of the new technology's resources, top management should be induced to fund AIT.

COM refers to the pressure an organization feels to adopt technology for its survival in the competition and functioning practices [20]. To solve the problems found when using new technology, STV must be implemented continuously during the implementation process. Assisting technology vendors in implementing and utilizing IS innovations are referred to as "STV" [19]. Additionally, STV is proved to be an important predictor of technology adoption. Hospitals adopted HRIS more readily when STV was considered. Based on the above literature, the following propositions were made:

H7: HRR moderates the relationship between PV and the Adoption of AI technology.
H8: TMS moderates the relationship between PV and the Adoption of AI technology.
H9: COM moderates the relationship between PV and the Adoption of AI technology.
H10: STV moderates the relationship between PV and Adoption of AI technology.

3.4 Moderating Role of "Stickiness (STK) to Traditional TA Methods" Between Adoption and Actual Usage of AI for TA

Human resources managers tend to stick to traditional recruiting and selection methods. Previously, the website was defined for attracting consumers' attention and retaining them. As described previously, STK is an intangible characteristic that stimulates individuals to stay on longer. [8] found that Stickiness to the traditional method moderated the relationship between ADP and ACU in digital payments. Traditional recruitment processes are preferred by HR managers, for they tend to use traditional recruiting methods.

Thus, based on the literature support, the following hypothesis was formulated.

H11: Stickiness (STK) moderates the relationship between adoption and actual usage of AI technology for TA.

3.5 Adoption (ADP) of AI and Actual Usage (ACU) of AIT for TA

Adoption means the usage of technology and the organizations have prepared to use and invest in the purchase of new technology, it becomes imperative to examine the relationship between the ADP and the actual ACU. ADP describes how an organization chooses new technology as per its needs. In this study, the actual usage refers to the assistance the organizations, the technology provides in achieving a set of goals for using it. Researchers have discussed the relationship between ADP and ACU in studies related to RFID and BI systems [12]. As such, AI and its potential to perform the functions of talent acquisition and save time and money by saving costs. Due to the benefits of AIT for HR managers, it is essential to investigate the adoption of AIT for TA which may affect the actual usage of AIT for TA. Thus, the following proposition was made based on the above literature:

H12: Adoption (ADP) of AI positively affects the Actual usage (ACU) of AI in TA.

4 Discussion and Conclusion

The primary objective of the current study was to propose a conceptual framework through the study and the relationship between independent, dependent, and moderating variables. This paper reviewed three technology adoption models used in organization settings such as VAM, TOE, and TFF. The pace of adoption of new technology depends on the perceived value derived from the technology itself for the organization. Perceived value can enhance by increasing the benefits or reducing the sacrifices as suggested by the VAM model [23]. For making AI adoption effective in talent acquisition, organizations should focus on increasing the usefulness, cost-effectiveness, and relative advantages of the technology concerning the traditional practices. Whereas, the factors may be considered as sacrifices such as security and privacy concerns, lack of trust, and resistance to innovation should be overcome to maximize perceived value. Additionally, the organizations also should focus on the availability of resources and skills required to determine the HR readiness to adopt the new technology. Moreover, the support of the top management and the pressure to become competitive in the industry would further trigger perceived value and adoption of AI technology with the collaborative efforts of AI vendors.

References

1. Acharjya, B., Das, S.: Adoption of e-learning during the COVID-19 pandemic: the moderating role of age and gender. Int. J. Web-Based Learn. Teach. Technol. (IJWLTT) **17**(2), 1–14 (2022)
2. Das, S., Acharjya, B.: Understanding organisational effectiveness through sustainable human relations approach: the role of empowerment climate in selected industrial establishments. Int. J. Syst. Dyn. Appl. (IJSDA) **10**(2), 33–52 (2021)
3. Dhamija, P., Bag, S.: Role of artificial intelligence in operations environment: a review and bibliometric analysis. TQM J. (2020). https://doi.org/10.1108/TQM-10-2019-0243
4. Hogg, P.: Artificial intelligence: HR friend or foe? Strateg. HR Rev. **18**(2), 47–51 (2019)
5. Sivathanu, B.: Adoption of digital payment systems in the era of demonetization in India: an empirical study. J. Sci. Technol. Policy Manag. **10**(1), 143–171 (2019)
6. Bafna, P., Shirwaikar, S., Pramod, D.: Task recommender system using semantic clustering to identify the right personnel. VINE J. Inf. Knowl. Manag. Syst. **49**(2) (2019)
7. Pathomphatthaphan, S., Das, S., Virat, K.: A conceptual approach to HRM practices influencing organizational outcomes and shared values to society: the mediating role of employee outcomes and organizational culture. Organ. Cult. Int. J. **21**(2), 19–33 (2021). https://doi.org/10.18848/2327-8013/CGP/v21i02/19-33
8. Oliveira, T., Martins, R., Sarker, S., Thomas, M., Popovic, A.: Understanding SaaS adoption: the moderating impact of the environment context. Int. J. Inf. Manag. **49**, 1–12 (2019)
9. Bersin, J.: Talent trends technology disruptions for 2018 (2018). https://www.isaconnection.org/assets/documents/2018BersinHRTechDisruptionsReport.pdf
10. Eubanks, B.: Artificial Intelligence for HR: Use AI to Support and Develop a Successful Workforce, 1 edn. Kogan Page Ltd., London, NY (2018)
11. Leong, C.: Technology and recruiting 101: how it works and where it's going. Strateg. HR Rev. **17**(1), 50–52 (2018)
12. Puklavec, B., Oliveira, T., Popovič, A.: Understanding the determinants of business intelligence system adoption stages an empirical study of SMEs. Ind. Manag. Data Syst. **118**(1), 236–261 (2018)
13. Howard, M.C., Rose, J.C.: Refining and extending task–technology fit theory: the creation of two task–technology fit scales and empirical clarification of the construct. Inf. Manag. (December), 1 (2018)
14. Scholz, T.M.: Big Data in Organizations and the Role of Human Resource Management: A Complex Systems Theory-Based Conceptualization. Peter Lang International Academic Publishers, Frankfurt (2017)
15. Khemthong, S., Roberts, L.M.: Adoption of internet and web technology for hotel marketing: a study of hotels in Thailand. J. Law Gov. **1**(2), 51–70 (2006)

16. Depietro, R., Wiarda, E., Fleischer, M.: The context for change: organization, technology, and environment. The processes of technological innovation. Process. Technol. Innov. **199**, 151–175 (1990)
17. Rogers, E.M.: Diffusion of Innovations, 5th ed. Free Press, New York (2003)
18. Premkumar, G., Roberts, M.: Adoption of new information technologies in rural small businesses. Omega **27**, 467–484 (1999)
19. Jia, Q., Guo, Y., Barnes, S.J.: E2.0 post-adoption: extending the IS continuance model based on the technology-organization-environment framework. In: Proceedings of the International Conference on Electronic Business (ICEB), No. September, pp. 695–707 (2016)
20. Ajzen, I.: The theory of planned behavior. Organ. Behav. Hum. Decis. Process. **50**, 179–211 (1991)
21. Hameed, M.A., Counsel, S., Swift, S.: Establishing relationships between innovation characteristics and innovation adoption in organizations: a meta-analysis approach. Inf. Manag. **49**(5), 218–232 (2012)
22. Sureshkrishna, G., Simanchala, D.: Occupational stress and job satisfaction: a study of automotive industry in India. Int. J. Mech. Eng. Technol. (IJMET) **9**(2), 690–697 (2018)
23. Hossain, M.A., Standing, C., Chan, C.: The development and validation of a two-staged adoption model of RFID technology in livestock businesses. Inf. Technol. People **30**(4), 785–808 (2017)
24. Bughin, J., Manyika, J., Woetzel, J.: Jobs lost, jobs gained: workforce transitions in a time of automation, McKinsey Global Institute, no. December, pp. 1–160 (2017)
25. Goodhue, D.L., Thompson, R.: Task-technology fit and individual performance. MIS Q. **19**(2), 213–236 (1995)
26. Oliveira, T., Martins, M.F.: Literature review of information technology adoption models at firm level, no. January (2011)
27. Davis, F.D., Davis, F.: Perceived usefulness, perceived ease of use, and user acceptance of Information Technology. MIS Q. **13**(3), 319–340 (1989)
28. Eze, S., Awa, H., Okoye, J., Emecheta, B., Anazodo, R.: Determinant factors of information communication technology (ICT) adoption by government-owned universities in Nigeria: a qualitative approach. J. Enterp. Inf. Manag. **26**(4), 427–443 (2013)

OntoRecipe: An Ontology Focussed Semantic Strategy for Recipe Recommendation

Siddhant Singh[1] and Gerard Deepak[2(✉)]

[1] Department of Computer Science and Engineering, SRM Institute of Science and Technology, Ramapuram, Chennai, India
[2] Department of Computer Science and Engineering, National Institute of Technology, Tiruchirappalli, India
gerard.deepak.christuni@gmail.com

Abstract. In present-day times there is a need for a recipe recommendation system to cater to the demanding culinary needs of several users around the globe. This paper proposes an ontology focused recipe recommendation approach by incorporating semantic models. Its deep learning infused using RNN classification. The dataset uses here is from Food.com Recipes and Interaction which contains 280 k+ recipes and 700 k+ recipe reviews which covers 18 years of user interaction and uploads. The dataset based on it increases a good number of auxiliary knowledge and uses pair of approaches like SemantoSim and Kullback-Leibler divergence under the Shuffled Frog-Leaping algorithm a knowledge-centric recipe is recommended. It shows that the proposed structure performs better than the other baseline approaches in terms of the F-Measure and False Discovery Rate having values 95.03 and 0.06.

Keywords: Domain-based ontology · SemantoSim · Kullback-Leibler divergence · SFLA · Recipe recommendation

1 Introduction

Many factors have been identified that influence a person's well-being. Actual exercise, rest, sustenance, heredity, contamination, and other external elements are all examples. Because nutrition is one of the most changeable aspects of our life, it's not surprising that small adjustments may have a big impact. Additionally, because our diets have strong cultural links, it's feasible to identify a wide range of cuisines from all over the world. Each's most prevalent component are strongly connected to geographical factors such as climate. This has a huge influence on the availability of each item utilised in regional recipes. With people becoming more accustomed to dealing with technology in their daily lives, one way to overcome this issue and assist people in making healthier choices is to build health-aware food recommender systems. One of the most difficult difficulties for such a system is providing accurate and tailored suggestions to its consumers.

S. Motahhir and B. Bossoufi (Eds.): ICDTA 2022, LNNS 454, pp. 21–33, 2022.
https://doi.org/10.1007/978-3-031-01942-5_3

Motivation: There A wide range of dietary products is recognised to have a favourable influence on health, particularly in the prevention of disease. Knowing which components have the greatest concentrations may help with illness treatment and prevention. Furthermore, by combining these components into attractive and cheap meals, it is possible to encourage a change in the population's dietary habits. In addition to the two preceding reasons, quickness of preparation is essential in a society where fast-food consumption is on the increase. Also, in a world where everyone is linked, there is a demand for recipes that are multi-regional and multi-cuisine in nature. There was a need for an ontology and knowledge-driven approach for a better food recipe recommendation system.

Contribution: In this paper, an ontology and query-driven in combined with the knowledge-centric approach are incorporated to provide multi-regional and multi-cuisine recipe recommendation to the user using a SemantoSim, KL divergence along with Shuffled Frog-Leaping algorithm (SFLA), with the results of lowest FDR of 0.06 it is evident that this proposed approach works better than other baseline models, and background information is infused into them using a modelled thesaurus and real-time web data from data Wikidata. An ontology is created and is subjected to semantic alignment and KL divergence. The micro fragments obtained are deep learning infused using a Recurrent neural network which further classifies and recommends recipes to the user. This proposed approach has recorded an F-Measure of 95.30.

Organization: The following is how the paper's body is organised. The preceding Sect. 2 depicts the Related Work already. The Proposed System Architecture is explained in Sect. 3. Section 4 goes through the Implementation and Performance Evaluation. The paper is concluded in Sect. 5.

2 Related Work

Nilesh et al., [1] offered a mechanism for proposing Indian cuisine based on cuisine and preferred food item matching. Ye-Jin et al., [2] proposed a system is based on web-based recipe recommendations and is a chatbot system. Du et al., [3] have proposed strategies for embedding word location information are proposed in this research. The goal is to specify the boundaries of Chinese named things so that they may be recognised more effectively. It researches to validate the usefulness of the word position information embedding approach in completing named entity recognition tasks. Dhungana et al., [4] recommended based on the clue words by using WordNet model for determining the proper sense of a polysemy word. Nouns, verbs, adjectives, and adverbs can all be clue words. Elbeltaget al., [5] studied five evolutionary-based search techniques. Each method is given a brief explanation, as well as pseudocode to help in its implementation. The PSO technique outperformed other algorithms in terms of success rate and solution quality. Thakkar et al., [6] focused on combining predictions from UbCF and IbCF with the goal of fusion. Li. et al., [7] proposed a system based on spark clusters is utilised to cope with enormous amounts of data from mobile applications and create a hybrid recommendation algorithm that combines content-based filtering and collaborative filtering to enhance recipe recommendations. A Yajima et al., [8] studied the goals to create a mechanism for recommending "easy" cooking recipes by evaluating recipe experiences and keeping user conditions in mind, and then to construct a recommendation system utilising the recommended technique. Stahl et al., [9] created semantic food descriptions based on the LanguaL thesaurus to suit two national food composition databases. Sanjo et al., [10] discussed a ground-breaking project to recommend seasonal meals based on learning from monthly view data First, we train a regression model on a set of recipe texts and their monthly views to generate seasonal scores that indicate the importance of various time scales for every cuisine. Wei et al., [11] developed a model using Kullback-Leibler divergence, and unlike most previous approaches, and systematically detect different pathways' regulations. By computing the distance of probability distributions from distinct phenotype gene expression data, our analytic model discovers the final significantly divergent pathway. Yue et al., [12] propose to study the parameters of SFLA that are investigated. The paper compares SFLA to the genetic algorithm (GA) and particle swarm optimization using the testing function (PSO). SFLA surpasses GA and PSO in terms of astringency and global search capabilities. In [13–27] several knowledge-centric models in support of the proposed literature have been depicted.

3 Proposed System Architecture

The architecture of the proposed Ontology-driven recipe recommendation system is depicted in Fig. 1. The proposed architecture is ontology focused by incorporating semantic models. It is infused with deep learning as RNN which is used as a classification framework. The input query is subjected to pre-processing. Tokenization, Lemmatization, Stop Word Removal, Named Entity Recognition [3], and Word Sense Disambiguation [4] are all part of the pre-processing procedure. As a tokenizer, negative space and special characters are used for tokenization. Individual words are achieved by tokenization based on white space and ideograms. Lemmatization is the process of obtaining the core form of a phrase from its derivational form by deriving the base term from the syntax of words. Stop words are a group of words that are often used in a language. Stop words are commonly used in Text Mining and Natural Language Processing (NLP) to remove keywords that are overly frequently used to the point of containing no important data. Element Recognition [3], also known as component affirmation, component division, and component extraction, is a sub-task of data retrieval. It refers to the process of locating and categorising identified components in text into preset groups such as persons, associations, areas, periods, and so on. Word Sense Disambiguation [4] is described in natural language processing (NLP) as the capacity to discern which meaning of a word is triggered by its use in each context. Lexical ambiguity, whether syntactic or semantic, is one of the first issues that every NLP system faces. Word's syntactic ambiguity can be resolved with high-accuracy part-of-speech (POS) taggers. WSD, on the other hand, is concerned with the problem of resolving semantic ambiguity.

At another step user query yields, individual query words or query terms these query terms are quite shallow in nature and content must be known although NER ensures that content can be determined to an extent, query words are incorporated to add background knowledge. Queries need background knowledge that is to be added from Wikidata

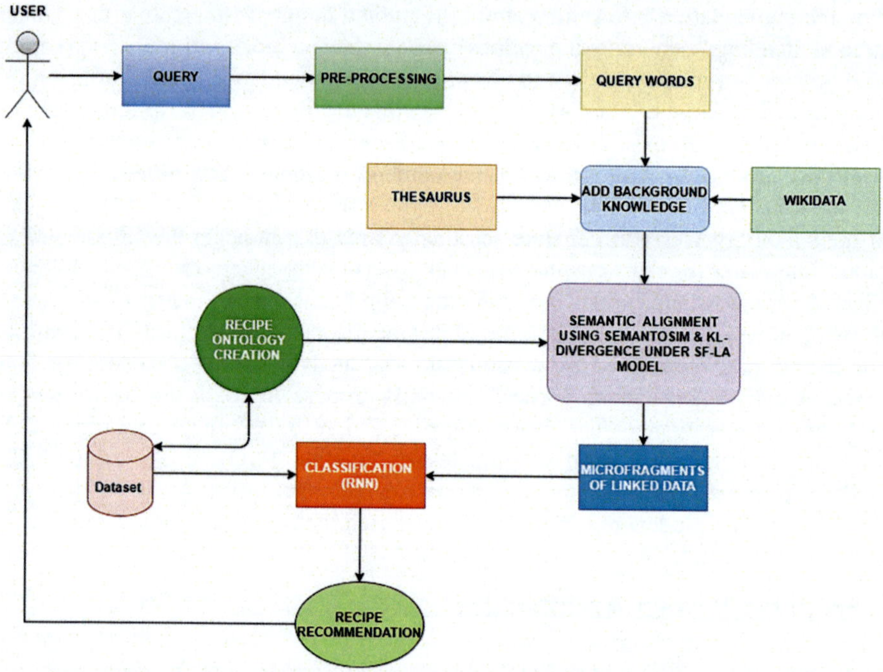

Fig. 1. Architecture diagram of the proposed system

API. Wikidata, a knowledge repository that includes multiple web articles that highlight elements such as concepts, taxonomies, data on a certain topic, and so on, is one of the academic forms integrated into the paradigm. The information is presented in the form of an attribute as well as a value. The Wikidata API is used to retrieve relevant knowledge sources connected to the content and to give supplementary knowledge. SPARQL i.e., standard query language protocol for linked open data, its endpoints are written to obtain data from Wikidata. All these entities are linked i.e., relevant to the query obtained from the structure of the World Wide Web which is consolidated as background knowledge.

Thesaurus is modelled and comprises several eBooks of recipes, different user blogs and user descriptions of YouTube of several recipes. This knowledge is categorised based on cuisines and subcategorised based on Regional Cuisines that consists of Indonesian, Turkish, Thai, Spanish, Moroccan, Japanese, Indian, Italian, French and Chinese. Also, the state of food whether the food is served hot or cold is considered. Entities are also labelled in other subcategories such as food is veg or nonveg, cooking time, Type of meal i.e., starters, main course, dessert and whether the food item is solid, or liquid is also incorporated. All of these are taken in form of keywords in thesaurus it is formulated and considered as a hash table in this case which is a basic data structure that contains key as the recipe name and value is the link to the NoSQL database and the entire recipe and its several versions are available. The beauty of incorporating a thesaurus is that the key comprises the title which can easily be parsed and can be added to the background knowledge as query words. Already modelled thesaurus is also considered for adding background knowledge to the approach. Since thesaurus and Wikidata gave been used, a large amount of background knowledge is added to the dataset which comprises of which contains 280 k + recipes and 700 k + recipe reviews which covers 18 years of user interaction and uploads.

The Ontology of the recipe is created using Protege. Protege is applied on documents reasoned and ontology for several cuisines have been modelled. Onto Collab is used for OWA ontology generation directly from World Wide Web and is generated for all the terms and categories available in the dataset. The ontology modelled is a categorical upper ontology with a minimum detailed implementation. The reason is that detailed implementation is not required only the recipe title, the recipe category, as well as recipe ingredients, are modelled up in the ontology. Details are available in the eBooks which have been summarised and loaded into the thesaurus. The ontology activities are subjected to semantic alignment using the SemantoSim measures long with KL divergence under the Shuffled Frog-Leaping algorithm. The Semantic Relationship between query words and thoughts is determined initially. The semantic similarity of the description logics in the Hash Table of corresponding concepts is then estimated, and the matching scenario and context are derived. Only semantically related conceptions and description logics are clustered together and placed in ascending order of semantic similarity. Finally, constraints in the form of representation logics are established between concepts and individuals given the unique hash value. This removes the need for branching paths in keeping boundaries between ideas and people.

$$\text{SemantoSim} = \frac{pmi(j,k) + p(j,k)\log([p(j,k)])}{[p(j).p(k)] + \log([p(y,x)]} \tag{1}$$

The SemantoSim measure is a normalised semantic similarity metric that evolved from the Pointwise Mutual Information measure. If the question contains two words, they are matched as follows: (j, k). Permutations of the three search words are analyzed once there are three inquiry phrases. Semantic similarity between the term j and its most closely linked semantically relevant term is considered while searching for a single phrase. The search words that have been specially adapted and stemmed are the search terms. The semantic similarity of two words is determined by the SemantoSim algorithm (j, k). The equation is used to determine the PMI (j, k) (1). The probability of the phrase x co-occurring with y is given by the equation p. (j, k). The probability of the word y appearing with x is determined by p. (j, k). The probabilities p(j) and p(k) show whether the variables x and y exist. The foundations of KL Divergence can be traced back to information theory. The primary goal of information theory is to quantify the volume of information in evidence. The most important measure in information theory is entropy, usually abbreviated as HH. A stochastic solution's complexity is defined as Eq. (2).

$$H = -\sum\nolimits_{xi=1}^{A} p(j_{xi}).\log p(j_{xi}) \tag{2}$$

What entropy does not tell us is the best encoding strategy to use to achieve this compression. Although optimal information encoding is an intriguing issue, it is not required for understanding KL divergence. The important thing about Entropy is that it allows us to quantify how much information is in our data merely by knowing the theoretical bottom constraint on the number of bits required. Now that we've quantified this, we'd like to see how much information is lost when we replace our observed distribution with a parameterized approximation. Kullback-Leibler Divergence is only a simple change to our entropy formula. In addition to our probability distribution pp, we have our approximation distribution qq. The difference in log values for each is then examined in Eq. (3).

$$D_{KLD}\,(p\|q) = \sum\nolimits_{xi=1}^{A} p(j_{xi})\,.(\log(p(j_{xi}) - \log q\,(j_{xi})) \tag{3}$$

We can calculate how much information is lost when we approximate one distribution with another using KL divergence. The Shuffled Frog Leaping Algorithm (SFLA) is a newly popular systematic methodology that is based on the evolutionary computation proliferation of a colony of frogs seeking of most readily accessible food. It was presented in 2003. In SFLA, the populace is comprised of a group of segmented frogs (solutions) (memeplexes). Diverse frog cultures are characterized by different memeplexes, each of which undertakes a local search (memes). Each memeplex's frogs go through memetic evolution, which means they engage in a local exploration of solution space utilising strategies that enhance meme transmission among local people. Information is relayed between memeplexes via a shuffling mechanism after a certain number of recursive development stages. Local exploration and stagger methods are varied until the required convergence threshold is attained.

The following is a description of the SFLA: Consider that the original population is made up of frogs that were produced S at random. A frog is represented j as in M dimensional issues (variables) as $Q_i = (Q_{i1}, Q_{i2} \ldots. Q_{iM})$. After that, the frogs are listed in order based on their fitness. The populace is then split into X memeplexes, each of which comprises Y frogs. (For example, S = X x Y). The very first frog enters the first memeplex, the second frog visits the second memeplex, the frog moves to the memeplex, the frog returns to the very first memeplex, the second frog then visits the memeplex, X visits the Xth memeplex, and X + 1 returns to the very first memeplex. H_b and H_w parameters are being used to determine which frogs perform best and worst inside each memeplex. The frog only with the best overall fitness is named H_g. In this scenario, a method like PSO is employed to boost H_w in each small group (not all frogs).

$$T = rand() \times (H_b - H_w) \tag{4}$$

$$H_{new} = H_w + T,\ -T_{max} \le T \le T_{max} \tag{5}$$

where is a rand() randomised integer between 0 and 1. $-T_{max}$ is the greatest change in the location of a frog. If a better solution H_{new} is found, it substitutes the frog with minimum metric H_w; otherwise, the computations are again processed of (4) and (5) by substituting H_b and H_w in (4). If no progress is made in this situation, a new solution is picked at random to replace it. Within local search periods, the memeplex is reordered and refreshed, and all memeplexes are mixed to share knowledge before being reassigned for the next search operation. In the proposed architecture the SemantoSim is considered of the threshold of 0.5 and KL divergence with step deviation of 0.25 which is used as an objective function for shuffled frog leap algorithm. The SFLA ensures that the best and most reliable solution set is generated from the initial list of solutions generated, so as a result highly relevant and the much-seasoned recipe for the terms in the given dataset are obtained. At the end of semantic alignment between the recipe ontology using the SFLA algorithm, micro fragments of linked data are obtained. The micro fragments are further fed into RNN for further classification. RNNs are a type of artificial neural network wherein the constituent interconnections create a graph structure that runs across a sequence of nodes. It is essentially a series of neural network pieces connected like a chain. Each is conveying a message to a successor. RNNs are intended to utilise sequential data when the current step is related to the prior stages in some way. This makes them excellent for time-based applications (music, time-series data) and natural language processing. RNNs work exceptionally well in situations where sequential information is significant since if sequential information is not employed, the meaning or grammar may be misunderstood or wrong. Image captioning, language modelling, and machine translation are some of the applications. So, the dataset is classified on micro fragments of linked data and features are automatically sent to RNN classification and the recipe is recommended to the user. The classifier instances are once again compared with the initial query word for SemantoSim measure, at this point the threshold is 0.75 for the recommendation so and the recipe is recommended using re-ranking the increasing order of the SemantoSim measures and recommended to the user.

4 Implementation and Performance Evaluation

Windows 10 operating system was used to design and build the proposed framework configuration. The development was performed on an Intel Core i7 9th Gen CPU with 16 GB RAM. NLTK, Re, Scikit-learn, and Matplotlib are Python packages used for data pre-processing and displaying parameters such as F-Measure of baseline model and other approaches. Onto Collab was used for Ontology Python's deep learning API known as Kera's, as well as TensorFlow, for the deployment and testing of the Term Based Ontology Model using the Food.com Recipes and Interaction collection on which methods were performed The recommended approach algorithm comprises a domain-centric ontology, an information resource, and datasets as inputs, but also the ontology and query words. The data collection, in this case, is Food.com Recipes and Interaction, which contains 280 k + recipes and 700 k + recipe reviews spanning 18 years of user interaction and uploads and is used for model implementation and construction. Individual query words or query phrases are returned by user queries; The Wikidata API is used to access relevant knowledge sources that are related to the content and contain auxiliary knowledge. Entities are also labelled by subcategories such as cooking time, meal type, and whether the food item is solid or liquid. The suggested architecture considers SemantoSim with a threshold of 0.5 and KL divergence with a step deviation of 0.25 as the goal function for the shuffling frog leap algorithm. The SFLA guarantees that the best and most reliable solution set is created from the original list of generated solutions, resulting in highly relevant and well-seasoned recipes for the words in the provided dataset. Micro fragments of connected data are generated after semantic alignment between the recipe ontology and the SFLA algorithm. The micro pieces are then put into RNN for classification. As a result, the dataset is categorised based on micro pieces of connected data, features are automatically submitted to RNN classification, and the recipe is proposed to the user. The classifier instances are compared again with the initial query word for the SemantoSim measure; at this point, the threshold is 0.75 for the recommendation, and the recipe is given to the user by re-ranking the ascending order of the SemantoSim measures.

The Precision, Recall, Accuracy, F-Measure, False Discovery Rate (FDR) are considered as performance metrics. The accuracy of the proposed framework vs. the number of entities is shown in Fig. 2 in contrast to the baseline model and several permutations of approaches for recipe recommendations. In Fig. 2 It is evident that the proposed OntoRecipe Recommendation system is baselined with RRBI, CSRS, SVM + KNN, Naïve Bayes along with Collaborative Filtering and Fuzzy C-Means Clustering along with SVM. The reason for baselining the proposed approach with these models is that these approaches are quite distinct by mainly covering all the possibilities which are existing in contrast with the proposed ontology recommendation approach.

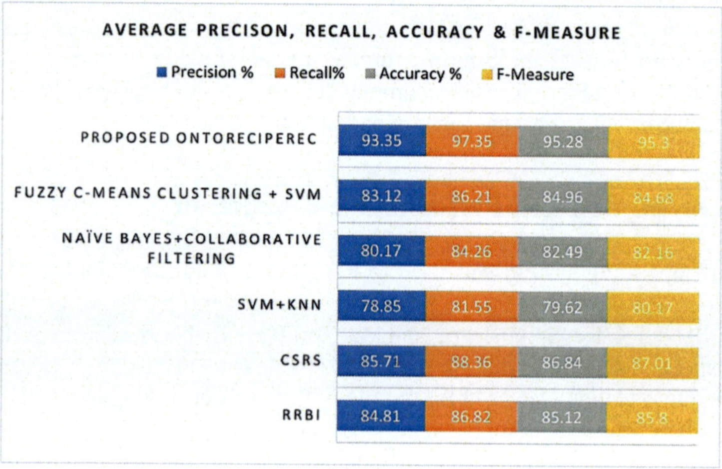

Fig. 2. Comparison of evaluation metrics of the proposed ontorecipe recommendation system.

Figure 2 shows that RRBI has Average Precision, Recall, Accuracy, and F-Measure values of 84.81,86.82,85.12,85.8. Similarly, CSRS produces average precision, recall, accuracy, and F-Measure values of 85.71, 88.36, 86.84, and 87.01. The SVM + KNN combination yields average precision, recall, accuracy, and F-Measure values of 78.85,81.55,79.62, and 80.17. Average Precision, Recall, Accuracy, and F-Measure for Nave Bayes + Collaborative Filtering are 80.17,84.26,82.49,82.16. Furthermore, Fuzzy C-Means Clustering + SVM yields Average Precision, Recall, Accuracy, and F-Measure values of 83.12, 86.21, 84.96, 84.68. However, the proposed method produces Average Precision, Recall, Accuracy, and F-Measure values of 93.35, 97.35, 95.28, 95.3, respectively. According to Fig. 2, the suggested system has the best performance metrics when compared to other baseline models.

As a result, depicted from Fig. 2 and Fig. 3 RRBI [2] yields a low range of Precision, Recall, F-Measure as it only combines the word to vector with cosine similarity nodes. Strategic language process schemes mainly word to vectorization with cosine similarities has been incorporated as a result there is no separate classification or clustering algorithm with vectorization which results in high FDR values [1] also yields a lower range of Precision, Recall, F-Measure and higher FDR because one-hot encoding is performed with vectorization, cosine similarity is mainly used for matching. The key or term based alone is followed since the approach only relies on term based matching and one hot coding for vectorization, it does not perform advantageously. However, in SVM + KNN the nearest neighbouring algorithm is incorporated, it is a combination of classifiers, as both SVM and KNN are naïve so as a result, It yields the lowest Precision, Recall, F-Measure, and highest FDR value. Naive Bayes and Collaborative Filtering is combined, again collaborative filtering is very unreliable because of item-item matchmaking which is based on item matrix. The items here in case recipe are being rated from the result as every item must be rated, so a major chance of repetition and there can be biases as well, As a result, biased recommendations are created, and the Naive Bayes classifier, which is a standard classifier used in conjunction with collaborative filtering, does not provide appropriate Precision, Recall, F-Measure, and FDR values. However, when Fuzzy C-Means clustering is combined slight improvement with the result is observed. As it gives a higher value of Precision, Recall, F-Measure SVM is a binary linear classifier and Fuzzy C-Means is a clustering algorithm.

It requires vectorization which makes the approach mathematically complex. At last, the proposed approach yields the highest Precision, Recall, F-Measure in comparison with baseline models mainly because it is ontology-driven as well as query driven. Ontologies which are generated from the dataset based on the increase of the good amount of semantic network to add auxiliary knowledge into the approach and the sematic alignment is quite strengthened using SemantoSim and KL-divergence using pair of approaches. Static similarity measures under SFLA helps in refining the initial population set into a final set of recommendable fragments. Moreover, the incorporation of RNN which is a deep learning classifier ensures with a high density of auxiliary knowledge that the recommendation of the recipe is up to the mark. While modelled thesaurus and Wikidata API ensures that background knowledge is consolidated in the initial query words. This amalgamation of auxiliary knowledge from modelled thesaurus as well as existing real-world knowledge ensures that the population is high with the density of knowledge associated with the query words and as a result, the proposed approach is much better than other baselined models. In Fig. 3. The column graph is plotted to demonstrate the comparison of False Discovery Rate of Proposed Approach and other baseline models. RBI, CSRS, SVM + KNN, Nave Bayes + Collaborative Filtering, Fuzzy C-Means Clustering + SVM, and Proposed Approach have FDR values of 0.15, 0.14,0.21, 0.19, 0.16, 0.06, respectively. The graph clearly shows that the proposed has the lowest FDR when compared to various baseline models and modifications. In the given approach an ontology-driven, and query-driven ontology is generated from the dataset based on the increase of goof amount of Semantic Network to add auxiliary knowledge moreover semantic alignment is strengthened with the use of SemantoSim and KL-Divergence under the SFLA algorithm which helps in refining population set into the final recommendable set and incorporation of RNN ensures that appropriate recommendation is given to the user.

The Precision Percentage of the Proposed Approach and the Baseline Models is shown in Fig. 4. We can see that the Proposed Approach is baselined using RRBI

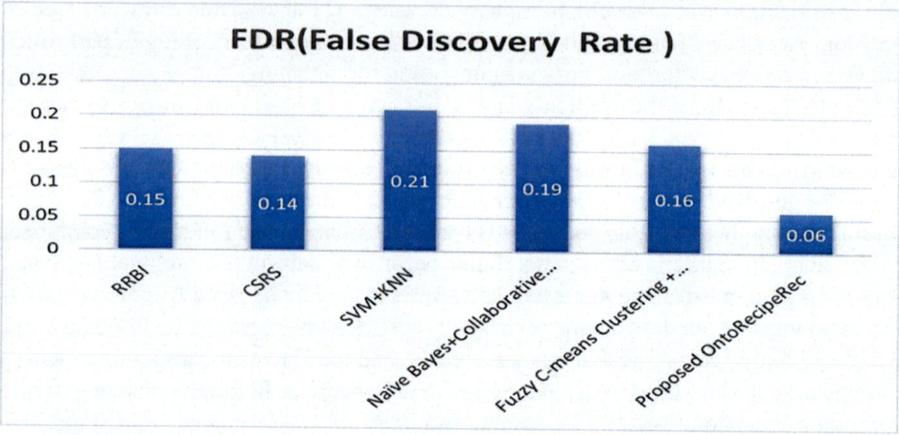

Fig. 3. FDR analysis of the proposed method concerning the existing paradigm models

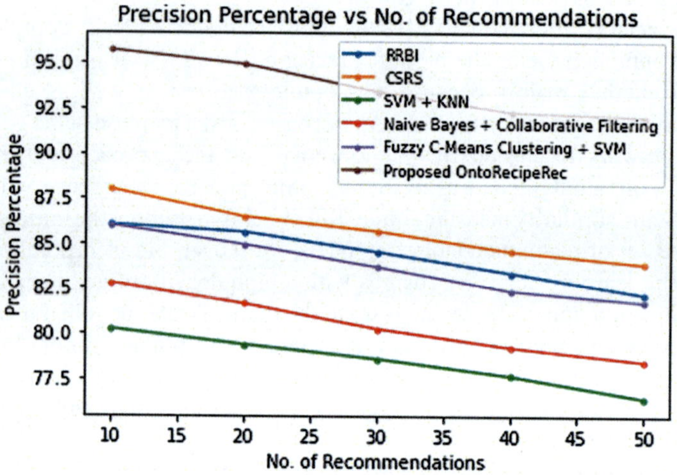

Fig. 4. Precision percentage of the proposed approach and existing paradigm models

[2], CSRS [1], SVM + KNN, Nave Bayes + Collaborative Filtering, Fuzzy C-Means Clustering + SVM, and the Proposed Approach. Analysing the graph's trend, the values of all techniques decline in general as the number of instances or variables that hold value rises, although the proposed technique is always larger than the existing mentioned approaches. When the precision per cent is measured in the number of terms of instances or entities, the maximum accuracy obtained is 95.69. Even when the quantity of variables is raised to 20, the accuracy percentage of the recommended system is 94.91. When the quantity of objects is raised to 50, the accuracy % increases to 93.42, 92.28, 92.04 when compared to less performing mentioned approaches.

5 Conclusions

Food preparation recipes are vital tools for everyone who wants to create nutritious and delicious meals at home. Even though many applications and websites feature large recipe compilations, individualized recommendations are rare. While most include some fundamental nutritional info such as calories, fat, and so on, they do not provide links to the food and nutrition composition libraries that are required. In this paper, a novel approach of Ontology-based Recipe Recommendation System which is query-driven along with the knowledge-centric approach that is incorporated by modelled thesaurus which contains multi-regional and different categories of food recipes. Because the suggested system is ontology and query-driven, and its semantic network is enlarged through the usage of SemantoSim and KL-Divergence, which are injected with deep learning in the form of RNN classifiers, the precision of 93.35% is greater when compared to other standard low performing mentioned models. Our findings suggest that the recommendation system has the computational power to handle vast amounts of recipe data.

References

1. Nilesh, N., Kumari, M., Hazarika, P., Raman, V.: Recommendation of Indian cuisine recipes based on ingredients. In: 2019 IEEE 35th International Conference on Data Engineering Workshops (ICDEW), pp. 96–99 (2019)
2. Moratanch, N., Gopalan, C.: A survey on extractive text summarization, pp. 1–6 (2017)
3. Du, Y., Zhao, W.: Named entity recognition method with word position. In: 2020 International Workshop on Electronic Communication and Artificial Intelligence (IWECAI), pp. 154–159 (2020)
4. Dhungana, U.R., Shakya, S., Baral, K., Sharma, B.: Word Sense Disambiguation using WSD specific WordNet of polysemy words. In: Proceedings of the 2015 IEEE 9th International Conference on Semantic Computing (IEEE ICSC 2015), pp. 148–152 (2015)
5. Elbeltagi, E., Hegazy, T., Grierson, D.: Comparison among five evolutionary-based optimization algorithms. Adv. Eng. Inform. **19**(1), 43–53 (2005)
6. Thakkar, P., Varma, K., Ukani, V., Mankad, S., Tanwar, S.: Combining user-based and item-based collaborative filtering using machine learning. In: Satapathy, S.C., Joshi, A. (eds.) Information and Communication Technology for Intelligent Systems. SIST, vol. 107, pp. 173–180. Springer, Singapore (2019). https://doi.org/10.1007/978-981-13-1747-7_17
7. Li, Z., Hu, J., Shen, J., Xu, Y.: A scalable recipe recommendation system for mobile application. In: 2016 3rd International Conference on Information Science and Control Engineering (ICISCE), pp. 91–94 (2016)
8. Yajima, A., Kobayashi, I.: Easy cooking recipe recommendation considering user's conditions. In: 2009 IEEE/WIC/ACM International Joint Conference on Web Intelligence and Intelligent Agent Technology, pp. 13–16 (2009)
9. Stahl, C., Gateau, B., Ferrini, K.: Experiments on the localisation of cooking recipes content using semantic food descriptions. In: 2020 15th International Workshop on Semantic and Social Media Adaptation and Personalization (SMA), pp. 1–5 (2020)
10. Sanjo, S., Katsurai, M.: Towards recommending diverse seasonal cooking recipes: a preliminary study based on monthly view data. In: 2017 IEEE International Symposium on Signal Processing and Information Technology (ISSPIT), pp. 306–310 (2017)
11. Wei, H., Zheng, H., Xu, Y.: A signalling pathway analysis model based on Kullback-Leibler divergence. In: 2015 International Symposium on Bioelectronics and Bioinformatics (ISBB), pp. 124–127 (2015)

12. Yue, M., Hu, T., Guo, B., Guo, X.: The research base on memetic meta-heuristic shuffled frog-leaping algorithm. In: 2009 2nd International Conference on Power Electronics and Intelligent Transportation System (PEITS), pp. 117–120 (2009)
13. Deepak, G., Kasaraneni, D.: Ontocommerce: an ontology focused semantic framework for personalised product recommendation for user targeted e-commerce. Int. J. Comput.-Aided Eng. Technol. **11**(4–5), 449–466 (2019)
14. Deepak, G., Rooban, S., Santhanavijayan, A.: A knowledge-centric hybridized approach for crime classification incorporating deep bi-LSTM neural network. Multimed. Tools Appl. 1–25 (2021)
15. Deepak, G., Santhanavijayan, A.: UQSCM-RFD: a query–knowledge interfacing approach for diversified query recommendation in semantic search based on river flow dynamics and dynamic user interaction. Neural Comput. Appl. 1–25 (2021)
16. Krishnan, N., Deepak, G.: Towards a novel framework for trust driven web URL recommendation incorporating semantic alignment and recurrent neural network. In: 2021 7th International Conference on Web Research (ICWR), pp. 232–237. IEEE (2021)
17. Roopak, N., Deepak, G.: OntoKnowNHS: ontology driven knowledge centric novel hybridised semantic scheme for image recommendation using knowledge graph. In: Villazón-Terrazas, B., Ortiz-Rodríguez, F., Tiwari, S., Goyal, A., Jabbar, M. (eds.) Knowledge Graphs and Semantic Web. KGSWC 2021. CCIS, vol. 1459, pp. 138–152. Springer, Cham (2021). https://doi.org/10.1007/978-3-030-91305-2_11
18. Ojha, R., Deepak, G.: Metadata driven semantically aware medical query expansion. In: Villazón-Terrazas, B., Ortiz-Rodríguez, F., Tiwari, S., Goyal, A., Jabbar, M. (eds.) Knowledge Graphs and Semantic Web. KGSWC 2021. CCIS, vol. 1459, pp. 223–233. Springer, Cham (2021). https://doi.org/10.1007/978-3-030-91305-2_17
19. Surya, D., Deepak, G.: USWSBS: user-centric sensor and web service search for IoT application using bagging and sunflower optimization. In: Noor, A., Sen, A., Trivedi, G. (eds.) Proceedings of Emerging Trends and Technologies on Intelligent Systems . ETTIS 2021. AISC, vol. 1371, pp. 349–359. Springer, Singapore (2022). https://doi.org/10.1007/978-981-16-3097-2_29
20. Arulmozhivarman, M., Deepak, G.: OWLW: ontology focused user centric architecture for web service recommendation based on LSTM and whale optimization. In: Musleh Al-Sartawi, A.M., Razzaque, A., Kamal, M.M. (eds.) Artificial Intelligence Systems and the Internet of Things in the Digital Era. EAMMIS 2021. LNNS, vol. 239, pp. 334–344. Springer, Cham (2021). https://doi.org/10.1007/978-3-030-77246-8_32
21. Adithya, V., Deepak, G., Santhanavijayan, A.: HCODF: hybrid cognitive ontology driven framework for socially relevant news validation. In: Motahhir, S., Bossoufi, B. (eds.) Digital Technologies and Applications. ICDTA 2021. Lecture Notes in Networks and Systems, vol. 211, pp. 731–739. Springer, Cham (2021). https://doi.org/10.1007/978-3-030-73882-2_66
22. Rithish, H., Deepak, G., Santhanavijayan, A.: Automated assessment of question quality on online community forums. In: Motahhir, S., Bossoufi, B. (eds.) Digital Technologies and Applications. ICDTA 2021. LNNS, vol. 211, pp. 791–800. Springer, Cham (2021). https://doi.org/10.1007/978-3-030-73882-2_72
23. Aditya, S., Muhil Aditya, P., Deepak, G., Santhanavijayan, A.: IIMDR: intelligence integration model for document retrieval. In: Motahhir, S., Bossoufi, B. (eds.) Digital Technologies and Applications. ICDTA 2021. LNNS, vol. 211, pp. 707–717. Springer, Cham (2021). https://doi.org/10.1007/978-3-030-73882-2_64
24. Tiwari, S., Al-Aswadi, F.N., Gaurav, D.: Recent trends in knowledge graphs: theory and practice. Soft. Comput. **25**(13), 8337–8355 (2021). https://doi.org/10.1007/s00500-021-05756-8
25. Abhishek, K., Pratihar, V., Shandilya, S.K., Tiwari, S., Ranjan, V.K., Tripathi, S.: An intelligent approach for mining knowledge graphs of online news. Int. J. Comput. Appl. 1–9 (2021)

26. Deepak, G., Santhanavijayan, A.: OntoBestFit: a best-fit occurrence estimation strategy for RDF driven faceted semantic search. Comput. Commun. **160**, 284–298 (2020)
27. Manoj, N., Deepak, G.: ODFWR: an ontology driven framework for web service recommendation. In: Shukla, S., Unal, A., Kureethara, J.V., Mishra, D.K., Han, D.S. (eds.) Data Science and Security. LNNS, vol. 290, pp. 150–158. Springer, Singapore (2021). https://doi.org/10.1007/978-981-16-4486-3_16
28. Roopak, N., Deepak, G.: KnowGen: A knowledge generation approach for tag recommendation using ontology and honey bee algorithm. In: European, Asian, Middle Eastern, North African Conference on Management & Information Systems, pp. 345–357 (March 2021)

Prediction of Water Quality Using Artificial Intelligence (AI) and Statistical Approach

Chaimae Zai(✉), Chaymae El Mechal, Najiba El Amrani El Idrissi, and Hicham Ghennioui

Signals, Systems and Components Laboratory (SSCL), Faculty of Science and Technology, University Sidi Mohamed Ben Abdellah, Fez, Morocco
{chaimae.zai,najiba.elamrani,hicham.ghennioui}@usmba.ac.ma

Abstract. Water is a valuable resource on the earth. To be consumed safely, it must be treated. But the increasing pollution of the reserves makes this operation more and more delicate. Accordingly, water quality modeling and prediction have become critical in tackling pollution. The goal of this study is to use an advanced artificial intelligence (AI) system to estimate the water quality index (WQI) for the surface. We aim through this research to generate a non-supervised machine learning Model that can predict the value of target variable inferred from available data (training data), as a soft computing technique: Decision Trees for Classification was used to predict the potability of water. It provides a good performance in this study. A big dataset, collected monthly over a period of 5 years, was employed in the modeling process. The data were splited into sets, one for model building (training dataset) and the other for model validation (testing dataset) to avoid over-learn. Statistical assessment indicators and visual evaluation Metrics estimated the prediction capacity of the developed model. The outcome demonstrates that DTC was opportune for the forecast of the potability indice and would be functional for water purification organizations in the management of divers water supply systems ($R^2 = 72\%$, RMSE = 0.41). The models developed in this study could serve as a foundation for decision-making that will aid in the maintenance and improvement of water supply system management.

Keywords: Surface water quality · Parametric study · Regression analysis · Artificial intelligence

1 Introduction

Water pollution is a huge problem that endangers human health, agriculture, and ecosystems. Therefore, it is critical to model surface water quality utilizing soft computing techniques as this could help identify the most effective ways of managing water resources. Even though there has been a lot of research on water quality evaluations based on deterministic and stochastic approaches, there is relatively little research on the use of machine learning techniques such as non-parametric supervised learning methods, especially in developing nations in Africa.

S. Motahhir and B. Bossoufi (Eds.): ICDTA 2022, LNNS 454, pp. 34–42, 2022.
https://doi.org/10.1007/978-3-031-01942-5_4

As globalization is moving to the fourth industrial transition, Digital technology, the internet of things (IoT), artificial intelligence (AI), and other technologies are becoming increasingly important to innovate and acquire the capabilities to rapidly adapt to changing circumstances. More crucially, African emerging countries like Morocco have yet to implement and verify machine learning's efficacy in water supply system management. As a result of this research, practitioners will be more aware of the robustness of the methodologies used.

According to the literature review, the water quality index (WQI) is commonly determined by the complex assessment of multiple WQ characteristics such as biochemical oxygen demand, dissolved oxygen, and total suspended solids. The WQI is a basic index that is used to identify and effectively solve water pollution issues [1] [2]. Then, effective steps to improve water quality will be implemented, resulting in improved human and ecosystem health. The WQI necessitates a lengthy conversion to estimate the sub-index [3], which requires a significant amount of time and effort to incorporate several equations to estimate the final index. Furthermore, the problem also carries the risk of accidental errors in estimating the index [4, 5].

To forecast the total dissolved solids of aquifers, the authors in Ref. [6] employed tree models of machine learning which are, support vector machines (SVMs), Fuzzy Interface System (ANFIS), and artificial neural network (ANN) models. These models were trained using cat swarm optimization (CSO), moth flam optimization (MFO), particle swarm optimization (PSO), shark algorithm (SA), grey wolf optimization (GWO), and gravitational search algorithm (GSA). The acquired results reveal that the ANFIS-MFO and ANFIS-CSO models outperformed the other models, with the ANFIS-MFO further reducing the RMSE by roughly 3% and 7%, respectively, when compared to the ANN-MFO and SVM-MFO.

The study presented in Ref. [7] used an ANFIS model for prediction of total dissolved solids (TDS), with $Mg2^{+}$, $HCO3^{-}$, and SO4 2^{-} as input parameters, the result affirms that the ANFIS model had an acceptable accuracy in the prediction of TDS with a root mean squared error (RMSE) of 16.7 ppm.

The research in Ref. [8] produced ANN models to predict WQI as the nonlinear autoregressive neural network (NARNET) and long short-term memory (LSTM). A NARNET model has the following properties: There are 12 hidden layers, 1: 8 delays, and 12 epochs. The LSTM model, on the other hand, comprises a total of 200 hidden layers, 150 epochs, and delays of [1, 3, 4, 7]. Based on the obtained R (Pearson's correlation coefficient approach) with regression coefficients; RNARNET = 96:17% and RLSTM = 94:21%, the NARNET model performed slightly better than the LSTM model.

Other models of machine learning algorithms were employed in Ref. [9] for estimating the WQI, they used the ANN and a Multi Linear Regression (MLR) as the conventional linear model. In the verification phase, they discovered that the ANN showed a significant improvement in performance accuracy over MLR, up to 10%.

Another study examined ANN, MLR, and gene expression programming (GEP) for simulating monthly total dissolved solids (TDS). The proposed GEP model had successfully taken the effect of all the modeling inputs. They employed four metrics to assess the models' performance: NSE, R2, MAE, and RMSE, which were determined to be 0.96, 96%, 6.58, 7.10 for model training data and 0.99, 99%, 1.38, 1.57 for testing data, respectively. For more explanation, refer to [10].

In terms of performance, GEP outclasses other models (ANN and LRM) during the training as well as the testing phase. The littlest RMSE values were also achieved by GEP, which underlined its overwhelming dominance. In terms of the ANN model's performance, it demonstrated high prediction capabilities during training but its performance was reduced on the testing dataset. This result was being approved by [11], in their study. They found that the hidden layer's total number of nodes and the transfers functions affect ANN model learning. For this purpose, The number of nodes should be chosen carefully, as a greater number can lead to an over-fitting, while a lesser number may not allow for effective data gathering, therefore the model could not perfectly predict the target variable.

The authors applied in [12] three different kinds of advanced machine learning approaches to predict TDS as water quality, which are GPR, BPNN, and PCR models, the efficiency and accuracy of each model was assessed by measuring the Evaluation Metrics; R^2, RMSE, MAE, Nash, and ENS. The results indicated that GPR and BPNN models had the lowest error in predicting compared to the PCR model. Respectively, gave an ENS of (0.967, 0.915, 0.874).

2 Study Area and Materials

2.1 Study Area

About 3270 data points obtained between 2016 and 2020, were used in this study. The data was collected from the Sebou basin, one of the most important basins in the Moroccan kingdom. Figure 1 depicts the study area's location and geological context. The area was selected because of the heavy- amount of pollution of water distribution networks from mining activities [13]. With an area of about 40,000 km^2 which currently

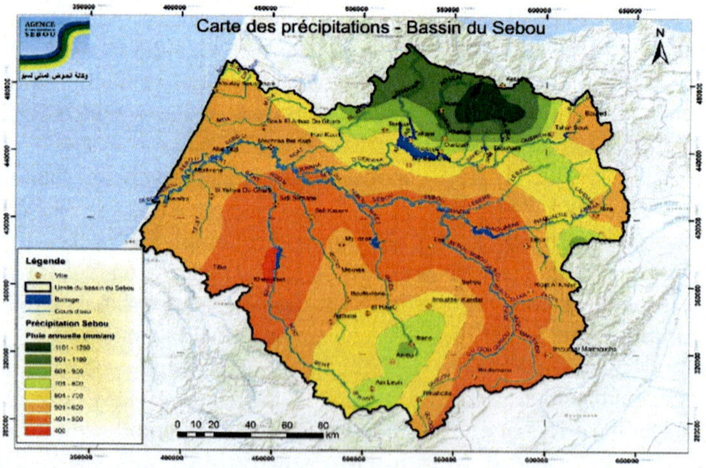

Fig. 1. The study area is depicted on a map [14].

has a population of nearly 6.2 million inhabitants (2004 census), 49% of them live in urban areas and 51% in rural areas. It's home to an agricultural and industrial economy that contributes significantly to the national economy.

Water is a breeding ground for viruses, parasites, bacteria, and insect vectors diseases and due to the growth of population and industrialization, the water resources become more polluted. Waste industrial, human wastes, automobile wastes, agricultural runoff from farmlands containing chemical factors, and other wastes can affect negatively the quality of the water.

2.2 Data Description

In this investigation, data measurements for the Sebou basin monitoring station were used, where water samples were collected and monitored using various methodologies. Since 2016, Data has been acquired with the time interval to carry out an efficient prediction process using this time-series data that includes parameters and their measurements along. The table below indicates Basic statistical analysis metrics for input parameters applied in this model. The potability of water was considered as the target parameter of the entire modeling because its value is affected by features parameters. The prediction model was built using the remaining 9 parameters as input parameters (Table 1).

Table 1. Descriptive table of statistical summary of parameters.

	PH	°TH	Solids	NH2Cl	SO2	TDS	TOC	TTHM	Turbidity
Count	2800	3122	3246	28761	2495	3276	3276	3114	3176
Mean	7.080	197.3	22014	7.122	333	426	14.24	66.39	3.966
Std	1.494	32.87	8768	1.583	41.4	80.8	3.303	16.17	0.780
Min	0.000	47.43	320	0.352	129	181	2.200	0.737	1.450
25%	5.393	176.8	15666	6.127	307	365	12.06	55.83	3.438
50%	6.926	196.9	20927	7.130	333	421.8	14.21	66.69	3.959
75%	6.001	216.6	27332	8.114	359	481.7	16.55	77.31	4.500
Max	17.00	323.1	61227	13.12	481	753.3	28.30	124.8	6.731

2.3 Methodology and Performance Evaluation of Models

Model of ML.

Decision Trees (DTs).

Modeling methods are the best alternatives for reducing the experimental burden for the estimation of water quality parameters. Using computer-aided models to depict features of water are useful in saving time, cost and making suitable predictions. In this study, we employed **DTs,** a non-parametric supervised learning method for classification and regression. The purpose is to develop a model that predicts the value of the target variable, based on a set of sequences of decision rules deduced from the training data. After training, the Decision Tree makes decision based on certain conditions through a graphical representation of all relevant input parameters. It chooses the root variable based on entropy and then looks at the If the set of predicted values of the other parameters. It shapes a top-down diagram that adjusts all of the parameter nodes and projects them based on different parameter values (Fig. 2).

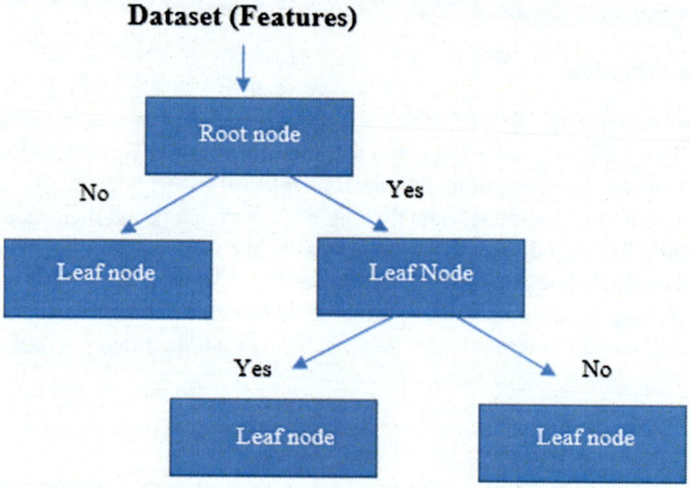

Fig. 2. The architecture of the DTs model.

Model performances measurement.

Generally, when working on classification tasks we should refer to some metrics and tools to get numerical values that show whether a model still needs training or is ready to be implemented. In this context, we choose to assess the network performance with three metrics, pursuant correlation coefficient (R2), which denotes the rate of alliance between Y_i and $\hat{Y}_i$; Root Mean Square Error (RMSE), which is preferred in many iterative predictions and optimization schemes. The Mathematical expressions for these measures are given below:

$$RMSE = \sqrt{\frac{\sum_{i=1}^{n} \left(Y_i - \hat{Y}_i\right)^2}{n}} \tag{1}$$

$$R^2 = \frac{\sum_{i=1}^{n} \left(Y_i - \hat{Y}_i\right)\left(Y_i - \overline{y}_i\right)}{\sqrt{\sum_{i=1}^{n} \left(Y_i - \hat{Y}_i\right)^2 \sum_{i=1}^{n} \left(Y_i - \overline{y}_i\right)^2}} \tag{2}$$

Where n is the number of the observations, $\boldsymbol{Y_i}$ the real outcome, $\boldsymbol{\hat{Y}_i}$ the predicted outcome and the $\overline{y}_i$ represents the variable's average.

Methodology.
In this study, we have implemented a water quality prediction using a machine-learning technique. The followed methodology is depicted in Figure below (Fig. 3);

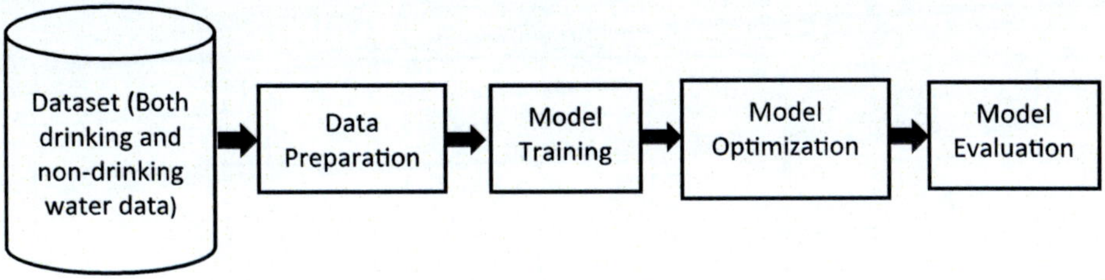

Fig. 3. Schematic representation of the resume methodology.

Data preprocessing.
Data cleaning is a critically important step in any machine-learning project, therefore we must apply some methods like statistical functions to acquire the smallest relevant subset that qualifies high precision and to remove inaccurate data, and then to identify data cleaning operations that we may want to perform. The data used was cleaned by using a **Dataframe.fillna**, **correlation heatmap,** and **box plot analysis**, as explained below.

Firstly, we used the **Dataframe filna** from Pandas library to replace the 'NaN' in the data frame, we have then used the **correlation heatmap** to quickly check correlation columns by visualizing the correlation matrix as a heatmap, and thereafter optimize algorithm performances (Fig. 4).

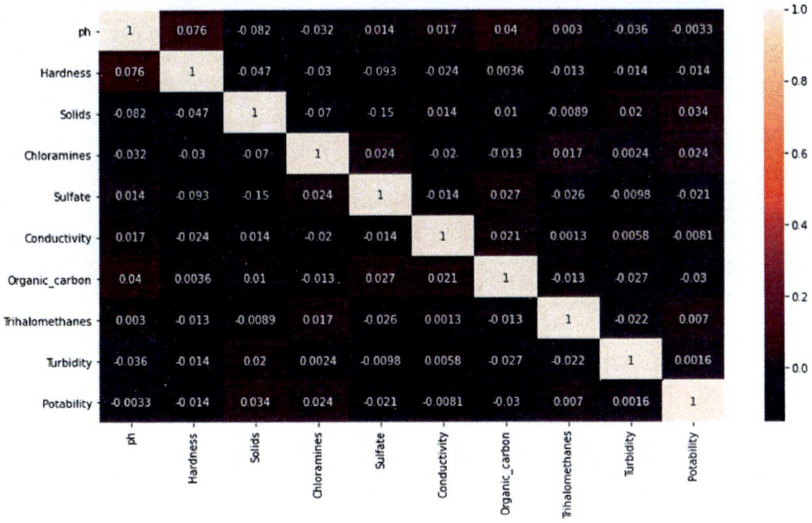

Fig. 4. Correlation heatmap of dataset.

For outlier detection, we have used the **box plot analysis** since most of the factors were sufficiently variable, and a boxplot gives an intuitive depiction for determining outliers. Most parameters were found to lay outside the box, indicating that outliers were normal (Fig. 5).

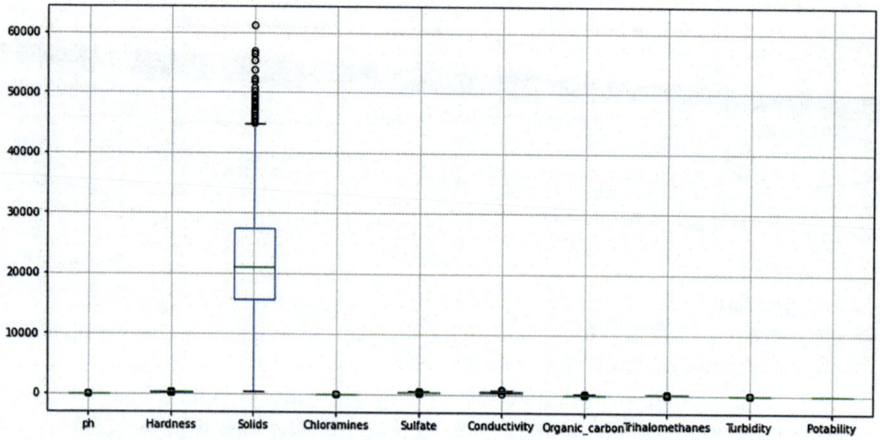

Fig. 5. Outlier detection using box plot analysis.

An example of the normalized distribution used in the present study is shown in Fig. 6. It describes the Q-value charts for two water quality parameters.

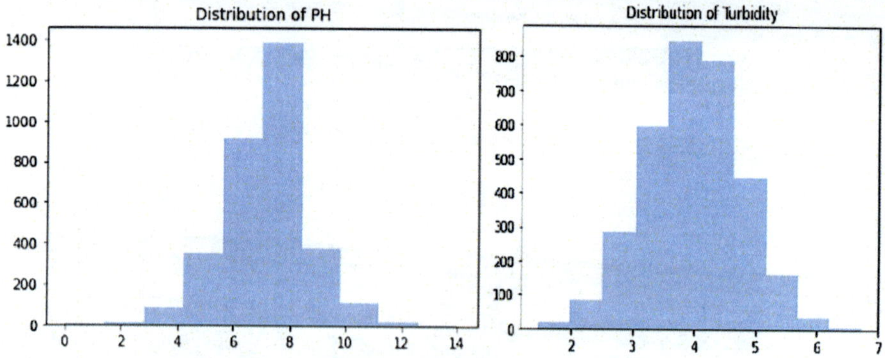

Fig. 6. Q-value normalization of data.

Model training.
After having prepared the Data, we applied the machine-learning model; except that before projecting the Decision Tree Classifier the supplied data must be split in order to train the model, test it with a certain part of the data and check the accuracy system response to substantiate the model's performance.

We have used the Accuracy score, with the following formula:

$$Accuracy\left(Y_i, \hat{Y}_i\right) = \frac{1}{N}\sum_{i=1}^{n} 1\left(Y_i - \hat{Y}_i\right) = 0.5 \tag{3}$$

The function returns the subset accuracy in multilabel classification. The subset accuracy is 1.0 if the set of projected labels perfectly matches the true set of labels; otherwise, it is 0.0.

Hyperparameters.
In order to increase system performance, we added **Cross-Validation**. it splits the data into k subdivisions and switches over all of them, with k-1 fragment useable for the training dataset and the last subset serving for testing, the process will be projected in the same way; until we reach iteration k as a test subdivision data, this function (**CV**) secured an efficient split. Given the iterations, this is typically computationally expensive, but our study employs a tiny dataset, which is typically the case with quality, therefore cross-validation is very appropriate for this problem. We did cross-validation on the data after dividing it into k = 6 subsets. Because the entire training set has 545 items, we made sure that each fold subset, including the test set, had at least 100 samples.

3 Result and Discussion

Before moving on to the results. we will evince the performance indicators used to rate the precision of the used machine learning algorithms. To evaluate our network, two metrics were followed in this research. First, R squared error (RSE), also known as the coefficient of determination, and often denoted as R2, in this study we have found $R^2 = 72\%$ and RMSE = 0.41 which denotes excellent performances in the proposed models for forecasting IWQ showing to other models that have been proposed in the literature. We recommend that future research should include the prototype of the data acquisition and the implementation in an electronic card such as Raspberry Pi Furthermore, the network performance can be improved by tuning the network's hyper-parameters.

4 Conclusion

According to a review of the literature, each machine-learning model has its own strengths and weaknesses, and its general behavior is influenced by many parameters. In the context of water quality and According to our results, the reason for the suitability of the Decision Tree model is the use of simple linear equations that provide the concept of visualization Data to estimate the quality of water. As a result, the findings of this work imply that sophisticated techniques and machine learning approaches are suitable for indicators of water quality prediction. Furthermore, The model established in this work might be used to help purification organizations makes better decisions in the future.

References

1. Tyagi, S., Sharma, B., Singh, P., Dobhal, R.: Water quality assessment in terms of water quality index. Am. J. Water Resour. **1**, 34–38 (2020). https://doi.org/10.12691/ajwr-1-3-3
2. Kumar, A., Dua, A.: Water quality index for assessment of water quality of river ravi at madhopur (INDIA) (2009)
3. Gazzaz, N.M., Yusoff, M.K., Aris, A.Z., Juahir, H., Ramli, M.F.: Artificial neural network modeling of the water quality index for Kinta River (Malaysia) using water quality variables as predictors. Mar. Pollut. Bull. **64**, 2409–2420 (2012). https://doi.org/10.1016/j.marpolbul.2012.08.005
4. Pham, Q.B., et al.: Application of soft computing to predict water quality in wetland. Environ. Sci. Pollut. Res. **28**(1), 185–200 (2020). https://doi.org/10.1007/s11356-020-10344-8
5. Mohammadpour, R., Shaharuddin, S., Chang, C.K., Zakaria, N.A., Ghani, A.A., Chan, N.W.: Prediction of water quality index in constructed wetlands using support vector machine. Environ. Sci. Pollut. Res. **22**(8), 6208–6219 (2014). https://doi.org/10.1007/s11356-014-3806-7
6. Banadkooki, F.B., Ehteram, M., Panahi, F., Sammen, S.S., Othman, F.B., EL-Shafie, A.: Estimation of total dissolved solids (TDS) using new hybrid machine learning models. J. Hydrol. **587** (2020). https://doi.org/10.1016/j.jhydrol.2020.124989
7. Shah, M.I., et al.: Modeling surface water quality using the adaptive neuro-fuzzy inference system aided by input optimization. Sustainability (Switzerland). **13** (2021). https://doi.org/10.3390/su13084576
8. Aldhyani, T.H.H., Al-Yaari, M., Alkahtani, H., Maashi, M.: Water quality prediction using artificial intelligence algorithms. Appl. Bionics Biomech. **2020** (2020). https://doi.org/10.1155/2020/6659314
9. Gaya, M.S., et al.: Estimation of water quality index using artificial intelligence approaches and multi-linear regression. IAES Int. J. Artif. Intell. **9**, 126–134 (2020). https://doi.org/10.11591/ijai.v9.i1.pp126-134
10. Izhar Shah, M., Alaloul, W.S., Alqahtani, A., Aldrees, A., Ali Musarat, M., Javed, M.F.: Predictive modeling approach for surface water quality: development and comparison of machine learning models. Sustainability (Switzerland). **13** (2021). https://doi.org/10.3390/su13147515
11. Emamgholizadeh, S., Kashi, H., Marofpoor, I., Zalaghi, E.: Prediction of water quality parameters of Karoon River (Iran) by artificial intelligence-based models. Int. J. Environ. Sci. Technol. **11**(3), 645–656 (2013). https://doi.org/10.1007/s13762-013-0378-x
12. Ewusi, A., Ahenkorah, I., Aikins, D.: Modelling of total dissolved solids in water supply systems using regression and supervised machine learning approaches. Appl. Water Sci. **11**(2), 1–16 (2021). https://doi.org/10.1007/s13201-020-01352-7
13. QUALITE DES EAUX
14. Eaux de surface–ABHS. http://www.abhsebou.ma/presentation-du-bassin/eaux-de-surface/

Vocal Parameters Analysis for Amazigh Phonemes Recognition System

Ahmed Ouhnini[1](✉), Brahim Aksasse[2], and Mohammed Ouanan[2]

[1] Faculty of Sciences and Technics, Moulay Ismail University, Errachidia, Morocco
ouhnini.ah@gmail.com

[2] MAIS Laboratory, Faculty of Science, Moulay Ismail University, Meknes, Morocco
b.akssasse@fste.umi.ac.ma, m.ouanan@umi.ac.ma

Abstract. Several studies in the development of Automatic Speech Recognition (ASR) systems have been yet presented and well studied for many languages. However, the particularities of the Amazigh Language due to its specific acoustic features such as the emphasis on Amazigh consonants pose many problems for developing its automatic speech recognition system. In this paper, we deal with the spoken recognition problem in Amazigh Language. We consider here a particular case of the problem, focusing on the phonemes formants frequencies which are a principal variable in specifying the phonetic substance of discourse sounds. We propose an approach allowing the vocal parameters analysis of phonetic units, as well as a recognition algorithm by analyzing the formants.

Keywords: Automatic speech recognition · Amazigh language · Artificial intelligence · Corpus · Formants frequency · Phoneme duration

1 Introduction

In order to improve human-machine (HM) communication, Automatic Speech Recognition (ASR) is used to analyze the human voice picked up by a microphone and transcribe it into machine-readable text. ASR is an active research area since a long time. For a long time, the formant frequencies have been considered as a principal variable when it comes to determining the phonetic content of speech sounds. A complex process from the production of speech to the determination of spelling, background noise, and sometimes significant differences according to accents and dialects within the same language, the inability to deal with bilingualism, a common situation in some countries, as well as very high computational needs, making real-time recognition very difficult [1]. We consider here a specific instance of the issue, focusing on the phonemes formants frequencies, which are a critical factor in determining the phonetic substance of discourse sounds [14]. We believe that this work will demonstrate the potential of vocal parameters analysis in phoneme recognition.

We propose an approach allowing the vocal parameters analysis of phonetic units, as well as a recognition algorithm by analyzing the formants.

S. Motahhir and B. Bossoufi (Eds.): ICDTA 2022, LNNS 454, pp. 43–53, 2022.
https://doi.org/10.1007/978-3-031-01942-5_5

The rest of the paper is as follows; in section two, we present some characteristics of the human voice in general and for the Amazigh language in particular. Further, in section three, we give some technological tools used for speech modeling followed by the proposed method for phoneme recognition. Furthermore, section four presents some results and finally, section five concludes the paper.

2 Phonetics and Speech Characteristics

2.1 The Human Hearing and Sound

Sound is a wave disseminating through a material medium, as little varieties pressure. It is detectable by the human ear when the frequency is between 20 Hz and 20 kHz. Nevertheless, it is determined that the phonetic information is less than 10 kHz [15]. When it comes to a phone call, frequencies over 3.5 kHz are cut, however, the discussion stays comprehensible, while some high-pitched phonemes are intense, similar to [S] are in effect gravely delivered.

From a physical point of view, it is normal in signal handling to simply treat amplitude and shift stage spectra [16]. This is due to the manner in which sound signs are sifted, we basically change the sufficiency range on the grounds that the ear is minimal delicate to stage distortions. This legitimizes us just only be keen on the accompanying applications module.

2.2 The Voice of Human

The voice of humans results from the interaction among breath and the diverse phonatory organs joint effort. In voiced phonemes case, firstly the sound is delivered by the vocal cords vibration [12]. Depending on the cavities it passes through, it is manipulated differently, mainly the pharynx and mouth. These cavities act as resonators which reinforce for some phonemes the frequencies equivalent to their resonance frequency. These reinforced frequencies are referred to as "formants", and they are the ones that phonologists attempt to identify in a spectrogram in order to identify the pronounced phonemes. It's worth noting that the nasal cavity attenuates the formants of the so-called nasalized phonemes, and this is why we speak of the phenomenon of anti-resonance [2].

These phonatory organs are not fixed, their shapes vary according to the articulation mode specific to a specific phoneme, and their reverberation frequencies, hence the appearance of various formants for every phoneme. Practically, we consider the first four formants, the initial two being the most identified with the mode of articulation. In our study, we choose to select four of them, designated by F1, F2, F3, and F4.

The following figure is a representation of the first three formants (F1, F2, F3) overlayed on the spectrogram of the voice signal «he took holidays» where there is an alternation of voiced and unvoiced sounds [13]. A formant structure is presented, in the voiced case (Fig. 1).

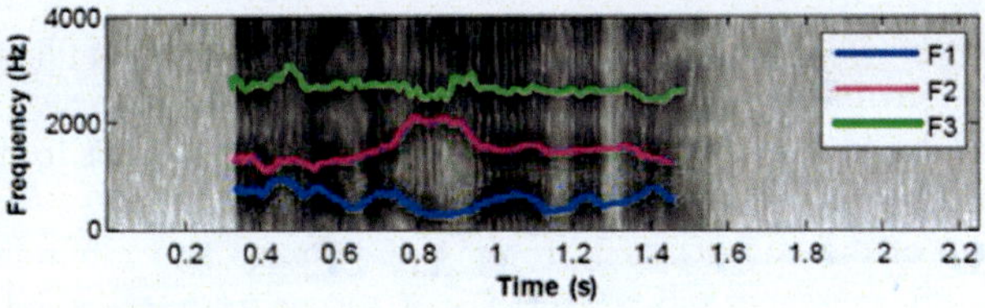

Fig. 1. Representation of the first three formants F1, F2 and F3.

2.3 Phonology and Phonetics

The phonetics is the study of phonemes, the last option being regularly characterized as the littlest particular phonetic unit. The opposition of the words bread and bath, for instance, demonstrates that [p] and [b] are phonemes. They are divided into several classes and subclasses, the most important of which are "vowels" and "consonants" [17].

Vowels relatively are long-duration sounds, characterized by a free flow and a great constancy of their frequency properties over time: the formants on a spectrogram are horizontal in the absence of overly marked prosodic elements. Vowels are frequently classified using the vowel trapezius, replicated in Fig. 2 below, which represents their position in the plane defined by the first two formants and reflects the position of the language during articulation.

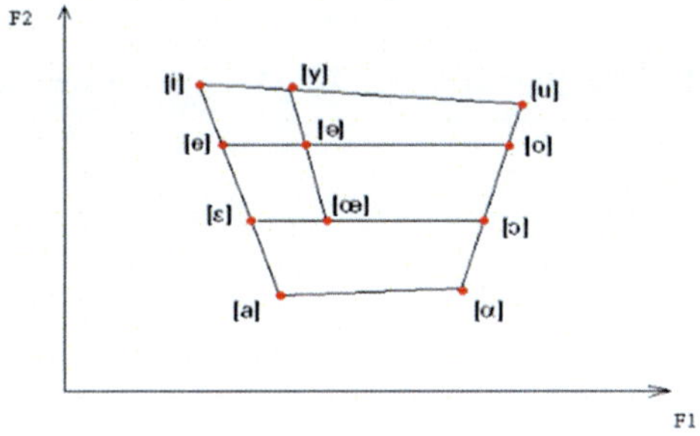

Fig. 2. The vowels classification vocal trapezium.

On the other hand, the consonants are phonemes, when articulated, encounter an obstacle (lips vowels for labial, teeth for dental, closing the palate for[k], etc.). They are typically much shorter than vowels and have a greater range of variation over time. They can be both noisy and sonorous. They only present formants in the latter case.

2.4 Frequency Characteristics of the Speech Signal

The bandwidth of the speech signal is much higher than 4 kHz, that of the telephone, and it contains all of the information required to understand the human voice [10].

The speed at which the vocal cords open and close during the phonation process, produces a vibration of a variable pitch called the fundamental frequency whose value is closely related to the size of the person's phonatory system. it varies from 80 to 600 Hz for a voice depending on age and gender.

The spectrum of the vocal signal presents maximums and minimums that correspond to the resonances and anti-resonances of the vocal tract called formants and anti-formants.

The spectrogram is a three-dimensional representation in which time is represented on the X-axis, frequency is represented on the Y-axis, and the level of each frequency on the Z-axis (symbolized by the grey level). To obtain it, an FFT (Fast Fourier Transform) with a sliding window is performed on the signal [11].

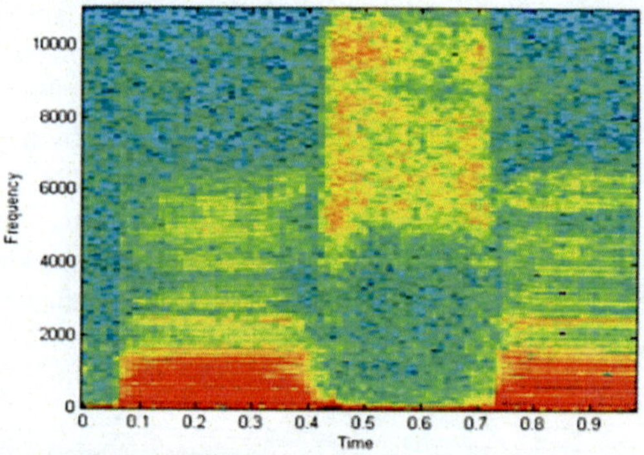

Fig. 3. Analysis of the voice signal of the syllable (asa).

2.5 The "Amazigh" Language

The Amazigh language, also known as Berber or Tamazight, is one of humanity's oldest languages. It now encompasses the northern part of Africa, extending from the Red Sea to the Canary Islands, and from Niger in the Sahara to the Mediterranean Sea. Due to historical, geographical, and sociolinguistic factors, this language is divided into three major regional varieties in Morocco, depending on the area and the communities: Tarifite in the north, Tamazight in Central Morocco, and the south-east, and Tachelhite in the south-west and the High Atlas. Even though the fact that Amazigh is spoken by 50% of the Moroccan population, the language has been exclusively reserved for familial and informal domains: Boukous (1995) [3]. However, thanks to the royal benevolence, this language has recently become institutionalized and integrated into the Moroccan educational system.

The official Moroccan Amazigh alphabetical system, known since February 2003 as Tifinaghe-IRCAM, is being implemented in Moroccan school curricula and is used in Amazigh historical research. This system contains the following alphabet [4] (Table 1):

Table 1. Amazigh language alphabet.

27 consonants	labials	ⴼ, ⴱ, ⵎ
	dentals	ⵜ, ⴷ, ⴹ, ⵟ, ⵏ, ⵔ, ⵕ, ⵍ
	alveolars	ⵙ, ⵣ, ⵚ, ⵥ
	palatals	ⵛ, ⵊ
	velar	ⴽ, ⴳ
	labiovelars	ⴽⵯ, ⴳⵯ
	uvulars	ⵇ, ⵅ, ⵖ
	pharyngeals	ⵄ, ⵃ
	laryngeal	ⵀ
2 semi-consonants	ⵢ and ⵡ	
4 vowels	full vowels	ⴰ, ⵉ, ⵓ
	neutral vowel (or schwa)	ⴻ which has a rather unique status in Amazigh phonology.

The words correct spelling, in Latin characters, very close to phonetic transcription, accurately reflects the pronunciation, including vowels and twinned.

Example: /illa/—> «illa» —> «il est», «il existe». Mekhlouf et al. [5] (Fig. 4).

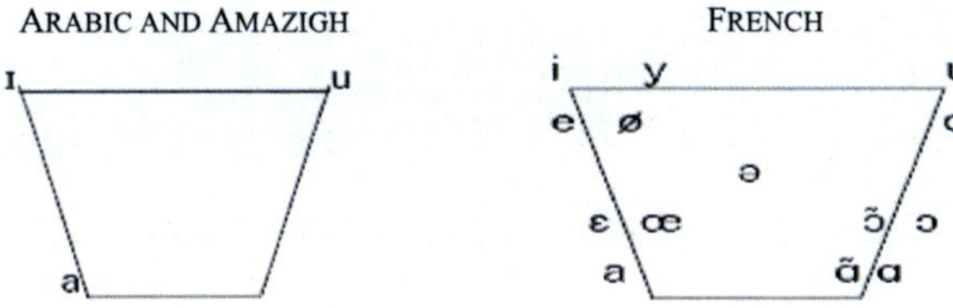

Fig. 4. Comparison of the Arabic, the Amazigh, and the French vocal systems.

The Amazigh language differs in pronunciation from region to region.

3 Technologies and Methods

3.1 Tools

PRAAT [6] is an open source computer software used by phoneticians and researchers to determine different phonetic characteristics of speech. It is a versatile program for analyzing and reconstructing acoustic speech signals. It analyzes speech, performs synthesis, manipulation, labeling and segmentation, graphics, and has a variety of other functionalities [7].

PRAAT was used to record and analyze the wav files in order to obtain all of the parameters presented in this work.

3.2 Methodology of Measurement

Pitch and four formants are the cornerstones of acoustic voice signal measurement and are frequently used as perceived voice quality indices [8] (Table 2 and Fig. 5).

Table 2. Amazigh language vowels were used in this study.

Tifinaghe	English transcription	Latin correspondence	Arabic transcription	IPA
ⴰ	YA	A	يا	Æ
ⵉ	YI	I	يي	I
ⵓ	YO	U	يو	ʊ

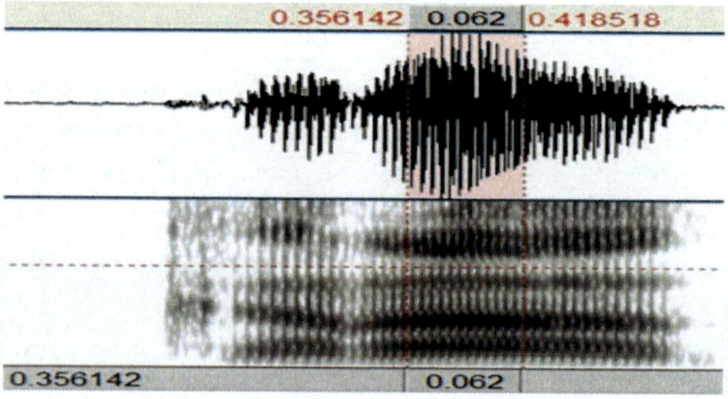

Fig. 5. Manual choice of vowel A from waveform and spectrogram of Krad digits

Corpus Preparation. Ten adults (5 males and 5 females) were selected from a voice database of Moroccan Amazigh native speakers come from all over the country, with no particular geographical allocation. The subjects were divided into groups based on their ages. The women's ages ranged from 23 to 50 years, with a mean of 35 years. The men age range from 25 to 50 years old, with a mean age of 36 years. The sequence was repeated ten times by all speakers. The total number of words analyzed is (10 speakers * 8 words * 10 repetitions), which gives us 800 files to analyze.

Our goal is to study the important voice parameters in order to analyze the pronounced consonants, semi-consonants, and vowels. The vowels A, I, and U were extracted manually from the spectrograms of Krad, Tanmirt, and Ayur. The consonants R, T, and Z are extracted from the words Aghrum, Attas, and Tazalit respectively. The following table shows more details about the database.

Table 3. Recording parameters used for the creation of the Amazigh phonemes corpus.

Parameter	Value
Sampling rate	22.05 kHz
Quantization	16 bits
Duration	2 s/syllabe
Wave format	Mono, wav
Corpus	10 Amazigh words
Speaker	10 (5 males + 5 females)
Accent	Moroccan Tamazight

Materials. We used a microphone and a laptop with 8 GB of RAM and a 2.5 GHz Intel Core i7 CPU for this study. In our experience, the operating system used is Windows 10 LTSB. The microphone was put among 4 and 10 cm from the person's mouth in a quiet room. We use the parameters mentioned in Table 3 above to record the wav file.

Pitch. The perceived fundamental of sound is well known as pitch closely related to frequency, which can be associated with the fundamental frequency of vocal cord vibration, allowing tone frequency perception. Along with loudness and quality, It is one of the most important auditory characteristics of sounds [9]. The command 'Get Pitch' was used to extract pitch, with the analysis parameters set to: pitch floor, 75 Hz, and pitch ceiling, 500 Hz.

Formants. The vocal signal spectrum presents maximums and minimums that correspond to vocal tract resonances and anti-resonance known as formants and anti-formants.

Generally, the frequency of the first three formants varies from 200 to 900 Hz for F1, from 500 to 2500 Hz for F2, and from 1500 to 3500 Hz for F3.

In this study, the first four formants (F1, F2, F3, and F4) of the vowels and consonants in the words were extracted to compare the formers for phonemes.

We choose formant analysis to recognize vowels and consonants. Before recognizing formants, a preprocessing treatment is necessary to be performed in order to bring them to light more clearly [18].

In fact, we observed that the magnitude decreases at a rate of 6 dB per octave in the spectrogram (Fig. 3) [19].

To get over this attenuation, which makes detecting the last formants difficult, the signal of voice $v(n)$ is accentuated by calculating the magnitude $v'(n)$. The greater α, the greater the magnitude at high frequency is enhanced.

Furthermore, the formants are obscured by different effects. The voice signal (t) is regularly demonstrated as the result of the convolution product between the source signal (t) caused by vocal cord vibration, and the signal $h(t)$ caused by the various resonators.

The homomorphic processing permits, the elimination of the source's impact after calculating an amount called "cepstrum". To carry out the deconvolution, the signal is divided into pieces and the windowed Fourier transform is applied to each piece. At that point we have (ω) and we take the module's logarithm (phase doesn't make any difference) to get to an additive representation $\ln|(\omega)|$. It then, at that point performs inverse

Fourier transform, giving the cepstrum c(t′). The source's contribution corresponds to high quéfrences, and that of the conduit to low quéfrences. As a result, the cutoff frequency from where one has further than the source's contribution must be determined. practically, finding the quefrency from where the cepstrum includes further peaks while remaining comparatively low is sufficient. This is referred to as liftering. Following that, it is possible to recompute h(t), but in order to obtain the "corrected" spectrogram, we are satisfied with H(ω).

The diagram below illustrates this preprocessing (Fig. 6).

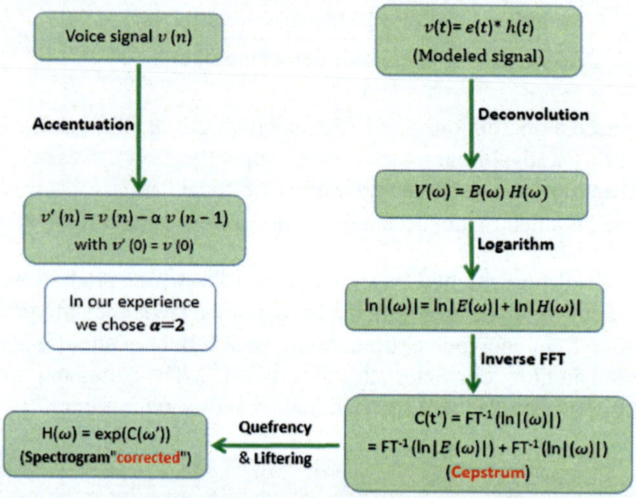

Fig. 6. Parameters preprocessing flowchart

Once the corrected spectrum has been obtained, we look for each formant within a frequency range that has been designated as finely as much as possible in order to have a sufficient chance of finding the formant that corresponds to the most significant average magnitude.

The formant values for a man's voice (in Hertz) are estimated and given in Tables 4 and 5 below. The extreme values for each formant have been indicated. It ought to be noted that these values can differ greatly depending on the individual and within the context of the discourse (Table 5).

Table 4. The estimated vowel formant value for a man's voice.

Voyelle	F1	F2	F3	F4
[i]	250	2250	2980	3280
[e]	420	2050	2630	3340
[a]	760	1450	2590	3280

Table 5. The estimated consonant formant value for a man's voice

Consonne	F1	F2	F3	F4
[m]	300	1300	2300	2770
[n]	350	1050	2300	3470
[l]	360	1700	2500	3300
[r]	550	1300	2300	2700

When the formant values have been determined and in order to compute similarity, we will compute the distance between each formant value stored in memory. Formant distance is purely a Euclidean distance; between A and B phonemes, this distance is given by:

$$d(A, B) = \sqrt{\sum_{i=1}^{4} \left(F_A^i - F_B^i\right)^2} \tag{1}$$

4 Results Discussion

The phoneme separation algorithm performs admirably as long as:

- The silences at the start and end have been removed;
- During the recording, the speaker does not blow into the microphone, which brings the signal to saturation level and results in very large peaks that the algorithm interprets as a transition;
- We are limited to a maximum of six phonemes (after that, the algorithm no longer distinguishes power variations and fluctuation between two phonemes).

The acoustic signal is highly variable from one speaker to another, which makes the problem of speech recognition very difficult.

This system achieves a performance rate of 75%, it depends on several phenomena such as location, recording materials, and the articulatory problems of the speakers.

5 Conclusion

In this paper, we have considered the problem of automatic speech recognition for the Amazigh language. We have particularly tried to distinguish between consonants and vowels of the Amazigh's voice signal. It is currently possible to make an Amazigh corpus to join those really accessible. This methodology gives a portal to the social and economic development of Moroccan Amazigh society.

This approach to speech recognition has allowed us to identify certain phonetic and spectral characteristics of the Amazigh's speech signal and to implement solutions to exploit them. However, it can be improved in the context of future work. If the software, upon arrival, does not allow to identify precisely each phoneme, it highlights the proximity of a good many of them. A more accurate determination of the problem's parameters (coefficient of emphasis, cepstrum frequency cut, and so on.), as well as a greater precision in analysis of trajectories and formant transitions and an improved consideration counting of prosody in the discourse, are all possible ways to obtain better results. Furthermore, for a language such as Amazigh, speakers must be proficient not only in the language itself, however in the utilization of computer resources.

References

1. Juang, B.H., Rabiner, L.R.: Automatic Speech Recognition: A Brief History of the Technology Development, p. 24. Georgia Institute of Technology, Atlanta and Rutgers University and the University of California, Santa Barbara (2004)
2. Samudravijaya, K.: Modeling Natural Language for Automatic Speech Recognition. Tata Institue of Fundamental Research, Mumbai, India (2008)
3. Boukous, A.: Phonologie de l'Amazighe. El Maarif Al Jadida, Rabat, Maroc (2009)
4. Ataa Allah, F., Boulaknadel, S.: Amazigh search engine: tifinaghe character based approach. In: Interntional Conference on Information and Knowledge Engineering, p. 5 (2010)
5. Makhlouf, M., Legros, D., Marin, B.: Influence de la langue maternelle kabyle et arabe sur l'apprentissage de l'orthographe française. Université Mouloud Mammeri, Tizi Ouzou et IUFM de Créteil (2006)
6. Kumbhakarn, M., Sathe-Pathak, B.: Analysis of emotional state of a person and its effect on speech features using PRAAT software. In: 2015 IEEE International Conference on Computing Communication Control and Automation (ICCUBEA), pp. 763–767 (2015)
7. Kreimana, J., Gerrattb, B.R.: Perception of ape-riodicity in pathological voice. J. Acoust. Soc. Am. **117**(4), 2201–2211 (2005)
8. Gerhard, D.: Pitch extraction and fundamental frequency: history and current techniques. Technical report. Department of Computer Science, University of Regina, Regina (2003)
9. Bonzi, E., Grad, G., Maggi, A., et al.: Study of the characteristic parameters of the normal voices of Argentinian speakers. Pap. Phys. **6**, 060002 (2014)
10. Dufour, R.: Transcription automatique da la parole spontanée. Informatique (cs). Université du Maine, Français (2010)
11. Agrawal, M., Raikwar, T.: Speech recognition using signal processing techniques. Int. J. Eng. Innov. Technol. **5**(8), 4 (2016). https://doi.org/10.17605/osf.io/zab7g
12. El Ouahabi, S., Atounti, M., Bellouki, M.: Toward an automatic speech recognition system for Amazigh-tarifit language. Int. J. Speech Technol. **22**(2), 421–432 (2019). https://doi.org/10.1007/s10772-019-09617-6
13. Dave, N.: Feature extraction methods LPC, PLP and MFCC in speech recognition. Int. J. Adv. Res. Eng. Technol. **1**(6), 5 (2013)
14. Satori, H., ElHaoussi, F.: Investigation Amazigh speech recognition using CMU tools. Int. J. Speech Technol. **17**(3), 235–243 (2014). https://doi.org/10.1007/s10772-014-9223-y
15. Hosni, H., Sakka, Z., Kachouri, A., Samet, M.: Étude de la Paramétrisation RASTA PLP envue de la Reconnaissance Automatique de la Parole Arabe. In: 5th IEEE International Conference on Sciences of Electronic Technologies of Information and Telecommunications, Tunisia, p. 7 (2009)

16. Ghazi, A.E., Daoui, C., Idrissi, N., Fakir, M., Bouikhalene, B.: Système de reconnaissance automatique de la parole Amazigh. In: Revue Méditerranéenne des Télécommunications, vol. 1, p. 7 (2011)
17. Abenaou, A., Ataa Allah, F., Nsiri, B.: Vers un système de reconnaissance automatique de la parole en amazighe basé sur les transformations orthogonales paramétrables. Asinag **9**, 133–145 (2014)
18. Gendrot, C., Adda-Decker, M.: Analyses formantiques automatiques de voyelles orales: évidence de la réduction vocalique en langues française et allemande. In: MIDL, Paris, p. 6 (2004)
19. Xiong, W., Wu, L., Alleva, F., Droppo, J., Huang, X.: Stolcke, A.: The Microsoft 2017 conversational speech recognition system. In: 2018 IEEE International Conference on Acoustics, Speech and Signal Processing (ICASSP), pp. 5934–5938 (2018). https://doi.org/10.1109/ICASSP.2018.8461870

Intelligent IoT Platform for Precocious Detection of Late Blight and TYLCV Tomato Disease in Morocco

Moad Essabbar(✉), Sara Bakkali, and Hajar Saikouk

Euromed Research Center, INSA Euro-Mediterranean, Euromed University of Fes, Fes, Morocco
{m.essabbar,s.bakkali,h.saikouk}@insa.ueuromed.org

Abstract. The agricultural sector is a key backbone of Morocco's economy. Agriculture has played a major role in the economy through agricultural exports and job creation, thus improving and maintaining good farming practices is critical in ensuring great agricultural yields. Tomatoes are a major product in Moroccan agricultural exports (51%). However, tomato plants are prone to diseases that can significantly reduce production volumes. The two main diseases to which tomato plants in Morocco are exposed are late blight and TYLCV (Tomato yellow leaf curl virus). The objective of this paper is to present an intelligent IoT platform that uses convolutional neural networks to ensure precocious detection of these two diseases, and thus limit the damage they could cause. The proposed convolutional neural networks model allows a 98% accuracy.

Keywords: Tomato diseases · Convolutional neural networks · IoT

1 Introduction

In Morocco, agriculture accounts for 14% of GDP, this number reflects the importance of this sector in the country's economy. Agricultural activity in Morocco increased sharply by 20.5% during the first quarter of 2021 [1] thanks to the Green Morocco Plan. One of the pillars of Moroccan agricultural exports is tomatoes. Indeed, according to statistics announced by the Ministry of Agriculture [2], the area devoted to tomato plants in 2021 is about 9,235 Ha. Tomatoes are a major product in Moroccan agricultural exports (51%) [3] and during the agricultural export season 2020–2021, tomato exports have registered an increase by 3% to reach a volume of 117,400 tonnes [4]. Looking closely at these numbers, we can see the importance of this product in the Moroccan agricultural sector. This importance has motivated us to pay more attention in this work to the difficulties which threaten tomato plants in Morocco. Indeed, tomato plants are exposed to several diseases that can severely reduce harvests. Among these diseases, we have chosen to focus in this article on two that most affect Moroccan tomato plants. The first, which is the most common, is Late blight which is a cryptogamic disease caused by a fungus called Phytophthora infestans and which can spread in the form of a meteoric epidemic because of its explosive epidemiological nature. This fungus

S. Motahhir and B. Bossoufi (Eds.): ICDTA 2022, LNNS 454, pp. 54–64, 2022.
https://doi.org/10.1007/978-3-031-01942-5_6

attacks all aerial parts of tomato plants (leaves, stems, and fruits) and quickly causes necrosis and drying out of the plant. The second disease, which is the most dangerous and which can lead to great losses if not controlled, is TYLCV which is a virus from the group of geminiviruses, transmitted exclusively by the species of whitefly Bemisia tabaci, this virus causes dwarfism of the plant especially when the infection is early, and therefore the infected plant does not produce any more fruits. So, to be able to minimize the damage caused by these two diseases our work introduces a new IoT platform using an Artificial Intelligence algorithm for precocious detection of late blight and TYLCV in tomato plants. This platform is based on the use of a network of cameras scanning the entire plant, these cameras regularly send images of the tomato plants to a server. At the server level, a convolutional neural network algorithm is run to process the received images and alert the plant manager if the disease is detected. The algorithm is based on the use of a convolutional neural network model. To be able to implement it, the first step of data processing is the segmentation and the augmentation of the collected image, the second step is the choice of the model's architecture, then the network training and the evaluation of the accuracy level to validate the model. The rest of the article is organized as follows: in Sect. 2 we briefly present the two diseases that we consider in our work. In Sect. 3, we present some related works which treat early detection of late blight and TYLCV in general. Section 4 is devoted to the presentation of the IoT platform architecture and convolutional neural networks model. Section 5 discusses the results obtained. And finally, a conclusion and possible perspectives.

2 Context and Related Work

2.1 Context

In Morocco, the main tomato production regions are Souss Massa, El Oualidia, El Jadida, and Casablanca for early crops and late-season cultivation. Most greenhouses are located on the coast. According to the Ministry of Agriculture and Maritime Fisheries, the area occupied by tomatoes during the 2014–2015 campaign is 17,000 ha. As mentioned before tomato's implantations are exposed to several diseases that cause huge losses which can in some cases go as far as a complete destruction of the tomato plant. In this paper, we choose to focus on two main diseases: Late blight that is one of the most widespread in Morocco, and TYLCV which is one of the most dangerous diseases. In this section, we present briefly these two diseases.

Late Blight

Late Blight attacks are rife all over the world and are very common. This cryptogamic disease can lead to great losses if left unchecked. Late Blight, caused by the infesting Phytophthora [5], can spread as a rapid epidemic because of its explosive epidemiological nature. Infestation with Phytophthora infestans often occurs late in the season. It is favored by a humid climate and moderate temperatures.

- Dissemination and biology of late blight
  Phytophthora infestans are preserved in the soil. Its oospores (sexual spores) are disseminated mainly by wind and rain, sometimes over great distances: several hundred

meters. Late Blight can also be spread through infected material. Seeds from diseased fruits are the source of contamination of future plants. Late Blight develops at temperatures between 3 and 25 °C but its sporulation is optimal between 16 and 22 °C. The fungus thrives exclusively at very high humidity levels: above 90%. Cool nights followed by hot days with high humidity promote its extension. On the contrary, dry air and temperatures above 30 °C slow down its development. It only takes 2 h of water on the tomato leaves to trigger the infection.

- Symptoms
  The first symptoms appear on the leaves and leaflets. Large spots, first yellowish and then light brown, appear and then rapidly necrotize. The outline of the spots is often pale and irregular. Sometimes the underside of the leaf blades has a white down consisting of sporangiophores and sporangia [6]. If the conditions are favorable, the damage spreads rapidly to the whole plant, which necroses and dries up entirely. As mentioned earlier, late blight has a strong epidemic character. It is therefore essential to react very quickly to the first symptoms that appear on the leaves. Thus, an IoT platform and CNN model use a leaves image for the precocious detection of Late Blight.

TYLCV

It is a disease caused by the Tomato Yellow Leaf Curl Virus (TYLCV), a virus of the geminivirus group, transmitted exclusively by the white fly species Bemisia tabaci.

- Symptoms
  Symptoms of TYLCV change little between tomato cultivars, but generally infected leaves are smaller, deformed, curled upwards with yellowish spots. The leaves become increasingly deformed and exhibit marginal yellowing which may extend to the entire leaf. Infected plants are stunted and young plants become severely stunted.
  The most significant effect of TYLCV is the abscission of flowers. Less than one flower in ten develops a fruit after infection with the virus, which considerably reduces the production yield.
- Transmission mode
  Transmission of TYLCV by the whitefly Bemisia tabaci occurs in the persistent mode. The virus is localized at the level of the phloem, in-depth. It can only be acquired by prolonged bites requiring several minutes. Once ingested with the sap, the virus passes through the insect's digestive tract to go into the hemolymph of the vector through a selective membrane thus explaining the degree of specificity between the virus and its vector. Then, the virus will be transported to the salivary glands to be transmitted to other plants during the feeding bites.

In Morocco, the disease due to TYLCV was first reported during the 1996/1997 season in El Jadida in the northwest of the country, during surveys carried out by the Plant Protection Service. In 1998, the disease spread to other tomato-growing areas including Agadir in the southwest and Berkane in the northeast. In 1999 and 2000, in some localities, 50% of farmers had to abandon their plantations in open fields or greenhouses [7]. Since that date, Morocco has launched a disease control program which has given very satisfactory results [8]. However, given the seriousness of this disease, the early detection that is proposed in this article remains a very effective way to avoid the risk of new waves like that of the early 2000s.

2.2 Related Works

Precocious detection of late blight and TYLCV diseases by using AI or IoT in tomato implantations in general or in tomato implantations specifically were treated in several research works. The objective of this section is to present some research works that try to respond to challenges of this context. Work in [9] uses a CNN model to detect tomato diseases by presenting two different deep architectures for detecting the type of infection in tomato leaves. The first architecture applies residual learning to learn significant features for classification. The second architecture applies attention mechanism on top of the residual deep network. For classifying leaf images of tomato plant diseases and pests, authors of [10] use well-known pre-trained convolutional neural network (CNN) models AlexNet [11], GoogLeNet [12] and ResNet-50 as feature extractors. In addition, they propose a deep learning model that concatenate deep features extracted from 3 CNN models. And these deep features were used to train the support vector machine classifier. Work in [13] presents also an advanced classification model which detects and classifies tomato leaf disease. A training dataset consisting of 450 images is used and image features are extracted using several models and kNN is applied for the classification. In [14], researchers develop a deep neural networks model for automatic late blight lesion recognition and severity quantification based on field imagery of diverse potato genotypes. Authors of [15] present also a deep learning-based approach to detect the early and late blight diseases in potato by analyzing the visual interpretation of the leaf of several potato crops. Paper in [16] tends to find a relationship between weather parameters and manifestation of Late Blight using Artificial Neural Network (ANN) in potato. Also, works in [17] presents a detailed review on Convolutional Neural Network (CNN) applied to plant leaf disease classification in general. Studies concerning the use of AI for detecting implantation diseases in Morocco are very limited, we can find in [18] a proposition of model to early diagnosis of the wheat plant in Moroccan region based on CNN. This model allows a rapid recognition of contaminated plants at early stages of appearance in order to provide the necessary treatment and limit propagation damage. In [19] and [20], authors present a classification of Tomato Diseases using CNN model. To fill this lack of studies, we propose in this paper a platform IoT that uses a CNN model to detect two major tomato diseases in Morocco which is late blight and TYLCV.

3 Platform Architecture and Operation

3.1 Architecture

The objective of the proposed platform is to implement cameras in tomato fields for collecting periodically images of tomato plants, these collected images are treated by

Raspberry Pi Card. Then, Raspberry uses a communication technology (wifi, 4G or Lora) to send images to a central server via Internet. In this server, an IoT platform and CNN model are implemented for detecting and deciding if a plant is infected by Late blight or TYLCV diseases. The general architecture is presented on Fig. 1.

3.2 Operation

As presented in Fig. 1 our platform is mainly based on the following units:

- **Image acquisition:** The idea is to place multiple cameras in a specific position within the tomato field so as to cover the whole field, each camera takes regularly pictures of the tomato plants in the area it covers. Each camera is connected to a Raspberry Pi Card that saves and treats images and periodically sends them to the central server.

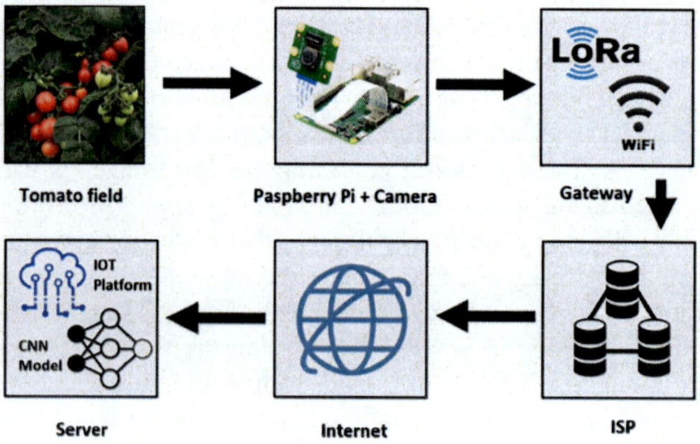

Fig. 1. Platform architecture.

- **Data transfer between Raspberry and the central server:** Images treated by the Raspberry Pi Card are sent to a gateway using communication technology. The role of the gateway is to establish communication between the card and the Internet Service Provider (ISP) network. It is possible to deploy at the gateways various communication technologies that exist on the market: short-range technologies like Zigbee, Z-Wave, WiFi, etc., and long-range technologies like Sigfox, LoRa, LTEM, NBIoT, 3G/4G, etc. Once the data arrives, at the ISP's backbone, internet routes it to the central server.
- **The central server:** To be able to make a decision based on collected images, it is necessary to develop and implement an IoT platform. This platform consists of an application to ensure, on the one hand, the reception of the data collected (the requests) and on the other hand, the display of results (after the execution of the AI algorithm) to the implantation supervisor, as well as an AI algorithm based on CNN model (as described in the next section) for detecting the infected plants.

4 CNN Model

4.1 Data Set

The first step is the preparation of the dataset. First, an open-source database was selected. This database contains thousands of tomato leaves images infected by several diseases. For our model, we focused on three types of images: healthy leaves, late blight leaves and TYLCV leaves. Figure 2 presents an example of each type.

Thus, for training phase we used 1926 healthy images, 1851 late blight images and 1961 TYLCV images. And for validation we used 481 healthy images, 463 late blight images, 490 TYLCV images. The next step is the data augmentation.

Actually, to adapt the original dataset (Fig. 3) to tomato field conditions specially for Morocco weather conditions we applied several operations to images.

Fig. 2. Example of dataset images

Fig. 3. Original image

In each operation we have arbitrarily varied parameters which allows us to generate new images and increase the size of our dataset which will allow us to better train our CNN model afterwards as shown in Fig. 4. The applied operations are as follows:

- **Brightness:** Image brightness can be increased by either darkening or brightening images, or both. The objective is to allow the model to generalize through images formed on different lighting levels that correspond to different weather conditions in the field, including an overcast sky or a partially covered sky or a completely clear sky.

- **Rotation:** By varying the rotation angle arbitrarily from 0 to 40 degrees, we generated new images which will allow us to train our model to recognize the tomato leaves whatever their position in the field, as captured by the raspberry camera.
- **Zoom:** Random zoom levels were applied to the images for feeding the CNN.
- **Mirroring:** The last operation is Mirroring along the vertical axis.

Fig. 4. Data augmentation to match real world data

4.2 Software Tools

For programming our CNN model, we use Python as a programming language. Python has simple syntax and has many AI libraries that facilitate the programming [18]. For our model we used "Keras" library on python. This library uses tensors to do the various deep leaning calculations which are mathematical objects generalizing scalars, vectors and matrices to higher dimensions. To implement our model, we used the Kaggle workspace which is adapted to integration and management of large data, using 16 GB for storage space, 16GB for the CPU RAM and 16 GB for the GPU RAM.

4.3 Convolutional Neural Network Model

Before convolutional neural networks, the most commonly used automatic classification methods for image classification were algorithms: k-Nearest Neighbors, k-NN, Support Vector Machine, SVM and Random Forest, RF.

Image classification has seen a major improvement of performance, thanks to the rise of convolutional neural networks. These are based on artificial neurons. An artificial neuron is a set of mathematical operations. First of all, a weight and a bias are applied in an affine way to an input value; in image analysis this is the value of a pixel. Then, an activation function is applied to the intermediate result to represent the data in the data space of that function. Often, this activation function is non-linear, because it allows to represent complex data where linear combination does not work.

The architecture of a convolutional neural network is formed by a succession of processing blocks to extract the characteristics discriminating the image class from the others. A treatment block consists of one to several:

- convolutional layers (CONV) which process the data of a receiving field;
- correction layers, often misused "ReLU" in reference to the activation function.
- pooling layers (POOL), which allows data compressing by reducing the size of the intermediate image (often by downsampling).

The processing blocks are linked to the final layers of the network which perform the classification of the image and the calculation of the prediction – target error:

- "fully connected" (FC) layer, which is a perceptron-type layer;
- **loss layer (LOSS).**

The implementation of CNNs requires a lot of data to make artificial neural networks work at their best. To improve the quality of learning a model, it is possible to pool knowledge from one model to another. This is called transfer learning, which consists of using an existing model to solve different but related problems. Basically, it helps to build on what has been learned in one task and improve generalization in another. This is ensured by using the pre-trained weights of the model. In this article, we use the pre-trained weights from the VGG19 model, modify the output layer, and solve a classification problem on our dataset. The VGG19 model is a deep CNN used to classify images using 16 convolution layers. The VGG19 architecture is presented in Fig. 5.

The model only requires a specific pre-processing which consists of subtracting the average RGB value, calculated over the training set, from each pixel. During the training of the model, the input of the first convolution layer is an RGB image of size 224 × 224.These layers filter the image by keeping only discriminating information such as

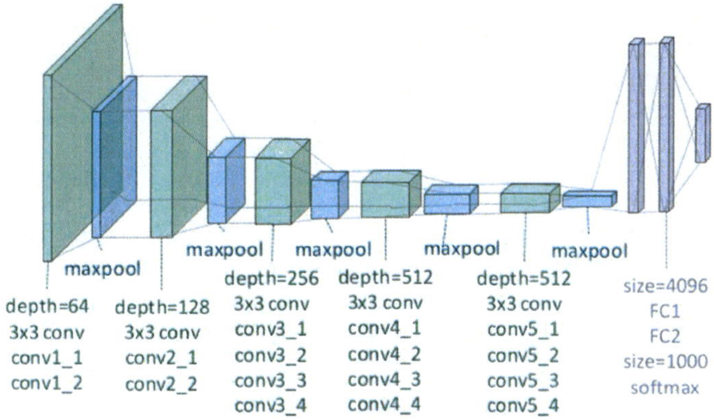

Fig. 5. VGG19 architecture [21]

atypical geometric shapes. These convolutional layers are accompanied by the Max-Pooling layer, each of size 2 × 2, to reduce the size of the filters during training. At the output of the convolution and pooling layers, we have 3 layers of Fully-Connected neurons. The first two are composed of 4096 neurons and the last of 1000 neurons with a softmax activation function to determine the class of the image.

4.4 Results

The suggested classification method (by transfer learning) performs well in detecting and classifying the dataset of tomato plant leaf diseases. In addition to a healthy type, plant illnesses are categorized into two other classes representing Late Blight and TYLCV disease. These datasets went through several steps, including picture resizing and splitting into training and testing images, augmentation, as well as classification using a VGG19 backed transfer learning neural network. Figure 6 shows the training and validation loss diagrams for each step time. Increasing training and validation accuracy leads to a reduction in training and validation loss. The classification accuracy achieved by the proposed network is 98% which is a significant increase over the past related works as shown on Table 1.

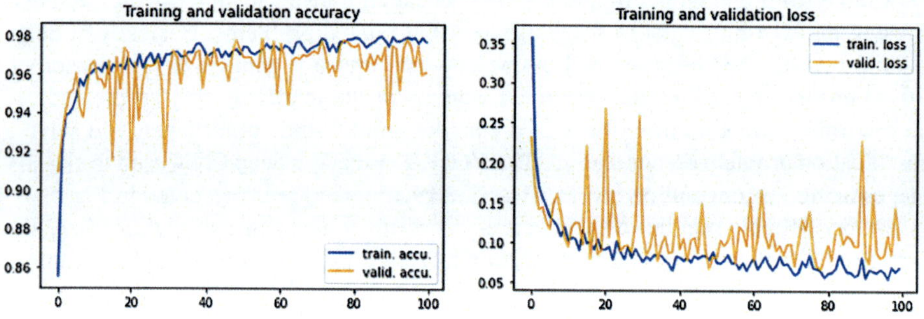

Fig. 6. Accuracy and loss variations

Table 1. Related work accuracy

Authors	Algorithm	Accuracy
Santosh Adhikari, Santosh et al. [20]	CNN	89%
Prajwala et al. [12]	LeNet	94%–95%
Solemane Coulibaly et al. [22]	VGG16	95%
Robert G. de Luna et al. [11]	AlexNet	95.75%

In addition, the confusion matrix method was used to assess the classification accuracy for each class. The confusion matrix approach compares network output and target output, then computes classification and error percentage for each class in a test dataset, describing the neural network model's performance, as illustrated by Fig. 7.

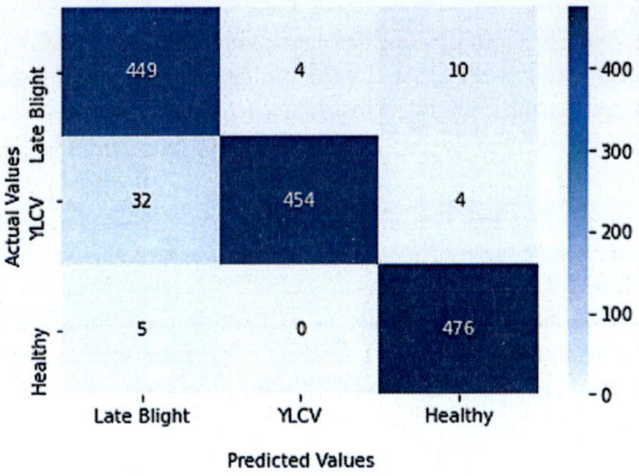

Fig. 7. Confusion matrix

5 Conclusion

In this paper we present a solution that can strongly contribute to precocious detection of two major and dangerous tomato diseases in Morocco: late blight and TYLCV. Indeed, an early detection of this diseases can significantly reduce the damages that they can cause. Actually, the proposed IoT platform offers a permanent monitoring of the tomato plants, detects the slightest sign of these diseases on the leaves, and alerts the responsible of the plantation at very early phase so he can act and do the necessary treatments before the infection spreads to other plants. This precocious detection is ensured thanks to the CNN model we proposed. The model allows a prediction of the two diseases at 98% which is an excellent rate. In future works, the system will be further developed for detecting more diseases of other types of plants in function of Moroccan agricultural needs. We are also considering a future collaboration with the Ministry of Agriculture for implementing our platform in a typical tomato farms.

References

1. Agrimaroc: Maroc: l'activité agricole en hausse de 20,5% au t1. https://www.agrimaroc.ma/maroc-activite-agricole-hausse/. Accessed 21 July 2021
2. Agence pour le Développement Agricole. Principales Réalisations du Plan Maroc Vert. https://www.ada.gov.ma/fr/principales-realisations-du-plan-maroc-vert. Accessed 2021
3. Harbouze, R., Pellissier, J.-P., Rolland, J.-P., Khechimi, W.: Rapport de synthèse sur l'agriculture au Maroc. CIHEAM-IAMM, p. 104 (2019). ffhal-02137637v2f
4. Campagne agricole 2020–2021: l'essentiel du communiqué du ministère de l'agriculture. http://mapecology.ma/actualites/185503/. Accessed 25 Oct 2020
5. Blancard, D. http://ephytia.inra.fr/fr/C/5086/Tomate-Mildiou-Phytophthora-infestans. Accessed 18 Apr 2013
6. Blancard, D.: INRAE. http://ephytia.inra.fr/fr/C/21433/Di-gno-Leg-Sporanges. Accessed 23 Sept 2015
7. Berger, P., Maroc, L.: Le virus qui dévaste les tomates. Réussir fruits et legumes, n° 194 01/03/2001, pp. 101–102(2) (2001)

8. Appui à la lutte contre le virus des feuilles jaunes en cuillère (TYLCV) de la tomate et son vecteur (Maroc). http://www.secheresse.info/spip.php?article38953. Accessed 31 May 2020
9. Karthik, R., et al.: Attention embedded residual CNN for disease detection in tomato leaves. Appl. Soft Comput. **86**, 105933 (2020). ISSN 1568-4946. https://doi.org/10.1016/j.asoc.2019.105933
10. Altuntaş, Y., Kocamaz, F.: Deep feature extraction for detection of tomato plant diseases and pests based on leaf images. Celal Bayar Univ. J. Sci. **17**(2), 145–157 (2021). https://doi.org/10.18466/cbayarfbe.812375
11. De Luna, R.G., Dadios, E.P., Bandala, A.A.: Automated image capturing system for deep learning-based tomato plant leaf disease detection and recognition. In: Proceedings/TENCON, vol. 2018, pp. 1414–1419 (2019). https://doi.org/10.1109/TENCON.2018.8650088
12. Tm, P., Pranathi, A., Saiashritha, K., Chittaragi, N.B., Koolagudi, S.G.: Tomato leaf disease detection using convolutional neural networks. In: The 11th International Conference on Contemporary Computing, Noida, 2–4 August 2018, pp. 1–5 (2018). https://doi.org/10.1109/IC3.2018.8530532
13. Batool, A., et al.: Classification and identification of tomato leaf disease using deep neural network. In: 2020 International Conference on Engineering and Emerging Technologies (ICEET), pp. 1–6 (2020). https://doi.org/10.1109/ICEET48479.2020.9048207
14. Gao, J., et al.: Automatic late blight lesion recognition and severity quantification based on field imagery of diverse potato genotypes by deep learning. Knowl.-Based Syst. **214**, 106723 (2021). ISSN 0950-7051. https://doi.org/10.1016/j.knosys.2020.106723
15. Agarwal, M., Sinha, A., Gupta, S.K., Mishra, D., Mishra, R.: Potato crop disease classification using convolutional neural network. In: Somani, A.K., Shekhawat, R.S., Mundra, A., Srivastava, S., Verma, V.K. (eds.) Smart Systems and IoT: Innovations in Computing. SIST, vol. 141, pp. 391–400. Springer, Singapore (2020). https://doi.org/10.1007/978-981-13-8406-6_37
16. Sharma, P., et al.: Prediction of potato late blight disease based upon weather parameters using artificial neural network approach. In: 2018 9th International Conference on Computing, Communication and Networking Technologies (ICCCNT), pp. 1–13 (2018). https://doi.org/10.1109/ICCCNT.2018.8494024
17. Lu, J., Tan, L., Jiang, H.: Review on convolutional neural network (CNN) applied to plant leaf disease classification. Agriculture **11**, 707 (2021). https://doi.org/10.3390/agriculture11080707
18. Aitelkadi, K., Chtaina, N., Bakouri, S., Belebrik, M.: Early detection of wheat diseases in morocco by convolutional neutral network. In: Motahhir, S., Bossoufi, B. (eds.) ICDTA 2021. LNNS, vol. 211, pp. 25–35. Springer, Cham (2021). https://doi.org/10.1007/978-3-030-73882-2_3
19. Brahimi, M., et al.: Deep learning for tomato diseases: classification and symptoms visualization. Appl. Artif. Intell. **31**(4), 299–315 (2017). https://doi.org/10.1080/08839514.2017.1315516
20. Adhikari, S., Saban Kumar, K.C., Balkumari, L., Shrestha, B., Baiju, B.: Tomato plant diseases detection system using image processing. In: 1st KEC Conference on Engineering and Technology, Lalitpur, vol. 1, pp. 81–86 (2018)
21. Karkare, P.: First steps—deep learning using Python and Keras (2019). https://meduim.com/x8-the-ai-community/first-steps-deep-learning-using-python-and-keras-3ae6bcb37d0e
22. Coulibaly, S., Kamsu-Foguem, B., Kamissoko, D., Traore, D.: Deep neural networks with transfer learning in millet crop images. Comput. Ind. **108**, 115–120 (2019). https://doi.org/10.1016/j.compind.2019.02.003

Forecasting Heating and Cooling Energy Consumption by Seasonal ARIMA Models

Lamya Lairgi[1](✉), Rachid Lagtayi[2,3], Yassir Lairgi[4], Abdelmajid Daya[1], Rabie Elotmani[5], Ahmed Khouya[3], and Mohammed Touzani[1]

[1] Department of Physics, Laboratory M3ER, FSTE, Moulay Ismail University, Meknes, Errachidia, Morocco
l.lairgi@edu.umi.ac.ma

[2] Department of Sciences, Industrial and Civil Technologies, Team of Sciences and Advanced Technologies, National School of Applied Sciences, Abdelmalek Essaadi University, 2121 Tetouan, Morocco

[3] Department of Electrical and Industrial Engineering, National School of Applied Sciences, Abdelmalek Essaadi University, 1818 Tangier, Morocco

[4] Department of Data Science, MINES School Saint-Étienne France, Saint-Etienne, France

[5] Sciences for Energy Laboratory LabSIPE, ENSAJ, Chouaib Doukkali University, 1166 El Jadida, Morocco

Abstract. The main purpose of this paper is to predict energy consumption of heating/cooling for the building's sector based on Seasonal Auto Regressive Integrated Moving Average SARIMA model for a tertial building located in Errachidia city in Morocco. The data of heating and cooling consumed energy have been collected during the last five years from January 2016 to December 2020 by using ECOTECT ANALYSIS software, considering the climate conditions plus the occupants' behaviors. After fitting and testing SARIMA model, the analysis of the Mean Absolute Error **MAE**, Mean Square Error **MSE**, Root Mean Square Error **RMSE** and the Scatter Index (SI) show that the seasonal ARIMA approach gives optimal results for both cooling and heating energy consumption forecasting. Finally, this developed time series SARIMA model can be applied as an efficient tool to forecast the energy needs to ensure a long-term development and energy management in the building sector.

Keywords: Cooling energy consumption · Heating energy consumption · Forecasting · Seasonal ARIMA

1 Introduction

In recent decades, the rate of energy consumption of tertiary sector has been increased in Morocco which represents 33% [1] of the total consumed energy. While, due to the increased demand of energy, the efficient energy management of the limited energy resources our planet (oil, gas, electricity, etc.) is also needed. A new trend is brightening up towards energy efficiency in buildings' sector to identify and forecast various

S. Motahhir and B. Bossoufi (Eds.): ICDTA 2022, LNNS 454, pp. 65–74, 2022.
https://doi.org/10.1007/978-3-031-01942-5_7

energy consumption by using Artificial Intelligence (AI) models [2–5] and deep machine learning tools [6–8].

Furthermore, the optimization of buildings' design, the identification's process of all parameters that substantially affect building energy consumption is required, Such as, climate [9, 10], building shape [11], orientation and envelope materials [12]. Thus, accurately predicting of building's energy needs is a challenging goal. Hence, the Artificial Intelligence (AI) is used to develop models such us SARIMA models that simulate the human inference processes and forecast new events from prior acquired data.

In the previous literature review, many research works have explored the application of SARIMA models to forecast various features. Ebhuoma et al. [13] have examined SARIMA $(0,1,1)(0,1,1)_{12}$ models to predict monthly malaria infections in KwaZulu-Natal, South Africa. It can provide reliable information to the relevant authority to act proactively. Shuojiang et al. [14] have proposed an hybrid SARIMA-SVR approach to forecast the needs of the aviation industry, and they can achieve better accuracy in this empirical study. Peng et al. [15] have studied the temperature values in Nanjing city, China, between 1951 and 2017 by applying SARIMA AI models. Moreover, they have used the data from 1951 until 2014 as a training part, while the data between 2015 and 2017 were used as a testing set section. Based on the obtained results, this proposed research approach gives a perfect forecasting accuracy.

In this paper, the SARIMA models are used to predict the cooling and the heating energy demand in Morocco for a tertial building and the forecasting performance parameters, such as Mean Absolute Error **MAE**, Mean Square Error **MSE**, Root Mean Square Error RMSE were used to confirm the approach's accuracy.

2 Data and Method

2.1 Building's Data Description

In this study, the used datasets are the monthly consumed energy for heating/cooling of a tertial building located in Errachidia city, Morocco during the last five years from January 2016 to December 2020 using ECOTECT ANALYSIS software. The simulation is done by considering the climate conditions of the Errachidia city and the occupant's behaviours.

Between 2016–2019, the data are utilised for training the SARIMA model, while the data between 2019–2020 are applied as a testing set for it.

The data description is presented on the Table 1, and the data visualisation is plotted on the Fig. 1.

Table 1. Data description

	Heating energy (kwh)	Cooling energy (kwh)
Count	60.000	60.000
Mean	2306.600	1831.316

(continued)

Table 1. (*continued*)

	Heating energy (kwh)	Cooling energy (kwh)
Std	2518.803	2731.626
Min	0.000	0.000
25%	0.000	0.000
50%	982.000	46.500
75%	4213.000	2953.750

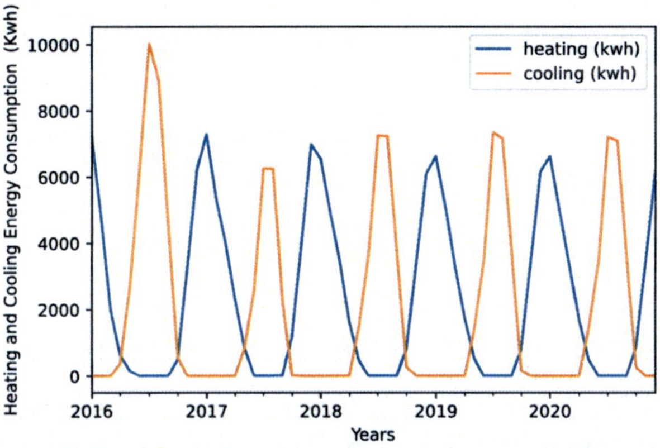

Fig. 1. Heating and cooling energy consumption from January 2016 to December 2020

2.2 SARIMA Models

The SARIMA artificial intelligence models combine non-seasonal and seasonal components, and can be defined as SARIMA (p, d, q) (P, D, Q) s, where P presents the orders of the seasonal AR, D is the seasonal differencing, Q is the seasonal MA sections of the model, and s is the length of the seasonal period. While (p, d, q) refer to ARIMA parameters.

- The AR process calculates for the prior observed values up to a specified maximum lag, plus an error term.
- The path of differencing is referred to as the integration part that accounts for stabilization of the data by deleting seasonality or trend
- The MA process accounts for the previous error terms, making the prediction easier.

The algebraic expression of the SARIMA model is given in the equation (1) as follow [16]:

$$\varphi(B)\phi\left(B^{s}\right)(1-B)^{d}\left(1-B^{s}\right)^{D}X_{t}=\theta(B)\Theta\left(B^{s}\right)\varepsilon_{t}+\theta_{0} \tag{1}$$

where:

- X_t is the data series.
- $\varphi(B) = 1 - \varphi_1 B - \varphi_2 B^2 - \cdots - \varphi_p B^p$ is the order p of AR term.
- $\theta(B) = 1 - \theta_1 B - \theta_2 B^2 - \cdots - \theta_p B^p$ is the order q of MA term.
- $\phi(B^s) = 1 - \phi_1 B^s - \phi_2 B^{2s} - \cdots - \phi_P B^{Ps}$ is the order P of seasonal AR term.
- $\Theta(B^s) = 1 - \Theta_1 B^s - \Theta_2 B^{2s} - \cdots - \Theta_Q B^{Qs}$ is the order Q of MA term.
- ε_t is the disturbance or error term at time t.
- B is the backward shift operator.
- φ is the Coefficient of non-seasonal autoregressive.
- θ is the coefficient of non-seasonal moving average.
- ϕ is the coefficient of seasonal autoregressive.
- Θ is the coefficient of seasonal moving average.

3 Results and Discussion

In this paper, a SARIMA (p, d, q) (P, D, Q)$_{12}$ model was built based on monthly cooling and heating energy consumption data of a tertial building located in Errachidia city, Morocco, and the Datasets were collected between January 2016 and December 2020 and simulated by using ECOTECT ANALYSIS software as the following steps below.

Step 1: The first step is the stationarity check of the data. The variance, the mean, and the Autocorrelation Functions (ACF) of the time series data should be steady over time. If the series has a trend or a seasonal component, then the stationarity is not satisfied. we should transform it by differencing.

Thus, we have tested the data stationarity by using dickey fuller ADF tool. It belongs to a group of tests named 'Unit Root Test'. The existence of a unit root refers to the time series is non-stationary. Thus, unit roots' number in the series indicates that the number of differencing operations needed to make the series stationary. The statistical hypothesises under this test are:

- H0: The series has a unit root.
- H1: The series hasn't a unit root, the data series is stationary.

The data is steady if the p-value is lower than 0.05, if not the data is not stationary. The results show that the cooling and heating data are not stationary, and it accept the null hypothesis H0 (Table 2).

Table 2. Data Dickey Fuller test 1

	Heating Dickey Fuller test	Cooling Dickey Fuller test
ADF test statistic	−2.615	−2.341
p-value	0.089	0.158
Stationarity result	The heating data is non-stationary	The cooling data is non-stationary

Therefore, the time series is not stationary, and it is Seasonal. We need to transform the data to be stationary by differencing. and it will not accept the null hypothesis H0.

So, this first stage serves to define the required number of differencing to realise the the stationarity of the series, whereas this number is the orders of both the non-seasonal (d = 1) and the seasonal (D = 1) differences (Table 3).

Table 3. Data Dickey Fuller test 2

	Heating Dickey Fuller test	Cooling Dickey Fuller test
ADF test statistic	−3.028	−5.495
p-value	0.032	2.126. 10^{-6}
Stationarity result	The heating data is now stationary	The cooling data is now stationary

Step 2: In this step, we should determine the orders (p, q, P, Q) of the possible SARIMA models based on the Autocorrelation Function ACF and the Partial Autocorrelation Function PACF by analysing in which lags we have the significant spikes (Figs. 2 and 3).

The figure below shows that the ACF plot has significant spikes at lags 1 and 2 for the cooling data which means that we can choose p = 1 or p = 2. While the PACF plot has significant spikes at lags 1 and 2, consequently, the q can take the values 1 or 2.

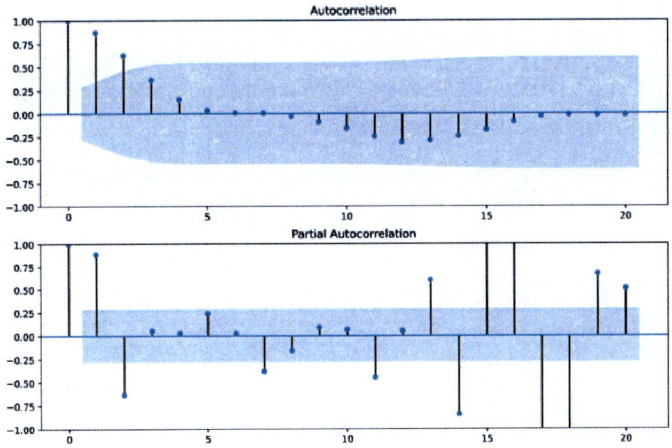

Fig. 2. ACF and PACF of monthly cooling data

The ACF and the PACF for the heating data shows in the Fig. 3 would suggest an AR (2) and a MA (2).

However, The ACF and PACF are not always adequately to identify the best values of (p, d, P, D) for SARIMA models. Thus, considering the Akaike Information Criterion **AIC** is the optimal method to search for the best parameters p, d, q, P, D and Q of SARIMA models. The model with the lowest AIC would be the best model that fits the data.

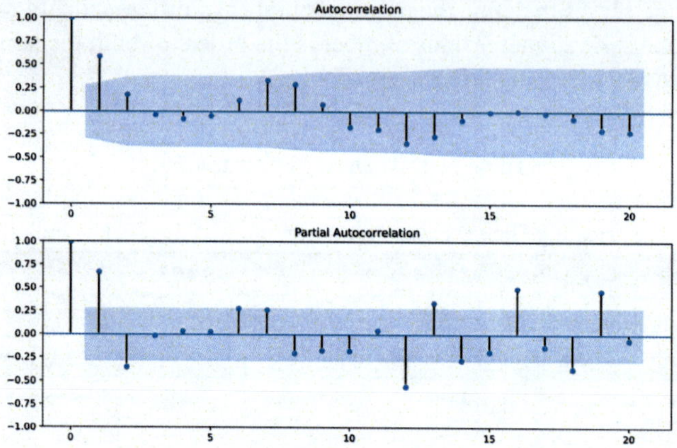

Fig. 3. ACF and PACF of monthly heating data

For our data, s = 12 for both cooling and heating series because their monthly datasets have a period of 12. The AIC parameters of the chosen models are presented in Tables 4 and 5.

According to Table 4, SARIMA (0,1,1) (2,1,0) $_{12}$ illustrates the lowest AIC value for the cooling series. Consequently, this model is good enough to forecast the cooling energy consumption.

Table 4. AIC values for the cooling consumption

(p, d, q) (P, D, Q)$_S$ parameters	AIC value
(0,1,0) (0,1,0)$_{12}$	551.684
(1,1,0) (1,1,0)$_{12}$	527.469
(1,1,0) (0,1,0)$_{12}$	542.738
(1,1,0) (2,1,0)$_{12}$	524.954
(0,1,0) (2,1,0)$_{12}$	529.433
(2,1,0) (2,1,0)$_{12}$	525.008
(1,1,1) (2,1,0)$_{12}$	526.035
(0,1,1) (2,1,0)$_{12}$	524.625
(0,1,1) (1,1,0)$_{12}$	527.712

For the heating consumption the best SARIMA model is SARIMA (0,1,0) (0,1,0)$_{12}$, it shows the lowest AIC = 534.420 (Table 5).

Step3: In this step, and after the selection of the optimal SARIMA model. We could predict the heating and the cooling energy consumption (Figs. 4 and 5).

Step 4: A test should be done to verify the prediction precision by calculating the Mean Square Error **MSE** "Eq. (2)", Root Mean Square Error **RMSE** "Eq. (3)", Mean

Table 5. AIC values for the heating consumption

(p, d, q) (P, D, Q)s parameters	AIC value
(0,1,0) (0,1,0)$_{12}$	534.420
(1,1,0) (1,1,0)$_{12}$	536.432
(0,1,0) (1,1,0)$_{12}$	534.513
(1,1,0) (0,1,0)$_{12}$	536.335
(0,1,1) (0,1,0)$_{12}$	536.204

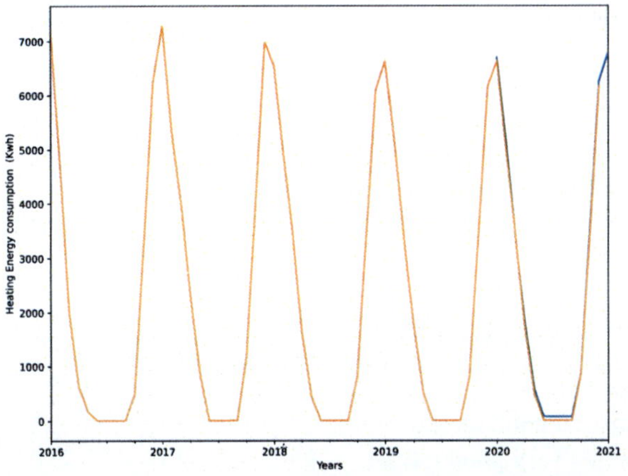

Fig. 4. The real heating values and the forecasted ones

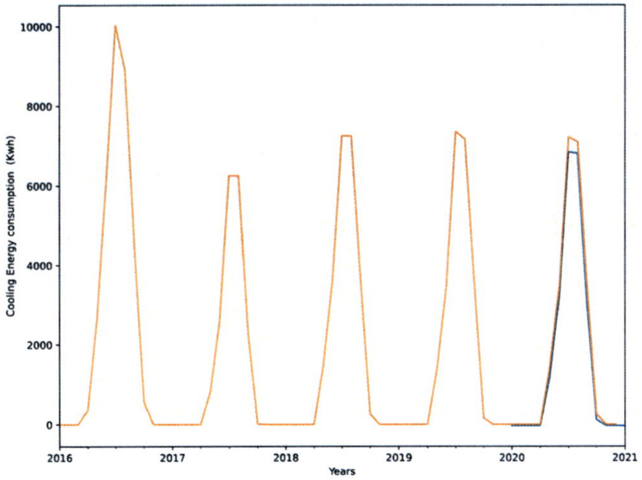

Fig. 5. The real cooling values and the forecasted ones

Absolute Error **MAE** "Eq. (4)".

$$\mathrm{MSE} = \frac{1}{N}\sum_{i=1}^{i=N} (yi - y')^2 \tag{2}$$

$$\mathrm{RMSE} = \sqrt{MSE} \tag{3}$$

$$\mathrm{MAE} = \frac{1}{N}\sum_{i=1}^{i=N} |yi - y'| \tag{4}$$

where:

- y' is the forecasted value.
- yi is the present value.
- N is the observations' number.

According to the above discussion and errors' values, RMSE = 86.77 for the cooling and 115.62 for the heating because of the biggest values of the data. Thus, we should calculate the Scatter Index "Eq. (5)", which is computed by dividing the RMSE by the average observed value.

$$\mathrm{SI} = \left(\frac{RMSE}{average\ observed\ value}\right) \tag{5}$$

- SI < 10% refers to a good model
- SI < 5% refers to a very good model

The selected SARIMA model for the cooling shows a SI = 4%. While the SI = 7% for the heating, which prove the high accuracy of this approach to predict the energy consumption in the tertial buildings (Table 6).

Table 6. Statistical evaluation results of the SARIMA models

	SARIMA models	MSE	RMSE	SI	MAE
Cooling	$(0,1,1)\ (2,1,0)_{12}$	7529.360	86.770	0.040	59.600
Heating	$(0,1,0)\ (0,1,0)_{12}$	13368.200	115.620	0.070	108.800

4 Conclusion

In this paper, we have used the SARIMA artificial intelligence model to forecast the consumed energy of heating/cooling for a tertial building in Errachidia, Morocco. The results show that's the SARIMA $(0,1,1)\ (2,1,0)_{12}$ model gives optimal outcomes for the cooling energy consumption, While the SARIMA $(0,1,0)\ (0,1,0)_{12}$ model presents the lowest errors analysis for the heating energy consumption. Finally, the application of the time series SARIMA model is a good solution to forecast the consumed energy of buildings' sector and that will be helpful to propose the appropriate energy management and operational systems of buildings to ensure energy savings and costs reduction.

References

1. Barradi, T.: UN APERÇU DE LA SITUATION DE L'EFFICACITÉ ÉNERGÉTIQUE DES MÉNAGES AU MAROC, RABAT (2019)
2. Mitkov, A., Noorzad, N., Gabrovska-Evstatieva, K., Mihailov, N.: Forecasting the energy consumption in Afghanistan with the ARIMA model. In: 2019 16th Conference on Electrical Machines, Drives and Power Systems, ELMA 2019 - Proceedings, no. June, pp. 1–4 (2019). https://doi.org/10.1109/ELMA.2019.8771680
3. Chou, J.S., Bui, D.K.: Modeling heating and cooling loads by artificial intelligence for energy-efficient building design. Energy Build. **82**(2014), 437–446 (2014). https://doi.org/10.1016/j.enbuild.2014.07.036
4. He, Y., Lin, B.: Forecasting China's total energy demand and its structure using ADL-MIDAS model. Energy **151**(April 2017), 420–429 (2018). https://doi.org/10.1016/j.energy.2018.03.067
5. Silva, J., Praça, I., Pinto, T., Vale, Z.: Energy consumption forecasting using ensemble learning algorithms. In: HerreraViedma, E., Vale, Z., Nielsen, P., Martin Del Rey, A., Casado Vara, R. (eds.) DCAI 2019. AISC, vol. 1004, pp. 5–13. Springer, Cham (2020). https://doi.org/10.1007/978-3-030-23946-6_1
6. Wang, Z., Srinivasan, R.S.: A review of artificial intelligence based building energy use prediction: contrasting the capabilities of single and ensemble prediction models. Renew. Sustain. Energy Rev. **75**(September 2015), 796–808 (2017). https://doi.org/10.1016/j.rser.2016.10.079
7. Camara, A., Feixing, W., Xiuqin, L.: Energy consumption forecasting using seasonal ARIMA with artificial neural networks models. Int. J. Bus. Manag. **11**(5), 231 (2016). https://doi.org/10.5539/ijbm.v11n5p231
8. Gao, G., Li, J., Wen, Y.: DeepComfort: energy-efficient thermal comfort control in buildings via reinforcement learning. IEEE Internet Things J. **7**(9), 8472–8484 (2020). https://doi.org/10.1109/JIOT.2020.2992117
9. Lairgi, L., Lagtayi, R., Daya, A., Elotmani, R., Touzani, M.: The impact of arid climate on the indoor thermal comfort in the South-East of Morocco. Int. J. Photoenergy **2021** (2021). https://doi.org/10.1155/2021/5554629
10. Lagtayi, R., Lairgi, L., Khouya, A., Daya, A.: Acoustic and thermal correlation for a building' s envelope in a Mediterranean climate in Morocco, **11**(3) (2021)
11. Pathirana, S., Rodrigo, A., Halwatura, R.: Effect of building shape, orientation, window to wall ratios and zones on energy efficiency and thermal comfort of naturally ventilated houses in tropical climate. Int. J. Energy Environ. Eng. **10**(1), 107–120 (2019). https://doi.org/10.1007/s40095-018-0295-3
12. Lairgi, L., Daya, A., Elotmani, R., Touzani, M.: Contribution to the study of the energy needs for a building in the city of errachidia influence of orientation and local materials. Key Eng. Mater. **820**, 18–28 (2019). https://doi.org/10.4028/www.scientific.net/KEM.820.18
13. Ebhuoma, O., Gebreslasie, M., Magubane, L.: A seasonal autoregressive integrated moving average (SARIMA) forecasting model to predict monthly Malaria cases in Kwazulu-Natal, South Africa. South Afr. Med. J. **108**(7), 573–578 (2018). https://doi.org/10.7196/SAMJ.2018.v108i7.12885

14. Xu, S., Chan, H.K., Zhang, T.: Forecasting the demand of the aviation industry using hybrid time series SARIMA-SVR approach. Transp. Res. Part E Logist. Transp. Rev. **122**(October 2018), 169–180 (2019). https://doi.org/10.1016/j.tre.2018.12.005
15. Chen, P., Niu, A., Liu, D., Jiang, W., Ma, B.: Time series forecasting of temperatures using SARIMA: an example from Nanjing. IOP Conf. Ser. Mater. Sci. Eng. **394**(5) (2018). https://doi.org/10.1088/1757-899X/394/5/052024
16. Brockwell, P.J., Davis, R.A.: Time Series: Theory and Methods. Springer, New York (1991). https://doi.org/10.1007/978-1-4419-0320-4

Predicting the Residential Energy Consumption in Morocco Based on Time Series Forecasting Models

Imane Hammou Ou Ali(✉), Mohammed Jamii, Mohammed Ouassaid, and Mohamed Maaroufi

Engineering for Smart and Sustainable Systems Research CenterMohammadia School of Engineers, Mohammed V University in Rabat, Rabat, Morocco
imanehammou@research.emi.ac.ma

Abstract. Forecasting the electricity demand is crucial for efficient planning, management, and conservation towards intelligent power systems. This study investigates the yearly electricity consumption of the Moroccan residential sector from 1990 to 2020. Two different time series forecasting methods are implemented and the best one is selected to predict the electricity consumption for the coming years. The performance of the Holts Linear Trend (HLT) method is compared to those of the Autoregressive Integrated Moving Average (ARIMA) Model. The outcome analysis shows that the similarities in prediction were high; meanwhile, the HLT performs better than ARIMA and predicts the data with more accuracy; lowest Mean Absolute Percentage Error (MAPE), Root Mean Square Error (RMSE), and Mean Square Error (MSE) compared to ARIMA model. The forecasted results based on HLT model show that the Moroccan residential electricity consumption will remain increasing by an average annual rate of 2.4% in the following 10 years. In 2030, Morocco will consume around 1253 Ktoe of electricity in the residential sector. This paper has policy implications that can be deployed for optimal planning in the electricity sector.

Keywords: Electricity consumption forecast · The Moroccan residential sector · ARIMA · HLT

1 Introduction

Power generation capacity requires high capital intensive, it takes a long period of time in construction, and the commodity cannot be stored in vast quantity. Moreover, there is a high level of uncertainty involved in planning due to the changing demands [1]. Owing of this particularity in the electricity sector, it is not an exaggeration to affirm that demand forecasting performs a crucial role in the integrated electricity planning process [2]. Reliable predictions mitigate any imbalance between the supply and the demand, which can compromise the socioeconomic well-being of a population of a developing country like Morocco, which have to review its way of consumption.

S. Motahhir and B. Bossoufi (Eds.): ICDTA 2022, LNNS 454, pp. 75–84, 2022.
https://doi.org/10.1007/978-3-031-01942-5_8

Electricity demand in Morocco has recognized an exponential growth because of industrialization, economic evolution, and increasing living standards. Figure 1 represents the annual electricity consumption per sector, according to the International Energy Agency [3]. As it can be seen from this graph, the residential sector is considered as one of the most electrical consuming sectors, reason why attentions are oriented to forecast the electricity consumption in this sector. The purpose of this study is to predict the Moroccan residential electricity consumption and provide a clear visibility on the electricity demand for the next ten coming years. The main contributions of the proposed research work can be listed as follows:

- Two time series forecasting methods, namely; HLT and ARIMA are deployed to predict the residential electricity consumption for the coming years.
- The efficiency of both models was evaluated in terms of prediction accuracy and the obtained results show that the HLT model extensively decreases the error rate when compared to the ARIMA model.
- HLT method can be considered as an effective technique to forecast the electricity demand and a reliable way to take strategical planning.

The rest of this paper is organized as follows, after introduction in Sect. 1, a short review about the existing forecasting research is presented in Sect. 2. The methodological analysis is explained in Sect. 3. Section 4 is reserved for the forecasting results and discussion of each model. Finally, Sect. 5 concludes the work.

2 Related Works

Forecasting the electricity demand is one of the elementary issues of the modern world. It is divided into three types depending on the time period in which the forecast is performed: short-term interval, medium-term interval, and long-term interval [4]. In literature, a considerable number of publications have been conducted in the long-term energy prediction. The work reported in [5] investigates three forecasting methods, namely; ARIMA, Exponential smoothing, and Temporal causality modeling to predict the energy demand in Morocco. In the same context, the authors in [6] forecast the Moroccan electric network based on ARIMA model and the obtained results was com pared to the official forecasts of ONEE (Office National de l'Eau et d'Electrcite). Other work in [7] proposes the prophet model and Holt-winters to forecast the peak load in Kuwait in the time period between 2020 and 2030. The findings show that the Prophet model provides more accurate prediction than Holt-Winters. A Hybrid method combing Vector Autoregressive model and Gray model is used in [8] to forecast the electricity consumption of the residential sector in Cameroon. The results obtained from the proposed model prove that the electricity needs by 2025 are estimated at 2641.632 GWh. Furthermore, Gray model is applied in [9] to forecast the building energy consumption in China. Depending on the obtained results, the building energy consumption shows an increase trend which indicates that the existing energy consumption problems in the building area have not been effectively solved. ARIMA model is also established in [10] to forecast the energy consumption in turkey for the next 25 years. The findings demonstrate that Turkey's need for oil, coal, renewable energy, and natural gas is raising which will result in energy shortage in the future. LSTM network is proposed in [11] to predict the periodic energy consumption. The obtained results show that the proposed technic has better prediction performances with respect to other methods such as ARMA, Propagation Neural Network (BPNN), and Autoregressive Fractional Integrated Moving Average model (ARFIMA). More than one variable can be exploited in forecasting the energy consumption. The work reported in [12] adopts a Multiple Linear Regression (MLR) model and a correlative approach to predict the residential energy demand. The outcomes demonstrate that correlative models based on a link Residential Energy Consumption (REC) to Gross Domestic Product (GDP) and the urbanization rate outperforms MLR models. Another work in [13] applies a multiple regression model and three Convolutional Neural Networks (CNNs) with the aim to forecast Florida's electricity usage. Economic, social, and climate data were gathered and introduced into the models. The performance of univariant, multichannel, and multi-head ANNs along with the MLR are developed. The performance metrics reveal the superiority of the multichannel and multi-head CNNs in contrast to the model based on the MLR. Another study in [14] displays three models of multiple regressions to forecast the energy consumption in the city of Bogota in Colombia based on the six socio-economic strata. The first model is a simple multiple linear regression, the second one has an economic construction of econometric, however, the third one is a double logarithm economic regression. The results show that the second model outperforms the linear regression model and the double logarithm in terms of precision.

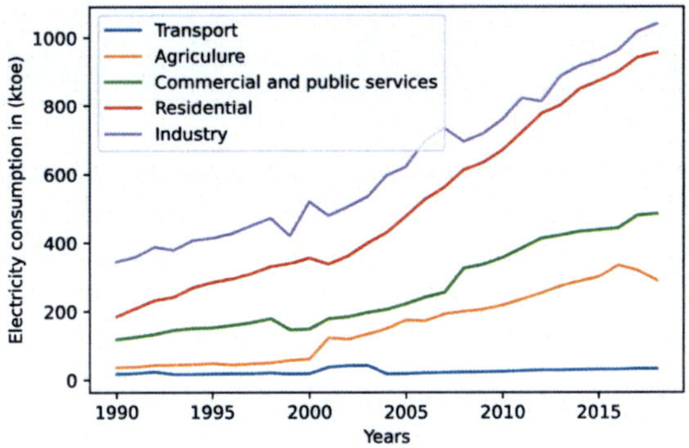

Fig. 1. Moroccan electricity consumption per sector [3].

3 The Methodology Analysis

Forecasting data involves taking models to fit historical data and use them to forecast future findings. Time series forecasting methods are one of the most applied data science techniques for prediction. There are a large number of techniques available for time series forecasting [15]. Among the most widely used approaches for time series forecasting ARIMA model and HLT model. Hereinafter the modeling of each technique.

3.1 ARIMA Model

ARIMA is a univariate model that tries to depict a single variable as an Autoregressive Integrated Moving Average process. It mixes three types of processes, the involvement of all of them are identified by the notation ARIMA (p, d, q). AR(p) denotes the Autoregressive process and p its order. I(d) refers to the Integration of the process and d its degree. MA(q) represents the Moving Average process and q its order. ARIMA (p, d, q) model is generally structured using the following equations [16]:

$$\nabla^d F_t = \mu + \frac{\gamma(D)}{\psi(D)}\varepsilon_t \tag{1}$$

and

$$\nabla^d = (1 - D)^d \tag{2}$$

$$\gamma(D) = 1 - \gamma_1 D - \gamma_2 D^2 - \cdots - \gamma_p D^p \tag{3}$$

$$\psi(D) = 1 - \psi_1 D - \psi_2 D^2 - \cdots - \psi_q D^q \tag{4}$$

where, μ represents the medium value of the series F_t, D is the delay factor, ∇^d is the d-th difference operator, $\gamma(D)$ represents the autoregressive polynomial of order p, $\psi(D)$ is the polynomial of the coefficients of the moving average of order q and ε_t is the estimated residual at each time period.

3.2 HLT Model

HLT method represents one of the most important time-series forecasting models. It is extended from the simple exponential smoothing to provide predictions of data with a trend [17]. Three equations are used in the process: the first and second equations stand for the level and the trend. They are both involved to get the forecast function in the third expression. Hence, the HLT model is organized as the following formation [18]:

$$L(t) = \alpha Y(t) + (1 - \alpha)(L(t - 1) + B(t - 1)) \tag{5}$$

$$B(t) = \beta(L(t) - L(t - 1)) + (1 - \beta)B(t - 1) \tag{6}$$

$$F(t + k) = L(t) + kB(t) \tag{7}$$

where $L(t)$ denotes the level of the series at time t, α is the smoothing parameter for the level it must be $0 \leq \alpha \leq 1$, $B(t)$ is the trend of the series at time t, β is the trend smoothing factor. It must be $0 \leq \beta \leq 1$. $F(t + k)$ is the k-step-ahead forecast.

3.3 Forecasting Performance Index

Electricity forecasting involves the estimation of the future electricity demand. The difference between the estimated values and the actual ones is the prediction error. Three commonly forecasting accuracy tests are used to estimate the forecasting performance of the proposed models. They are defined as below:

Mean Square Error (MSE):

$$MSE = \frac{1}{N}\sum\nolimits_{t=1}^{N}(Y_t - F_t)^2 \tag{8}$$

Mean Absolute Percentage Error (MAPE):

$$MAPE = \frac{1}{N}\sum\nolimits_{t=1}^{N}\left|\frac{Y_t - F_t}{Y_t}\right| \times 100 \tag{9}$$

Root Mean Square Error (RMSE):

$$RMSE = \sqrt{\frac{\sum_{t=1}^{N}(Y_t - F_t)^2}{N}} \tag{10}$$

where F_t and Y_t are the t-th predicted and measured values respectively. N is the number of years of the forecast period. The model's performance depends on the estimated error; smaller values for MSE, MAPE, and RMSE, indicate that the mode's performance is better.

3.4 Case Study

The historical data investigated in this paper was collected from IEA [3] from 1990 to 2020. Figure 2 displays the energy consumption expressed in Ktoe of the residential sector in Morocco. In order to make successful predictions, a systematical forecasting process is followed as explained in Fig. 3. First, we collect the data, then we move to build the forecasting model, next we proceed to the diagnostic checking to select the best model. Finally, we make the future predictions.

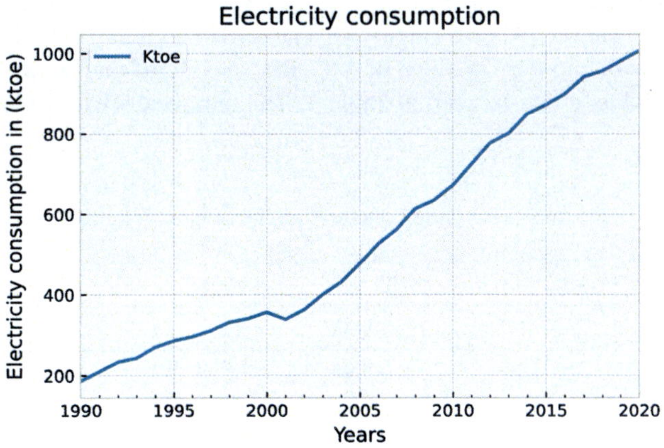

Fig. 2. The electricity consumption profile for the Moroccan residential sector [3].

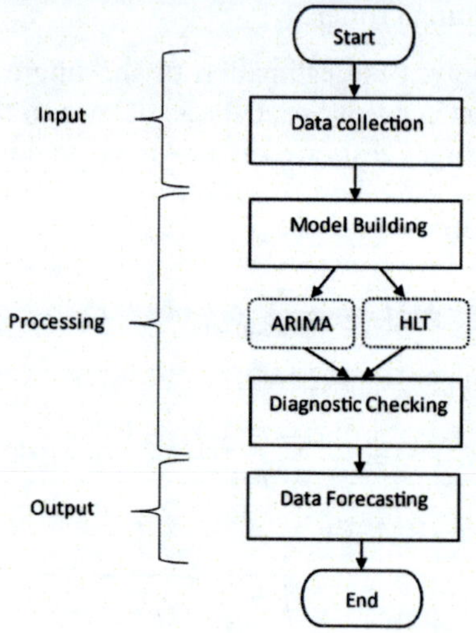

Fig. 3. Schematic of the process.

4 Results and Discussion

The study handled so far aims to assess two time series forecasting methods, namely; HLT and ARIMA to forecast the energy consumption of the Moroccan residential sector. The obtained results were exanimated through the forecasting performance index described in Sect. 3 and conducted to the following results.

For ARIMA Model, first we perform an Augmented Dickey Fuller (ADF) test since the Moroccan residential electricity consumption is a non-stationary time series. The results get from the test prove that the value of ADF is not significant, but it performs significant at the first difference. As a result, our series is stationary in the first difference. Thus, we implement a first-order differential for the electricity consumption time series. In order to adopt the best ARIMA, selection criteria like Log Likelihood, Akaike Information Criterion (AIC), Bayesian Information Criterion (BIC) and Hannan–Quinn Information Criterion (HQIC) are employed. The ARIMA (1, 1, 1) model is chosen as the optimal model due to the low value of AIC, BIC and HQIC and the maximum value of Log Likelihood as it can be seen in Table 1. The parameters of ARIMA (1,1,1) are listed in Table 2.

Table 1. Selection of best fitted ARIMA model of electricity consumption.

Model	AIC	BIC	HQIC	LogLikelihood
ARIMA (1,1,1)	253.747*	257.950*	255.092*	−123.873*
ARIMA (1,1,2)	255.325	260.930	257.118	−123.663
ARIMA (1,1,3)	257.324	264.330	259.566	−123.662

Table 2. Parameters related to ARIMA model.

Term	Coefficient	Standard error	Test statistic	Probability
AR (1)	0.9761	0.047	20.878	0.000
MA (1)	−0.6096	0.178	−3.402	0.001

A comparison between the actual data gets from the data set and the predicted data based on ARIMA model is presented in Fig. 4. It is noted that the two graphs match well. The predicted data follows the shape of the data graph, however small error is detected. To see whether the obtained results are good enough or not, we need to perform other forecasting model.

In case of fitting the HLT for forecasting the Moroccan residential electricity consumption, the auto-optimization approach is used to automatically find the optimized values for parameters α and **β** that provides the lowest error. Table 3 displays the estimate values of the parameters related the HLT model. To validate the effectiveness of the forecasting model, the annual data set which includes the amount of the electricity consumption in the Moroccan residential sector was selected to build the model and test the predicted results. The way to prove whether the method is accurate, a comparison with actual consumption data is needed. Figure 5 illustrates the comparison between the actual data and the predicted data using HLT. As it can be seen, the forecasting graph results in following the pattern of the actual data which demonstrates the quality of the forecasting model. However, small difference is detected.

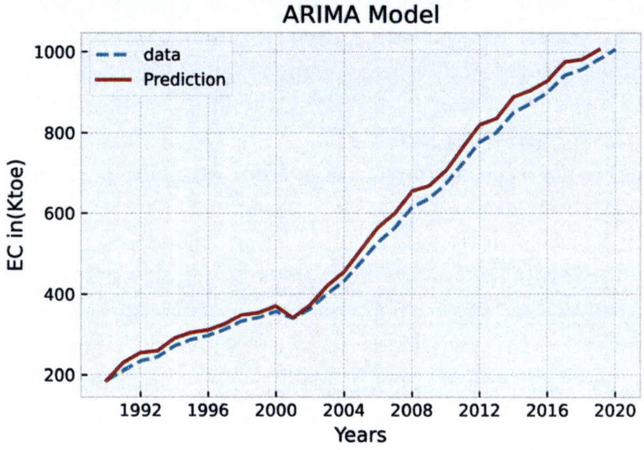

Fig. 4. Comparison between the actual data and the predicted data using ARIMA (1,1,1) Model.

As it can be seen from the Figs. 4 and 5 the prediction graphs generated by ARIMA Model and HLT model fit the data graph effectively but with a small difference. At the sight, we cannot distinguish which model accurate better. However, the difference between the predicted values and the actual data must be measured.

Table 3. Parameters related to HLT model.

Parameter	α	β	α_0	β_0	AIC	BIC	AICC
Value	0.88	0.47	164.5	21.2	172.345	178.081	175.845

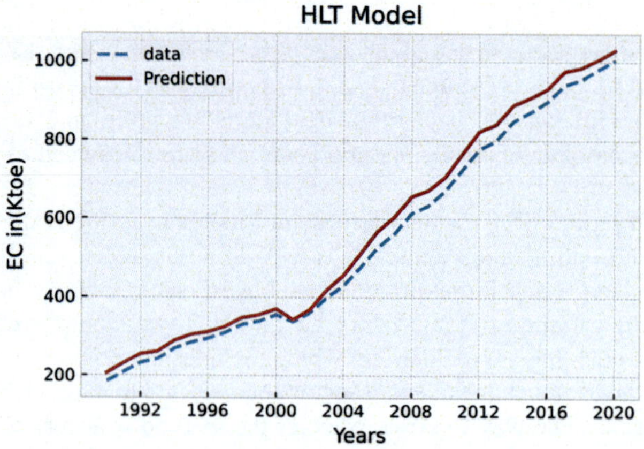

Fig. 5. Comparison between the actual data and the predicted data using HLT Model.

To assess the efficiency of each forecasting model, we need to calculate the difference between the predicted values and the actual observed values. For the models presented, we employ the commonly used measurement metrics such as: MSE, RMSE, and MAPE. Generally, the best model in the one with the lowest errors. The predicted values must be close to the actual ones to get a better forecasting result. As it can be seen from Table 4, the proposed HLT performs better than ARIMA with the minimum value of MSE of 853.52, RMSE of 29.21 and MAPE of 5.55%.

As HLT model is proved to be powerful in forecasting the electricity consumption in contrast to the ARIMA model. Thus, HLT model will be employed in this work to forecast the Moroccan electricity consumption in the residential sector for the coming 10 years. Figure 6 presents the forecasting values of the electricity consumption until 2030. It is quite clear that the electricity consumption will tend to increase in a positive direction. The Moroccan residential electricity consumption will reach 1253 Ktoe in 2030. Therefore, attention must be paid to manage that increasing consumption at the residential side.

This knowledge can be exploited in the decision-making process such as energy control in the residential sector. Some recommendations can be listed as follows:

- Encourage the householders to consume less by proposing a discount percentage on price per kWh.
- Allow reductions on the acquisition of electrical home devices and environmentally friendly products that provide low consumption.

- Support the integration of renewable energy sources in homes for householder self-consumption.
- Establish a power limit not to be exceeded and enforce penalties in case of non-compliance by the householders.
- Require some building norms that seek significantly to reduce the electricity consumption.

Table 4. Comparison between model according to the measures of accuracy.

Models	MSE	MAPE	RMSE
ARIMA (1,1,1)	1324.675	70.76%	36.39
HLT	853.52	5.55%	29.21

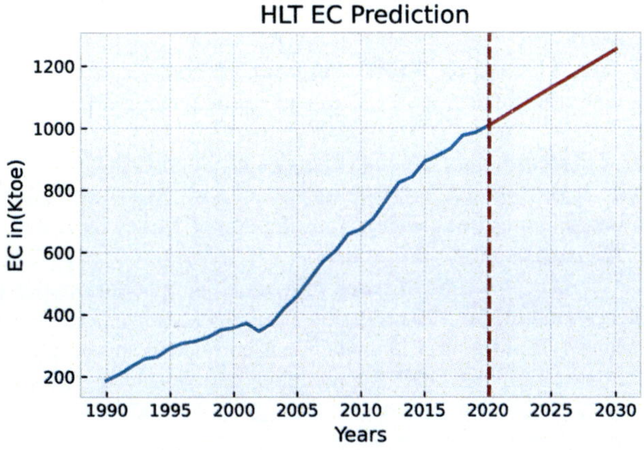

Fig. 6. Electricity consumption prediction using HLT.

5 Conclusion

The work reported in this study aims to predict the electricity consumption of the Moroccan residential sector despite the lack of the historical data. Two time series are studied and their efficiency was verified under the forecasting performance metrics. Both methods provide good results, however the HLT outperforms the ARIMA model with a minimum value of MAPE, MSE, and RMSE. Furthermore, the HLT method is proved to be simple, fast, and computationally inexpensive. The obtained forecasted results based on the HLT method reveal that the Moroccan residential electricity consumption will continue to increase during the coming years until it reaches 1253 Ktoe in 2030. This knowledge must be exploited by the Moroccan government in the decision-making process, in order to match the production to the demand.

References

1. Current challenges in Energy. https://www.bbvaopenmind.com/en/articles/current-challenges-in-energy/. Accessed 21 Nov 2021
2. Debnath, K.B., Mourshed, M.: Forecasting methods in energy planning models. Renew. Sustain. Energy Rev. **88**, 297–325 (2018)
3. International Energy Agency. https://www.iea.org/countries/morocco. Accessed 21 Nov 2021
4. Ghalehkhondabi, I., Ardjmand, E., Weckman, G.R., Young, W.A.: An overview of energy demand forecasting methods published in 2005–2015. Energy Syst. **8**(2), 411–447 (2016). https://doi.org/10.1007/s12667-016-0203-y
5. Nafil, A., Bouzi, M., Anoune, K., Ettalabi, N.: Comparative study of forecasting methods for energy demand in Morocco. Energy Rep. **6**, 523–536 (2020)
6. Citroen, N., Ouassaid, M., Maaroufi, M.: Long term electricity demand forecasting using autoregressive integrated moving average model: case study of Morocco. In: 2015 International Conference on Electrical and Information Technologies (ICEIT), pp. 59–64. IEEE, March 2015
7. Almazrouee, A.I., Almeshal, A.M., Almutairi, A.S., Alenezi, M.R., Alhajeri, S.N.: Long-term forecasting of electrical loads in Kuwait using prophet and holt-winters models. Appl. Sci. **10**(16), 5627 (2020)
8. Guefano, S., Tamba, J.G., Azong, T.E.W., Monkam, L.: Forecast of electricity consumption in the Cameroonian residential sector by Grey and vector autoregressive models. Energy **214**, 118791 (2021)
9. Dun, M., Wu, L.: Forecasting the building energy consumption in China using Grey model. Environ. Process. **7**(3), 1009–1022 (2020). https://doi.org/10.1007/s40710-020-00438-3
10. Ozturk, S., Ozturk, F.: Forecasting energy consumption of Turkey by Arima model. J. Asian Sci. Res. **8**(2), 52 (2018)
11. Wang, J.Q., Du, Y., Wang, J.: LSTM based long-term energy consumption prediction with periodicity. Energy **197**, 117197 (2020)
12. Haouraji, C., Mounir, B., Mounir, I., Farchi, A.: A correlative approach, combining energy consumption, urbanization and GDP, for modeling and forecasting Morocco's residential energy consumption. Int. J. Energy Environ. Eng. **11**(1), 163–176 (2020). https://doi.org/10.1007/s40095-020-00336-2
13. Elkamel, M., Schleider, L., Pasiliao, E.L., Diabat, A., Zheng, Q.P.: Long-term electricity demand prediction via socioeconomic factors a machine learning approach with Florida as a case study. Energies **13**(15), 3996 (2020)
14. Peña-Guzmán, C., Rey, J.: Forecasting residential electric power consumption for Bogotá Colombia using regression models. Energy Rep. **6**, 561–566 (2020)
15. Deb, C., Zhang, F., Yang, J., Lee, S.E., Shah, K.W.: A review on time series forecasting techniques for building energy consumption. Renew. Sustain. Energy Rev. **74**, 902–924 (2017)
16. Desbois, D.: Une introduction à la méthodologie de Box et Jenkins: l'utilisation de modèles ARIMA avec SPSS. La revue MODULAD (2005)
17. Charles, C.H.: Forecasting seasonals and trends by exponentially weighted moving averages. Int. J. Forecast. **20**(1), 5–10 (2004)
18. Box, G.E., Jenkins, G.M., Reinsel, G.C., Ljung, G.M.: Time Series Analysis: Forecasting and Control. Wiley, New York (2015)

Generation and Analysis of Residential Microgrid Database

Sonia Moussa[1(✉)], Houda Ben Attia Sethom[1,2], Afef Bennani Ben Abdelghani[1,3], and Ilhem Slama-Belkhodja[1]

[1] Laboratoire Des Systèmes Electriques, Ecole Nationale d'Ingénieurs de Tunis, Université de Tunis El Manar, LR11ES15, 1002 Tunis, Tunisie
sonia.moussa@enit.utm.tn

[2] Ecole Nationale d'Ingénieurs de Carthage, Université de Carthage, Charguia II, 2035 Tunis, Tunisie

[3] Institut National Des Sciences Appliquées Et de Technologie, Centre Urbain Nord, Université de Carthage, 1080 Tunis, Tunisie

Abstract. Nowadays, the high level of solar PV rooftop penetration leads to the increase of residential microgrids and brings new challenges mainly to the Energy Management Systems (EMS). To investigate such issues, a Microgrid Platform (MGP) has been developed in the National Engineers School of Tunis (ENIT). As all residential microgrids, MGP provides a large amount of data from various types of equipment and devices. Each of these data has its format and sampling period and is accessible via its dedicated portal requiring authentication. Hence, the analysis and exploitation of such microgrid data are tedious. In this paper, we propose the design of a centralized database that can be used for Microgrid analysis and EMS development. The data formats are standardized according to the chosen database management system (DBMS), namely MySQL. Depending on the user's needs, unified sampling periods for all the microgrid data are implemented. In addition, access to the proposed database is designed to be possible from a unique portal.

Keywords: Microgrid · Database · Data pre-processing

1 Introduction

Residential microgrids are a promising concept to cope with the impact of the increasing penetration of rooftop photovoltaic (PV) in distribution networks [1, 2]. They include distributed renewable energy generation, especially PV, distributed energy storage systems, based on batteries, and residential loads. Microgrids operate in either grid-connected or islanded mode [3, 4]. Microgrids and their different components are nowadays equipped with new technologies (e.g. internet-connected sensors) to ensure the cohesive operation of the overall system [5, 6]. Hence, data information exchange on the operating point of each microgrid component is necessary for energy management system (EMS) deployment [7, 8], or energy trading solution development [9]. A significant measured

S. Motahhir and B. Bossoufi (Eds.): ICDTA 2022, LNNS 454, pp. 85–96, 2022.
https://doi.org/10.1007/978-3-031-01942-5_9

data amount is then generated by microgrids. Such data are essential to investigate EMS solutions to face the energy supply balance issue and to manage overproduction or curtailment of variable and uncertain energy generation and demand.

Many studies and research works have been reported in the literature, focusing on data-driven models and control applied to microgrids. Some authors detail their specific case study with their platform dataset, like at Fort Collins [10], Colorado, USA, at Kimen [11], Taiwan, and at university building of FERIT Osijek [12], Croatia. Other authors have addressed data-driven challenges solutions investigations, related to high-level renewable energy sources penetration into utility grid [13], renewable energy integration in remote islanded microgrid [14], impacts of electric vehicle functionalities in wind energy-powered utility grid [15], and the flexibility test system for the studies of variable renewable energy resources [16]. These studies highlight the importance of microgrids data.

The National Engineers School of Tunis (ENIT) has developed a microgrid platform (MGP) to conduct its investigations on new technologies and new microgrid control concepts. The MGP components are of various types with different technologies, such as solar inverters, battery inverters, weather-station, and energy meters. The data produced are scattered and must be first processed to be readily useable for analysis. Therefore, this paper deals with ENIT Microgrid Platform's database generation and data analysis. A centralized database gathering MGP dataset is developed by standardizing the data formats according to the chosen database management system (DBMS), namely MySQL, then by synchronizing data. It only collects the needed data and allows their analysis and display with no need to connect to the different devices' portals. This database will be used to establish a data-driven MGP model for EMS development. This paper describes the generation of such a database from MGP's dataset.

2 Microgrid Platform Description

The Microgrid Platform (MGP) is located in QehnA Laboratory of the National Engineers School of Tunis. This platform includes two microgrids that have been added in parallel to the conventional electrical installation. The two microgrids are named Pla-NeTE microgrid and SMARTNESS microgrid, respectively. The whole forms the MGP.

Each microgrid is supplied by its own power distribution line. In that case, the laboratory is equipped with three distribution lines (Fig. 1): a single-phase one for SMARTNESS, a three-phase four-wires one for Pla-NeTE, and the already existing one, named "Load" line. This latter feed loads like lighting, air conditioners, computers, etc. Pla-NeTE microgrid is operating with individual self-consumption mode and SMARTNESS one with collective self-consumption concept. SMARTNESS microgrid comprises a central system with battery inverter that connects the microgrid to the SMARTNESS line distribution and a hybrid solar inverter that create a local line distribution with four Points of Connection (PoC), corresponding each to a single-phase house connection. Each of these houses, designed by H1 to H4, is equipped with a hybrid solar inverter from Goodwe (GW1 to GW4 in Fig. 2). The central system data are collected in JUA portal (www.dev.jua.solar) and each Goodwe inverter data are stored in Goodwe portal (www.semsportal.com).

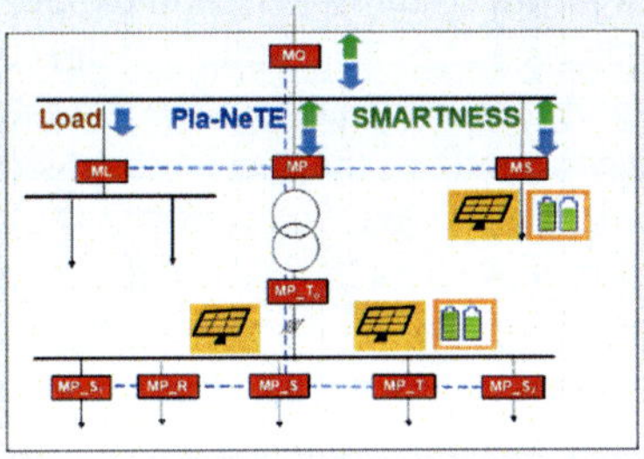

Fig. 1. Meters provided by SACEM Industries company and implemented in the Lab.

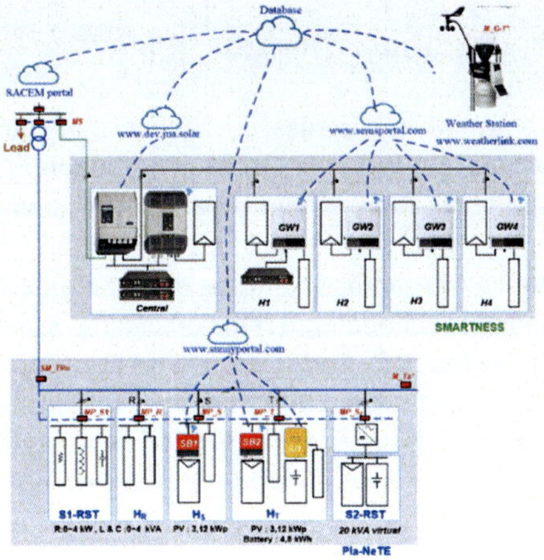

Fig. 2. Studied microgrid platform.

Pla-NeTE microgrid has five PoC. Two of them are three-phase ones, (S1-RST and S2-RST in Fig. 2) and the three others are single-phase. To each single-phase R, S, and T is connected what corresponds to a house designed by H_R, H_S, and H_T, respectively. The first one, H_R comprises only loads, while both H_S and H_T are equipped with single-phase solar inverters from SMA company (SB1 and SB2 in Fig. 2). H_T is also equipped with a battery system composed of a battery inverter from SMA and lead-acid batteries from Assad Battery company (SI1 in Fig. 2). The SMA inverter data are collected in the cloud of this company: www.sunnyportal.com.

Smart meters from SACEM Industries company are installed to supervise power flow, between QehnA laboratory and ENIT school (Meter MQ in Fig. 1) and power flow through each distribution line, SMARTNESS, Pla-NeTE, and Load (meters MS, MP, and ML, respectively in Fig. 1). Each PoC of Pla-NeTE is also equipped with a meter: MP-S1 and MP-S2 for each of the three-phase PoC and MP-R, MP-S, and MP-T for H_R, H_S, and H_T, respectively. The SACEM meters data are collected in the research team server at ENIT school.

In addition, a weather station from DAVIS Instruments company provides several data such as solar radiation and ambient temperature. Figure 2 highlights different portals that user has to manage for data analysis. It is to note that the connection to the aforementioned portals requires user authentication. In other words, the access to the overall information goes through five different user logins. Moreover, the login provided is unique and cannot be shared to tierce users for security reasons. Hence, the objective of centralizing data to a single database. This paper focuses especially on the generation of the Pla-NeTE platform's database since the same approach will be applied to the SMARTNESS platform.

3 Database Design

As previously discussed, each piece of equipment provides a set of data that are whether related to weather or electricity production. The weather station provides data that can be of integer, float, or text type with timestamp given in two columns as date and time. The *sunnyportal* gives the data related to the average production of the inverters in float type, with a timestamp including only the time variable. As for the data from *semsportal*, the datatype encountered is the timestamp provided in two columns, the float for the electrical measured data, and text for working status of the inverters. And finally, *SACEM portal* offers data related to the energy exchange of the different Lab's PoC. The datatype is all float with a timestamp provided in a single column.

For the new database establishment, our focus will be on the energy and power exchanged on the MGP since this database is primarily aimed to investigate an energy management system. Hence, only the solar radiation and the ambient temperature will be used from *weatherlink* portal. Likewise, only active power, imported and exported energy, voltage, current of inverters, and state of charge of batteries will be considered from the other portals.

Each of the microgrid data acquisition portals has its own data acquisition period. Indeed, *sunnyportal* acquires data every 15 min, whereas the weather station offers the possibility of acquiring data in steps of 1, 5, 10, 15, 30, 60, or 120 min. However, the choice of the weather station acquisition step is configured once for all. On the other hand, the data from *SACEM portal* can be downloaded in steps of 1, 5, 10, 15, 20, 30, 60 min, or 1 day or month according to the user's choice. Since the data will be synchronized, the data acquisition time step is chosen identical to the one of sunnyportal for the sake of implementation easiness, i.e., 15 min.

A relational database model is adopted to simplify the different categories of data classification and store them efficiently. Indeed, a relational database has a simple structure in contrast to a hierarchical database. The data are organized in tables and views can be created to display data from different tables. The concept of primary and foreign keys that can relate some tables to each other makes the data non-repetitive, enabling to avoid data duplication. In addition, relational database offers easy access to data through a simple query, and also flexibility with the facility of increasing the incoming amount of data, or its deleting or updating. Furthermore, its high-security level makes it possible to define privileges according to the users. Additionally, the feasibility of future modification and expansion is easy since the insertion, updating, or deleting of records are straightforward thanks to data tables.

To implement the proposed database, MySQL relational database management system is used since it is free and open source. It uses a Structured Query Language (SQL) to communicate with the database. The size of the database is then estimated over a day, a month, and a year, so that to predict the storage needed for its web hosting. The database size estimation is presented in Table 1. It amounts to 992 kilobytes/day. Thus, the annual data is estimated to be 354 megabytes. Since the views are virtual tables, they do not occupy storage space. Only the table sizes matter during the estimation of the database's total size. Tables with an acquisition period of 15 min have a relatively small size.

Table 1. Database size estimation.

Platform		Table	Size/day	Size/month	Size/year
MGP	Pla-NeTe	dsunny	16 kB	480 kB	5.7 MB
		dsacem	16 kB	480 kB	5.7 MB
	SMARTNESS	dgw1	224 kB	6720 kB	79.8 MB
		dgw2	224 kB	6720 kB	79.8 MB
		dgw3	224 kB	6720 kB	79.8 MB
		dgw4	224 kB	6720 kB	79.8 MB
		dsacemmgp	16 kB	480 kB	5.7 MB
	Weather Station	dweather	48 kB	1440 kB	17.1 MB
Total			992 kB	29.06 MB	354 MB

However, the multitude of data from the weather station makes its size more important. The same goes for the data size of the SMARTNESS platform tables, which not only have numerous data but are also characterized by an acquisition period of 1 min.

The database will be hosted in the Lab's server and with access granted through a link with single authentication for each user with a defined privilege.

4 Data Pre-processing

Data pre-processing is performed by targeting the necessary data among the various data provided by the portals, followed by their extraction and formatting, as well as their synchronization. For the case of Pla-NeTE platform, some original data obtained from the weather station, sunnyportal.com, and SACEM portal are reported in Tables 2, 3, and 4, respectively.

Table 2. Weather station data.

Column	Type	Comments
Date	Date	
Time	Time	
TempOut	Float	Outside temperature [°C]
SolarRad	Float	Solar radiation [W/m^2]

The weather station provides data downloadable in text file format whereas sunnyportal and SACEM portal in CSV format. While the weather station data gives timestamp data in two columns, the sunnyportal does not provide the user with the date data whereas *SACEM portal* affords it in a single column. Therefore, without formatting the timestamp, the data synchronization will be made difficult.

Table 3. Data from sunnyportal

Column	Type	Comments
Time	Time	
InstallationPower	Float	Average value of the installation total production [kW]
InverterSPower	Float	Average value of house S inverter [kW]
InverterTPower	Float	Average value of house T inverter [kW]

Table 4. Data from SACEM portal

Column	Type	Comments
Date/Time	DateTime	
SYST-RST-1.A-PIII	Float	S1-RST active power [kW]
SYST-RST-1.Energie-Active-Exporte	Float	S1-RST exported active energy [kWh]
SYST-RST-1.Energie-Active-Importe	Float	S1-RST imported active energy [kWh]
House-R.A-PIII	Float	House R active power [kW]
House-R.Energie-Active-Exporte	Float	House R exported active energy [kWh]
House-R.Energie-Active-Importe	Float	House R imported active energy [kWh]

Furthermore, given the various instrumentations, the data types and formats also differ, as can be seen in Tables 2, 3 and 4. In addition, due to the format of the file in which the data are obtained, some data must be converted in appropriate format type recognized by the used DBMS.

4.1 Data Extraction and Formatting

As aforementioned previously, the data to be imported into the database must have a format recognized by the DBMS to be used. For our case, MySQL will be used as DBMS. Hence, decimal numbers must be written using *dot*, and the date format should be as "yyyy-mm-dd" while the time format as "hh:mm:ss". For the case of the data from *sunnyportal*, for instance, an example of downloaded file is depicted in Fig. 3.

	A	B	C	D
1		ENIT 2017 / P	SB 3000TL-21	SB 3000TL-21 639 / Puissance / Valeurs moyennes [kW]
2	00:15	0	0	0
3	00:30	0	0	0
4	00:45	0	0	0
5	01:00	0	0	0
6	01:15	0	0	0
7	01:30	0	0	0
			. . .	
26	06:15	0	0	0
27	06:30	0	0	0
28	06:45	0	0	0
29	07:00	0	0	0
30	07:15	0	0	0
31	07:30	0,07	0,03	0,03
32	07:45	0,16	0,09	0,07
33	08:00	0,36	0,27	0,09
34	08:15	0,58	0,46	0,12
35	08:30	1,2	0,82	0,38
36	08:45	2,32	1,17	1,15
37	09:00	2,94	1,47	1,48
38	09:15	3,19	1,59	1,6
39	09:30	3,46	1,72	1,74
40	09:45	3,7	1,84	1,86

Fig. 3. Example of data download file from sunnyportal.

Since the decimal numbers in this file are written with a *comma*, these data must be formatted. The same goes for the data from *SACEM portal*, where the decimal numbers are written with a *comma*. As for the data weather station, no formatting is necessary for the number. However, the timestamp must be formatted for all files since their format corresponds to french timestamp one.

The data can then be imported to the database after meeting the required datatype format.

4.2 Data Synchronization

Once the data has been imported into the database, the grouping of all these data according to the timestamp is then performed. Authors refer to this as synchronization. The latter is achieved by a join request of the different database tables according to the identical timestamp criterion of each table, by creating a view. This created view corresponds to a virtual table that takes up very little storage space since it does not store real data. Thus, the size of the database corresponds only to the size of the data stored in the tables.

5 Proposed Database

As the data are extracted from different web portals and dataloggers, the datasets from each portal are stored in different tables that constitute the data warehouse. In other words, for PlaNe-TE data, for example, the warehouse data includes 2 tables storing data from SACEM portal (*dsacem*) and *sunnyportal* (*dsunny*), as shown in Fig. 4. The data from the weather station (*dweather*) are stored in Weather warehouse.

Fig. 4. Entity-relationship diagram for the proposed database

Then, to fetch information or gather data for reporting or analysis purposes, SQL queries are applied. Hence, any set of data under different categories can be inserted in and retrieved from the database without altering the existing data. The entity-relationship diagram corresponding to this model is shown in Fig. 4. In Pla-NeTE views, *all_data* view allows access to the recording of all data from Pla-NeTE and Weather warehouses.

The views *s1_rst*, *s2_rst*, and *houser* display the data corresponding to *S1-RST, S2-RST*, and H_R from Fig. 2, respectively. Since they do not have photovoltaic production, only data from *dsacem* are used here. The views *houses* and *houset*, however, include synchronized data from *dsunny*, *dsacem*, and *dweather* since they include photovoltaic production.

A graphical illustration of some variables from *houses* view is shown in Fig. 5.

In this figure, the solar radiation from the weather station, the average power at the output of the inverter from sunnyportal, and the injected power to the grid measured by SACEM's smart meter are presented. Given that the weather station has been recently acquired, its fixation on the rooftop has not yet been done. For now, it operates as a mobile station that can be moved out of the lab when performing tests. For this day's test, the station was out from 9 am to 3 pm. During this period, the solar radiation's curve shape and the one of the power injected into the grid from SACEM meters are correlated. However, the shape of the power from sunnyportal (inverterSPower) is smoother and slightly delayed compared to the others. This can be explained by the fact that the output power of the inverter provided by sunnyportal corresponds to an average value over the 15 min step acquisition while the others are instantaneous values.

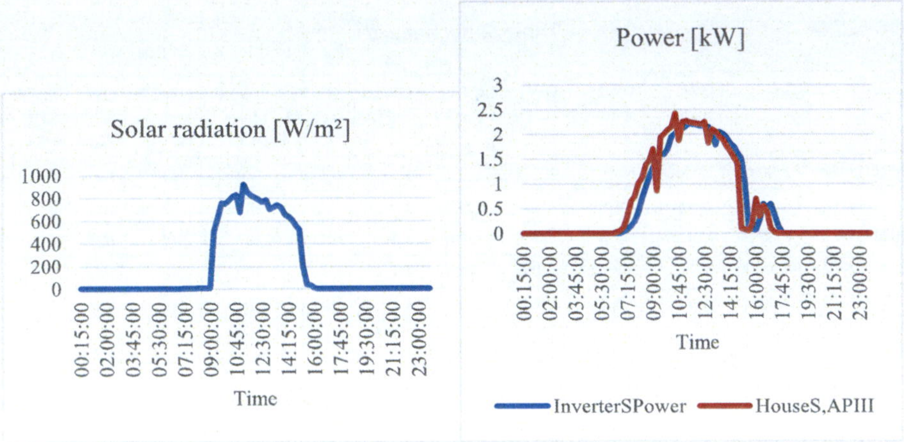

Fig. 5. House S' graphical representation of solar radiation, inverter power, and active power injected into the grid

A data sample from two inverters of SMARTNESS microgrid view is presented in Table 5. In this table, the negative values of power correspond to energy absorption while the positive ones correspond to energy injection. Although the data from each inverter is acquired every 1 min, the choice to treat data according to the desired period is facilitated thanks to the database. For instance, for the case of the data in Table 5 for instance, the period is set to 15 min.

Table 5. H1, H2's data sample from SMARTNESS microgrid

Date	Time	GW1 status	GW2 status	GW1 Ua [V]	GW2 Ua [V]	GW1 I AC [A]	GW2 I AC [A]	GW1 P Meter [W]	GW2 P Meter [W]	GW1 SOC [%]	GW2 SOC [%}
2021-11-12	05:00:00	Normal	Wait	226.4	226.8	0.2	0	−23	−12	98	0
2021-11-12	05:15:00	Normal	Wait	225.9	226.6	0.1	0	−22	−12	98	0
2021-11-12	05:30:00	Normal	Wait	225.3	225.5	0.1	0	−26	−12	98	0
2021-11-12	05:45:00	Normal	Wait	225.3	226.2	0.1	0	−24	−12	98	0
2021-11-12	06:00:00	Normal	Wait	225.2	225.5	0.1	0	−21	−13	98	0
2021-11-12	06:15:00	Normal	Wait	223.7	224.4	0.1	0	−23	−13	98	0
2021-11-12	06:30:00	Normal	Wait	224.2	224.5	0.1	0	−25	−12	98	0
2021-11-12	06:45:00	Normal	Normal	225.1	225.2	0.2	0	−60	0	98	0
2021-11-12	07:00:00	Normal	Normal	222.5	222.9	0.2	0	−57	−12	98	0
2021-11-12	07:15:00	Normal	Normal	223.7	224.1	0.1	0.5	−59	30	98	0
2021-11-12	07:30:00	Normal	Normal	223.7	224.7	1	0.5	189	40	100	0
2021-11-12	07:45:00	Normal	Normal	222.1	222.2	1	0.5	188	35	100	0

(continued)

Table 5. (*continued*)

Date	Time	GW1 status	GW2 status	GW1 Ua [V]	GW2 Ua [V]	GW1 I AC [A]	GW2 I AC [A]	GW1 P Meter [W]	GW2 P Meter [W]	GW1 SOC [%]	GW2 SOC [%}
2021-11-12	08:00:00	Normal	Normal	222.3	222.2	1	0.5	180	61	100	0
2021-11-12	08:15:00	Normal	Normal	219.9	220.8	1	0.5	181	68	100	0
2021-11-12	08:30:00	Normal	Normal	221.3	223.6	1.1	0.5	207	47	100	0
2021-11-12	08:45:00	Normal	Normal	220.6	221.5	1.1	0.5	200	63	100	0
2021-11-12	09:00:00	Normal	Normal	221.8	223.2	1.1	0.5	220	49	100	0

As aforementioned in Sect. 2, SMARTNESS microgrid has been implemented for the collective self-consumption purpose study. H1 includes battery storage while H2 does not, which explains the second house's state of charge (SOC) to always be null. As for the exchanged power between the inverters and the central, their energy management system has yet to be analyzed. An assumption on the inverters' power absorption during the night might be the power drawn by the inverters from the central to power their communication system as well as their energy management system board.

6 Conclusion

In summary, this paper presents the database generation and data analysis of a grid-connected residential microgrid platform. This platform consists of several distributed sources and loads, which are instrumented to provide their data via different web portals. As a result, the platform data are fragmented, making their access and analysis difficult. To overcome these issues, a centralized database, gathering all data from the web portals has been set up using MySQL as relational database management system. The choice of adopting relational database was argued by the easiness implementation of relational database while ensuring various advantages such as simplicity, data integrity, flexibility, and future modification feasibility and expansion without altering the ancient data. The objective of gathering and synchronizing the data from the different web portals in a single database has been achieved. The presented results show the correlation of the data from the different portals.

However, since the microgrid platform has been developed to investigate the new energy technologies as well as the new control concepts, the events related to the tests' condition to be carried out must be added to the database. Therefore, in future works, authors plan to extend the database by including tables recording the events describing the type and conditions of tests to be performed.

Acknowledgment. This work was supported by the Tunisian Ministry of Higher Education and Scientific Research under Grant LSE-ENIT-LR 11ES15 and funded in part by PAQ-Collabora (PAR & I-Tk) program and in part by NAS and USAID under the USAID Prime Award Number AID-OAA-A-11-00012. Any opinions, findings, conclusions, or recommendations expressed in this article are those of the authors alone and do not necessarily reflect the views of USAID or NAS.

References

1. Alam, M., Al-Ismail, F.S., Salem, A., Abido, M.A.: High-level penetration of renewable energy sources into grid utility: challenges and solutions. IEEE Access **8**, 190277–190299 (2020). https://doi.org/10.1109/ACCESS.2020.3031481
2. Raval, D.Y., Pandya, S.N.: Investigation on impact of rooftop solar system on LV distribution network. In: Mehta, A., Rawat, A., Chauhan, P. (eds.) Advances in Electric Power and Energy Infrastructure. LNEE, vol. 608, pp. 187–198. Springer, Singapore (2020). https://doi.org/10.1007/978-981-15-0206-4_16
3. Shen, W., et al.: A comprehensive review of variable renewable energy levelized cost of electricity. Renew. Sustain. Energy Rev. **133**, 110301 (2020). https://doi.org/10.1016/j.rser.2020.110301
4. Singh, S., Kumar, S., Kalla, U.K., Chandra, A., Saad, M.: Optimization of Rooftop PV system deployment for LV distribution network. In: 2021 International Conference on Sustainable Energy and Future Electric Transportation (SEFET), pp. 1–8 (2021). https://doi.org/10.1109/SeFet48154.2021.9375688
5. Jiang, J., Peyghami, S., Coates, C., Blaabjerg, F.: A comprehensive study on reliability performance of photovoltaic-battery-based microgrids under different energy management strategies. J. Energy Storage **43**, 103051 (2021). https://doi.org/10.1016/j.est.2021.103051
6. Marino, C.A., Marufuzzaman, M.: A microgrid energy management system based on chance-constrained stochastic optimization and big data analytics. Comput. Ind. Eng. **143**, 106392 (2020). https://doi.org/10.1016/j.cie.2020.106392
7. Hassan, M.U., Rehmani, M.H., Chen, J.: DEAL: differentially private auction for blockchain-based microgrids energy trading. IEEE Trans. Serv. Comput. **13**(2), 263–275 (2020). https://doi.org/10.1109/TSC.2019.2947471
8. Zhu, L., Yan, Z., Lee, W.-J., Yang, X., Fu, Y., Cao, W.: Direct load control in microgrids to enhance the performance of integrated resources planning. IEEE Trans. Ind. Appl. **51**(5), 3553–3560 (2015). https://doi.org/10.1109/TIA.2015.2413960
9. Chen, Q., Xia, M., Zhou, Y., Cai, H., Wu, J., Zhang, H.: Optimal planning for partially self-sufficient microgrid with limited annual electricity exchange with distribution grid. IEEE Access **7**, 123505–123520 (2019). https://doi.org/10.1109/ACCESS.2019.2936762
10. Panwar, M., Zimmerle, D., Suryanarayanan, S.: Data analysis and visualization for electric microgrids: a case study on the FortZED RDSI microgrid. In: 2013 IEEE Green Technologies Conference (GreenTech), pp. 330–337, April 2013. https://doi.org/10.1109/GreenTech.2013.57
11. Liu, Y.-J., Chen, S.-I.: Development of a simulation-based electricity analysis environment for a community microgrid system. In: 2017 International Conference on Applied System Innovation (ICASI), pp. 1461–1464, May 2017. doi:https://doi.org/10.1109/ICASI.2017.7988192
12. Marguš, H., Žnidarec, M., Šljivac, D., Stojkov, M.: Performance analysis of upgraded university building of FERIT Osijek microgrid achieving nearly zero energy standard based on real measurement data. In: 2019 International Conference on Smart Energy Systems and Technologies (SEST), pp. 1–6, September 2019. https://doi.org/10.1109/SEST.2019.8849138
13. Sun, X., Jiang, B., He, X.: Database query optimization based on distributed photovoltaic power generation. In: 2018 2nd IEEE Advanced Information Management, Communicates, Electronic and Automation Control Conference (IMCEC), pp. 2382–2386, May 2018. https://doi.org/10.1109/IMCEC.2018.8469696
14. Holdmann, G.P., Wies, R.W., Vandermeer, J.B.: Renewable energy integration in Alaska's remote islanded microgrids: economic drivers, technical strategies, technological niche development, and policy implications. Proc. IEEE **107**(9), 1820–1837 (2019). https://doi.org/10.1109/JPROC.2019.2932755

15. Salama, H.S., Said, S.M., Aly, M., Vokony, I., Hartmann, B.: Studying impacts of electric vehicle functionalities in wind energy-powered utility grids with energy storage device. IEEE Access **9**, 45754–45769 (2021). https://doi.org/10.1109/ACCESS.2021.3066877
16. Li, H., Lu, Z., Qiao, Y., Zhang, B., Lin, Y.: The flexibility test system for studies of variable renewable energy resources. IEEE Trans. Power Syst. **36**(2), 1526–1536 (2021). https://doi.org/10.1109/TPWRS.2020.3019983

Technological and Entrepreneurial Factors as Drivers of Private Equity Funding in Morocco

Mimoun Benali[1] and Jawad El Ghalfiki[2](✉)

[1] Finance and Economics at National School of Business and Management, Sidi Mohamed Ben Abdellah University Fez, Fez, Morocco
mimoun.benali@usmba.ac.ma
[2] National School of Business and Management, Sidi Mohamed Ben Abdellah University Fez, Fez, Morocco
jawad.elghalfiki@usmba.ac.ma, j.elghalfiki@gmail.com

Abstract. This paper investigates technological and entrepreneurial factors as drivers of private equity fundraising in Morocco. Based on ARDL analysis, the results of this research indicate that, in the short run, only productivity has a significant and negative effect on fundraising. Then, patents have no significant impact in the long run, while the number of new businesses, productivity, research, and development has a considerable effect. Finally, these findings demonstrated that the government must enhance the entrepreneurial ecosystem to promote private equity investments.

Keywords: ARDL · Entrepreneurship · Private equity · Technology

1 Introduction

The promotion of innovation is among the objectives of the 2020 finance bill, with - among other things - the "strengthening of mechanisms for the valorization of scientific research and innovation results" [1]. Startups, especially high-growth ones, are essential sources of innovation. For this reason, Moroccan policymakers are working to create effective ecosystems where innovative startups can thrive and grow. One of the pillars of this ecosystem is financing. Indeed, access to finance is a crucial factor in the creation [2, 4], survival [5], and growth [6] of firms, especially for small [7] and innovative firms [8]. Innovative firms face many challenges related to information asymmetry and communication noise between innovative business founders and traditional financing systems. One of the critical challenges young innovative companies face is the cautious behavior of some funders. This caution often results in the eviction of young innovative companies from traditional financing channels [9]. As a result, private equity as a source of innovation [10] is at the core of this innovative financing dynamic in Morocco.

Unemployment remains a significant economic challenge for Morocco. Given the unstable, dual, and inefficient nature of the Moroccan labor market, it can be seen as a considerable constraint to Morocco's emergence. The result is a misallocation and underutilization of the workforce. Nevertheless, there are opportunities to strengthen

S. Motahhir and B. Bossoufi (Eds.): ICDTA 2022, LNNS 454, pp. 97–106, 2022.
https://doi.org/10.1007/978-3-031-01942-5_10

the Moroccan labor market and increase its participation in economic growth. Private equity investments are likely to contribute to job creation by stimulating entrepreneurship and contributing to business growth.

Established literature on the drivers of private equity investment focuses principally on venture capital investment determinants. There is a lack of studies on the determinants of private equity investment more generally, and they tend to focus on developed economies, particularly in Europe and the United States. In addition, there is limited interest in studying the determinants of private equity in developing and African markets. That presents a gap in the existing literature. There seems to be an enormous scope of research to explore the determinants of private equity investments (in its entirety) in developing markets (Moroccan case).

This paper aims to contribute to the literature on the determinants of private equity. We enrich the existing literature to highlight the role of technological and entrepreneurial factors in Morocco's private equity dynamics.

2 Moroccan Private Equity Market: A Strong Growth Trend

Moroccan Private Equity Association (MPEA) considers that private equity is a financial activity that consists of a professional investor entering, for a determined time, in the capital of companies not listed on the stock market, in need of equity. Private equity supports the company at all stages (startups, growth, transmission, or recapitalization). The dynamics of the private equity market are essentially measured by the amounts raised and invested by private equity funds and companies.

The Moroccan private equity market has experienced significant growth since the mid-1990s. Investments made by private equity investors have grown rapidly and regularly. The cumulative amount of capital invested has risen from 1,808 million Moroccan Dirham (MAD) in 2008 to 8.5 billion MAD in 2019.

Funds raised, which are more cyclical, have seen a much more uneven trend: the high point in 2013 was followed by a period of decline, then a recovery in activity from 2017. Over the long term, there is a significant evolution. While the total amount of capital raised totaled 6.2 billion MAD in 2008, this amount reached 20.8 billion MAD in 2019.

3 Technological and Entrepreneurial Factors as Drivers of Private Equity: Literature Review

A country needs to have a genuine research and development culture To promote a private equity market. Indeed, there is a strong correlation between research and development spending and the number of patents, on the one hand, and private equity investment, on the other. Thus, Groh and Wallmeroth [11] conclude in their study that the level of entrepreneurship interacts with the R&D capital stock, technological opportunities, and the number of patents. We will discuss determinants from four aspects: research and development, entrepreneurial dynamics, innovation capacity, and the level of human capital.

3.1 Research and Development

According to the literature review, high levels of research and development (R&D) spending are associated with a positive impact on the level of private equity transactions (particularly venture capital and development capital). The argument is that a high level of research and development enhances the ability of firms to create new products or services and therefore increases the opportunities for private equity investment.

Gompers and Lerner [12], who examined venture capital transactions in the United States over 1972–1994, found that university and industry R&D spending in the United States in the mid-1990s had a substantial and positive impact on venture capital investment period. Schertler's [13] research on venture capital deals in fourteen European countries over the period 1988–2000 shows the presence of a positive and significant correlation between the level of R&D spending and the level of venture capital growth. Another study, that of Romain and Van Pottelsberghe [14], conducted on a larger panel of venture capital operations in sixteen OECD countries over the period 1990–2000, also shows a positive impact of R&D on venture capital. It underlines the role of the R&D growth rate as an indicator of a country's research dynamics and its important and beneficial influence on venture capital activities. Da Rin obtains a similar result and al [15], Schröder [16], Felix and al. [17], and Groh and Wallmeroth [11], although there are some exceptions in this regard e.g., Bonin and Alkan [18].

3.2 Entrepreneurial Environment

Gompers and Lerner [19] were among the first to establish an empirical relationship between entrepreneurial dynamics and the level of available venture capital. In this perspective, Romain and van Pottelsberghe [20] point out that any economic system is supposed to guarantee a minimum level of entrepreneurship and entrepreneurial activity to stimulate the demand for venture capital.

Previous research also points out that personal relationships are more critical in regions with more traditional cultures and that social capital is more helpful in making private equity funds available to entrepreneurs in these regions than in others [21]. Informal institutions determine individuals' risk aversion attitudes toward entrepreneurship and influence entrepreneurs' desire to develop business and seek help from external investors.

3.3 Innovation Capacity

The fundamental characteristic of the entrepreneur is his ability to innovate [22]. Innovation is the multi-step process by which organizations transform ideas into new or improved products, services, or processes to advance, compete, and successfully differentiate themselves in their markets [23]. A sustained level of innovation and technological improvement helps to stimulate economic growth and generate more investment opportunities.

Innovations generally lead to more patent applications and startups. That, in turn, can increase the demand for private equity. Therefore, the number of patent applications can measure technological innovations in a country. In addition, as many research findings reveal, there is a positive relationship between the number of patents registered and the dynamics of private equity [13, 16, 20, 24].

In addition, technological opportunities can encourage entrepreneurship and stimulate venture capital investments [12]. Venture capital investments focus on startups exploiting scientific discoveries and converting inventions and innovations into products, goods, and services [25].

3.4 Human Capital and Productivity

The quality of education of the population is a crucial determinant of a country's economic growth rate [26]. It is generally recognized that countries benefit from investing in public education: by augmenting spending on education, the dexterities of employees increase, and, consequently, these increased skills promote economic growth.

Farag and al [27] have addressed this issue and studied the private equity markets in Hungary, the Czech Republic, and Poland, compared to Germany. The most significant handicap in these countries is the lack of managerial skills to manage private equity-backed companies. These results confirm the findings of Bliss [28], Karsai, and al. [29] and Klonowski [30] for Central and Eastern European Countries and with Groh and Liechtenstein [31] for a broader panel of emerging economies.

4 Data and Methodology

This section discusses the empirical methodology. It presents variables used and the method of data processing.

4.1 Variable and Description

We examine the variables included in the empirical analysis model to assess their impact on the Moroccan private equity market dynamics.

4.2 Dependent Variable

As suggested by Precup [32], in our case, the private equity proxy is measured by the total amount of capital raised in Morocco. This choice is consistent with Gompers and Lerner [12], who analyzed the determinants of private equity fundraising, as well as Poterba [14], Jeng, and Wells [33] (Table 1).

Table 1. Dependent variable and data source

Variable	Abbreviation	Description	Source
Fundraising	FONLEV	The annual amount of funds raised by private equity funds in billions of MAD	MPEA

4.3 Independent Variables

We retained four dependent variables to assess the Moroccan private equity market's technological and entrepreneurial drivers (Table 2).

Table 2. Independent variables and data source

Variable	Abbreviation	Description	Source
Research and development	R_D	Research and development expenditure as a percentage of GDP	World Bank
Patents	BREVINV	Patent applications by residents	World Bank
Number of new businesses	ENTRCRE	The number of new businesses registered	World Bank
Productivity	PRODTE	Level of production factors (current purchasing power parities)	Penn World Table

4.4 Empirical Model Estimation Method: ARDL Approach

Autoregressive Distributed Lag (ARDL) allows, on the one hand, to test long-term relationships by using the bounds test on series that are not integrated of the same order and, on the other hand, to obtain better estimates on small samples [34].

Therefore, ARDL offers the ability to address long-term dynamics and short-term adjustments simultaneously. For this reason, we use this approach to explore the impact of technological and entrepreneurial factors on the dynamics of private equity in Morocco.

Given the dependent variable "*Yt*" and the independent variable "X*t*," in its general form, an ARDL model is described as follows:

$$Y_t = \varphi + a_1 Y_{t-1} + \cdots + a_p Y_{t-p} + b_0 \mathrm{X}_t + \cdots + b_q \mathrm{X}_{t-q} + e_t.$$

With: $e_t \sim$ ii*d* (0, σ): error term; "b_0" reflects the short-term effect of X_t on Y_t.

5 Results and Interpretations

All variables are either integrated of order 0 or 1, which confirms using the ARDL model for our modeling (Table 3).

We used the "Variance Inflation Factor" (VIF) to eliminate variables with a VIF more significant than 5 (Table 4).

The results of the cointegration bounds test are given in Table 5. The test statistic is well above their critical values to conclude cointegration between the variables.

Table 3. Dickey-Fuller unit root test

BRVINV	−2.077748	−4.125899***	I(0)
ENTRCRE	−1.574059	−3.936522***	I(1)
PRODVTE	−0.645888	−10.99346	I(1)
R_D	−5.547367***		I(0)

Robust standard errors in parentheses *** $p < 0.01$

Table 4. Multicollinearity test

Variable	VIF
BRVINV	1.598681
ENTRCRE	2.181299
PRODVTE	2.255473
R_D	1.012515

Table 5. Cointegration bounds test

Test statistic	Value	Significance	I(0)	I(1)
Asymptotic: n = 1000				
F-statistic	19.35356	10%	2.2	3.09
K	4	5%	2.56	3.49
		2,5%	2.88	3.87
		1%	3.29	4.37

Table 6. Long-term relationship

Variable	Coefficient	Std. error	t-statistic	Prob.
D(BRVINV)	0.431346	0.884777	0.487520	0.6432
D(PRODVTE)	−9347.642	2266.643	−4.124002	0.0062
CointEq(−1)*	−1.045681	0.071668	−14.59072	0.0000
R-squared	0.955458	Mean dependent		52.78571
Adjusted R-squared	0.947360	S.D. dependent		759.2794
S.E. of regression	174.2048	Akaike info criterion		13.34575
Sum squared residual	333820.2	Schwarz criterion		13.48269
Log-likelihood	−90.42024	Hannan-Quinn criterion		13.33307
Durbin-Watson stat	3.163325			

In the long term, there is a cointegration relationship between the variables. Apart from the variable BRVINV, which has no significant effect, ENTRCRE, PRODVTE, and R_D considerably impact the long run. The impact of ENTRCRE and R_D is positive, while that of PRODVTE is negative (Table 6).

The error correction term is significant and negative, indicating the existence of an error correction mechanism. The value of our error correction term suggests that an imbalance on an exogenous variable is corrected after 1/1.04 year (about one year). In the short run, only PRODVTE has a significant and negative effect on fundraising.

In the following table, we use the model results to determine the unidirectional causality of the variables. Apart from the variable BRINV, there is a causal relationship between the other technological and entrepreneurial variables and fundraising (Table 7).

Table 7. Causality test

Variable	F-statistic	df	Probability	Results
BRVINV	2.803925	(2, 6)	0.1381	BRVINV does not cause the fundraising
ENTRCRE	13.39588	(1, 6)	0.0106	ENTRCRE cause fundraising
PRODVTE	10.24481	(2, 6)	0.0116	PRODVTE cause fundraising
R_D	8.828312	(1, 6)	0.0249	R_D cause fundraising

The p-value of our significativity test is 0.6090 > 5%. Therefore, we can conclude that our errors follow a normal distribution (Table 8).

Table 8. Normality test of the residues

Mean	4.16e−12
Median	−27.73227
Maximum	361.1224
Minimum	−257.6729
Std.Dev	160.2451
Skewness	0.642799
Kurtosis	3.217097
Jrque-Bera	0.991604
Probability	0.609082

The p-value of our autocorrelation test is 0.0511 > 5%, so we can conclude that our errors are not auto-correlated (Table 9).

The P-value of the heteroscedasticity test (Breusch-Pagan-Godfrey) is more than 5%. We can conclude that our errors are homoscedastic (Table 10).

Table 9. Autocorrelation test

F-statistic	Prob. F(18, 3)
2.992408	0.1014

Table 10. Heteroscedasticity test (Breusch-Pagan-Godfrey)

F-statistic	Prob. F(18, 3)
2.992408	0.1014

6 Conclusion

The cost of doing business in Africa, by global standards, can be high. That is generally due to a complex entrepreneurial environment, characterized by increased political risks, bureaucratic burdens, underdeveloped legal systems, inadequate and unreliable infrastructure, an insufficient supply of qualified professionals of local origin [35]. Despite institutional reforms implemented by the government in Morocco, much remains to be done in this area [36].

The lengthy paperwork and related expenses are perceived as a concern for equity investors in Morocco [37]. Procedures on tax declaration, records, registrations, and approvals are tedious and in dire need of relief. Administrative burdens hinder the profitability of investments and deter the creation of new businesses. Therefore, policymakers are called upon to reduce the number of administrative procedures perceived as unnecessary and improve their efficiency to stimulate the entrepreneurial culture and reduce costs and risks. There is also a need to improve access to information for entrepreneurs and investors to help them understand the processes and requirements.

One of the main challenges facing the private equity market in Morocco is the lack of local talent, i.e., the lack of management expertise on the ground. Indeed, private equity involves an extensive network and technical skills that the Moroccan government must deploy to achieve investors' desired results [35]. That suggests that the government should improve higher education quality and strengthen the link between the education system and the labour market. That starts with developing a national program to develop entrepreneurship in educational institutions. Secondly, establishing training platforms for entrepreneurs allows access to a database and a list of experts/coaches to upgrade project leaders on various topics (finance, marketing, communication, business plans, and others). Finally, strengthening the participation of private operators in educational programs (Universities and training institutes) and the stimulation of creativity and imagination through the multiplication of initiation actions to the principles of innovation and creativity in centres dedicated to youth (cultural centers, centers of artistic initiation, libraries, museums, and others) [37].

Finally, encouraging the use of technology and digital innovation in Morocco is expected to positively impact the dynamism of Moroccan companies and, therefore, on investment in private equity. The Moroccan government has a vital role as a catalyst in developing innovation. It is a matter of increasing R&D research spending, strengthening the institutional structure of digital innovation policy, improving the private sector's contribution to the financing of digital innovation, and fostering the spirit of innovation in companies.

References

1. EL Yamlahi, I.: Le financement de l'innovation par le Capital-Risque: Cas du Maroc, Rev. Econ. Kap. **15** (2019)
2. Cassar, G.: The financing of business startups. J. Bus. Ventur. **19**(2), 261–283 (2004)
3. Popov, A., Roosenboom, P.: Venture capital and new business creation. J. Bank. Finance **37**(12), 4695–4710 (2013)
4. Kim, D.-H., Lin, S.-C., Chen, T.-C.: Financial structure, firm size and industry growth. Int. Rev. Econ. Finance, **41**, 23–39 (2016)
5. Tsoukas, S.: Firm survival and financial development: evidence from a panel of emerging Asian economies. J. Bank. Finance **35**(7), 1736–1752 (2011)
6. Rahaman, M.M.: Access to financing and firm growth. J. Bank. Finance **35**(3), 709–723 (2011)
7. Beck, T., Demirguc-Kunt, A.: Small and medium-size enterprises: access to finance as a growth constraint. J. Bank. Finance **30**(11), 2931–2943 (2006)
8. Lee, N., Sameen, H., Cowling, M.: Access to finance for innovative SMEs since the financial crisis. Res. Policy **44**(2), 370–380 (2015)
9. Diamane, M., Koubaa, S.: Financement des entreprises innovantes au Maroc Etat des lieux. Rev. L'entrepreneuriat L'innovation **1**(3) (2017)
10. Lerner, J., Sorensen, M., Strömberg, P.: Private equity and long-run investment: the case of innovation. J. Finance **66**(2), 445–477 (2011)
11. Groh, A.P., Wallmeroth, J.: Determinants of venture capital investments in emerging markets. Emerg. Mark. Rev. **29**, 104–132 (2016)
12. Gompers, P., Lerner, J.: What drives venture capital fundraising? (1999)
13. Schertler, A.: Driving forces of venture capital investments in Europe: a dynamic panel data analysis, no. 1172 (2003)
14. Poterba, J.M.: Venture capital and capital gains taxation. Tax Policy Econ. **3**, 47–67 (1989)
15. Rin, D., Marco, G.N., Sembenelli, A.: Public policy and the creation of active venture capital markets. J. Publ. Econ. **90**, 8–9 (2006)
16. Schröder, C.: Financial system and innovations: determinants of early-stage venture capital in Europe. Financ. Mark. Integr. Growth, 127–158 (2011)
17. Félix, E.G.S., Pires, C.P., Gulamhussen, M.A.: The determinants of venture capital in Europe - evidence across countries. J. Financ. Serv. Re-Search, **44**(3), 259–279 (2013)
18. Bonini, S., Alkan, S.: The political and legal determinants of venture capital investments around the world. Small Bus. Econ. **39**(4), 997–1016 (2014)
19. Gompers, P., Lerner, J.: An analysis of compensation in the US venture capital partnership. J. Financ. Econ. **51**(1), 3–44 (1999)
20. Romain, A., Pottelsberghe, B.V.: The determinants of venture capital: a panel data analysis of 16 OECD countries. Res. Inst. Manag. Sci. Work. Pap. **5** (2004)
21. Batjargal, B., Liu, M.: Entrepreneurs' access to private equity in China: the role of social capital. Organ. Sci. **15**(2), 159–172 (2004)
22. Schumpeter, J.A.: Bus. Cycles **1** (1939)
23. Bareghheh, A., Rowley, J., Sambrook, S.: Towards a multidisciplinary definition of innovation. Manag. Decis. (2009)

24. Guler, I., Guillén, M.F.: Institutions and the internationalization of US venture capital firms. J. Int. Bus. Stud. **41**(2), 185–205 (2010)
25. Kortum, S., Lerner, J.: Does venture capital spur innovation?. Emerald Group Publishing Limited (2001)
26. Stevens, P., Weale, M.: Education and Economic Growth, vol. 13, p. 2013. National Institute of Economic and Social Research, London (2003)
27. Farag, H.: Contracting, monitoring, and exiting venture investments in transitioning economies: a comparative analysis of Eastern European and German markets. Venture Cap. Int. J. Entrep. Finance **6**(4), 257–282 (2004)
28. Bliss, R.T.: A venture capital model for transitioning economies: the case of Poland. Venture Cap. Int. J. Entrep. Finance **1**(3), 241–257 (1999)
29. Karsai, J.: Screening and valuing venture capital investments: evidence from Hungary, Poland and Slovakia. Entrep. Reg. Dev. **10**(3), 203–224 (1998)
30. Klonowski, D.: Venture capital as a method of financing enterprise development in Central and Eastern Europe. Int. J. Emerg. Mark. (2006)
31. Groh, A.P., Liechtenstein, H.: International allocation determinants for institutional investments in venture capital and private equity limited partnerships. Int. J. Bank. Account. Finance **16**(3), 2–3 (2011)
32. Precup, M.: The future of private equity in Europe-the determinants across countries. Roman. J. Eur. Aff. **15**, 72 (2015)
33. Jeng, L.A., Wells, P.C.: The determinants of venture capital funding: evidence across countries. J. Corp. Finance **6**(3), 241–289 (2000)
34. Narayan, P., Smyth, R.: Trade liberalization and economic growth in Fiji. An empirical assessment using the ARDL approach, pp. 96–115 (2005)
35. Babarinde, O.: The private equity market in Africa: trends, opportunities, challenges, and impact. J. Priv. Equity **16**(1), 56–73 (2012)
36. Benali, M., El Ghalfiki, J.: Access to venture capital in Africa: the role of public institutions and corporate governance. Int. J. Bus. Perform. Manag. **22**(2/3), 22 (2021)
37. M.P.E.A.: Diagnostic: capital amorçage et risque au Maroc, p. 27 (2018). https://amic.org.ma/publications/. Accessed 27 Aug 2021

Literature Review Around: Sustainable Urban Logistics

Fadwa Farchi[1(✉)], Badr Touzi[2], Chayma Farchi[1], and Charif Mabrouki[1]

[1] Hassan 1er University, B.P: 577, Casablanca Road, 26000 Settat, Morocco
f.farchi@uhp.ac.ma
[2] Mohammed V University, B.P: 8007.N.U, 100000 Rabat, Morocco

Abstract. Currently, we observe large numbers associated with internal urban migration exacerbating the environmental urban scenario, which is the outcome of various prior circumstances, and this is why urbanization is becoming more widespread. This is true of internet shopping, globalization, and damaging industrial and human resources, to name a few examples. The key solutions are primarily involved with urban mobility, the involvement of the relevant actors, trip optimization, the usage of lanes and roads, and vehicle selection. One of the most significant foundations of the notion of "urban logistics" is sustainability, which is one of the main parts of flow management. Indeed, the ultimate goal extends beyond the delivery of commodities, promoting the environment as a vital component of urban logistics. We will attempt to solve the following problem: Where are we in terms of urban logistics? To address this issue, we used the following strategy: Presentation of the fundamental principles, selection of articles to be used in the study, identification of the major axes, actors, and clarification of the most important axis.

Keywords: Sustainable urban logistics · Literature review · Bibliographic study

1 Introduction

The future vision in the urban logistic context or the "Smart city" is a framework for urban development strategies that incorporate multiple levels of development. The state of the world's atmosphere is constantly worsening., and all stakeholders face a problem in recognizing this negative equilibrium [1–4]. Industrials are the first to be singled out, due to internal production or machining, both axis have extremely high pollution levels. Because of the complexities of urban mobility and traffic, urban logistics is included in this green strategy [5, 6].

In particular, city logistics as a concept aims to optimize freight transportation management on the one hand while reducing undesirable impacts on the other (non-respect of the security of goods & people, air or noise pollution and congestion), indeed, this subject relies on many expectations and realities [7, 8].

The offered solutions are varied, but they are not a subject for any correct transfer from one representation to another. As a result, city logistics need to be able to respond

S. Motahhir and B. Bossoufi (Eds.): ICDTA 2022, LNNS 454, pp. 107–119, 2022.
https://doi.org/10.1007/978-3-031-01942-5_11

to a variety of contradictory restrictions concerning the economic, the environment, and urban planning, also, most importantly, a better logistics organization [9, 10].

To do this we fare using a simple plan. The structure of our paper is as follows: In the next paragraph we carry out a bibliographical study, we explain the method of selecting the articles that make up our database as well as analyzing the data of this latter. In Sect. 3 we introduce urban logistics, its definitions, its pillars as well as the existing distribution methods and types of transports. We conclude the article with a summary of the most important axes by giving a clear idea of the possible solutions that can be adopted.

2 Bibliographic Studies

At this level we will show how we built our database. and this, based on a method of sorting and selection which will be explained later. Also, we will analyze the results of the data collected according to the results of other researchers and our point of view.

2.1 Methodology

This first section shows the methodology we followed and how we did build and select our database.

Analysis of Methodologies and Approaches

We came across a variety of research and analysis tools and methodologies while constructing our database. The most common method is literature reviews, which are followed by case studies reviews. Mass studies are important as well.

We also found other techniques that were encountered only once and were therefore ignored at the graph level. Here are some examples of symbolic types:

This treatment is limited to hundreds of articles retained in our database (Fig. 1).

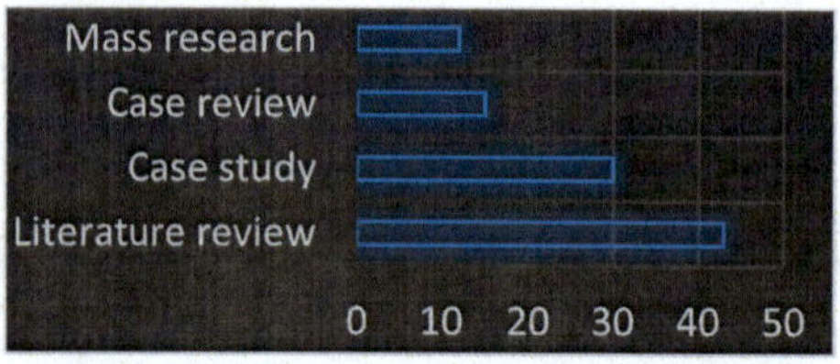

This treatment is limited to hundreds of articles retained in our database.

Fig. 1. Percentage of methodologies

The next section explains the methodology we built this database, as well as the choice and sorting criteria we used.

Selection of Articles

We conducted a synthesis research on last-mile logistics in order to build our database. To visualize the issue in its whole while widening our area of reflection. The latter relies primarily on a bibliographic study. This theoretical approach necessitates a wide range of readings in order to broaden our understanding of terms like the ecologically sustainable modes of transportation and last mile coordination.

The final objective of this critical documentary is to address the following issue: Urban logistics: where are we? To solve this problem, we used the largest search engines: Science Direct, Elsevier and Google Scholar to compile the largest number of articles published between 2015 and 2021 which deal with our subject. Initially, we tried to constitute a total sample of 546 articles centered on urban logistics in order to identify the key terms of the theme. Secondly, we sorted the articles according to a few key terms (freight transport, urban logistics, smart city, sustainable development) to eliminate those that did not answer the questions. A sample of 100 articles was chosen. The majority of articles -71%- discuss the relevance of sustainability and viability (Fig. 2).

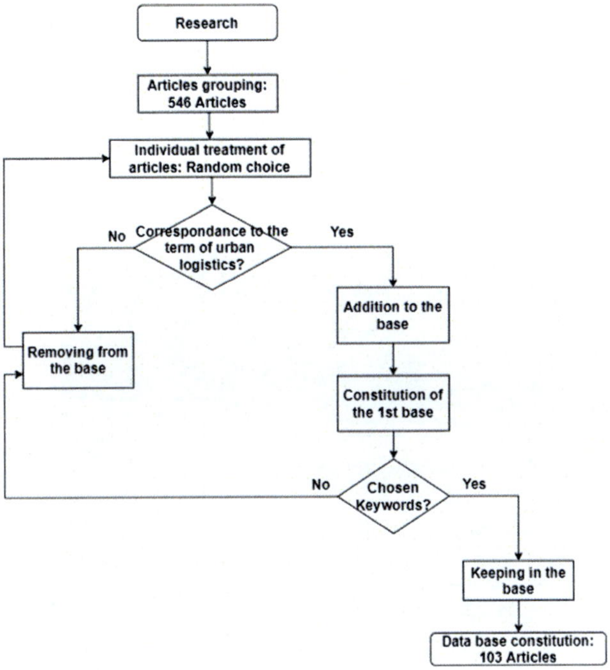

Fig. 2. Flowchart selection of articles

2.2 Analyzing Data

The analysis of the selected papers is described in the following section.

Our database contains publications published between 2015 and 2021. With a share of 54%, we arranged our collection mostly by publications published in 2020 (Fig. 3).

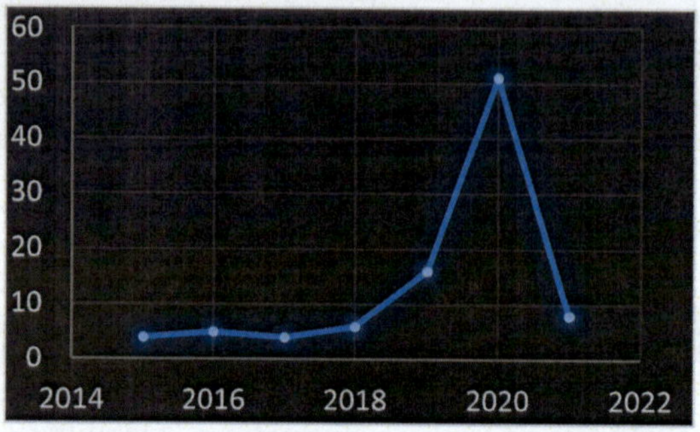

Fig. 3. Years chart

The topic of urban logistics has piqued the interest of researchers all around the world. China, on the other hand, continues to be the country with the most urban logistics publications. Unlike many countries, China is one of the few countries, if not the only one, that did not alter its searching practices or include it in scientist publications (Fig. 4 and Table 1).

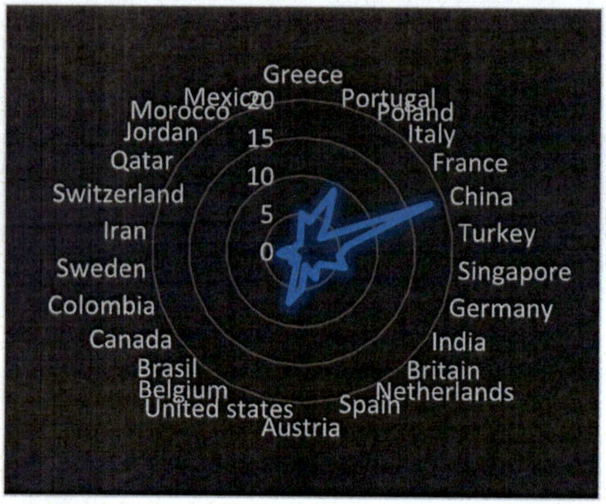

Fig. 4. Locations chart

More than two thirds of the articles analyze urban logistics in a broad way by adopting a large view, whereas contributions that are focused on a specific domain and geared toward a representative framework make up only a small percentage of the total.

3 Urban Logistics

As long as we are talking about urban logistics, it is important to define it. In this section, we will discuss urban logistics definitions, parts, pillars, distribution and transport types.

Table 1. Proportion per field

Fields	Proportion
Omnichannel B2C transport	69.6%
General	8.7%
E-commerce	6.4%
FMGC industry	2.1%
Construction	2.2%
Logistics 4.0	2.5%
Uncertain markets	2%
Smart logistics zones	2%
Intermodal terminals	2.3%
Cold chain	2.2%

3.1 Definitions

The term "last mile" was coined in the telecommunications industry. According to [11] last mile logistics is now allocated to the last link or mile in the process of distribution, that is typically regarded the most demanding and inefficient component of the value and supply chain [12].

To help clarify the idea of urban logistics, we've gathered different author perspectives on the topic [13] (Table 2):

Table 2. Authors' definitions around "Urban Logistics"

Author	Definition
The Organization for Economic Cooperation and Development	Freight distribution and delivery to the consumer in cities and suburbs (not only to retail but also by other activity areas such as manufacture), including return-reverse logistics for consumed, damaged, defective or non-compliant products, in the form of clean waste
Taylor Michael and Coll	At the city or metropolitan area level, logistics is a system of managing the movement of products and data flows. "In a market economy, it is the process where private enterprises' logistics and transportation activities are optimized by using pushed and advanced information systems for the management of congestion, energy resources, traffic and security in agglomerations"

(continued)

Table 2. (*continued*)

Author	Definition
Dablanc	A service that contributes to the efficient management of goods flow in a city environment
Pipame	Transport of goods through densely populated areas, their treatment in these places, the distribution to the final recipient and the management of reverse flows (damaged products, products to be recycled and waste) As a result, it is the "last link" in the supply chain, the most visible for end consumers but also one of the most demanding in terms of values (about a quarter of the cost of the total supply chain). Thus, the transport of goods is more difficult and complex in urban areas, where a large part of consumption is concentrated
Taniguchi et al.	Within the framework of a market economy, the process of fully optimizing private enterprises' logistics and transportation activities in metropolitan regions by giving importance to certain elements such as the traffic environment, energy usage, and congestion
Durand and Senkel	Urban logistics concerns 2 types of flows; • Flows related to urban management, primarily generated by waste transportation, specific needs of public services, removals, home deliveries, postal services, and hospitals; • Flows related to end-user supply (i.e. purchasing trips but also new delivery practices at home or near the point of consumption)

"Urban logistics" has a lot of different definitions. It does, however, have some limits. Many elements are required such as viability, optimal consumption, passengers' transport, commodities and information flow management and means of transportation are all crucial concepts to include in the description.

Like traditional logistics, which is concerned with the management of worldwide movements connected with industry, urban logistics is a subset of the branch that deals with the same elements in the city. Specifically, urban logistics is "the application and assistance of conventional logistics content with the goal of maximizing the long-term management of good, people, and information in the urban context: urbanized or peripheral areas, main cities and inter-city areas also". The latter masters the construction of urban infrastructures, selection of vehicles, coordination between actors and sustainable thinking.

3.2 Pillars of Urban Logistics

The actual functionality and goal of urban logistics vary greatly from one author to the next by depending on a variety of axes. The solutions are based on the environmental and infrastructural conditions of each urban region.

For the two cities of Barcelona and Valencia, [14] find that the usage of E-cycles and the installation of transhipment stations or micro-distribution platforms perform well. [15, 16] for example, recommend cyclical logistics or cargo bikes. According to [17] the selection of logistics areas should be based on a multi-criteria analysis that includes everything from primary selection to logistics area size. [18] looked at the SMART planning process for freight transportation from three perspectives: the desk, the Living Lab, and the modeling approach (Fig. 5).

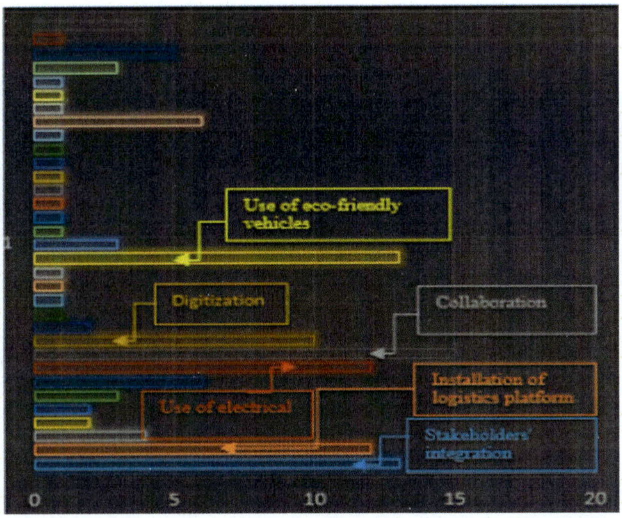

Fig. 5. Pillars of urban logistics

Customer satisfaction assessment will need to be automated, according to [19]. Similarly, [20, 21] agree that smart techniques must be included into urban logistics.

[22] associates urban logistics with the concept of "Smart City," at a time when [23] link sustainable urban logistics to proper policy assessment.

According to [24, 25], clean last-mile transportation is dependent on a vast range of soft measures, operator collaboration, and stakeholder involvement.

[26] prefer to group passenger and freight transportation in an energy optimization framework, however [27] believe that grouping the first and last kilometers is preferable.

The utilization of underground or rail urban logistics is gaining traction in China [28]. [29] argue that integrating electric mobility is critical for reducing energy consumption and urban pollution.

3.3 Urban Logistics' Distribution Systems

We determine from our analysis that freight can be transported from the factory to the final customer by one of these three modes of urban delivery systems [30, 34]:

- Scooters, cargo bikes, bicycles, and Bullitt-bikes, etc. are the most efficacious forasmuch as they can access to specific narrow locations without causing danger to people's lives.
- Electric means of transport: Transportation systems that use new computer and telecommunication technology to make them able of reacting and adapting themselves to real-world situations. – E-tricycles, electric vans, and bicycles.
- Common transport: This refers to the capability of combining both man and freight inside one same mode of transport – Metro, trains, buses are all available (Table 3).

Table 3. Types of transport means analysis

Type of transport	Advantages	Disadvantages
Cyclic transport	Can be used in confined spaces Doesn't put passengers in danger Can perform a variety of alternations	Reduced capacities Non-respectful for driver's safety Requires specific infrastructure Limited velocity and speed
Intelligent transportation	Adaptable to delivery's requirements Vehicles continuously tracked	Not appropriate for all cultures Vehicles that necessitate frequent and costly maintenance Problem with battery charging
Public transport	Delivery based on public transportation routes Use of current public transportation	Several back-and-forth distributions Containers for commodities are required Vehicles that are well-equipped for the type of the materials being conveyed

It is necessary to write down that the best settlement is a ternary combination that prioritizes green thoughts (safety and environment), affordability or cost, and efficiency.

To accurately select the most appropriate mode of transportation, it is necessary to remind that these sorts of distribution systems can implicate the customer/consumer who can either have it delivered or pick up their own goods based on three options as well [14–29, 35–40]:

- The HYBRID system: the product is first transferred to an intermediate agent, who keeps it until the consumer arrives at the intermediate site to pick it up.
- The PULL system: The consumer is the most important link in the delivery process. He is responsible for collecting goods by his own assets.
- The PUSH system: The customers are not concerned or embroiled by the delivery process; the good is delivered to the consumer immediately by the industrials or by a third-party transportation company (Table 4).

Table 4. Systems of distribution analysis

System	Advantages	Disadvantages
Hybrid	Recovering the items based on the availability of the consumer Companies profit from a partial reduction in transportation expenses	Consumer aversion Price & value of intermediate storage
Pull	Recovering the items based on the availability of the consumer Companies gain from a whole reduction in transportation costs	Costs of additional transportation Use of route that isn't optimal
Push	Transport delegated to a freight forwarder	The existence of several distributions Manufacturers will have more work to do

A variety of factors go into deciding on a distribution system. For starters, it has to do with the form of transportation used to convey commodities. Second, the involved parties' culture and opposition have a significant impact on system selection: fewer-priced solutions can be refused by the customer. Thirdly, the expenses caused by the mobility of consumers are never borne by the carrier, but always create an unreasonable burden if the customer has to travel long distances.

3.4 Urban Logistics' Segments

A body of interconnections and dependencies are used to characterize urban logistics activity. The goal of segmenting urban logistics sections is to separate the various actors so that each can be identified and the expectations and restrictions of each link can be understood. City logistics is composed of a variety of subcomponents that we summarize as Fig. 6.

The main benefit of staggering is the differentiation of units, which allows to choose at what level we should interfere in order to maximize the entire urban logistics activity [41, 46]:

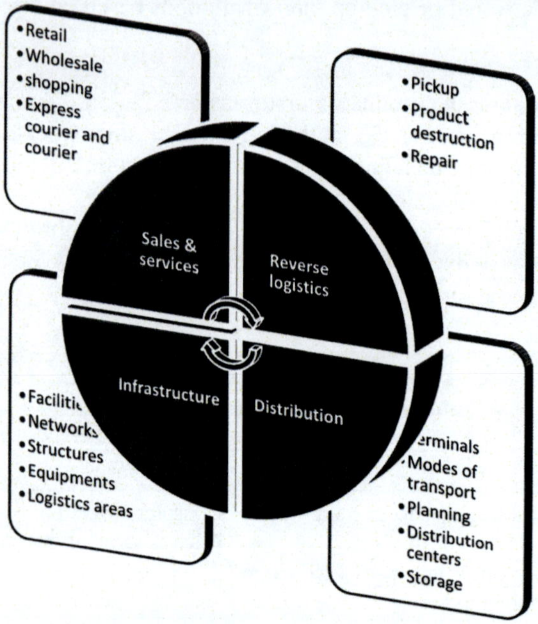

Fig. 6. Segments of urban logistics

4 Conclusion

An optimum freight distribution procedure in urban logistics necessitates the coordination of numerous components. Private and public players must work together and pool their resources in a sustainable logic.

The mode of transportation should be selected in accordance with the urban context in question, which is determined by the volume of traffic, the type of the city, the infrastructure and the anticipated investment cost. Establishment of urban logistics platforms is a factor to consider as well, because it creates recovery hubs and breaks up the monotony of journeys.

The future of city logistics in the use of technology, especially industry 4.0. In the few coming years, the supply chain will be digitalized and able to act autonomously without human aid.

References

1. Gonzalez-Feliu, J., Semet, F., Routhier, J.- L.: Sustainable Urban Logistics: Concepts, Methods and Information Systems, pp. 113–143 (2014). https://doi.org/10.1007/978-3-642-31788-0_7
2. Strale, M.: Sustainable Urban Logistics: What are we Talking About? (2019). https://doi.org/10.1016/j.tra.2019.10.002
3. Semanjski, I., Gautama, S.: A collaborative stakeholder decision-making approach for sustainable urban logistics (2019). https://doi.org/10.3390/su11010234

4. Taniguchi, E., Thompson, R.G., Yamada, T.: New opportunities and challenges for city logistics (2016). https://doi.org/10.1016/j.trpro.2016.02.004
5. Akgun, E.Z., Monios, J., Rye, T., Fonzone, A.: Influences on urban freight transport policy choice by local authorities (2019). https://doi.org/10.1016/j.tranpol.2019.01.009
6. Anand, N., Quak, H., Duin, R.V., Tavasszy, L.: City logistics modeling efforts: trends and gaps - a review (2012). https://doi.org/10.1016/j.sbspro.2012.03.094
7. Behnke, M., Kirschstein, T.: The impact of path selection on GHG emissions in city logistics (2017). https://doi.org/10.1016/j.tre.2017.08.011
8. Kawamura, K., Sriraj, P.S.: Building freight-friendly environment (2016). https://doi.org/10.1016/j.trpro.2016.02.052
9. Flaming, H., Wolff, J.: Impacts of planning and policy strategies on freight flows in urban areas (2016). https://doi.org/10.1016/j.trpro.2016.02.013
10. Matusiewicz, M.: Towards sustainable urban logistics: creating sustainable urban freight transport on the example of a limited accessibility zone in Gdansk (2019). https://doi.org/10.3390/su11143879
11. Macharis, C., Melo, S.: City distribution and urban freight transport (2011). https://doi.org/10.4337/9780857932754. (2011)
12. Kull, T.J., Boyer, K., Calantone, R.: Last-mile supply chain efficiency: an analysis of learning curves in online ordering. Int. J. Oper. Prod. Manag. **27**(4), 409–434 (2007). https://doi.org/10.1108/01443570710736985
13. Dündar, H., Ömürgönülşen, M., Soysal, M.: A review on sustainable urban vehicle routing. J. Clean. Prod. **285**, 125444 (2021). https://doi.org/10.1016/j.jclepro.2020.125444
14. Navarro, C., RocaRiu, M., Furió, S., Estrada, M.: Designing new models for energy efficiency in urban freight transport for smart cities and its application to the Spanish case. Transp. Res. Procedia **12**, 314–324 (2016). https://doi.org/10.1016/j.trpro.2016.02.068
15. Schliwa, G., Armitage, R., Aziz, S., Evans, J., Rhoades, J.: Sustainable city logistics — making cargo cycles viable for urban freight transport. Res. Transp. Bus. Manag. **15**, 50–57 (2015). https://doi.org/10.1016/j.rtbm.2015.02.001
16. Nürnberg, M.: Analysis of using cargo bikes in urban logistics on the example of Stargard. Transp. Res. Procedia **39**, 360–369 (2019). https://doi.org/10.1016/j.trpro.2019.06.038
17. Ruesch, M., Todesco, P., Hegi, P.: A positive planning approach to secure logistics facilities in urban areas. Transp. Res. Procedia **46**, 69–76 (2020). https://doi.org/10.1016/j.trpro.2020.03.165
18. Gatta, V., Marcucci, E., Le Pira, M.: Smart urban freight planning process: integrating desk, living lab and modelling approaches in decision-making. Eur. Transp. Res. Rev. **9**(3), 1–11 (2017). https://doi.org/10.1007/s12544-017-0245-9
19. Tian, Z., Zhong, R.Y., VatankhahBarenji, A., Wang, Y.T., Li, Z., Rong, Y.: A blockchain-based evaluation approach for customer delivery satisfaction in sustainable urban logistics. Int. J. Prod. Res. **59**(7), 2229–2249 (2020). https://doi.org/10.1080/00207543.2020.1809733
20. Giret, A.: Smart and sustainable urban logistic applications aided by intelligent techniques. SOCA **13**(3), 185–186 (2019). https://doi.org/10.1007/s11761-019-00271-z
21. Drop, N., Garlińska, D.: Evaluation of intelligent transport systems used in urban agglomerations and intercity roads by professional truck drivers. Sustainability **13**(5), 2935 (2021). https://doi.org/10.3390/su13052935
22. Kauf, S.: City logistics – a strategic element of sustainable urban development. Transp. Res. Procedia **16**, 158–164 (2016). https://doi.org/10.1016/j.trpro.2016.11.016
23. Morfoulaki, M., Kotoula, K., Stathacopoulos, A., Mikiki, F., Aifadopoulou, G.: Evaluation of specific policy measures to promote sustainable urban logistics in small-medium sized cities: the case of Serres, Greece. Transp. Res. Procedia **12**, 667–678 (2016). https://doi.org/10.1016/j.trpro.2016.02.020

24. Cossu, P.: Clean last mile transport and logistics management for smart and efficient local governments in Europe. Transp. Res. Procedia **14**, 1523–1532 (2016). https://doi.org/10.1016/j.trpro.2016.05.117
25. Lagorio, A., Pinto, R., Golini, R.: Urban logistics ecosystem: a system of system framework for stakeholders in urban freight transport projects. IFAC-PapersOnLine **50**(1), 7284–7289 (2017). https://doi.org/10.1016/j.ifacol.2017.08.1402
26. Pimentel, C., Alvelos, F.: Integrated urban freight logistics combining passenger and freight flows – mathematical model proposal. Transp. Res. Procedia **30**, 80–89 (2018). https://doi.org/10.1016/j.trpro.2018.09.010
27. Bergmann, F.M., Wagner, S.M., Winkenbach, M.: Integrating first-mile pickup and last-mile delivery on shared vehicle routes for efficient urban e-commerce distribution. Transp. Res. Part B Methodol. **131**, 26–62 (2020). https://doi.org/10.1016/j.trb.2019.09.013
28. Yue, S., Wang, Y., Zhao, M., Liu, Y.: Research on the feasibility and distribution mode of urban logistics distribution by rail transit. J. Phys. Conf. Ser. **1827**(1), 012101 (2021). https://doi.org/10.1088/1742-6596/1827/1/012101
29. Ruggieri, R., Ruggeri, M., Vinci, G., Poponi, S.: Electric mobility in a smart city: European overview. Energies **14**(2), 315 (2021). https://doi.org/10.3390/en14020315
30. Szymczyk, K., Kadłubek, M.: Challenges in general cargo distribution strategy in urban logistics – comparative analysis of the biggest logistics operators in EU. Transp. Res. Procedia **39**, 525–533 (2019). https://doi.org/10.1016/j.trpro.2019.06.054
31. Tahami, H., Rabadi, G., Haouari, M.: Exact approaches for routing capacitated electric vehicles. Transp. Res. Part E Logist. Transp. Rev. **144**, 102126 (2020). https://doi.org/10.1016/j.tre.2020.102126
32. Crainic, T.G., Gendreau, M., Jemai, L.: Planning hyperconnected, urban logistics systems. Transp. Res. Procedia **47**, 35–42 (2020). https://doi.org/10.1016/j.trpro.2020.03.070
33. Nathanail, E., Gogas, M., Adamos, G.: Smart interconnections of interurban and urban freight transport towards achieving sustainable city logistics. Transportation Research Procedia **14**, 983–992 (2016). https://doi.org/10.1016/j.trpro.2016.05.078
34. Zhang, Z., Tafreshian, A., Masoud, N.: Modular transit: using autonomy and modularity to improve performance in public transportation. Transp. Res. Part E Logist. Transp. Rev. **141**, 102033 (2020). https://doi.org/10.1016/j.tre.2020.102033
35. Farvaresh, H., Shahmansouri, S.: A coalition structure algorithm for large-scale collaborative pickup and delivery problem. Comput. Ind. Eng. **149**, 106737 (2020). https://doi.org/10.1016/j.cie.2020.106737
36. Kumar, S., Bharj, R.S.: Solar hybrid e-cargo rickshaw for urban transportation demand in India. Transp. Res. Procedia **48**, 1998–2005 (2020). https://doi.org/10.1016/j.trpro.2020.08.228
37. Lin, Y.H., Wang, Y., He, D., Lee, L.H.: Last-mile delivery: optimal locker location under multinomial logit choice model. Transp. Res. Part E Logist. Transp. Rev. **142**, 102059 (2020). https://doi.org/10.1016/j.tre.2020.102059
38. Jiang, B., Fan, Z.-P.: Optimal allocation of shared parking slots considering parking unpunctuality under a platform-based management approach. Transp. Res. Part E Logist. Transp. Rev. **142**, 102062 (2020). https://doi.org/10.1016/j.tre.2020.102062
39. Quak, H., Lindholm, M., Tavasszy, L., Browne, M.: From freight partnerships to city logistics living labs – giving meaning to the elusive concept of living labs. Transp. Res. Procedia **12**, 461–473 (2016). https://doi.org/10.1016/j.trpro.2016.02.080
40. Russo, F., Comi, A.: Behavioural simulation of urban goods transport and logistics: the integrated choices of end consumers. Transp. Res. Procedia **46**, 165–172 (2020). https://doi.org/10.1016/j.trpro.2020.03.177

41. Schöder, D., Ding, F., Campos, J.K.: The impact of e-commerce development on urban logistics sustainability. Open J. Soc. Sci. **04**(03), 1–6 (2016). https://doi.org/10.4236/jss.2016.43001
42. Wolbertus, R., Jansen, S., Kroesen, M.: Stakeholders' perspectives on future electric vehicle charging infrastructure developments. Futures **123**, 102610 (2020). https://doi.org/10.1016/j.futures.2020.102610
43. Bhasker, A., Sarmah, S.P., Kim, T.: Collaborative last-mile delivery and pick-up in city logistics. Int. J. Logist. Syst. Manag. **34**(4), 533 (2019). https://doi.org/10.1504/ijlsm.2019.103518
44. de Gusmão, A.C.S., Ribeiro, P.C.M.: Guidelines for the efficiency of urban goods distribution: the Brazilian case. Case Stud. Transp. Policy **8**(4), 1478–1488 (2020). https://doi.org/10.1016/j.cstp.2020.10.013
45. Karakikes, I., Nathanail, E.: Simulation techniques for evaluating smart logistics solutions for sustainable urban distribution. Procedia Eng. **178**, 569–578 (2017). https://doi.org/10.1016/j.proeng.2017.01.110
46. He, P., He, Y., Xu, H.: Buy-online-and-deliver-from-store strategy for a dual-channel supply chain considering retailer's location advantage. Transp. Res. Part E Logist. Transp. Rev. **144**, 102127 (2020). https://doi.org/10.1016/j.tre.2020.102127

Simulation and Forecasting the Covid-19 Pandemic in the Rabat Region

Khalid Qbouche[1(✉)] and Khadija Rhoulami[2]

[1] LRIT Associated Unit to CNRST (URAC °29), Faculty of Science, Mohammed V University in Rabat, 4 Av.Ibn Battouta B.P. 1014 RP, 10006 Rabat, Morocco
qbouchekhalid@gmail.com

[2] DESTEC, FLSHR Mohammed V University in Rabat, Rabat, Morocco
rhoualmikhadija@yahoo.fr

Abstract. There have been 19 pandemics since December 2019 in Wuhan, China. It then spread to all of the world's major cities and capitals. As a result, the World Health Organization declared it a worldwide pandemic. It also impacts some industries such as industry, trade, the economy, and public transit. In this study, we utilized the SIR (Susceptible, Infectious, Recovered) model to forecast the status of Covid-19 in the Rabat region, and the model obtained was compared with the current epidemic situation in the region. Furthermore, we have simulated the propagation of the Covid-19 virus by the Multi-Agent System Gama-Platform.

Keywords: Covid-19 · SIR model · Gama-platform and Multi-agent system

1 Introduction

The pandemic of Covid 19 has spread to many countries around the globe. The virus causes havoc on the country's health, political strategies, and working methods. This virus can be spread by direct contact between individuals (individual-individual) through secretions, saliva, and droplets released during breathing, coughing, or sneezing. Indirect contact is another way to spread the virus (individual-objects - individual) when secretions reach food and surfaces. On 02/03/2020, the first Moroccan case was reported. The Moroccan authorities declared a health emergency on March 20, after only 12 cases had been confirmed to stop the spread of the virus. According to [3], Morocco took quick and practical steps to increase its number of intensive care beds to 1,640 to 3,000 beds. It also has units for respirators that can be used to treat severe cases. They are adequate for some instances but not enough to meet demand. To predict the evolution of the virus, one must develop models and analyzes, as shown in Fig. 1.

The fast spread of the coronavirus (Covid-19) has turned it into a worldwide menace that has affected nearly every country on the planet. As nations reach the height of infection, it is envisaged that they would return to a new normal under various circumstances of cohabitation in order to mitigate the economic impacts of the whole or partial closure of corporations, institutions, stores, and other businesses. In such cases, employing mathematical models to estimate the danger of Covid-19 in various regions

S. Motahhir and B. Bossoufi (Eds.): ICDTA 2022, LNNS 454, pp. 120–130, 2022.
https://doi.org/10.1007/978-3-031-01942-5_12

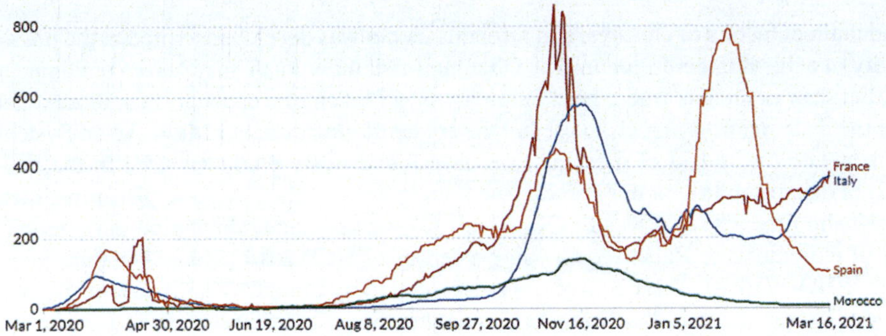

Fig. 1. Total number of cases reported in Morocco compared to other countries (France, Italy, and Spain).

is a critical tool for authorities to make educated judgments. Agent-based modeling, on the other hand, is a relatively recent way to modeling large systems made up of agents whose behavior is governed by basic rules. Agent-based techniques, in contrast to standard mathematical models (which consider a homogeneous population), simulate people with specific traits and provide more realistic results.

In our work, we made a mathematical model predict the epidemic situation of the covid-19 virus in the region of Rabat, and we compared it with the actual situation in the same region. Then we simulated the propagation of this virus by Multi-Agent System Gama [12].

The remainder of this paper is structured following the order of sections and subsections. Section 2 gives a short discussion of previous mathematics models for predicting epidemic situations and models that exist Multi-Agents System. Section 3 will detail the model SIR for predicting the actual situation and model Multi-Agent System to simulate the propagation of this epidemic in the Rabat region. In Sect. 4, we will describe our experiments.

2 Related Works

We discuss related work on the Covid-19 epidemic predictor implementation model and simulated mobility with a multi-agent system in this part:

Alanazi Saad and Kamruzzaman [4] established a method for predicting COVID-19 utilizing the SIR and machine learning for competent health care and KSA citizens' well-being. Saulo Bastos and Daniel Cajueiro [5] used Brazilian data to simulate and estimate the COVID-19 pandemic's route in Brazil, utilizing two variations of the SIR model to account for the impact of containment measures. Also, Silva, Adilson [10] applied the SIR model to simulate the propagation dynamics of COVID-19 on the Cape Verde Islands. Vergara, Edmundo & Leon, Ronald & More, Juliana [6] They studied the evolution of COVID-19 in each of the regions of Peru by the epidemiological model SIR, and for the parameters of this model, they based on the data of the regional director of health of Peru. Bayraktar, Erhan & Cohen, Asaf & Nellis [9] They developed a macro-SIR model that takes into account behavior-dependent transmission rates, remote workers, and indirect factors of closures, and also, this model was developed to predict the possibility of achieving herd immunity, so that high-risk individuals may leave containment earlier than in models where herd immunity is not taken into account. In addition, that for the mathematical models to predict the epidemic situation. For Multi-Agent System to simulate the spread of the virus, we have found many platforms used in the context of daily mobility; in this context, we found topics worked by these platforms, such as Khalid Qbouche and Khadija Rhoulami [7] worked about simulation daily mobility of individuals in Rabat region using a census of HCP 2014 [3] by Gama platform. The WHO-WHERE-WHEN (3W) model, proposed by Smolak, Kamil, Rohm, Witold, Knop, Krzysztof, and Sila- Nowicka, Katarzyna [8], is an enhanced privacy-protective mobility modelling approach for synthetic mobility data creation.

3 Methods

3.1 The SIR Model (Susceptible-Infected-Recovered) to Predict the Epidemic Situation in the Rabat Region

They are a lot of mathematical models used to predict epidemic propagation. Furthermore, modelling or numerical modelling makes it possible to build a set of mathematical functions to represent a phenomenon; among these models, SIR (Susceptible, Infectious, Recovered) model [5]:

- S: "Susceptible": The agent is healthy;
- I: "Infectious": The agent is contagious. We are talking about a period of infection;
- R: "Recovered": The agent has recovered or is removed (dead).

Between these sections, there are transitions. Susceptible people are identified by removing individuals who have been proven as viral carriers by hospital testing from the overall population. The contact rate, β defines the disease velocity in the population as you progress from susceptible to infected cases. The shift from the S-state to the I-state is never deterministic, but always stochastic. As a result, the multiplication of rate, probability, and population number is included in β. The fundamental model is as follows (Fig. 2):

Fig. 2. Structure of model SIR.

Also, the model can be described as:

$$\frac{ds}{dt} = -\beta is, \frac{di}{dt} = -\beta is - \gamma i, \frac{dr}{dt} = -\gamma i \tag{1}$$

The ratios of Susceptible, Infected, and Recovered persons in the population are represented by S, I, and R, respectively. Normally, differential equations are used to address the problem. Agent-based methodologies in networks, on the other hand, can depict the nature of viral propagation. And, based on the time t, the total number of persons equals the sum of the individuals in each compartment, so:

$$N(t) = S(t) + I(t) + R(t) \tag{2}$$

- N: total population.
- S(t): number of people susceptible on day t.
- I(t): number of people infected on day t.
- R(t): number of people recovered on day t.

3.2 The Gama Platform Multi-agent System Model to Simulate Propagation Covid-19 in the Rabat Region

Agent-based modelling relies on the simulation of actions and interactions of autonomous agents to evaluate their effects on the system. It is often used to predict the projections that we will obtain given a complex phenomenon. In our paper, we used the Gama platform for modelling and simulating the spread of the covid-19 virus in the Rabat region; it was used to simulate the movement of people in a geographical area. It combines explicit multi-agent simulations with GIS data management, multi-level modelling, and the ability to create BDI and reactive agents. As a result, its agent-oriented language GAML makes it practical for prototyping and automated simulation generation. Besides this, we used this software to simulate daily mobility in the Rabat region based on the 2014 HCP survey [6].

Agent-based Modeling and Simulation: When a researcher wishes to study a particular phenomenon, he or she has several research strategies. He can collect data on this phenomenon and study it with statistical tools. He can conduct "real" experiments in the laboratory "experiments in the laboratory. Finally, he can try to recreate the phenomenon with the help of a computer simulation. Computer simulation is a scientific approach involving an artificial reproduction, called a model, of a natural phenomenon that one wishes to study [7]. Simulation is used when the analytical approach proposed by the statistical or mathematical models is insufficient or mathematical models are insufficient. Moreover, when one wishes to imitate the mechanisms of an actual process that are difficult to reproduce experimentally or to put into the equation or to put into the equation or when one wishes to develop a theory, and when one does not, we do not have a solid enough knowledge about it. Agent-based modelling has developed at the same time as the notions of distributed algorithms and artificial intelligence [8] from these roots, (Distributed Artificial Intelligence) DAI. Moreover, MAS (Multi-Agent Systems) was born, allowing the study using computer tools, the emerging structures of their virtual interactions.

Geospatial Data: To draw our study area, we have collected the data of type shp which contained all the buildings with their type either residential or industries of the region of the flap under the format of files, and other files containing the roads of the region were extracted from OpenStreetMap [11]. We can add the people to whom it is attached with their buildings of residence from these files. For this, we can visualize the displacements of the people on our zone of study by the system multi-agents Gama Platform, such as we see in Fig. 7.

4 Experimental Evaluation

4.1 Implementation of the SIR Model

The objective of this section is to implement the SIR model and verify that it is capable of predicting the situation of the spread of the covid-19 virus and comparing the results obtained with the accurate statistical data in the Rabat region. The mathematical models do not take into account all the parameters on which the epidemic depends, but they give an idea on the propagation shortly; for our case, we applied the SIR model on the rabat region of Morocco, we took the initial values of the base of the Model, N = 6000, $I_0 = 20$, $S_0 = 5980$, $\beta = 0.2$ and $\gamma = 0.05$. The result of the Model is shown in the following Fig. 3; python developed it.

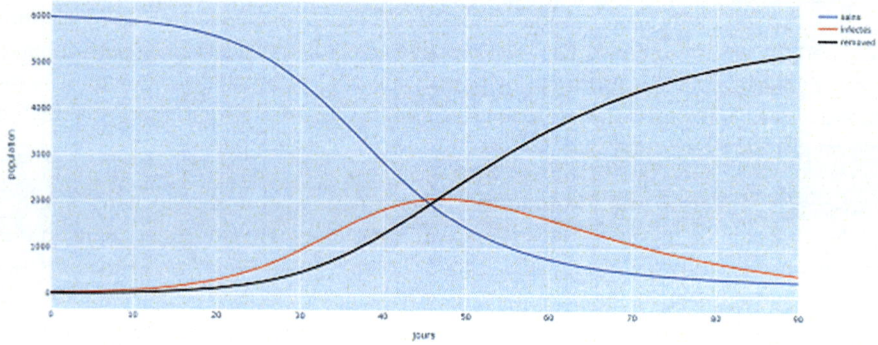

Fig. 3. Schematic representation, differential equations, and plot for the basic (Susceptible, Infectious, and Recovered) SIR model.

The figure above shows the overall result of the model SIR, but we are interested in the curve result of the infected person's figure (Fig. 4).

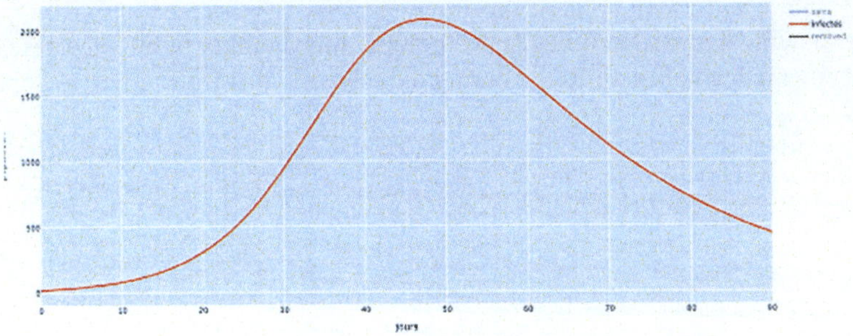

Fig. 4. The curve of infected persons by SIR model.

4.2 The Data Used to Validate SIR Model

From the daily covid-19 bulletin [10], we can to extract the number of people is affected each day in the region of Rabat. We recorded the number of people infected every day for three months, from July to September 2021, to trace the curve of people infected with the covid-19 virus in the Rabat region. The following Fig. 5 shows the number of people infected on the day.

Days	number of infections per day in the Rabat region
01-july	102
02-july	144
03-july	184
04-july	102
05-july	52
06-july	163
07-july	180
08-july	231
09-july	189
10-july	218
11-july	125
12-july	128
13-july	187
14-july	322
15-july	268

Fig. 5. Extract of data of infected people per day in Rabat region.

The following figure represents the histogram of the number of infected cases of the covid-19 virus in the region of Rabat during the three months from July 1 to September 27 in 2021 (Fig. 6).

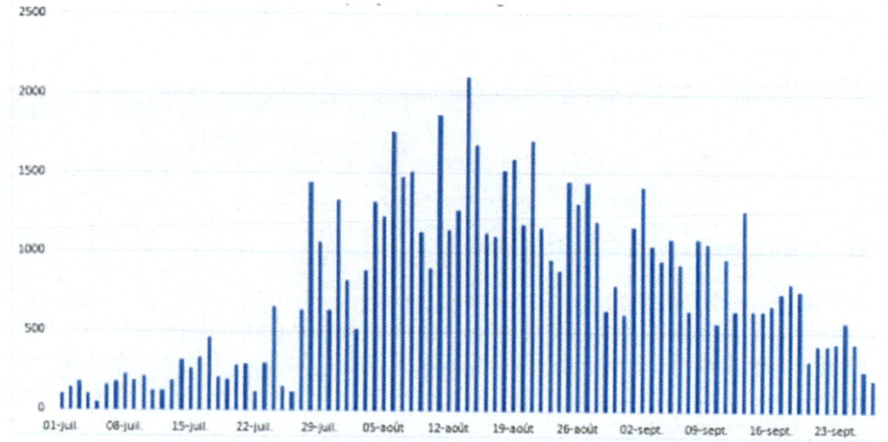

Fig. 6. Daily epidemic histogram of new cases in Rabat region from July 1 to September 27 in 2021.

From the epidemic bulletin [10], we have a demographic distribution of infected people (by age, sex…), as illustrated in the following Fig. 7.

Variable		Effective	%
	Total Cases	483	100,0
	Severely	73	15
Sexe			
	Feminine	262	54,2
	Male	196	40,6
	not specified	20	5,2
Age			
	Median(Q1-Q3)ans	41 (29-55)	
	<18 ans	9	1,9
	18-44 ans	235	48,7
	45-64 ans	119	24,6
	65-74 ans	43	8,9
	>74 ans	21	4,3
	not specified	56	11,6

Fig. 7. Demographic characteristics of people are infected in the Rabat region.

From these demographic characteristics, we have the category of people between 18 and 44 years old was predominant (48.7%), followed by the age group (45–64 years) (24.6%). The sex ratio M/F was 0.75. In addition, we have 38% of the patients (137 cases) with adverse reactions had comorbidity as a contributing factor: diabetes for 43 patients, arterial hypertension for 34, heart disease for 12, asthma for 10, drug addiction for six and renal insufficiency for four patients, knowing that a patient may have one to three associated pathologies.

4.3 Validation of the SIR Model with Actual Statistical Data from the Rabat Region for Three months

In this, the SIR model developed by Python language was compared with the actual statistical data of people infected with the covid-19 virus in the region of Rabat. The following Fig. 8 shows that the SIR model can predict the situation of covid-19 virus propagation during three months based on variables such as infection rate, withdrawal rate, and the number of infected people at the beginning of the epidemic.β Infection rate, γ the withdrawal rate, and the number of infected persons at the epidemic's beginning.

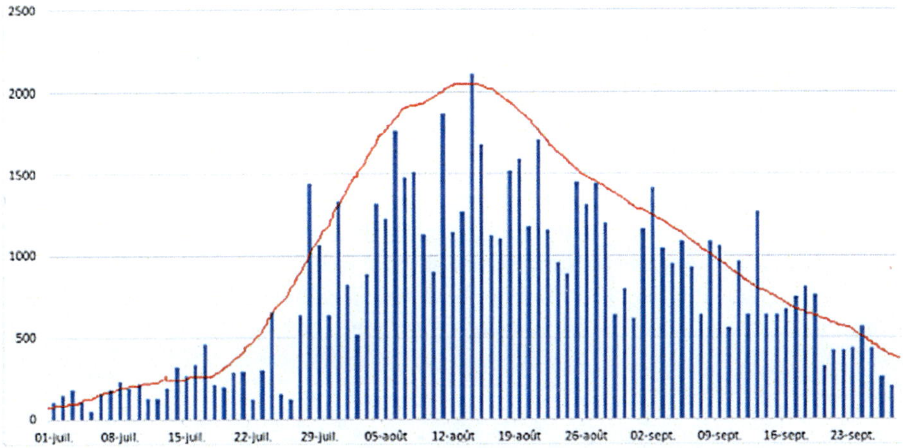

Fig. 8. Comparison between the predictive model and the real situation in the Rabat region.

In the Fig. 8 represents a comparison between the predictive model for infected people represented by the red color and the actual distribution of infected people represented by the histograms.

4.4 Simulation of the Spread of the Virus by the Multi-agent System Gama

Realistic City "case Rabat region": We utilized a building shapefile based on the Rabat area map. The buildings are divided into kinds, such as house, industry, office, school, stores, grocery, coffee, restaurant, and public park, and we present them by type. In our multi-agent system, we have added 800 people to the map of the Rabat region to show how the virus is spread. Among this number, there are 50 people infected by the covid-19 virus, which is indicated by the red color, and the green color does not infect the people. In addition, the people are initialized in a building (their residence) and go to work (place of work represented by the buildings of type industry).We illustrate that in the following Fig. 9.

Fig. 9. A realistic population sample of the Rabat region.

Simulation: Agents report to work between 6 and 9 a.m when the simulation begins. They come to a halt between 4 and 8 p.m and might then travel to a leisurely destination or return home. The user has control over all of these settings. We could illustrate these agents in the following Fig. 10, where we have indicated the agents carrying the virus by the red color and the people not infected by the green color.

Fig. 10. The map of the Rabat region with the agents and their health status.

We based on the simulation parameters, such as the distance of infection that has the value 1 m and can be this distance valid to infect if the person was infected that is interconnected with it was infected. It was proposed by the world health organization [9]. In addition, after executing our model with these parameters, we obtained the result of the propagation of the virus covid-19 in the Rabat region based on the multi-agent system that can be seen in the following Fig. 12. Moreover, Fig. 11 represents a statistical diagram that shows the evolution of the spread of the covid-19 virus.

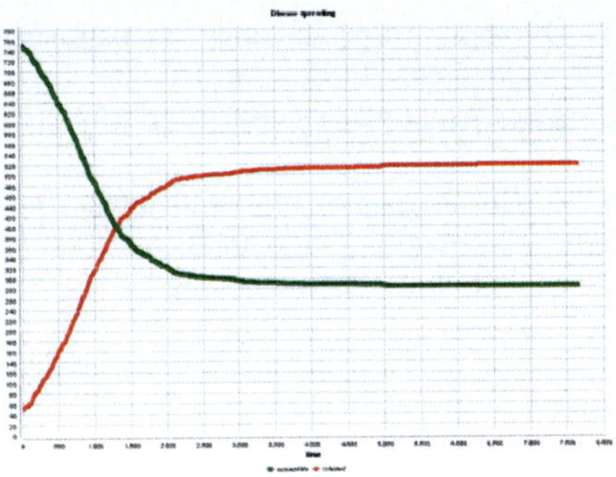

Fig. 11. Epidemic spread in Rabat region

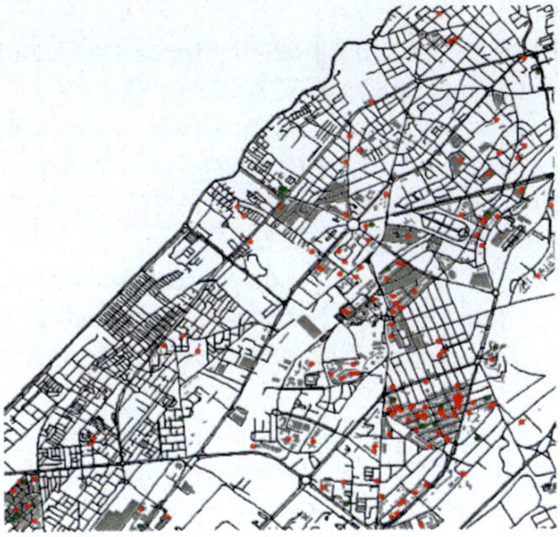

Fig. 12. The simulation of the spread of the coronavirus in the rabat region.

5 Conclusion

In this paper, we have demonstrated the use of the mathematical Model SIR to predict the state of the epidemic in the region of Rabat in the Kingdom of Morocco, and a simulation of the virus's propagation using the multi-agent system Gama platform is in this article. In addition, we have compared the result obtained with the actual situation epidemic of the Rabat region. Besides this, we have simulated the propagation of virus covid-19 using the multi-agent system Gama for the same region based on the parameters (the distance of infection, the working hours between 9 am and 4 pm. The multi-agent system we have taken is based on three layers of social interactions: home, workplace, employment and public transport. Each layer has its characteristics based on how individuals engage in that activity daily. The propagation is carried out using an agent-based approach in which weights reflect the infection likelihood, which changes depending on the layers and groups with which the agent interacts, such as illustrated in the figure Fig. 16, is scaled by a term that regulates the infection possibilities. Shortly, we will work to develop model intelligence by Artificial intelligence algorithms for predicting the situation of epidemic.

Acknowledgement. This research is carried out as a part of the project Plateforme logicielle d'intégration de stratégies d'immunisation contre la pandémie COVID-19 funded by the grant of the Hassan II Academy of Sciences and Technology of Morocco.

References

1. Bacaër, N.: A Short History of Mathematical Population Dynamics (2011). https://doi.org/10.1007/978-0-85729-115-8
2. Gabriel, J.-P. Harpe, P. : Daniel Bernoulli, pionnier mathematical models in medicine (2010)
3. Institutional site of the High Commission for Planning of the Kingdom of Morocco. https://www.hcp.ma/

4. Hamer, W.H.: Epidemic diseases in England – the evidence of variability and of persistency of type. Lancet **1**, 733–739 (1906)
5. Marques, J.A.L., Gois, F.N.B., Xavier-Neto, J., Fong, S.J.: Epidemiology compartmental models—SIR, SEIR, and SEIR with intervention. In: Predictive Models for Decision Support in the COVID-19 Crisis. SAST, pp. 15–39. Springer, Cham (2021). https://doi.org/10.1007/978-3-030-61913-8_2
6. Qbouche, K., Rhoulami, K.: Simulation daily mobility in rabat region. In: The 4th International Conference on Networking Information Systems and Security. (NISS2021), April 01, 02, 2021, KENITRA, AA, Morocco, p. 6. ACM, New York (2021). https://doi.org/10.1145/3454127.3454128
7. Drogoul, A.: From multi-agent simulation to collective problem solving of problems. PhD thesis, Thesis at University of Paris VI (1993)
8. McCarthy, J., Hayes, P.: Some Philosophical Problems from the Standpoint of Artificial Intelligence. Stanford University USA (1968)
9. Official website of the World Health Organization. https://www.who.int/
10. The official portal of the Corona virus in Morocco. http://www.covidmaroc.ma
11. Haklay, M., Weber, P.: Openstreetmap: User-generated street maps. IEEE Pervasive Comput. **7**(4), 12–18 (2008)
12. Official website of the Gama Platform. https://gama-platform.org/wiki/Home

New ETL Process for a Smart Approach of Data Migration from Relational System to MongoDB System

Abdelhak Erraji[1](✉), Abderrahim Maizate[2], and Mohamed Ouzzif[2]

[1] RITM ESTC Laboratory, Hassan II University, ENSEM, Casablanca, Morocco
erraji.abdelhak@gmail.com

[2] RITM ESTC Laboratory, Hassan II University, ESTC, Casablanca, Morocco

Abstract. Today, data is experiencing a real explosion in terms of quantity and nature. They also play a very important role in reading the past, managing the present, and planning for the future. To this end, organizations consider them a treasure and are always looking for a veritable way to manage and exploit them. And since the old data management system has some weaknesses compared to Bigdata, and designers have designed systems called NoSQL to overcome these weaknesses, there is a critical need for migrating data to the new NoSQL system, to keep their old data and take advantage of the power of NoSQL systems. To meet this need, several approaches have been developed by researchers to ensure this migration. The problem is that these approaches transform data, structures, or both. These approaches at best mimic a relational database with its limitations and constraints in another NoSQL environment, which loses much of its efficiency when applying relational processing in the migration result database. In this article, we will develop a smart approach that aims to migrate the three essential parts of relational databases: the first is data, the second is structure, and the third is the semantic component, using an ETL to be defined. The solution respects the analytical part of the relational systems that we have developed and the transformations of the structure and semantics of the relational system towards the NoSQL system, to develop our model of our ETL, which consists of three phases: data extraction, transforming this data, and uploading it to the destination system.

Keywords: Migration database · NoSQL database · Relational database · ETL · BigData

1 Introduction

Relational Database Management Systems (RDBMS) appeared in the 1970s as a real revolution in data storage and management. As everyone seeks to overcome the limits of binary languages, it assembles languages or basic languages. At this time, the hardware that is used in data management is very weak, and the databases are small and poorly designed, with major problems such as the representation of data on disk, storage optimization, verification of data integrity, and restitution of complete information. These

S. Motahhir and B. Bossoufi (Eds.): ICDTA 2022, LNNS 454, pp. 131–140, 2022.
https://doi.org/10.1007/978-3-031-01942-5_13

limits are exceeded at present by the relational model and its systems, which process data on both physical and logical levels. Subsequently, the use of RDBMS continues to emerge in all sectors at an accelerated and continuous pace, causing the creation of strong communication, from which it produces an explosion in the amount of data. In turn, it will continue to be developed in the future. On the one hand, this data is currently doubling in volume every two years [1], and it is estimated that in 2025, an amount of data in the order of 163 trillion gigabytes will be created [2]. It will give rise to what is called "BigData", which plays a very important role in analytical systems. On the other hand, the implication of these data in the management of the sectors has proved its interest through a precise description of the past. Furthermore, its explanation, the good control of the present and the good planning of the future, lets us analyze the behavior of customers, competitors, and markets over time. This situation gives the RDBMS the main role and a real opportunity to develop and expand with the appearance of decision-making systems, artificial intelligence, management, and governance support systems. At this stage, the quality and quantity of data form a real force for the existence and evolution of organizations. In addition to that, the use and exploitation of this data have given rise to a variety of analysis and study methods, tools for extracting and exploiting this data, and making relevant decisions that will play a mandatory role in increasing the profits of these organizations. At this point, exploiting these data methods becomes an essential asset in the evolution of all sectors of life [3–5], by creating a strong link between data management and the evolution of these organizations. Managing and using this huge amount of data requires a powerful database management system capable of storing, processing, and analyzing this data. The old relational database management systems, which have proved their effectiveness for small and medium databases, present several limitations concerning the management and exploitation of big data. On the contrary, the hardware becomes more powerful. These limits are due to the constraints and the rules applied by this system, which create strengths and weaknesses at the same time. To overcome these limits, database designers have designed new systems called NoSQL (Not Only SQL), benefiting from new data representation technologies that are simple, flexible, efficient, and powerful. To put it differently, NoSQL systems make it possible to implement the notion of horizontal scalability, which allows a set of database servers to be used either to replicate or to subdivide data. Besides, it aims to always increase data availability, broadening the data container to be able to store more data, increasing the processing speed, and boosting the power of the entire system. Consequently, this results in the additional resources of the different database servers and their use in this complex data container [6–11]. To reap these benefits, organizations want to adopt NoSQL systems and keep their old data accumulated for years in RDBMS at the same time. This data for these organizations represents a real treasure, describing their past, their positions with the market and their competitors, as well as customer needs, market studies, sector studies, the control of their current functioning, and future opportunities. Based on this situation, these organizations will need to adopt the new data management system without losing their accumulated data, resulting in a real need to migrate their old data from the relational system to the NoSQL system.

Researchers have also developed several migration approaches [12–15]. As we notice, all of these approaches transform relational databases from relational systems to

NoSQL systems. It will create a new NoSQL database as the transformation of our basic database. However, not only do these transforms affect their data or their structure, or both. According to careful analysis, we can see that these transformations either waste part of the database data offered by the semantics of the relational system, or they put the same architecture and the same relational design back into the NoSQL system through their transformation. The latter situation will recreate a relational database in a NoSQL environment, which is contradictory to the objective of system change, which aims to exceed the limits of the relational model philosophy.

As a solution to this problem, we are offering a complete migration approach that aims to transform the structure of the database with its semantic aspect. Then, we migrate the data from the relational system to our target NoSQL system according to the first stage of migration. In a previous work [in press], we developed the transformation of the structural and semantic parts of our approach, and in this article, we will define the different parts of our ETL, whose objective is the migration of data following the models and structures of the first phase. At this stage, the question that arises is: how will the proposed ETL use the previous data and the first phase to define its operation?

2 Related Work

Currently, scientific research rarely discusses how to effectively validate the best way to migrate data from RDBMS to NoSQL databases to ensure the quality of the data being migrated. Therefore, those needs become a major concern and an interesting research topic. According to [19], data has become an essential hub for information processing. The authors present a framework, named NoSQLayer, based on two parts: a migration module and a mapping module. The first one is a set of methods enabling seamless migration between RDBMS, and the second one provides a persistence layer to process database requests and translate and execute requests in any RDBMS. In 2016, the authors of the article [20] present an automatic transformation of a multidimensional conceptual model into two NoSQL databases: column-oriented and document-oriented models, and to validate the transformation, they implemented four data warehouses using Cassandra (as a column-oriented NoSQL system) and MongoDB (as a document-oriented NoSQL system). The results show that MongoDB with hierarchical transformation is more suitable when dealing with OLAP queries. In the same year (2016), the authors of Article [21] proposed a flexible and highly modularized data adapter for hybrid database systems by using a general SQL layer that accepts queries from application services. It also controls query flow during database transformation. Therefore, they can provide a seamless mechanism to use RDBMS and NoSQL databases at the same time. NoSQL databases are designed to solve data processing problems in volume [22]. Cloud computing also enables the database as a service model to manage large volumes of user-generated data using NoSQL data repositories. Authors in [23] have presented a NoSQL data migration framework to foster data portability across cloud-based heterogeneous NoSQL data repositories; results show that the document-based database supports efficient migration of data, whereas the graph database ensures effective management of large volumes of data [24]. In 2017, the authors of the article [25] proposed a meta-model-based data merging approach, which allows simultaneous querying of data from heterogeneous database systems. Results of database queries are translated into JSON objects, and data merging is done by concatenating separate JSON files, so the result can be represented by any data visualization source (tables, Excel, CSV files, etc.). In addition, Big Data has also become a crucial issue and one of the most important technologies in the modern world [26]. The authors have presented a study by proposing a document-oriented data model for big data and then applying this model to migrate relational database applications to NoSQL.

All of these approaches transform relational databases into NoSQL systems from the point of view of data or structures only or both. Now, a relational database has the semantic aspect, which defines how we construct the entities, the tables, and their structures, and how we can generate new information from this structure and the stored data.

3 Relational Database and NoSQL Database Overview

3.1 Relational Database Analyzing

To define the framework of a complete migration, it's necessary to specify all the constituents of the relational system so that the approach can define its transformation rules. In this study, we present, in the following table, an exhaustive list of the constituents of the relational system to migrate to the NoSQL system (Table 1):

Table 1. Components of relational database system and theirs corresponding

Natures	Elements of RDBMS	In MongoDB or Mongoose
Data	Data	Exist
Structure data	Table	Redeploy
	Name of table	Redeploy
	Column	Redeploy
	Name of column	Redeploy
	Data type	Exist
	Check constraint	Redeploy
	Unique constraint	Exist
	Not null constraint	Exist
	Primary key	Exist
	Index	Exist
	Foreign key	Redeploy
	Relational Ship	To migrate
Semantic data	Dependence between fields	Exist

(*continued*)

Table 1. (*continued*)

Natures	Elements of RDBMS	In MongoDB or Mongoose
	View (stored select query)	To migrate
	Function & procedure	To migrate
	Trigger to control data in lines	Not supported and without added value
	Structure definition queries	Not supported and without added value
	CRUD operation	Exist
	Join operation	To migrate
	Usual functions	Redeploy

3.2 NoSQL Databases

NoSQL database management systems are distinguished from relational DBMSs (RDBMS), particularly by their performance. Records are no longer tabular but usually in the form of collections managed by a key/value system. They make it possible not to restrict the base to its schema. The system is simplified and responds to the problems of big data thanks to its scalability principle (it is possible to maintain performance by increasing the number of servers). It also allows greater portability on different platforms (Android, Windows, and Linux) while using fewer machine resources (memory and processor). NoSQL eliminates a certain structure in advance. We are no longer constrained by tables and data whose type is specified. It is also possible to distribute the data more easily by adding new servers. These databases also meet the needs for rapid change, especially for developers who do not have the exact structure of their data during the various phases of development [20].

The NoSQL movement represents a set of technologies offering new architectures for databases. These new types of databases have been stipulated to align with the current technological evolution, especially related to the concepts of Big Data and Cloud Computing. On one hand, NoSQL databases allow one to make the system much more powerful, especially with the sharp increase in data volumes, and on the other hand, they also make the system more resistant to failures by taking advantage of the possibilities of Cloud Computing. This type of relational system implements the BASE principles, as presented by [27]:

- **Basically Available.**
- **Soft State.**
- **Eventually Consistent.**

4 Proposed Approach

4.1 The Overall Architecture of Our Migration Approach

After watchful reading of the migration approaches proposed in the research works cited in the related work above, we were able to discuss these approaches, which neglect a considerable amount of data during their migration and transformation from relational systems to another NoSQL. These approaches did not set a framework for transformation and migration.

In our approach, which we explained in detail in a previous paper [in press], which we called the “TMSDRDND Approach” (Transformation and Migration Structure and Data from Relational Database to NoSQL Database), we will respect a specific framework during the transformation and migration of relational databases to big data databases of NoSQL nature. The “TMSDRDND Approach” is a model-based that describes relational systems and NoSQL systems according to their meta-model for complete migration. We note that at this level, our approach treats different constituents of the relational system and that NoSQL databases, which are considered Big Data.

To define the “TMSDRDND Approach” we will go through two stages: The first, in the “TSRSNLayer” (Transformation of the Structure from Relational System to NoSQL System) layer, aims to transform the structure data and semantic data into a NoSQL model according to a set of rules, and the second, in the “MDRSNLayer” (Migrating Data from Relational System to NoSQL System) layer, consists of using an ETL, which will begin its processing by extracting the data according to the models proposed, then it performs the necessary transformations according to the transformation rules of the layer “TSRSNLayer”. In Fig. 1 we present the architecture of our approach:

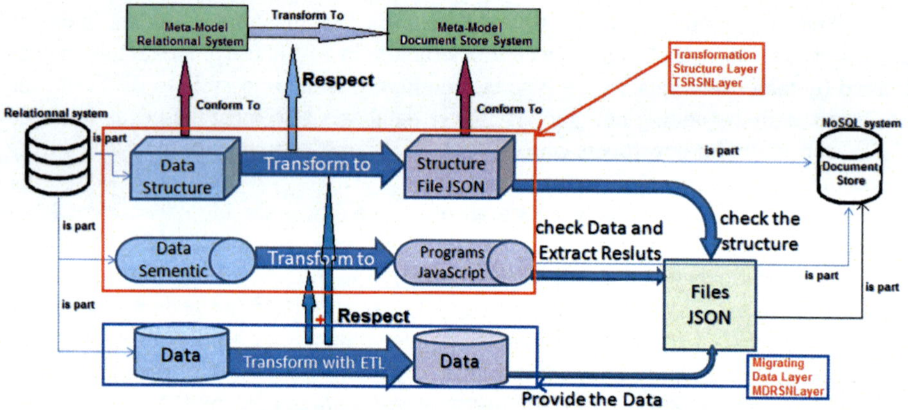

Fig. 1. The Architecture of the “TMSDRDND approach ”

In our study, we focus on the MDRSNLayer, which uses an ETL (Extract, Transform and Load) to migrate data according to the first layer.

4.2 ETL System for Data Migration in TMSDRDND Approach

To use an ETL, we must extract data from the relational systems according to their structures to start the transformation process according to the models and file the results of the first transformation of structure and semantic data in the first step of our global approach. This transformation respects the destination system and will be finished by loading data into the destination system. In Fig. 2 we present the architecture of our ETL:

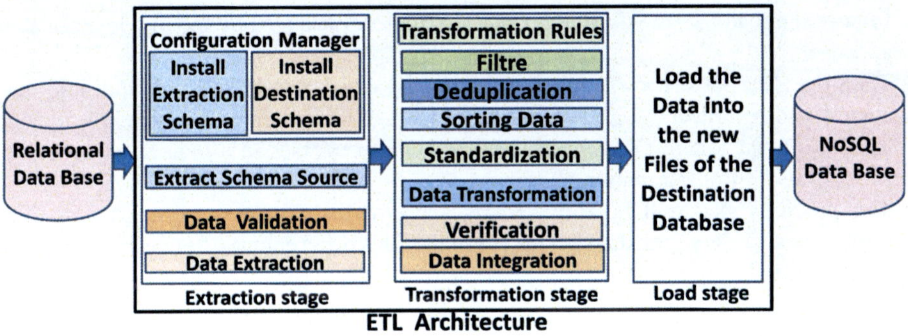

Fig. 2. ETL architecture data migration in TMSDRDND approach

This architecture shows that our ETL has three stages:

The first stage is "Data Extraction", which will start with:

- The extraction program is instantiated, then configured as an extraction engine by importing the two diagrams: the diagram of the extraction process, which begins with the precision of the data server source and the establishment of a connection with it, then the preparation of objects, agents, and interfaces required to begin this phase, and finally, the installation of the diagram of the destination system, which results from the phase of the transformation of the structure and the semantic component to developing the additional elements for the various diagrams and documents of the destination system.
- After this import, the system loads the diagram and the model of the various components of the source, as well as the prototypes and references of the components, to begin the validation stage.
- Then one passes through a validation of the models, of the data structures, and of the processing to avoid or minimize the errors introduced during the transformation.
- Finally, we start the data extraction, and we load them into our objects with precision following the validation of the models, to put each data point in its exact location so as not to lose the data and to succeed in transforming all the data.

The second stage is "Data transformation", which will start with:
In this transformation phase, ETL begins with the following steps:

- Import and load the transformation rules and filters produced by the first layer of the approach.
- Then the ETL performs a duplication of the extracted data to put each item into the corresponding object, which represents the start of its transformation process.
- Then we must sort to classify these objects and processes in a logical sequence to succeed in the transformation phase.
- The previous step will be followed by a standardization step to avoid semantic errors, especially in the JavaScript language that differentiates between lowercase and uppercase.
- Afterward, it is the moment to begin the general transformation.
- The previous step will be followed by verification and final data integration to prepare them for data loading.

This last stage is "Data Load":
At this stage, ETL loads the data in the final state into the destination system by the effect of the import. This step sets up the latest version of the database in MongoDB so that it is operational and able to verify the various constraints, rules, triggers, and other components that make the database very rich and powerful.

5 Conclusion and Perspective

The purpose of this article is to propose a new approach, called the "TMSDRDND Approach", for the complete migration of a relational database to a NoSQL database in a distributed environment. This approach solves many problems in the previous similar approaches and has two phases: the first will transform the structure and semantic data from the relational system into two files in the NoSQL system, materialized by the layer called "TSRSNLayer" and according to many transformation rules; the second phase uses an ETL tool, which we develop according to the first frames defined by the first layer, to extract, transform, and load data. This last step represents the second layer, called "MDRSNLayer";

Within the framework of this article, we have succeeded in presenting the different axes and articulations of this ETL within the framework of the definition of the mechanism of our ETL. To design this interesting tool, we proposed their architecture and different stages and components, explaining the role of each component to achieve a satisfactory and error-free approach.

In future work, we will present an implementation of this ETL to implement our global approach, by a heuristic definition, to evaluate the "TMSDRDND Approach".

References

1. The digital universe of opportunities. https://www.emc.com/leadership/digitaluniverse/2014iview/executive-summary.htm

2. Data is eating the world. https://whatsthebigdata.com/2017/04/18/idc-163-trillion-gigabytesof-data-will-be-created-in-2025/
3. Sokolova, M.V., Gómez, F.J., Borisoglebskaya, L.N.: Migration from an SQL to a hybrid SQL/NoSQL data model. J. Manag. Anal. **7**(1), 1–11 (2020)
4. de Oliveira, V.F., de Oliveira Pessoa, M.A., Junqueira, F., Miyagi, P.E.: SQL and NoSQL Databases in the Context of Industry 4.0 (2021)
5. Gamero, D., Dugenske, A., Kurfess, T., Saldana, C., Fu, K.: SQL and NoSQL databases for cyber physical production systems in internet of things for manufacturing (IoTfM). In: International Manufacturing Science and Engineering Conference, vol. 85079, p. V002T07A014. American Society of Mechanical Engineers (2021
6. Dai, J.: SQL to NoSQL: what to do and how. In: IOP Conference Series: Earth and Environmental Science, vol. 234, no. 1, p. 012080. IOP Publishing (2019
7. Chang, M.-L.E., Chua, H.: SQL and NoSQL database comparison: from performance perspective in supporting semi-structured data. In: Arai, K., Kapoor, S., Bhatia, R. (eds.) Advances in Information and Communication Networks: Proceedings of the 2018 Future of Information and Communication Conference (FICC), Vol. 1, pp. 294–310. Springer International Publishing, Cham (2019). https://doi.org/10.1007/978-3-030-03402-3_20
8. Flores, A., Ramírez, S., Toasa, R., Vargas, J., Urvina-Barrionuevo, R., Lavin, J.M.: Performance evaluation of NoSQL and SQL queries in response time for the E-government. In: 2018 International Conference on eDemocracy & eGovernment (ICEDEG), pp. 257–262. IEEE (2018
9. Reetishwaree, S., Hurbungs, V.: Evaluating the performance of SQL and NoSQL databases in an IoT environment. In: 2020 3rd International Conference on Emerging Trends in Electrical, Electronic and Communications Engineering (ELECOM), pp. 229–234. IEEE (2020)
10. Sahatqija, K., Ajdari, J., Zenuni, X., Raufi, B., Ismaili, F.: Comparison between relational and NOSQL databases. In: 2018 41st International Convention on Information and Communication Technology, Electronics and Microelectronics (MIPRO), pp. 0216–0221. IEEE (2018
11. Mukherjee, S.: The battle between NoSQL Databases and RDBMS. Available at SSRN 3393986 (2019)
12. Sokolova, M.V., Gómez, F.J., Borisoglebskaya, L.N.: Migration from an SQL to a hybrid SQL/NoSQL data model. J. Manag. Anal. **7**(1), 1–11 (2020)
13. Ramzan, S., Bajwa, I.S., Ramzan, B., Anwar, W.: Intelligent data engineering for migration to NoSQL based secure environments. IEEE Access **7**, 69042–69057 (2019)
14. Hanine, M., Bendarag, A., Boutkhoum, O.: Data migration methodology from relational to NoSQL databases. World Acad. Sci. Eng. Technol. Int. J. Comput. Electr. Autom. Control Inf. Eng. **9**(12), 2369–2373 (2016)
15. Stanescu, L., Brezovan, M., Burdescu, D.D.: An algorithm for mapping the relational databases to mongodb-a case study. Int. J. Comput. Sci. Appl. **14**(1), 1–16 (2017)
16. Sharma, K., Attar, V.: Generalized big data test framework for ETL migration. In: 2016 International Conference on Computing, Analytics and Security Trends (CAST), pp. 528–532. IEEE (2016)
17. Hu, H., Wen, Y., Chua, T.S., Li, X.: Toward scalable systems for big data analytics: a technology tutorial. IEEE Access **2**, 652–687 (2014)
18. DB-Engines - Knowledge Base of Relational and NoSQL Database Management Systems (2015). https://db-engines.com/en/
19. Rocha, L., Vale, F., Cirilo, E., Barbosa, D., Mourão, F.: A framework for migrating relational datasets to NoSQL. Procedia Comput. Sci. **51**, 2593–2602 (2015)
20. Yangui, R., Nabli, A., Gargouri, F.: Automatic transformation of data warehouse schema to NoSQL data base: comparative study. Procedia Comput. Sci. **96**, 255–264 (2016)

21. Liao, Y.T., et al.: Data adapter for querying and transformation between SQL and NoSQL database. Futur. Gener. Comput. Syst. **65**, 111–121 (2016)
22. Bansel, A., González-Vélez, H., Chis, A.E.: Cloud-based NoSQL data migration. In: 2016 24th Euromicro International Conference on Parallel, Distributed, and Network-Based Processing (PDP), pp. 224–231. IEEE (2016)
23. Atzeni, P., Bugiotti, F., Cabibbo, L., Torlone, R.: Data modeling in the NoSQL world. Comput. Stand. Interf. **67**, 103149 (2020)
24. Vathy-Fogarassy, Á., Hugyák, T.: Uniform data access platform for SQL and NoSQL database systems. Inf. Syst. **69**, 93–105 (2017)
25. Hamouda, S., Zainol, Z.: Document-oriented data schema for relational database migration to NoSQL. In: 2017 International Conference on Big Data Innovations and Applications (innovate-data), pp. 43–50. IEEE (2017)
26. Bante, P.M., Rajeswari, K.: Big data analytics using hadoop map reduce framework and data migration process. In: 2017 International Conference on Computing, Communication, Control and Automation (ICCUBEA), pp. 1–5. IEEE (2017)
27. https://www.dummies.com/programming/big-data/hadoop/acid-versus-base-data-stores/

Innovation Activities of Ethiopian Enterprises: Current Issues and Recommendations

Le Lan Anh[1](✉) and Menelik Samuel[2]

[1] VASS_ Institute of Americas Studies, Hanoi, Vietnam
lelananh84.vias@gmail.com

[2] Technologies and Innovation Institute, Addis Ababa, Ethiopia

Abstract. This article examines the findings of Ethiopia's second national innovation survey, which took place in 2019. The conclusions of this study analyze the status of technological and non-technology innovations in Ethiopian firms, as well as indicate the hurdles to innovation. The findings of this study demonstrate that only 19.4% of enterprises were innovative, based on a list of 300,000 legally registered business-oriented corporations gathered from various government ministries and organizations. Product innovations were the most common followed by process innovations; around % of innovative businesses introduced technology improvements (product and/or process innovations). The inventive enterprises introduced 686 inventions, with 56% being product advances and 78.6% being technology improvements.

Keywords: Ethiopian enterprises · Technological innovation · Innovation

1 Introduction

With increasing global competition and the quickly spreading of knowledge, the future of many businesses depends upon their ability to innovate [9]. New technologies and procedures are developed as a result of innovation, allowing economies to take advantage of previously unattainable or unavailable opportunities. A given set of inputs, allows organizations to minimize manufacturing costs or boost productivity. The studies conducted by Aminreza et al. [2] and Davila et al. [3] indicated the importance of innovation in enterprises: to improve quality of product, create new markets, reduce labor costs, environmental protection; improve production processes and materials, and replace products or services. In addition, Elshamy, H. [4] pointed out the efforts for stimulating new products should favor the strategies of market expansion by firms, through, for instance, an expansionary demand policy, organization of new markets and targeted procurement. Therefore, innovation has for many decades been subject to thorough analysis and research. It would also eliminate or delay the phenomenon of the diminishing return that plagues developing nations' traditional resource-based production operations. As a result, the expanding populations of developing countries would have more production and employment options.

S. Motahhir and B. Bossoufi (Eds.): ICDTA 2022, LNNS 454, pp. 141–148, 2022.
https://doi.org/10.1007/978-3-031-01942-5_14

Ethiopia's economy has grown by double digits for more than a decade. Despite the tremendous growth of the economy, the government claims that it is not due to technological advancement. As a result, the government's lofty objective of restructuring the economy and becoming a middle-income country by 2025, fuelled mostly by public sector investment [7], may not be met. In 2012, the Ethiopian government issued a national science, technology, and innovation policy to coordinate and organize the country's knowledge generation and innovation efforts, recognizing the limits in the framework that underpins the present economic growth process.

Ethiopia's STI policy demonstrates the government's apparent preference for foreign technology transfer over new technology development. On the other side, national innovation initiatives are claimed to foster the adoption and adaptation of foreign technologies. While developing countries have the option of adopting industrialized countries' technologies, their capacity to properly utilize or exploit such technologies is limited by their own technological or absorption capacities [1, 5, 6]. Furthermore, the socio-economic settings in which foreign technologies are developed may differ significantly from those in technology-importing countries.

The study focuses on themes such as the rate of innovation in both technology and non-technology domains, as well as the variables that promote or restrict innovation. It provides Ethiopian business executives and officials with important insights and policy recommendations. The survey results will also serve as a starting point for further research into Ethiopia's business innovation difficulties.

2 Methodology

2.1 Method of Data Collection

The data collection instruments were made in compliance with the guidelines in the Oslo manual (3rd eds.). The survey employed a typical self–administered questionnaire based on the handbook, plus a few factors adjusted to the country's circumstances.

2.2 Target Population

The survey's target group was businesses, which included both private and public businesses (whether publicly listed or not) (corporations owned by the government).

Small, medium, and large enterprises in the manufacturing, construction, service, and mining sectors are included in the survey. A list of 300,000 legally registered business-oriented businesses was provided by various government ministries and agencies. The main entities where business registration data was sought were the Federal Small and Medium Manufacturing Industry Development Agency, the Ministry of Industry, the Ministry of Trade, the Ministry of Construction, the Ministry of Mines, Petroleum and Natural Gas, and the Privatization and Public Enterprise Supervising Agency.

2.3 Sampling Technique and Sample Size

To select firms, stratified sampling was used, first by size and then by sector of economic activity. The primary criterion for the stratification of target firms was firm size. The size of a company is usually determined by the number of employees or the turnover rate. Firm size is typically defined by the number of employees or the annual revenue. Using both of these criteria at the same time may result in a firm being classified into different size groups. As a result, this survey only used the number of employees to categorize firms into different size groups.

The existing national company size classification system was utilized in this study, which categorizes businesses like small, medium, or big depending on the number of employees they have, which is fewer than 49, 50–250, or greater than 250. The sampling frame includes a total of 21,548 business businesses with at least 10 employees, as recommended by the Oslo guideline. The questionnaire was sent to a total of 1047 commercial enterprises at random. Proportional weights were employed to distribute the sample enterprises across size and sector groups after the overall sample size was determined.

2.4 Data Collection Instruments and Field Work

A self-administered standardized survey questionnaire created under the Oslo manual's recommendations was the primary data gathering instrument. To strengthen the accuracy of the data acquired, the selected business organizations' managers, finance directors, and human resource directors each individually responded to questions concerning their respective authorities and tasks.

A total of 50 convenient sample businesses were used to pilot test the poll. The necessary alterations and corrections were made based on the findings of the sample test.

2.5 Method of Data Analysis

Descriptive data analyses were used to explain, present, or summarize the quantitative data gathered from the survey questionnaire in a comprehensible way. Traditional and convenient visualization approaches such as tables, charts, and figures were used to illustrate the quantitative measurement of innovation results.

3 Results and Discussions

3.1 Current Status of Technological Innovations and Non-technological

Product, process, organizational, and marketing innovation are the four major types of innovation. Product innovation is defined as the introduction of a new or significantly improved good or service in terms of technical specifications, components, and materials, including software, user-friendliness, or other functional features, as illustrated in Fig. 1. Product innovation has been implemented by approximately 83% of innovative firms. Process innovation, which includes new or significantly improved production or delivery methods, changes in techniques, equipment, and/or software, and changes in techniques, equipment, and/or software aimed at lowering production or delivery unit costs, was implemented by approximately 60% of the innovative firms. Marketing innovation was used by 49.3% of the innovative businesses (new or significant changes in marketing methods involving product design or packaging, product placement, product promotion, or pricing).

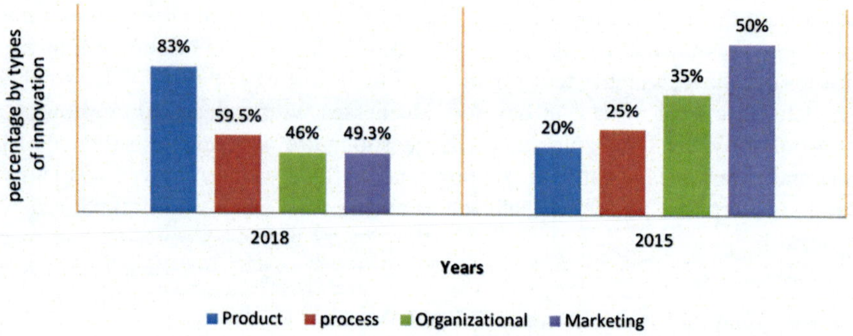

Fig. 1. Rate by type of innovation

In addition, nearly half of the innovative firms have implemented marketing organizational innovations. Contrary to the findings of the 2018 innovation survey, product and marketing innovations were the least and most prevalent types of innovations in the 2015 survey, respectively. In the 2015 survey, approximately 20% and 50% of the innovative firms introduced product and marketing innovations, respectively. In comparison to the innovative firms in the 2015 survey, the results indicate that a large proportion of the innovative firms in the 2018 survey were able to introduce innovations in more than one dimension. It is also possible that only the most innovative firms, capable of introducing innovation in multiple dimensions, have remained innovative recently. This is also evident in Fig. 2, which depicts the degree of conjointness of various types of innovations.

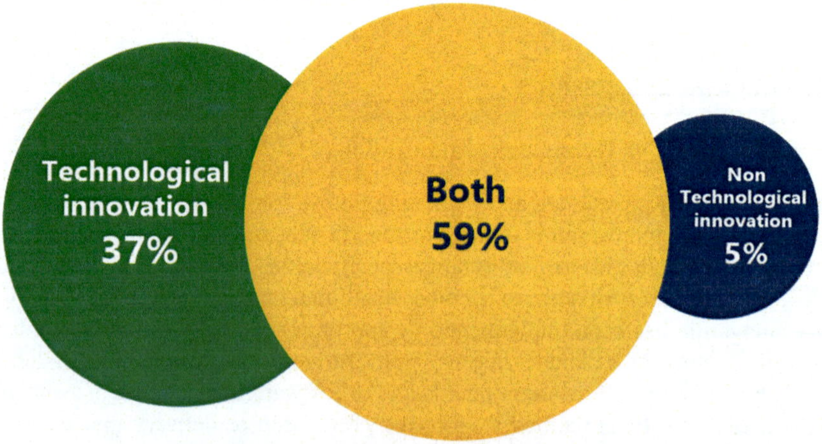

Fig. 2. Conjointness of innovations

According to Fig. 2, 96% of innovative firms (combined product and process innovations) introduced technological innovations in the current survey, with 37% introducing exclusively technological innovations. Only 5% of innovative organizations introduced only non-technological types of innovation, even though 64% of innovative firms introduced non-technological innovations. Firms that introduced non-technical innovations have a much bigger percentage of the market than those that introduced technology innovations. At least one technology and one non-technological type of innovation were introduced by 59% of innovative companies.

The differences in the structure or content of innovation between the 2015 and 2018 surveys could be linked to variances in the goals of innovation, which will be influenced by the economy and the issues that businesses confront. Indeed, a comprehensive inquiry into the causes of this transition is required. Nonetheless, the survey results show that in 2015, the primary purpose of innovation was to solve marketing or distributional challenges, whereas, in 2018, the focus was on production or technical issues.

3.2 Factors Encouraging and Inhibiting Innovation

Numerous factors can either stimulate or impede innovation in one way or another.

- *Factors encouraging innovation*

Many factors motivate the innovation of Ethiopian enterprises. The innovation drive is important in driving the innovation process and demonstrating progress in a company's operations. The primary goal of an enterprise is to generate more profits, either by increasing revenue, market share or reducing costs. In addition, companies may need to comply with social responsibility regulations and expectations.

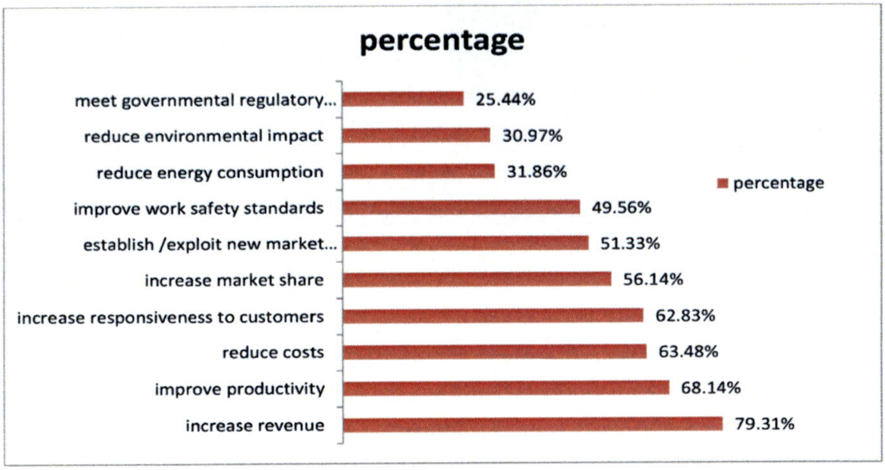

Fig. 3. Factors encouraging innovation

Figure 3 shows that the majority of innovative firms (79.31%) are motivated to innovate to increase revenue, and about 68% of them also innovate to improve their productivity. This is related to revenue. Only a small percentage of innovative enterprises (25.44%) are motivated by the requirement to meet Government regulations.

- *Factors inhibiting innovation*

Some several causes or obstructions may prohibit firms from inventing, just as there are variables that urge them to do so. The key reasons why businesses do not innovate, as well as the accompanying hurdles to innovation, are discussed in this section. It also goes over the effects of numerous variables on innovation, as measured by a binary logistic regression model.

There are compelling reasons that prevent business enterprises from innovating, just as there are compelling reasons that encourage firms to innovate. This could be due to the enterprises' unwillingness to face challenges, or it could be due to their lack of interest in innovation. As shown in Fig. 4, the vast majority of non-innovators considered innovating but were prevented from doing so due to a variety of factors. These vexing factors can be both internal and external. On the other hand, 38% of them do not appear to be interested in innovating. The following figure (Fig. 5) depicts multiple impediments to innovation for those who consider it but face numerous impediments.

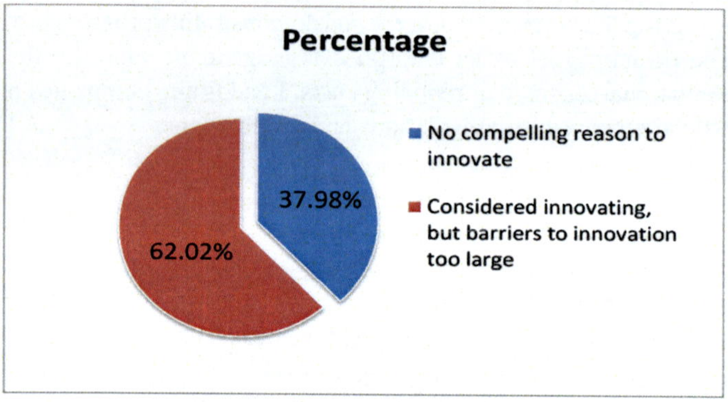

Fig. 4. The reason for non-innovation

According to the data in Fig. 5, the main barriers to enterprise innovation are economic factors such as a lack of internal finance, a lack of credit, and a lack of private equity. When it comes to internal factors, the lack of skilled employees should be highlighted. Other external factors such as difficulties in obtaining government grants and subsidies for innovation, a lack of collaboration partners, and excessive competition are barriers to innovation.

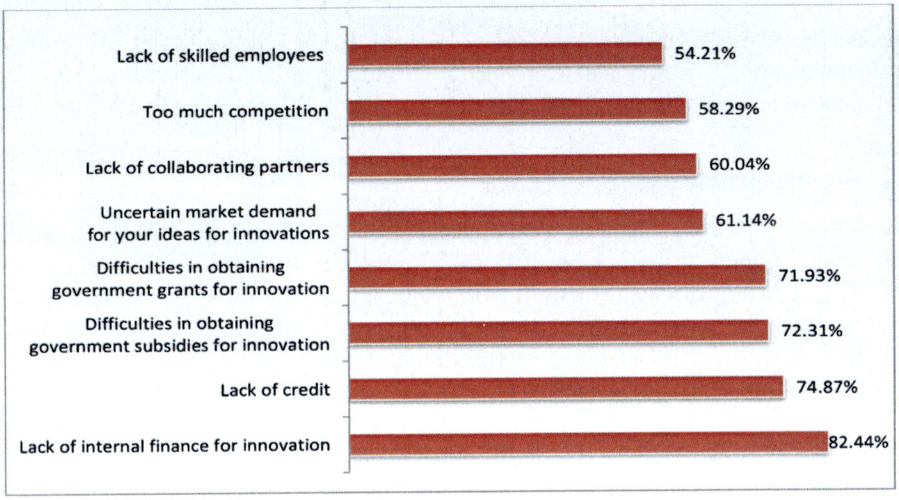

Fig. 5. Barriers to innovation

4 Conclusion and Recommendations

4.1 Conclusion

According to the survey results, only 19.4% of enterprises were innovative. Product innovations were the most common type of innovation, followed by process innovations; approximately 96% of innovative enterprises introduced technological innovations (product and/or process innovations). Six hundred and eighty-six innovations were introduced by the innovative firms, with 56% being product innovations and 78.6% being technological innovations. The average percentage of enterprises that innovated was 4.64%.

The percentage of innovation in the 2018 study, which covered the years 2017 and 2018, was significantly lower than the 59.7% rate shown in the 2015 national innovation survey, which covered the years 2012–2014. The rate is also lower than in certain other African countries, such as Kenya, where the rate of innovation was 45% from 2012 to 2015, and Ghana, where the rate was 72.5% from 2008 to 2010 [8]. Ethiopia's acute shortage of foreign exchange revenues and political instability during the last few years could be blamed for the reduction in innovation rates. These circumstances may have an immediate influence on innovation by limiting the funding allocated to it, or they may have an indirect impact by lowering the demand for or need for it.

According to the findings of the survey, Ethiopian businesses not only have a low rate of innovation but also have limited uniqueness in their ideas. The bulk of the innovative firms (56%) or 11% of the sample firms introduced novel advances. Only about a third of the innovative enterprises introduced unique inventions to their markets, and only about third introduced novel innovations to the country. Despite this, only about 3% of businesses were able to introduce fresh innovations to the globe. Although the prevalence of "new-to-the-firm" innovations is a symptom of a general lack of originality in innovation, delivering a new product or business method to the world would be a commendable achievement.

4.2 Recommendations

The following recommendations were created based on the findings of this Ethiopian national innovation survey done in 2011 E.C. for the reference years of 2009 E.C. and 2010 E.C. The proposals may be useful to a wide range of stakeholders, including governments, businesses, universities, research institutes, and others interested in Ethiopia's growth.

The majority of the companies polled said they didn't have enough money to invest in innovation. As a result, the government should consider implementing high-tech innovation-boosting programs and incentive packages. The government may, for example, create an innovation fund and provide tax incentives and subsidies to the innovative enterprise.

The government should seek to strengthen the country's cyberinfrastructure and internet access in partnership with other private parties. Collaborations for innovation must be encouraged as soon as possible. This could include collaboration between universities, corporations, and government agencies to build collaborative R&D centers. Firms could share facilities, people resources, experiences, and contracting of each other's inventions even if they conduct distinct R&D efforts. Coordinators could be the government or government agencies, such as the ministries of innovation and technology, research and higher education, and industry, among others. Either of these ministries or other relevant authorities should take the lead in gathering diverse stakeholders to debate and plan the formation of cooperation for knowledge production and sharing, innovation, and technology transfer in the country.

References

1. Abramovitz, M.: Thinking About Growth: and Other Essays on Economic Growth and Welfare. Cambridge University Press, Cambridge (1989)
2. Aminreza, K., Maryam, R., Lagha, A.M.: Barriers to innovation among Iranian SMEs. Asian J. Bus. Manag. **5**(2), 251–265 (2011)
3. Davila, T., Epstein, M.J., Shelton, R.: Making Innovation Work: How to Manage it, Measure it, and Profit From it. Wharton School Publishing, Upper Saddle River (2006)
4. Elshamy, H.: The impact of firm size on innovative activity: an analysis based on Egyptian firm data. Int. J. Bus. Technopreneurship **5**, 397–406 (2015)
5. Gerschenkron, A.: Economic Backwardness in Historical Perspective: A Book of Essays. Harvard University Press, Cambridge (1962)
6. Lall, S.: Technological capabilities and industrialization. World Dev. **20**(2), 165–186 (1992)
7. Moller, L.C.: Ethiopia's Great Run: The Growth Acceleration and How to Pace it. World Bank Group, Washington, DC (2015)
8. NEPAD. The NEPAD Strategic Plan for the period 2014–2017. African Union Development Agency (2014)
9. Talegeta, S.: Innovation and barriers to innovation: small and medium enterprises in Addis Ababa. J. Small Bus. Entrepreneurship Dev. **2**(1), 84 (2014)

A Trace-Based System for the Categorization of Actors in a Social Network

Samra Sabeg[1,2](✉) and Brahim Belgroun[1,2]

[1] Department of Computer Science, The University Abbes Laghrour of Khenchela, Khenchela, Algeria
samra.sabeg@univ-khenchela.dz, b_belgroune@esi.dz
[2] ICOSI Laboratory, The University Abbes Laghrour of Khenchela, Khenchela, Algeria

Abstract. Facebook presents the largest social network on the Internet. Users' interaction traces seem to provide a good basis for analyzing this environment. Several methods and approaches have proposed to treat different questions (categorization, social relations detection, behavior evaluation, etc.). However, the existing methods cannot satisfy users' requirements.

In this paper, we propose a new approach to detect the social pole in Facebook. Our system measures the effectiveness of a user in the social network using the technics of the Trace-Based System. It aims to collect and analyze user' interaction traces over the Facebook platform then transform them into indicators to evaluate a list of categorization metrics.

Finally, we evaluate our approach with a case study. We suggest a list of metrics and evaluate each metric by a list of indicators. Every indicator has calculated from a list of traces (observed elements).

Keywords: Social networks · Trace · Indicators · Metrics · Trace-based system

1 Introduction

Recently, Online Social Networks represent a new way of communicating between actors (people, organizations, companies) such as Facebook and Twitter. They allow millions of users to participate in several groups through different engagements (interactions, comments, and sharing). During the wide use of these virtual environments, actions, events, and interactions of users leave a digital trace that has recorded to analyze the activities of a user (or group of users) and to understand its behavior when using computer systems [1]. We point out some of the studies in the literature addressed that exploit traces recorded during interactions to analyze social networks in several fields. In [2], the authors present an analysis of a regional Facebook dataset, with attention given to the correlation among user connectivity, activity, and similarity.

An automated process in multi-agent systems is presented in [3], which collects data from users' activity in social networks and analyzes the social behavior at different levels in online events. In [4], authors design a model based on traces of human activities to calculate indicators in E-learning. In [5], a system is proposed for classifying user activities in network traffic using supervised and unsupervised learning. It uses the

S. Motahhir and B. Bossoufi (Eds.): ICDTA 2022, LNNS 454, pp. 149–157, 2022.
https://doi.org/10.1007/978-3-031-01942-5_15

behavior of users over the network to classify the underlying user activity, taking into consideration all of the traffic generated by the user within a given time. Authors in [6] proposed a framework for intrusion detection in an informal social environment based on users' traces to improve their learning. In [7] authors aim to analyze the impact of Facebook in the teaching-learning process in financial mathematics education.

Several social networks analysis approaches have been proposed, they have still based on user' traces, as was shown in [8]. Authors prove that any social media network users' personality is predicted based on their posts and status by using a digital footprint left by users.

Moreover, even the traces offer an excellent foundation to analyze social networks. They were not always easy to exploit them directly with the available tools due to the difficulty in their collection, processing, and handling in a simple manner. Therefore, the existing approaches cannot fulfill researchers' requirements. We still need methods that exploit the traces of interactions in social networks. In this paper, we have proposed a system that uses user traces to detect the social pole. Its objective is to collect and analyze generated traces in Facebook and transform them into indicators (computed Traces) to evaluate a list of metrics. A vector of three binary elements has been calculated to represent each metric and indicate its category (strong, medium, low). Finally, we classify actors of Facebook using these metrics.

Our paper is organized into different sections: we will present several works based on analysis of traces in social networks in the second section. The third section is devoted to modeled trace-based systems. Section 4 will explain our contribution. Section 5 introduces an example that serves as a running example for illustrating our approach. Before concluding and presenting the future directions of this work, we discuss the obtained results in Sect. 6.

2 Related Work

Exploiting users' interactions traces to analyze a social network has been outlined by different studies. In this research, we adopted the research described in [9]. The authors have developed a system that uses learners' traces to evaluate the Knowledge Level. It uses a list of criteria described by calculated indicators from a set of Traces in E-learning environments. In [10], the authors proposed an approach to handle action traces in Technology Enhanced Learning TEL systems. In [11], the authors proposed a software platform to define a model of traces to calculate indicators on the activity in the system. In [12], the authors developed a model for predicting peoples' activity indicators using fuzzy logic, without making rigid calculations but based on traces collected from the Moodle Learning Management System (LMS). In [13], the authors provided an in-depth comparative analysis of the proposed approaches to grouping social network actors into communities of interest.

3 Modeled Trace-Based Systems

TBS is one of the most recent systems. It used to analyze users' interactions [14–16]. It collects traces from their sources and writes them according to the model of Trace.

Next, it transforms them into indicators. Each indicator is a variable (digital, semantic, or graphical form) that represents some characteristics of a given individual (person, web page, phenomenon, etc.) [17]. Figure 1 depicts the core elements of a trace-based system.

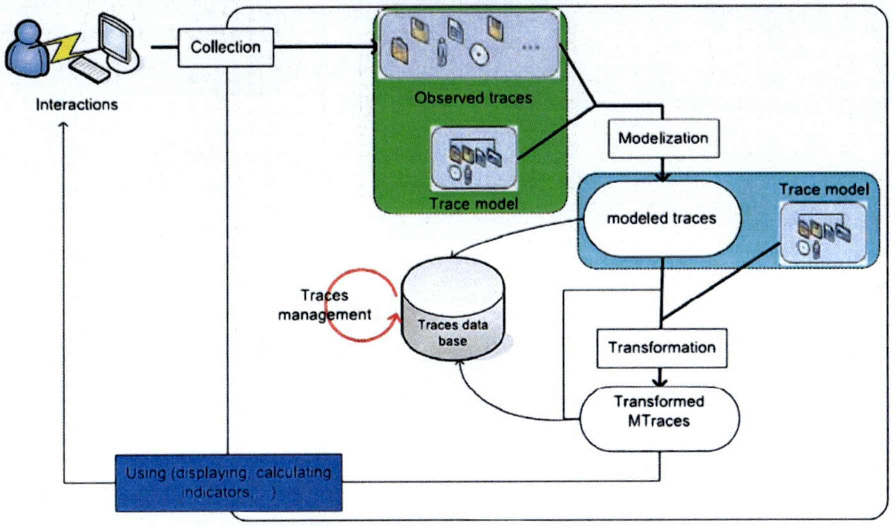

Fig. 1. Core elements of a trace-based system.

4 Contribution

We present our approach to detect a social pole in social networks. Our system analyzes traces in a social network to calculate indicators. They will have used to evaluate each metric. For example, engagement and engagement rate are indicators calculated from different traces in Facebook to evaluate page appreciation metric.

The detection system evaluates each metric by a global vector of three binary elements that indicate its category (strong, medium, low). All global vectors will be gathered in a global matrix to be analyzed by our detection system to categorize actors. Our approach aims to help designers measure the effectiveness of pages in the social network. The detection process has applied to one actor or a group of actors. Our approach determines the strengths and weaknesses of each actor on several occasions. Therefore, the first step towards our purpose is to suggest a list of metrics. Then we evaluate each metric by four indicators; each indicator has calculated from a list of traces. Three models have been proposed to model traces, indicators, and metrics.

4.1 Proposed Models

Before presenting the process of social pole detection, we will propose three models: model of trace, model of indicator, and finally, model of metric. Figure 2 presents the proposed model of Trace; the other models are not presented in this paper due to the length limitation.

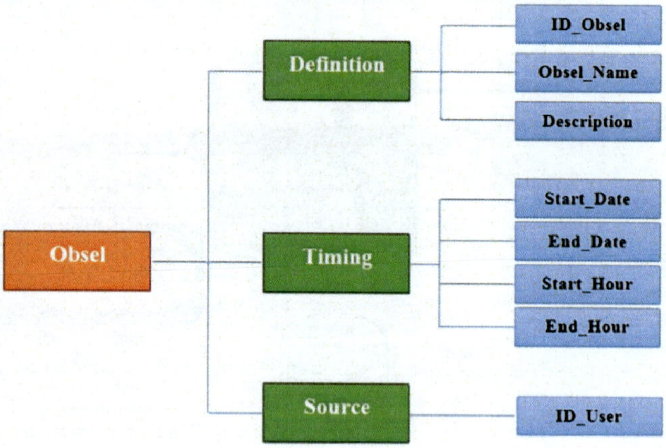

Fig. 2. Trace model

4.2 Description of TBS_DSP Architecture

We present our TBS_DSP (Trace-Based System for the Detection of Social Pole) architecture in three levels: trace processing, detection of the social pole, and capitalization. It uses the three models described above. Figure 3 shows the overall architecture of TBS_DSP.

Trace Processing Process. The trace processing process consists of producing indicators through four sub-processes: (A) filtering the collected traces, (B) models traces according to a trace model, (C) Calculating indicators, and (D) Capitalization.

Detection Process. This process detects the social pole. It consists of two parts: evaluate the metrics by vectors and detection.

Evaluate the Metrics. We evaluated metrics using a list of calculated indicators: R1, R2, G1, and G2. These metrics will have classified according to the mechanism shown in Fig. 4, (R1, R2) are two indicators calculated from traces, (G1, G2) are the medium of (R1, R2) respectively.

For example, engagement and engagement rate are two indicators calculated for each post of an actor.

A vector of three binary elements will describe each metric. It determines its category. One of their elements must have the value "1" to indicate its class, and the rest of the elements will have evaluated as "0". For instance, the vector (0, 0, 1) represents a metric of the low category.

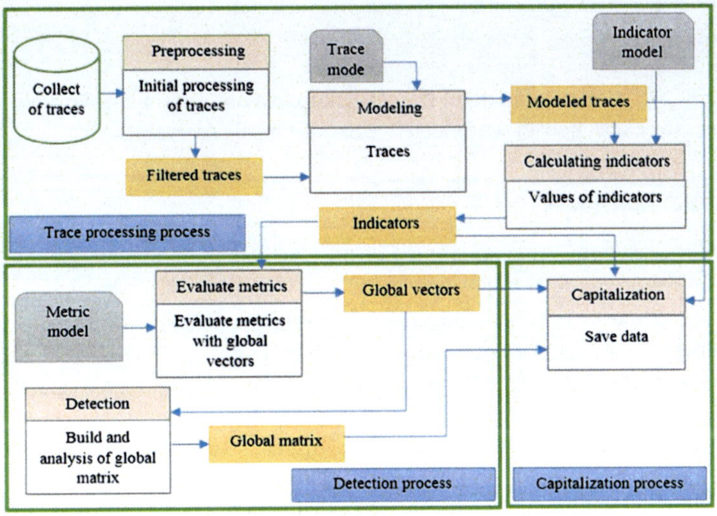

Fig. 3. TBS_DSP architecture

The global vector for each metric will have calculated from all vectors of different posts. Equation (1) explains how to calculate the global vector. V_{global} (v_1, v_2, v_3):

$$V_i = \frac{Nb_{Post \in class(i)}}{Nb_{Post}} \tag{1}$$

All global vectors of different metrics will have collected in a global matrix created by the detection process. This latter examines this global matrix to categorize and classify actors.

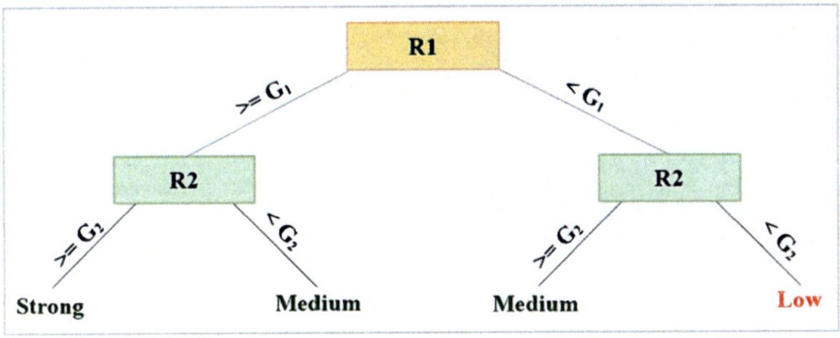

Fig. 4. Classification of metric

The process of capitalization is a shared process in our architecture; it is ready at all times to save all intermediate processing.

5 Case Study

In this section, we present a practical framework to test our approach. We have proposed four metrics; for each metric, we selected a list of Obsels to evaluate it.

5.1 Proposed Indicators

We used five indicators to represent the metrics that are:

1. The scope of a publication defines the number of people who have seen a post.
2. Engagement is the number of people who clicked anywhere on post
3. Engagement rate is the percentage of people who liked, commented on, shared post, or clicked on it after seeing it
4. Click-through rate is the ratio between the number of views of a post and the number of clicks received.
5. Conversion rate is defined as the ratio between the number of people who have carried out a given action (buy a product, fill out a contact form) and the total number of people affected by the campaign.

5.2 Proposed Metrics

To implement our TBS_DSP, we have proposed four (04) metrics. An Obsel list will have required to represent them as (likes, shares, clicks, comments, etc.). Figure 5 details the needed elements for evaluating a metric.

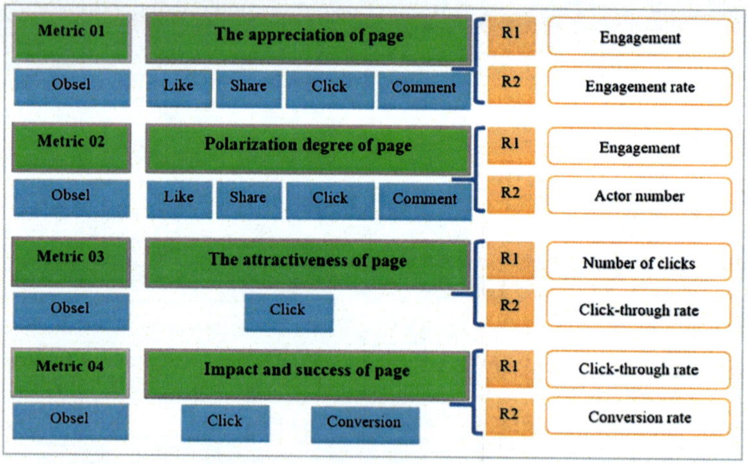

Fig. 5. Proposed metrics to measure the effectiveness of page

For instance, engagement and engagement rate (indicators) are two crucial indicators to estimate page appreciation (metric). Figure 6 clarifies the adopted way to measure the appreciation metric. The result will be a global vector that determines the category of

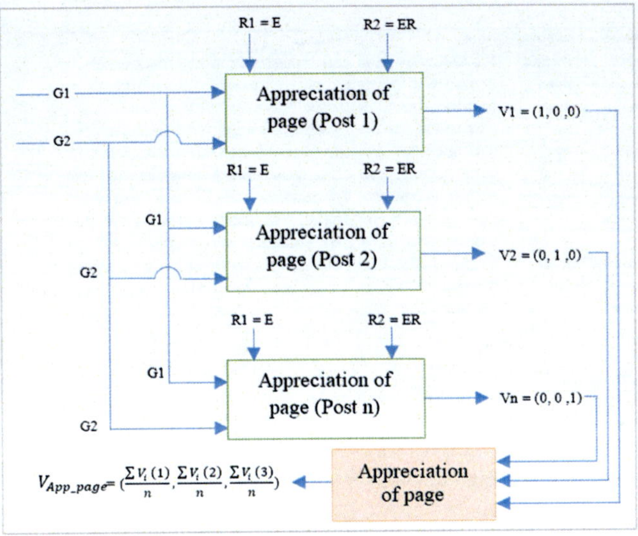

$$V_{App\_page} = \left(\frac{\sum V_i(1)}{n}, \frac{\sum V_i(2)}{n}, \frac{\sum V_i(3)}{n}\right)$$

Fig. 6. How to calculate the appreciation metric

metric. Similarly, we will evaluate the rest of the metrics. Figure 7 appears an appreciation metric for each actor.

Each metric has described as a row in the global matrix. Figure 8 shows the matrix of all metrics. This matrix will be the input of our detection system.

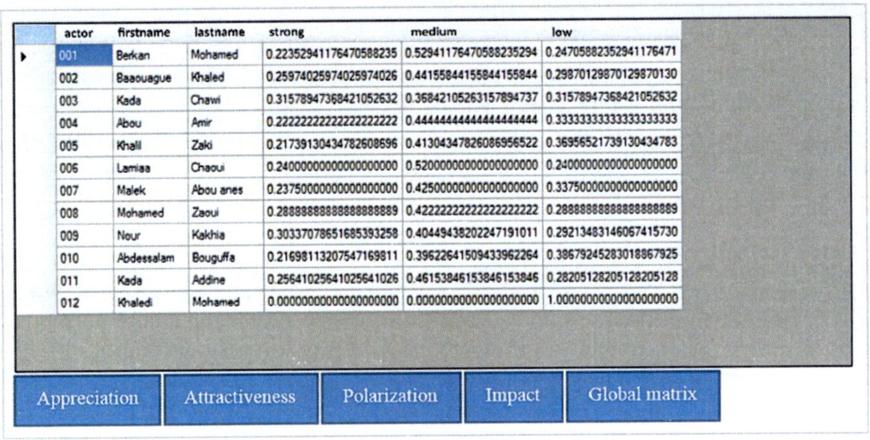

actor	firstname	lastname	strong	medium	low
001	Berkan	Mohamed	0.22352941176470588235	0.52941176470588235294	0.24705882352941176471
002	Baaouague	Khaled	0.25974025974025974026	0.44155844155844155844	0.29870129870129870130
003	Kada	Chawi	0.31578947368421052632	0.36842105263157894737	0.31578947368421052632
004	Abou	Amir	0.22222222222222222222	0.44444444444444444444	0.33333333333333333333
005	Khalil	Zaki	0.21739130434782608696	0.41304347826086956522	0.36956521739130434783
006	Lamiaa	Chaoui	0.24000000000000000000	0.52000000000000000000	0.24000000000000000000
007	Malek	Abou anes	0.23750000000000000000	0.42500000000000000000	0.33750000000000000000
008	Mohamed	Zaoui	0.28888888888888888889	0.42222222222222222222	0.28888888888888888889
009	Nour	Kakhia	0.30337078651685393258	0.40449438202247191011	0.29213483146067415730
010	Abdessalam	Bouguffa	0.21698113207547169811	0.39622641509433962264	0.38679245283018867925
011	Kada	Addine	0.25641025641025641026	0.46153846153846153846	0.28205128205128205128
012	Khaledi	Mohamed	0.00000000000000000000	0.00000000000000000000	1.00000000000000000000

Fig. 7. Result of the appreciation metric

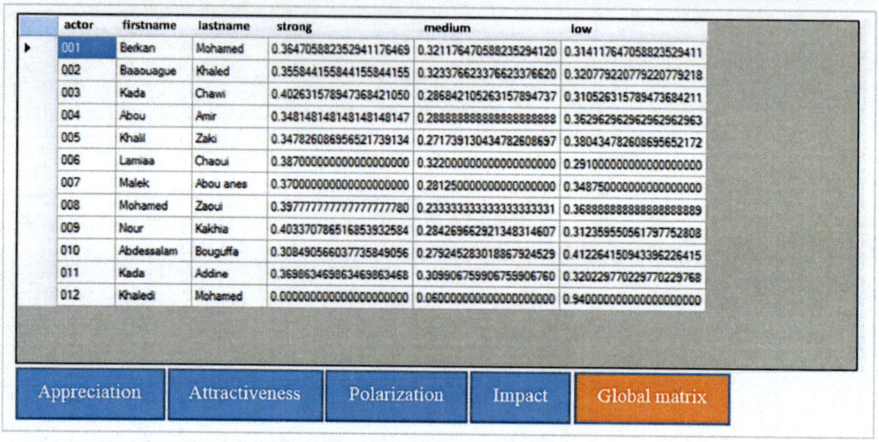

actor	firstname	lastname	strong	medium	low
001	Berkan	Mohamed	0.364705882352941176469	0.321176470588235294120	0.314117647058823529411
002	Baaouague	Khaled	0.355844155844155844155	0.323376623376623376620	0.320779220779220779218
003	Kada	Chawi	0.402631578947368421050	0.286842105263157894737	0.310526315789473684211
004	Abou	Amir	0.348148148148148148147	0.288888888888888888888	0.362962962962962962963
005	Khalil	Zaki	0.347826086956521739134	0.271739130434782608697	0.380434782608695652172
006	Lamiaa	Chaoui	0.387000000000000000000	0.322000000000000000000	0.291000000000000000000
007	Malek	Abou anes	0.370000000000000000000	0.281250000000000000000	0.348750000000000000000
008	Mohamed	Zaoui	0.397777777777777777780	0.233333333333333333331	0.368888888888888888889
009	Nour	Kakhia	0.403370786516853932584	0.284269662921348314607	0.312359550561797752808
010	Abdessalam	Bouguffa	0.308490566037735849056	0.279245283018867924529	0.412264150943396226415
011	Kada	Addine	0.369863469863469863468	0.309906759906755906760	0.320229770229770229768
012	Khaledi	Mohamed	0.000000000000000000000	0.060000000000000000000	0.940000000000000000000

Fig. 8. Global matrix of all metrics

6 Discussion

Our research aims to exploit interactions traces over the Facebook environment to detect social pole. Compared to previous studies presented in [18] and [19], our process is improved when we use the concept of vectors to evaluate metrics. This process of detection relies on the users' actions and interactions. Furthermore, the representation of the metrics by vectors gives the possibility to compare actors between them. The system can focus on the most popular post (of an actor); it can send guidance messages to actors who are not active (with low class). Hence, the representation of metrics with calculated indicators (computed from traces) is more efficient in detecting a social pole. Comparing actors over different periods allows us to concentrate precisely on the weaknesses of one actor compared to other actors. Using a Trace-Based System helps designers know the strengths and weaknesses of the page. It allows actors to improve the quality of their publications.

7 Conclusion and Perspectives

In this study, we proposed a system based on traces to detect the social pole in the Facebook environment. It analyses digital Traces (actions, events, and interactions) to compute the indicators and evaluate a list of metrics. A vector of three binary elements represents each metric and defines its category (strong, medium, low). For instance, V(0, 0, 1) is a vector of the Low class. All vectors representing metrics will have grouped in the global matrix that will be the input of our detection system to compare actors between them and categorize actors in the social network platform. To test our approach Four (04) metrics have been proposed. We used five indicators to represent them.

We hope in future work to enhance the effectiveness of the system. We desire to endow the system with other indicators and metrics to enrich the database with new and more significant traces.

References

1. Djouad, T., Mille, A., Reffay, C.: Ingénierie des indicateurs d'activités à partir de traces modélisées pour un environnement informatique d'apprentissage humain. Sci. Technol. l'Inform. Commun. l'Éduc. Format. **16**, 1–28 (2012)
2. Nguyen, K., Tran, D.A.: An analysis of activities in Facebook. In: IEEE Consumer Communications and Networking Conference (CCNC) (2011)
3. del Val, E., Martínez, C., Botti, V.: Analyzing users' activity in online social networks over time through a multi-agent framework. Soft. Comput. **20**(11), 4331–4345 (2016). https://doi.org/10.1007/s00500-016-2301-0
4. Djouad, T., Mille, A.: Observing and understanding an on-line learning activity: a model-based approach for activity indicator engineering. Technol. Knowl. Learn. **23**(1), 41–64 (2017). https://doi.org/10.1007/s10758-017-9337-9
5. Labayen, V., Magaña, E., Morató, D., Izal, M.: Online classification of user activities using machine learning on network traffic. Comput. Netw. **181**, 107557 (2020)
6. Zedadra, A., Zedadra, O., Kebabi, A.: SM-ILE: a framework for intrusion detection in an informal social environment based on users' traces. Int. J. Innov. Learn. **28**(3), 368 (2020)
7. Salas-Rueda, R.A.: Analysis of Facebook in the teaching-learning process about mathematics through data science. CJLT/RCAT **47**(2) (2021)
8. Valanarasu, R.: Comparative analysis for personality prediction by digital footprints in social media. J. Inf. Technol. Digit. World **03**(02), 77–91 (2021)
9. Brahim, B., Lotfi, A.: A traces based system helping to assess knowledge level in e-learning system. J. King Saud Univ. – Comput. Inf. Sci. (2018)
10. Djouad, T., Mille, A.: Trace based system in TEL systems: theory and practice. In: Amine, A., Mouhoub, M., Ait Mohamed, O., Djebbar, B. (eds.) Computational Intelligence and Its Applications. CIIA 2018. IFIP Advances in Information and Communication Technology, vol. 522, pp. 257–266. Springer, Cham (2018). https://doi.org/10.1007/978-3-319-89743-1_23
11. Casado, R., Guin, N., Champin, P.A., Lefevre, M.: kTBSLA: une plateforme d'analyse de traces fondée sur une modélisation sémantique des traces, Font-Romeu, France. hal-01637548 (2017)
12. Chehbi, S., Fri, C., Elouahbi, R.: Learners' activity indicators prediction in e-learning using fuzzy logic. Int. J. Adv. Comput. Sci. Appl. (IJACSA) **11**(12), 2020 (2020)
13. Chouchani, N., Abed, M.: Online social network analysis: detection of communities of interest. J. Intell. Inf. Syst. **54**, 5–21 (2018)
14. Djouad, T., Settouti, L.S., Reffay, C.: Un systeme a base de traces pour la modelisation et l'elaboration d'indicateurs d'activites éducatives individuelles et collectives. Mise à l'épreuve Sur Moodle. Tsi. **29**(6) 1–21 (2010). X(May 2014)
15. Champin, P.-A., Mille, A., Prié, Y.: Vers des traces numériques comme objets informatiques de premier niveau. Intellect. Assoc. Pour Recherche Cogn. **59**, 2013 (2013)
16. Brahami, M., Cherbal, M., Settouti, L.: Système à base de traces pour les EIAH (2008)
17. Dimitrakopoulou, A., Petrou, A., Martinez, A., Marcos, A., Kollias, V., Jermann, P.: State of the art of interaction analysis for metacognitive support diagnosis. Inform. Soc. Technol. Netw. **2**, 1–117 (2006). <hal-00190146>
18. Benyounes, B.: Évaluation de la fonctionnalité et l'utilisabilité de la plateforme d'apprentissage Moodle: une approche du génie logiciel. Université du Québec à Montréal (2009)
19. Fusaro, M.: Rapport du comité institutionnel sur les plates-formes d'apprentissage en ligne (2005)

Ranking Task in RAS: A Comparative Study of Learning to Rank Algorithms and Interleaving Methods

Sara Lasri(✉) and El Habib Nfaoui

LISAC Laboratory, Faculty of Sciences Dhar El Mahraz, Sidi Mohamed Ben Abdellah University, Fez, Morocco
{sara.lasri,elhabib.nfaoui}@usmba.ac.ma

Abstract. Relational Aggregated Search (RAS) is a paradigm that refers to a set of techniques for retrieving and aggregating information nuggets based on their relationships. The performance of a relational aggregated search is dedicated to how to retrieve an information nugget, and, how to present this information to users to fulfill their information need. Recent RAS research has focused only on information retrieval strategies. As consequence, there is a critical need to research strategies for aggregating and ranking information nuggets. In this paper, we focus on the recent issue, namely answering the question of how information is shown to the users. In addition, what are the most effective approaches for the ranking task? We use the ranked results from RAS. These ranked results include different content types (images, videos, news, maps, blogs, groups, and books, etc.). Thus, the focal challenge in the vertical results aggregation task is how to rank these heterogeneous results to obtain pertinent information. Our experimental results of the application of learning to rank algorithms and interleaving methods are shown in this study. From the evaluation results, we can affirm that LambdaMART proves its performance compared with the other; LambdaRank, MART, ListNet, and RankNet. Finally, we apply the interleaving approaches to satisfy the user's need by providing appropriate, organized, and most relevant results. Our experimental study proves that VI-TDA performs better than TDA in all steps of our tests.

Keywords: Relational aggregated search · Ranking task · Learning to rank algorithms · Interleaved methods

1 Introduction

Modern web search engines mainly use vertical search, by combining results from different sources (e.g. web, news, or media: image, video, blog) and displaying them in groups to provide a better user experience, what is called aggregated search. Therefore, we can define the aggregated search by trying to fulfill the requirements of users by collecting information from multiple verticals and displaying it on a single result page [1].

S. Motahhir and B. Bossoufi (Eds.): ICDTA 2022, LNNS 454, pp. 158–168, 2022.
https://doi.org/10.1007/978-3-031-01942-5_16

Query dispatching (QD), nugget retrieval (ND), and result aggregation are the three key components of an aggregated search system (RA) [1]. The focal benefit of the aggregated search system is that it adds two additional phases to traditional retrieval, known as query dispatching and result aggregation, while nugget retrieval serves as an information collector [2].

- **Query dispatching** includes the actions that come before query matching, such as initial query interpretation (query analysis) and other actions that rely heavily on the query and knowledge of the collections [1].
- **Nugget retrieval** is a crucial phase, with a query as the input and a set of relevant information nuggets as the output. Different information nuggets, such as images, videos, news items, etc., can be retrieved and related [1]
- **Result aggregation** in this step, we can involve different methods of putting the relevant and coherent information nuggets together [1] (Fig. 1)

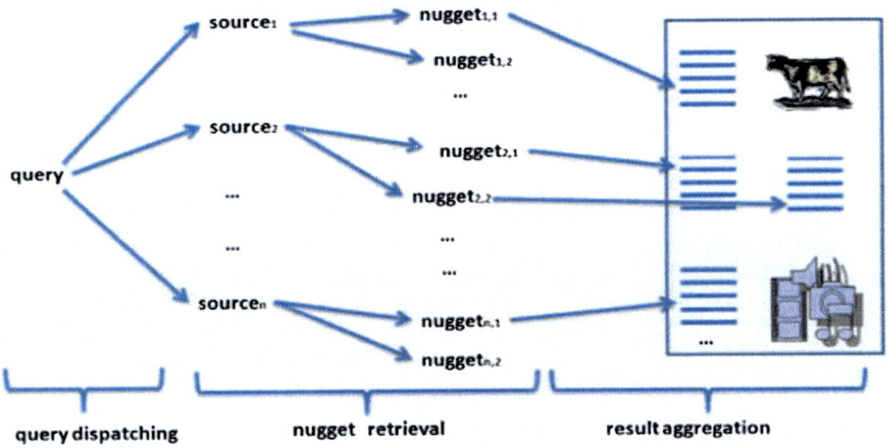

Fig. 1. A general framework for aggregated search [1]

The two study directions in aggregated search are cross-vertical aggregated search (CVAS) and relational aggregated search (RAS) [3]. The first is an example of a multi-source search, which means CVAS is the task of searching and collecting information from vertical search and Web search [3]. The second one is based on the hypothesis that the information can be divided into information nuggets; In addition, the relational aggregated search should uncover dependencies/relations among various keywords of the query [2]. The components of the general framework for aggregated search are simple to identify in cross-vertical aggregated search (query dispatching, nugget retrieval, and result aggregation). The same components (query dispatching, nugget retrieval, and result aggregation) may be found in relational aggregated search, but with the addition of a relations layer [3]. In this work, we focus on the relation-aggregated search, especially how the result aggregation is presented to the users. Moreover, what is the efficient technique to provide a relevant result? The remainder of this paper is structured as follows, in the following section, we review the related works concerning the Relation aggregated search and learning to rank approaches. Section 3, presents the learning to rank algorithms and the interleaving methods used in our study. Section 4 provides our experimental results and our discussion. Finally, the paper is ultimately summarized in the final section (Figs. 2 and 3).

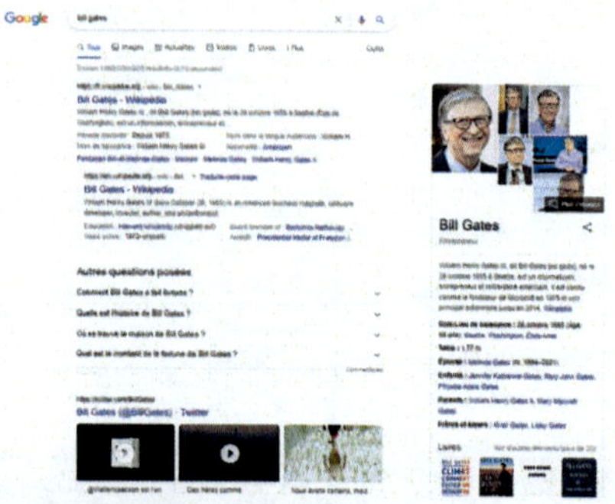

Fig. 2. Example of CVAS: Yahoo results for the query "bill gates"

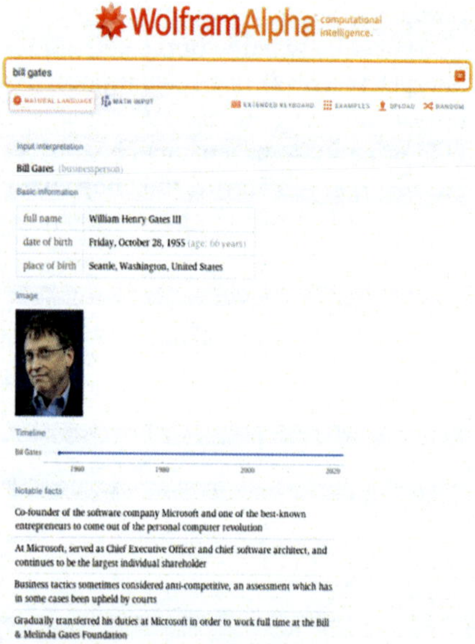

Fig. 3. Example of RAS: Wolfram Alpha results for the query "bill gates

2 Related Works

The aggregated search aims to find and assemble relevant documents from a variety of verticals and combine them with normal online search results on a single result page [4].

2.1 Relational Aggregated Search

The earliest work on aggregated search presented [1, 3] Relational aggregated search as a method of retrieving information based on nuggets of information and their relationships. In [1], The purpose of the relational aggregated search is to produce consistent results that may be combined into a single page. In [3], RAS decomposes into four sub-tasks, Vertical Selection (VS), Nuggets Retrieval (NR), Relation retrieval (RR), and Vertical Results Presentation (RP).

According to [3, 5], the Vertical Selection (VS) is considered the initial step in the relational aggregated search process after receiving the user's query. It focuses on predicting which verticals are relevant to the user query.

According to [1], Nuggets Retrieval (NR) enables for the selection of relevant information. For relational aggregated search, we can retrieve and relate any information nugget, such as images, videos, news items, In terms of vertical results presentation or aggregation, the goal is to deliver a result that includes a diversity of content from several verticals [1].

2.2 Ranking Tasks

According to [1, 3, 5], the performance of a relational aggregated search system is determined not only by the results retrieved from the selected verticals (nugget retrieval) but also by how these results are presented to the users. The items retrieved from relevant verticals are different (web pages, videos, images, blogs, etc.) and are represented using different features. That considered the main challenge related to the vertical results aggregation task. In aggregated search, there were two main ranking strategies: ranking by item and ranking by block [3, 5]. For the first, ranking by item requires ranking each result individually. In block ranking, however, the results are ranked in groups of similar results from the same source. The authors of [6], suggest a machine-learning architecture that can help major search engines rank search results in blocks. They offer and compare three general ranking approaches: classification, voting, and learning to rank.

2.3 Learning to Rank Approach

In [7], the main challenge of the search engine is to find efficient and relevant results and present them to the users. In this work [8], the authors define the learning to Rank (LTR) as a strategy that uses supervised or semi-supervised machine learning techniques to solve the problem of ranking systems. LTR is a task that involves automatically creating a ranking model from training data so that it can classify new objects according to their degrees of relevance, preference, or importance. In [8], The pointwise, pairwise, and listwise approaches are the three learning-to-rank methods.

3 LTR Algorithms and Interleaving Methods for Ranking the Vertical Results

In our research, we adopt the Learning to Rank approach, for ranking the vertical results such as (images, videos, news, maps, blogs, groups, and books, etc.). The main challenge is how to rank these different types of results. As a result, our research proposes that the vertical results be presented in textual form by using the features extracted from the text (web pages, News, Books, Forums, etc…). However, for no textual verticals (image, video), we should exploit their textual description to obtain significant results (Fig. 4).

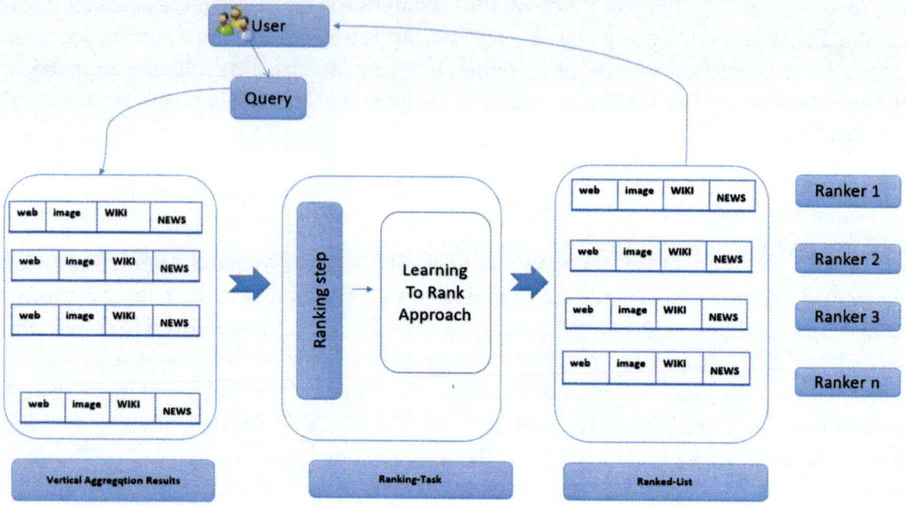

Fig. 4. Learning to rank algorithm in vertical result

3.1 Learning to Rank Algorithm

In this study, we use the learning to rank algorithms naming RankNet, LambdaMART, LambdaRank, and MART ListNet.

[9] for **RankNet** is proposed by [10], is a pairwise ranking algorithm, employs a loss function and gradient descent as an algorithm to train a neural network model. Feature vectors are computed for each query/document pair, in our case Feature vectors are computed for each query/cluster pair. According to [9], **LambdaRank** is a ranking algorithm based on RankNet it also uses neural nets, it tries to solve the cost function problem by working with implicit cost functions. It considers employing Gradient Descent to optimize the evaluation result and tries to directly define and utilize the gradient function of the evaluation result. From [9] and [11], the **MART algorithm** (Multiple Additive Regression Trees), or also known as Gradient boosted regression tree, is a class of boosting algorithms that may be viewed as performing gradient descent in function space, using regression trees. In [9], **LambdaMART** is described as a method for merging MART and LambdaRank, using Boosting and the Lambda function in LambdaRank, it employs the MART (Multiple Additive Regression Trees) algorithms to learn a boosted regression tree as a ranking model. However, LambdaMART improves its result by using gradient boosted decision trees with a cost function derived from LambdaRank to order any ranking situation. According to [12], **ListNet** is proposed by authors as a method for optimizing the listwise loss function based on top one probability, with Neural Network as a model and Gradient Descent as an optimization algorithm. The difference between ListNet and RankNet is the former uses cluster lists as instances while the latter uses cluster pairs as instances and the former utilizes a listwise loss function while the latter utilizes a pairwise loss function.

3.2 Interleaving Methods Example

In our study, we aim to improve and provide richer results that are more relevant. As a result, we use and demonstrate interleaving approaches to evaluate two retrieval functions, RA and RB, as well as a paired comparison test to examine ranks [13]. Instead of showing RA results to certain users and RB results to the rest, both result lists are prepared for each query, and users are given a randomized combination of the two. In the literature, several interleaved approaches have been proposed, especially Team-Draft Interleaving and Vertical-Aware Interleaving [14].

In [14], the Team-Draft Interleaving strategy is based on the concept of teams competing in a friendly team sport. Each cluster corresponded to a player in this method, and the rankings A and B signified the captains' favorable ranks.

This work [15] presents the vertical-aware team-draft interleaving (VA-TDI) method that extends the TDI algorithm over multiple vertical blocks. The VA-TDI method's basic point is to force vertical clusters to be grouped.

4 Experiment Results

4.1 Data

In our work, we use the results obtained from the RAS [16] system. These results are composed of several clusters of vertical search results and each cluster's elements are similar to each other and "dissimilar" to elements from other clusters.

4.2 Algorithms

This system uses this manner to give its result, so for each query, the RAS system generates a vertical aggregation result because the main goal of RAS is to get and provide relevant results. Next, we need to rank these results by using the LTR algorithms (LambdaMART, LambdaRank, MART, ListNet, and RankNet). After applying the LTR algorithms, we propose to use the interleaving approach (TDI, VI-TDI) for merging two ranking lists and present an interleaved list.

4.3 Traditional Metrics

Traditional metrics concentrate the topic relevance of a query q to its results [17]. They evaluate the relevance of a ranked list L, ignoring the type (vertical). In [18] there were three performance measures used for the evaluation of the learning to rank algorithms' performance NDCG (Normalized Discounted Cumulative Gain), Expected Reciprocal Rank. Following the application of Learning to Rank Algorithms, we must compare the algorithms' performance and detect the most reliable algorithm for ranking. In this case, we need to use not only the traditional metrics but also the aggregated metrics.

4.4 AS Metrics

Several AS metrics have been suggested by the researchers to analyze The three main components of AS systems [17] vertical selection (VS), item selection (IS), and result presentation. The AS measures are designed to balance four important AS properties (VS, IS, RP, and VD); however, certain metrics, such as ASDCG, include all four (VS IS, RP, and VD) [17].

4.5 Results and Discussions

In this section, we present our results and discuss the effectiveness of each LTR algorithms. Our study applies the LTR algorithms for ranking the results obtained from a relational aggregated search. In the first step, we use the traditional and aggregated metrics to discover the best and the power of LTR algorithms. Next, we apply the interleaving methods on the ranked list generated by the LTR algorithms to find the optimal and appropriate results.

After ranking the results obtained from a RAS system by LTR algorithms, we calculate different operations; Normalized Discounted Cumulative Gain, Expected Reciprocal Rank, and Mean Reciprocal Rank. Then, we present their average and values in Fig. 5. We are constant that all tested algorithms (LambdaMART, LambdaRank, MART, ListNet, and RankNet) achieve significant results. Furthermore, LambdaMART gets the highest values with the three metrics, followed by the MART algorithm is very similar to LambdaMART, it proved successful results.

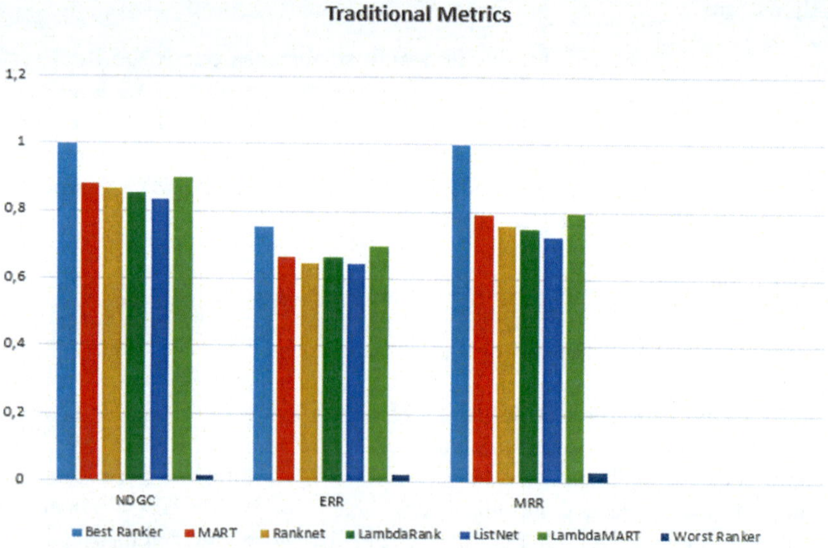

Fig. 5. Average traditional metrics of different LTR algorithms

Then we measure the AS metrics to evaluate each key component (VS, IS, RP, and VD) of AS systems, and we calculate this metric; ASDCG that combines all four elements. Their average and the values are obtained and illustrated in Fig. 6. From these results, we can remark LambdaMART obtains the top-ranking results.

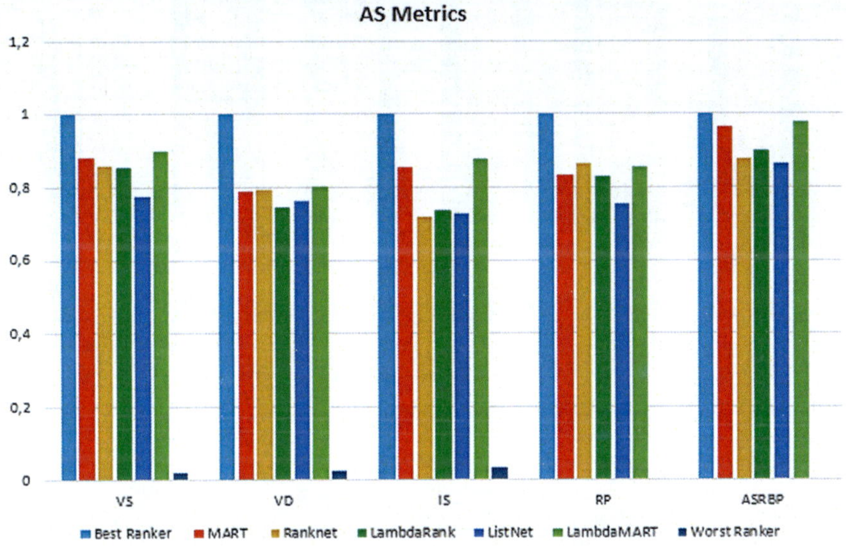

Fig. 6. Average AS metrics of different LTR

In our research, we propose to apply the interleaving methods to the ranking results generated by LTR algorithms. We use two principal approaches, Team-Draft Interleaving and vertical-aware team-draft interleaving these methods take two ranked lists as an input and generate an interleaved list. In this study, we take:

- Ranking A is a result generated by MART algorithm.
- Ranking B is a result generated by LambdaMART algorithm
- Ranking D is a result generated by RankNet algorithm.

We continue the step of evaluation of our results by evaluating the effectiveness of the interleaved list by measuring NDCG, and a specific metric to aggregated search, such as (ASRBP, VS, VD, IS, RP). The obtained results are Presented and drawn in Figs. 7 and 8 respectively. After analyzing all obtained results, we can conclude that the vertical-aware team-draft interleaving is performed well than Team-Draft interleaving especially when we use AS metrics.

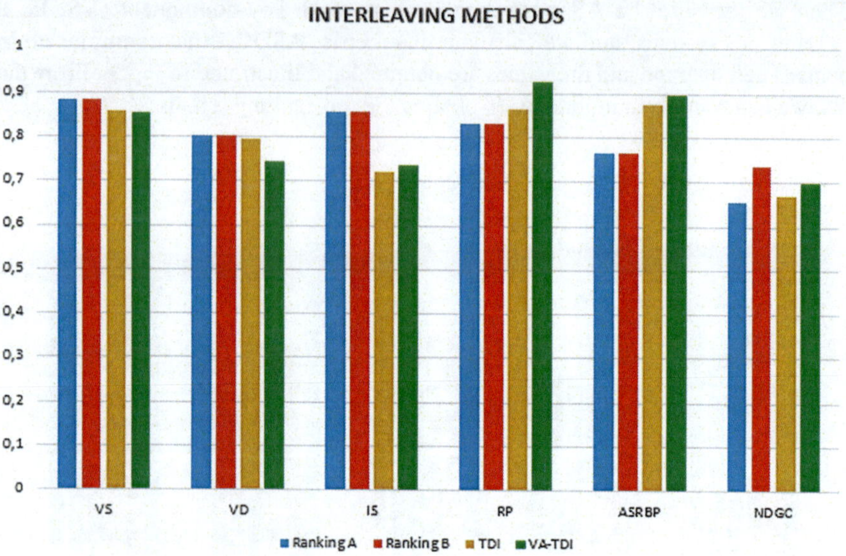

Fig. 7. Average AS metrics of different interleaving methods

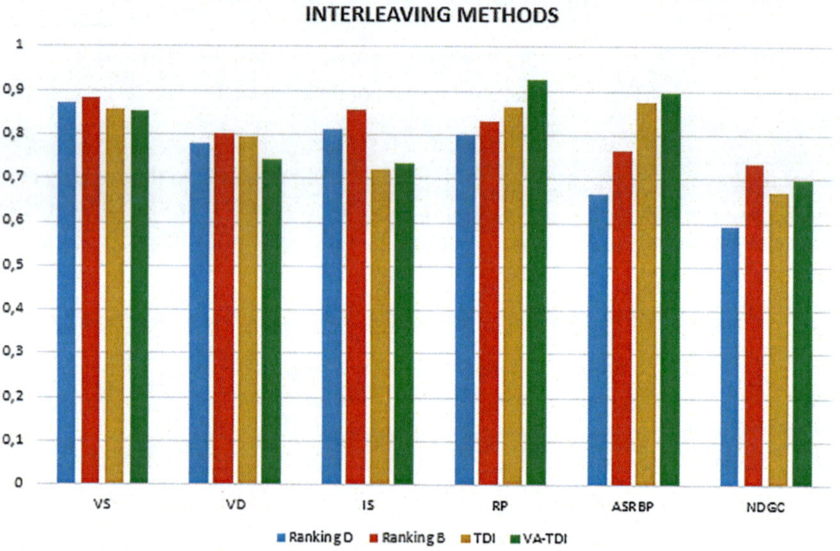

Fig. 8. Average AS metrics of different interleaving methods

5 Conclusion

In this research, we focused on the ranking task on Relational aggregated search to present an adequate result to users and satisfy their needs. For this reason, we applied different learning to rank algorithms (LambdaMART, LambdaRank, MART, ListNet, and RankNet) on different types of results (news, video, image, books, etc.) that obtained from the RAS system. The proposed result is presented in textual form. We took the features extracted from the text (web pages, News, Books, Forums, etc.) but for no textual verticals (image, video), we took their textual description. Following the results, we discover that LambdaMART is more efficient than (LambdaRank, MART, ListNet, and RankNet). To achieve powerful results in our research, we used the interleaving methods hence; we discovered that VA-TDI produces important results.

References

1. Kopliku, A., Pinel-Sauvagnat, K., Boughanem, M.: Aggregated search: a new information retrieval paradigm. ACM Comput. Surv. **46**(3), 1–31 (2014). https://doi.org/10.1145/2523817
2. Lalmas, M.: Aggregated search. In: Melucci, M., Baeza-Yates, R. (eds.) Advanced Topics in Information Retrieval. The Information Retrieval Series, vol. 33, pp. 109–123. Springer, Heidelberg (2011). https://doi.org/10.1007/978-3-642-20946-8_5
3. Kopliku, A.: Approaches to implement and evaluate aggregated search. Ph.D. thesis, Université de Toulouse, Université Toulouse III-Paul Sabatier (2011)
4. Shokouhi, M., Si, L.: Federated search. FNT Inf. Retrieval **5**(1), 1–102 (2011). https://doi.org/10.1561/1500000010
5. Yan, L.T.: Learning to Rank for Information Retrieval. Microsoft Research Asia, Sigma Center, Beijing (2009)
6. Arguello, J., Diaz, F., Callan, J.: Learning to aggregate vertical results into web search results. In: Proceedings of the 20th ACM International Conference on Information and Knowledge Management - CIKM 2011, Glasgow, Scotland, UK, p. 201 (2011). https://doi.org/10.1145/2063576.2063611
7. Li, H.: Learning to rank for information retrieval and natural language processing. Synth. Lect. Hum. Lang. Technol. **4**(1), 1–113 (2011). https://doi.org/10.2200/S00348ED1V01Y201104HLT012
8. Liu, T.-Y.: Learning to rank for information retrieval. FNT Inf. Retrieval **3**(3), 225–331 (2007). https://doi.org/10.1561/1500000016
9. Burges, C.J.C.: From RankNet to LambdaRank to LambdaMART: an overview, 19 (2010)
10. Burges, C., et al.: Learning to rank using gradient descent. In: Proceedings of the 22nd International Conference on Machine Learning - ICML 2005, Bonn, Germany, pp. 89–96 (2005). https://doi.org/10.1145/1102351.1102363
11. Ding, R.: Literature survey for Learning to rank. Computer and Information Science Department University of Delaware, p. 6 (2009)
12. Marek Modrý, Bc.: Learning to rank algorithms. Ph.D. thesis, Faculty of Electrical Engineering, Department of Cybernetics, Czech Technical University in Prague (2014)
13. Lasri, S., Nfaoui, E.H.: Aggregated search and interleaving methods: a survey. In: Proceedings of the International Conference on Big Data and Advanced Wireless Technologies - BDAW 2016, Blagoevgrad, Bulgaria, pp. 1–9 (2016). https://doi.org/10.1145/3010089.3010098
14. Kharitonov, E., Macdonald, C., Serdyukov, P., Ounis, I.: Generalized team draft interleaving. In: Proceedings of the 24th ACM International on Conference on Information and Knowledge Management, Melbourne Australia, October 2015, pp. 773–782 (2015). https://doi.org/10.1145/2806416.2806477
15. Chuklin, A., Schuth, A., Hofmann, K., Serdyukov, P., de Rijke, M.: Evaluating aggregated search using interleaving. In: Proceedings of the 22nd ACM International Conference on Information and Knowledge Management - CIKM 2013, San Francisco, California, USA, pp. 669–678 (2013). https://doi.org/10.1145/2505515.2505698

16. Achsas, S., Nfaoui, E.H.: Improving relational aggregated search from big data sources using stacked autoencoders. Cogn. Syst. Res. **51**, 61–71 (2018). https://doi.org/10.1016/j.cogsys.2018.05.002
17. Zhou, K., Lalmas, M., Sakai, T., Cummins, R., Jose, J.M.: On the reliability and intuitiveness of aggregated search metrics. In: Proceedings of the 22nd ACM International Conference on Information and Knowledge Management - CIKM 2013, San Francisco, California, USA, pp. 689–698 (2013). https://doi.org/10.1145/2505515.2505691
18. Schuth, A., Hofmann, K., Radlinski, F.: Predicting search satisfaction metrics with interleaved comparisons. In: Proceedings of the 38th International ACM SIGIR Conference on Research and Development in Information Retrieval, Santiago Chile, August 2015, pp. 463–472 (2015). https://doi.org/10.1145/2766462.2767695

Used Car Price Prediction with Pyspark

Maida Ahtesham(✉) and Javairia Zulfiqar(✉)

Department of Computer Science and Software Engineering, Jinnah University for Women, Karachi, Pakistan

maidaahtesham@gmail.com, javariaZulfiqar@yahoo.com

Abstract. Price prediction has always been a challenging task as there are several factors involved that tend to cause variation in prices. There is similar case with car price prediction. It is a difficult task to decide either a used car actually justifies the posted price. Factors such as mileage, model, year influence the car prices. The paper aims to develop machine learning models which include linear regression, gradient boosting algorithm, and decision tree in order to predict used car prices based on its features. In order to obtain results, we made an implementation using Apache Spark, which is a big data tool. Among several machine learning libraries used such as pandas and others, we have also preferred Apache spark for implementation. The dataset of PakWheels is utilized which originally have 56186 instances and 16 attributes. A considerable amount of work has been done on this research domain; however, very little amount of work has been done using pyspark. Our proposed models predicted accuracy with linear regression of about 50% while Gradient boosting algorithm obtained 89% and 86% accuracy with decision tree regression.

Keywords: Linear regression · Prediction · Machine learning · Preprocessing · GradientBoost

1 Introduction

The increased prices of brand-new cars and financial instability have turned several customers' attention towards the preference of used cars [1]. It is one of the reasons that the market for the second-hand car continues to grow in recent times. This expansion has turned into a business for both buyer and seller. Most individuals prefer buying used cars due to their considerable prices and tend to sell them again after a certain time of usage in order to gain some profit. But in all such scenarios, it is often a complex task to conclude that either the used car is worth the posted price. There are several attributes including mileage, model, year, Engine Type, Engine Capacity, and many others that affect car prices [2]. Since the prices are subjected to constant fluctuation and hence it requires an evaluation model to predict the prices.

Machine learning models have ever been a remarkable choice for prediction purposes and by training statistical models for predicting prices, one can get a certain estimate of the price. In recent times, due to the growing trend of big data, machine learning is considered the most useful approach as it sustains the ability to make predictions more

S. Motahhir and B. Bossoufi (Eds.): ICDTA 2022, LNNS 454, pp. 169–179, 2022.
https://doi.org/10.1007/978-3-031-01942-5_17

accurately on the basis of their attributes. The number of studies that have explored this problem has also justified the prediction capability of the machine learning approach [3–6]. However, most of the research studies conducted have not utilized spark to address their problems. Apache Spark is known for its easy-to-use, general and fast engine for big data processing. It has built-in modules for machine learning, streaming, graph processing, and SQL. Predicting the used car price requires a system or model which can be built by machine learning. In this paper we have built, optimize and evaluate a number of machine learning models which can contribute to predicting the prices of used cars based on the features. Machine Learning Models such as linear regression, Gradient boosting algorithms, and decision trees are applied to the structure of the Pak Wheel data set which is obtained from Kaggle [7]. The data set contains 56186 instances and 16 attributes. The features include Name, mileage, color, body type, and many others. These models can be used by both buyers and sellers for the true prediction of used cars. Results associated with three models are calculated based on root mean square and accuracy. It has been noticed that the accuracy of Gradient boosting algorithm and decision trees is better than linear regression.

The paper is structured as, Sect. 2 is for background and related work whereas, Sect. 3 is based on Methodology. Section 4 discusses the overall implementation and work and Sect. 5 is for results and discussion.

2 Background and Related Work

Prediction is considered one of the popular approaches that help to improve business processes and also supports decision-making. In machine learning, prediction models are developed to support decisions for pricing. A large number of studies have presented that such state-of-the-art strategies for pricing can help to enhance the effectiveness of operations that are customer-centric. Similarly, the research study carried out by Stefan Lessmann et al. (2017) [3], has focused on the second-hand market and also developed empirical models in order to forecast resale prices. The data set used in the research is provided by a leading German car manufacturer and it contains about 450,000 observations taken from true sale prices of six different car models in the second-hand market. Resale price in the research study is estimated by using several different attributes such as mileage, car age, engine type, model year, etc. The results are provided on basis of three research questions which aim to evaluate the comparative performance of several alternative forecasting methods. The results of the research study depict that the market value of used cars depends on several different factors and the forecasting accuracy can be achieved by incorporating such factors into price-prediction models.

Prediction techniques applied by machine learning can be regarded as either deductive or inductive. Deductive learning is concerned with the research that is carried out based on existing facts and knowledge in order to deduce new knowledge and facts. Whereas, inductive machine learning refers to developing new computer programs by finding rules and patterns in some new dataset that has never been explored before. Research conducted by Noor and Jan (2017) [6], has used the deductive approach of multiple linear regression as it is supposed to create new values based on existing values. It has used multiple linear regression as there is more than one independent variable. The initial dataset recorded in research was about two thousand records that were obtained from PakWheels. The research study has only utilized those cars which have price details. Features such as version, city, engine capacity, mileage, cars model, engine type, power steering, alloy rims are included. After preprocessing the dataset, it is fed to a regression model to estimate price. The linear model explains the percentage response variation in a variable by R-square. The R-sq obtained in the research is about 98.61%. Representation of high percentage shows how well the model has predicted the response for new input records.

A comparative study is conducted by Monburinon et al. (2018) [8], using multiple linear regression, random forest regression and gradient Boosted regression trees in order to build a model that predicts the price of a used car. The data set used for the research is taken from an e-commerce website. The objective behind the study is to search predictive model that is best or worthy for predicting used car prices. The data set that contained about 371, 528 observations initially required some tweaking and engineering. The regression-based methods have proved to be reliable in supervised machine learning to predict continuous variables. Among three models, Gradient boosted regression trees obtain the best performance among others while having 0.28 mean absolute error. In comparison to this, Random Forest was second-best having a mean absolute error of 0.35 and the multiple linear regression model had a relatively large mean absolute error of about 0.55.

Anamika Das Mou et al. (2021) [9], applied four best algorithms including SVM, Naïve bayes, K nearest neighbor and Random Forest and analyzed the accuracy of different algorithms that predict the probability of purchasing car. Data is collected from different shops in Bangladesh and from social media. Algorithms are evaluated on basis of precision-recall, accuracy measurement and execution time of algorithms. They got highest 86% of accuracy in SVM than KNN, random forest, naïve bayes. The study has mainly focused on customers reviews by using cosine similarity to analyse customer reviews. On the basis of previous related work, it can be justifying that none of the work has been done using pyspark while working with three models i.e., Linear regression, decision trees and Gradient boosting algorithm.

3 Methodology

3.1 Data Exploration

The dataset of PakWheels used in this research is taken from Kaggle. It is a massive data set that has about 56186 entries along with 16 features as given in Table 1. The data set might contain null, unhandled, and unwanted values that need to be checked and verified before training the dataset for model development. The data set contains the Car Name along with its different features such as Model Year, Mileage, Engine Type, etc. The Feature description of the data set is given in Table 1.

3.2 Data Pre-processing

Data Cleaning

It is one of the most crucial steps that must be performed before the implementation

Table 1. Feature description of actual dataset

S. no.	Name	Data type
1.	AdNo	Numerical
2.	Name	Categorical
3.	Price	Numerical
4..	Model year	Numerical
5	Location	Categorical
6.	Mileage	Numerical
7.	Registered city	Categorical
8.	Transmission	Categorical
9.	EngineCapacity	Numerical
10.	EngineType	Categorical
11.	Color	Categorical
12.	Assembly	Categorical
13.	BodyType	Categorical
14.	Features	Categorical
15.	URL	Categorical
16.	LastUpdated	Categorical

phase. It refers to identifying and correcting errors in the data set that might negatively affect the predicting model. It includes all kinds of activities that help in detecting and repairing errors. In this procedure, we removed the rows that holds missing values and also dropped the features or attributes that might not be effective. Only those features are kept that are most important and strongly correlated. After data preprocessing, we were left with 13 features.

3.3 Separating Categorical and Numerical Data

The dataset we are using contains both categorical and numerical values as described in Table 1. The numerical features are separated from the categorical ones including Name, location, Transmission, EngineType and others.

3.4 Conversion of Categorical Data into Numerical Data

Most machine learning algorithms are unable to handle categorical data unless we convert them to numerical values to make it more machine understandable. The Dataset obtained after preprocessing is described in Table 2.

Table 2. Feature description of dataset after preprocessing

S. no.	Name	Data type
1.	Name	Integer
2.	Price	Integer
3.	Model year	Integer
4.	Location	Integer
5.	Mileage	Integer
6.	Registered city	Integer
7.	Transmission	Integer
8.	EngineCapacity	Integer
9.	EngineType	Integer
10.	Color	Integer
11.	Assembly	Integer
12.	BodyType	Integer
13.	Features	Integer

4 Implementation

4.1 Reading Data

The Study has been performed using pyspark which is a spark library that helps in running python applications while utilizing Apache Spark Capabilities. The data set file is loaded into the Spark Data frame. While reading a CSV file into a data frame, spark operates in eager mode, and all of the data is loaded into the memory before execution.

4.2 Feature Selection

It is one of the most substantial steps in machine learning. It is a process of reducing the number of features or input variables while developing a predictive model. When number of variables are reduced, it helps in minimizing the computational cost and improving the performance of the model. The feature selection process is done by correlation matrix. It helps in determining the correlation between independent variables and target variables. The correlation values range from -1 to 1. When the values are close to 1, it indicates a strong correlation between the independent and target variable. We have determined the correlation by plotting the correlation matrix as shown in Fig. 1 and also by displaying the correlation coefficient as depicted in Fig. 2. The features selected from this procedure include Name, Model Year, EngineType, Engine Capacity, Assembly, and Body Type.

4.3 Transformation Using Vector Assembler

Vector Assembler in pyspark is a transformer that combines a given list of columns into a single vector column. It combines raw features or features that have been generated using

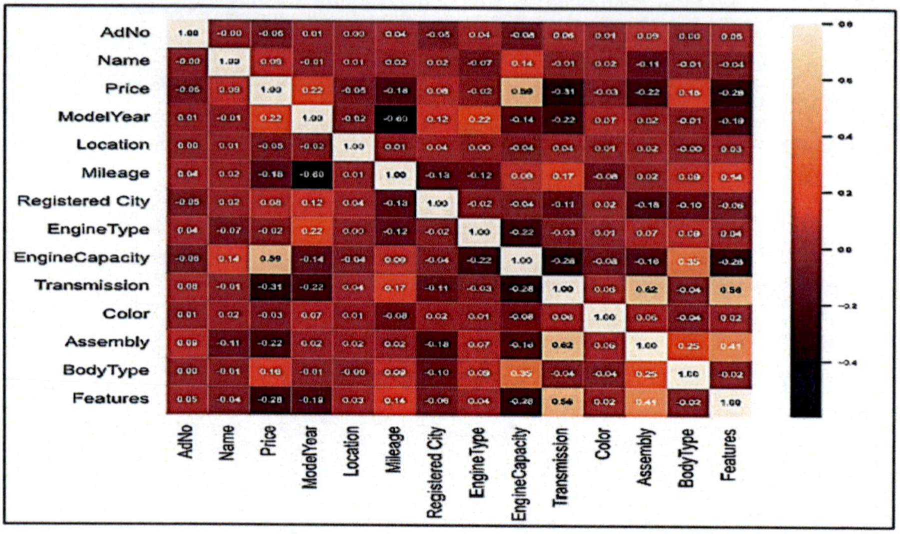

Fig. 1. Correlation matrix in between independent and target variable

```
Correlation to Price for  Name 0.09481727648288188
Correlation to Price for  Price 1.0
Correlation to Price for  ModelYear 0.22594720502852123
Correlation to Price for  Location -0.05124683566297722
Correlation to Price for  Mileage -0.18981985857254946
Correlation to Price for  Registered City 0.09525112956321838
Correlation to Price for  EngineType -0.015654391800120297
Correlation to Price for  EngineCapacity 0.618304888733702
Correlation to Price for  Transmission -0.32765236773775536
Correlation to Price for  Color -0.027123027022060573
Correlation to Price for  Assembly -0.23278522627117643
Correlation to Price for  BodyType 0.16260315203937656
Correlation to Price for  Features -0.2905157928944784
```

Fig. 2. Correlation coefficient for price.

different transformers into a single feature vector to train ML models such as logistic regression and decision trees. The study has also implemented a vector assembler to prepare data for machine learning, here we needed just two variables Price and features. Figure 3 shows details of the transformed data selection using vector assembler.

4.4 Linear Regression

In this work, the first algorithm we have used is Linear Regression to prepare a model that predicts used car prices. Linear Regression in machine learning is about searching a linear relationship among a target and one or more prediction variables. The relationship among two variables is supposed to be deterministic if one variable can be accurately expressed by the other. Here, we have estimated Root Mean Square Error, R-Squared, and Mean Absolute Error by training the model. Mean Square Error refers to the difference between predicted and actual values. Our linear regression model has obtained MSE of

```
+--------------------+------+
|           features_| Price|
+--------------------+------+
|[1841.0,2019.0,11...|104000|
|[4187.0,1998.0,70...|110000|
|[5490.0,2019.0,11...|111000|
|[4188.0,1990.0,15...|120000|
|[4286.0,1992.0,19...|120000|
|[5894.0,2017.0,97...|123000|
|[5440.0,1998.0,12...|123000|
|[3486.0,1993.0,15...|130000|
|[4500.0,1990.0,12...|130000|
|[4450.0,1993.0,12...|135000|
|[2090.0,1996.0,10...|135000|
|[288.0,1996.0,100...|140000|
|[4502.0,1992.0,10...|140000|
|[4285.0,1991.0,15...|145000|
|[4285.0,1991.0,12...|145000|
|[4295.0,1990.0,12...|145000|
|[4295.0,1990.0,30...|145000|
|[289.0,1992.0,265...|150000|
|[294.0,1993.0,100...|150000|
|[4296.0,1991.0,28...|150000|
+--------------------+------+
only showing top 20 rows
```

Fig. 3. Transformation using vector assembler.

about 3990188155075.87 and R Squared value of about 0.5162 which means 51% of the variability in price can be explained using our model. Table 3 describes the price and prediction made by our linear regression model and Table 4 describes the percentage difference between Actual and predicted prices. It helps us validate how well our model has predicted the prices of used cars.

Table 3. Predictions made by linear regression model

Predictions	Price	Features
2361650.977522105	5700000	[6.0, 2017.0, 12000…]
3248811.522700906	6175000	[16.0, 2018.0, 5000…]
2619491.239606887	6500000	[17.0, 2019.0, 4000…]
3754163.552587092	5800000	[19.0, 2018.0, 2600…]
3702608.681743592	8000000	[28.0, 2017.0, 1100…]

Table 4. Percentage difference between price and prediction

S. no.	Price	Prediction	Difference	Percentage difference%
1	5700000	2361651.0	3338349.0	41.410240
2	6175000	3248811.5	2926188.5	31.051008
3	6500000	2619491.2	3880508.8	42.551813
4	5800000	3754163.5	2045836.5	21.413036
5	8000000	3702608.8	4297391.2	36.721651

To obtain a better visualization among predicted and actual prices obtained through our model, Fig. 4 plots the difference among predicted and Actual Prices.

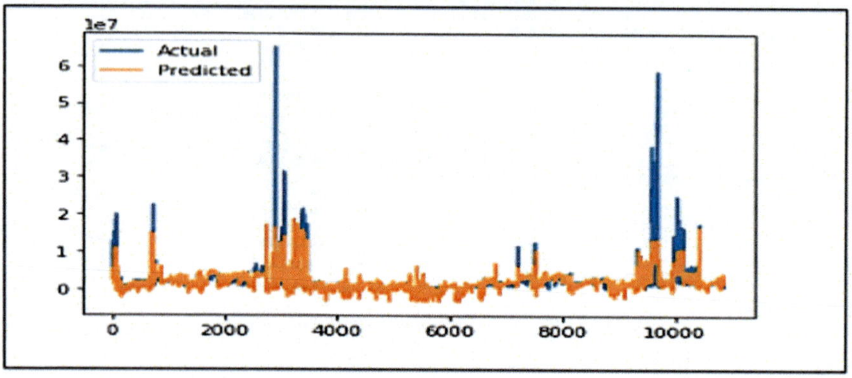

Fig. 4. Plotting the predicted vs actual prices.

4.5 Decision Tree Regression

The paper has also used decision tree regression to build a model that can predict the prices of used cars. It is a nonparametric supervised learning method that is utilized in both regression and classification problems. It observes features of an object and trains the model in the structure of the tree to predict data in the future. The predictions made by our model are shown in Table 5. The model obtained a Root Mean Square Error of 1090548.7135263914 along with an R squared value of 0.861. Table 6 describes the percentage difference among price and predicted values obtained from the model. It can be observed from the percentage difference and predicted values that this model obtained better results than linear regression since the percentage difference is much less than the previous model.

Table 5. Predictions made by decision tree regression model.

Predictions	Price	Features
5581071.428571428	5700000	[6.0, 2017.0, 12000...]
5581071.428571428	6175000	[16.0, 2018.0, 5000...]
5581071.428571428	6500000	[17.0, 2019.0, 4000...]
8665953.216374269	5800000	[19.0, 2018.0, 2600...]
8665953.216374269	8000000	[28.0, 2017.0, 1100...]

Table 6. Percentage difference between price and prediction from decision tree regression

S. no.	Price	Prediction	Difference	Percentage difference%
1	5700000	5581071.5	118928.5	1.054231
2	6175000	5581071.5	593928.5	5.052100
3	6500000	5581071.5	918928.5	7.606349
4	5800000	8665953.0	−2865953.0	−19.811712
5	8000000	8665953.0	−665953.0	−3.995889

4.6 Gradient Boosted Tree Regression

The final algorithm implemented in the paper to obtain the prediction of the car sale price is Gradient boosted Tree Regression. It is considered one of the most effective machine learning models. Table 7 proves that Gradient Boosted Tree regression is efficient in getting better results and predictions among the rest of the models. The model obtained a Root Mean Square Error of 942280 and R square of 0.89672 (Table 9).

Table 7. Predictions made by gradient boosted tree regression model.

Predictions	Price	Features
5600137.333494508	5700000	[6.0, 2017.0, 12000…]
5660300.955942343	6175000	[16.0, 2018.0, 5000…]
5640094.844199583	6500000	[17.0, 2019.0, 4000…]
8042417.027370915	5800000	[19.0, 2018.0, 2600…]
1.2781927940069327E7	8000000	[28.0, 2017.0, 1100…]

Table 8. Percentage difference between price & prediction on gradient boosted tree regression

S. no	Price	Prediction	Difference	Percentage difference%
1	5700000	5600137.5	99862.5	0.883728
2	6175000	5660301.0	514699.0	4.348846
3	6500000	5640095.0	859905.0	7.083182
4	5800000	8042417.0	−2242417.0	−16.199606
5	8000000	12781928.0	4781928.0	−23.010031

5 Results and Discussion

The primary objective of the research study is to compare the accuracy while building different models to predict used car prices. The research uses linear regression, Decision

Trees and Gradient Boosted Tree Regression for this purpose. Results of the predicting models are compared based on Root Mean square error, mean absolute error, and R Square. The R square value in the linear model is considered as the response variable in the model and high R-square values mean the model has better fitness and has resulted in accurate results. Table 9 provides the summary of results obtained from all three models.

We compare the prediction accuracy of the proposed prediction models, As shown in Table 9, Decision Tree Regression has performed better as compared to Linear regression as the R square value has increased from 0.51 to 0.86. Whereas, the gradient boosted regression has performed much better than the rest of the two models. Gradient boosted regression is considered to perform better as compared to the rest of the predicting models. Considering the better predicting accuracy, it is perceived that Gradient boosted regression could be of great use for used car price prediction.

Table 9. Summary of results obtained

Models	RMSE	MAE	R2
Linear regression	3990188155075.87	830266.906	0.5162
Decision trees regression	1090548.713	495499.582	0.861
Gradient boosted tree regression	942279.590	375078.94	0.89672

6 Conclusion

The study has evaluated comparative performances of models including linear regression, Decision Tree Regression, and Gradient Boost tree regression to predict used car prices. The car price has a strong dependence on its variety of factors or attributes, and it helps to yield forecast accuracy for used car prices. The results of the research study show that the linear regression method has performed significantly less than the rest of the methods, which makes the model less preferable for price prediction. On the contrary, it is observed that linear regression has been extensively used in literature for prediction. In the future, it is expected to make predictions comprised of KNN and other models with improved results.

References

1. Jayadeera, T.P., Jayamanne, D.J.: Fair price prediction system for used cars in Sri Lanka using machine learning and robotic process automation. In: 2019 International Conference On Business Innovation (ICOBI), Colombo (2019)
2. Pal, N., Arora, P., Kohli, P., Sundararaman, D., Palakurthy, S.S.: How much is my car worth? A methodology for predicting used cars' prices using random forest. In: Arai, K., Kapoor, S., Bhatia, R. (eds.) FICC 2018. AISC, vol. 886, pp. 413–422. Springer, Cham (2019). https://doi.org/10.1007/978-3-030-03402-3_28

3. Lessman, S., VoB, S.: Car resale price forecasting: the impact of regression method, private information, and heterogeneity on forecast accuracy. Int. J. Forecast. **33**, 864–877 (2017)
4. Monburinon, N., Chertchom, P., Kaewkiriya, T., Rungpheung, S., Buya, S., Boonpou, P.: Prediction of prices for used car by using regression models. In: 2018 5th International Conference on Business and Industrial Research (ICBIR), Bankok (2018)
5. Zhang, F., Yang, J., Guo, Y., Gu, H.: Multi-source heterogeneous and XBOOST vehicle sales forecasting model. In: MacIntyre, J., Zhao, J., Ma, X. (eds.) SPIOT 2020. AISC, vol. 1283, pp. 340–347. Springer, Cham (2021). https://doi.org/10.1007/978-3-030-62746-1_50
6. Noor, K., Jan, S.: Vehicle price prediction system using machine learning techniques. Int. J. Comput. Appl. **167**(9), 0975–8887 (2017)
7. Kaggle. https://www.kaggle.com/spideysloth/pakwheels-cars-dataset
8. Monburinon, N., Chertchom, P., Kaewkiriya, T., Rungpheung, S., Buya, S., Boonpou, P.: Prediction of prices for used car by using regression models. In: 2018 5th International Conference on Business and Industrial Research (ICBIR), Bangkok (2018)
9. Mou, D., Saha, P.K., Nisher, S.A., Saha, A.: A comprehensive study of machine learning algorithms for predicting car purchase based on customers demands. In: 2021 International Conference on Information and Communication Technology for Sustainable Development (ICICT4SD), Dhaka (2021)

Convolutional Neural Networks: Architecture Optimization and Regularization

El Houssaine Hssayni[1(✉)], Nour-Eddine Joudar[2], and Mohamed Ettaouil[1]

[1] Faculty of Science and Technology of Fez, University Sidi Mohamed Ben Abdellah, Fez, Morocco
elhoussaine.hssayni@usmba.ac.ma
[2] ENSAM, Mohammed V University in Rabat, Rabat, Morocco

Abstract. Since its invention, convolutional neural network (CNN) has revolutionized the artificial intelligence field. Furthermore, its development has become a very active research area. This is due to its unavoidable achievements in most real world applications. However, the determination of an optimal CNN architecture is still unclear until now. Furthermore, the choice of CNN architectures previously proposed is based only on experiments. To deal with this issue, we suggest, in this work, a mathematical modelling for CNN Architecture Optimization and training in terms of a mixed integer nonlinear optimization problem with linear constraints. We adopt the genetic algorithm to solve the resulting model. To further evaluate our suggested mathematical modeling, we conduct various experiments on two benchmark datasets. Numerical results confirm that our proposal achieves improvement in terms of optimization and classification performance.

Keywords: Binary optimization · Deep neural networks · Architecture optimization · Convolutional neural networks · Regularization

1 Introduction

Recently, Convolutional neural network (CNN), one of the competitive deep learning models, which originally introduced for document recognition [1], has imported relevant enhancements to various areas and applications in the field of intelligence artificial. These applications include image classification [2–5], time series prediction [6], and many other applications.

Unlike the classical machine learning models, CNN covers the features extraction part that is responsible to extract relevant characteristics within the data. This part makes CNN very powerful by integrating the data pretreatment in the network training. Generally, the features extraction is achieved using a sequence of two layer types: convolutional and pooling. The features extracted in the first part are then classified using a multiplayer perceptron. This latter represents the second part in CNN (Fig. 1).

However, successful CNNs are deeper and composed of large number of layers, yet parameters. Therefore, training a successful CNN can produce some undesirable effects such as, the overfitting [7] and parameters redundancy [8]. Furthermore, due to

S. Motahhir and B. Bossoufi (Eds.): ICDTA 2022, LNNS 454, pp. 180–189, 2022.
https://doi.org/10.1007/978-3-031-01942-5_18

the lack of theoretical analysis serves to find an optimal CNN architecture, most CNN architectures previously proposed are based on experiments without a mathematical justification. To deal with these issues, considerable efforts have been made. Among these efforts, the combination of multilevel and multiscale with Bayesian optimization in order to optimize some CNN hyper-parameters [9]. Another important algorithm named psoCNN introduced by Junior et al. [10]. This latter allows searching architecture using PSO optimizer. psoCNN focused only to approximate layers number, which limits its performance. A different approach to CNN architecture design is presented by Louati et al. [11]. In this approach, the authors invested generic bi-level EA CODBA to solve a bi-level optimization represented CNN topology choice.

Despite all the previously proposed works the CNN architecture optimization still unclear and needs more considerations. In this context and motivating by the important role of the choice of the optimal CNN architecture, we propose an optimization model in order to control the contribution of each CNN parameter. For this purpose, we associate to each weight a decision variable to control its impact in the network. The obtained model is a nonlinear optimization problem with mixed variables and linear constraints. The combinatorial aspect makes the resolution of our model by exact methods not feasible. In this context, the genetic algorithm is one of the efficient meta-heuristics that handle the constraints in combinatorial problems. For this purpose, this efficient evolutionary algorithm is adopted to solve the obtained decision model. In the last part of this work, we report numerical results, which show that our model can import improvements to convolutional neural networks in terms of classification performance and architecture optimization.

The rest of this paper is structured as follows: In the next section, we introduce the functioning and the general architecture of CNN. In Sect. 3, we present the modeling steps of the optimization model. Section 4 describe the experiments and reports numerical results that demonstrate the efficiency of the proposal. Section 5 covers a conclusion and perspectives of this work.

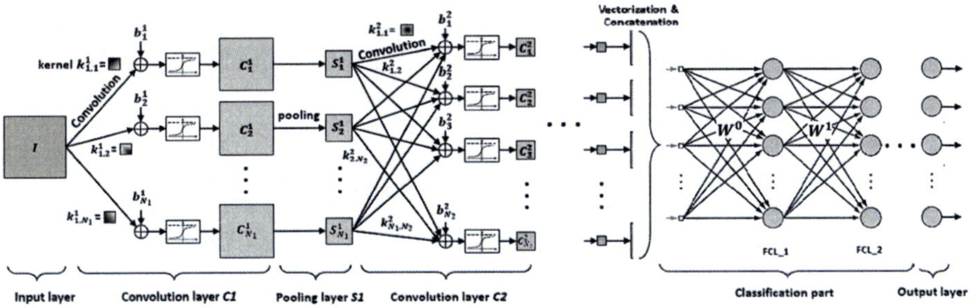

Fig. 1. A standard convolutional neural network architecture

2 Convolutional Neural Networks

Unlike the classical machine learning models, CNN has the particularity of the features extraction part that is responsible to extract relevant characteristics within the data. This part makes CNN very powerful by integrating the data pretreatment in the network training. Generally, the features extraction is achieved using a sequence of two layer types: convolutional and pooling. The features extracted in the first part are then classified using a multiplayer perceptron. This latter represents the second part in CNN (as shown in Fig. 1). In this section, we describe each CNN component as well as, the functioning mechanisms.

2.1 Convolution Layer

The convolution is one of the principal mechanisms in a CNN. It is responsible to extract relevant features within input feature maps. This extraction is accomplished using the following expression:

$$C_l^k = f\left(\sum\nolimits_{j=1}^{N_{k-1}} K_{j,l}^k * X_j^{k-1}\right) \quad (1)$$

where f represents the used activation function in the convolutional layer and $*$ refers to the discrete convolution product.

2.2 Pooling Layer

The pooling (subsampling) mechanism applies to the feature maps generated from the previous convolution layer. It is another relevant component of CNN based on dimension reduction. In fact, it allows transforming inputs to a reduced dimension feature maps by the use of a pooling operation. These operations include max-pooling [12] and average pooling [13]. We next give the definition of these two pooling methods.

- Max Pooling [12] is a successful pooling method, which allows reducing feature maps dimension by taking the largest activation in each pooling region as expressed in Eq. 2.

  $$s_k = \max_{a_i \in R_k} a_i \quad for \quad k = 1, \ldots, K \quad (2)$$

  This method is a very powerful subsampling technique achieved better performance especially in extracting local characteristics.
- Average Pooling [13] is another reducing dimension method that produces the average of activations in each pooling region. This is accomplished using the following expression (Eq. 3).

  $$s_k = \frac{1}{\#R_k} \sum\nolimits_{a_i \in R_k} a_i \quad for \quad k = 1, \ldots, K \quad (3)$$

  The use of average operation allows to smooth subsampling regions; as a result, this method is useful for extracting global characteristics.

Where $\{a_1, a_2, .., a_{\#R_k}\}$ denotes the kth subsampling region.

2.3 Batch Normalization Layer

Data normalization is a successful process widely used during the learning. This process allows data to be comparable with respect to its characteristics. However, during training, changing of network parameters generate a change in the network activations distribution. For this reason, various approaches have been developed in order to improving the convergence during training. Among these approaches, we highlight batch normalization, which is a successful method that greatly solved the mentioned issue. It consists of normalizing on mean and in variance the layers outputs of the network; mathematically, let us consider $B = \{X_1, X_2, \ldots, X_d\}$ a mini batch of size d, and $(y_1, y_2, \ldots, y_d)$ their linear transformations correspondents. Batch normalization $BN_{\gamma,\beta}$ is presented by the following expression:

$$BN_{\gamma,\beta} : \{X_1, X_2 \ldots, X_d\} \rightarrow \{Y_1, Y_2 \ldots, Y_d\}$$
$$X_i \rightarrow Y_i = \gamma \cdot \hat{X}_i + \beta$$

where $\hat{X}_i = \frac{X_i - \mu_B}{\sqrt{\sigma_B^2 + \varepsilon}}$ is the normalized value of X_i and

$$\left(\mu_B, \sigma_B^2\right) = \left(\frac{1}{m}\sum\nolimits_{i=1}^{m} X_i, \frac{1}{m}\sum\nolimits_{i=1}^{m} (X_i - \mu_B)^2\right)$$

present the (mean, variance) of X.

2.4 Classification Layers

Once we achieve the last layer in the features extraction, all the feature maps are converted to a one dimension (1D) array, which, in turn, is presented as an input to a classifier composed of fully connected layers where the outputs are given as follows:

$$Z_j^k = f\left(\sum\nolimits_{i=1}^{n_{k-1}} \left(v_{ji}^k \cdot W_{ji}^k\right) \times Z_i^{k-1}\right) \quad (4)$$

The Softmax function is then used to calculate the network output,

$$Y = softmax\left(Z^{N_{FC}}\right)$$

where N_{FC} is the number of hidden layers.

3 Proposed CNN Architecture Optimization Model

After presenting the necessary background related to our work, we describe in this section, the steps of our suggested CNN architecture optimization, from modelling to resolution.

3.1 Objective Function

Our approach for CNN architecture optimization is based on controlling the contribution of each connection between two neurons in the classification part of CNN. Indeed, each connection does not contribute will be removed in the network. This control is achieved by associating a decision variable v_{ji}^k to each connection W_{ji}^k. This variable is defined as:

$$v_{ji}^k = \begin{cases} 0 \text{ if the weight } \mathrm{W_{ji}^k} \text{ does not contribute} \\ 1 \text{ else} \end{cases} \quad (5)$$

Thus, integrating this binary variable allows controlling the influence of weights in the network. Since, the process of propagation in the classification layers becomes:

$$Z_j^k = f\left(\sum\nolimits_{i=1}^{n_{k-1}} \left(v_{ji}^k \cdot W_{ji}^k\right) \times Z_i^{k-1}\right) \quad (6)$$

The suggested binary variable V allows controlling the contribution of each weight in fully connected layers at the learning process. In fact, removing connections that have not an important role in the network generate at the end an optimal architecture with no loss in the performance.

In presence of binary variables V, and due to the success of Cross-Entropy loss in the classification task, we suggest the following objective function:

$$E(\mathbb{W}, V) = -\sum\nolimits_{j=1}^{m_c} t_j \log(p_j) \quad (7)$$

where p_j, m_c and t_j are respectively the Softmax probability, the class number and the truth label of the ith class.

3.2 Constraints

Integrating binary variables V in fully connected layers within the classification part of CNN necessitates specific consideration. Indeed, to prevent the total elimination of connections between neurons in the current and the next layers, it is necessary to keep at least one connection for each neuron. Thus, this constraint can be formulated as follow:

$$\sum\nolimits_{i=1}^{n_{h-1}} v_{i,j}^h \geq 1 \quad (8)$$

for $j = 1, \ldots, n_l$ and $h = 1, \ldots, N_{FC}$.

The CNN parameters $\mathbb{W}$ represent an union of two tensors K and W. Such that K presents the CNN variables within the features extraction part.

$$K = ((K_{i,j}^l)_{p,q}) \begin{array}{l} l = 1, \ldots, N_{conv} \\ i = 1, \ldots, N_k \\ j = 1, \ldots, N_{k-1} \\ p, q = 1, \ldots, D \end{array}$$

Each element of this tensor is a real variable $(K_{i,j}^k)_{p,q} \in \mathbb{R}$.

The second parameters tensor W composed of CNN variables within fully connected layers:

$$W = (W_{i,j}^k)_{\substack{k=1,\ldots,N_{FC}\\ i=1,\ldots,n_k\\ j=1,\ldots,n_{k-1}}}$$

such that $W_{i,j}^k \in \mathbb{R}$.

3.3 Solving the Suggested Optimization Model

Since our model in a mixed integer nonlinear problem, the resolution using exact methods is not feasible. For this reason, we propose solving it in two phases. Firstly, we exploit an optimizer based on gradient descent such as Adam [14] to train the CNN. The obtained solution $\mathbb{W}^*$ is then used to produce a new optimization problem $\min_V E(\mathbb{W}^*, V)$. Based on the success of genetic algorithm in handling the constraints in combinatorial problems, we suggest using this efficient evolutionary optimizer to solve our binary model.

4 Experiments

In this section, we will evaluate the efficiency of our proposal using numerical results. For this purpose, we build three CNN models (CNN1, CNN2, and CNN3), where each one is composed of a sequence of convolutional and pooling layers produce at the end relevant features of input data. The obtained features are then classified using a sequence of fully connected layers ended with a soft-max layer. The numbers of features maps and convolutional layers in each model are reported in Table 1. These models are used to classify two benchmark datasets that are described in the next paragraph.

Table 1. Proposed CNN models

CNN1	CNN2	CNN3
cnv $-5 \times 5(92 * 92 * 32)$	cnv $-5 \times 5(92 * 92 * 64)$	cnv $-3 \times 3(32 * 32 * 64)$
Max-pool $(46 * 46 * 32)$	Max-pool $(46 * 46 * 32)$	Max-pool $(16 * 16 * 64)$
cnv $-3 \times 3(44 * 44 * 64)$	cnv $-3 \times 3(44 * 44 * 72)$	cnv $-3 \times 3(16 * 16 * 96)$
Max-pool $(22 * 22 * 64)$	Max-pool $(22 * 22 * 72)$	Max-pool $(8 * 8 * 96)$
FC-1 1024	cnv $-3 \times 3(20 * 20 * 96)$	cnv $-3 \times 3(8 * 8 * 1024)$
FC-2 512	Max-pool $(10 * 10 * 96)$	Max-pool $(4 * 4 * 1024)$
Output 5	FC-1 1048	FC-1 2048
	FC-2 964	FC-2 1024
	Output 5	Output 10

4.1 Description of Datasets

- **SVHN** is a challenging image dataset generated from house number in Google [15]. The images constituted this dataset are RGB color images with 32×32 pixels; each image presents a house number. This dataset is composed of $73, 257$ and $26, 032$ images for training and testing respectively.
- **NORB** (NYU Object Recognition Benchmark) widely utilized for the performance evaluation of a suggested deep learning approach in three-dimensional object recognition [16]. It composed of $24, 300$ training images and the same number for testing divided into five categories.

4.2 Evaluation Metrics

In all case studied, fore metrics are exploited to measure the classification performance. These metrics are described as follow:

Accuracy is one of the most widely used evaluation metrics. It represents the proportion of examples correctly classified. It is expressed by the formula:

$$Ac = \frac{N_c}{N_d}$$

Such that N_d and N_c represent, respectively, the total examples number and the number of correctly classified examples.

Precision represents the proportion of correctly classified examples among all examples in a class. It is expressed as:

$$P_r = \frac{N_c^c}{N_d^c}$$

Such that N_d^c, N_c^c represent the restrictions of N_d and N_c on each class.

The recall is another metrics that represents the proportion of correctly classified examples among all examples classified in the corresponding class. This is given by the following expression:

$$R_c = \frac{N_c^c}{K_d^c}$$

where K_d^c represents the total of examples classified as belonging to class c.

F1-Score represents the harmonic average of precision and recall. For a class c, F1-Score is expressed as:

$$F_1 - S = \frac{2R_c P_r}{P_r + R_c}$$

4.3 Results and Discussion

In the first level, we show the efficiency of our model in terms of parameters optimization. For this purpose, the suggested CNN1 and CNN2 are used to classify the NORB dataset, while the classification of SVHN is evaluated using CNN3. For all these model, the Adam optimizer is exploited as the learning algorithm. The proposed optimization model is utilized to remove unnecessary weights. Figure 2 illustrates the comparison of parameters number in the classification part of suggested CNNs and their correspondent optimized CNN*. As shown in this figure, the number of parameters is considerably reduced for each model. For instance, the number of parameters in CNN3 is reduced from 32247813 to 23795084. This means that 8452729 parameters in CNN3 are redundant and can be removed in the network, which is a very important reduction.

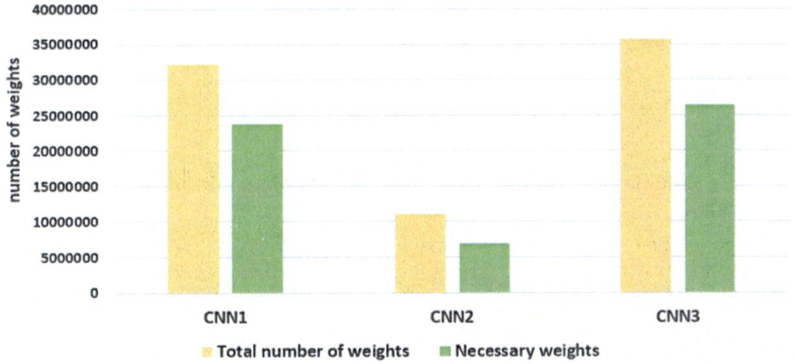

Fig. 2. Weights number in CNN and correspondent optimized CNN for several proposed models

Table 2. Evaluation metrics obtained using suggested CNNs and their correspondent optimized architectures (CNN*)

Evaluation metrics	NORB				SVHN	
	CNN1	CNN1*	CNN2	CNN2*	CNN3	CNN3*
Accuracy	0.8103	0.8107	0.8243	0.8251	0.8940	0.8940
F1 score	0.8103	0.8106	0.8243	0.8250	0.8938	0.8941
Precision	0.8104	0.8107	0.8244	0.8251	0.8937	0.8942
Recall	0.8102	0.8105	0.8242	0.8249	0.8939	0.8940

Regarding the classification performance, we compare the evaluation metrics described above, of the suggested CNNs and their correspondent optimized architectures obtained using the proposed optimization model. As reported from Table 2, the network with optimized architecture (CNN*) is capable to engender remarkable results in comparison with standard CNN. In fact, in all studied cases, the evaluation metrics generated using the optimized architecture are greater than those produced using CNN without optimization. Thus, the imported numerical results illustrate the effectiveness of our proposal regarding the classification performance.

5 Conclusion

In this work, a novel CNN architecture optimization model is proposed. The suggested model allowed associating a decision variable to each parameter in fully connected layer in order to control its influence on the network. The parameters that have not contribute are eliminated. This integration of binary variables allows generating a mixed variables optimization model. Based on the success of genetic algorithm in solving a such problems, we have suggested adopting GA to solve our model. Experiments demonstrate that our proposal can import enhancements to CNN in terms of classification performance and architecture optimization.

References

1. LeCun, Y., Bottou, L., Bengio, Y., Haffner, P.: Gradient-based learning applied to document recognition. Proc. IEEE **86**, 2278–2324 (1998)
2. Hssayni, E., Ettaouil, M.: New approach to face recognition using co-occurrence matrix and bayesian neural networks. In: 2020 IEEE 6th International Conference on Optimization and Applications (ICOA), pp. 1–5 (2020)
3. Hssayni, E., Ettaouil, M.: A novel pooling method for regularization of deep neural networks. In: 2020 International Conference on Intelligent Systems and Computer Vision (ISCV), pp. 1–6 (2020)
4. Hssayni, E., Ettaouil, M.: Regularization of deep neural networks with average pooling dropout. J. Adv. Res. Dyn. Control Syst. **12**(04-sp), 1720–1726 (2020)
5. Hssayni, E., Ettaouil, M.: Generalization ability augmentation and regularization of deep neural networks using pooling. Int. J. Tech. Phys. Probl. Eng. **13**(3), 1–6 (2021)
6. Chen, W., Shi, K.: A deep learning framework for time series classification using relative position matrix and convolutional neural network. Neurocomputing **359**, 384–394 (2019)
7. Krizhevsky, A., Sutskever, I., Hinton, G.E.: Imagenet classification with deep convolutional neural networks. Adv. Neural. Inf. Process. Syst. **25**, 1097–1105 (2012)
8. Hssayni, E.H., Joudar, N.-E., Ettaouil, M.: KRR-CNN: kernels redundancy reduction in convolutional neural networks. Neural Comput. Appl. **34**(3), 2443–2454 (2021)
9. Bai, C., Huang, L., Pan, X., et al.: Optimization of deep convolutional neural network for large scale image retrieval. Neurocomputing **303**, 60–67 (2018)
10. Junior, F.E.F., Yen, G.G.: Particle swarm optimization of deep neural networks architectures for image classification. Swarm Evol. Comput. **49**, 62–74 (2019)
11. Louati, H., Bechikh, S., Louati, A., et al.: Deep convolutional neural network architecture design as a bi-level optimization problem. Neurocomputing **439**, 44–62 (2021)
12. Ranzato, M., Boureau, Y.-L., LeCun, Y.: Sparse feature learning for deep belief networks. Adv. Neural. Inf. Process. Syst. **20**, 1185–1192 (2007)
13. Le Cun, Y., Boser, B., Denker, J.S., et al.: Handwritten digit recognition with a back-propagation network. In: Proceedings of the 2nd International Conference on Neural Information Processing Systems, pp. 396–404 (1989)
14. Hoseini, F., Shahbahrami, A., Bayat, P.: AdaptAhead optimization algorithm for learning deep CNN applied to MRI segmentation. J. Digit. Imaging **32**, 105–115 (2019)

15. Netzer, Y., Wang, T., Coates, A., Bissacco, A., Wu, B., Ng, A.Y.: Reading digits in natural images with unsupervised feature learning (2011)
16. LeCun, Y., Huang, F.J., Bottou, L.: Learning methods for generic object recognition with invariance to pose and lighting. In: Proceedings of the 2004 IEEE Computer Society Conference on Computer Vision and Pattern Recognition, 2004 (CVPR 2004), vol. 2, pp. II-104. IEEE (2004)

Cross Virtual Machine Security: A Survey

Khalid Zirari[1(✉)], Hamza Kamal Idrissi[1], Hicham Bensaid[1], Abdeslam En-Nouaary[1], and Ahmed El-Yahyaoui[2]

[1] CEDOC 2TI Institut National des Postes et Télécommunications INPT, Rabat, Morocco
{zirari.khalid,kamalidrissi,bensaid,abdeslam}@inpt.ac.ma
[2] Mohammed V University in Rabat, Rabat, Morocco
a.elyahyaoui@um5r.ac.ma

Abstract. Cloud computing security has emerged as the most pressing problem on the globe. Moreover, hosting several machines in a single physical environment offer more flexibility and performance, especially with the emergence of modern concepts such as smart cities, IoT, and robotics. The benefits and advantages of virtualization technology are enormous. On the other hand, sharing resources such as network, CPU, and RAM by several tenants cause security vulnerability related to virtualization components or side-channel attacks that challenge those gains. The present paper highlights the virtualization technologies and their security issues and survey the critical vulnerabilities related to side-channel attacks.

Keywords: Cross virtual machine security · Cloud computing · Virtualization · Hypervisor security · Cache-based side-channel attacks

1 Introduction

Virtualization is the core of cloud computing. IBM first developed this technology in 1960. It became more mature by developing advanced techniques and software that permit multiple machines to share computer resources and coexist on the same hardware to ensure dynamic resource provisioning, scalability, and service delivery [1]. However, diverse new security concerns have emerged because of the extensive use of virtualization that may challenge tenants' data privacy due to the physical colocation that still influences the virtualized infrastructure's security since the exchange of data or metadata circulates in shared hardware [2].

Despite the security mechanism, control, and partial isolation granted by all hypervisors in the market, the co-residency affects virtual tenants' isolation and privacy. Some of the famous isolation violations arise from side-channel attacks that target the transition of sensitive information in the different levels of the system by using a hidden channel [3]. The attackers may analyze the physical signals (power consumption, heat, laser, radiation) variations or exploit tasks executions time in the shared memory, such as the CPU caches. However, cloud service providers have developed several security mechanisms and concepts like containers [4] and Unikernels [5] to protect the customers' information from data leakage [6]. However, it remains insufficient to meet these challenges.

S. Motahhir and B. Bossoufi (Eds.): ICDTA 2022, LNNS 454, pp. 190–199, 2022.
https://doi.org/10.1007/978-3-031-01942-5_19

The shared resources between virtual machines lead to different security problems. Therefore, cloud providers, virtualization components, and logical isolation may cause potential vulnerabilities. Cache-based side-channel attacks are considered a real threat to cloud tenants. These attacks are based either on timing variations while accessing processor caches or on the collection of encryption traces. These traces may be used to create cache hits and misses profiles. Hence, enabling the capability of an attacker to deduce information about the secret key [7]. Furthermore, many mitigation methods suggested against side-channel attacks lead to performance degradation and a little overhead [8]. On the other hand, some researchers have discovered approaches to detect cache side-channel attacks by recording system calls made by virtual machines operating on the same hypervisor employing specific software or modifying the hardware [9, 10].

The organization of this paper is as follows: Sect. 2 presents the virtualization technology's timeline and its benefits. Section 3 discusses the architectural security model and the privilege levels according to the technology used. Sections 4 and 5, respectively, presented a comprehensive study on virtualization vulnerabilities and side-channel attacks. Finally, the conclusion of the article and some future directions are described in Sect. 6.

2 Virtualization Background

IBM Corporation invented virtualization in the 1960s, primarily to split the giant mainframe computer into multiple logical instances and operate them on a single physical hardware host. In 1998, VMware introduced an ×86 virtualization solution by creating a software-based solution. On November 13, 2005, Intel released hardware virtualization extensions (Intel® VT), often known as "hardware-assisted virtualization." These modifications were introduced to the CPU to facilitate and improve the efficiency of hypervisor activities. Therefore, company customers and IT developers will profit from virtualization since it allows them to operate various operating systems and applications simultaneously on a single physical platform by allocating the physical resources (memory, CPU, storage, and network) to the appropriate virtual machines.

Virtualized architectures and virtualization methods provide an extra execution layer, requiring administration and security protection. This layer contains many components that have a part in the virtualization process, among them:

- The hypervisor: is a software program used to handle numerous operating systems and manage system resources like processor memory and storage
- The virtual machines (VM): is a total computer environment abstraction by combining memory processor and I/O components virtualization of a computer
- Virtual machine monitor (VMM): is an application component of the hypervisor that keeps track of activities carried out by virtual machines

Nowadays, virtualization technology is essential for cloud service models, and 77% of on-premises server workloads are virtualized, according to the Spiceworks 2020 state of virtualization technology study [11].

Due to the progression of the COVID-19 pandemic, many organizations have adopted virtualization. This technology is becoming essential and improves new digital workplaces, such as desktop virtualization, allowing users to access virtual desktops from any connected device (Fig. 1).

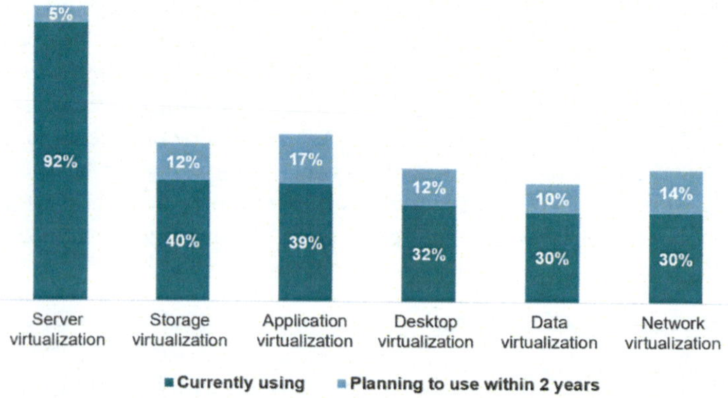

Fig. 1. Business adoption of virtualization technologies

Virtualization is utilized to meet client security, control, cost, scalability, and speed. It may influence cloud provider selection. It also allows cloud customers to quickly start and stop their resources, which is helpful in specific applications. However, several risks and vulnerabilities exist that an attacker may use to breach security and privacy measures in cloud computing environments.

3 Virtualization and Protection Rings

Understanding the ×86 security ring model of the processor is necessary to understand the virtualization security aspects. The ×86 is designed to specify the privilege level of instructions running inside the CPU. There are four privilege levels named rings 0 through 3 provided by the ×86 architecture, and the OS kernels often use Ring 0 privileges which are the highest elite level of all rings.

In the virtual environment, guest VMs can only control hardware resources via the hypervisor or directly through the host operating system. Moreover, the transition between user mode and kernel mode is achieved depending on the type of virtualization used. The following section describes each type and the transition method (Fig. 2).

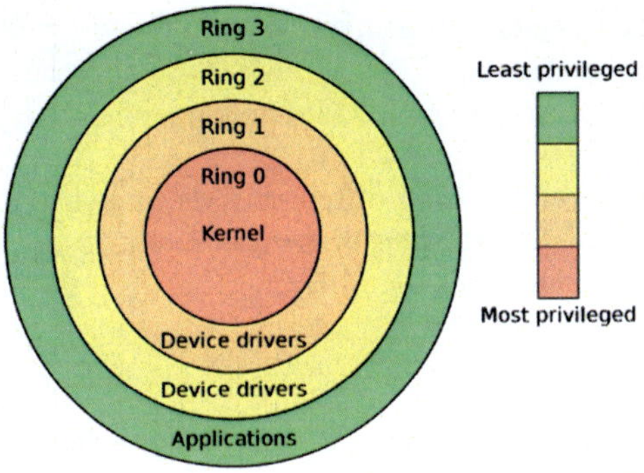

Fig. 2. Ring security model

3.1 Full Virtualization

In this technique, users might run several isolated guest operating systems simultaneously, and the VMM emulates hardware (BIOS, Memory, I/O) to isolate an unmodified guest OS.

Binary translation and direct execution are used to achieve full virtualization and ensure maximum compatibility with hardware. In other words, the hypervisor provides CPU emulation to control and modify privileged and protected CPU operations made by the unmodified guest operating system kernels.

The full virtualization approach provides the highest security and isolation for virtual machines and the compatibility with different operating systems by using software products like VMware, Hyper-V, or KVM [12]. However, the performance is impacted due to the trap and emulate technique.

3.2 Paravirtualization

In opposition to full virtualization, the operating guest OS in paravirtualization must be modified and adapted to replace privileged instruction execution with hypercalls and allow direct communication with the hypervisor. In other words, the transition between the user mode and the OS is assured through this interface. Also, to avoid the emulation of hardware features.

This technique improves performance and efficiency due to direct communication, but the compatibility with the unmodified operating system is limited.

3.3 Hardware-assisted Virtualization

In 2006, Intel developed various CPU extensions like the Intel VT-X technology that provides architectural support for virtualization and enables processor virtualization to enhance speed and minimize memory overhead [13].

The two previous techniques use the hypervisor in the virtualization process, which loads the VMM. In the hardware-assisted virtualization approach, the devices may be assigned directly to VMs. As a result, the VMM no longer needs to provide device/driver emulation.

In this approach, instructions from different VMs in a shared CPU operate in the non-root mode while the hypervisor runs in the root mode.

The following table summarizes the difference between the existing virtualization technique (Table 1):

Table 1. Comparison of the virtualization techniques

	Full virtualization	Paravirtualization	Hardware-assisted virtualization
Technique	Binary translation and direct execution	Hypercalls	Root and non-root mode
Performance	Good	Medium	Good
Guest modification	Unmodified Guest OS	Modified Guest OS	Unmodified Guest OS
Compatibility	Excellent	poor	Excellent
Utilized by	VMware, Microsoft	VMware, Xen	VMware, Microsoft, Xen

4 Security Vulnerabilities and Risks Categories of Virtualization Technology

4.1 Colocation Issues

Different tenants co-locate multiple virtual machines on the same host in cloud computing, which is common in the public cloud. If a tenant's virtual resources are co-located on the same physical system, they could be exploited, and many challenges can arise due to colocation.

The shared resource such as memory, CPU, and storage in cloud computing may influence the security of connected VM and lead to data leakage by recovering sensitive data using a developed technique like side-channel attacks. Moreover, the cloud service provider holds different customer instances using the same application code. Therefore, the data transition in the same environment leads to information leakage and produces a service disruption.

4.2 Hypervisor

The Hypervisors is the core of virtual architecture and one of the significant components in cloud computing that run in a high privilege. It manages and maintains the virtual machine's security and checks any privileged instruction sent by the guest to ensure the authenticity of the instruction before its execution.

The assessment of vulnerability exposed by different researchers [14, 15] demonstrate that the hypervisor is an attractive attack surface regardless of the variety of VMM existing in the market (VMware, Xen, KVM, Hyper-V), among them:

- **VM Escape**: This vulnerability may cause a denial of service due to the injection of a malicious program by exploiting a VM in the VMM
- **Privilege escalation:** It can happen by obtaining the root privilege and full access to the host OS
- **VM-Hypervisor Crosstalks:** The communication pipe provided by the hypervisor and the resources exposed to the VM may cause software vulnerabilities [16].

Furthermore, the hypervisor employs a virtual switch by using many physical networks, allowing connection between VM in the same host or into the external network. However, Son Duc Nguyen confirms that the hypervisor can be exploited using the fuzzing approach's denial of service in the virtual switch [17].

4.3 Virtual Machine Migration

Earlier, physical machines suffered from hardware faults that impacted production. Nevertheless, with the arrival of virtualization technology, cloud providers manage hardware-related failures by live migrating the afflicted virtual machines to another host. The migration process is one of the significant savers of client applications by ensuring the continuity of customer services. However, the transfer duration depends on VM size and network bandwidth and leads to downtime due to the state of the VM during the migration process.

The migration capability employed to relocate a virtual machine from one host to another helps achieve fault tolerance, load balancing, and power efficiency indispensable to cloud computing requirements. The benefits of VM migration are numerous. Nevertheless, the process may be vulnerable and presents serious security issues that can be summarized as follows [18]:

- Using a network as a transmission channel may expose data to an attacker during the migration. The source and destination platforms should establish a secure cryptographic path to ensure the confidentiality and integrity of data.
- Before starting the migration process, the destination host should be identified and trusted to avoid man-in-the-middle attacks [19].
- The start of an unauthorized migration call or the transfer of unneeded VM at the same time may overload the network resources and cause a Denial-of-Service

5 Side Channel Attacks

Cloud service providers often adopt virtualization to improve their computer resources efficiency. Still, it produces high-security risks due to shared resources and co-residency of VMs. Therefore, exploiting the environmental vulnerability to obtain sensitive information using techniques such as side-channel attacks.

By monitoring and analyzing the power consummation during an execution, magnetics radiation, or any other environmental parameters, the attackers can correlate the results of each action to form an idea of the security mechanisms adopted, especially the secret key used [20] (Fig. 3).

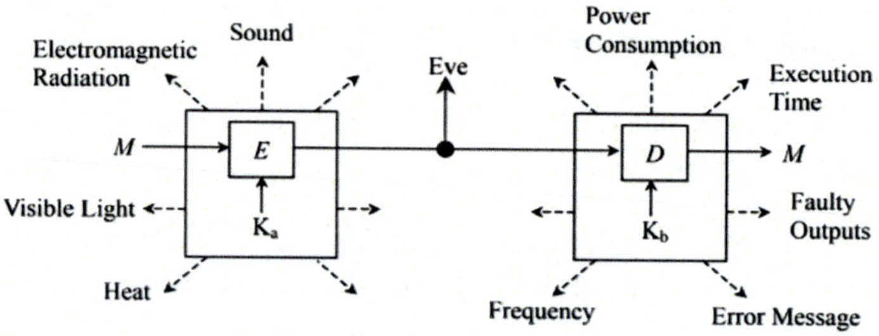

Fig. 3. The block view of side-channel attack [21]

5.1 Side-channel Vulnerability assessment

Vulnerability assessment is a crucial component of risk analysis in computer science. We analyze vulnerability reports in the CVE Common Vulnerabilities and Exposures database hosted at MITRE (a Public and open structure for information-security and vulnerability rapports) [22].

The study of 170 CVEs until the end of 2020 demonstrates that in 2007, just one attack was identified while the number of side-channel attacks increased years after years, as shown in the diagram below. The challenge of this type of attack currently brings the community's attention due to the immense exploit of the technology (Edge Computing, Artificial intelligence, Robotics) (Fig. 4).

Fig. 4. Evolution of side-channel vulnerabilities

Each CVE has a unique ID that simplifies the sharing of vulnerability and its details, including the affected products. The analysis of these IDs allows us to classify and trace the dependencies of each menace. Figure 5 reports that information exposure is the main existing weakness that permits unauthorized persons to access sensitive information, including private personal, geographic, or contact data like CVE-2020-6521 attack.

Observable Discrepancy weakness (Information Exposure Through Discrepancy) is listed second. It refers to the analysis of the behavior of communication and queries that may provide an unauthorized actor with knowledge about the product's functionality or internal status, followed by cryptographic issues that may challenge data confidentiality and integrity via exploiting encryption libraries.

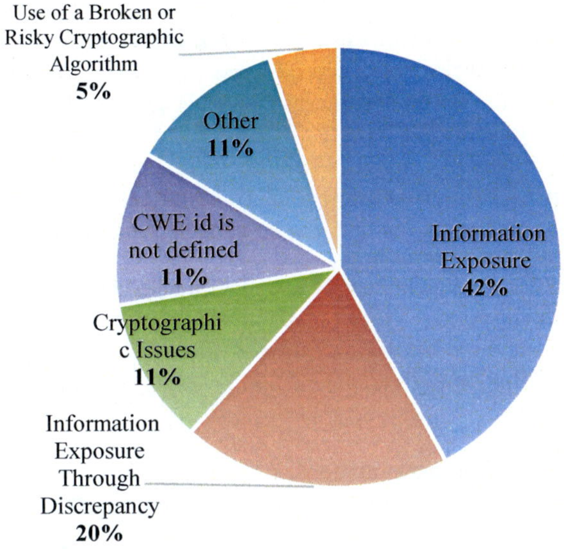

Fig. 5. CWE of side-channel attacks

5.2 Cache-Based Side-Channel Attacks

The CPU cache is a kind of storage with a smaller capacity and fast memory, positioned between CPU and DRAM with speed access differences. It is modeled by a hierarchy architecture of three levels, L1, L2, and L3.

When the processor requires data, it first checks the L1 cache; if the requested information is not reached, searches are continued in L2 than L3 until it has been found; which signifies a cache hit. Otherwise, The CPU extracts information from RAM and caches it for the subsequent request, which means a cache miss [23].

Caches have been a popular target of side-channel attacks due to the differences in tasks execution time between the cache hit and the cache miss. At the same time, the co-residency of virtual machines in cloud computing may boost the impact due to the shared hardware between the tenants. In other words, the L3 cache of the host is shared among the vCPUs of the guest machine, the monitoring and the analysis of cache state lead to extract private information.

5.3 Cache Side-Channel Attack Category

We can classify the cache-based side-channel attacks into three categories:

Access-Driven Attack: In access-driven assaults, the attacker runs detective software on his machine and the victim's machine to get information about cache sets. Simultaneously, it monitors the instruction cache and data cache and learns about the cryptographic process being executed to get the secret key. One of the most well-known examples of this attack is the Prime and Probe attacks.

The term Prime in this attack refers to determining how long it takes to read data from a cache set's memory pages. In this case, the hacker employs a process to overwrite the data in a cache with their own. Then the process will remain for a predetermined period before allowing the victim to reaccess the local cache. After the predestined time, the procedure moves to the probe level [24]. Currently, the cache lines are manipulated, and the attacker supervises its content. When the victim accesses the cache line, a cache miss occurs. In this case, the attacker will analyze and calculate the time variation to obtain the data from memory. Conclusively, he gets the sensitive information needed.

Time-Driven Attack: An attacker uses a correlation between a cryptographic transaction and a victim's cache misses. Moreover, the time required to access memory depends on the cache's status and the key value. Consequently, various inputs may have varied execution durations; this timing variation could be analyzed and exploited to identify the encryption key.

Trace-Driven Attack: The cache hits and misses are observed with extreme care in this type. The attacker inspects the victim's cache activity and searches for any link within an encryption technique and cache lines visited. However, these methods need a lot of high rate event monitoring, which is typically only possible with high-end technology [25].

6 Conclusion

In this study, we reviewed several virtualization techniques and design options in addition to the security concerns of each approach and an in-depth enumeration of the known attacks and vulnerability of side-channel in a virtual environment.

To reduce arisen security issues, much focus should be considered regarding co-residency and cache-based side-channel attacks due to the untrusted transition of data and metadata in the shared components, which can be an exciting axis in future research.

References

1. Saeed, A.: Cross-VM Network Attacks and their Countermeasures within Cloud Computing Environments DEDICATION (2020)
2. Sierra-Arriaga, F., Branco, R., Lee, B.: Security issues and challenges for virtualization technologies. ACM Comput. Surv. **53**, 1–37 (2020)

3. Tsalis, N., Vasilellis, E., Mentzelioti, D., Apostolopoulos, T.: A taxonomy of side channel attacks on critical infrastructures and relevant systems. In: Gritzalis, D., Theocharidou, M., Stergiopoulos, G. (eds.) Critical Infrastructure Security and Resilience. ASTSA, pp. 283–313. Springer, Cham (2019). https://doi.org/10.1007/978-3-030-00024-0_15
4. Sultan, S., Ahmad, I., Dimitriou, T.: Container security: issues, challenges, and the road ahead. IEEE Access **7**, 52976–52996 (2019)
5. Talbot, J., et al.: A Security Perspective on Unikernels (2019)
6. Gupta, I., Singh, A.K.: An integrated approach for data leaker detection in cloud environment. J. Inf. Sci. Eng. **36**, 993–1005 (2020)
7. Ahmadi, M.M., Khalid, F., Shafique, M.: Side-Channel Attacks on RISC-V Processors: Current Progress, Challenges, and Opportunities (2021)
8. Liu, L., Wang, A., Zang, W.Y., Yu, M., Xiao, M., Chen, S.: Shuffler: mitigate cross-VM side-channel attacks via hypervisor scheduling. In: Beyah, R., Chang, B., Li, Y., Zhu, S. (eds.) SecureComm 2018. LNICSSITE, vol. 254, pp. 491–511. Springer, Cham (2018). https://doi.org/10.1007/978-3-030-01701-9_27
9. Ainapure, B.S.: A Novel Approach to Prevent Cloud Infrastructure against Cache Attacks (2021)
10. Sangeetha, G., Sumathi, G.: An optimistic technique to detect cache based side channel attacks in cloud. Peer-to-Peer Networking Appl. **14**(4), 2473–2486 (2020). https://doi.org/10.1007/s12083-020-00996-1
11. Spiceworks. https://www.spiceworks.com/marketing/reports/state-of-virtualization. Accessed 15 Nov 2021
12. Tofiq Ahmed, B.: Virtualization mechanisms and its tools: A comprehensive survey Biometric View project (2020)
13. Asvija, B., Eswari, R., Bijoy, M.B.: Security in hardware assisted virtualization for cloud computing–State of the art issues and challenges. Comput. Netw. **151**, 68–92 (2019)
14. Thongthua, A., Ngamsuriyaroj, S.: Assessment of hypervisor vulnerabilities. In: Proceedings - International Conference on Cloud Computing Research and Innovation 2016, ICCCRI 2016, pp. 71–77. Institute of Electrical and Electronics Engineers Inc. (2016)
15. Security aspects of virtualization about ENISA (2017)
16. Compastié, M., Badonnel, R., Festor, O., He, R.: From virtualization security issues to cloud protection opportunities: An in-depth analysis of system virtualization models. Comput. Secur. **97**, 101905 (2020)
17. Nguyen, S.D., Mimura, M., Tanaka, H.: SVTester: finding DoS vulnerabilities of virtual switches. J. Inf. Process. **29**, 581–591 (2021)
18. Choudhary, A., Govil, M.C., Singh, G., Awasthi, L.K., Pilli, E.S., Kapil, D.: A critical survey of live virtual machine migration techniques (2017)
19. Vinay Deep, S., Gowda, P.S.J., Nagendra, H., Dhanalakshmi, M.: Secure Protected Live Migration for Energy Efficient Techniques (2020)
20. Mandal, J.K., Mukhopadhyay, S., Roy, A.: Lecture Notes in Networks and Systems 137
21. Zhou, Y., Feng, D.: Side-Channel Attacks: Ten Years After Its Publication and the Impacts on Cryptographic Module Security Testing (2005)
22. MITRE. https://cve.mitre.org/cgi-bin/cvekey.cgi?keyword=side+channel. Accessed 15 Nov 2021
23. International Conference on Advances in Computing, Communication Control and Networking (ICACCCN). IEEE (2018)
24. Su, C., Zeng, Q.: Survey of CPU Cache-Based Side-Channel Attacks: Systematic Analysis, Security Models, and Countermeasures (2021)
25. Jaamoun, A., Hiscock, T., di Natale, G., Jaamoum, A.: Scramble Cache: An Efficient Cache Architecture for Randomized Set Permutation (2021)

Adversarial Attacks on Machine Learning Systems

Safae Laatyaoui(✉) and Mohammed Saber

SmartICT Lab, Mohammed First University, Oujda, ENSAO, Oujda, Morocco
{s.laatyaoui,m.saber}@ump.ac.ma

Abstract. Due to their impressive performance, Machine Learning Systems; including Deep Learning networks; are widely used in several application domains including sensitive areas like security applications, autonomous vehicles, medicines, weapon manufacturing and other areas which require high robustness and safety. However, these systems are vulnerable to adversarial attacks. This vulnerability has been investigated by researchers to improve the robustness of Machine Learning models and consequently increase confidence in their use. For this purpose, various attacks against these systems and defenses appeared and were classified. In this work, our objective is to provide a comprehensive review in the Adversarial Machine Learning field of research. Especially; to achieve our goal; we have detailed the Fast Gradient Sign Method as well as its variants, we have cited several adversarial attacks and their generated adversarial examples, we have also explained the similarity constraint and metrics that measure it in addition to other interesting notions like the adversary's 3D model.

Keywords: Adversarial Machine Learning · Deep learning · Adversarial attack · Adversarial example · Adversary's 3D model

1 Introduction

Machine Learning systems; those of Deep Learning and others; achieve performance that approaches human performance. They even exceeded them in some areas. These systems that we find everywhere in our daily life, especially in entertainment areas, are also used in sensitive areas such as anomaly detection, medicine, drones and autonomous cars, which require robustness and security. However, these systems are vulnerable.

The error may be acceptable if an incorrect translation is made. However, if we make the wrong diagnosis of a patient by considering him to be ill when he is not, this can traumatize him. And if a sick patient is diagnosed as being breast, this can postpone the start of the treatment and cause serious damage. So how can we trust, especially in sensitive areas, a system that has vulnerabilities? Moreover, for example, to secure banking transaction systems, anomaly detection systems based on Machine Learning are widely used: How can an unsecured system secure another system?

Especially, Machine Learning models are often vulnerable to adversarial manipulation of their input intended to cause incorrect classification [1]. These inputs are called adversarial examples and we talk about adversarial attacks [2].

S. Motahhir and B. Bossoufi (Eds.): ICDTA 2022, LNNS 454, pp. 200–208, 2022.
https://doi.org/10.1007/978-3-031-01942-5_20

Researchers try to understand the cause of these vulnerabilities and increase the robustness of these systems to increase confidence in the use of Machine Learning. For this purpose, various attacks and defenses appeared and were classified. Defenses try to improve the model robustness and attacks give as tools to evaluate the model safety. Adversarial Attacks concern a lot of models and areas of Machine Learning: There are attacks generated against conventional Machine Learning models and others against Graph Neural Networks (GNN) [3] and Vision Transformers (VIT) [4]. There are generated adversarial examples in various application areas such as speech recognition [5] and physical world [6] and various attacks against different types of learning as unsupervised learning, semi-supervised learning and reinforcement learning.

In this paper, we aim to provide an overview of this research area. Our goal is to focus on the attacks and present them in a way that is easy to understand while trying to answer questions posed by beginners in this research area. Therefore, our examples and illustrations are simplified and usually based on image classification.

After listing used notations (Sect. 2), we first present the notion of adversarial attacks providing a mathematical formulation, discussing the similarity constraint presented by three norms and giving a graphical illustration (Sect. 3). We then tried to explain the Fast Gradient Sign method (Sect. 4) before moving on to presenting the adversary's 3D model: (i) Adversary's Goal, (ii) Adversary's Knowledge and (iii) Adversary's Capability (Sect. 5). Finally, we have cited (i) variants of the Fast Gradient Sign Method in addition to (ii) two other attacks.

2 Notations

The following Table 1 lists the notation which will be used in the subsequent sections.

Table 1. Notations.

Notations	Description
x	Victim data sample
x'	Perturbed data sample
x_i	The ith coordinate of x vector. $i \in [1,m]$
δ	Perturbation
y	Sample x 's truth label
t	Target label
F	ML model
θ	Parameters of model F

3 Adversarial Attack

"In computer security, an attack against a secure system means an attempt to make it do something other than what it was intended to do".[1]

Machine Learning systems are also vulnerable to attacks: ML systems can be fooled by perturbing input samples that may be imperceptible to the human eye but can force the system to produce incorrect outputs: It is an adversarial attack and the perturbed input is an adversarial example [2].

3.1 Mathematical Formulation

To understand adversarial attacks, let's see them mathematically:

The goal of the attacker of F is to find a small δ to add to x so that the output will not be y [7].

For x find δ under similarity constraints such that (see Eq. 1):

$$F(x+\delta) \neq y \tag{1}$$

3.2 Similarity Constraints

In the previous subsection, we have seen that similarity constraints are required to generate an adversarial example. These constraints are required so that the adversarial example does not deviate too much from the original input and the perturbation is imperceptible. So, why do we need the perturbation to be imperceptible?

To understand this, we will take an example: To launch a practical adversarial attack against a license plate recognition (LPR) system based on deep learning that has been widely used in public transport such as car parking management and electronic toll collection, we need to handle the License Plate. However, any covering or scrawling to license plates is prohibited by law. In this case, the success of the attack will also be measured by the level of imperceptibility [8].

The similarity constraint may be translated by a norm of δ that must be less than or equal to an ε previously defined. Thus, the attacker goal will be to find δ such that (see Eq. 2):

$$F(x+\delta) \neq y;\ ||\delta|| \leq \varepsilon \tag{2}$$

3.3 The Norm Metric

To quantify similarity, the definition of adversarial examples uses a distance metric. The question is: What distance to use? Sure, no distance metric is a perfect measure of human perceptual similarity but here we will name three widely used distance metrics in the literature for generating adversarial examples.

[1] *Andrew Ng, Coursera, AI for Everyone, Lecture 4.4: Adversarial attacks on AI.*

L_0 Distance. It is the number of the x coordinates which have been changed (see Eq. 3):

$$x_i \neq x_I' \tag{3}$$

L_2 Distance. It is the Standard Euclidean Distance between x and x'.

L_∞ Distance. It is the maximum change applied to all the coordinates (see Eq. 4):

$$||x - x'||_\infty = max\left(|x_1 - x_1'|, |x_2 - x_2'|, \ldots, |x_m - x_m'|\right) \tag{4}$$

3.4 Graphical Illustration

To understand adversarial attacks let's take this basic example of a binary classification with linear boundary decision (see Fig. 1):

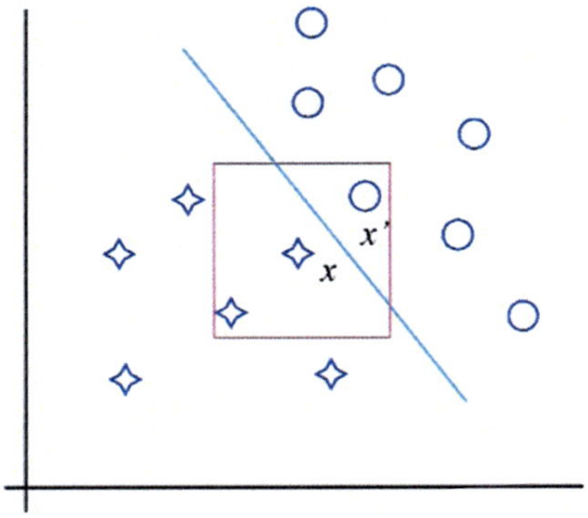

Fig. 1. An adversarial attack illustration.

x is part of the class on the left. We are looking for δ so that x' will be part of the class on the right. δ is small to respect similarity constraints represented by the square.

4 The Fast Gradient Sign Method

Remember that the training goal of the clean model is to find θ that minimize the loss function of F (see Eq. 5):

$$min_\theta \, Loss\,(F(x;\ \theta),\ y) \tag{5}$$

The minimization function during the training phase allows new inputs to be correctly predicted as much as possible. The attacker wants to have a wrong label result for x', so his goal is to maximize the loss function. Since he handles the inputs and not the parameters of the model, then the maximization will be applied on x and not θ. With respect to the similarity constraints, the attacker goal will be (see Eq. 6):

$$max_\delta \, Loss\,(F(x + \delta;\ \theta), y);\ \|\delta\|_\infty \leq \varepsilon \tag{6}$$

To resolve this maximization problem, the Gradient Ascent Method was used as follow [7] (see Eq. 7):

$$x + \delta = x + \varepsilon.sign\left(\nabla_x Loss\left(F(x;\ \theta), y\right)\right) \tag{7}$$

The chosen δ must maximize the loss function with respect to the correct class. So, the idea is to find the correct direction to move x. This step is presented by (see Eq. 8):

$$sign\left(\nabla_x Loss(F(x;\ \theta), y)\right) \tag{8}$$

The sign function has a vector for input and return a vector result that content as values –1, 0 or 1. It return 1 if the input is strict positive, -1 if the input is strict negative and 0 if the input is null. For example, (see Eq. 9):

$$sign\left((-3,\ 5,\ 3,\ 0,\ -1)\right) = (-1,\ 1,\ 1,\ 0,\ -1) \tag{9}$$

So, the idea is that we calculate the gradient of the loss function for each input feature and if the augmentation of the feature's value will augment the loss function value so we will augment it, if the augmentation of the feature's value will reduce the loss function value so we will reduce the feature's value to augment the loss function value and if the change of the feature's value does not impact the loss function value so we do not modify it. In this way, we try to intersect with the decision boundary and move to the other class (see Fig. 2).

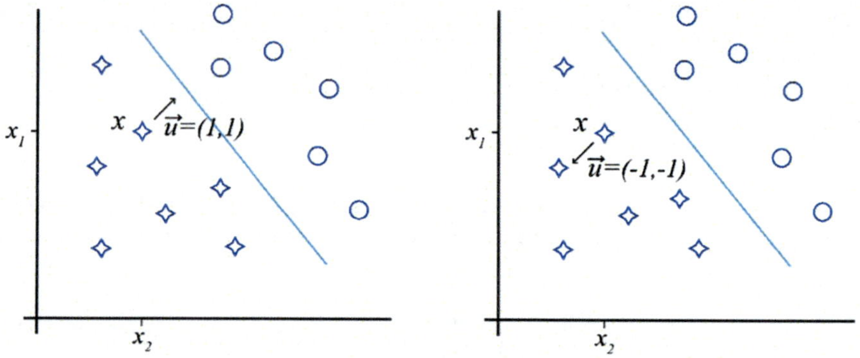

Fig. 2. On the left, we move in the direction of the gradient sign. On the right, we move in another direction.

In other terms, in this first step, the adversary evaluates the impact, on the output, of changing each input feature positively or negatively.

Now, one needs an ε that intersects the boundary decision. ε is a hyper-parameter to be chosen without forgetting that it must be small to meet the similarity constraints.

This method is "fast" because the algorithm is one step: It does not require an iterative procedure to compute adversarial examples.

5 Threat Model

To be able to defend a model against adversarial attacks, we need to make assumptions about the threat model: The goal of the attacker, his knowledge about the model and how can he manipulate data to reach his goal.

In other terms, we talk about Adversary's 3D model [9]: Adversary's Goal, Adversary's Knowledge and Adversary's capability.

5.1 Adversary's Goal

The adversary goal can be defined by the desired Security Violation (Integrity Violation, Availability Violation or Privacy Violation), Attack Specificity and Error Specificity [9].

Especially in a deep learning classification system, to violate the classifier integrity, the adversary tries to give an input x' that has an incorrect output $\neq y$. The nature of the incorrectness represents the adversary goal [10]:

- Confidence reduction: The adversary reduces the confidence of the correct output class.
- Misclassification: The adversary changes the prediction by altering the output classification to any other class.
- Targeted misclassification: The adversary changes the prediction by altering the output classification to a specific target class.
- Source/target misclassification: For a specific input, the adversary forces the output classification to be a specific target class.

5.2 Adversary's Knowledge

To defend a model against attacks, we make assumptions about the adversary's knowledge of the model:

- Training Data: Is the model trained in a public or private dataset?
- Feature Space: Are we using a statistic method to define the features or is the analysis more personalized?
- Learning Algorithm: What is the type of learning algorithm used for training? SVN, DNN, decision tree… and what are its parameters?

The model F can be defined by: Training Data Distribution D, Features Space x, Learning Algorithm f and Parameters θ [9]. We talk then about:

- White Box (WB) Attack or Perfect Knowledge (PK): When the attacker knows all this previous information about the target model.
- Gray Box (GB) Attack or Semi-White Box Attack or Limited Knowledge (LK): When the attacker has limited knowledge of the inner configuration of the model. He then trains a surrogate model for producing adversarial examples in a white-box setting and next he feeds these adversarial examples to the target model. This is possible due to the transferability property: The generated adversarial examples in a substitute model are also very likely to mislead a target model.

- Black Box (BB) Attack or Zero Knowledge (ZK): When the inner configuration of the model is unavailable to the attacker. In this case, adversaries can feed the input data to the box and observe the outputs of the model to exploit the input-output relationship and identify its weakness [11]. Also, in this case, the attacker can use a surrogate model and the transferability property. That is possible because one knows a minimum of information about the model [9].

5.3 Adversary's Capability

The attack is a poisoning attack when the attacker can insert or modify adversarial samples into training data. For example, when the model is trained online. The attack is an evasion attack when the attacker has not the ability to manipulate training data but can craft some malicious samples that evade detection by the model. In this second case, the attacker does not impact the model but evade it [11].

In addition, some practical constraints limit the attacker's capabilities. For example: When an attacker sends a spam email, this email must remain readable by the receiver and when an attacker sends malware to antivirus, the malware must keep its intrusive functionality. These constraints can be formalized as the number of samples that can be added to the data and the feature space that can be modified [12].

6 Other Attacks

In this section, we will present a non-exclusive list of attack methods.

6.1 Variants of FGSM

One Step Target Class Method. FGSM as seen in Sect. 3 is a White Box untargeted attack. But can be extended to a White Box targeted attack algorithm: Instead of minimizing the probability of predicting the correct class *P(y |x),* the attacker will try to maximize the probability of predicting the target class *P(t |x'* [13].

For a Neural Network instead of increasing the value of the loss function with respect to the true class, the attacker will decreachoosingse the loss function with respect to the target class by descendant the gradient instead of ascending it. The adversarial example will be defined by (see Eq. 10):

$$x' = x - \varepsilon . sign\left(\nabla_x Loss\left(F(x;\theta),\ t\right)\right) \tag{10}$$

One Step Least-Likely Class Method. Instead of choosing the target class t randomly from the incorrect ones, the attacker can choose the least-likely class predicted by the model and defined by [14] (see Eq. 11):

$$t_{LL} = argmin_t \left\{P(t/x)\right\} \tag{11}$$

Since the least-likely class predicted by a well-trained model is the less similar class to the correct class, the attack that chose it as a target class results in more interesting mistakes. In this case, the adversarial example is defined by (see Eq. 12):

$$x' = x - \varepsilon . sign\left(\nabla_x Loss\left(F(x;\theta), t_{LL}\right)\right) \tag{12}$$

Basic Iterative Method (BIM). This method tries to do as the FGSM but in multiple steps. The goal is to find a better maximum of the loss function in many iterations [14] (see Eq. 13):

$$x_0 = x;\ x_{t+1} = Clip_{x,\varepsilon}(x_t + \alpha.\,sign\ (\nabla_x Loss\,(F(x_t;\ \theta), y)) \tag{13}$$

The *Clip* function ensure that *x'* does not go out of the ε-neighbor of *x* (see Eq. 14):

$$x' : ||x' - x||_\infty \le \varepsilon \tag{14}$$

The step size alpha is relatively small and the number of iterations guarantees that the perturbation can reach the border.

This method is also named Projected Gradient Descent (PGD) if the initial x_0 is different from *x* and is randomly chosen in the ε-neighbor of *x* [11] (see Eq. 15):

$$x_0 : ||x_0 - x||_\infty \le \varepsilon \tag{15}$$

Iterative Least-Likely Class Method. As the FGSM is generalized to One Step Target Least-Likely Class Method, the BIM is also generalized as bellow [14] (see Eq. 16):

$$x_0 = x;\ x_{t+1} = Clip_{x,\varepsilon}(x_t - \alpha.\,sign(\nabla_x Loss\,(F(x_t;\ \theta), t_{LL})) \tag{16}$$

6.2 Szegedy's L-BFGS Attack

To generate adversarial examples for this attack, the optimization problem was formulated at beginning as [2] (see Eq. 17):

$$min||\,\delta\,||_{2;}\ F\,(x + \delta) = t; x + \delta \in [0, 1]^m \tag{17}$$

In other words, they try to find the closest adversarial example $x + \delta$ to x under L2 norm that is classified as *t* class target. This optimization problem can be very hard to resolve, it was approximated and resolved using limited memory BFGS. This attack was applied to the MNIST dataset and ImageNet dataset.

6.3 One Pixel Attack

For this attack, the used norm is L_0: Only one pixel was changed to generate adversarial examples. The attack was applied to a public dataset using famous Neural Networks architectures and had interesting results [15].

7 Conclusion

In this paper, we have tried to present a comprehensive review on adversarial attacks on Machine Learning Systems. First, we defined adversarial attacks and gave their mathematical formulation then we explained the similarity constraint. We have also detailed the FGSM method and presented the threat model. Finally, we have listed variants of the FGSM method in addition to other attack methods. In this work, we did not review metrics of evaluating the robustness of ML models against adversary attacks and the existing defense mechanisms with their efficiencies. We left this to future work.

References

1. Dalvi, N., Domingos, P., Sanghai, S., Verma, D., et al.: Adversarial classification. In: Proceedings of the tenth ACM SIGKDD international conference on Knowledge discovery and data mining, pp. 99–108. ACM (2004)
2. Szegedy, C., et al.: Intriguing properties of neural networks. In: 2nd International Conference on Learning Representations, ICLR 2014 - Conference Track Proceedings (2014)
3. Zügner, D., Borchert, O., Akbarnejad, A., Günnemann, S.: Adversarial attacks on graph neural networks: perturbations and their paterns. ACM Trans. Knowl. Discov. Data **14**(5), 3394520 (2020)
4. Hu, H., Lu, X., Zhang, X., Zhang, T., Sun, G.: Inheritance attention matrix-based universal adversarial perturbations on vision transformers. IEEE Signal Process. Lett. **28**, 1923–1927 (2021)
5. Kwon, H., Kim, Y., Yoon, H., Choi, D.: Selective audio adversarial example in evasion attack on speech recognition system. IEEE Trans. Inf. Forensics Secur. **15**(8747397), 526–538 (2020)
6. Eykholt, K., et al.: Robust physical-world attacks on deep learning visual classification. In: Proceedings of the IEEE Computer Society Conference on Computer Vision and Pattern Recognition, vol. 8578273, pp. 1625–1634 (2018)
7. Goodfellow, I.J., Shlens, J., Szegedy, C.: Explaining and harnessing adversarial examples. In: 3rd International Conference on Learning Representations, ICLR 2015 - Conference Track Proceedings (2015)
8. Zha, M., Meng, G., Lin, C., Zhou, Z., Chen, K.: RoLMA: a practical adversarial attack against deep learning-based LPR systems. In: Liu, Z., Yung, M. (eds.) Inscrypt 2019. LNCS, vol. 12020, pp. 101–117. Springer, Cham (2020). https://doi.org/10.1007/978-3-030-42921-8_6
9. Biggio, B., Roli, F.: Wild Patterns: ten years after the rise of adversarial machine learning. Pattern Recogn. **84**, 317–331(2018)
10. Papernot, N., Mcdaniel, P., Jha, S., Fredrikson, M., Celik, Z.B., Swami, A.: The limitations of deep learning in adversarial settings. In: Proceedings - 2016 IEEE European Symposium on Security and Privacy, EURO S and P 2016.7467366, pp. 372–387 (2016)
11. Xu, H., et al.: Adversarial attacks and defenses in images, graphs and text: a review. Int. J. Autom. Comput. **17**(2), 151-178 (2020).https://doi.org/10.1007/s11633-019-1211-x
12. Biggio, B., Fumera, G., Roli, F.: Security evaluation of pattern classifiers under attack. IEEE Trans. Knowl. Data Eng. **26**(4), 6494573, 984–996 (2014)
13. Kurakin, A., Goodfellow, I.J., Bengio, S.: Adversarial machine learning at scale.In: 5th International Conference on Learning Representations, ICLR 2017 - Conference Track Proceedings (2017)
14. Kurakin, A., Goodfellow, I.J., Bengio, S.: Adversarial examples in the physical world. In: 5th International Conference on Learning Representations, ICLR 2017 - Workshop Track Proceedings (2019)
15. Su, J., Vargas, D.V., Sakurai, K.: One pixel attack for fooling deep neural networks. IEEE Trans. Evol. Comput. **23**(5), 8601309, 828–841(2019)

DF E-Commerce: A Deep Learning Integrated Ontology Driven Model for E-Commerce Product Recommendation for Improved Machine Intelligence

Sanjeev Vijayakumar and Gerard Deepak(✉)

Department of Computer Science and Engineering, National Institute of Technology, Tiruchirappalli, India
gerard.deepak.christuni@gmail.com

Abstract. A product recommendation system is the critical component of any functioning E-commerce industry. It is crucial for these corporations to present their application with a recommendation system capable of furnishing products according to the desires of their customers by considering an array of factors. This paper proposes a unique approach that integrates textual data and data comprised of images by hybridizing both Machine Learning and Deep Learning strategies, thereby emphasizing user search queries and product images impartially. Distinct stages are presented for intermediate image set generation prior to user search preferences and term set generation related to the user query. A customized web-crawler along with search queries and click rates are used respectively. The architecture utilizes a nature-inspired Dolphin's echolocation metaheuristic algorithm, with objective functions scilicet SemantoSim, Normalized Google Distance (NGD), and Shannon's entropy to compute semantic similarity and semantic deviants. Experimentations indicated when compared with baseline models that the proposed DF E-commerce outperformed the baseline models with an accuracy of 97.3% with an FDR of 0.04.

Keywords: Dolphin's echolocation · NGD · Semantic similarity · SemantoSim · Shannon's Entropy

1 Introduction

The accumulation of data with a lot of heterogeneity on the Web in the recent era has increased the density of data. Multimedia content like images and videos has also exponentially grown, and there is a need to retrieve these multimedia contents. The upcoming 3rd generation of internet services, Web 3.0 or Semantic Web, has compelled a need for semantically inclined retrieval mechanisms. The advancement of the internet and incessant multiplication of hand-held devices has revolutionized the E-commerce sector. Mobile E-Commerce [1] and E-commerce in the fashion industry [2] are dependent on business models, and semantic recommendation systems have boomed in the recent

S. Motahhir and B. Bossoufi (Eds.): ICDTA 2022, LNNS 454, pp. 209–218, 2022.
https://doi.org/10.1007/978-3-031-01942-5_21

era. At present, there are multiple recommendation systems employed in E-commerce applications. Each recommendation system is unique in its own way; some emphasize on user query, while some give more weightage to the feature vectors of the product image. Owing to this, the accuracy of the results depends on a wide range of factors. However, a Web 3.0 compliant recommendation system that gives equal significance to all variables at stake, including semantic similarity measures, is vital.

Contribution: This paper proposes a Dual Feature E-commerce recommendation system that hybridizes Logistic Regression and Convolutional Neural Network for user-query and Product Images, respectively. The initial stage involves accumulating the database prior to any user search queries by utilizing a web-crawler to prioritize the top products trending and match it with the product database by computing semantic similarities using the Nature-inspired Dolphin's echolocation algorithm and generating the Intermediate Image Dataset. Compiling a Term set based on user queries and click rates by utilizing Jaccard similarity and Semantic similar measure is carried out. Experimentations are conducted on the Myntra All Products Dataset, and results are obtained with high accuracy and precision percentage, with low FDR.

Organization: Following this section, Sect. 2 compiles the relevant research previously performed under this domain. Section 3 depicts the proposed system architecture, Sect. 4 summarizes the implementation of the classifier, along with comparison and result, Sect. 5 concludes the study.

2 Related Works

Yan et al., [1] presented an upgraded Apriori algorithm to devoid information overload in a mobile e-commerce application. Hyunwoo et al., [2] proposed a system called K-RecSys that reckons customer's season variation of interests and motive to replace previously preferred items with a superior one. Yiyang et al., [3] proposed a recommendation system based on the classification that utilizes knowledge discovery to deal with semi-structured data. Shakila et al., [4] proposed a recommendation system capable of bridging the semantic gap for product recommendation. Rahul et al., [5] proposed a recommendation system based on model-based filtering. Chen [6] proposed a recommendation model that utilizes a text-matching algorithm for goods searching. Tessy et al., [7] proposed a hybrid-classifier that recommends products by intaking the description of products and users' search history based on content-based filtering and collaborative filtering methods. Mahesh et al., [8] proposed a system that employs a knowledge graph that recognizes relationships among user history and product data. In [9–20], several models supporting the proposed literature and the evaluations are depicted. Wu et al., [21] proposed a model known as SAPR, where a deep neural network architecture is modeled on a list-wise basis considering anonymous users and gathering data like the number of clicks and number of views to generate a ranking model. Sejal D et al., [22] proposed a model known as ACSIR where visual and textual data are collaborated to fill the void of semanticity. Xiao et al., [23] proposed a model known as EHPCD, where a bond between user click preferences and purchase history is mined to generate recommendations using the historical purchase frequency matrix. Bandyopadhyay et al., [24] proposed a model known as PCA+ K-means, where word to vector conversion is performed along with dimensionality reduction using PCA and classification using K-means.

3 Proposed System Architecture

The proposed system architecture involves two phases. Phase-1 comprises extracting product images, image labels, and descriptions from the dataset, classifying them, and reordering them based on semantic similarities computed between the image labels and product description to facilitate coherent recommendations. The intermediate image dataset generated is presented to the user. Phase-2 involves obtaining search keywords from the user, which is subjected to pre-processing and queried from the dataset by semantic similarity. In the motive of training a model with a large and diverse dataset, the "Myntra all product" dataset has been used for this study. The reason for the inclusion of this dataset, is its varied attribute. As depicted in Fig. 1, Phase-1 considers a variegated attribute of the dataset. It is subjected to pre-processing using the natural language toolkit of python involving Tokenization, Lemmatization, Stop Word Removal, and Named Entity Recognition (NER).

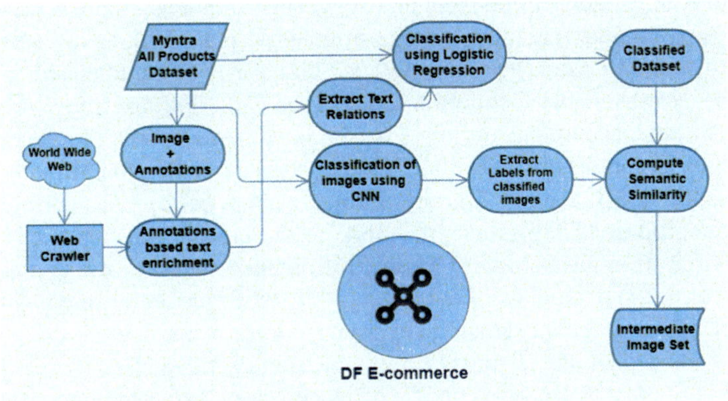

Fig. 1. Phase-1 of proposed system architecture

The attributes of interest, namely Product image, product description, and product label, are extracted from the pre-processed data. The product label, product description collectively known as the annotations and the product image are enriched by employing a customized web-crawler. They are matched based on annotations and the Product image set is indexed based on semantic similarities. Further, the data enriched based on annotations is classified using a Convolutional neural network (CNN), a deep learning algorithm. The CNN uses feature vectors from product images and is classified using the handcrafted feature. The product annotations are input as training data, and output is supplied as Product labels to help train the data. Thus, the annotation labels extracted are stored as Extracted Labels. Simultaneously, the dataset prior to any alterations by the web-crawler is compared for text relations from the annotations and finally passed to the Logistic regression algorithm. Since logistic regression is a machine learning algorithm, the dataset is classified based on product annotations. The necessity of a pair of classifiers, namely the CNN classifier and the Logistic regression classifier, is to achieve perspective on the distinct features of the available dataset, i.e. product images, product description, and product label, and be precise in image recommendations. Succeeding to classification using CNN and Logistic regression, the classified dataset and the term labels obtained respectively are compared for semantic similarities to produce the intermediate image dataset.

A nature-inspired meta-heuristic algorithm has been utilized to minimize the semantic gap between user desired products and recommended products. The Dolphin's echolocation algorithm is the most suitable one based on trials with various algorithms. The inspiration lies with bottlenose dolphins and their method to search for their target. A bottlenose dolphin can generate sounds in the form of clicks whose frequency is higher than that used for communication and varies with different species. As soon as the dolphin receives feedback from its target, it produces another click. The time-lapse between the clicks assists the dolphin in predicting the target distance from its current position. Hence initially, the dolphin emits sound in all directions hoping to find its target and as it approaches its target, the sound frequency is focussed in a particular direction. Similar to the methodology of dolphin's echolocation, an optimization algorithm can be developed to rank products effectively and suggest recommendations in E-commerce. In the algorithm, random initiation of the position of dolphins is done in a search space. Later PP (Predefined Probability) is computed, and a loop is commenced to calculate the fitness of each available location in search space and optimize the fitness in each iteration. In the end, accumulative fitness is computed corresponding to the best fitness value. Utilizing Dolphin's echolocation algorithm along with SemantoSim, Normalized Google Distance, and Shannon's Entropy as objective functions proves to be effective in the case of our proposed DF E-commerce model. The SemantoSim measure is a sematic measure devised from the Pointwise Mutual Information measure and is a normalized semantic measure. The study from our input query, when passed to it, consists only of a single term or word, then a semantically relevant term is considered along with it. If it consists of more than one term, then a permutation of the queried words is stacked. As depicted in Eq. (1), Normalized google distance – a semantic similarity computation measure that outputs the number of matches for a set of keywords. In NGD, keywords with a similar sense are clustered together, whereas the distinct meanings are spread farther apart. Equation (2) depicted below is utilized to compute the NGD factor, where the variable 'N' is the number of websites searched multiplied with the average of search terms. 'f(x)' and 'f(y)' signifies total positive matches for a search query 'x' and 'y' respectively. f(x, y) depicts the total number of web pages where both 'x' and 'y' exist. NGD factor near zero is preferred for similar terms and more than one for different terms. Shannon's entropy – Information of an event, also known as surprisal, is the prime concept in Shannon's entropy. An unlikely outcome may associate itself with an immense amount of information, whereas a likely outcome does not supply much information.

For example, a heavily biased coin on heads probably results in heads; hence it provides zero information or entropy. The self-information should serve two properties, it should lie between zero to one, and the additivity of independent events must be ensured. Hence it is represented by $I(p_i)$. Further, the entropy H(x) is computed by employing Eqs. (3) and (4).

$$\mathrm{SemantoSim}(x, y) = \frac{\mathrm{pmi}(x, y) + p(x, y)\log[p(x, y)]}{[p(x)p(y)] + \log[p(y, x)]} \tag{1}$$

$$NGD(x, y) = \frac{\max\{\log(fx),\quad \log(fy)\} - \log f(x, y)}{\log N - \min\{\log fx, \log fy\}} \tag{2}$$

$$I(p_{i)} = -\log(p_i) \tag{3}$$

$$H(p(x)) = \int p(x)\log(p(x))dx \tag{4}$$

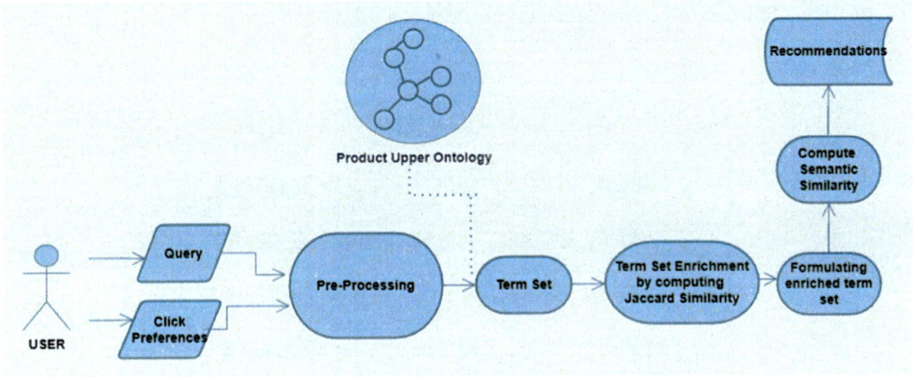

Fig. 2. Phase-2 of proposed system architecture

Sequel to computing the semantic similarity, the dataset is ranked, and finally, the intermediate image set is supplied to the user. As depicted in Fig. 2, Phase-2 deals with providing products based on user input query or search keywords. User input query keyword and user click preferences are extracted and subject to pre-processing. Similar to the phase 1 approach. The result of pre-processing yields the term set of data. The term set extracted is shallow in general scenarios; hence it is enriched by the upper ontology dataset. The upper ontology dataset is modeled by paradigm scaling of major classification metadata from e-commerce websites such as Myntra, Flipkart, and Amazon. Succeeding this step, the Jaccard index is computed between the user input query and each element of the term set and is depicted as Eq. (5). The threshold is set as 0.75. So data with a Jaccard index less than 0.75 are discarded from the term set to neglect the least matched products and further re-organized in decreasing order of indices. Succeeding this, the term set is subjected to compute semantic similarity. The methods involved in obtaining the intermediate image dataset are employed here, SemantoSim measure to determine the closeness of products, Normalized google distance to compare google search engine hits, and Shannon's entropy to compute the entropy of the obtained data. According to the ranked sequence of suggested products, the user is supplied with the corresponding list based on his input query.

$$Jaccard(String1, String2) = \frac{String1 \cap \mathrm{String2}}{String1 \cup \mathrm{String2}} \tag{5}$$

4 Implementation and Result Analysis

The proposed model is implemented in Python 3.9.0 in Google Collaboratory, powered by iPython. The proposed Dual Feature E-Commerce (DF E-commerce) and the baseline models used to compare the said model are implemented on a Windows 10 Operating System installed holding an i7 8th Gen Processor supported by 16 GB RAM. The Dataset is fabricated using 7856 queries provided to 1141 users. Each user was given around 50–60 queries and was asked to use their best product recommending approach. The top 15% of their recommendations were taken into consideration.

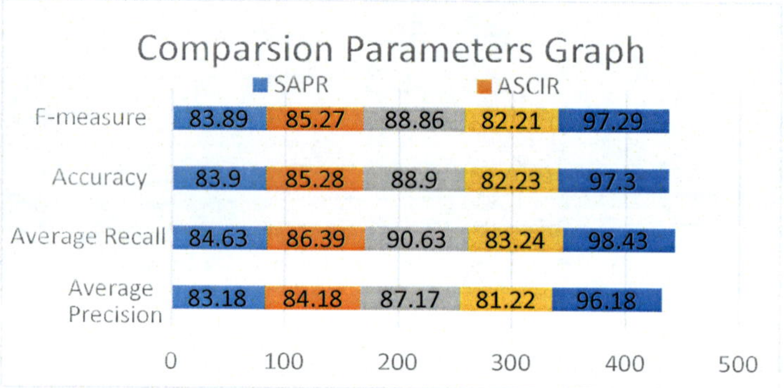

Fig. 3. F-Measure, Accuracy, Average Recall, and average precision comparison with baseline models

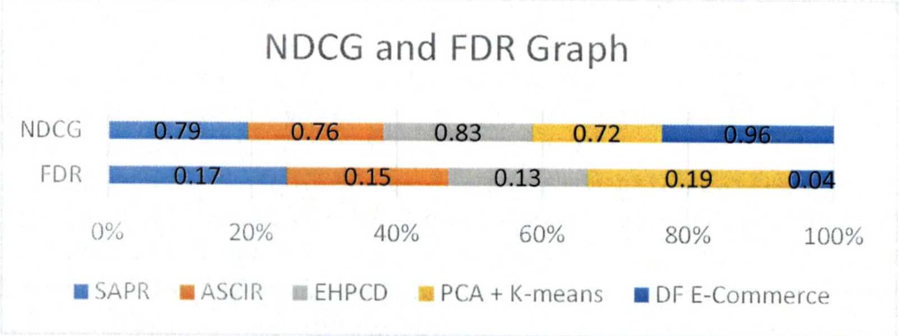

Fig. 4. NDCG and FDR comparison with baseline models

A total of 4 baseline models were used to compare the performance of DF E-Commerce, scilicet SAPR [21] ACSIR [22], EHPCD [23], and PCA with K-means clustering [24]. In this study, Precision, Accuracy, Recall, F-measure, False Discovery Rate (FDR), and Normalized Discounted Cumulative Gain (nDCG) are used as potential metrics. Standard Formulations for Precision, Recall, Accuracy, F-Measure, FDR, and nDCG have been used for evaluations. The baseline papers are known to have the following performance metrics. SAPR produces a precision of 83.18%, recall of 84.63%, an accuracy of 93.9%, F-measure of 83.89%, FDR of 0.17, and NDCG of 0.79. ACSIR produces a precision of 84.18%, recall of 86.39%, an accuracy of 85.28%, F-measure of 85.27%, FDR of 0.15, and NDCG of 0.76. EHPCD produces a precision of 87.17%, recall of 90.63%, an accuracy of 88.90%, F-measure of 88.86%, FDR of 0.13, and NDCG of 0.83. PCA+ K-means produces a precision of 81.22%, recall of 83.24%, an accuracy of 82.23%, F-measure of 82.21%, FDR of 0.19, and NDCG of 0.72. The proposed DF E-commerce model produces a precision of 96.18%, recall of 98.43%, an accuracy of 97.3%, F-measure of 97.29%, FDR of 0.04, and NDCG of 0.96. In SAPR, a deep neural network architecture is modeled on a list-wise base considering anonymous users. Here, activities such as the number of clicks and views are gathered, and a ranking model is generated to provide outputs. The vision of supplying recommendations to anonymous users, has reduced the quality of said recommendations. The prime reason for average values of precision%, recall%, F-measure, a very high value of FDR, and meager value of NDCG is because of its session-wise product recommendation approach, where information is embedded based on individual sessions.

The session-aware data is relatively sparse i.e. deficient in terms of density, which can be strenuous when training a deep learning classifier despite the enormous volume of data being fed. Hence in DF E-commerce, session-wise knowledge such as click rates and views are embedded into the model to facilitate the existence of word-embeddings. In ACSIR, visual and textual data are collaborated to fill the void of semanticity. The initial result set is generated using test-based user search of input query, and suggested results are computed using Pair-wise image cosine similarity. However, a lot of emphasis is given to image feature extraction using ANOVA cosine similarity, a classic measure of classification. Hence, certain hidden or uncovered entities, oblique to the approach fail to reveal. One of the presumed rationales could be the utilization of ANOVA similarity, which makes a stringent window leading to ambiguous results. Thus, the recommendation produced by ACSIR can be either be highly relevant to the user query or positively deviate from it. In EHPCD, the historical purchase frequency matrix is utilized. The bond between user click preferences and purchase history is mined to generate recommendations that could improve recommendation quality. The incorporation of click stream recommendation ensures a stream of concurrent knowledge is passed to the classifier. However, developing the frequency matrix, which comprises high session-based knowledge requires ranking product items. Hence, this approach would involve high integration of user knowledge to rank each item. As a result, the probability of bias in the recommendations is high, and deviation of output from the query term is imminent. In PCA+ K-means, PCA is utilized to reduce the dimensionality of product features and customer preferences in terms of product details. It also requires word to vector conversion to categorize similar words. PCA ensures adequate data density and reduces high relevance to query terms. Hence, the integration of PCA with the K-means clustering algorithm provides the applicability of recommendations to users. But since PCA handles scantiness of data, it could lead to extreme trimming. Thus, a very sparse and incoherent recommendation is susceptible, which is not vital. The proposed DF E-Commerce classifier is proposed to overcome the setbacks faced by the baseline models. From Fig. 3, Fig. 4, and Fig. 5, it can be seen that the proposed model carries the highest precision%, recall%, accuracy%, NDCG value, and a very low FDR. Offering equitable significance to the dual features (User query and Product Images) is vital for the performance. Coalescing Text features obtained from a web crawler to image features annexes the classifier of auxiliary knowledge from the real world. Enabling CNN for product images and Logistic Regression for user queries improves the performance manifold. Further, entity enrichment of dataset is performed using Google's knowledge-based API, a web crawler that obtains real-world data from the World Wide Web.

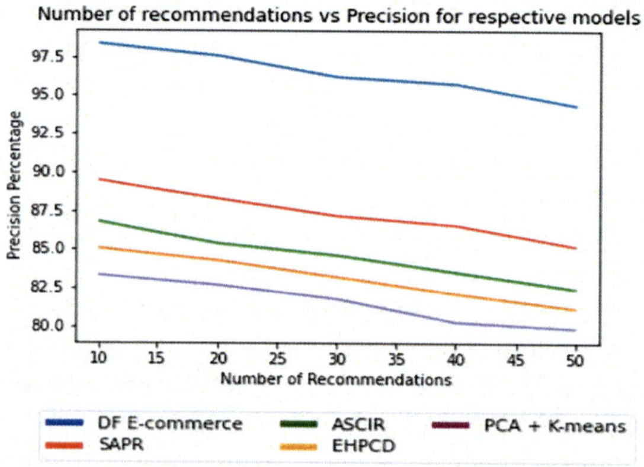

Fig. 5. Precision percentage vs. no. of recommendations

The data obtained is fed into the framework in a subgraph based on the Resource Description Framework (RDF), ensuring a high density of relevant knowledge. Finally, a virtual semantic network is built to classify text using Logistic Regression efficiently. Incorporating product-upper ontology promptly to the approach ensures taxonomy over the result. As a result, a distinctive reasoning scheme is established that recommends best-in-class products concerning user queries, current user clicks, user interests. Moreover, the inclusiveness of user clicks to the classifier facilitates implicit personalization of the product. Utilizing the Dolphin's echolocation algorithm for optimization with objective functions as SemantoSim, Normalized Google's Distance (NGD), and Shannon's entropy ensures that the optimized set of recommendable entities is semantically cohesive and similar. The amalgamation of SemantoSim and Normalized Google's distance and Shannon's entropy (responsible for information content measure), which are contrasting semantic similarity measures, ensures a practical relevance.

5 Conclusions

The recommendation system is an essential part of any E-commerce industry. The models proposed before DF E-commerce generally are not diverse and can focus on a single parameter. Hence the Dual Feature E-commerce was designed to give equal weightage on the user search query and product image. The proposed model proves to be superior when compared to its baseline models by a very significant amount. The inclusion of Dolphin's echolocation algorithm for optimization increases the accuracy by a remarkable quantity. The paper can be concluded by saying that the utilization of the Myntra All Product Dataset has made the recommendation system test under diverse conditions. The model becomes compliant to Semantic Web as the semantic similarity measures in a data rich environment. Hence, the model can be utilized under various environments to give an accuracy percentage of 97.3. As a future scope, explainability in the architecture of the DF E-commerce can be provisioned by further statistical and heuristic hybridizations and eliminating learning schemes.

References s

1. Guo, Y., Wang, M., Li, X.: Application of an improved Apriori algorithm in a mobile e-commerce recommendation system. Ind. Manag. Data Syst. (2017)
2. Hwangbo, H., Kim, Y.S., Cha, K.J.: Recommendation system development for fashion retail e-commerce. Electron. Commer. Res. Appl. **28**, 94–101 (2018)
3. Zhang, Y., Jiao, J.R.: An associative classification-based recommendation system for personalization in B2C e-commerce applications. Expert Syst. Appl. **33**(2), 357–367 (2007)
4. Shaikh, S., Rathi, S., Janrao, P.: Recommendation system in E-commerce websites: a graph based approached. In: 2017 IEEE 7th International Advance Computing Conference (IACC), pp. 931–934. IEEE (2017)
5. Gaikwad, R.S., Udmale, S.S., Sambhe, V.K.: E-commerce recommendation system using improved probabilistic model. In: Mishra, D., Nayak, M., Joshi, A. (eds.) Information and Communication Technology for Sustainable Development. LNNS, vol. 10, pp. 277–284. Springer, Singapore (2018). https://doi.org/10.1007/978-981-10-3920-1_28
6. Chen, H.: Personalized recommendation system of e-commerce based on big data analysis. J. Interdiscip. Math. **21**(5), 1243–1247 (2018)
7. Badriyah, T., Wijayanto, E.T., Syarif, I., Kristalina, P.: A hybrid recommendation system for E-commerce based on product description and user profile. In: 2017 Seventh International Conference on Innovative Computing Technology, pp. 95–100 (2017)
8. Singh, M.K., Rishi, O.P.: Event driven recommendation system for E-commerce using knowledge based collaborative filtering technique. Scalable Comput. Pract. Exp. **21**(3), 369–378 (2020)
9. Surya, D., Deepak, G., Santhanavijayan, A.: KSTAR: a knowledge based approach for socially relevant term aggregation for web page recommendation. In: Motahhir, S., Bossoufi, B. (eds.) Digital Technologies and Applications. ICDTA 2021. LNNS, vol. 211, pp. 555–564. Springer, Cham (2021). https://doi.org/10.1007/978-3-030-73882-2_50
10. Deepak, G., Ahmed, A., Skanda, B.: An intelligent inventive system for personalised webpage recommendation based on ontology semantics. Int. J. Intell. Syst. Technol. Appl. **18**(1–2), 115–132 (2019)

11. Varghese, L., Deepak, G., Santhanavijayan, A.: A fuzzy ontology driven integrated IoT approach for home automation. In: Motahhir, S., Bossoufi, B. (eds.) Digital Technologies and Applications. ICDTA 2021. LNNS, vol. 211, pp. 271–277. Springer, Cham (2021). https://doi.org/10.1007/978-3-030-73882-2_25
12. Aditya, S., Muhil Aditya, P., Deepak, G., Santhanavijayan, A. (2021). IIMDR: intelligence integration model for document retrieval. In: Motahhir, S., Bossoufi, B. (eds.) Digital Technologies and Applications. ICDTA 2021. LNNS, vol. 211, pp. 707–717. Springer, Cham (2021). https://doi.org/10.1007/978-3-030-73882-2_64
13. Manaswini, S., Deepak, G., Santhanavijayan, A.: Knowledge driven paradigm for anomaly detection from tweets using gated recurrent units. In: Motahhir, S., Bossoufi, B. (eds.) Digital Technologies and Applications. ICDTA 2021. LNNS, vol. 211, pp. 145–154. Springer, Cham (2021). https://doi.org/10.1007/978-3-030-73882-2_14
14. Deepak, G., Priyadarshini, J.S.: Personalized and enhanced hybridized semantic algorithm for web image retrieval incorporating ontology classification, strategic query expansion, and content-based analysis. Comput. Electr. Eng. **72**, 14–25 (2018)
15. Deepak, G., Shwetha, B.N., Pushpa, C.N., Thriveni, J., Venugopal, K.R.: A hybridized semantic trust-based framework for personalized web page recommendation. Int. J. Comput. Appl. **42**(8), 729–739 (2020)
16. Deepak, G., Kasaraneni, D.: OntoCommerce: an ontology focused semantic framework for personalised product recommendation for user targeted e-commerce. Int. J. Comput. Aided Eng. Technol. **11**(4–5), 449–466 (2019)
17. Deepak, G., Santhanavijayan, A.: OntoBestFit: a best-fit occurrence estimation strategy for RDF driven faceted semantic search. Comput. Commun. **160**, 284–298 (2020)
18. Deepak, G., Santhanavijayan, A.: UQSCM-RFD: a query–knowledge interfacing approach for diversified query recommendation in semantic search based on river flow dynamics and dynamic user interaction. Neural Comput. Appl. 1–25 (2021)
19. Krishnan, N., Deepak, G.: Towards a novel framework for trust driven web URL recommendation incorporating semantic alignment and recurrent neural network. In: 2021 7th International Conference on Web Research (ICWR), pp. 232–237 (2021)
20. Roopak, N., Deepak, G.: OntoKnowNHS: ontology driven knowledge centric novel hybridised semantic scheme for image recommendation using knowledge graph. In: Villazón-Terrazas, B., Ortiz-Rodríguez, F., Tiwari, S., Goyal, A., Jabbar, M. (eds.) Knowledge Graphs and Semantic Web. KGSWC 2021. CCIS, vol. 1459, pp. 138–152. Springer, Cham (2021). https://doi.org/10.1007/978-3-030-91305-2_11
21. Wu, C., Yan, M.: Session-aware information embedding for e-commerce product recommendation. In: Proceedings of the 2017 ACM on Conference on Information and Knowledge Management, pp. 2379–2382 (2017)
22. Sejal, D., Ganeshsingh, T., Venugopal, K.R., Iyengar, S.S., Patnaik, L.M.: Image recommendation based on ANOVA cosine similarity. Procedia Comput. Sci. **89**, 562–567 (2016)
23. Xiao, Y., Ezeife, C.I.: E-commerce product recommendation using historical purchases and clickstream data. In: Ordonez, C., Bellatreche, L. (eds.) Big Data Analytics and Knowledge Discovery. DaWaK 2018. LNCS, vol. 11031, pp. 70–82. Springer, Cham (2018). https://doi.org/10.1007/978-3-319-98539-8_6
24. Bandyopadhyay, S., Thakur, S.S., Mandal, J.K.: Product recommendation for e-commerce business by applying principal component analysis (PCA) and K-means clustering: benefit for the society. Innov. Syst. Softw. Eng. **17**(1), 45–52 (2021)

From Sketch to Simulation in Virtual Laboratory Using Deep Learning

Fatima Cheddi[1(✉)], Mohammed Serrhini[2], and Tahiri Ahmed[1]

[1] ACSA Laboratory, Mohammed I University in OUJDA, Oujda, Morocco
Fa.cheddi@ump.ac.ma
[2] LaRI Laboratory, Mohammed I University in OUJDA, Oujda, Morocco

Abstract. Virtual labs are simulated learning environment softwares that allow students to complete laboratory experiments online or offline to explore concepts and theories without stepping into a physical science lab. Developing such a software code is a long process and a hard task to achieve that requires a skilled expert developer. Human free-hand sketches have been studied in various fields including sketch recognition, synthesis, sketch-based image retrieval, etc. This paper will present a new sketch paradigm to generate a simulation from a Hand sketch. We propose a novel method that uses Convolutional Neural Networks (CNNs), Recurrent Neural Networks (RNN) combined with a Laboratory as a Service (LaaS) in order to transform a user-drawing sketch into a simulation code. Our approach is based on the use of both SSD (Single Shut Detector) and Mobile-Net as feature extractions for object identification in the sketch, called corresponding web services, which eventually regenerate a simulation software. Our method has been trained and evaluated using the "Quick, Draw" dataset. Experimental results have demonstrated that the approach has achieved over 80% of accuracy.

Keywords: Sketch to code · Virtual laboratories · Simulated learning environments · User interface · Simulation · CNN

1 Introduction

Nowadays, online laboratories (remote and virtual) play an important role in Engineering Education, especially in applied sciences Wherein experimentation simulation is an essential operation.

A virtual laboratory is a piece of software where students interact with a system through Graphical User Interfaces (GUIs) such as 2D Drawables elements, interactive sketches, buttons, sliders, and input fields. Moreover, sketching is an intuitive and powerful representation method for people to present their ideas. Despite its relatively concise and abstract character, sketch making is effective in a variety of situations like those of communication and design. Even sketching a wireframe is the principal means to outline and diagram user interfaces in the development of computer-based design tools [1]. With the advances in science and technology, the virtual laboratory has evolved and has gotten much attention from researchers; many solutions were provided by users to

S. Motahhir and B. Bossoufi (Eds.): ICDTA 2022, LNNS 454, pp. 219–228, 2022.
https://doi.org/10.1007/978-3-031-01942-5_22

build and implement virtual laboratories. These solutions include Ejss [2], LabVIEW [3], Simulink [4]: Adobe Flash, javaApplet, and others. However, adopting such a solution is a time-consuming process that requires skilled specialists in programming who are willing to put high technical effort in working with the suggested software.

Recently, Deep Learning has demonstrated its proficiency in different domains of Visual Recognition [5], like that of Facial Point Detection [6], House Numbers Digit Classification [7], Multi-digit Number, object detection on the real images, and freehand image (sketch). Nevertheless, free-hand sketches are fundamentally different from natural images; In particular, Sketches are typically abstract and stylized, depending on the user's artistic ability, as they contain the most basic details only. On the contrary, real images contain a lot of characteristics such as background, colors…

Our contribution is a novel method based on a Convolutional and Recurrent neural network that is connected with the Laboratory as a service in order to generate a simulation code from the sketched image as an input. To achieve this, our work is divided into four essential steps. The first one is the recognition method based on conventional and Recurrent neural network technology that is used for identifying and detecting objects (Elements, texts) that exist in the input image drawn by the user. We have chosen Single Shot Detector (SSD) [8] with (MobileNet) [9] for computer vision and RNN for optical character recognition. The second step is processing which is based on the paradigm Laboratory As a Service (LAAS) [10] and modular components [11]. The final step is the regeneration of the experimentation simulation.

This paper is structured as follows:

In Sect. 2 related works, we present an overview of sketch-based deep learning that exists in the field. Section 3 describes the proposed approach. In Sect. 4, we elaborate on an application used for testing our approach. Finally, Sect. 5 is devoted to a conclusion that suggests some areas of future work.

2 Related Works

The automatic generation of codes from sketch images using Machine learning algorithms is a new field of research. Some related works in the transformation of sketches into codes are as follows. Beltramelli [12] proposed a new method named Pix2code. This model can generate a code from Screenshots for three platforms: web, iOS, and Android. It uses Convolutional Neural Network (CNN) and Recurrent Neural Network (RNN) to convert the raw input image into a learned representation. It also performs language processing on the textual description of the matching image. Another application that deals with the same subject was proposed by Nguyen et al. It is named REMAUI [13]; It automatically transforms conceptual drawings into codes for iOS and Android mobile devices. REMAUI merges between computer vision algorithms and Optical Character Recognition (OCR) [14] to classify items from the input image. It builds a new application that looks exactly like one of the input screenshots. In 2019, Alexander Robinson introduced a method named sketch2code [15] for generating websites from designs. The author of this paper provides two approaches for automating this interaction: one is based on classic computer vision and the other is grounded on a novel use of deep learning segmentation networks. In 2019, another application called Sketch2Code [16] was proposed. This work presented a new approach that utilizes a Deep Neural Network which aims at converting a sketch into its corresponding User Interface. Sketch2Code converts a hand-drawn image on a white surface into a JavaScript Object Notation (JSON). In 2018, a framework called Sketch-Based Physics Tutoring Systems [17] was proposed to offer animation support for a variety of physics topics. The objective was to create animations from a student's written response to a physics problem, giving visual feedback in the context of the problem.

3 Methodology "Sketch to Simulation"

3.1 Pipeline of the Proposed Architecture

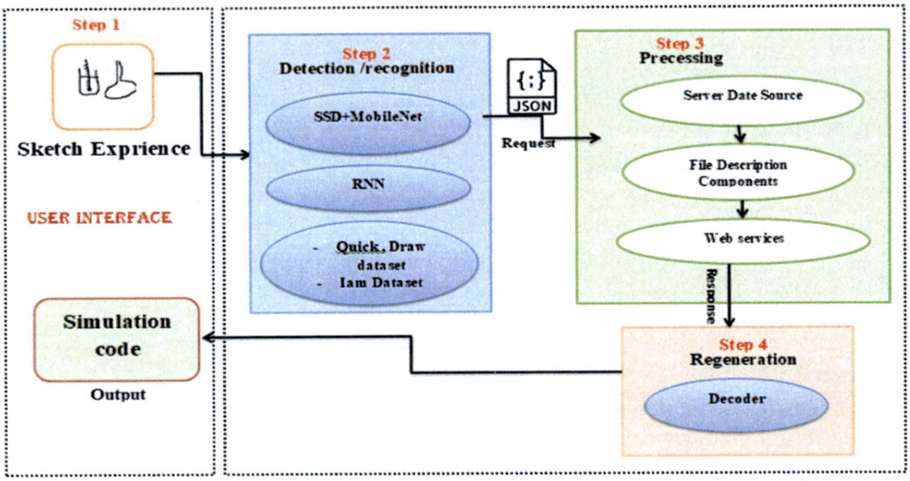

Fig. 1. System architecture

Our contribution "sketch to simulation" is a novel method based on CNN and RNN combined with LaaS allowing us to transform a sketch as input into a simulation code. To achieve this goal, we divided our solution into four steps. The pipeline of our proposed model is given in Fig. 1. The first step is a white interface where the user draws an experience. The aim of the second step named Detection/Recognition is to understand the input sketch and deduce the present items (Elements, positions, texts, values…). The third step which is processing is responsible for providing all services called. The last step regeneration represents a decoder that builds an output simulation code. Our model takes a sketch as an input. Firstly, the input sketch will go to block the recognition which consists of three sub-parts. The goal of the first one which is based on a convolutional neural network is to extract and identify the elements present. The second step concerns optical character recognition (OCR). Here, we adopted a recurrent neural network for inferring different kinds of hand-written elements such as text, numbers, …. And finally, we use the result of both previous parts to generate a file that contains a description of the input sketch. In this case, we choose a JSON format. The output of the previous

Block "recognition" is considered as an input to a third step Processing. In this step, a process request-response method is enabled between the "input file" and the virtual laboratory which is based on the Laboratory as a service (LaaS) [10, 11] and the smart device laboratory paradigms for implementing laboratories as independent Elements and developing them as sets of services to be utilized with abstract level and virtualization. And finally, a decoder will generate an output simulation code.

3.2 The First Step: Input Image

An input image is a sketch drawn by users on a white background using a mouse and a stylus of a graphic tablet. this sketch consists of one or more objects and descriptions of the sketch experimentation such as values, functions.... The input sketch has the dimensions 64 * 64.

3.3 The Second Step: Recognition

A sketch can be contained by both hand-drawn images and handwritten texts. For this reason, we propose a two-branch neural network that combines computer vision and character optic recognition. It should be mentioned that two neural networks have been trained independently. While working on the first branch, we used a convolutional object detector SSD [8] with pre-trained Model "MobileNet" [9] for computer vision. Therefore, in the second branch, we utilize Recurrent Neural Network layers with CTC for Optical Character Recognition. Finally, we concatenate their results to obtain a merging representation.

Computer Vision
The first branch, shown in Fig. 2, is composed of two main parts: feature extraction and object detection. In this work, we used for the first one MobileNet as a feature extractor and Single Shot Detector SSD as a convolutional object detector.

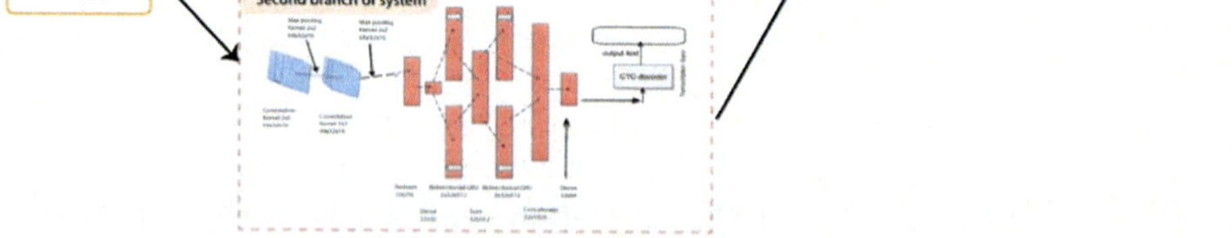

Fig. 2. The framework of proposed network of two-branch architecture. The first branch aims at recoganizing shapes using CNN. The second branch concerns optical character recognition (OCR) Using RNN+CTC. Afterwards, the results merge to identify each element with their properties

- SSD [8] was created by Wei Liu in 2016 based on a convolutional neural network. The object detector divides the output region of bounding boxes into a set of default frames depending on aspect ratios and scales per feature map position. The network produces a collection of bounding boxes with fixed size and confidence scores for the presence of each object class instance in bounding boxes, followed by a non-maximum suppression function to output the final detections. To increase accuracy, SSD introduces the following key features: a) convolutional filters predictors of object classes for detection b) separation of filters and default boxes to manage on aspect ratios, and c) multi-scale feature maps for object detection.
- MobileNet [9] is a model which is developed be used in mobile and embedded vision systems. It is developed to enhance the real-time performance of deep learning. MobileNets is based on depth-separable convolutions. The basic architecture of MobileNets is based on depth-wise layers and point-wise layers. The depth-wise refers to deep convolutional layers using 3×3 kernels, while the point-wise refers to convolutional layers using 1×1 kernel. The depthwise and pointwise convolutions are separate layers. MobileNet generally has 28 layers and 4.2 million parameters.

Optical Character Recognition: The network (second branch) as shown in Fig. 3 begins with two convolution layers, followed by maximum pooling. The convolution stride is 1 and the max-pooling stride is 2. The first convolution layer employs 32 convolution filters to construct the convolution output matrix's 32 dimensions. The result is scaled to be a two-dimensional matrix after the second max pooling. The dimension of this matrix will be reduced by a dense layer in a direction other than the temporal direction. We can see that the network's time dimension has been kept since the beginning; it has just been reduced. A second bidirectional GRU layer and a bidirectional GRU recurrent neuron layer with the sum of the two output matrices will follow. This time, for output matrix concatenation, we'll employ 51230 GRU neurons in each direction of the sequence. In the end, the network employs the CTC output layer. This layer is nothing more than a dense layer that has been matched to our needs.

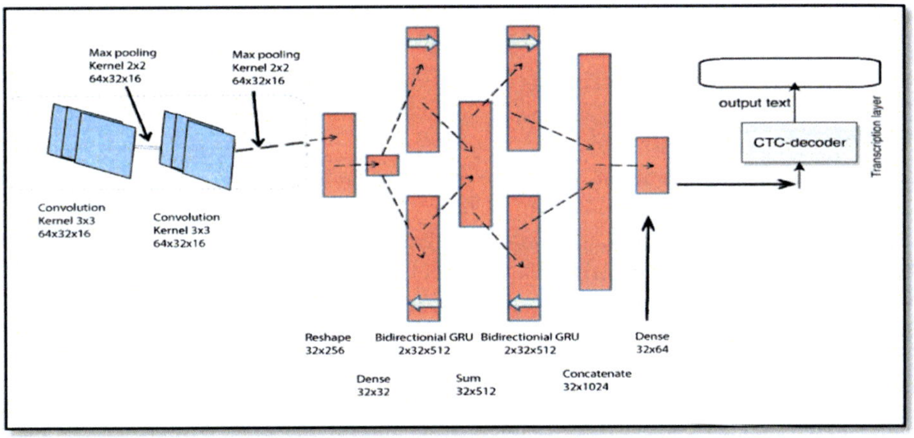

Fig. 3. The neural network of OCR

Merging Branch: This part presents a merging process that is used to concatenate the results of two previous steps computer vision and OCR to structure an output representation features of input sketch. The output is a set of information that represents the notations of the sketch. In this case, we used JavaScript object notation (JSON) format.

3.4 The Third Step: Processing

In this section, we present the third Step which is processing and we explain the communication process between the generated file by the previous block and the laboratory provider. The main idea is that the adoption of a Deconstruction Modular laboratory paradigms [11] allows separating all functionalities from each other, and Laboratory as a service LaaS aims at converting laboratories into a set of services. This process is built through an API that is responsible for the transfer of the parameters by means of sending the request and receiving responses. The virtual laboratory provider searches for the "services description file" in a database using a request defined according to the information that exists in (file.json). All the services provided by the laboratory are composite in the "service description file". Moreover, LAAS is exposed as a set of web services RESTful HTTP; each service represents a virtual laboratory element. The internal specifics of these web services are masked to the final service consumer. The output of this block is a set of customized services according to the user's needs. Each service is a table describing all configurations that the smart element provides. Each element is identified by the ID, the full name, and the description of all functions.

3.5 The Fourth Step: Regeneration

The automatic generator is a decoder that generates a simulation code using the configuration obtained from the webserver. For this contribution, we deployed a generator based on the NodeJS server. For the connection between the client and the server, we have utilized WebSocket. Since we are using web services, we have chosen HTML/javascript as an output code, which can be executed on any browser. It's worth mentioning that the method doesn't generate HTML directly; instead, they employ a domain-specific language (DSL) to represent the structure of the functions.

4 The Experimental Setup

4.1 Datasets

Our proposed method is trained, validated, evaluated on two datasets:

- Quick, Draw! [18]: this dataset is created by Google publicly which includes Over 50 million photos in 345 categories. Each drawing is represented as a sequence of line vectors in one dataset, and each picture is represented as a 28×28 gray scale matrix in another.
- *Iam* [19]: This dataset is composed of over 1500 scanned text pages handwritten by 657 different writers, submitted by 657 authors. this dataset is divided into three parts: the training set which has 747 documents, the validation set which contains 116 documents, and the test set which has 336 documents.

4.2 The Subjects and Experiment Procedure

In this work, we conducted our experiments with 20 students to obtain the performance of our proposed system. We recruited these participants (10 female and 10 male) from the specialized technician level. For our informal evaluation, all participants had studied physics. Each participant took 30–45 min to complete the experiment tasks. First, students were given an explication about the system's functionality. All participants were invited to provide three examples of each experience.

4.3 Performance Evaluation Metrics

To evaluate the performance of our proposed model, four metrics: Recall (1), Precision (2), accuracy (3), and F1-score (4) were used in this study. To calculate these metrics for the evaluation, we will extract the values of True Positive (TP), False Positive (FP), False Negative (FN), and True Negative (TN) of the classes (calorimeter, pendulum, …).

$$\text{Recall} = \text{TP}/\text{TP} + \text{FN} \tag{1}$$

$$\text{Precision} = \text{TP}/\text{TP} + \text{FP} \tag{2}$$

$$\text{Accuracy} = \text{TP} + \text{TN}/\text{TP} + \text{TN} + \text{FP} + \text{F} \tag{3}$$

$$\text{F1-Score} = 2 * (\text{Precision} * \text{Recall}) \,/\, (\text{Precision} + \text{Recall}) \tag{4}$$

4.4 Experiment Results and Discussion

Qualitative Results
To test and validate our previous architecture, we have developed our first system for some basic experiments in the physics domain (thermometer, calorimeter, rotation event …). Table 1 presents some simulations that the system can generate.

Quantitative Results
The proposed system implements deep learning-based SSD and MobileNet for the regeneration of a simulation from a sketch image. We evaluate The performance of our system in terms of Precision, Recall, Accuracy, and F-score. Table 1 illustrates the detection results. The average of the values received from 90 samples was used to determine each value. To measure performance, we used two different sketching methods. The results show that the methods have achieved a mean accuracy of over 90. It is seen that the system shows significant accuracy of 93.33. The proposed method achieves a significant result in terms of F-score, which counts both precision and recall. we noticed that the use of the tablet stylus technique has achieved better results compared to the other technique of using the mouse (Table 2).

Table 1. Presentation of some simulations generated by our system

Title	Experience sketch	Generated simulation	Description
spring	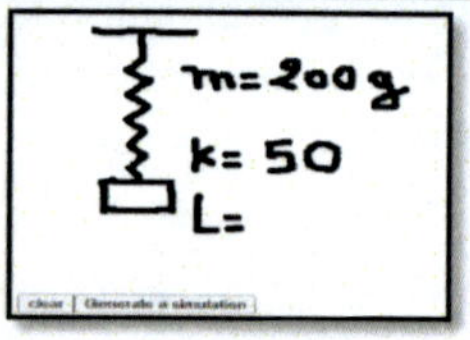	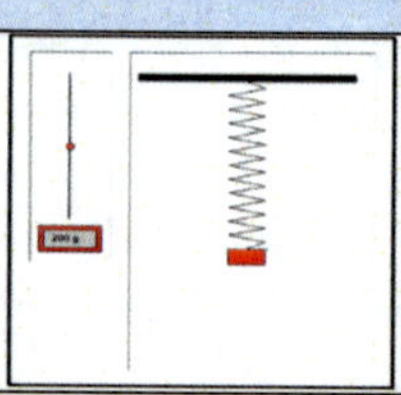	Spring is characterized by L and K and the mass m. The position of an object of mass m. Attached to its other ends M and assumed to be punctual at this point. Is identified by its obscure x (T). At the instant t = 0 the object is moved away from a distance a downwards a> 0 and released without initial speed.
calorimeter	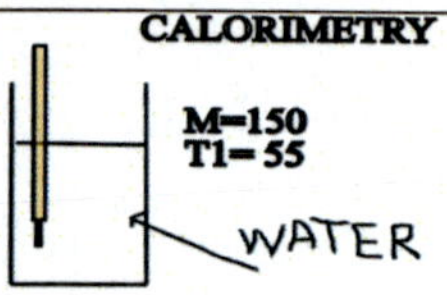	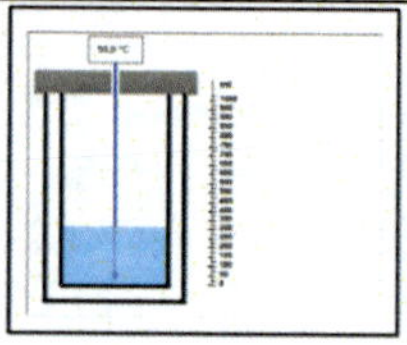	This manipulation is to measure the specific heat or heat capacity of the element and its accessories, of a liquid. For this purpose, a mass of cold water (approximately 50 ml) is poured into the calorimeter, of water value k. Stirred and the temperature t1 is recorded. After we add A mass of hot water (approximately 60 ml) of temperature t2 of between 55 and 60 ° C.
Pendulum	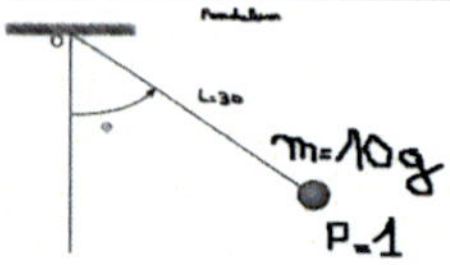	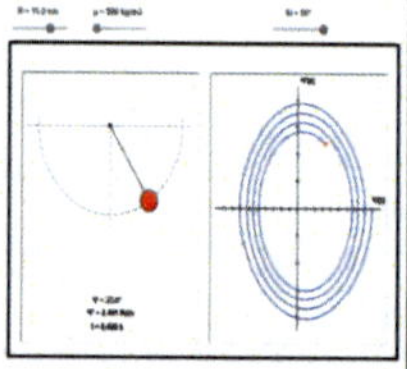	The pendulum consists of a point mass m with a weight P at the final of a rod without a mass of length L, which can pivot freely around its upper end. The question is to determine the natural frequency of oscillation and influence of this pendulum. The mass position is identified by the angle φ between the vertical and the direction of the wire.
Trajectory & acceleration	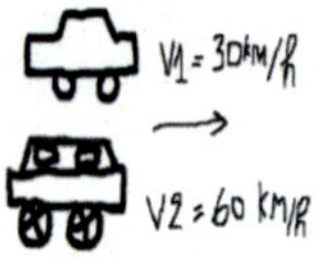		This animation illustrates the following concepts: trajectory, rectilinear translation, average speed, acceleration. The speed of movement is specified by the user in the sketch

Table 2. Evaluation performance

	Simulations	Precision%	Recall%	Accuracy%	F1-Score %
Use the mouse On Pc	Spring	96,00	82,76	80,00	88,89
	calorimeter	92,86	92,86	86,67	92,86
	pendulum	96,30	89,66	86,67	92,86
	movement of Wheel	96,55	96,55	93,33	96,55
	Rotation of objects	96,67	98.54	96,67	98,31
	Ball	89,29	92,59	83,33	90,91
using the stylus of the graphic tablet	Spring	96,55	96,55	93,33	96,55
	calorimeter	89,47	94,44	85,00	91,89
	pendulum	98,28	96,61	95,00	97,44
	movement of Wheel)	98,31	98,31	96,67	98,31
	Rotation of objects	94,92	98,25	93,33	96,55
	Ball	99,16	99,16	98,33	99,16

Discussion

Our contribution deals with the recognition of shapes, texts, and numeric values. The system utilizes the text and values to provide more information in order to make a better representation of the sketch. The processing step is used to add meaning and context to identify elements, their properties, associated values for each parameter. It is also used to provide the desired service. Moreover, our prototype is deconstructed and moduled for a) the Creation of customized simulations b) adding more services and improvements to each step in isolation. Our approach is the first contribution in this field that applies deep learning to generate an interactive simulation code. The efficiency and power of our approach stem from the combination methods between sketch analysis (recognition/detection of shapes, optical character recognition) and the use of the laboratory as a service. It should be mentioned that this approach is applicable for all areas such as (chemistry, optics, electricity…) These results are interpreted for the following reasons:

- The bad recognition of the values affects the configuration of the experience.
- The bad detection of the elements in the sketch affects the regeneration of the inappropriate services with the need of the user. It generates another element that does not exist in the sketch; especially when there are elements that resemble each other in appearance.
- CSS Color of our simulation is not customized in our prototype. We have used the default colors.

5 Conclusions and Future Work

In this work, we proposed a novel architecture called "sketch to simulation" based on CNN and RNN combined with allowing to generate a simulation in a virtual laboratory from a sketch of an experiment as an input. Our approach was trained, tested, evaluated on relative datasets quick, draw! and IAM. Our results illustrate that the proposed approach can be employed to achieve important performance for the regeneration of a code from the sketch. The future work is to create a dataset that concerns only scientific labs and also, we try to evolve our model to prolong to create a complete software based on our dataset and to create online laboratories and regenerate interfaces web-based on the intelligent screenshot.

References

1. Landay, J.A., Myers, B.A.: Interactive sketching for the early stages of user interface design. In: Proceedings of the SIGCHI Conference on Human Factors in Computing Systems, pp. 43–50. ACM Press Addison-Wesley Publishing Co. (1995)
2. de la Torre, L., Sanchez, J., Andrade, T.F., Restivo, M.T.: Easy creation and deployment of javascript remote labs with EjsS and moodle. In: 13th International Conference on Remote Engineering and Virtual Instrumentation (REV), pp. 260–261. IEEE (2016)
3. Wang, Z., Shang, Y., Liu, J., Wu, X.: A LabVIEW based automatic test system for sieving chips. Measurement **46**, 402–410 (2013)

4. Hearst, M.A., Gross, M.D., Landay, J.A., Stahovich, T.F.: Sketching intelligent systems. IEEE Intell. Syst. **13**(3), 10–19 (1998)
5. Lecun, Y., Kavukcuoglu, K., Farabet, C.: Convolutional networks and applications in vision. In: Proceedings of IEEE International Symposium, vol. 20, 10, pp. 253–256 (2010)
6. Sun, Y., Wang, X., Tang, X.: Deep convolutional network cascade for facial point detection. In: IEEE Conference on (CVPR), pp. 3476–3483. IEEE (2013)
7. Lecun, P.S.: Convolutional neural networks applied to housenumbers digit classification. In: 21st International Conference on (ICPR), pp. 3288–3291. IEEE (2012)
8. Liu, W., et al.: SSD: Single shot multibox detector. In: Leibe, B., Matas, J., Sebe, N., Welling, M. (eds.) ECCV 2016. LNCS, vol. 9905, pp. 21–37. Springer, Cham (2016). https://doi.org/10.1007/978-3-319-46448-0_2
9. Howard, A.G., et al.: MobileNets: efficient convolutional neural networks for mobile vision applications (2017)
10. Tawfik, M., et al.: Laboratory as a Service (LaaS): a novel paradigm for developing and implementing modular remote laboratories. Int. J. Online Eng. **10**(4), 13–21 (2014)
11. Tobarra, L., et al.: Creation of customized remote laboratories using deconstruction. IEEE Revista Iberoamericana de Tecnologias del Aprendizaje **10**(2), 69–76 (2015)
12. Beltramelli, T.: pix2code: generating code from a graphical user interface screenshot. In: Proceedings of the ACM SIGCHI Symposium on Engineering Interactive Computing Systems, p. 3. ACM (2018)
13. Nguyen, T.A., Csallner, C.: Reverse engineering mobile application user interfaces with remaui. In: Automated Software Engineering, 30th IEEE/ACM International Conference, pp. 248–259 (2015)
14. Nagy, G., Nartker, T.A., Rice, S.V.: Optical character recognition: an illustrated guide to the frontier. In: Document Recognition and Retrieval VII, vol. 3967, pp. 58–70. International Society for Optics and Photonics (1999)
15. Robinson, A.: sketch2code: Generating a website from a paper mockup. arXiv:1905.13750v1 [cs.CV], 9 May 2019
16. Jain, V., Agrawal, P., Banga, S., Kapoor, R., Gulyani, S.: Sketch2Code: transformation of sketches to UI in real-time using deep neural network. arXiv:1910.08930v1 (2019)
17. Cheema, S., LaViola, J.J., Jr.: Using animation to enrich learning experience in sketch-based physics tutoring systems. In: Kapros, E., Koutsombogera, M. (eds.) Designing for the User Experience in Learning Systems. HIS, pp. 201–227. Springer, Cham (2018). https://doi.org/10.1007/978-3-319-94794-5_10
18. Jongejan, J., Rowley, H., Kawashima, T., Kim, J., Fox-Gieg, N.: The Quick, Draw! - A.I. Experiment (2016). https://quickdraw.withgoogle.com/
19. Marti, U.-V., Bunke, H.: The IAM-database: an English sentence database for offline handwriting recognition. Int. J. Doc. Anal. Recogn. **5**(1), 39–46 (2002)

Unsupervised Deep Learning Model Based on Autoencoders for Cancer Classification

Chahrazed Mediani[1(✉)] and Yassamina Mediani[2]

[1] LRSD Laboratory, Ferhat Abbas Sétif-1-University, 19000 Sétif, Algeria
chahrazed.mediani@univ-setif.dz
[2] MISC Laboratory, Abdelhamid Mehri-Constantine2 University, Constantine, Algeria
yassamina.mediani@univ-constantine2.dz

Abstract. Genomics and proteomics activities are the fundamental controllers of the living processes of any organism. They carry out most of the work, perform most life functions, and make up the majority of cellular structures. Moreover, their abnormal behaviors are the driving factors of many genetic diseases. Therefore, understanding these activities has become a fundamental mission of large-scale genomics research. However, this study field has some limitations, such as the sensitivity, the high dimensional, and the large scale of data. That is why it is a challenging application area of deep learning. However, this last is more effective when one wants to make sense of large and complex data sets. This paper uses a deep learning method based on autoencoders to solve a classification problem, which is cancer detection using two distinct annotated data sets: the microRNA data set and the reverse phase proteins array. The results were promising and proved that our deep integrative model was superior in the Cancer Type and the PFI score classification and prediction with high accuracy.

Keywords: Machine learning · Deep learning · Autoencoders · Cancer · miRNA · RPPA · TCGA

1 Introduction

Today, medical technology advances have provided explosive volumes of biological, physiological, and clinical data [1]. They have accelerated the accumulation of massive amounts of omics data from multiple sources, such as the genome, epi-genome, transcriptome, proteome, and metabolome. For example, in the stratification of cancer patients based on their DNA methylation, miRNA expression, and protein expression measurements, the number of variables can be substantially higher than the number of samples (thousands of variables measured on just a few hundred patients) [2]. It is the era of massive data, including billions or trillions of data points. Learning from these data facilitates understanding the human body, health, and disease [3]. However, it has a certain risk level because this data is sensitive information, and a small mistake can be paid with life.

Genomics and proteomics activities controlled the living processes of any human organism. In the recent decade, high-throughput molecular data have provided abundant

S. Motahhir and B. Bossoufi (Eds.): ICDTA 2022, LNNS 454, pp. 229–238, 2022.
https://doi.org/10.1007/978-3-031-01942-5_23

information about the whole genome and has become an important research direction to study genomics from these data by deploying computational tools. However, traditional computational learning methodologies often have many limitations when dealing with high-throughput genomic data sets because of the large scale of data with a very high dimensional and sensitivity.

Deep learning (DL) and machine learning that uses neural networks to extract new features from input data are promising approaches now being applied in the genomics field. Genomics is a challenging application area of deep learning that entails unique challenges compared to other fields. However, the DL is most effective when one wants to make sense of large and complex data sets. Deep learning methods are likely to become more significant to genomics as more large datasets become available. DL has been developed from artificial neural networks. It refers to a collection of new techniques that, together, have demonstrated breakthrough gains over existing machine learning algorithms across several fields [4]. It takes rows features from extremely large, annotated data set, such as a collection of genomes, and uses them for creating a predictive tool based on these patterns.

The autoencoder is one of the DL algorithms that have shown great promise in extracting features and learning patterns to complex and large data. It is defined as an unsupervised learning neural network used to learn data unsupervised coded. Its aim is the dimensionality reduction of data representation. Autoencoder reconstructs its input by applying the principle of back-propagation. Moreover, it usually represents data through multiple hidden layers such that the target values are set as close to inputs. It aims to learn efficient data coding in an unsupervised environment. The main task of autoencoders is to identify regular data and outliers or anomalies. A deep autoencoder is composed of significant steps. The first one is the encoding of the input to identify latent feature representation. Then, the second step is decoding the latent feature representation to recreate the input vector. After, it calculates the loss by comparing the input and the output vectors and uses the back-propagation to reduce the reconstruction error and update the weights, which are updated based on how much they are responsible for the error [5, 6].

This paper investigates deep learning methods and their applications. The main focus has been classifying cancer using Deep Autoencoders (DAE) and analyzing the performance. To further investigate the performance and capability of the DAE, two data sets consisting of miRNAs and RPPAs of human cancer samples were used. This work investigated whether DAE can be trained to integrate genomics data sets to help cancer diagnosis and classification. In this model, we have applied unsupervised deep learning based on autoencoders. We have also explored the integration of multi-source genomic data sets, dimensionality reduction, and powerful performances in handling non-spatial data sets. In the following, we will state our problem and present existing work related to machine and deep learning in cancer gene expression analysis. In Sect. 3, we detail the data collection and preprocessing. In Sect. 4, We illustrate the proposed model and a set of experimentation. Finally, We conclude before advancing prospects for this work.

2 Deep Learning and Cancer Data Analysis

The big advances in biological and medical technologies have provided explosive volumes of biological, physiological, and clinical data, such as genomics, protein sequences, and different medical images. It is the era of big data, and learning from this data facilitates understanding the human body, health, and disease. However, transforming big biomedical data into valuable knowledge has been one of the most critical challenges in bioinformatics. Many studies for predicting cancer prognosis risk have used single omics data. The gene expression measured by microarray is among the most frequently used data [7, 8]. With the recent advance of next-generation sequencing techniques, other types of genomic data have been used, like DNA methylation [9], miRNA, and copy number variation (CNV) [10]. Each type of omics data only represents a single view for patients. A global view of patients in genomics is obtained by integrating multiple omics data from the same patient [11, 12]. Many studies have used The Cancer Genome Atlas (TCGA), which provides more than ten thousand samples over 33 cancer types [13, 14]. The integrated analysis based on multiple omics data is used to analyze cancer prognosis comprehensively.

Chen et al. [15] have used three data sets: the GEO expression data, the GTEx expression data, and the Genomes expression data. They have presented their deep learning method for gene expression inference D-GEX and two methods, linear and k-nearest neighbor regression. Furthermore, they showed the predictive performances of the three methods on both the GEO data and the GTEx data. They concluded that deep learning provides a better model than linear regression for gene expression inference.

Chaudhary et al. [16] have employed a deep learning-based model to identify and integrate multi-omics features linked to the differential survival of hepatocellular carcinoma (HCC) patients by using clinical information, RNA-seq data, miRNA-seq data, and DNA methylation data from TCGA data. They have chosen the autoencoder framework to implement DL for multi-omics integration. They have used the reduced new features to cluster the samples using the K-means clustering algorithm, and they have built a supervised classification model using the SVM algorithm. Finally, they have compared the performances of the deep learning framework with two alternative approaches: the PCA analysis and the K-means procedure. The results were clear and satisfactory to the success of the study.

Hu et al. [17] have developed a deep learning method to diagnose malignant mesothelioma automatically (MM). They have designed and implemented a diagnostic model based on stacked sparse autoencoder (SSAE) algorithms, and they have used the Genetic Algorithm (GA) and the ReliefF algorithm to select high discriminative features. To diagnose the model for MM and evaluate its performance, they have applied three different algorithms, with or without feature selection, on the same training data set. The comparison of their algorithms performances had shown that the SSAE and GA + SSAE models exhibited the highest overall performance; the accuracy was 100%, and it was for SSAE+ ReliefF 96.90%.

Wong et al. [18] have presented a trained deep learning model to discover prognostic gene features in glioblastoma for patients' survival, using tumor gene expression data from TCGA, Microarray data, and clinical data. They built a deep multilayer perceptron network to predict patient survival risk. They compared the prediction performance of the deep learning method with other machine learning methods. As a result, they have concluded that the deep learning survival prediction model learned genes that are strongly related to glioblastoma stem cells and treatment-resistant, which may be helpful to inform patient therapy.

Zeng et al. [19] have used autoencoders' neural network strategy to create features reduced for similarity measurement and computing correlation between cancer and reference tissue expression. Similarly, they have performed principal component analysis (PCA) of these data sets using the RNA-Seq data. As a result, the autoencoder performed the best in predicting cancers among the three methods.

3 Data Collection and Preprocessing

We collected two distinct annotated data sets from The Cancer Genome Atlas TCGA[1] (Table 1): the TCGA-RPPA-pan-cancer and the TCGA-miRNA data sets. Both of them have been used for unsupervised training and supervised classification. The RPPA data set (date 09-01-2019) comprises 7789 samples collected from patients admitted in the United States hospitals with different cancer types.

- The RPPA data set was measured using Reverse Phase Protein Arrays, representing an innovative, precise technology that allows studying protein expression levels and protein modifications. This data set contains the protein expression series matrix and various patients' information such as patient SampleID or Barcode, tumor type, and different proteins.
- The second data set is miRNA (date 10-12-2020). It comprises 742 microRNAs, representing small endogenous non-coding RNAs that function as the universal specificity factors in post-transcriptional gene silencing, and 100 cancer samples extracted from 10824 individuals. All series matrixes were generated by using a microRNA profiling assay.

Table 1. Data description

Name	TCGA-RPPA	TCGA-miRNA
Description	Reverse phase protein arrays data set	microRNA expression data set
Use	Cancer classification	Cancer classification
Number of samples	7789 patients	10824 patients

The preprocessing steps contain:

- **Data curation**: the first step was processing, organizing, and cleaning our genetic data. We filtered our two data sets by deleting patients with a missing PFI.1 value:

[1] https://gdc.cancer.gov/about-data/publications/pancanatlas.

the progression-free interval event. The PFI.1 value is set to one for a patient having a new tumor event; zero for censored otherwise. This last has been extracted from the supplementary file TCGA-CDR-SupplementalTableS1, which contains supplementary patient information such as patients Barcode, cancer type, PFI.1, DSS. The second step was the elimination of the genes in the miRNA data set and proteins in the RPPA data set with a lot of missing values. We deleted those having zeros in more than 30% of the rows. Table 2 shows what the data sets look like after deleting the patients without the characteristic of PFI.1.

- **Imputation of missing values**: We replaced the missing mRNA and RPPA expression values with the mean score in this step. After that, we applied the Log transformation on the whole date. Then, we normalized our data by using the Z score normalization technique.
- **Splitting the data set**: we split the data into two random samples, 80% for training and 20% for testing.

Table 2 displays the data set description before and after preprocessing.

Table 2. PFI data description

Data set	Number of all the primary samples	Number of samples having PFI.1 value	Deleted values
RPPA data set	7789	7140	679
miRNA data set	10824	7140	3714
Labels data set	11160	7140	4050

4 Proposed Architecture

We have applied unsupervised deep learning based on auto-encoders in the proposed architecture to integrate two multilevel genomics data: the short non-coding regions miRNA and proteins RPPA (Fig. 1). Our architecture is based on three phases: (1) the dimensionality reduction process. We have used deep auto-encoders to reduce the miRNA data's dimension, omit the noisiness of the inputs, and reduce the error cost in the second phase. (2) We have integrated the reduced data obtained from the first phase with the RPPA data set by using another autoencoder to produce new features representation. (3) The classification phase in which the new features space and clinical data have been applied as a training set to train a supervised classifier to predict prognosis and diagnosis outcomes.

4.1 Dimensionality Reduction of miRNA Data

The tackled dimensionality reduction problem is formulated as follows: Given a matrix X of NxM dimensions, where N represents the number of samples and M represents the number of features. Each element X_{ij} represents the value of the non-coding region miRNA *j* corresponding to sample *i*. Since we applied an unsupervised dimensionality reduction, we ignored the labels of samples. In this analytical phase, the first encoder takes the high dimensional matrix *X* of range *M* as the input vector E_1. Then, it progresses E_1 through successive layers until it reaches the bottleneck layer. This last represents the latent space F_1 of range *K*, where $K < M$.

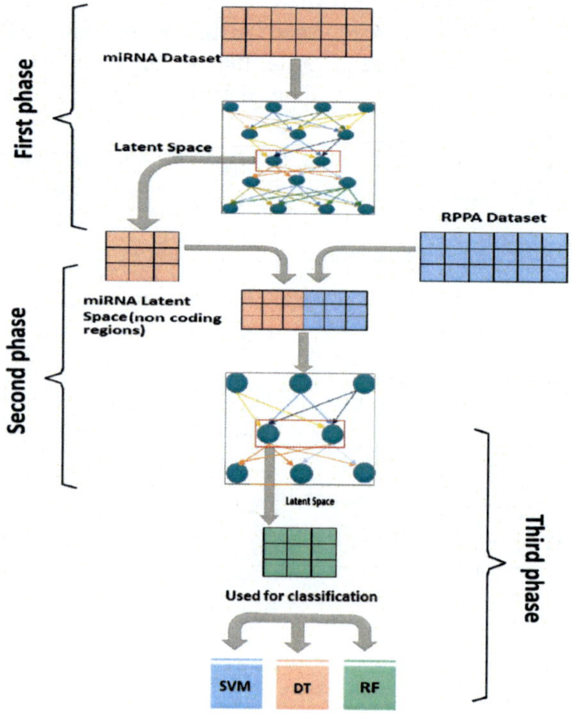

Fig. 1. Analytical architecture for multilevel genomic data integration using deep autoencoders.

The bottleneck layer fed its output to the decoder, which produces a new reconstructed output E_1^0 of range *K*. However, our main objective is not the new output; instead, it is the latent space F1, the new reduced features representation. The consistency of the model and mainly the F_1 features vector have been evaluated through the reconstruction loss error, where we compared the similarity between the input vector E_1 and the reconstructed output vector E_1^0. In our case, we have applied the *mean_absolute_error* (Eq. 1) to evaluate the consistency of the F_1 vector.

$$MAE = \frac{1}{n}\sum\nolimits_{i=1}^{n}|x_i - x| \quad (1)$$

So, the dimensionality reduction deep autoencoder, which is used to train the whole of miRNA samples, is built as follows: Three fully connected layers to represent the encoder (291, 200,100), the fully connected bottleneck layer comprises 50 nodes representing the reduced space F_1, and three fully connected layers to represent the decoder (100, 200, 291). We have used the nonlinear *Softplus* and *Relu* activation functions to update the layers' weights.

4.2 Integrative Autoencoder

We integrated the two data sets in this phase: the extracted features vector F1 from the first phase with the RPPA data set matrix. We have taken the two data sets as one input matrix X0 of range NxL where L = T + K in which T is the number of proteins in the RPPA data set and K is the new space generated from the first phase. Following the same steps in phase one, we fed the whole features vector of X0 to the encoder E2. Moreover, a new latent features space F2 is generated in the bottleneck layer. It integrates the two types of data sets. Equally, F2 is fed to a decoder D2 to reconstruct the input features vector. To evaluate the consistency of the model, we have used the measurement mean_abselute_error loss function (Eq. 1).

The second Autoenceder E_2 is built as follows: one fully connected layer representing each of the encoder and the decoder, the bottleneck fully connected layer composed of 50 nodes further representing the reduced space F_1, which will be used in the third phase. We have used the nonlinear Softplus and Relu activation functions to update the layers' weights.

4.3 Classification and Overall Evaluation

In this supervised classification phase, we have used the selected features vector F2 obtained from the second phase, and we have trained supervised classifiers, namely support vector machines (SVM), random forest (RF), and decision trees (DT). In addition, we tested the deep model's discrimination consistency of our analytical model in this step. First, we selected and integrated the miRNA and RPPA data sets. Then, we compared the classifiers outcomes trained on the features provided by our proposal with those, which are provided by famous unsupervised dimensionality reduction and features integration models. Principal Component Analysis (PCA) and Kernel Principal Component Analysis (KPCA) methods are used. In the first phase, each of these methods was used to perform the dimensionality reduction and extract a new miRNA space with 50 features. And then, the new features have been added to the RPPA data set. The second phase is furthermore executed to extract the second vector of features. Finally, we assessed the performance of our classifiers by using hold-out cross-validation and splitting our data set to 80% of samples for the training and 20% for the testing sample. To evaluate the performance of the classifiers along with the dimensionality reduction and integrative phase, we trained all the classifiers to predict the diagnostic of a prognosis outcome, mainly Cancer type and Progression-free interval 1 PFI.1.

5 Experimental Results

We present our experiments and evaluation steps undertaken in this model to test the effectiveness of our proposed model. We have used MiRNA and RPPA data sets proposed in The Cancer Genome Atlas TCGA. We have deployed the Keras open-source deep learning framework with Tensorflow backend to build and train the unsupervised deep learning model based on autoencoders. Our experiments were run on a DELL Intel(CORE i5) with Processor Intel(R) Core(TM) i5-4310U CPU @ 2.00 GHz 2.60 GHz and RAM 4.00 GB.

In our model, we have applied three training phases, which are the unsupervised phase, the integrative phase, and the supervised training phase. In the first phase model, we have trained the first autoencoder E1 on the whole miRNA data set. After the unsupervised training, the second autoencoder *E*2 is trained to extract the most relevant features in the new data sets miRNA and RPPA. As a result, we found our models converged to their best performances following the parameters illustrated in Table 3. Figure (Fig. 2) shows the training loss function of the two phases.

Table 3. Phase 1 and phase 2 parameters. E1: represents the first phase and E2 the second.

Parameters	E1 values	E2 values
Epochs	100	80
Batch size	150	250
Optimizer	Adam	
Loss function	Mae	
Activation function	Relu-Softplus	
Reconstruction error	**0.666**	**0.622**

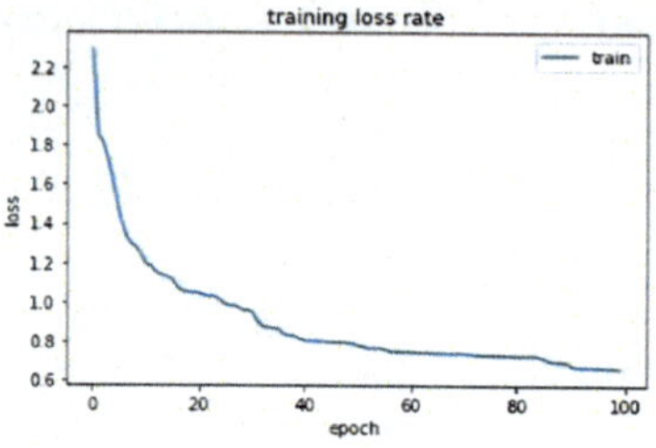

(a) Training loss model of the first phase

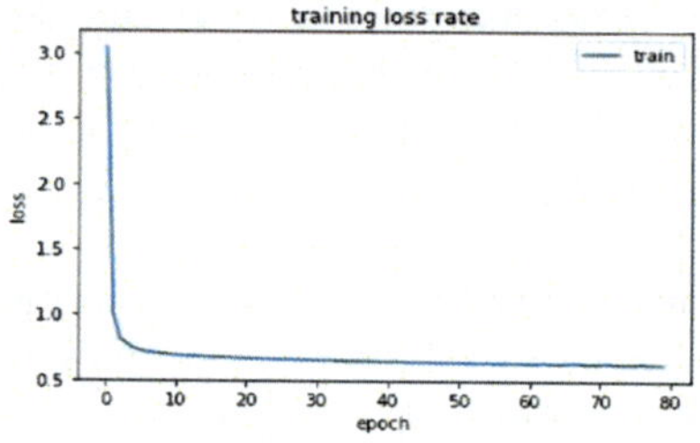

(b) Training loss model of the second phase

Fig. 2. Training loss model

5.1 Evaluation and Results Discussion

We used the accuracy metric to evaluate the overall performance of the models, where the rows represent the classifiers, and the columns represent the features extraction models. The results prove that our integrative model was superior in predicting the Cancer Type and the PFI score. Our analytical feature representation learning reached an accuracy of 94% in predicting cancer type, where training on the SVM classifier compared with a range of [63%–89%] in PCA and KPCA trained classifiers (Fig. 3 left). The DT/PCA classifier overcame our model with a difference of 0.02. Also, the RF/PCA classifier scored the same accuracy rate as our proposal. In the PFI prediction, our proposal model conquered all the other compared models with an accuracy of 71% (Fig. 3 right). However, we can enhance this score by collecting PFI class2 samples.

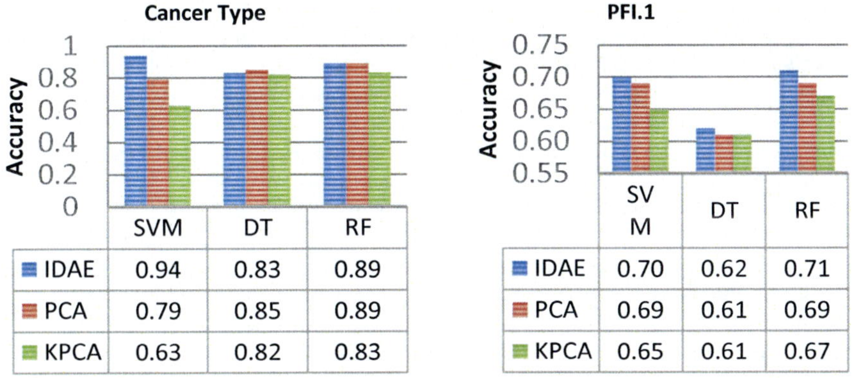

Fig. 3. The predicting model accuracy of the Cancer Type (left) and the PFI score (right).

6 Conclusion

This paper has applied unsupervised deep learning based on autoencoders for features selection and dimensionality reduction to integrate multilevel genomic miRNA and RPPA (the reverse phase proteins array) data sets. To solve the classification problem of cancer detection, we have used the two distinct annotated data sets.

First, we have explained our methodology by giving an overview of the data sets' cleaning and preprocessing. Then, we have described our deep autoencoder proposed architecture passing through its three phases. Finally, we have given an overview of our experiments and discussed our results. The results were promising and proved that our deep integrative model was superior in Cancer Type and the PFI score classification and prediction with high accuracy. Much work is still to be done. This approach for extracting information from miRNA and RPPA data sets can give researchers a direction for investigation and help them diagnose and treat cancer.

In future works, we could include more neurons (layer) in our contribution algorithms to assess which of them produces the most biologically accurate representation. Since in our work, we dealt with the two last genomic levels: the short non-coding regions miRNA and proteins RPPA, we could integrate the first level, which is the level of the genes, using deep learning methods. Finally, we could perform the classification based on more than PFI and cancer type. We could use the cancer progression as a classifier, the age, and the gender of the patient.

References

1. Cao, C., et al.: Deep learning and its applications in biomedicine. Genom. Proteom. Bioinform. **16**(1), 17–32 (2018)
2. Mirza, B., Wang, W., Wang, J., Choi, H., Chung, N.C., Ping, P.: Machine learning and integrative analysis of biomedical big data. Genes **10**(2), 87 (2019)
3. Tan, L., Ong, K.: The impact of medical technology on healthcare today. Hong Kong J. Emerg. Med. **9**(4), 231–236 (2002)
4. Ching, T., et al.: Opportunities and obstacles for deep learning in biology and medicine. J. R. Soc. Interface **15**(141), 20170387 (2018)
5. Deng, L., Yu, D.: Deep learning: methods and applications. Found. Trends R Sig. Process. **7**(3–4), 197–387 (2014)
6. Goodfellow, I., Bengio, Y., Courville, A.: Deep learning. MIT Press, Cambridge (2016)
7. Hao, J., Kim, Y., Kim, T.-K., Kang, M.: PASNet: pathway-associated sparse deep neural network for prognosis prediction from high-throughput data. BMC Bioinform. **19**, 510 (2018)
8. Sun, D., Wang, M., Li, A.: A multimodal deep neural network for human breast cancer prognosis prediction by integrating multi-dimensional data. IEEE/ACM Trans. Comput. Biol. Bioinform. **16**, 841–850 (2019)
9. Liu, B., Liu, Y., Pan, X., Li, M., Yang, S., Cheng Li, S.: DNA methylation markers for pan-cancer prediction by deep learning. Genes (Basel) **10**(10), 778 (2019)
10. Li, Y., Wu, F.X., Ngom, A.: A review on machine learning principles for multi-view biological data integration. Brief Bioinform. **19**(2), 325–340 (2018)
11. Chai, H., et al.: Integrating multi-omics data with deep learning for predicting cancer prognosis. In: Computers in Biology and Medicine, vol. 134, July 2021
12. Kim, M., Tagkopoulos, I.: Data integration and predictive modeling methods for multi-omics datasets. Mol. Omics **14**(1), 8–25 (2018)
13. Huang, Z., et al.: SALMON: survival analysis learning with multi-omics neural networks on breast cancer. Front. Genet. **10**, 166 (2019)
14. Chaudhary, K., Poirion, O.B., Lu, L., Garmire, L.X.: Deep learning-based multi-omics integration robustly predicts survival in liver cancer. Clin. Cancer Res. **24**, 1248–1259 (2017)
15. Chen, Y., Li, Y., Narayan, R., Subramanian, A., Xie, X.: Gene expression inference with deep learning. Bioinformatics **32**(12), 1832–1839 (2016)
16. Chaudhary, K., Poirion, O.B., Lu, L., Garmire, L.X.: Deep learning-based multi-omics integration robustly predicts survival in liver cancer. Clin. Cancer Res. **24**(6), 1248–1259 (2018)
17. Hu, X., Yu, Z.: Diagnosis of mesothelioma with deep learning. Oncol. Lett. **17**(2), 1483–1490 (2019)
18. Wong, K.K., Rostomily, R., Wong, S.T.: Prognostic gene discovery in glioblastoma patients using deep learning. Cancers **11**(1), 53 (2019)
19. Zeng, W.Z., Glicksberg, B.S., Li, Y., Chen, B.: Selecting precise reference normal tissue samples for cancer research using a deep learning approach. BMC Med. Genom. **12**(1), 21 (2019)

Internet of Things, Blockchain and Security and Network Technology

Fighting Fake News Propagation Using Blockchain in Journalism: A Systematic Literature Review

Meryem Chabini[1(✉)], Khadija Sabiri[2], Abdessadek Aaroud[1], and Khalid Akodadi[1]

[1] Laboratory LAROSERI, Department of Computer Science, Faculty of Science, Chouaib Doukkali University, El Jadida, Morocco
meryemchabini123@gmail.com, Aaroud.a@ucd.ac.ma

[2] Fraunhofer Portugal AICOS, Rua Alfredo Allen, 455/461, 4200-135 Porto, Portugal
Khadija.sabiri@aicos.fraunhofer.pt

Abstract. The challenge facing journalism these days in the Internet environment is greater than ever. The public no longer believes in traditional media in light of fake news and infected news. This is possibly due to the emergence of many diverse sources of conflicting information via social media. The central intention of this article is to present a review of previously conducted studies related to the use of the blockchain to fight against the proliferation of fake news to acquire knowledge on the results obtained in this context. The literature review includes articles published between 2017 and 2021. This review shows that despite the published research opportunities, there is not a standard deployable framework in the journalism ecosystem.

Keywords: Blockchain · Fake news · Journalism · Literature review

1 Introduction

In the era of the Internet, the media are witnessing a huge change and have placed traditional media in the face of great challenges. This may be since the Internet is an open space that does not respect the restrictions to which traditional media are subject.

Nowadays, social media has replaced traditional news sources and has become the most popular news-sharing site in recent years. However, not all information disseminated on social media is accurate and some of it is from the unverified source.

Fake News then is a big problem in today's world, especially when it comes to the field of journalism, whether online or offline.

It is undeniable that the issue of Fake News is real and it is not a new phenomenon but its impact on the world and people is damaging and, in some cases, fatal. There have been many cases that have had unintended and even tragic consequences. We have cited some heavily reported fake news that has had unintended and even fatal outcomes:

(1) Fake news about the 2016 US presidential election, which inevitably misleads voters' decisions and influences election results. She showed the impact of digital transformation on democracy and politics [2].

S. Motahhir and B. Bossoufi (Eds.): ICDTA 2022, LNNS 454, pp. 241–255, 2022.
https://doi.org/10.1007/978-3-031-01942-5_24

(2) Fake news led to the lynching of 2 bikes in India in 2017 [3].
(3) Fake news caused riots in the Indian state of Kolkata in 2017 [4].
(4) Fake news led Dylann Roof to murder nine people [5].
(5) Fake news which caused a loss of about Rs. 500 crores (RM 286,087,500.00) for Kalyan Jewelers [6].

The list doesn't end there; fake news has caused several deaths in countries like India, the United States, Kenya, and some countries, it has sparked violent protests and riots. The impact of fake news in the future is something that needs to be addressed as we do not know how damaging fake news can be and can have a devastating and lasting impact on society if left unchecked [7].

Journalism is in a state of considerable flux. The ambition of the journalism industry is to be the arbiter of information. This makes transparency in journalism currently a key element in exercising serious and rigorous journalism. If the emerging Blockchain technology can store almost any type of data that requires to be secured and can be viewed and edited by many people, then it is a potential solution to many problems. The journalism industry is one of the case studies for securing the confidence of its readers in the face of fake news and infected news.

The central intention of this article is to present a systematic review of previously conducted studies related to the use of blockchain for the prevention of Fake News and this had shown that the journalism industry has not been well explored in this context.

In this review, the research questions investigated are as follows:

RQ1: How can we trust the media sources or the authenticity of the news published in the media?
RQ2: Can Blockchain protect journalism and push against threats of censorship and Fake News proliferation?
RQ3: How Blockchain can preserve the credibility of media content?
RQ4: How to assess the credibility of a media in journalism?

The rest of this paper is organized as follows. In Sect. 2, we describe the blockchain technology and its implications on journalism; Sect. 3 shows methodology adopted for the literature review; Sect. 4 details the articles we established for this review; Sect. 5 discusses the applicability of blockchain in the field of journalism and comparison of the discussed studies with the four RQs. Finally, Sect. 6 provides some conclusions and future works.

2 Blockchain and Its Implications in Journalism

A blockchain is a form of distributed and immutable ledger technology for storing and transmitting information in a transparent, secure, and working without a central control body or intermediary institution.

Blockchain technology is in the early stages of development, and the blockchain definition recognized by the industry has not been formed. However, the financial industry is seen as a primary user of the concept of blockchain. This is not only because the best-known application of this technology is the bitcoin cryptocurrency. Bitcoin is a cryptocurrency based on a distributed design.

Blockchain as a technology in the media industry is in its development phase. Even today, there are many efforts to apply blockchain to media in several ways; for example, Steem, which pursues blockchain-based SNS (Social Networking Service) systems without censorship; system Po.et, which attempts to secure the ownership of digital creative assets through a blockchain in the internet environment; and the civil blockchain-based journalism platform using the Ethereum token-based encryption currency called CVL.

There are two main features of blockchain technology that are important for journalism implementation:

- Secure, verified, and reliable information exchange via the blockchain in real-time which makes them accessible to all concerned member parties.
- Possibility of verification and automatic execution of agreed transactions when certain requirements are met through smart contracts.

Based on these main features of blockchain, the areas of implementation of its use in journalism are developing in various directions. Some of the most significant current implementation areas in journalism: tracking the origin of the media content with one click, proving the authenticity of a news item, reducing the risk of fake news proliferation, direct tracking the editorial processes of information, and verification of the veracity and credibility of news.

3 Methodology

This study aims to review the published literature to discover articles related to our research context. To do this, a systematic literature review was done. We search for publications in this context using related keywords such as the following: Blockchain AND "Fake News". The Scopus, Web of Science, and PubMed databases were used to find articles describing the use of blockchain against the proliferation of Fake News. Figure 1 shows the flowchart for the selection of studies.

Initially, 110 articles were obtained from the string query (TITLE-ABS-KEY (Blockchain and "Fake News")) of which 58 potentially relevant articles are identified and 39 articles are excluded by studying the titles and summaries because most of them were irrelevant. In addition, a representation of the selection funnel of initially found results has been made in terms of the following characteristics: year (from 2017); keyword ('Blockchain', 'Fake News'); Subject Area (Computer Science); Document Type (Conference Paper, Article, Conference Review); Language (English).

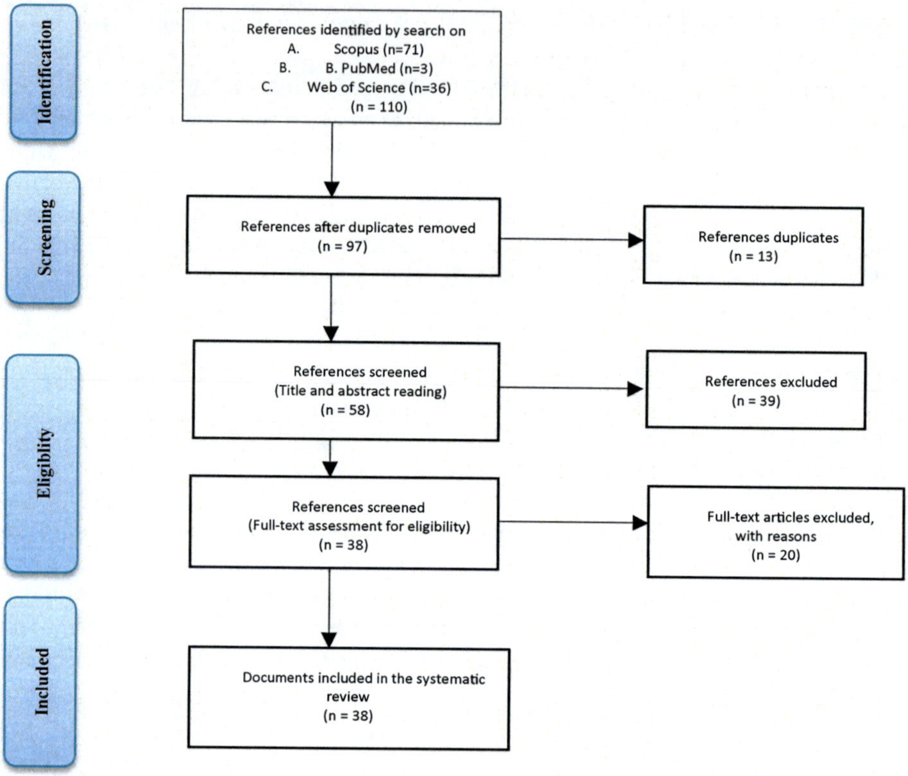

Fig. 1. Flow chart for selection of articles

4 Literature Review

Major blockchain research to combat fake news began in 2017. We can see that there is an exponential increase with some 12 papers in 2019 compared to 2018 published papers, indicating an explosive growth trend in this area of research which is on the escalate for the years 2020 and 2021 (see Fig. 2).

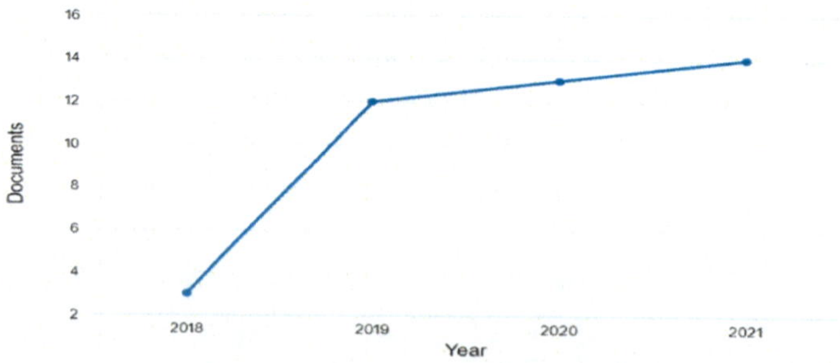

Fig. 2. Documents by year classification.

In total, 38 documents are included in our review. But we applied a criterion for choosing high-quality articles in our review of the literature by applying a criterion based on the number of citations for each of them. We made a limitation on the number of citations in descending order up to the number 5.

Table 1 shows the distribution of selected articles by publishers, publication year, and publication types.

Table 1. Distribution of selected articles by publishers.

Publisher	Paper reference	Year	Publication type	Name of Journal/Conference
Elsevier	[7]	2020	Journal	Information Processing and Management
IEEE	[8]	2019	Journal	IT Professional
	[9]	2019	Conference Paper	Communication and Network Security, CNS 2019
	[10]	2019	Conference Paper	7th International Conference on Smart Computing & Communications, ICSCC 2019
	[11]	2018	Conference Paper	17th International Conference IEEE/ ACIS on Computer and Information Science
Springer	[12]	2019	Book Chapters	Communications in Computer and Information Science
	[13]	2019	Book Chapters	Advances in Intelligent Systems and Computing
Mary Ann Liebert	[14]	2017	Journal	Big Data
MDPI	[15]	2019	Journal	Symmetry

The details of the included research are organized in a tabular (see Table 2) form showing the details of the work performed.

Table 2. Literature survey of articles selected.

Paper reference	Year	Aim of studies	Type of studies	Details of work done	Limitation
[7]	2020	Verify the integrity and authenticity of the news	A protocol	The authors proposed a blockchain solution adapted to IoFMT (Internet of Fake Media Things) incorporated with a gamification component. Specifically, the proposed approach uses the concepts of a custom proof of authority consensus algorithm, as well as a weighted ranking algorithm, serving as an incentive mechanism in the gamification component to determine the integrity of false news	The proposed proof of authority protocol does not include the public of news being published in the validation process since social media plays a key role in disseminating information and this must be taken into account

(*continued*)

Table 2. (*continued*)

Paper reference	Year	Aim of studies	Type of studies	Details of work done	Limitation
[8]	2019		A framework	In this article, the authors have offered a high-level overview of a blockchain-based framework for the prevention and mitigation of fake news. The proposed framework architecture has three components: 1) the publisher management protocol; 2) The smart contract for news; 3) The news blockchain	• They have not developed a real prototype for the fake news problem • It is important to integrate contextual knowledge to verify the integrity of the news

(*continued*)

Table 2. (*continued*)

Paper reference	Year	Aim of studies	Type of studies	Details of work done	Limitation
[9]	2019		A prototype	In this article, a prototype blockchain system that aims to limit the spread of fake news in social networks	• The authors did not provide a criterion for selecting the stakeholders including making their prototype effective • The authors did not specify the type of news taken for their system • The authors did not describe a deployable and feasible solution to show the effectiveness of the blockchain to fight fake news in the ecosystem of social networks

(*continued*)

Table 2. (*continued*)

Paper reference	Year	Aim of studies	Type of studies	Details of work done	Limitation
[10]	2019		An algorithm	The authors of this article have proposed a method to detect fake news on social media as disseminating fake news via social media, which is a huge problem. The idea is that they integrated social media into a blockchain in such a way that random users (including journalists) act as news validators with the assignment of an accuracy value for the news issued. The average of these values will be the authenticity of this news	Sometimes, if the minors (e.g. news portals, journalists, etc.) are under political influence, they may consider false news that has been published in support of their political party to be valid. In this case, their verification will no longer be transparent and trustworthy
[12]	2019		An algorithm	The authors proposed an architecture that uses a centralized blockchain on the fake news detection process. They used data mining as a consensus algorithm to authenticate information posted on social networks	Requires a large number of sources to ensure the efficiency of the proposed architecture

(*continued*)

Table 2. (*continued*)

Paper reference	Year	Aim of studies	Type of studies	Details of work done	Limitation
[14]	2017	Proving and tracking the provenance of media content	A prototype	In this article, the authors proposed a prototype based on blockchain technology to prove the origins of captured digital media. The goal is to prove the validity of media resources used in fake news with an application named Provenator, which is particularly capable of recording metadata on digital media at the prototype level	They only took as a type of media (the images) in their prototype
[11]	2018		A blockchain platform	This article follows news based on distributed storage, decentralization, and other blockchain features, with consensus algorithms and smart technology	The authors did not specify the source of the information in their work
[13]	2019	Check the credibility of news content	A case study	This study aims to provide an overview of the implementation of blockchain technology and social networks to verify the credibility of content and prevent false information on social networks	The authors did not clarify in their study how to check the credibility of the news content as they spoke already mentioned in their article

(*continued*)

Table 2. (*continued*)

Paper reference	Year	Aim of studies	Type of studies	Details of work done	Limitation
[15]	2019	Determine the veracity of the source	A personalized journalism platform (protocol+ a personalized recommendation system)	The authors of this article have proposed a model of journalism with sharing spaces based on a hybrid blockchain and which aims to accomplish three things: • The delivery of articles presenting a sharing value • The distribution of roles of personalized agenda settings • And finally, the creation of a public space, where public opinions are collected	• They have not integrated trusted media agencies into their model • They have not implemented the provenance of the articles in their solution

5 Discussion

As the literature review shows, blockchain technology, being a decentralized ledger technology, promises to bring transparency and trust to this new "post-truth" world by enabling features such as smart contracts, decentralized consensus, and tamper-proof authentication. Indeed, blockchain in the media industry creates bold possibilities.

Table 3 shows an evaluation of performance at the implementation level between the different studies.

As the table shows, most of the approaches used the Ethereum blockchain with anonymous or private transactions.

Currently, journalism is facing unprecedented assaults due to attempts at censorship and the proliferation of fake news and infected news around the world. To fend off these threats, the potential of blockchain technology in journalism could make it more credible, maintain the integrity of journalism, and provide information vital to the public interest. Blockchain in journalism can be an effective means, not only to fight against "Fake News", but also to guarantee and protect the intellectual property of authors dues to the inviolability and traceability of the information provided by this technology.

Table 3. Evaluation between different studies

Paper reference	Blockchain platform	Type of deployment	Smart contract	Stakeholders
[7]	Hyperledger fabric	Confidential	Yes	Press agencies
[8]	–	–	Yes	News media organizations
[9]	–	Private	Yes	Social Networks users
[10]	Ethereum	Anonymous	Yes	Information portals, journalists…
[12]	–	–	**Yes**	–
[14]	Ethereum	Anonymous	Yes	Content creators
[11]	–	–	Yes	News media
[13]	–	–	–	Social media users
[15]	–	Hybrid	No	Journalists, news consumers

In the traditional news process, a professional(journalist) creates the content which is then edited by media editors. Afterward, the content is checked according to the standards carried out and subsequently validated. A broadcast takes place after validation on media platforms until the content is viewed and shared by end-users. These phases are illustrated in Fig. 3.

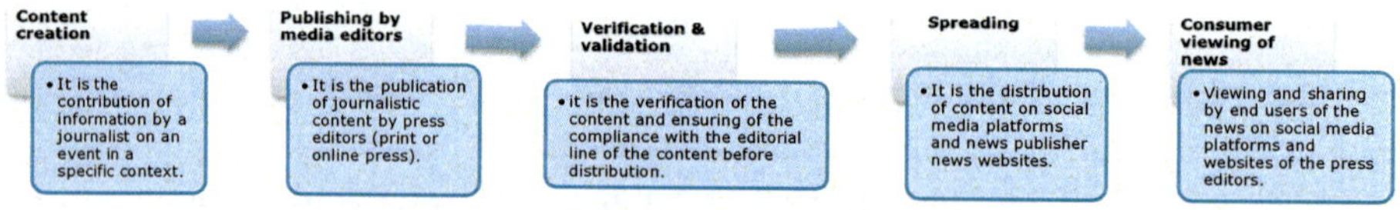

Fig. 3. Workflow of the news cycle

The subject of fake news detection has aroused great interest from researchers as we discussed in our review of the literature. They looked at the problem from different angles.

Below we have made a comparison table between the studies taken in our literature review to formulate our research contribution. We took the different phases of the news cycle as a comparison metric as shown in Table 4.

This table shows that few researchers have controlled the creation, editing, verification, publication, and amplification stages. Applying the principle of journalistic transparency to the different phases allows readers to corroborate information by directly

Table 4. Comparison between different studies

Paper reference	Content creation	Publishing by media editors	Verification & Validation	Spreading	Consumer viewing of news
[7]	✖	✖	✓	✖	✖
[8]	✖	✖	✖	✓	✖
[9]	✖	✖	✖	✖	✓
[10]	✖	✖	✖	✖	✓
[12]	✖	✖	✖	✖	✓
[14]	✓	✖	✖	✖	✖
[11]	✖	✖	✖	✖	✓
[13]	✖	✖	✖	✖	✓
[15]	✖	✓	✓	✖	✖

following the editorial processes and easily to the source of the information and verifying its veracity and credibility. So much so that the credibility of a news story in journalism is currently a key element to exercise serious and rigorous journalism.

Figure 4 shows the weaknesses encountered for each news cycle.

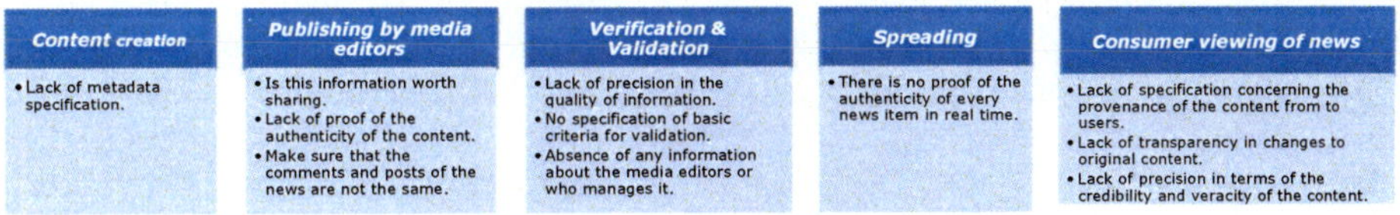

Fig. 4. Flow weakness of the news cycle

During our systematic review, we defined research questions. Below is the set of questions raised with answers for each question.

RQ1: How can we trust the media sources or the authenticity of the news published in the media?

- Should look for the source of the topic, or in other words, the media that published it, to check whether this medium has a good reputation, whether it is a newspaper, a website, a radio station, or a television station.
- Do a search on the Internet for the name of the journalist who prepared the topic that interests you, to know the nature of the topics he usually writes about (it is necessary to be suspicious of news that does not come from specific sources or people, and this may be justified in some cases, but if this is not explained in the news, it cannot be trusted).

- It is necessary to be careful of news that was not published on any other site. Important, shocking, and unusual news will have more than one source; it must be doubted if it is limited to one site.
- Verify the date of the news, as some news forgers resort to republishing old news. This news may be true when it is published, but it becomes false if it is republished to mislead.

RQ2: Can Blockchain protect journalism and push against threats of censorship and Fake News proliferation?

Journalism faces unprecedented assaults due to attempts at censorship and the proliferation of fake news around the world. Blockchain can help fend off these threats thanks to the immutability and decentralized verifiability offered by blockchain as well as the fact that blockchain makes it possible to trace the origin and flow of data, content, and information from creation to amplification through perpetual distribution, this will help much more easily access and verify the authenticity of multimedia content, preserve recordings and maintain the integrity of journalism.

RQ3: How Blockchain can preserve the credibility of media content?

The challenge facing journalism these days is the loss of trust of readers. The blockchain makes it possible to guarantee and protect the intellectual property of authors to the inviolability and traceability of the information provided by this technology.

RQ4: How to assess the credibility of a media in journalism?

The big challenge for journalism today is to preserve the credibility of media content. It is more important than ever to know how to distinguish a credible media: see if the information serves the public interest, the author cites and clearly identifies his sources of information and finally the author mentions in his media a clear distinction between the two notions (facts and opinions).

6 Conclusion and Future Works

The crisis of public confidence in mainstream journalism declined precipitously and is part of the explanation for the current popularity of fake news.

Strong of its promises of independence and decentralization, the blockchain remains very obscure for the public to fight against the "Fake News" of journalism and also to encourage the public to show a healthy skepticism towards the media online as a whole.

In this article, we tried to spend a literature review on the implementation of the blockchain to fight against the proliferation of fake news in journalism that highlighted its popularity of it as a field of research.

The existing solutions are not commensurate with the enormity of this problem. As a future working perspective, we plan to work on this lack to propose a feasible prototype for the journalism ecosystem with emphasis on the news cycle namely content production, dissemination, verification and validation, and finally consumer viewing of news.

References

1. US Election Analysis 2016: Media, Voters and the Campaign. https://www.abc.net.au/news/2016-11-14/fake-news-would-have-influenced-us-election-experts-say/8024660
2. India WhatsApp killings: Why mobs are lynching outsiders over fake videos. https://globalnews.ca/news/4333499/india-whatsapp-lynchings-child-kidnappers-fake-news/
3. A Facebook post was all it took to undo decades of communal harmony in a small east Indian town. https://qz.com/india/1030653/west-bengal-violence-how-a-facebook-post-broke-the-decades-long-communal-peace-of-a-west-bengal-town/
4. How fake news les Dylann Roof to murder nine people. https://theundefeated.com/features/how-fake-news-led-to-dylann-roof-to-murder-nine-peo-ple/
5. Kalyan Jewellers moves court after fake news on YouTube causes Rs 500-crore loss. https://www.indiatoday.in/india/story/kalyan-jewellers-moves-court-after-fake-news-on-youtube-causes-rs-500-crore-loss-1282400-2018-07-11
6. Khan, S.A., Alkawaz, M.H., Zangana, H.M.: The use and abuse of social media for spreading fake news. In: 2019 IEEE International Conference on Automatic Control and Intelligent Systems, I2CACIS, 2019-Proceedings, pp. 145–148. IEEE, Malaysia (2019)
7. Chen, Q., Srivastava, G., Parizi, R.M., Aloqaily, M.: An incentive-aware blockchain-based solution for internet of fake media things. Inf. Process. Manage. **57**(6), 102370 (2020)
8. Qayyum, A., Qadir, J., Janjua, M.U., Sher, F.: Using blockchain to rein in the new post-truth world and check the spread of fake news. IT Professional **21**(4), 16–24 (2019)
9. Saad, M., Ahmad, A., Mohaisen, A.: Fighting fake news propagation with blockchains. In: 2019 IEEE Conference on Communications and Network Security, CNS, pp. 1–4. IEEE, Washington (2019)
10. Paul, S., Joy, J.I., Sarker, S., Ahmed, S., Das, A.K.: Fake news detection in social media using blockchain. In: 2019 7th International Conference on Smart Computing & Communications, ICSCC, pp. 1–5. IEEE (2019)
11. Shang, W., Liu, M., Lin, W., Jia, M.: Tracing the source of news based on blockchain. In: 2018 IEEE/ACIS 17th International Conference on Computer and Information Science, ICIS, pp. 377–381. IEEE, Singapore (2018)
12. Ochoa, I.S., de Mello, G., Silva, L.A., Gomes, A.J.P., Fernandes, A.M.R., Leithardt, V.R.Q.: FakeChain: a blockchain architecture to ensure trust in social media networks. In: Piattini, M., Rupino da Cunha, P., García Rodríguez de Guzmán, I., Pérez-Castillo, R. (eds.) QUATIC 2019. CCIS, vol. 1010, pp. 105–118. Springer, Cham (2019). https://doi.org/10.1007/978-3-030-29238-6_8
13. Jing, T.W., Murugesan, R.K.: A theoretical framework to build trust and prevent fake news in social media using blockchain. In: Saaed, F., Gazem, N., Mohammed, F., Busalim, A. (eds.) Recent Trends in Data Science and Soft Computing, IRICT. AISC, vol. 843, pp. 955–962. Springer, Cham (2018)
14. Huckle, S., White, M.: Fake news: a technological approach to proving the origins of content, using blockchains. Big Data **5**(4), 356–371 (2017)
15. Kim, B., Yoon, Y.: Journalism model based on blockchain with sharing space. Symmetry **11**(1), 19 (2019)

Vulnerabilities in Android OS: Challenges and Mitigation Techniques

Nazish Nouman(✉), Zain Noreen, and Fouzia Naz

Mohammad Ali Jinnah University, Karachi, Pakistan
{nazish,zain.noreen,fouzia.naz}@jinnah.edu

Abstract. Android is one of the most popular operating systems for smartphones. Although Android OS includes many salient features which include its open-source nature, and its affinity to be customizable but its security remains debatable. Android uses SE-Linux for its security model, which can be leveraged and easily manipulated if ignored. The rise in IoT, smart gadgets, and wearable devices that also use Android as their operating system can't be ignored. Over the period many vulnerabilities have been discovered and exploited. This research paper focuses on some recent and most popular security vulnerabilities found in the android mobile operating system. The key contribution of this research includes the identification of mitigation techniques to overcome the challenges faced by its user The security of android devices can be significantly improved If mitigation techniques are applied effectively.

Keywords: Android · Security · Vulnerability · Hacking · Operating system

1 Introduction

Android has been one of the most used operating systems for smartphones and portable devices for more than a decade. With over 2 billion smartphones shipped with Android OS during 2018, [1, 2] and continued to lead the position of mobile operating system market with 74.13% till December 2019. The operating system is a target for malicious attacks [3] as well as being the target for many hackers who want to cause chaos in this overgrowing segment. Google Inc. the developer of Android OS, had started a bug bounty program to thwart bugs and various vulnerabilities found in the mobile operating system [3]. In 2015, Google started monthly security updates for Android [4] after a study in Cambridge University [5] found that up to 87% of Android smartphones were susceptible to the latest known security vulnerabilities which prompted Google to act [4]. These updates usually take time for various OEM manufacturers which use Android as their smartphones' operating system due to the various customizations and changes they do to the software, which is allowed due to Android's open-source nature. Ron Amadeo of Ars Technical has stated that although Google is taking security on Android seriously, it was not designed to be this way, as Android came to the market with zero market share, and to circumvent that, Google let these issues slide under the radar so that the OS could be spread and used by more OEM manufacturers.

S. Motahhir and B. Bossoufi (Eds.): ICDTA 2022, LNNS 454, pp. 256–266, 2022.
https://doi.org/10.1007/978-3-031-01942-5_25

From these bounties and various security organizations, such as Zimperium and Exodus Intelligence are working in tandem with Google to research and find new bugs and vulnerabilities within the mobile operating system and to patch them before hackers and various organizations take advantage of them [4, 5].

This paper aims to highlight the major vulnerability that surfaced during the last few years. Furthermore, the paper presents easy and probable solutions for a user to deal with android vulnerabilities in the best possible way.

This paper is divided as follows. Section 2 provides a detailed literature review about various android operating system security issues and weaknesses. Section 3 examines the common android vulnerabilities along with mitigation techniques. Section 4 provides a conclusion of the paper.

2 Literature Review

The author of this paper [4], has focused on the timely updates that are supposed to be delivered to the millions of smartphones that currently run Android. In a nutshell, around 50 OEM manufacturers currently sell mobile phones with Android as their Operating System, but few supports and update their respective models of smartphones for more than a year [5]. With Android being updated monthly to provide Security updates to their wide and large userbase, the various smartphones manufacturers must update their models on time. While the researchers used a FUM security metric to give the various devices in their corpus a score based on how quickly their devices receive updates, and how often these manufacturers provide updates when a serious vulnerability or flaw is discovered [6]. Smartphones are usually sold with 12–24 months contracts, but their study showed that on average, smartphones receive 1.26 updates per year, which is astonishing. Smartphones, being the most used piece of technology in the last decade has been left unpatched for long periods even though they are used the most by us during the day. According to their findings, Google with their Nexus and Pixel line of devices has a better score than most of the other OEM manufacturers with LG coming second in their findings. According to [7], the Android operating system implements an advanced security model. It is modeled after application-oriented mandatory access control and sandboxing. This allows the application access to only those privileges that were assigned to it at installation time. In recent versions of Android, as the security issues were raised on Android, the access control was modified to include runtime access control in which a user can change the access to privileges and data of the smartphone that the app can access. The applications are run within their sandbox environment that is separate within the confines of the operating system. In their research, they were able to confirm that because of unpatched vulnerabilities, they were able to access privileges that the users did not give to the application at install time, as well as the privileges the application did not even show while it was being installed through the app store. This was achieved through advanced attack techniques, namely code injection, return-oriented programming, and ROP without returns. They were able to escalate permissions within the applications, which allows test applications to bypass transitive privilege use, which allows it to bypass restrictions imposed by their respective sandboxes. In this paper, it's not discussed as an implementation bug, but as a fundamental flaw in the way, Android OS was coded.

Many layers of Android's core applications framework, the kernel framework, as well as the library framework, is discussed [8]. Any operating system is an amalgamation of various modules working in conjunction with each other. With that in mind, it's imperative to realize that Android in its current state is susceptible to various forms of exploitations and techniques by which its architecture can be leveraged to be used in a bad way. Android is built over the Linux kernel, which depends on the vast open-source community of the Linux foundation. The kernel is the bridge between the hardware aspect and the software aspect of the whole experience. With the sample space, the research was based on, the Application framework layer was found to consist of most of the vulnerabilities. In this layer, the exploits occur mostly through the browser, which attackers use to execute arbitrary code that provides unauthorized access to the smartphones' private resources. It also allows man-in-the-middle attacks which allows monitoring of user activities on the browser. There are flaws in the External Drivers that Android uses which include Qualcomm, and PowerVR, which are one of the largest vendors of smartphones parts in the world. According to [9], as these vendors use their drivers with Android OS, they are susceptible to various vulnerabilities such as Levitator exploit which allowed privilege escalation through kernel memory corruption caused by a graphics vendor driver.

The author of this paper [10] has focused on the lifecycle of how bugs and vulnerabilities are fixed within Android. Their study indicated that out of the Device Analyzer study mentioned in their report, around 19300 devices were tested, with 271 distinct OEM manufacturers. Such a big network of devices and companies makes it hard for the main developer of Android, Google to fix these bugs and vulnerabilities. With that said, even network operators and carriers modify their devices that run Android to suit their needs. Android being an open-source project, around 176 different organizations are responsible for running this machine. Keeping that in mind, the research takes a deeper look at how APIs or Application Programming Interfaces are manipulated, and vulnerabilities are exploited using that protocol. The designers of that API have certain protocols and sequences when they design them, but exploiters manipulate the software to run them as they want them to, basically breaking the intended security model. Fixing API vulnerabilities is hard as it requires keeping past versions of Android in mind. As Google and application developers normally need to keep the backward compatibility aspect of their applications, fixing exploits is hard. As various applications still are updated to accommodate the newer versions of the OS, the older versions are left vulnerable, which allows exploits and vulnerabilities to be made and found. The vulnerability the research focused on was the JavaScript vulnerability that allowed exploiters to run instances of Android applications on a browser within the smartphone. The vulnerability was fixed in 2 weeks, but more than 5.2 years later, the patch is only available to 65% of all smartphones running Android OS, which is alarming.

[11] discussed the vulnerabilities caused by some messaging design decisions in the Android Operating System. The research shows how a malware application is made to leverage those vulnerabilities. As many mobile operators and carriers use the traditional SMS protocol to transfer and share credits, it is easy for attackers to illegally hijack someone else's credits and use them for their use. The permission system of Android,

the broadcast and message-sending subsystem causes complete control of infected applications to send, receive and hide messages within the operating system. As the application can also hide messages, when telecom operators send acknowledgment messages to users, they are oblivious to the fact that their phone is being misused because the operating system gives a 3rd party application that much control over the core messaging service. As much of this malware is disguised as a legitimate 3rd party messenger for Android, it can be downloaded millions of times from the app store. Even if these applications target 1% of their users, and take $1 from each, it could be a huge sum of thousands of dollars that the malware can steal as most users will not bother if a dollar is deducted from their account only. In their testing, they also made the application available to the official Google Play Store for Android, but it was quickly taken off as Google's own antimalware was able to detect the flagged application. Many trojans were caught being operated in China and the UK, which were able to steal millions of dollars within a short period. [12–14]. Their proposed solution to this problem was to notify the user when the phone or the application prompts to send any message without the user's acknowledgment. Applications should not be given explicit permissions to read and write messages at install time, as this permission should be given at the user's behest in real-time. As well as a user should be able to remove the application's permission when an SMS has been sent, which includes one-time authentication applications such as WhatsApp, Facebook, and Telegram et cetera.

[15] inspects the friendliness of Android OS as a software, how it is easy for new users to adopt the software in their daily lives. But it does have some glaring issues, like allowing users to sideload applications that could be disguised as malware. In their research, there are many vulnerabilities discussed such as Code Execution, which allows malicious code to be run in a user's device without their knowledge or permission, which was patched in 18 months. Another vulnerability caused smartphones to become virtually dead, as it rendered the phone unable to make calls, texts with a dead screen. This vulnerability was known as Overflow. Another vulnerability discussed in the paper was Memory Corruption, which exploited the RAM of an Android smartphone, and could be used to read and write data within the active processes of the CPU. The research also proposed an architecture that could be implemented that could mitigate most of the vulnerabilities and provide an extra layer of security already implemented in the Operating system, which includes patching vulnerabilities in applications within the cloud. Generating keys made by the developer of the application will further strengthen the applications allowing no external modifications to be committed inside the application once it's built.

In this paper [16] the architecture of Android OS is criticized. As Android uses Linux kernel as their main form of security protection, many loopholes can be exploited as the Linux kernel was not developed as a mobile Operating System, nor was Linux made at the time Android was in existence. This leads to various discrepancies within the way Android works as an operating system. Due to this, Android has been the target of various exploiters and hackers who want to leverage the way Android works and functions. Another vulnerability found in the Operating system was Gain Privilege, which allows the attacker to gain access or control of those resources and functions which are normally protected from normal applications. After getting these permissions, the attacker could virtually own the smartphone and do unauthorized actions. Remote Code Execution vulnerability allowed attackers to execute commands and code on a target device with their own choice, even on the media server resource in the Android OS. Denial of Service vulnerability is a very common form of attacking networking resources and networks with the help of a large, infected computer network, but in Android OS, this pertains to rendering a smartphone dead or bricked due to blocking core services and resources from running on the infected device which causes pain in emergencies. The research also proposed a Security Framework that would do static and dynamic checks on applications that are installed and run from the smartphone. This would mitigate the install time permissions issue during which the applications gain permissions without giving the user any prompts, dynamic analysis includes inspecting applications that are running in real-time and raising flags or blocking them when they do anything suspicious.

3 Latest Android Vulnerabilities and Possible Mitigation Techniques

3.1 Stagefright

Stagefright is an umbrella term given to a group of vulnerabilities found in the Android Stagefright Media Library. This vulnerability was found in April 2015 [17] by Joshua Drake of the Zimperium security firm which was used to send malware through MMS messages to target Android devices. This vulnerability was dangerous because it did not require any user intervention to trigger the exploits it targeted. It used Privilege Escalation, and Remote Code Execution vulnerability, but with a remote aspect that could leave users from being targeted remotely if they had an MMS service with their carriers. This vulnerability left almost all Android smartphones running Android version 2.2 to the latest version to be affected. Another extension of this vulnerability, dubbed Stagefright 2.0 was discovered in October 2015 by Zimperium. This vulnerability used the media containers for music and video files, MP3, and MP4. As these two containers are the most popular within the industry, this bug was thought to be affecting more than 1 billion Android smartphones. According to [18], the vulnerabilities were patched within days of being public, as the security firm that led to their discovery, Zimperium, also gave the patches which eventually fixed the vulnerabilities to Google. The CVE-ID given to these vulnerabilities was CVE-2015–1538-39, CVE-2015–3824, CVE-2015–3826-29, and CVE-2015–3864.

Solution: One of the mitigation strategies against stagefright is to block MMS from unknown senders. another possible solution is to block automatically retrieving MMS from any sender as it is possible to infect messages from known senders. Users can also block MMS services from carriers and telecom providers. For Stagefright 2.0, as far as MP3 and MP4 containers are concerned, one thing that can be done is to stop downloading content from untrusted parties.

3.2 Krack

KRACK, short for 'Key Reinstallation Attack', is one of the vulnerabilities that led to the eventual demise of the security protocol for Wi-Fi, that is WPA2. This vulnerability was discovered in 2016 [19], by Belgian researchers. This flaw left virtually 41% of Android devices vulnerable [20]. The flaw was not discovered in the software, but rather in the implementation of the protocol itself. As Android OS uses the same implementation of Wi-Fi protocol as Linux does, which is wpa_supplicant, it was easier for attackers to nullify the security provided by the system as it could be manipulated to an all zero-encryption key, basically leading to man-in-the-middle attacks quite easily. Although the mitigation and patches were released by vendors and Google quickly, it was still easy to work around the patches [21]. This led to the creation of a newer Wi-Fi protection protocol, WPA3 in January 2018. The CVE-ID given to these vulnerabilities was CVE-2017–11013[22].

Solution: To mitigate this issue, as this is considered a security and hardware vulnerability, one possible solution is to update all wireless peripherals that one uses at the work and home environment. As WPA3 solutions are more readily available in the market, updating routers, switches, and access points, and PC's receiver modules, as well as updating security credentials. Most Router manufacturers, like TP-Link, Tenda, etc. have released updated firmware that mitigates this vulnerability also.

3.3 Spectre and Meltdown

These two hardware vulnerabilities were found to be affecting the Intel x86 platform and the ARM family of microprocessors [23]. Meltdown vulnerability was discovered independently by Google employee Jann Horn [24]. Security Analysts did not believe the reports of these vulnerabilities as they were so severe. It was given the CVE-ID code of CVE-2017–5754. This vulnerability causes a rogue process to read data in the RAM from the device it is attacking.

This was an inherent design flaw in the architecture of the design CPUs. Patching this bug meant that every CPU within these devices would have to be replaced, which is not possible [25]. This led to creative fixes through software which has mitigated this issue from reoccurring. Another vulnerability, dubbed Spectre [24], was found days after Meltdown was discovered. Although both are characterized differently, they both have some similarities. Specter is a vulnerability that stems from those processors that can perform Branch Prediction, which is a feature on most modern microprocessors. Through branch prediction, processors can perform speculative processing, which depends on private data, which could be exposed if the exploits are made to leverage that vulnerability. Data could be read from other applications as well within the Android OS, whether access to that data was given to the exploited application itself. Spectre was given the CVE-ID of CVE-2017–5753 and CVE-2017–5715. An extension of this vulnerability was found which was dubbed Spectre V2 as well.

Solution: The basic solution for this vulnerability is to update the devices like smartphones, personal computers, etc. so it does not have that hardware vulnerability. If one is not able to update their hardware, the easiest form of mitigation is to update software to the last patches as the vulnerabilities are patched in Android, as well as most, if not all OEM manufacturers.

3.4 API Breaking

Another recent vulnerability found in Android, given the CVE-ID of CVE-2018–9489. This vulnerability was discovered by researchers of Nightwatch Cybersecurity [25], which affected Android's communication management. Android itself broadcasts data within to its applications and services, which includes Wi-Fi BSSID, name, IP addresses, DNS server information, and MAC addresses [26]. Although applications are programmed to listen to these broadcasts, rogue applications can be used to eavesdrop on smartphones. This is alarming since by knowing this data, users' location can easily be traced via geolocation. Most of the data that Android OS broadcasts, is available in database format on the internet, which could be used to trace the locations of smartphones that is, the users. These vulnerabilities were patched in the latest version of Android, 9.0 Pie, and 10.0 Android 10.

Solution: To update your phone to the latest version of Android is by far seems the only option to protect from this vulnerability.

3.5 Permanent Denial of Service (PDoS)

Another very recent and so far, the most critical vulnerability found in Android of CVE-2019–2231. This vulnerability was highlighted by Google [27], in its December 2019 edition of the android security bulletin. According to the research, a single malicious crafted message by a threat player or a hacker [28], can generate a permanent denial of service (DoS) on an Android device.[29] To perform this malicious act perpetrator doesn't require additional execution privileges. According to the official NIST National Vulnerability Database, the vulnerability in the "handleRun" of TextLine.java" could create a "possible application crash. This could lead to a permanent Denial of Service attack by a remote user on an android device and can brick a smartphone. This vulnerability affects Android 8.0, Android 8.1, Android 9, Android 10 versions Android 11 versions [30].

Solution: To update your phone to the latest version of Android is one of the possible solutions so far which may protect your device to get paralyzed by a crafted message from a hacker. But for this, the latest updates should be available for the device. The major obstruction here is the availability of these fixes depends on the device manufacturers (Table 1).

Table 1. Summary of android vulnerabilities and possible solutions.

No	Vulnerabilities	Year Discovered	Description	Mitigation Strategies Consequences
1	Stagefright	April 2015	It was used to send malware through MMS messages to target Android devices	- Block MMS from any senders - Disable automatic MMS retrieval in Messaging setup - Use a browser that is not vulnerable to Stagefright (for example, Firefox 38 +)
	Stagefright 2.0	October 2015	This vulnerability used the media containers for music and video files, MP3 and MP4	
2	KRACK	2016	One of the critical replay attacks (a type of exploitable flaw) on the Wi-Fi Protected Access protocol that secures Wi-Fi connections	update all wireless peripherals that one uses at the work and home environment
3	Meltdown	January 2018	This vulnerability causes a rogue process to read data in the RAM from the device it is attacking. This was an inherent design flaw in the architecture of the design CPUs	- Solution for this vulnerability is to update the device - Another possible solution is to update hardware; the easiest form of mitigation is to update your software to the last patches
4	Spectre	January 2018	Spectre is a vulnerability that stems from those processors that can perform Branch Prediction, which is a feature on most modern microprocessors	
5	API Breaking	August 2018	This vulnerability in android smartphones affects Android's communication management and can expose users' personal information	- To update your phone to the latest version of Android, which is 10.0 Android 10

(*continued*)

Table 1. (*continued*)

No	Vulnerabilities	Year Discovered	Description	Mitigation Strategies Consequences
6	Permanent DoS	December 2019	It allows a remote attacker to cause a permanent denial of service which may result in a permanent freeze or block of the smartphone	- One of the plausible solutions is to install the Android updates of December 2019, if they are not already installed and if they are available for your device

4 Research Directions and Future Scope

Android's market will continue to expand in near future. Due to Android's open-source nature and increasing market share there are high chances of growing vulnerabilities as compared to other popular operating systems. Android vulnerabilitie3s are increasing rapidly. For this purpose more efficient malware detection techniques are to be applied with less human intervention. Machine learning (ML) and Deep learning (DL) algorithms must be used for better malware detection. For this purpose, it is now become a necessity to improve malware datasets because existing datasets are not updated, and sufficienct. Researher should fous on designing the extensive,structured and standard datasets.

5 Conclusion

This research has emphasized the various vulnerabilities and bugs of the Android operating system discovered in recent years. Furthermore, this study intends to provide possible mitigation techniques against the existing vulnerabilities of the android system. Android is an open-source operating system, which makes it vulnerable as attackers do not have to guess the code that's running the entire system. With the increasing demand for Android devices, the security of the operating system must be ensured. This also depends on the timely security updates that most OEM manufacturers neglect to provide. Android now provides monthly security patches to every manufacturer. Now, this responsibility falls on corporate organizations to provide it on every smartphone model they sell publicly.

References

1. Smartphone unit shipments worldwide by operating system from 2016 to 2018 (in million units), Statista, September 2019. https://www.statista.com/statistics/309448/global-smartphone-shipments-forecast-operating-system/. Accessed Sept 2019
2. Possemato, A., Aonzo, S., Balzarotti, D., Fratantonio: Trust, but verify A longitudinal analysis of Android OEM compliance and customization. In: 2021 IEEE Symposium on Security and Privacy (SP), pp. 87–102. IEEE (2021)
3. Zakaria, S.N., Zolkipli, M.F.: Review on mobile attacks: operating system, threats, and solution. Borneo Int. J. eISSN 2636-9826 **4**(2), 8–16 (2021)

4. Thomas, D., Beresford, A., Rice, A.: Security Metrics for the Android System. SPSM (2015)
5. Google. An Update to Nexus Devices. Google, August 2015. https://android.googleblog.com/2015/08/an-update-to-nexus-devices.html. Accessed Dec 2018
6. Mahalakshmi, K., Kavitha, K.: A comparative study on customers satisfaction towards android operating system and iphone operating system in moblie phone. Annals of the Romanian Society for Cell Biology, pp. 12337–12344 (2021)
7. Davi, L., Dmitrienko, A., Sadeghi, A.-R., Winandy, M.: Privilege escalation attacks on android. In: Information Security: 13th International Conference, Boca Raton, FL., pp. 346–360 (2010)
8. Shewale, H., Patil, S., Deshmukh, V., Singh, P.: Analysis of android vulnerabilities and modern exploitation techniques. Ictact J. Commun. Technol. **5**(1), 863–867 (2014)
9. Robinson, S.: Strategy Analytics: Q1 2018 Smartphone Apps Processor Market Share: Chips with On-device Artificial Intelligence (AI) Grew Three-Fold. Business Wire, August 2018. https://www.businesswire.com/news/home/20180808005464/en/Strategy-Analytics-Q1-2018-Smartphone-Apps-Processor. Accessed Dec 2018
10. Thomas, D., Beresford, A., Coudray, T., Sutcliffe, T., Taylor, A.: The lifetime of android API vulnerabilities: case study on the JavaScript-to-Java interface. In: Christianson, Bruce, Švenda, Petr, Matyáš, Vashek, Malcolm, James, Stajano, Frank, Anderson, Jonathan (eds.) Security Protocols 2015. LNCS, vol. 9379, pp. 126–138. Springer, Cham (2015). https://doi.org/10.1007/978-3-319-26096-9_13
11. Hamandi, K., Chehab, A., Elhajj, I., Kayssi, A.: Android SMS malware: vulnerability and mitigation. In: International Conference on Advanced Networking and Applications (2013)
12. Russel, J.: SMS Payment Virus Identified in China, 500,000 Android Device Infected. The Next Web (2012). http://thenextweb.com/asia/2012/08/19/stealth-sms-paymentmalware-identified-chinese-app-stores-500000-android-devicesinfected/. Accessed Dec 2018
13. Osborne, C.: SMS malware firm ordered to compensate victims. ZDNet, September 2012. http://www.zdnet.com/sms-malware-firm-ordered-to-compensatevictims-7000003639/. Accessed Dec 2018
14. Traynor, P., Enck, W., McDaniel, P., Porta, T.: Exploiting open functionality in SMSCapable cellular networks. J. Comput. Secur. **16**(6), 713–742 (2008)
15. Joshi, J., Parekh, C.: Android smartphone vulnerabilities: a survey. In: International Conference on Advances in Computing, Communication, & Automation (2016)
16. Umasanker. Analysis of latest vulnerabilities in android. In: International Conference on Advances in Computing, Communications, and Informatics (2017)
17. Rundle, M.: 'Stagefright' Android bug is the 'worst ever discovered'. Wired UK, July 2015. https://www.wired.co.uk/article/stagefight-android-bug. Accessed Dec 2018
18. Mimoso, M.: Stagefright 2.0 Vulnerabilities Affect 1 Billion Android Devices. Threatpost Inc, October 2015. https://threatpost.com/stagefright-2-0-vulnerabilities-affect-1-billion-android-devices/114863/. Accessed Dec 2018
19. Cimpanu, C.: New KRACK Attack Breaks WPA2 WiFi Protocol. Bleeping Computer, October 2017. https://www.bleepingcomputer.com/news/security/new-krack-attack-breaks-wpa2-wifi-protocol/. Accessed Dec 2018
20. Warren, T.: 41 percent of Android phones are vulnerable to 'devastating' Wi-Fi attack. The Verge, October 2017. https://www.theverge.com/2017/10/16/16481252/wi-fi-hack-attack-android-wpa-2-details. Accessed Dec 2018
21. Chirgwin, R.: Man the harpoons: The KRACK-en reawakens in updated WPA2 attack. The Register, October 2018. https://www.theregister.co.uk/2018/10/05/krack_updated_wpa2_attack/. Accessed Dec 2018
22. Burke, S.: Wi-Fi Alliance® introduces Wi-Fi CERTIFIED WPA3™ security. WiFi.org, June 2018. https://www.wi-fi.org/news-events/newsroom/wi-fi-alliance-introduces-wi-fi-certified-wpa3-security. Accessed Dec 2018

23. Arm Security Updates. ARM Technologies. https://developer.arm.com/support/arm-security-updates. Accessed Dec 2018
24. Meltdown and Spectre. meltdownattack.com. https://meltdownattack.com/#faq-systems-meltdown. Accessed Dec 2018
25. Osborne, C.: Android 'API breaking' vulnerability leaks device data, allows user tracking. ZDNet, August 2018. https://www.zdnet.com/article/android-operating-system-vulnerability-leaks-device-data-allows-user-tracking/. Accessed Dec 2018
26. Popper, B.: Google announces over 2 billion monthly active devices on Android. The Verge, May 2017
27. https://www.forbes.com/sites/daveywinder/2019/12/07/google-confirms-critical-android-8-9-and-10-permanent-denial-of-service-threat/#693ebf0466fe
28. Omar, M., Mohammed, D., Nguyen, V., Dawson, M., Banisakher, M.: Android application security. In: Research Anthology on Securing Mobile Technologies and Applications 2021, pp. 610–625. IGI Global
29. Garg, S., Baliyan, N.: Comparative analysis of Android and iOS from security viewpoint. Comput. Sci. Rev. **40**, 100372 (2021)
30. Ditton, S., Tekeoglu, A., Bekiroglu, K., Srinivasan, S.: A proof-of-concept denial of service attack against bluetooth IoT devices. In: 2020 IEEE International Conference on Pervasive Computing and Communications Workshops (PerCom Workshops), pp. 1–6. IEEE (2020)

Blockchain for Smart City: Opportunities and Future Research Directions

Abdullah Al Mahfuj Shaan[1], Tasfia Nausheen[1], and Akm Bahalul Haque[2(✉)]

[1] Department of ECE, North South University, Dhaka, Bangladesh
{abdullah.mahfuj,tasfia.nausheen}@northsouth.edu
[2] Software Engineering, LUT University, Lappeenranta, Finland
bahalul.haque@lut.fi

Abstract. Blockchain for smart cities is a growing prospect and gaining attention from academic and industrical standpoint. The increased population of the world has become a challenge to provide state-of-the-art living standards and resources. Smart cities promote innovative healthcare, transportation, education, security, etc. Blockchain is characterized by facilitating transparency, reliability, data integrity, user access control, immutability, decentralization, audibility. Integrating Blockchain in smart cities has its challenges that require extensive research and development. Smart cities hve been in in motion for last few years in different parts of the world. Still, it has created data security, authentication, sustainability, and vulnerability issues, especially when smart cities are not adequately planned. The application of Blockchain has put forward a lot of improvements in the quality and sustainability of smart cities. In this paper, we have discussed the application of Blockchain across various innovative smart city application domains. Blockchain fundamentals have been discussed before to help the reader gain ideas about Blockchain. This paper also provides future research agendas for blockchain integration in smart cities. The agendas will provide the researchers with helpful research directions.

Keywords: Blockchain · Smart cities · Consensus algorithm · Scalability · Decentralization

1 Introduction

In recent years, smart cities have been considered the modern form of urbanization. In the last few decades, the explosion in population has forced people to develop and build bigger and better cities [1]. This massive load on the urban area has brought new technologies that have enabled scientists and researchers to fight back many social, economic, technical, and organizational problems [2]. The main idea of creating smart cities is to develop a sustainable environment with the help of recent technologies to create seamless operations and management for the citizens [2, 3]. Smart cities have a lot to offer to the quality of living and other amenities that include healthcare, education, and finance which can be developed and maintained with the help of blockchain integration.

S. Motahhir and B. Bossoufi (Eds.): ICDTA 2022, LNNS 454, pp. 267–275, 2022.
https://doi.org/10.1007/978-3-031-01942-5_26

Blockchain enables the creation of safe, secure, reliable, and transparent collection, retrieval, and processing of data inside smart cities infrastructure [2].

Smart cities have diverse application domains, such as smart economy, marketing, healthcare, transportation, financial and banking systems, supply chain management, power and energy, and data centers [3–5]. To be more precise, the smart city concept is a multidimensional area requiring an extension in technological innovations [5, 6]. Therefore, future research directions regarding application domains and smart city infrastructure are crucial for urbanization.

In this article, blockchain fundamentals have been discussed briefly to set the background of this research. Later, blockchain integration in current smart city applications has been portrayed in detail. Furthermore, future research directions have been outlined based on the existing knowledge.

The article's objective was to put forward the potential research areas of blockchains that need to be thoroughly looked upon and the current applications of blockchains in smart cities. The article has been divided into five sections. Section 2 presents the fundamentals of Blockchain and smart cities. Section 3 discusses and analyses the current applications of Blockchain in smart cities. And in the Sects. 4 and 5, future research directions have been discussed, and the article has been concluded.

2 Blockchain Fundamentals

Blockchain comprises blocks, distributed and decentralized ledger of transactions that manages records. The distributed ledger technology keeps the records of digital transactions unaltered and transparent [4].

2.1 Types of Blockchain

Based on control mechanism and authentication, four types are discussed further below.

- Public Blockchain is open where everyone can participate in making a transaction. This does not belong to any organization. Instead, it is available for participation and contribution, and the user's identity for privacy and security is well hidden. The consensus algorithm plays a vital role, without which it is tough to keep it consistent [2, 4].
- Private Blockchain, is for a specified group of people of an organization, unlike public Blockchain. Selected people can access through this platform other than some specified few. There are restrictions and special rules which make it more preventive and inaccessible for everyone other than the one who has the permission to access it. The commonly used private blockchains are Hyperledger, Hash Graph, Corda or so on [4, 6].
- Consortium Blockchain has the features of both public and private Blockchain apart from the feature of the public Blockchain that allows everyone to access. Here, specific entities are only allowed within the network, which partially reveals a decentralized architecture rather than showing it all [6].

- Hybrid Blockchain is also the combination of public and private Blockchain where its features include access with permission, modifications are done based on organization's requirement and soon. When a user wants to access, some specific rules have to be followed to enter the network in this Blockchain. This Blockchain creates enough space to verify and check a security and privacy breach and take immediate actions [2, 4, 6].

2.2 Consensus Algorithm

The consensus algorithm is used to validate a transaction in the Blockchain. This algorithm helps the participants reach a consensus for decision-making [4, 7]. Various algorithms exist in Blockchain technology which is briefly described in Table 1.

Table 1. Types of consensus algorithms

Consensus algorithms	
Proof of Work (PoW)	The algorithm confirms the transaction and creates a new block to the chain. A group of people competes against each other to complete the transaction on the network. Once the user successfully creates a valid block, he gets rewarded [4, 7]
Delegated Proof of Stake (DPoS)	It is a democratic consensus protocol where the valid blocks are chosen based on voting. The consensus algorithms in these blockchains have few nodes to validate blocks and authentication purposes [4, 7]
Proof of Capacity (PoC)	PoC is a solution where the nonce values are needed for unlocking the blocks using the space in the hard drive of the nodes within the blockchain network [4, 7]
Proof of Elapsed Time (PoET)	PoET is used to estimating the waiting time before creating a block by the nodes generating random numbers in order [4, 7]
Raft	A voting-based consensus relies on two operations: leader selection, which manages the transactions' orders. The other is log replication, which is activated when the leader accepts log entries from the clients, creating its version of the transaction log by showing the accepted log entries [4, 7]
Proof of Stake (PoS)	Proof of stake consensus algorithm is for blockchains that work by selecting validators in proportion to their quantity of stake [4, 7]

(*continued*)

Table 1. *(continued)*

Consensus algorithms	
Proof of Burn (PoB)	This algorithm focuses on verifiably burning down cryptocurrencies. It has two functions to initiate the burning process and find if the address is spendable or not and its verifiability [4, 7]
Proof of Authority (PoA)	This algorithm is a permissioned blockchain that creates new blocks that then exploits the criteria of majority votes and helps form a consensus and new blocks
Ripple	Ripple works in two nodes, server type, and client node. The task of the server node is to give consensus for block validation and authentication, and the client node task is to transfer the data [4, 7]
Practical Byzantine Fault Tolerance (PBFT)	The algorithm is used to make the nodes byzantine fault-proof. When the consensus safely reaches between two nodes across the distributed network where some nodes cause problems and create protection is this algorithm's aim [4, 7]

3 Applications of Blockchain in Smart Cities

Some crucial application domains of blockhain in a smart city environment are outlined as follows. Figure 1 comprehensively depicts the application and future research agendas.

3.1 Transportation

Blockchain can make the whole transportation system more secure transparent in data transmission and ensure data integrity. Blockchain can address various transportation systems' security and privacy issues [1]. The data coming from the network can be used to analyze and make roads and highways more efficient. Blockchain can facilitate data security and integrity by default, increasing trust among different smart city entities [1, 8, 9]. Furthermore, with the help of smart contracts, Blockchain can manage access control of the traffic network vehicles and ensure secure financial transactions if needed.

3.2 Education

As the world moves towards a more digitized version of the workforce, verifying the integrity of educational qualification becomes an integral part of the process. The employers have to go through a multi-staged approach to verify the academic certificates. Blockchain can be used to maintain a central ledger to track the educational records and verify them simultaneously. This will significantly impact both the applicants and employers [4, 11].

3.3 Healthcare

Blockchain can facilitate healthcare data integrity confidentiality, prevent data breaches and facilitate privacy protection [5]. Currently, there is ongoing research on various healthcare systems based on Blockchain. Using private Blockchain for application, an in-depth learning approach for more accuracy while detecting disease, and many authors also propose feature extraction of different conditions. In addition, Blockchain and smart contracts can ensure a reliable insurance mechanism, patient data storage, transfer, etc.

3.4 Power and Energy Management

Blockchain technology can be used as a decentralized method to address local energy production firms [8]. Energy trading can also be possible with the help of blockchain technology [1]. A decentralized technology will use tokens to enable peers to form the trading network in a more secure environment and negotiate the prices. Meter reading for electricity consumption is another sector where smart grid networks based on blockchain technology can be used to prevent tampering with the readings [13]. Blockchain can solve this problem by creating a secure platform to transmit and receive data over the internet [1].

3.5 Smart Infrastructure

Blockchain technology can be used as a decentralized method to address energy consumption, data integrity, data privacy, and secure information transaction. Smart infrastructure includes smart homes and smart buildings used for offices and other organizations. Blockchain can be integrated with all the infrastructures to facilitate decentralized and secure data storage and transaction [2].

4 Future Research Directions

Future efforts need to be aware of the advancement of blockchain technology in smart cities to make blockchain implementation more secure and reliable. Figure 1 portrays applications & potential research areas of Blockchain for smart cities.

4.1 Security in Smart City

In smart cities, preventing any sort of breach or malicious attack is the biggest concern because the confidential information of the dwellers is at stake. With such abundant information, connections, and networks, it becomes hard to maintain infrastructure from any cyber-attack or breaches. It has to be made sure all the data and information goes through secured connections and is end-to-end encrypted. For example, several cryptographic technologies, homomorphic encryption, helps maintain data integrity [1]. Again, various techniques exist to detect traces of malicious attacks or cyber-attacks. No matter what measures are taken, it isn't possible to make the infrastructure or the system attack or malicious free [1].

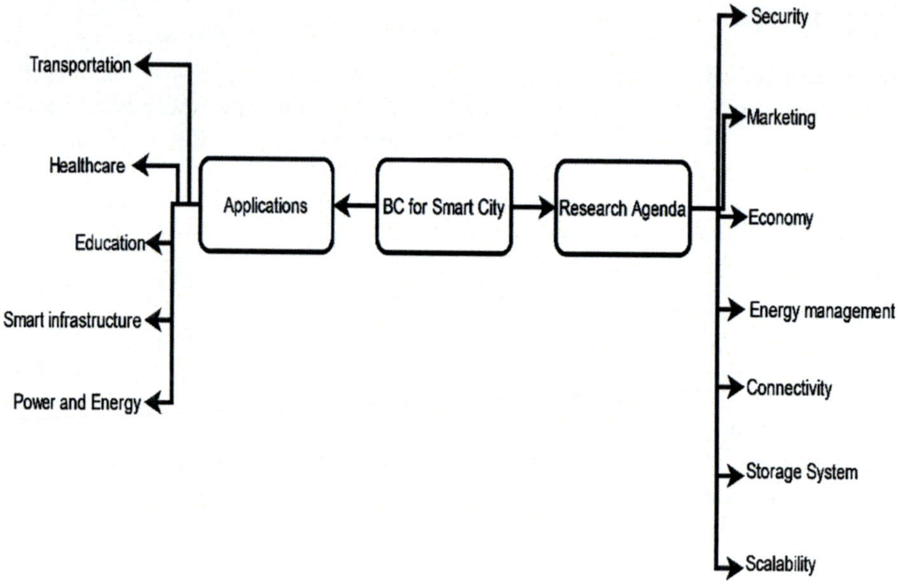

Fig. 1. A diagram that shows the applications & potential research directions of Blockchain for smart cities.

4.2 Marketing in Smart City

With the advent of technology, marketing has become much more manageable. The marketing in smart cities is mainly categorized into two groups. One group is smart city solution providers, and the other is the city authorities presenting on their behalf. Blockchain creates a sense of straightforwardness to the flexible chain resulting in trust in any situation and allowing the proper organization to succeed by reducing flexible Blockchain. It will take time for every organization to use blockchain technology for marketing. As usual, it's one-way communication that only conveys to the dwellers rather than helping increase the interaction [4].

4.3 Economy in Smart City

The infrastructural and design development of any smart city requires a boost in the economy [1]. Optimizing the budget allocation for any development purpose needs to be economically stable. Blockchain can provide economic management and deep learning techniques to predict future costs and optimize the budget. Economic policies can be regulated by decentralized blockchain methods that will give less allowance to the banks to maintain the economy. This makes the government more prudent in using cryptocurrencies and faces regularity issues. Therefore, ensuring the data integrity requires regulation, and it is a field where more experiments and research are required [1, 5].

4.4 Connectivity

Networks in a smart city are very costly and complex as they need to be deployed in a large-scale manner [14]. It is essential to understand common communication, data security, data storage, and ease of data sharing between peers and networks. The final target of any smart city is to have automated objects that can be deployed and run in any environment with an immutable and interoperable back-end structure that will allow them to interconnect the devices. Another research direction for connectivity in smart cities are data center networks. Network, storage and computational resources are connected via the internet that form the whole network structure. Blockchain can provide an easy solution as data center networks are centrally connected. Blockchain will allow them to manage privacy, make the storage system more secure, and solve the data integrity problem. Blockchain-based solutions for data center networks are cloud computing edge or fog computing [1]. Blockchain-based secure distributed systems can be designed to solve data center networks' security and privacy issues.

4.5 Scalability and Performance

Blockchain-based solutions for scalability and performance need to meet different business and government services requirements. PoW[1] consensus algorithms help to solve the problem of scalability, but it is resource-intensive and has low-performance issues because PoW is CPU intensive [2]. This makes it an unfit option for solving real-time problems. To have the economy up and without delay, platforms must handle a million transactions per second without any delay. This makes blockchain's scalability and performance sector in smart cities a concern and needs more research.

4.6 Storage Systems

Data storage that is scalable and can be incorporated using big data analytics is still in the research phase [15]. Integrating Blockchain in the system is impossible as the chains become very long and inefficient. The prominent rise in data generation has made it a big challenge. Using the consensus algorithm PoR[2] instead of PoW can create a decentralized token system [2]. In this method, the participant is required to contribute in the file storage and earn digital tokens as rewards.

Furthermore, a normal blockchain system requires each chain to process and maintain the transactions. This incident makes it a resource-constrained challenge [2]. Therefore, a smart city's data and storage system require intensive research and investigation.

4.7 Ransomware Issue

Ransomware attacks are getting very common these days. The payment for these ransomware transactions is sometimes accomplished with cryptocurrencies. The attack should be detected early on to tackle the threat. Pre-encryption algorithms can be used

[1] Proof of Work.

[2] Proof of Reputation.

to detect crypto-ransomwares very early [16]. Deep learning algorithms such as LSTM (Long Short-Term Memory) can be used to detect malware in a BitCoin System [17]. Bitcoin traceability research methods clustering heuristics can be used to automatically trace monetary flows and identify and compare the activity inside the Bitcoin Network [18].

4.8 Energy Consumption Issue of Blockchain

Blockchain energy consumption is related to consensus and redundant operations. PoW blockchain consumes more energy. However, the energy consumption is reduced for non-PoW blockchains [19]. Lightning network, where transactions are held off-chain by positioning channels and permitting transactions, helps to reduce energy consumption [20]. Another consensus algorithm, Delegated Proof of Stake, DPoS, is also a less energy-consuming algorithm as it requires less participation and is faster than PoS and PoW [21, 22].

5 Conclusion

What will a future blockchain-based smart city look like? If the research and development go as planned, the rewards will be higher than expected, and the current issues will be solved. It will become easier to picture a smart city that is Blockchain enabled. The technology will have more to offer the citizens, and it can save resources and save time to connect the entire economy with every other country in the world. Smart cities can improve the quality of living, but it also allows governments to make their country more secure. Integrating Blockchain in different sectors and promoting it would make infrastructures more sustainable and efficient. Smart cities will provide the citizens with additional facilities and amenities that rapid population and urbanization are currently unable to deliver on a large scale.

References

1. Haque, A.K.M.B., Bhushan, B., Dhiman, G.: Conceptualizing smart city applications: Requirements, architecture, security issues, and emerging trends. Expert Systems (2021). Published. https://doi.org/10.1111/exsy.12753
2. Bhushan, B., Khamparia, A., Sagayam, K.M., Sharma, S.K., Ahad, M.A., Debnath, N.C.: Blockchain for smart cities: a review of architectures, integration trends and future research directions. Sustain. Cities Soc. **61**, 102360 (2020). https://doi.org/10.1016/j.scs.2020.102360
3. Pranto, T.H., Noman, A.A., Rahaman, M., Haque, A.K., Islam, A.K., Rahman, R.M.: A blockchain, smart contract and data mining-based approach toward the betterment of e-commerce. Cybernet. Syst. 1–25 (2021). https://doi.org/10.1080/01969722.2021.2018545
4. Haque, A.K.M.B., Bhushan, B.: Blockchain in a nutshell. In: Advances in Data Mining and Database Management, pp. 124–143 (2021). https://doi.org/10.4018/978-1-7998-6694-7.ch009
5. Bagloee, S.A., Heshmati, M., Dia, H., Ghaderi, H., Pettit, C., Asadi, M.: Blockchain: the operating system of smart cities. Cities **112**, 103104 (2021). https://doi.org/10.1016/j.cities.2021.103104

6. Saxena, S., Bhushan, B., Ahad, M.A.: Blockchain based solutions to secure IoT: background, integration trends and a way forward. J. Netw. Comput. Appl. **181**, 103050 (2021). https://doi.org/10.1016/j.jnca.2021.103050
7. Mingxiao, D., Xiaofeng, M., Zhe, Z., Xiangwei, W., Qijun, C.: A review on consensus algorithm of blockchain. In: 2017 IEEE International Conference on Systems, Man, and Cybernetics (SMC), pp. 2567–2572 (2017). https://doi.org/10.1109/SMC.2017.8123011
8. Akter, R., Khandaker, M.J.H., Ahmed, S., Mugdho, M.M., Haque, A.B.: RFID based smart transportation system with android application. In: 2020 2nd International Conference on Innovative Mechanisms for Industry Applications (ICIMIA), pp. 614–619. IEEE, March 2020
9. Ghandour, A.G., Elhoseny, M., Hassanien, A.E.: Blockchains for smart cities: a survey. In: Hassanien, A.E., Elhoseny, M., Ahmed, S.H., Singh, A.K. (eds.) Security in Smart Cities: Models, Applications, and Challenges. LNITI, pp. 193–210. Springer, Cham (2019). https://doi.org/10.1007/978-3-030-01560-2_9
10. Gaved, M., Jones, A., Kukulska-Hulme, A., Scanlon, E.: A citizen-centred approach to education in the smart city: incidental language learning for supporting the inclusion of recent migrants. Int. J. Digit. Lit. Digit. Compet. (IJDLDC) **3**(4), 50–64 (2012)
11. Srivastava, G., Dhar, S., Dwivedi, A.D., Crichigno, J.: Blockchain education. In: 2019 IEEE Canadian Conference of Electrical and Computer Engineering (CCECE), pp. 1–5. IEEE, May 2019
12. Aymen, F., Mahmoudi, C.: A novel energy optimization approach for electrical vehicles in a smart city. Energies **12**(5), 929 (2019)
13. Kundu, D.: Blockchain and trust in a smart city. Environ. Urban. ASIA **10**(1), 31–43 (2019). https://doi.org/10.1177/0975425319832392
14. Kotevska, O., Kusne, A.G., Samarov, D.V., Lbath, A., Battou, A.: Dynamic network model for smart city data-loss resilience case study: city-to-city network for crime analytics. IEEE Access **5**, 20524–20535 (2017)
15. Strohbach, M., Ziekow, H., Gazis, V., Akiva, N.: Towards a big data analytics framework for IoT and smart city applications. In: Xhafa, F.., Barolli, L.., Barolli, A.., Papajorgji, P.. (eds.) Modeling and processing for next-generation big-data technologies. MOST, vol. 4, pp. 257–282. Springer, Cham (2015). https://doi.org/10.1007/978-3-319-09177-8_11
16. Kok, S., Abdullah, A., Jhanjhi, N.: Early detection of crypto-ransomware using pre-encryption detection algorithm. J. King Saud Univ.-Comput. Inf. Sci. (2020, in press)
17. Yazdinejad, A., HaddadPajouh, H., Dehghantanha, A., Parizi, R.M., Srivastava, G., Chen, M.-Y.: Cryptocurrency malware hunting: a deep Recurrent Neural Network approach. Appl. Soft Comput. J. **96**, 106630 (2020)
18. Paquet-Clouston, M., Haslhofer, B., Dupont, B.: Ransomware payments in the bitcoin ecosystem. J. Cybersecur. **5** (2018). https://doi.org/10.1093/cybsec/tyz003
19. Sedlmeir, J., Buhl, H.U., Fridgen, G., Keller, R.: The energy consumption of blockchain technology: beyond myth. Bus. Inf. Syst. Eng. **62**(6), 599–608 (2020). https://doi.org/10.1007/s12599-020-00656-x
20. Ray, S.: What is Delegated Proof-of-Stake, 15 April 2018. http://hackernoon.com/what-is-delegated-proof-of-stake-897a2f0558f9. Accessed 26 Dec 2021
21. Ghosh, E., Das, B.: A study on the issue of blockchain's energy consumption. In: Chakraborty, M., Chakrabarti, S., Balas, V.E. (eds.) eHaCON 2019. AISC, vol. 1065, pp. 63–75. Springer, Singapore (2020). https://doi.org/10.1007/978-981-15-0361-0_5
22. Haque, A.K.M., Rahman, M.: Blockchain technology: methodology, application and security issues. IJCSNS Int. J. Comput. Sci. Network Secur. **20**(2), 21–30 (2020)

Sensors Deployment in IoT Environment

Soukaina Bouarourou[1](✉), Abderrahim Zannou[1], Abdelhak Boulaalam[2], and El Habib Nfaoui[1]

[1] LISAC Laboratory, University Sidi Mohamed Ben Abdellah, Fez, Morocco
{soukaina.bouarourou,abderrahim.zannou, elhabib.nfaoui}@usmba.ac.ma
[2] LISA Laboratory, University Sidi Mohamed Ben Abdellah, Fez, Morocco
abdelhak.boulaalam@usmba.ac.ma

Abstract. The rapid growth of wireless sensor networks (WSNs) demand creates numerous Internet of Things (IoT) applications applied in several domains. On account of its wide-ranging application, WSNs that can monitor specific sensing range is the most adopted IoT platform. Therefore, the search for preferable placements and the optimal number of sensor node deployments can reduce computation, energy, cost, and communication overhead, which are the most factors that directly impact the coverage network. To address the problem, we propose a deployment and connectivity model bonded to sensor networks to ensure sensor deployment on a grid for target coverage while allowing connectivity. The model focuses on the k-coverage problem, which necessitates selecting a minimum subset of nodes such that at least k sensor nodes cover each point of a field of interest. The simulation results proved the proposed deployment model's efficacy and capability to reach maximum coverage with a minimum number of sensors.

Keywords: Sensing range · Coverage network · Target coverage · k-coverage

1 Introduction

Over the past few decades, the Internet of Things [1] has received important attention in large-scale applications [2] such as smart factories, smart buildings, smart devices, smartphones, Etc.The main idea of IoT is that it provides different objects or things as sensors, actuators, Radio-Frequency Identification (RFID), able to interconnect and exchange data continuously via the Internet to attain a common goal [3].

Nowadays, IoT application knows a lot of demand [4] for extended data collection for monitor and intruder detection in a specific area such as smart manufacturing, smart environment. Wireless sensor networks [5] are a framework favored to handle interests in formulating and managing the sensor nodes for controlling and monitoring a specific area. A wireless sensor network comprises multiple sensor nodes [6], which are usually deployed over remote space to sense information and forward the data wirelessly to the base station node.

There are several optimization problems: location and tracking, connectivity, quality of service, self-organization, maintenance, data process, security, energy efficiency, coverage, theoretical modeling, Etc. The most problematic and interesting tasks are energy

S. Motahhir and B. Bossoufi (Eds.): ICDTA 2022, LNNS 454, pp. 276–283, 2022.
https://doi.org/10.1007/978-3-031-01942-5_27

efficiency [7] and coverage problems [8]. In other words, propose protocols that maximize the network lifetime and reduce the energy consumption with desired connected coverage.

In WSN, sensor nodes are possibly deployed in deterministic or random places based on accessibility and the nature of the monitoring region. In random deployment, the coverage is not always guaranteed when the sensors are deployed in a dangerous area, and a few sensors are either densely or sparsely covered. Deterministic deployment is usually applied when the region details are known in advance and accessible [3–9], but it has an exceptional area where random deployment is the best option in many applications, as explained in [9]. Hence, the coverage can evaluate the capability of controlling every region's concern [10]. Some methods define the deployment regarding the studied area as a continuous surface [11], and others discretize the area into a group of discrete points [12].

In other words, there are three different classifications of coverage type: target coverage [13], area coverage [14], and barrier coverage [15]. The target coverage intends to track a set of monitoring points to monitor all targets continuously or deploy at least one sensor node all the time [16]. The area coverage aims to control an area of interest. The barrier coverage has a goal to test the highest support that crosses a sensor area.

Selecting a better placement with an optimal number of sensors for coverage monitoring of every position of a concern region is the most important problem of WSNs. This work provides a grid-based model for sensor node deployment for target coverage, ensuring connectivity. The proposed model assures a better placement to the sensor, based on clustering all targets of interest and present relation that allows selecting an optimal number of the sensor considering the connectivity.

The rest of this paper is structured as follows. Section 2 presents the literature reviews. The proposed deployment and connectivity model is explained in Sect. 3. In Sect. 4, the simulation results have explained the consistency of our model. In Sect. 5, we then conclude the whole paper.

2 Related Work

Coverage and connectivity are the major important, difficult problems in WSNs. The coverage in WSNs challenge is to monitor each field of interest, and the connectivity challenge is to forward all the collected data to a given destination.

In [17], a sequential hybrid approach is adopted to deal with the problem of target coverage and connectivity. The approach is instituted on three algorithms: the first algorithm locates the sensors to achieve coverage of all targets by allowing each sensor to one target. The second algorithm excludes redundancies from the deployment algorithm to decrease the number of sensors deployed. The inspired idea of the genetic algorithm with the third algorithm intends to generate a connected graph that offers the lowest path among the deployed sensors node and the base station. Although this approach has been proposed to ensure coverage with connectivity, it is still not guaranteed for various mobile nodes under uncertainty factors.

Another work [18] addressed coverage and connectivity problems in WSNs using a method finding the optimal deployment of the sensor node resulted in prolonging the network lifetime. They offered a queuing model based on virtual grid clustering, dynamic program, and a heuristic for the optimal problem. However, they do not design and consider optimizing energy consumption to assure network coverage.

In many industrial IoT applications, there are important research and studies that use intelligent algorithms to optimize energy [19–22]; the authors recommended the Bat algorithm to resolve the problem of resource allocation (RA) in the Internet of Things aiming to minimize the total cost of communication between gateways and resources. The algorithm's purpose in choosing the optimum monitoring sensor node in the best route resulted in reducing energy dissipation. The limitation of this approach is that the environment's intrusion on the nodes' communication signals, the wireless radio frequency (RF), and energy transmission in IoT sensor networks are not considered.

The abovementioned literature adopted determinist deployment. In an excepted situation where the deployment is infeasible, and the information environment is unknown, the random deployment is more practical from a moving vehicle [23]. In order to achieve full coverage and connectivity for random deployment, the appropriate network model is important in energy optimization.

The author in [16] suggested distributed data gathering algorithm that provides both shortest and multi-path routing for assuring energy-efficient and reliability of the network with static and mobile WSNs. However, the deployed sensor produces redundant data, which drives energy dissipation and blockage in the network.

The model [24] deploys the sensors node in both deterministic and random deployment in industrial IoT systems. It aims to determine the optimized number of sensors that maintain good coverage and determines conditions for regional connectivity. Although without suiting it into the wider context of data quality.

3 Our Proposal

In this contribution, we deal with sensors deployment in an IoT environment, where these devices are constrained in link failures, storage processing, energy consumption, Etc. The number of sensors should be minimized, and the number of covered targets should be maximized. The objective is to deploy sensors in a dynamic environment and create connectivity between them to gather data for the sink node.

3.1 Field of Interest Construction

The field of interest is the environment where the targets are found, as depicted in Fig. 1. A target in an environment is a set of points that should be monitored. These targets can be different, such as soil condition, pest infection, temperature, humidity, Etc.

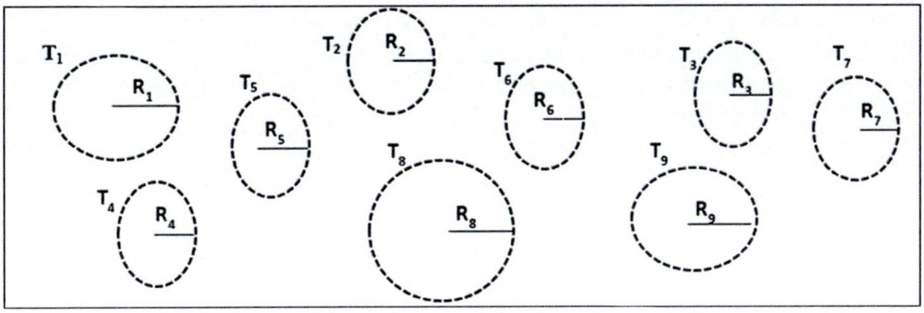

Fig. 1. The field of interest of different radius of targets.

This scenario considers the field of interest with two dimensions; it is a plan. The targets in the area of interest as form rectangle are constructed based on the extremums of circles. These targets are transformed in the form of circles with different radius. As the targets can have a different form, we construct their circles as described in Fig. 2.

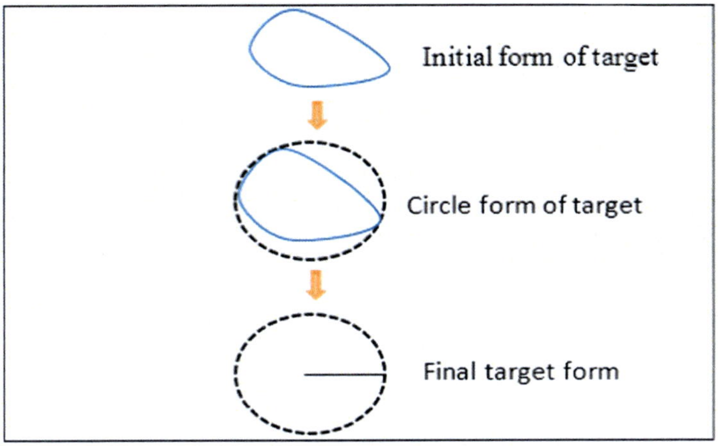

Fig. 2. The field of interest of different radius of targets.

In reality, the final targets do not have circle forms; therefore, we consider it a circle environment target to avoid ignoring some interesting area. Thus, the sensors should cover the entire target.

3.2 Grids Construction and Sensor Deployment

In this contribution, we consider that all the sensors are stationary, have the same communication range, and each sensor can collect all types of targets.

We construct a set of grids from the field of interest, where each grid can contain at least one target or many targets. A target can belong to one grid or more than one grid. In this situation, we focus on the size of grids and the communication range of sensors. To well study the field of interest environment, we calculate a set of parameters related to targets and sensors.

We measure the total surface area of targets A_T in the field of interest as follows.

$$A_T = \sum_{i=1}^{|T|} a^{(t_i)} \tag{1}$$

$$a^{(t_i)} = \pi R_i^2 \tag{2}$$

where, t_i is the target i, $|T|$ is the number of targets, $a^{(t_i)}$ is the surface area of a target disk, and R_i is the radius of target i.

We compute the targets ration as follows.

$$P = \frac{A_T}{F} \tag{3}$$

where, F is the surface area of the field of interest and $0 \prec P \leq 1$.

We determine the number of candidate sensors that can satisfy the field of interest, such that the surface area that sensors can cover approximately equals the surface of targets as given in the next formula.

$$S_T = \sum_{i=1}^{|N|} s_i = A_T \tag{4}$$

where, s_i is the surface range of sensor i, N is the number of candidate sensors, and S_T is the total surface range of all candidate sensors.

To determine the accurate points in the field of interest, the k-means clustering is used to achieve the given goal. The number of clusters k is determined as follows.

$$k = integer(|\mathrm{N}| \times \mathrm{P}) \tag{5}$$

We note that, more the value of p is higher, more the value of k is higher. Also, the targets are added to cluster if the next formula is verified.

$$d_{ij} = \sqrt{\sum_{h=1}^{2} (T_{i,h} - C_{j,i})^2} \leq \frac{R}{2} - \varepsilon_+ \tag{6}$$

where $T_{i,h}$ and $C_{j,i}$ are coordinate locations of target i and centroids of clusters respectively, R is range communication of sensors, and ε_+ is a random number that verifies

$$0 \prec \varepsilon_+ \leq \frac{R}{2}.$$

Finally, when the clustering in the field of interest is performed, the sensors will be deployed in centroids.

4 Simulation and Results

This simulation implements our clustering algorithm on a Dell 9650 computer with an Intel i7 CPU running at 2.80 GHz with Windows 8 (64-bit) and 6 GB of RAM using the Python language.

We realized five trials with different targets to validate our approach, as shown in Table 1. Also, to facilitate the implementation, we initialized the target with its surrounding circle. We calculate the targets ratio of targets for each trial, and the results show that there are different ratios from ratio to another.

Table 1. The trials number with the number of targets.

Trials	Number of targets	Targets ratio
Trial 1	10	72%
Trial 2	20	45%
Trial 3	30	55%
Trial 4	40	81%
Trial 5	50	63%

Applying our mechanism for sensors deployment, we measure the number of sensors disposed of in the area of interest and the coverage ratio among sensors, as presented in Table 2. As illustrated, the number of sensors increases with the number of targets; this result is evident. Besides, the coverage in the network achieved 100% in trials 1 and 4, which means the more the ratio of the target is higher, the more the coverage ratio increases. It indicates that our approach is more efficient in high targets ratio.

Table 2. Ratio coverage of sensors.

Trials	Number of required sensors	Coverage ratio
Trial 1	8	100%
Trial 2	15	64%
Trial 3	24	75%
Trial 4	33	100%
Trial 5	27	82%

To study the behavior when the coverage ratio is smaller than 100%, we measure the required number of sensors for coverage of 100% of targets considering our mechanism compared with the standard required number of sensors as displayed in Fig. 3. Our mechanism achieves 100% of coverage with a minimal number of sensors. However, we deploy the number of sensors more than the standard number of sensors.

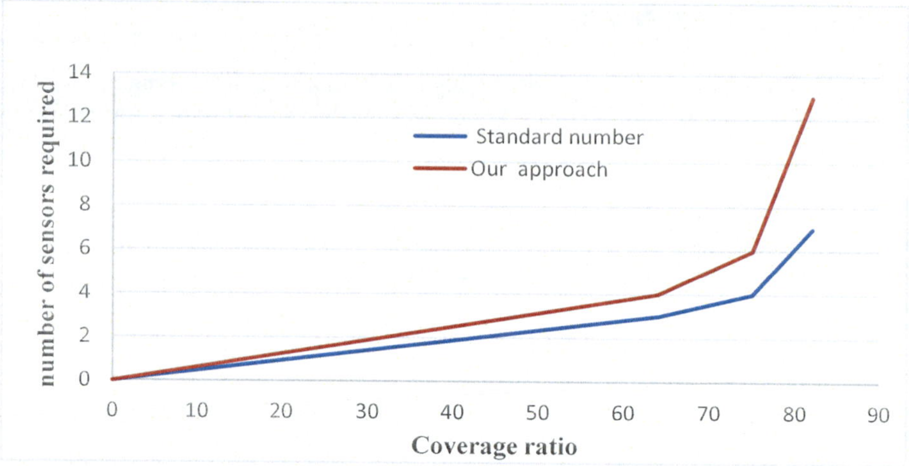

Fig. 3. The required number of sensors for coverage 100% of targets

5 Conclusion

This paper considered the k-coverage difficulty in WSNs to choose the least number of sensors for target coverage on a network considering connectivity. We have acquainted a model-inspired clustering of all targets of the area of interest that achieve sequentially an optimal number and a better position of sensor nodes that place the sensors in the centroid of the clusters.

References

1. Fagroud, F.Z., Ajallouda, L., Lahmar, E.H.B., Toumi, H., Zellou, A., ElFilali, S.: A brief survey on Internet of Things (IoT). In: Motahhir, S., Bossoufi, B. (eds.) ICDTA 2021. LNNS, vol. 211, pp. 335–344. Springer, Cham (2021). https://doi.org/10.1007/978-3-030-73882-2_31
2. Zannou, A., Boulaalam, A., Nfaoui, E.H.: A multi-layer architecture for services management in IoT. In: Ben Ahmed, M., Boudhir, A.A. (eds.) SCAMS 2017. LNNS, vol. 37, pp. 324–334. Springer, Cham (2018). https://doi.org/10.1007/978-3-319-74500-8_30
3. Ammari, H.M., Das, S.K.: Fault tolerance measures for large-scale wireless sensor networks. ACM Trans. Auton. Adapt. Syst. TAAS **4**(1), 1–28 (2009)
4. Zannou, A., Boulaalam, A., Nfaoui, E.H.: A Node Capability Classification in Internet of Things (2020). https://doi.org/10.1109/ISCV49265.2020.9204024
5. Zannou, A., Boulaalam, A., Nfaoui, E.H.: System Service Provider–Customer for IoT (SSPC-IoT). In: Bhateja, V., Satapathy, S.C., Satori, H. (eds.) Embedded Systems and Artificial Intelligence. AISC, vol. 1076, pp. 731–739. Springer, Singapore (2020). https://doi.org/10.1007/978-981-15-0947-6_70
6. Zannou, A., Boulaalam, A., Nfaoui, E.H.: Relevant node discovery and selection approach for the Internet of Things based on neural networks and ant colony optimization. Pervasive Mob. Comput. **70** (2021). https://doi.org/10.1016/j.pmcj.2020.101311

7. Zannou, A., Boulaalam, A., Nfaoui, E.H.: SIoT: a new strategy to improve the network lifetime with an efficient search process. Future Internet **13**(1), 1–23 (2021). https://doi.org/10.3390/fi13010004
8. Amutha, J., Sharma, S., Nagar, J.: WSN strategies based on sensors, deployment, sensing models, coverage and energy efficiency: review, approaches and open issues. Wirel. Pers. Commun. **111**(2), 1089–1115 (2020)
9. Balister, P., Kumar, S.: Random vs. deterministic deployment of sensors in the presence of failures and placement errors. In: IEEE INFOCOM 2009, pp. 2896–2900 (2009)
10. Huang, C.-F., Tseng, Y.-C.: A survey of solutions to the coverage problems in wireless sensor networks. J. Internet Technol. **6**(1), 1–8 (2005)
11. Njoya, A.N., et al.: Efficient scalable sensor node placement algorithm for fixed target coverage applications of wireless sensor networks. IET Wirel. Sens. Syst. **7**(2), 44–54 (2017)
12. Liu, X.: A deployment strategy for multiple types of requirements in wireless sensor networks. IEEE Trans. Cybern. **45**(10), 2364–2376 (2015)
13. Wang, J., Niu, C., Shen, R.: Priority-based target coverage in directional sensor networks using a genetic algorithm. Comput. Math. Appl. **57**(11–12), 1915–1922 (2009)
14. Carle, J., Simplot-Ryl, D.: Energy-efficient area monitoring for sensor networks. Computer **37**(2), 40–46 (2004)
15. Li, X.-Y., Wan, P.-J., Frieder, O.: Coverage in wireless ad hoc sensor networks. IEEE Trans. Comput. **52**(6), 753–763 (2003)
16. Biswas, S., Das, R., Chatterjee, P.: Energy-efficient connected target coverage in multi-hop wireless sensor networks. In: Bhattacharyya, S., Sen, S., Dutta, M., Biswas, P., Chattopadhyay, H. (eds.) Industry interactive innovations in science, engineering and technology. LNNS, vol. 11, pp. 411–421. Springer, Singapore (2018). https://doi.org/10.1007/978-981-10-3953-9_40
17. Njoya, A.N., et al.: Hybrid wireless sensors deployment scheme with connectivity and coverage maintaining in wireless sensor networks. Wirel. Pers. Commun. **112**(3), 1893–1917 (2020)
18. Al-Karaki, J.N., Gawanmeh, A.: The optimal deployment, coverage, and connectivity problems in wireless sensor networks: revisited. IEEE Access **5**, 18051–18065 (2017)
19. Sangaiah, A.K., Sadeghilalimi, M., Hosseinabadi, A.A.R., Zhang, W.: Energy consumption in point-coverage wireless sensor networks via bat algorithm. IEEE Access **7**, 180258–180269 (2019)
20. Sangaiah, A.K., Medhane, D.V., Bian, G.-B., Ghoneim, A., Alrashoud, M., Hossain, M.S.: Energy-aware green adversary model for cyberphysical security in industrial system. IEEE Trans. Ind. Inform. **16**(5), 3322–3329 (2019)
21. Sangaiah, A.K., Medhane, D.V., Han, T., Hossain, M.S., Muhammad, G.: Enforcing position-based confidentiality with machine learning paradigm through mobile edge computing in real-time industrial informatics. IEEE Trans. Ind. Inform. **15**(7), 4189–4196 (2019)
22. Sangaiah, A.K., Hosseinabadi, A.A.R., Shareh, M.B., Bozorgi Rad, S.Y., Zolfagharian, A., Chilamkurti, N.: IoT resource allocation and optimization based on heuristic algorithm. Sensors **20**(2), 539 (2020)
23. Liu, C., Wu, K., Xiao, Y., Sun, B.: Random coverage with guaranteed connectivity: joint scheduling for wireless sensor networks. IEEE Trans. Parallel Distrib. Syst. **17**(6), 562–575 (2006)
24. Dou, R., Nan, G.: Optimizing sensor network coverage and regional connectivity in industrial IoT systems. IEEE Syst. J. **11**(3), 1351–1360 (2015)
25. Bouarourou, S., Boulaalam, A., Nfaoui, E.H.: Services search techniques architecture for the Internet of Things. In: Ezziyyani, M. (ed.) AI2SD 2019. LNNS, vol. 92, pp. 226–236. Springer, Cham (2020). https://doi.org/10.1007/978-3-030-33103-0_23

Operating Modeling of Medium Range Protocols IoT

Sakina Elhadi(✉), Loubna Chhiba, Nawal Sael, and Abdelaziz Marzak

Laboratory of Modeling and Information Technology, Faculty of Sciences Ben M'SIK, University Hassan II, Casablanca, Morocco
elhadi.sakina1993@gmail.com

Abstract. The Internet of Things (IoT) is a hot topic. It considers a system of billions of physical objects, industrial machines, and billions of humans where these objects connected to the Internet. The Internet of Things based on existing communication techniques and protocols. The diversity of protocols responds to the multiplicity of billions of connected objects. There are different technologies, each of which has its own particularities in terms of a set of criteria. However, the absence of connectivity protocol results in a lack of interoperability between IoT objects and applications. In this article we answer the question of protocol operation, we will present the operating models of medium-range network protocols. These models ensure a deep and flexible conception of how network protocols work using UML modeling. We used the activity diagram.it permit to graphically represent the behavior of these protocols. In this paper, we focus on medium-range are Z-wave, Wi-Fi, and Bluetooth Low Energy (BLE).

Keywords: Internet of Things · IoT · Models · Protocols · Z-Wave · Wifi · Bluetooth · UML

1 Introduction

IoT is facilitating our personal and professional life. There are countless IoT applications from many areas, including health, industry, transportation, education, entertainment, etc. Thus, many devices, services and protocols have been created. The IoT system makes it possible to gather information from the physical environment using detectors such as the various sensors, to process this information locally and to transfer it via the Internet for a useful decision asset for the user thanks to a set technology integrated into this system while respecting the IoT architecture. According to our research, there is not a standard architecture for the IoT, but there is different proposed architecture according to the needs of organizations or developers. We find that there are four principal layers which are device, network, middleware and application [1]. Each has its main role in the success of it. We focus on the network layer. This layer ensures the connection of the sensors to each other and to the gateway to transmit the information collected via the Internet. This connection thanks to the use of different network protocols, namely Low Energy Bluetooth, Zigbee, WiFi. Each protocol has its characteristics, advantages,

S. Motahhir and B. Bossoufi (Eds.): ICDTA 2022, LNNS 454, pp. 284–293, 2022.
https://doi.org/10.1007/978-3-031-01942-5_28

and disadvantages. These protocols play a major role in the production and development of IoT, it allows different objects to connect to the internet and provide various useful services. This will help researchers and developers to choose one protocols. This paper is a continuation of our research [2–6]. For the organization of this article, section two introduces network protocols. In section three, we will provide the analytical study of IoT network protocols. In section four focuses on the proposed operating models for medium-range network protocols. For the fifth sections, we present the projection of the criteria on the proposed model. We will finish with the conclusion.

2 Network Protocols

Network protocols are one of the main communication techniques for the Internet of Things. The protocol is the language used to establish communication between the different connected objects. To be able to communicate with each other, two objects must therefore have the same communication language. There are several protocols on the scientific market. Just like Bluetooth, Wifi, Z-wave, which ensure that the devices communicate with each other. Each has its own peculiarities in terms of different criteria.

2.1 Z-Wave

Z-Wave is a wireless communication protocol developed by Zensys and Introduced in 1999 [7]. Mainly intended for home automation. Recently, it has been used in many IoT applications, including smart homes and small business areas. Z-Wave is a low power consumption MAC standard. It covers up to 100 m distance, point to point communication and is suitable for small messages. It uses radio frequencies to establish communications (800–900 MHz). It therefore allows two Z-Wave electronic components to discuss together to exchange information. It uses CSMA/CA for media access in addition to small ACK messages for reliable transmission. Z-wave follows a master/slave architecture, in which the master controls the slaves, sends them commands and manages the planning of the whole network. Z-wave is relatively secure, in both directions, used in a mesh network system [8].

2.2 WiFi

WiFi (Wireless Fidelity) is the most popular set of wireless communications protocols. It follows the IEEE 802.11n/a/ b/g standard for the wireless local area network (WLAN). As IEEE 802.11b/g uses the 2.4 GHz band and IEEE 802.11a uses the 5 GHz band. Furthermore, IEEE 802.11n uses the 2.4 and 5 GHz bands with MIMO (Multiple-Input Multiple-Output) mechanisms. It allows many data to transfer quickly. There are different WIFI standards for varying range and speed. The data rate can reach 11 to 867 Mbps. Its higher power consumption (–116 MA) due to its high data rate. WIFI therefore has a good loss range up to 100 m and an interior of around 35 m. In addition, it allows high-speed connections over distances of 20 to 100 m. Wi-Fi security at all levels and, in high security mode, Wi-Fi 2 (WPA2) 802.11i protected access and has no security option (open network). It is only suitable for devices connected to the mains where the power supply can be easily and frequently [9, 10].

2.3 Bluetooth Low Energy

Bluetooth Low Energy (BLE) or Smart Bluetooth is the widely used short-range communication standard in IoT. BLE is a wireless transmission technique created by Nokia in 2006 as an open standard based on Bluetooth. This technology appeared in 2010 with the release of version 4.0 of the Bluetooth Core Specification. BLE is a communications standard for the two-way exchange of data at very short distances using UHF radio waves on a frequency band of 2.4 GHz. BLE allows a speed of the same order of magnitude (1 Mb/s) for a power consumption 10 times less than original Bluetooth. Thus, its low energy can reach ten times less. Its access control uses contention-free MAC with low latency and fast transmission [11].

3 Analytical Study

We presented a comparative study network protocols [12]. we focus on medium range network protocols. This study evaluates these protocols in terms of capacities, main characteristics, and behaviors. This study is based on the criteria shown in the following Table 1.

Table 1. Comparative study.

Protocols criteria	Wifi	Z-Wave	BLE
Specification	Based on 802.11n	Z-Wave Alliance ZAD12837 / ITU-T G.995	Bluetooth Core Specification 4.0
Cost	Low	Effective	Low
Energy consumption	High	Low	Low
Network type	LAN WPAN/P2P	LAN	PAN
Topology	Star	Mesh	Star-bus
Power	Low-High	Very low	0.01 to 0.5 (use case dependent)
Network size	Medium	Large	Medium
Data rate	Up to 1.3Gbps	40 Gbps	1 Mbit/s
Frequency band	54 Mb/s	900 MHz (ISM)	2,4 GHz
Modulation Technique	BPSK, QPSK, OFDM, M-QAM	BFSK, GFSK	GFSK, CPFSK
Spread spectrum	MC-DSSS,CCK, OFDM	DSSS	FHSS
Range	Up to 100 m	30 km	50 m (160 ft)

(continued)

Table 1. (*continued*)

Protocols criteria	Wifi	Z-Wave	BLE
Security	WPA and WPA2	Data encryption	128-bit AES with Counter Mode CBC-MAC
Risk of data collusion	Low	–	High

4 Operating Models

In this part, we will present the operating models of the medium-range network protocols which are Low Energy Bluetooth, Z-wave, and WiFi. We use for modeling the activity diagrams. This diagram is divided into lanes. Each corresponding lane is responsible for creating a set of activities. Thus, this diagram represents the detailed process of operation of the protocols.

4.1 Z-Wave

Z-Wave is structured in three layers are physical, Mac layer and Application (Fig. 1).

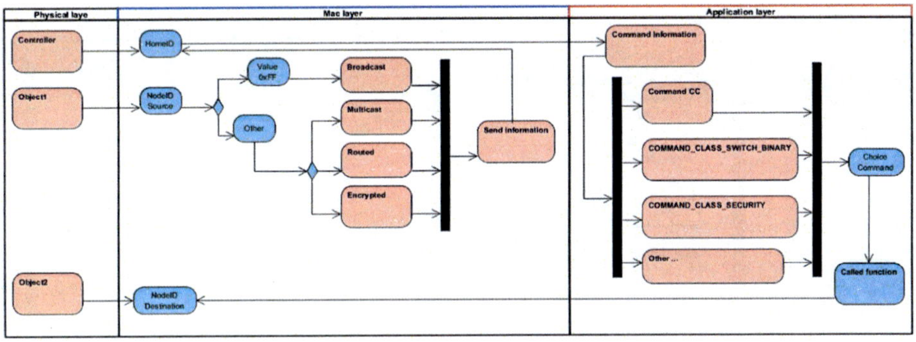

Fig. 1. Z-wave

For the physical layer, there are two types of equipment present on a Z-Wave network is one or more controllers, nodes (slaves) which represent the different modules of the network. A main controller allows up to 232 nodes to be incuded in a network. This network is represented by an identifier (HomeID) encoded in 32 bits. It is communicated during an inclusion phase to all equipment wish-ing to attach to the controller. Each device in the network is also assigned its own identifier (NodeId) coded on 8 bits. For the Mac layer, The HomeID, source and destination of a message are defined on the MAC layer. For the destination, the first NodeID is al-ways 0×01, and is automatically incremented each time a module is add. If the value is set to "0xFF", the frame is of the "broadcast" type and therefore sent to all the devices on the network. In addition, the content of the first fields of this layer may vary slightly for multicast, routed or encrypted frames.

The application layer provides information on the type of Z-Wave control class used, the function called and a list of parameters. Indeed, all communications are organized into com-mand classes (CC). These classes used to perform specific actions related to the functionalities offered by the equipment. About fifty different class types currently exist. Each class can then define one or more commands, with different parameters.

4.2 WiFi

WiFi is structured in two layers are the physical and the data link.

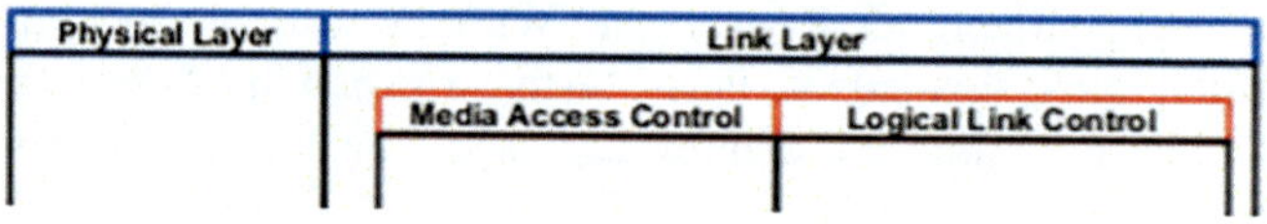

Fig. 2. General operating model WiFi

The physical layer defines the modulation of radio waves and the signaling characteristics for data transmission (Fig. 3).

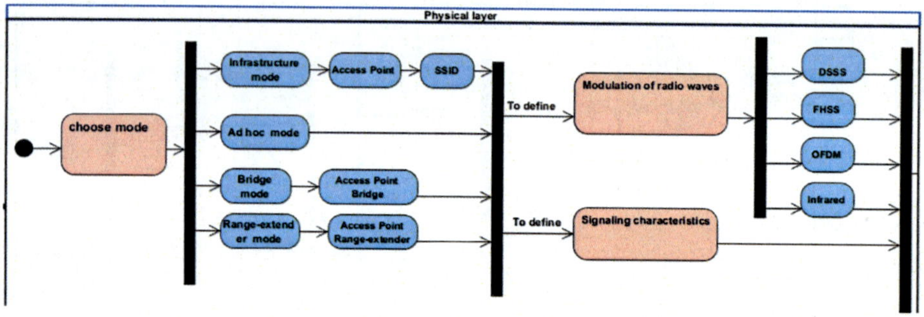

Fig. 3. Operating model of physical layer

The data link layer, consisting of two sub-layers: Logical Link Control (LLC) and Media Access Control (MAC). MAC layer defines the interface between the machine and the physical layer, as well modes of alternative transmission. Thus, it en-sures different functionality, which is clearly determined in the Fig. 2. LLC makes the MAC protocol more reliable through error control and flow control. It includes three types of transmission service. LLC type 1 is unconnected mode and datagram without acknowledgment. LLC type 2 is connected mode with acknowledgment. LLC type 3 is a mode not connected with acknowledgment (Fig. 4).

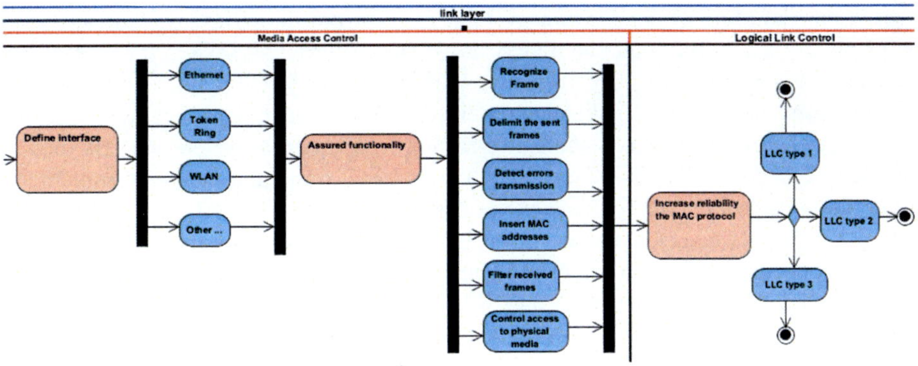

Fig. 4. Operating model the link layer

4.3 Bluetooth Low Energy (BLE)

BLE is structured in 3 layers are the controller, host 1 and application. The controller consists of three sub-layers are physical, Link and Host Con-troller Interface (HCI). The host layer consists of four sup-layers are Logical Link Control and Adaptation Layer Protocol (L2CAP), Attribute Protocol (ATT), Generic Attribute Profile (GAPP), Generic Access Profile (GAP) (Fig. 5).

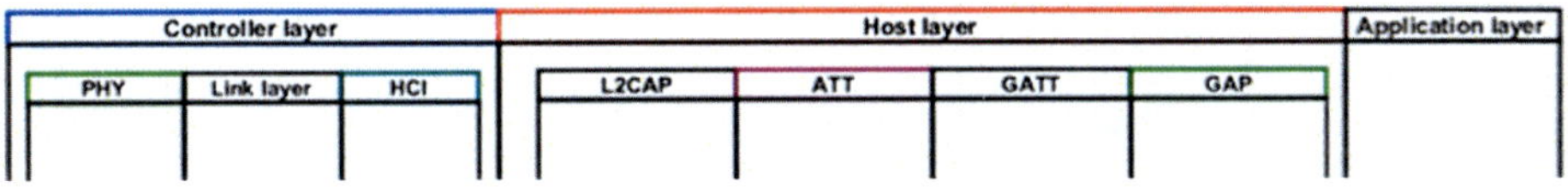

Fig. 5. General model BLE

For PHY layer defines the physical characteristics of the Bluetooth transmitter and receiver. Also, for the physical link. The baseband establishes the connection in a piconet.

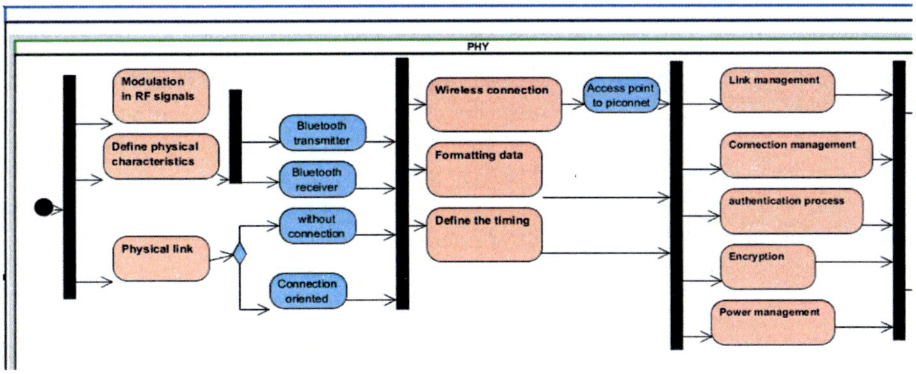

Fig. 6. The physical sub-layer

It pro-vides data formatting for transmission to and from the radio layer. In addition, PHY defines timing, framing, and flow control. Thus, it provides several functionalities shown in the (Fig. 6).

Link layer ensures the establishment of the connection and the management of packages. HCI is a standardized communication format between a host system and a Bluetooth controller. Indeed, the Bluetooth specification makes it possible to split its implementation into a controller, which manages the Physical and Link layers and a host, which manages the upper layers. This allows for interoperability between the different Bluetooth components. This communication can carry out between four types of transport: USB, UART 1 Wire, UART 3 Wires and SDIO. HCI allows both the configuration of its controller, the creation of a connection, and the reception and transmission of data on the physical layer (Fig. 7).

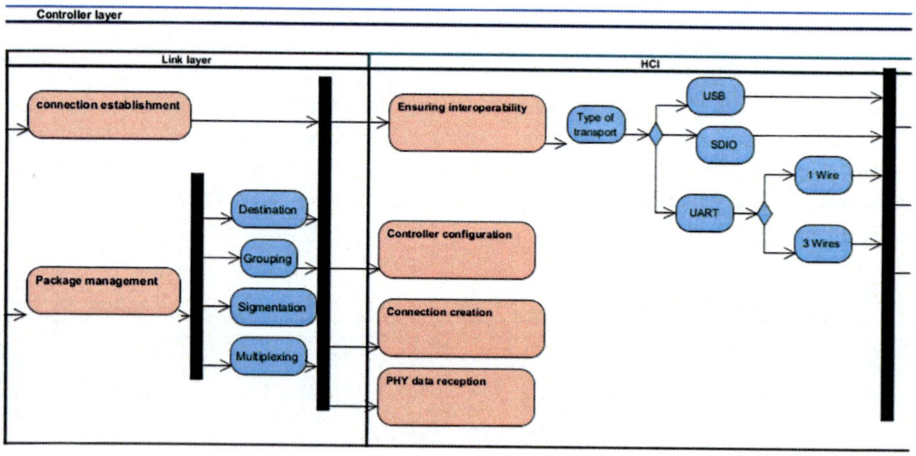

Fig. 7. Link and HCI sub-layers

L2CAP provides data flow control, fragmentation, and reassembly. It also manages the quality of service (QoS) to prioritize certain applications over others. L2CAP can see as the basis for creating new features for Bluetooth. The GATT relies on the ATT data format to provide a sort of database in the form of attributes that can be services or characters. The GATT takes care of ensuring the interaction of the system (the client and the server). Thus, it ensures the management of the data. GAP allows access to the configuration, advertising, scanning and connection functions of the controller. It also defines the role of the device: Broadcaster, Observer, Peripheral or Central. In BLE, GAP enforces the existence of GATT in the software stack of the Bluetooth device. Thus, any device called Blue-tooth Low Energy necessarily implements this profile. The application layer allows the interaction between user (Fig. 8).

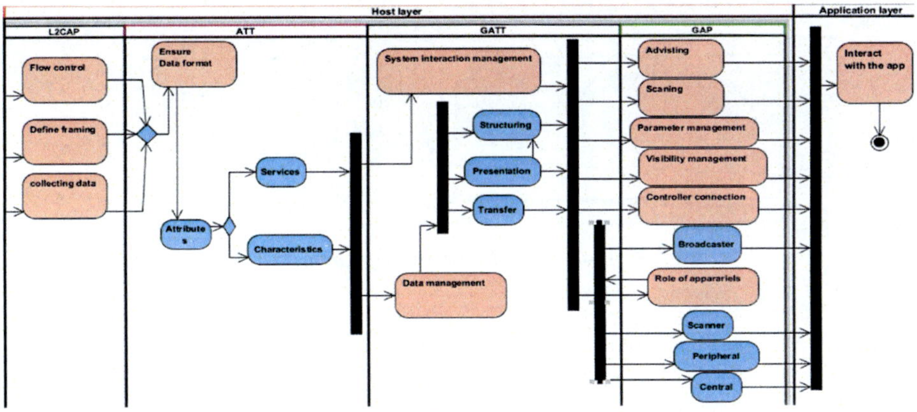

Fig. 8. L2CAP and ATT and GATT sub-layers and application layer

5 Projection of the Criteria on the Operating Models

After the detailed study of the protocol and the analysis of the operating models, we project the criteria of the protocols on this model. It allows us to categorize the criteria of the protocols thereafter (Figs. 9, 10 and 11).

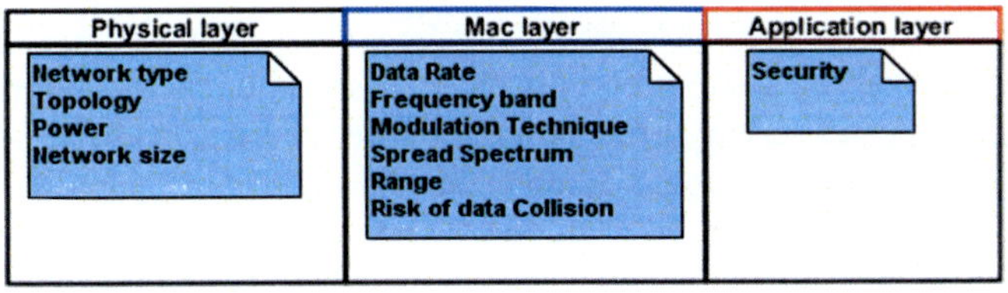

Fig. 9. Projection of criteria on the model Z-wave

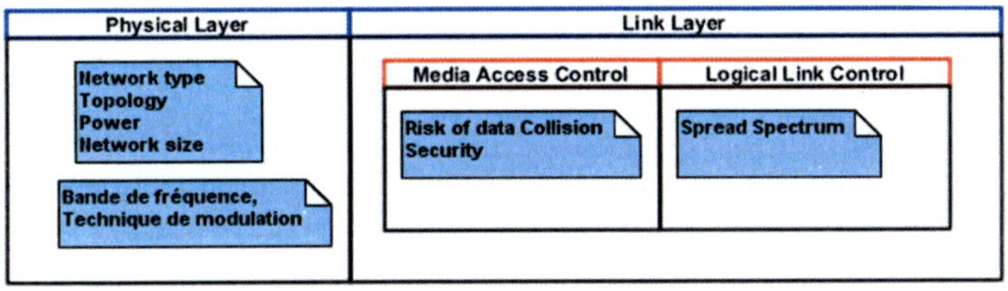

Fig. 10. Projection n of criteria on the model WiFi

From this projection, we classify the criteria in all service, namely:

- Specification service: which gives general information on the protocols. We cannot project this criterion on these models, then that 'there are not related to the operating model.

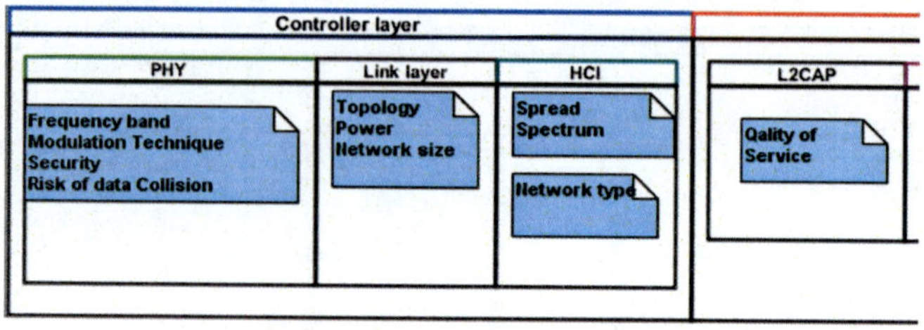

Fig. 11. Projection of criteria on the model BLE

- Network service: who assess the protocol based on network quality
- Data management assessment service: who to evaluate the protocol based on the given quality.

6 Conclusion

Wi-fi, Bluetooth, Z-wave radio communication protocols are becoming major for IoT. They are used extensively for medium-distance IoT applications and especially in the home. In this article, we provide five sections. After a general introduction, first we presented the protocol to medium range. Then we develop a comparative study based on criteria and measurements which allows the evaluation of these protocols. Subsequently, we propose models of the functioning of these protocols which learn and understand and detailed the functioning of these protocols using the activity diagram. Finally, we project the criteria which evaluate these protocols on the models of their functioning. These projections allowed us to classify these criteria to three services namely: Specification, Network service, Data management assessment service. These operating models and this projection and the other published study which have already been mentioned in the interlocution allow us to propose a guide for choosing network and application protocols and which is among the most common problems of the Internet of Objects. So, these studies will help to identify the priority of the protocol criteria to propose a protocol decision model and that will be our next article.

References

1. Glaroudis, D., Iossifides, A., Chatzimisios, P.: Survey, comparison and research challenges of IoT application protocols for smart farming. Comput. Netw. **168**, 107037 (2020). https://doi.org/10.1016/j.comnet.2019.107037
2. ELhadi, S., Marzak, A., Sael, N.: Operating models of application protocols. In: Proceedings of the 4th International Conference on Smart City Applications, Casablanca, Morocco, pp. 1-7 (2019). https://doi.org/10.1145/3368756.3369073
3. Elhadi, S., Marzak, A., Sael, N.: Functional modeling of IoT protocols. In: Ben Ahmed, M., Boudhir, A.A., Santos, D., El Aroussi, M., Karas, İR. (eds.) SCA 2019. LNITI, pp. 619–633. Springer, Cham (2020). https://doi.org/10.1007/978-3-030-37629-1_45
4. Elhadi, S., Marzak, A., Sael, N.: IoT Short-range network protocols: analytical study and operating models. EAI Endorsed Trans. Internet Things **7**(25), 25 (2021). https://doi.org/10.4108/eai.16-2-2021.168717

5. Operating models of Network protocols IoT: Short-range protocols|IEEE Conference Publication | IEEE Xplore. https://ieeexplore.ieee.org/abstract/document/9523646/ (consulté le 22 sept 2021)
6. Operating Models of Network Protocols IoT: Long-Range Protocols|SpringerLink. https://link.springer.com/chapter/https://doi.org/10.1007/978-3-030-66840-2_81. (consulté le 22 sept 2021)
7. Salman, T., Jain, R.: A Survey of Protocols and Standards for Internet of Things, Adv. Comput. Commun. (2017). https://doi.org/10.34048/2017.1.F3
8. Naidu, G.A., Kumar, J.: Wireless protocols: Wi-Fi SON, bluetooth, ZigBee, Z-Wave, and Wi-Fi. In: Saini, H.S., Singh, R.K., Kumar, G., Rather, G.M., Santhi, K. (eds.) Innovations in Electronics and Communication Engineering. LNNS, vol. 65, pp. 229–239. Springer, Singapore (2019). https://doi.org/10.1007/978-981-13-3765-9_24
9. Khanchuea, K., Siripokarpirom, R.: A multi-protocol IoT gateway and WiFi/BLE sensor nodes for smart home and building automation: design and implementation. In: 2019 10th International Conference of Information and Communication Technology for Embedded Systems (IC-ICTES), Bangkok, Thailand, mars, pp. 1–6 (2019). https://doi.org/10.1109/ICTEmSys.2019.8695968
10. Zemrane, H., Baddi, Y., Hasbi, A.: IoT Smart home ecosystem: architecture and communication protocols. In: 2019 International Conference of Computer Science and Renewable Energies (ICCSRE), Agadir, Morocco, pp. 1–8 (2019).https://doi.org/10.1109/ICCSRE.2019.8807652
11. Oliveira, G.M.B., et al.: Comparison between MQTT and websocket protocols for IoT applications using ESP8266. In: 2018 Workshop on Metrology for Industry 4.0 and IoT, avr. pp. 236–241 (2018). https://doi.org/10.1109/METROI4.2018.8428348
12. Elhadi, S., Marzak, A., Sael, N., Merzouk, S.: Comparative Study of IoT Protocols, SSRN Electron. J. (2018). https://doi.org/10.2139/ssrn.3186315

Improved IOT/SDN Architecture with the Concept of NFV

Jaadouni Hatim(✉), Saadi Chaimae, and Chaoui Habiba

Team of Data Analysis and Information Security, System Engineering Laboratory, National School of Applied Sciences, Ibn Tofail, Kénitra, Morocco
hatimjaadouni1@gmail.com

Abstract. For a standard software-defined Network (SDN), the controller class the flux gotten for the switch to using it in the network virtualization platform (NFV). we wanted to create a new IoT architecture forced with SDN while adding NFV to the equation. And then we see that the IoT is taking the development between the internet and the new connected devices in our everyday life to another level. Therefore, to be more effective and secure, it needs new approaches and new architecture designs to keep a good level of performance and security. Our paper presents a developed IoT architecture design with the use of SDN technology and also possibilities to use NFV and how it's going to all work!

Keywords: IOT · SDN · NFV · Architecture

1 Introduction

In distinction to ancient networks, software-defined network (SDN) [1] is here to manage plain information and make the network more programmable and adaptable [1]. SDN manages this information by creating a centralized controller and dispatching the information safely with the help of OpenFlow protocol [2–4]. Because the controller contains an international read of the network, SDN [1] is appropriate for controlling property. If network operates virtualization (NFV) modules [3] are to be deployed as SDN services [4], management may be forced to handle a variety of control messages and incoming packets from the information plane to classify or analyze packet contents. The old IOT platforms [5] and the used protocols aren't built to accommodate high levels of measurability, volume, or quality, let alone the higher-than-mentioned requirements. They are unstable and limited in their ability to meet these new requirements. Furthermore, with the massive numbers of connected things predicted by Gartner [6], it's getting hard to control the connected nodes, which generate an enormous number of information, all this and not having physical property & suppleness naturally owned and used inside the structure the. If the networks aren't ready, its influx of the connected environment (IoT), which generates massive amounts of traffic, could bring the network to a halt.

To accomplish these objectives, emerging techs [7, 8] such as SDN and network function virtualization (NFV) are now looked at as the new tech that can propose a more efficient solution to the equation. Software-Defined Network (SDN) in these past few

S. Motahhir and B. Bossoufi (Eds.): ICDTA 2022, LNNS 454, pp. 294–301, 2022.
https://doi.org/10.1007/978-3-031-01942-5_29

years had an excessive amount of attention from experts all over the world [1] all having a goal to make the IoT a safer place. The work presented in this paper will be organized as follows: Sect. 2 will explore mainly the background of SDN. Section 3 is a fallback to the work already done by other researchers and then Sect. 4 will discuss the proposed architecture of the IoT architecture including NFV and SDN and its details then Sect. 5 presents the full communication and how the new architecture works. The last section concludes the work and gives future scopes and an area for potential directions.

2 Background

The old network sensors tend to lock off computer code and functions produced by creators, making it difficult to experiment with new networks. The lack of flexibility in network management is exacerbated by the tightly connected management plane and knowledge plane. The organic process that leads to the necessity for a programmable network and, as a result, the separation of the management plane and the knowledge plane is known as SDN. McKeown et al. proposed OpenFlow to standardize communications between the centralized controller and, as a result, controlled switches in SDN [4]. The idea comes from past work in programmable networks, such as active networks, which allow packets to carry directives and network devices to die. When it comes to the separation of the management plane and thus the knowledge plane in programmable networks, frameworks such as the forwarding and management part separation (ForCES) framework were developed to standardize communications between multiple management and knowledge components at constant network intervals. OpenFlow followed, and its success aided in the widespread adoption of SDN. Because of the limited space, we tend to suggest readers to [4] for a comprehensive overview of Open-Flow development.

3 Related Work

Cisco, Qosmos, Juniper, Huawei, and Ericsson [7–11] all provide solutions for constructing service chains that include traffic categorization functionalities. We, on the other hand, are concentrating on expanding the data plane and OpenFlow protocol messages. The two-tier classification, enhanced OpenFlow protocol messages, and service chaining techniques are the main differences from previous efforts in terms of traffic categorization. Furthermore, past works do not fully reveal or evaluate its design. Curtis et al. [6] built the DevoFlow framework to maintain as many flows on the data plane as feasible and decrease controller traffic overheads, but that work has no bearing on traffic classification on the switch or NFV provisioning. Shin et al. [12] installed the FRESCO architecture on the controller to provide security services, but it did not lessen the controller's load and is irrelevant to NFV.

3.1 Proposed IoT Architecture Including NFV and SDN

In this part, we'll present the IOT structure with the addition of the centralized controller from the SDN technology and function virtualization on the NFV.

First, we'll light a spot on the original IOT design and how it works and then the SDN and NFV, finally we'll end with their combination in a proposed architecture.

3.2 IoT Design

Figure 1 shows an associate IoT design that consists of 3 layers in keeping with Al-Fuqaha,fAla, et alf [13] particularly these layers: Perception, network, and also application.

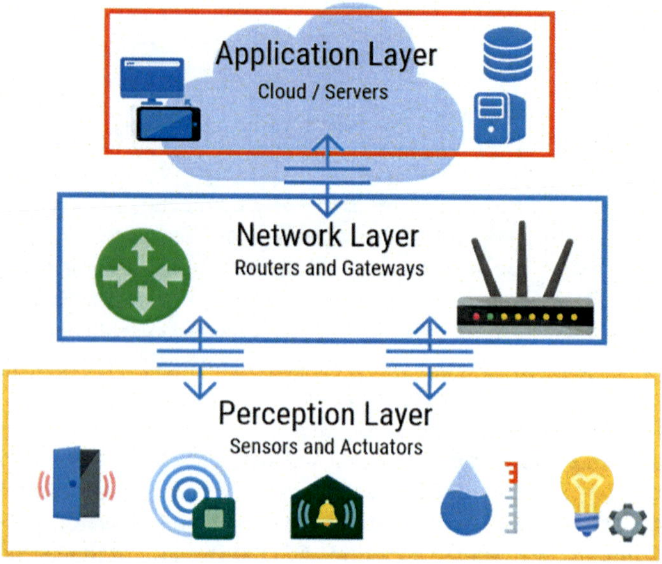

Fig. 1. IoT Architecture

1) Perception Layer: it contains the sensing layer and that's composed of varied reasonably physical with low energy use (like RFID, good meters, WSN, NFC, etc.) that permits communication with IOT. They sense associated collect knowledge from varied varieties of devices in an intelligent means. To summarize, the importance of this part is to identify and communicate between devices.
2) Network Layer: The information gathered by these devices must be disseminated and archived. The network layer is responsible for this. It connects these smart/intelligent items to other smart/intelligent objects. In addition, it is in charge of data transport. The network layer is responsible for establishing connections between smart objects, network devices, and servers. It's also how sensor data is distributed and analyzed.
3) Application Layer: This application layer is where the user interacts. It is in charge of offering software resources to the consumer. For example, with a smart home app, users may turn on a coffee machine by pressing a button in the app. The application layer is responsible for supplying application-specific resources to the consumer. It defines several IoT applications, including smart homes, smart cities, and smart health.

Making the IoT systems more secure presents different obstacles to deal with such as problems in the protection of data and privileges, unreliable communications, and hostile environments [14]. And as given in Table 1, there is plenty of related security problem in the perception layer. And that is caused by the easy way of access physical the sensors placed outside.

Hence, we can resume from this that in IoT, sensors, node, and material devices are the main attack surface for hackers.

Table 1. Table Current challenges in IoT security and the given solution [15].

Layer	Security challenges	Mitigation
Perception	Detection of the unusual sensor node The decision cryptography algorithms and key management algorithms to be used Anonymity of data and sender Device vulnerabilities	Deficiency detection algorithm, decentralized intrusion Detection system public key encryption due to the large-scale network Slot reservation protocol Access control, relief of asset consumption attacks
Network	Enabling IPSec; Communication with IPV6 node	Researching in the reasonablenessaofaIPv6 and IPSec for secure correspondence
Application	The computer-systems is more embedded and configurable	No solution or suggestion is proposed from this paper

3.3 SDN Background

The management plane (low-level configuration and options, etc.) is concerned with the proprietary software of outdated network parts (e.g., switches or routers) while setting them up. This implies that networks are difficult to dynamically adapt and administer. Software-Defined Networking (SDN), as defined in [14], attempts to address the issue by separating the administration and knowledge planes, allowing for a software-based efficient solution, as illustrated in Figure one. [15] presents a high-level bailiwick overview of SDN, which delves further, but it consists of three basic principles:

- Separation of the control and knowledge planes.
- Management is logically centralized.
- External applications have access to network resources and states that are abstracted.

The concept is here to make network management and the configurability easier within the installed software component, enabling all of the benefits that go along with it (such as dynamic control). You may monitor overall network performance and make dynamic configuration changes using a logically centralized controller (be it re-routing traffic, or ap-plying a replacement information measure rate limit to a greedy flow, etc.)

3.4 SDN Based Security Applications

It exists hardly any market for newly created solutions just to improve the existing SDN controllers, also there's no other Software Defined Network-based security addons or research or analysis!

Redware Flow of Defense: Redware [16] is a network DDoS attack prevention and cyber control application designed to detect and mitigate network-wide, multi-vector attacks. Defense Flow supports always-on/SmarTap and hosted customer protection use cases for service providers to provide the widest DDoS attack defense and detection coverage coupled with immediate attack mitigation.

Organizations may use Radware's DefenseFlow Software Defined Networking (SDN) solution to dynamically mitigate Network, Application, and APT assaults while leveraging the whole network capacity.

New security risks to telecommunications providers and businesses necessitate the quick deployment of security protection systems capable of detecting and preventing unforeseen assaults at unknown scales. Radware DefenseFlow is an SDN-based adaptive attack mitigation system that may drastically reduce security capability provisioning time while seamlessly integrating into a wide range of networks. DefenseFlow is a network-wide DDoS attack detection and cyber command and control tool that protects networks from known and new network assaults that jeopardize network resource availability. DefenseFlow is a software-defined network (SDN)/OpenFlow network that supports both classic NetFlow and new SDN/OpenFlow networks. DefenseFlow, when used in conjunction with DefensePro, allows carriers, ISPs, and cloud service providers to assure service availability by preventing network and application DDoS assaults.

4 The Extended Architecture

The approval of the Network Function Virtualization (NFV) in the IoT in combination with the concept of Software Defined Network (SDN) opens up ginormous possibilities for increasing its system potency and also the lightness of the system for the IoT applications. By integrating network services in the cloud and decoupling network administration and management performance from hardware, the infrastructure becomes more diverse and property, reducing the reliance of emerging wireless technologies on hardware. As illustrated in Fig. 2, we choose to offer an SDN with NFV design for IoT. This concept incorporates a virtualized approach to the IoT framework and expands on the SDN architecture for IoT. The design is based on the NFV architecture illustrated in Fig. 1, but with several changes to match the SDN concept. Due to NFV, the info layer consists of multiple network resources arranged into virtualized resources. We have a propensity to presume that all virtual network functions (VNF) are SDN-enabled in this work because we might have network functions virtualized without SDN capabilities. As a result, we tend to refer to virtual network components as VNFs that are SDN aware. We've created an SDN/NFV edge node to allow network operators to rapidly deploy new services, ensuring that IoT apps receive unique features such as ultra-low latency, customized user context (location information), and high data measurement, among other things. When extreme-real-time connections are necessary between IoT applications with a large number of devices, this is one of the most critical demands for the tactile web.

The applying layer, which contains IoT servers for various IoT applications, is at the top of the design. We even have 'Operations Support Systems' (OSS) and Business Support Systems (BSS), that encompass a variety of networks and their application used a service providing uses for a goal to manage the whole operations. Because of increasing the two big concepts of Software Defined Network (SDN) and Network Function Virtualization (NFV), which can run aggressively using to develop creative methods of taking care of operations and exploring new methods of it for IOT, the OSS/BSS is moving to a brand-new generation type. The associate operator's BSS/OSS is also linked to the fluid way to manage and deal with systems through normal interfaces.

Within the realm of quantifiability, a common example of highlighting the efficacy of the proposed design might be found. We tend to believe that using several gateways will resolve quantifiability issues, however, their square measure constraints are at the bottleneck due to overloaded knowledge flow when many IoT devices provide data at the same time. With NFV, we'll be able to execute our network functions on the cloud, which will provide Brobdingnagian benefits in terms of quantifiability. Because of the growing number of IoT devices, NFV will be used to move a range of IoT entrance functionality into the virtual realm, providing both users and repair operators with more options. The separation of network operations, let alone SDN, allows the cloud architecture to allocate resources automatically to meet the increased load. This also provides dynamic quantifiability by allowing users to scale the performance of virtual network operations in real-time.

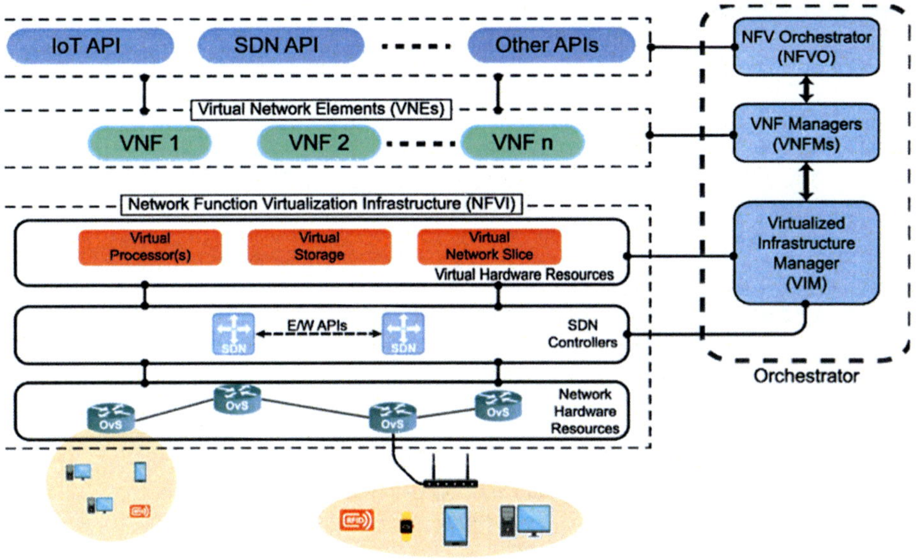

Fig. 2. SDN/IoT implemented with NFV

The area of quality management is another subject where the pertinence of the proposed technique might be highly beneficial. As previously stated, we have a proclivity to employ virtualized SDN- Gateways to replace traditional gateways. By virtualizing the IoT entry, NFV and SDN can make it easier to change a programmable SDN plane and maintain active session continuity for IoT devices. By allowing dynamic sensor connection to numerous networks, entry virtualization aids in the growth of IoT. This might allow smart devices to move from one area to another, with the SDN controller adjusting the flow tables in OpenFlow switches to take the device's quality into account.

5 Conclusion and Future Scope

SDN and NFV are the next steps to change the whole communication platform, and they may become the principal structure of big new companies soon. We've detailed when and in which place we need to use them in the IoT new architecture to address the difficulties of quantifiability and quality in IoT networks, a straightforward and generic SDN-IoT architecture using NFV was proposed. It is fair enough to be flexible and easy to use in IoT scenarios by virtualizing the IoT entranceway. We also see the market side of things by adding a software-defined network (SDN) with the use of NFV to existing infrastructure, and it's clear that the strategy provides new income potential inside the services value chain. We believe that combining both of those powerful concepts in the IoT infrastructure and architecture will provide intriguing opportunities as well as problems.

Acknowledgment. I'd like to thank Pr. SAADI Chaimae and Pr. CHAOUI Habiba for their insightful and helpful recommendations during the conception and implementation of this research project. their willingness to sacrifice so much of their time and knowledge has been much appreciated.

References

1. ONF. SDN Architecture Overview, 12 December 2013
2. ONF. OpenFlow Switch Specification version 1.4.0, 14 October 2013
3. ETSI. Network Functions Virtualization — Introductory White Paper. http://portal.etsi.org/NFV/NFV_White_Paper.pdf
4. Lara, A., Kolasani, A., Ramamurthy, B.: Network Innovation using OpenFlow: A Survey. IEEE Commun. Surv. Tutor. **16**(1), pp. 493–512 (2014). 1st Quarter
5. Quinn, P., et al.: Network service header. IETF Internet-Draft, draft-quinn-sfc-nsh-02.txt, 14 February 2014
6. Curtis, A.R., et al.: DevoFlow: scaling flow management for high-performance networks. In: ACM SIGCOMM, August 2011
7. Cisco. Enabling Service Chaining on Cisco Nexus 1000V Series. http://www.cisco.com/c/en/us/products/collateral/switches/nexus-1000v-switch-vmware-vsphere/white_paper_c11–716028.pdf
8. Qosmos. Service-Aware Network Architecture Based on SDN, NFV, and Network Intelligence. http://www.qosmos.com/wp-content/uploads/2014/01/Intel_Qosmos_SDN_NFV_329290–002US-secured.pdf

9. Juniper Networks. "Contrail Architecture". http://www.juniper.net/us/en/local/pdf/whitepapers/2000535-en.pdf
10. Huawei. Enabling Agile Service Chaining with Service Based Routing. http://www.huawei.com/ilink/en/download/HW_308622
11. John, W., et al.: Research directions in network service chaining. In: IEEE SDN for Future Networks and Services (SDN4FNS) (2013)
12. Shin, S., et al.: Fresco: modular composable security services for software-defined networks. Network & Distributed System Security Symposium (NDSS), February 2013
13. Al-Fuqaha, A., Guizani, M., Mohammadi, M., Aledhari, M., Ayyash, M.: Internet of things: a survey on enabling technologies, protocols, and applications. IEEE Commun. Surv. Tutor. **17**(4), 2347–2376 (2015)
14. Chen, K., et al.: Internet-of-Things security and vulnerabilities: taxonomy, challenges, and practice. J. Hardw. Syst. Secur. **2**, 97–110 (2018)
15. Yuchen Yang, H.Z., Wu, L., Yin, G., Li, L.: A survey on security and privacy issues in internet-of-things. In: 2015 10th International Conference Internet Technology Security Transaction, vol. 4, pp. 202–207 (2015)
16. Radware Ltd. "Defense Flow 2.0 - Flow-Based Detection, Technical report (2015)

Evaluating IoT-Based Healthcare Architecture with Quality Factors

Loubna Chhiba[1(✉)], Sakina Elhadi[1], Abdelaziz Marzak[1], and Mustapha Sidqui[2]

[1] Faculty of Science Ben M'sik, Laboratory of Technology of Information and Modeling, Hassan II University, Casablanca, Morocco
chhibaloubna@gmail.com

[2] Faculty of Dental Medicine of Casablanca, Hassan II University, Casablanca, Morocco

Abstract. The Internet of Things (IoT) has become a major technological revolution. Evaluating any IoT advancements comprehensively is critical to understand the conditions under which they can be more useful, as well as to assess the robustness and efficiency of IoT systems to validate them before their deployment in real life. The compliance of IoT architecture to quality is paramount to achieving users' satisfaction. Currently, we do not have a comprehensive set of guidelines to appraise and select the most suitable IoT platform architectures that meet relevant criteria. Considering the challenges associated with developing IoT healthcare applications. This paper proposes a reference platform as a quality guideline to help both researchers and practitioners who are developing and maintaining IoT platforms. We focus on the suggestion of quality factors to evaluate IoT-Based Healthcare Architecture. We adopted the design science research approach to develop a novel evaluation.

Keywords: Design science research · IoT platform · Architecture evaluation · Quality factors · Smart healthcare

1 Introduction

Many IT-based organizations have shown their interest in embarking on IoT platforms to implement and integrate the next generation of data-intensive Internet-based applications. IoT platforms provide the required infrastructure and core services for connectivity and managing a large amount of data, dealing with a large variety of smart healthcare, and improved scalability. From the perspective of an IoT platform user or consumer, who views an IoT platform as a black box, it becomes a challenging exercise to select one or more platforms that fulfill business requirements as each platform offers features that are common with other platforms and distinct features at different levels and costs. That is, reliable decision-making depends on an understanding of pertinent quality factors contributing to a quality IoT platform as well as a rigorous way to qualitatively/quantitatively measure these factors.

This understanding mitigates from a cursory selection and unforeseen contingencies which may result in the poor quality and maintenance overhead cost of applications that

S. Motahhir and B. Bossoufi (Eds.): ICDTA 2022, LNNS 454, pp. 302–311, 2022.
https://doi.org/10.1007/978-3-031-01942-5_30

are deployed on a platform. Furthermore, practitioners may need to compare and rank IoT platforms based on their specific set of requirements with different priorities.

The number of healthcare applications incorporating the concept of the Internet of Things (IoT) is increasing considerably [1–9]. The global market size for IoT-connected devices is expected to reach 41.6 billion by 2025. This increase cannot continue without high-quality assurance. There are some difficulties in testing IoT-Healthcare applications; system heterogeneity, mobility, connectivity, cost, and accessibility in real-time. Therefore, it is necessary to have differentiated quality standards and quality models. One of the most important methods of verifying the quality of the software is the quality assurance (QA) approach.

Quality compliance of the IoT-Healthcare architecture is essential to achieve user satisfaction. However, there are many challenges in developing and deploying this type of architecture, such as interoperability, availability, usability, and security.

The purpose of this paper is to introduce criteria to support the evaluation and comparison of IoT platform architectures. We compiled a list of commonly agreed criteria derived from the IoT literature together with measuring guidance to provide a basis for an in-depth evaluation of IoT platforms. We applied the Design Science Research (DSR) approach to achieving our research objective. The DSR relies on the theory-based development of a design artifact for real-world problems and an interplay between the design and use of the artifact. This article is structured as follows: Sect. 2 provides background on IoT platform architecture and the problem of IoT platform evaluation and selection. Section 3 presents the full description of the proposed criteria related to assessing the IoT platform, Sect. 4 describes the software reference platform for healthcare applications. Finally, this article concludes in Sect. 5.

2 Research Background

Typically, an IoT-based system constitutes a few key building elements, namely, people, applications, infrastructures, things, and platforms, as shown in Fig. 1. An IoT platform in such a harness is a key element. It is defined as an integrated middleware environment relying on enabling technologies such as cloud computing, big data analytics platforms, and cyber-physical systems to provide necessary infrastructure and core services to enable the development and integration of applications for different domains, such as smart healthcare and smart cities.

We view an IoT platform architecture from the perspective of software architecture that is defined in conventional software engineering, i.e., breaking the platform down into its high-level interconnected structure such as layers and software/hardware components [12]. Several architectural representations with a different recognition of layers exist in IoT literature; though when collectively viewed, they have intersections.

Cerina et al. [13] proposed a multi-level architecture leveraging fog computing and context awareness. The authors used the Field-Programmable Gate Array (FPGA) as a fog node. The architecture consists of four layers, namely, i) interaction layer, ii) mesh layer, iii) fog layer, and iv) cloud layer.

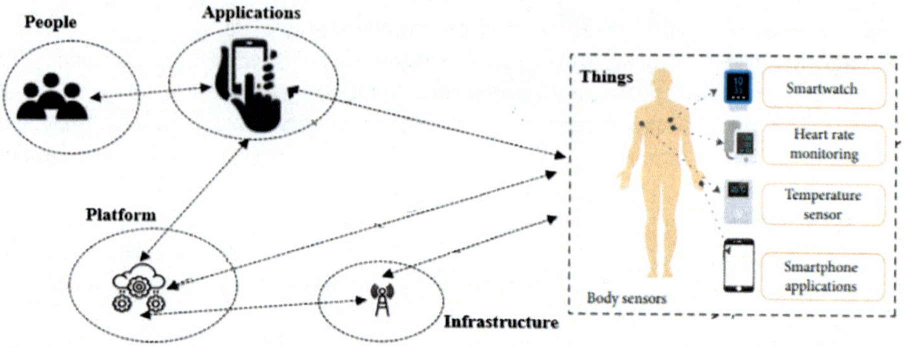

Fig. 1. Key elements of an IoT-based system

The authors suggested the removal of the mesh layer in the case of time-critical healthcare applications. Nandyala et al. [14] proposed a 4-tier architecture for real-time ubiquitous healthcare monitoring. The four tiers are. Namely, i) smart devices tier, ii) fog tier, iii) core tier, and iv) cloud tier. This architecture was heavily influenced by CISCO's fog architecture. The task of the core tier is to provide security services. Verma et al. [15] proposed a five-layer fog-based architecture for remote patient health monitoring. The layers are i) data acquisition layer, ii) event classification layer, iii) information mining layer, iv) decision making layer, and v) cloud storage layer. The event classification layer was implemented using a smart gateway at the fog layer. In line with these definitions, the architecture of an IoT platform may constitute the following layers:

- **The user interface layer** provides interactivity between users and platform services via providing dashboards, reports, message boards, 3D spaces, and 2D maps so that users can get access services based on the subject of interest and send queries.
- **The application layer**: This layer covers smart environmental spaces such as transportation cities, factories, and emergency health care. This layer enables users to receive data from smart objects, perform processing algorithms over the data, and send the results back to users or other objects.
- **The service layer**: This layer enables the processing of information possible through controlling the security, analytic, process modeling, and management devices. The most important feature of this layer is the business and the process rule engines. IoT brings connection and interconnection of objects and the system together presenting the information in the form of data, events, current location, and traffic data.
- **The data layer** (Middleware Layer): This layer keeps data from various data sources floating across the operational environment such as real-time data from sensors.
- **Physical layer** (Smart Device/Sensor Layer) provides hardware for data processing, storing, computing, and interconnectivity among data centers, servers, and smart objects via different network communication protocols.

3 The Proposed Criteria Related to Assessing IoT-Healthcare Platform

The criteria in this dimension examine the platform support of commonly expected non-functional requirements, i.e., quality factors. It is important to realize that these criteria are orthogonal to the functional requirements. For example, the security criteria are concerned across all layers of the IoT platform. These criteria are listed in Table 1. In the following, we describe the criteria the layers for which they are concerned in an architecture assessment exercise.

Table 1. MCs Criteria related to assessing IoT platform functions (UL: user interface layer, AL: application layer, SL: service layer, DL: data layer, PL: physical layer, p: related, ×: Not related).

Criterion	Leading evaluation question(s) for measuring criterion	Layer				
		UL	AL	SL	DL	PL
Security	Does the platform limit the data acquisition from users' devices and smart objects to keep the security and privacy of users?	✓	✓	✓	✓	✓
	Does the platform enable a user to define a customized smart object's discoverability at different levels? These levels can be, for example, public where an object can be discovered to all users, private where an object is discoverable by the object owner, and friends where the object is discoverable by a friend key provided by thE virtual object.	X	X	X	X	✓
	Does the platform define encryption/decryption mechanisms across layers?	✓	✓	✓	✓	✓
Interoperability	Does the platform provide integration mechanisms, protocols, or middleware such as HTTP, MQTT (message queuing telemetry transport which is a broker-based publishing/subscribing), an AMQP (advanced message queuing protocol) for integrating applications?	X	✓	X	X	X
	Does the platform use mechanisms to exchange and process heterogeneous events coming from multiple sources at different levels of the system?	X	✓	✓	✓	✓

(continued)

Table 1. (*continued*)

Criterion	Leading evaluation question(s) for measuring criterion	Layer				
		UL	AL	SL	DL	PL
	Does the platform have mechanisms, e.g., ontologies?	X	X	✓	X	X
	Does the platform provide mechanisms such as web service, virtual object, representational State Transfer/ReST to interact with other IoT platforms to call/use their services/data?	X	X	✓	X	X
	Does the platform provide mechanisms to identify heterogeneous resources in a network such as a sensor networks, IP networks?	X	X	X	X	✓
	Does the platform provide the necessary interfaces to facilitate invoking its services?	X	X	✓	X	X
	Does the platform provide open APIs to avoid vendor lock-in issues?	X	✓	✓	✓	✓
Scalability	Does the platform include mechanisms for effective processing algorithms?	X	✓	✓	X	X
Performance	Does the platform include mechanisms for effective processing algorithms?	X	✓	✓	X	X
Usability	Does the platform provide a suitable and effective user interfaces for interaction with presenting information to users?	✓	X	X	X	X
Mobility	Does the platform provide mechanisms for mobility protocols in the support of continuous user movement to different locations?	X	X	X	X	✓
Reusability	Does the platform have templates of objects/services to be used for other scenarios different from the default?	X	✓	✓	X	X
	Does the platform have the necessary interfaces to hide service incompatibilities and enable the reuse of calls by its services?	X	✓	✓	X	X

(*continued*)

Table 1. (*continued*)

Criterion	Leading evaluation question(s) for measuring criterion	Layer				
		UL	AL	SL	DL	PL
	Does the platform allow modifying and reusing its core functions/services for applications?	X	✓	✓	X	X
	Does the platform apply architecture design principles such as loose coupling, modularity, or layering to facilitate reuse?	X	✓	✓	X	X
	Does the platform have simple and small services instead of coarse grain and composite services?	X	✓	✓	X	X
	Can the platform operate across multiple connectivity protocols?	X	✓	✓	✓	✓
Availability	Does the platform provide component/data clustering or replication/distribution mechanisms to increase availability?	X	✓	✓	✓	✓
	Does the platform provide redundant physical links for the network layer?	X	X	X	X	✓
	Does the platform provide redundant nodes, e.g., routers allowing service discovery, for PL?	X	X	X	X	✓
Adaptability	Does the platform define mechanisms such as machine learning algorithms, rules, or policies to support the dynamic transformation from the current configuration to a new configuration or in the case of occurring disruptive events?	X	✓	✓	X	X
	Does the platform have mechanisms for incorporating upcoming functional, data, and objects to accommodate new platform configuration toward continuous improvement?	X	✓	✓	X	X

(*continued*)

Table 1. (*continued*)

Criterion	Leading evaluation question(s) for measuring criterion	Layer				
		UL	AL	SL	DL	PL
Maintainability	Does the platform simplify applying refinements, correcting faults to its components, or adding new functionalities/increments to the existing components via applying design principles such as low coupling, high cohesion, and separation of concerns?	X	✓	✓	X	X
	Does the platform define clear definitions of objects such as naming, icons, documentation/modeling, and interactions, as well as design principles such as modularity, less coupling, and separation of concerns, to make modifications simpler and less costly?	X	✓	✓	X	X
Efficiency	Does the platform define mechanisms to estimate energy consumption by its internal and external components?	X	✓	✓	X	✓
	Does the platform have data storage related to energy logs to store a history of energy consumption by objects?	X	X	X	✓	X
	Does the platform have data storage related to the environment to log the history of changes in the environments such as temperature, humidity, and the the number of people who are typically collected by objects?	X	X	X	✓	X
	Does the platform have mechanisms for service composition to resource constraints of involved smart objects such as batty power?	X	X	✓	X	X

4 A Software Reference Platform for IoT-Based Healthcare Applications

In this section, we propose a software reference platform, describing its elements and the relationship between them. From the criteria of IoT-based healthcare architecture, which were previously defined in Sect. 3, a reference IoT platform was designed for healthcare applications.

The reference platform, presented in Fig. 2, is organized into four layers according to its criteria. The criteria in the second dimension of the proposed reference platform examine the platform (architecture) support of commonly expected non-functional requirements, i.e., quality factors. It is important to realize that these criteria are orthogonal to the functional requirements. For example, the security criteria are concerned across the applications layer of the IoT platform. These criteria are listed in Table 1. In the following, we describe the criteria the layers for which they are concerned in an architecture assessment exercise.

The reference platform has been underpinned on a commonly grounded set of architecture non-functional criteria to which an ideal IoT platform is expected to adhere to. The rationale behind designing the framework has been to provide a basis for a fine-grained and in-depth evaluation and comparison of IoT platforms.

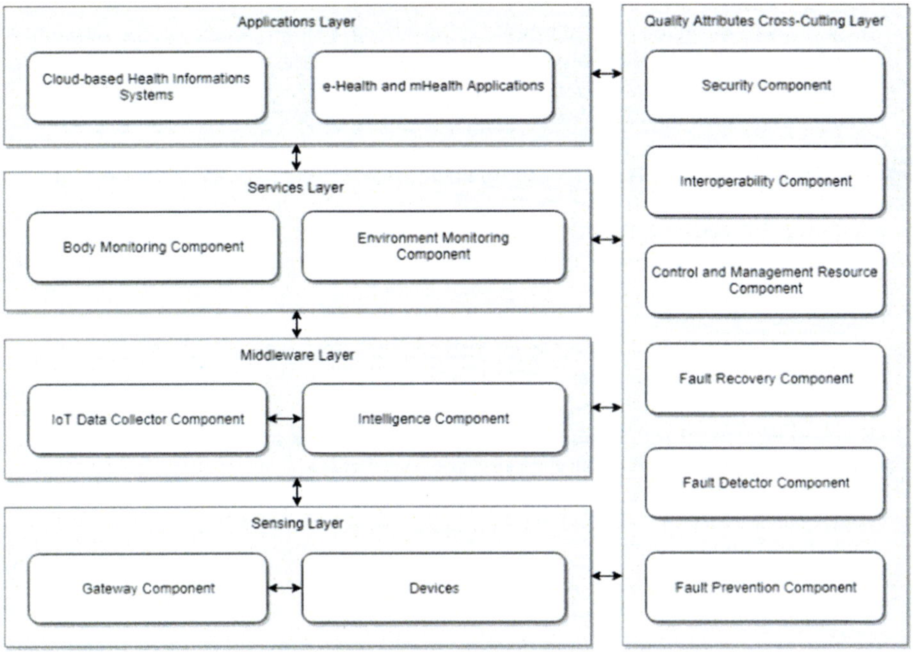

Fig. 2. Reference architecture

The reference is the first attempt to enable users to systematically assess different layers of platforms and can be viewed as a quality guideline for both researchers and practitioners who are developing and maintaining IoT platforms for healthcare applications.

5 Conclusion and Future Work

The growth in the number of IoT platform variants has raised the need for evaluation and comparison efforts, mostly to assist in appraising an existing platform architecture, selecting, or designing a new one aimed at satisfying specific quality criteria. In this context, this paper proposes a Criteria for IoT platform architecture evaluation, from these criteria a reference IoT platform was designed for healthcare applications.

In future work, we develop a novel evaluation framework consisting of criteria presented in this paper: functional- and non-functional-related criteria (quality factor). The reference will provide a basis for a fine-grained and in-depth evaluation and comparison of IoT platforms.

References

1. Ali, Z., Hossain, M.S., Muhammad, G., Sangaiah, A.K.: An intelligent healthcare system for detection and classification to discriminate vocal fold disorders. Futur. Gener. Comput. Syst. **85**, 19–28 (2018)
2. Yang, G., Xie, L., Mantysalo, M., et al.: A health-IoT platform based on the integration of intelligent packaging, unobtrusive bio-sensor, and intelligent medicine box. IEEE Trans. Industr. Inf. **10**(4), 2180–2191 (2014)
3. Yan, Y.: A home-based health information acquisition system. Health Inf. Sci. Syst. **1**, 12 (2013)
4. Khan, M., Han, K., Karthik, S.: Designing smart control systems based on internet of things and big data analytics. Wireless Pers. Commun. **99**(4), 1683–1697 (2018)
5. Nachankar, P.J.: IOT in agriculture. Decis. Making **1**(3) (2018)
6. Menon,V.G.: An IoT-enabled intelligent automobile system for smart cities. Internet of @ings, p. 100213 (2020)
7. Qin, E., Long, Y., Zhang, C., Huang, L.: Cloud computing and the internet of things: technology innovation in automobile service. In: Yamamoto, S. (ed.) HIMI 2013. LNCS, vol. 8017, pp. 173–180. Springer, Heidelberg (2013). https://doi.org/10.1007/978-3-642-39215-3_21
8. Froiz-Míguez, I., Fernández-Caramés, T.M., Fraga-Lamas, P., Castedo, L.: Design, implementation and practical evaluation of an IoT home automation system for fog computing applications based on MQTT and ZigBee-WiFi sensor nodes. Sensors **18**(8), 2660 (2018)
9. Mathew, P., Pillai, A., Palade, V.: Applications of IoT in healthcare. In: Sangaiah, A.K., Thangavelu, A., Meenakshi Sundaram, V. (eds.) Cognitive Computing for Big Data Systems over IoT. LNDECT, vol. 14, pp. 263–288. Springer, Cham (2018). https://doi.org/10.1007/978-3-319-70688-7_11
10. Fahmideh, M., Zowghi, D.: An exploration of IoT platform development. Inf. Syst. **87**, 101409 (2020)
11. Shin, D.-H.: Conceptualizing and measuring quality of experience of the internet of things: Exploring how quality is perceived by users Inf. Manag. **54**, 998–1011 (2017)
12. Bass, L., Clements, P., Kazman, R.: Software Architecture in Practice. Addison-Wesley Professional, Boston (2003)

13. Cerina, L., Notargiacomo, S., Paccanit, M.G., Santambrogio, M.D.: A fog-computing architecture for preventive healthcare and assisted living in smart ambients. In: 2017 IEEE 3rd International Forum on Research and Technologies for Society and Industry (RTSI) (2017)
14. Nandyala, C.S., Kim, H.-K.: From cloud to fog and iotbased real-time U-healthcare monitoring for smart homes and hospitals. Int. J. Smart Home **10**(2), 187–196 (2016)
15. Verma, P., Sood, S.K.: FogAssisted-IoT enabled patient health monitoring in smart homes. IEEE Internet Things J. **5**(3), 1789–1796 (2018)
16. Bichler, M.: Design science in information systems research. Wirtschaftsinformatik **48**(2), 133–135 (2006). https://doi.org/10.1007/s11576-006-0028-8
17. Kitchenham, B., Linkman, S., Law, D.: DESMET: a methodology for evaluating software engineering methods and tools. Comput. Control Eng. J. **8**, 120–126 (1997)
18. Taromirad, M., Ramsin, R.: An appraisal of existing evaluation frameworks for agile methodologies. In: Proceedings of the 15th Annual IEEE International Conference and Workshop on the Engineering of Computer Based Systems (ECBS 2008), Belfast, UK, 31 March–4 April 2008, pp. 418–427. IEEE, Piscataway (2008)

Terahertz Band Advances and Applications

Imane Halkhams[1] (✉) and Wafae El Hamdani[2]

[1] LSEED Laboratory, UPF, Fez, Morocco
imane.halkhams@usmba.ac.ma
[2] IASSE Laboratory, SMBA University, Fez, Morocco
w.elhamdani@ump.ac.ma

Abstract. The terahertz band, which is located between the microwave and the infrared bands, has shown countless assets that have allowed researchers to make huge advances in different sectors. In this paper, we try to present an overview of terahertz digital and analog applications in some areas such as communication, material characterization, and imaging. Terahertz communication systems deliver data rates of the order of TBps, allowing data-intensive applications to operate more quickly. Terahertz spectroscopy can detect chemicals or liquids that are difficult to identify with other techniques, using FTS (Fourier transform spectroscopy). Terahertz rays offer very high image resolution thanks to their wavelength. Applications include medical imaging, protein detection, the effects of radiation on biological processes, disease diagnosis and the analysis of the optical properties of human tissues. In the field of security, the detection of illicit or dangerous objects is done by short pulses lasers. Finally, TD non-destructive spectroscopy makes it possible to observe and preserve cultural heritage through short laser pulses. In the field of terahertz communication systems, we will present the simulation results of an active bandpass filter at the frequency 129 GHz. This filter has the particularity of being tunable over a 300 MHz band.

Keywords: Terahertz communications · Spectroscopy · Imaging · Sensing · Material characterization · Medical applications · Food industry · Security · Bandpass filter

1 Introduction

Terahertz (THz) waves, which are being exploited by researchers and industry, are revolutionising the world by offering a multitude of valuable applications. Hitherto confined to the laboratory, terahertz rays (also known as T-rays) have in recent years made their way into the industrial world [1]. With pandemic covid19, terahertz applications have experienced unprecedented growth in various fields [2].

With the emergence of new applications such as medical imaging [3], high-speed communication [4], spectroscopy [5] or satellite communications [6, 7], the need for higher and vacant frequencies is growing.

The frequency of terahertz waves ranges from 0.1 to 30 THz. These electromagnetic waves lie between the widely used microwave and infrared bands (Fig. 1). However, they

S. Motahhir and B. Bossoufi (Eds.): ICDTA 2022, LNNS 454, pp. 312–320, 2022.
https://doi.org/10.1007/978-3-031-01942-5_31

have very attractive physical characteristics, such as the strong penetrating power that allows them to penetrate materials such as wood, clothing and plastic [8]. And unlike infrared and X-rays, they are non-ionising. The monophotonic energy of tera rays is much lower than that of x-rays, and therefore have no visibly adverse effects on humans. Hence, it can be used for medical applications.

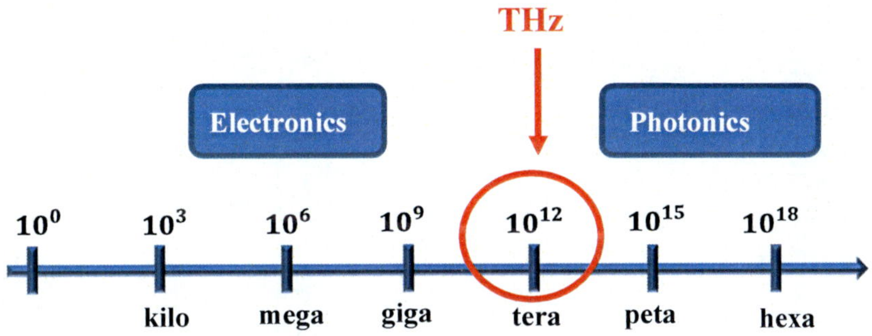

Fig. 1. Electromagnetic spectrum

Due to atmospheric attenuation, absorption and technical characteristics of the terahertz band, the researchers identified frequency channels and their possible bandwidths suitable for high-speed applications (Fig. 2) [9].

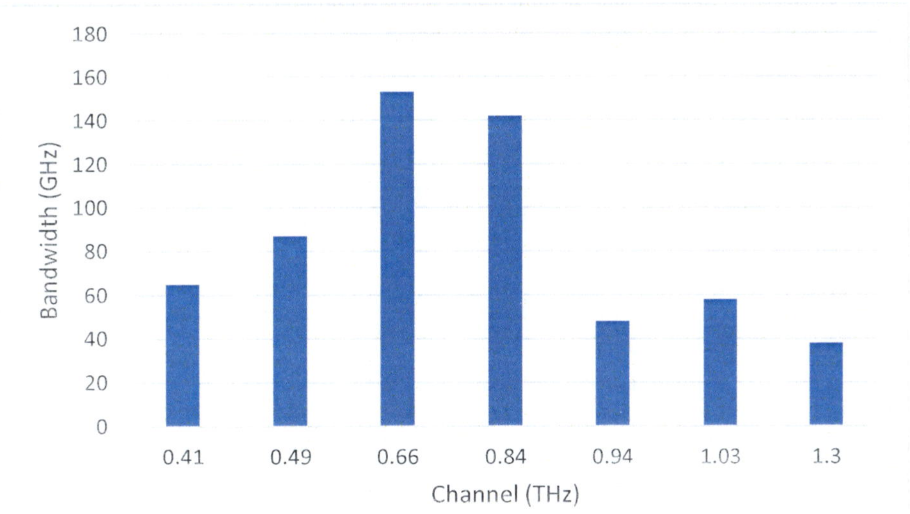

Fig. 2. Available bandwidths in various channels in terahertz band.

With the growing demand for available new frequency bands, plenty of applications deal with the terahertz band, taking advantage of the many assets of this frequency band (Fig. 3).

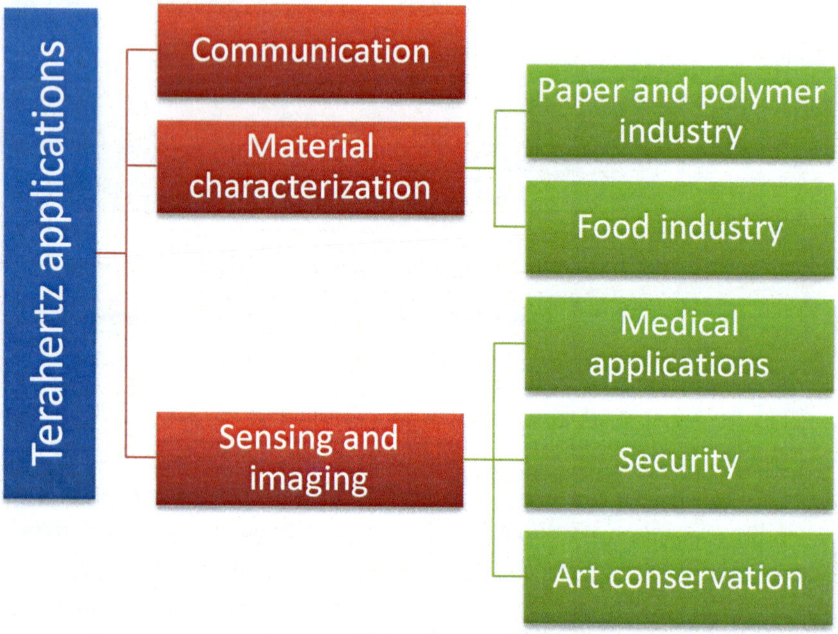

Fig. 3. Terahertz applications areas

In the following, we will describe the different applications cited in the figure. We will also present at the end the simulation results of an active bandpass filter to align with innovations in the terahertz band.

2 Terahertz Applications

2.1 Wireless Communication

Telecommunication systems that are heavily used by high-capacity applications and mobility need to migrate to new frequencies offering higher data rates. The terahertz band seems to be a good solution to these technical constraints. In this section, we will review the promising communication advances being deployed in the terahertz band.

Wireless communications use electromagnetic waves to achieve very high data rates, this is particularly possible above 100 GHz [10]. Data rates of 10, 20 or 100 Gbps have already been achieved. The maximum achieved so far is 115 Gbps on a 0.3 THz carrier at a distance of 110 m [11]. The study uses a single Schottky diode receiver which allows a wide bandwidth but requires the time-dependent phase specific to the terahertz band. To achieve this, the researchers used signal processing techniques to reconstruct the phase.

With regard to cellular communication, the design of 5G cells uses mainly small cells, which according to the ANFR (French National Frequency Agency) would significantly improve the quality of the network without significantly increasing the power of electromagnetic emissions. At least, more than two meters [12]. The antennas have undergone over time continuous improvements to accompany the evolution of systems (Fig. 4) [13].

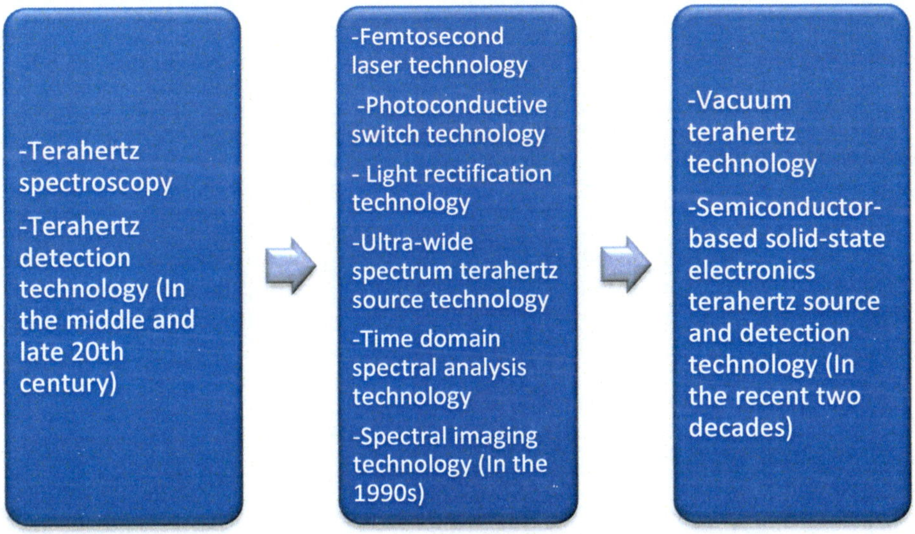

Fig. 4. Evolution of THz antennas

To take into account the technical requirements of the terahertz band, the terahertz antennas present a very delicate design challenge. These structures offer countless advantages compared to conventional antennas. Three design axes are currently being exploited by researchers (Fig. 5) [13].

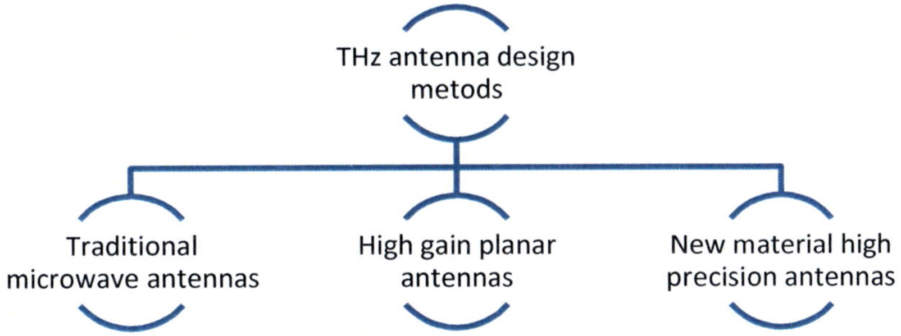

Fig. 5. THz antenna design methods

The use of a terahertz carrier to transport data can reach a TBps level. In addition, the waves will have better penetration, efficiency and anti-interference performance. Several searches have been launched in this direction with wide bandwidths and gains ranging from 5 to 7 dB [14, 15]. The work [16] evokes an antenna of up to 7.9 dB and a reflection coefficient of −21.36 operating on the 120 GHz frequency.

In addition to all the advantages mentioned compared to microwave antennas, terahertz antennas are more miniaturized due to their high frequency. The major challenge facing these promising antennas is the choice of manufacturing materials and process technology.

2.2 Material Characterization

Spectroscopy, which has long used the far-infrared FTS (Fourier transform spectroscopy), is currently based on terahertz technology to detect and control the properties of materials [17].

Terahertz spectroscopy covers two fields: the identification of substances from their terahertz transmission spectrum and that based on their optical and dielectric properties in the terahertz band. Spectrometers reported in the literature have different characteristics. These are summarized in Table 1 [18].

Table 1. THz spectrometer specifications

Spectrometer	Frequency range (THz)	Frequency resolution (Hz)
Fourier-Transform Far-Infrared	>0.3 (broadband)	10^9
Molecular gas laser	0.2–8.0 (tunable)	10^5
backward-wave oscillator	0.035–1.4 (tunable)	10^6
Photoconductive mixing	0.2–1.8 (tunable)	10^9
Time-domain THz	0.1–5 (broadband)	10^{10}

Spectroscopy has the advantage of interacting directly with the vibrational states and rotational movements of molecules. This allows to obtain a fingerprint of the molecule studied and complementary signatures compared to those observed in the infrared, visible or ultraviolet. For example, THz spectroscopy can detect chemicals that are difficult to identify with other techniques, such as hydrogen disulfide (H2S) and other hazardous substances such as explosives and gases or illegal substances such as drugs [19].

Spectroscopy is widely used in the food industry. It detects the presence of solid or soluble matter in food, of toxins, and of foreign bodies and defects, enabling safety, maintaining quality and reducing waste and resulting costs [20].

Terahertz spectroscopy allows to study of the structure of the polymer and to observe the dynamics of phase transitions. The most transparent material in the terahertz field is monocrystalline silicon [21].

2.3 Sensing and Imaging

Because THz wavelengths are lower than those of microwaves, THz radiation provides a spatial resolution of a few hundred microns. In addition, THz radiation can penetrate shallow into organic materials. This property is in high demand in some fields of medicine and biology. The medical applications of terahertz waves can cover several areas. As already mentioned, these radiations do not pose a danger to biological cells since they have low non-ionizing energy. Terahertz imaging can be used to check the integrity and thickness of tablets and confirm their 3D morphology, which provides an industrial application of THz sensing.

The applications also cover the identification of the structural state of proteins, the effects of radiation on biological processes, the diagnosis of diseases, as well as the analysis of the optical properties of human tissues. In dermatology, quantitative characterization makes it possible to evaluate the level of hydration of the skin and thus to compare or evaluate the effectiveness of cosmetic products to increase water retention in the skin. In oncology, terahertz technology can diagnose cancer earlier and improve biopsy [22].

In the field of security, several techniques are deployed to filter and detect illicit or threatening objects (weapons, explosives, etc.): X-rays and microwave rays. However, the danger of X-rays is still controversial. Moreover, these two techniques do not offer a chemical analysis of suspicious objects. For these reasons, the terahertz laser technique that generates short pulses, with its strong penetrating power, will allow to detect objects more accurately and faster [23].

Cultural heritage and artwork deteriorate over time due to materials ageing, weather, human intervention, etc. Non-destructive time-domain spectroscopy is a technique used to preserve this heritage, clean it, or repair it. The principle of THz time-domain imaging uses a THz pulse as a probe to observe cross-sectional images or/and sharp images of artworks. The particular advantage of THz spectroscopy is that it can detect internal structures of objects even if they are opaque. And this is thanks to THz rays that penetrate the majority of dielectrics [24, 25].

3 Example of Application

The filtering operation is an essential step for the proper functioning of transmission systems. Designing a filter in the terahertz band with good out-of-band rejection, good selectivity, and sufficient bandwidth is a great challenge for designers.

To contribute to the development of applications in the terahertz band, we have designed an active bandpass filter operating at the frequency of 129 GHz [26]. The filter uses the concept of active inductance to provide the centre frequency [27]. The simulation results of the S parameters are presented in Fig. 6.

The results obtained demonstrate a good band selection, good input and output reflection coefficients and a bandwidth of 400 MHz. Simulations are still in progress to reach higher frequencies.

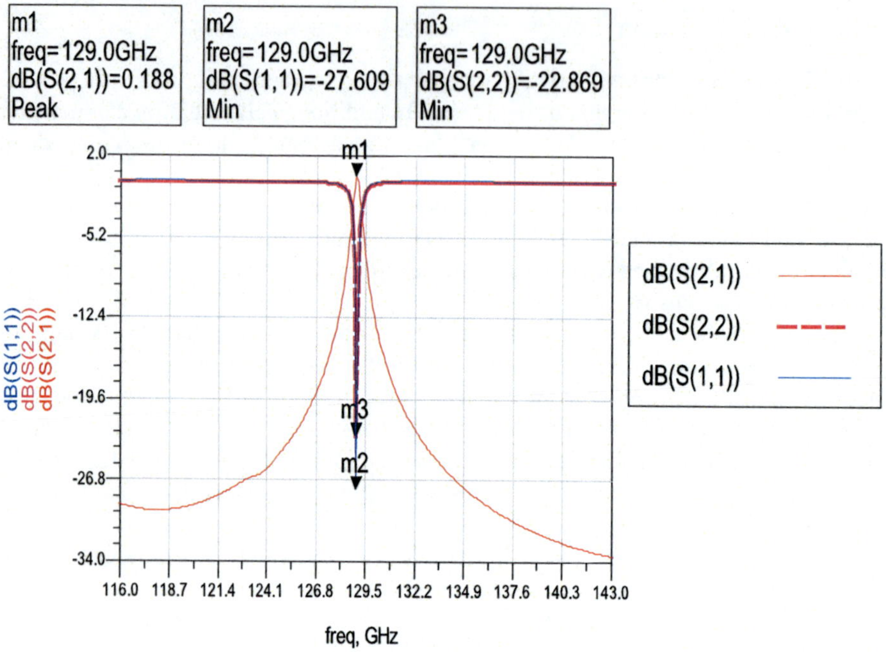

Fig. 6. The filter's S parameters

4 Conclusion

The terahertz frequency band holds great promise for multiple industrial, medical, food and cultural fields. We have demonstrated the capabilities of terahertz, namely: propagation speed, object penetration, absorption and high resolution. All these assets open up a large field of research for researchers to further develop the cited applications and take advantage of the opportunities of this band.

As far as we are concerned, we have tried to design an active filter that will select the terahertz frequencies. The filter is being improved to achieve higher frequencies and a wider bandwidth.

On a global scale, developments are still underway, and new advances promise to further explore this band by linking microwave benefits to photonic ones to produce new technologies that are faster, more powerful and less expensive.

References

1. Naftaly, M., Vieweg, N., Deninger, A.: Industrial applications of terahertz sensing: state of play. Sensors (Basel). **19**(19), 4203 (2019). https://doi.org/10.3390/s19194203.PMID:31569789;PMCID:PMC6806174
2. Mordor intelligence. https://www.mordorintelligence.com/industry-reports/terahertz-technologies-market. Accessed 07 Dec 2021

3. Mukherjee, M., Chatterjee, S.: THz Medical Imaging: Current Status and Future Outlook, Chapter 6. Banerjee, A., Chakraborty, B., Inokawa, H., Roy, J.N. (eds.) Terahertz Biomedical and Healthcare Technologies, pp. 113–125. Elsevier (2020). ISBN 9780128185568, https://doi.org/10.1016/B978-0-12-818556-8.00006-9
4. Wang, S., et al.: 26.8-m THz wireless transmission of probabilistic shaping 16-QAM-OFDM signals, APL Photonics (2020). https://doi.org/10.1063/5.0003998
5. Ghann, W., Uddin, J.: Terahertz (THz) Spectroscopy: A Cutting-Edge Technology, Terahertz Spectroscopy - A Cutting Edge Technology, Jamal Uddin, IntechOpen. (2017). https://doi.org/10.5772/67031
6. Civas, M., Akan, O.B.: Terahertz Wireless Communications in Space, ITU. J. Fut. Evol. Technol. **2**(7), 5 (2021)
7. Browne, J.: Modeling high-data-rate terahertz satellite links, modeling high-data-rate terahertz satellite links | microwaves & RF (mwrf.com). Accessed 07 Dec 2021
8. Vladimir, Y.: Fedorov and Stelios Tzortzakis: powerful terahertz waves from long-wavelength infrared laser filaments. Light. Sci. Appl. **9**, 186 (2020). https://doi.org/10.1038/s41377-020-00423-3
9. Bozanic, M., Sinha, S.: Methodologies for design of millimeter-wave and terahertz integrated circuits: an LNA case study. In: 2020 30th International Conference Radioelektronika (RADIOELEKTRONIKA). (2020) https://doi.org/10.1109/radioelektronika49387.2020.9092337
10. Nagatsuma, T.: Terahertz technologies: present and future. IEICE Electronics Express **8**(14), 1127–1142 (2011)
11. Harter, T., et al.: Generalized Kramers–Kronig receiver for coherent terahertz communications. Nature Photonics (2020). https://doi.org/10.1038/s41566-020-0675-0
12. ANFR: urban connectivity Technical report on small antenna experiments in Annecy (2017)
13. He, Y., Chen, Y., Zhang, L., Wong, S.-W., Chen, Z.N.: An overview of terahertz antennas. China Commun. **17**(7), 124–165 (2020). https://doi.org/10.23919/j.cc.2020.07.011
14. Liu, X., Leung, K.W., Yang, N.: Frequency reconfigurable filtering dielectric resonator antenna with harmonics suppression. IEEE Trans. Antennas Propagation **69**(6), 3224–3233 (2021). https://doi.org/10.1109/TAP.2020.3044387
15. Zheng, Z., Zhang, Y., Shi, L., Wu, L., Mao, J.-F.: An overview of probe-based millimeter-wave/terahertz far-field antenna measurement setups [Measurements Corner]. IEEE Antennas Propag. Mag. **63**(2), 63–118 (2021). https://doi.org/10.1109/MAP.2021.3054017
16. Didi, S.E., et al.: Study and design of the microstrip patch antenna operating at 120 GHz. In: El Ghzaoui, M., Das, S., Lenka, T.R., Biswas, A. (eds.) Terahertz Wireless Communication Components and System Technologies. Springer, Singapore (2022). https://doi.org/10.1007/978-981-16-9182-9_12
17. Naftaly, M., Miles, R.E.: Terahertz time-domain spectroscopy for material characterization. Proc. IEEE **95**(8), 1658–1665 (2007). https://doi.org/10.1109/JPROC.2007.898835
18. Wilke, I.: Terahertz spectroscopy applications. In: Worsfold, P., Poole, C., Townshend, A., Miró, M. (eds.) Encyclopedia of Analytical Science, 3rd edn. pp. 1–5. Academic Press (2017). ISBN 9780081019849. https://doi.org/10.1016/B978-0-08-101983-2.12094-8
19. Coutaz, J.L.: Optoélectronique térahertz. EDP Sci. 25–34 (2012)
20. Khushbu, S., Yashini, M., Rawson, A., Sunil, C.K.: Recent advances in terahertz time-domain spectroscopy and imaging techniques for automation in agriculture and food sector. Food Anal. Methods **15**, 498–526 (2021). https://doi.org/10.1007/s12161-021-02132-y
21. Hoshina, H., et al.: Terahertz Spectroscopy in Polymer Research: Assignment of Intermolecular Vibrational Modes and Structural Characterization of Poly(3-Hydroxybutyrate). IEEE Trans. Terahertz Sci. Technol. **3**, 248–258 (2013). https://doi.org/10.1109/TTHZ.2013.2253154

22. Pawar, S.Y., Sonawane, D.D., Erande, K.B., Derle, D.V.: Terahertz technology and its applications. Drug Invention Today **5**(2), 157–163 (2013). ISSN 0975–7619, https://doi.org/10.1016/j.dit.2013.03.009
23. Kemp, M., Taday, P., Cole, B., Cluff, J.A., Fitzgerald, A., Tribe, W.: Security applications of terahertz technology. In: Proceedings of SPIE - The International Society for Optical Engineering (2003). 5070. https://doi.org/10.1117/12.500491
24. Fukunaga, K.: Terahertz applications in art conservation. In: Handbook of Terahertz Technology for Imaging, Sensing and Communications, pp. 615–623 (2013). https://doi.org/10.1533/9780857096494.3.615
25. Krugener, K., et al.: On-site inspection of conservation works using THz TDS. In: 2019 44th International Conference on Infrared, Millimeter, and Terahertz Waves (IRMMW-THz). (2019). https://doi.org/10.1109/irmmw-thz.2019.8874079
26. Halkhams, I., El Hamdani, W., Mazer, S., El Bekkali, M., Fattah, M.: Terahertz band frequency tuning for a CMOS active filter. In: El Ghzaoui, M., Das, S., Lenka, T.R., Biswas, A. (eds.) Terahertz Wireless Communication Components and System Technologies. Springer, Singapore. (2022) https://doi.org/10.1007/978-981-16-9182-9_6
27. Halkhams, I., et al.: Design of a dual-band bandpass filter using active inductor principle. International Journal on Communications, Antenna and Propagation (IRECAP), **7**(4) (2017). https://doi.org/10.15866/irecap.v7i4.11791

Towards an Approach for Efficient Energy Management in WSN

Lamyaa Moulad[1](✉), Laila Mousaid[2], and Rachid Chaabita[1]

[1] Hassan II University, FSJES AC, BIGOFCF, Casablanca, Morocco
lamyaa.moulad@gmail.com

[2] Hassan II University, ENSEM, LRI/2IT Casablanca, Morocco

Abstract. Since they are endowed with small batteries and limited energy, it is expensive and difficult to change or charge the batteries of the sensors, especially if they are deployed in large numbers. The exhaustion of the sensor nodes' energy most often implies the malfunction of the entire network. Therefore, at the time of design, communication protocols for WSN should take into account the degree of energy consumption to provide optimal management of this resource. An effective model of conservation and Energy will be presented. In this work, our contribution focuses on the implementation of an integrated, efficient solution based on the integration of the following system: A new scheme for the management of energy consumption, based priority mechanism, using the AODV protocol, is proposed in this contribution. Most of the existing energy models only analyze the energy state of the communication module in the absence of studies of global energy consumption from the view of nodes.

However, by modeling the energy consumption related to different node components in different modes of operation and state transitions, our work proposes a new energy management model, based on the improvement of the reactive routing protocol AODV, under the NS simulator, providing energy management and load balancing mechanism for the entire network, applying a technique of prioritizing nodes with more energy resources and event triggering. This model showed good performance in terms of energy consumption, average time, and flow.

Keywords: Wireless sensor network · Energy consumption priority mechanism · Energy management · NS · AODV

1 Introduction

With limited energy resources, energy consumption management and energy modeling are important issues in the design and implementation of a wireless sensor network (WSN) [17]. Good knowledge of the sources of energy consumption in the WSN is the first step to reducing energy consumption. Therefore, it is not only important to know the quantity of energy that can give a harvesting system, but also the quantity of energy that will be consumed by a specific application running on a node and combining the two models to obtain a functional system from an energy point of view. Therefore, a specific energy model is required for the evaluation of communication protocols.

S. Motahhir and B. Bossoufi (Eds.): ICDTA 2022, LNNS 454, pp. 321–330, 2022.
https://doi.org/10.1007/978-3-031-01942-5_32

In the literature, many methods for optimizing energy consumption have been proposed, so there is still much research underway in this context for limited battery sensor networks. However, none of the proposed solutions apply universally. For example, if security applications require rapid reactivity, there are other applications, such as in agriculture where the delay is not as significant. However, energy-saving problems in wireless sensor networks [19] need to be addressed more systematically, taking into account application requirements.

Our goal in this work is to provide a model to increase the life of wireless sensor networks. In this sense, we propose a new approach to managing the transmission and routing of packets, to improve the AODV routing protocol (Ad-hoc On-demand Distance Vector) [14], based on the principle of priority allocation based on the energy of each node. We can provide more priority for nodes that have more energy. And as already mentioned, to protect the network against informal random use, resulting in the depletion of energy resources. This approach has shown a minimization of delays and energy expenses and increases throughput.

2 Problem and Motivation

Nowadays, WSN [18] is used in many sensitive and difficult areas such as medicine, the military… etc. In such networks, monitoring and energy management is very important to ensure protection against the misuse of this resource concerning the nodes that make up the wireless sensor network and against attacks by malicious nodes, such as the jamming of the linking layer which consists in finding a data packet and jamming it [7]. This type of attack can cause packet collisions during transmission, thus depleting the resources and confusion of nodes and why not stopping the functionality of WSN components.

Over the past few years, many approaches to optimizing energy consumption have been proposed, so there is still much research underway in this context for limited battery sensor networks. However, none of the proposed solutions apply universally. However, energy-saving issues in wireless sensor networks need to be addressed more systematically, taking into account application requirements.

Hence, the aim is to provide a robust and lightweight model to maximize the lifetime of wireless sensor networks. In this work, we will present a new approach to packet transmission and routing management, based on energy capacity priorities through the improvement of the Ad-hoc On-demand Distance Vector AODV [14] protocol of routing.

This experiment is tested under NS2 to compare the performance of AODV. We studied and compared the packet ratio, the differences in control packages, and the end-to-end delay of both AODV and the modified version (AODV Enhanced).

3 Related Works

Minimizing the energy consumption of wireless sensors or maximizing their life has always been a topic of debate in the research community. Numerous works have been published describing new algorithms, routing protocols, and architectures to maximize the life of the WSN. Already proposed [2, 3] energy-saving routing techniques for

WSN use routing protocols [3] to balance network-wide energy consumption or data aggregation techniques [4] that minimize the overall network traffic amount, efficient MAC protocols [2] and [1] that minimize radio activation time, etc....

These protocols and algorithms can be classified depending on the type of network structure, for flat network routing (Data-centric routing [2]), hierarchical network routing, and location-based routing [3]. Intanagonwiwat [5] has proposed a data-centric release (i.e., all communications for named data) and application-oriented to avoid unnecessary network layer routing operations to minimize energy consumption by shooing carefully good paths, caching, and processing data in the network. Yao in [3] proposed an improved data-centric protocol, called COUGAR, for an architecture that considers the network as a huge distributed database system. [2] Proposed a new protocol Energy-Aware Routing, similar to derived diffusion, differs as it sometimes uses suboptimal paths to benefit from energy benefits. In [12] Manjeshwar proposed a hierarchical routing protocol named TEEN (Threshold Sensitive Energy-Efficient Sensor Network Protocol) and another called APTEEN (Adaptive Threshold Sensitive Energy-Efficient Sensor Network Protocol) is adapted to time-critical applications. In [6] Byers proposed a concept of utility-based generic decision-making and attempt to measure the cost of each operation done by a sensor node, using heuristic evaluations for wireless sensor networks. In [6] the author modeled the behavior of the node and the energy consumed like a cyclic charge. In [7] Kansal proposed an energy management solution that executes tasks, based on the energy input and use model.

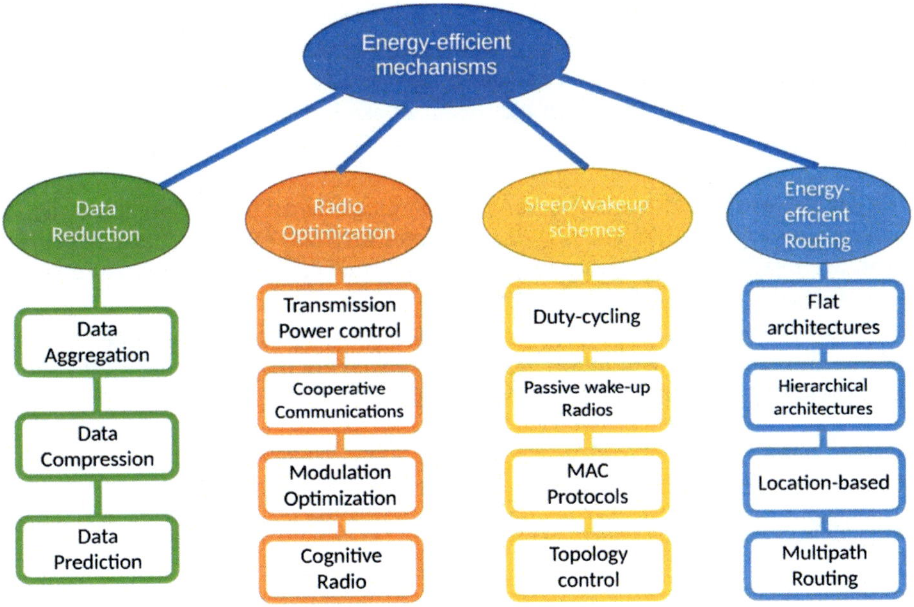

Fig. 1. Classification of Energy Efficiency Mechanisms, this figure provides a summary and details of the energy efficiency mechanisms classification.

Vigorito, Ganesan and Barto [8] also discuss the issue of managing the sensor node service cycle, based on past values for harvested energy. Their algorithm use adaptive control theory. In [9], author Zheng discusses the issue of data flow management in multi-jump networks using energy recovery. They gave a solution to define the data rates for each sensor node based on the expected traffic flow. The solution performs best in static topologies. So, there is a need to predict sensor node energy consumption according to its hardware. This allows network deployers to define how long the node will operate, and if the service life is not satisfactory to update the parameters to expand it. Many energy consumption solutions exist which follow the hardware platform (energy consumption/clock cycle) [10] and the physical radio layer [11]. [12] PixieOS is a Wireless sensor network operating system that uses the resource-sensitive model to increase efficiency (Fig. 1).

4 Our Proposed Approach

The use of the limited and small batteries capacity requires the use of energy-efficient protocols. These protocols must reduce the total amount of energy consumption in the network to transmit packets between each pair of neighboring sensor nodes.

The key idea is therefore to take advantage of the energy of all the nodes available in the network before its fall. For this, we proposed a priority routing scheme for load balancing. So, we developed a model to assess the energy consumption at the sensor nodes. The model captures the actual behavior of the WSN because it takes into account random and sometimes excessive access to a few specific nodes without the others, this will lead to the brutal depletion of their energy resources, especially if the nodes in question are the target of a targeted attack. We have shown that this schema is best suited for WSN, and that significant improvement can be made by our routing system based on improving the AODV protocol in terms of network life, time, and packet rate transmitted.

Primarily, Wireless sensor networks exist according to two types of routing protocols: proactive and reactive [13]. The first one preserves new lists of destinations and route appropriate by distributing routing tables through the network periodically. The second discover a route on-demand by inundating the network with route request packets. We have chosen the second type of protocol that is appropriate for the limited resources of the wireless sensor network. We choose AODV which does not handle routing information but is dependent on communication needs (Fig. 2).

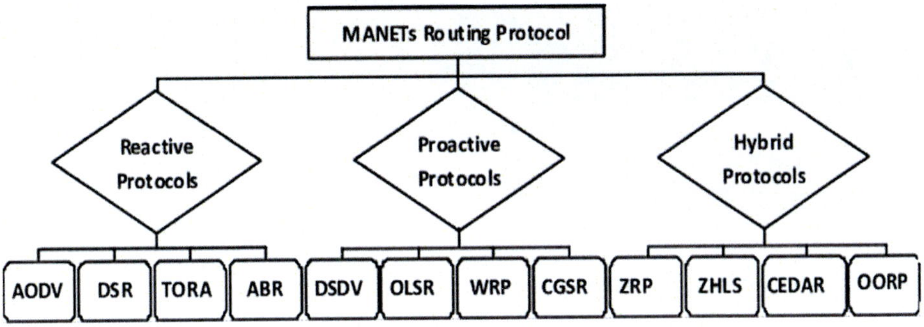

Fig. 2. Different types of routing protocols: proactive and reactive.

Based on the communication demand with its neighborhood, AODV broadcasts discovery packages when needed this routing protocol is adapted to the limited resources of the wireless sensor network and it does not maintain routing information. So, it is an extensively used reactive routing protocol that uses little energy as it searches for a path to the on-demand destination [13]. Contrary to other protocols, AODV does not periodically store destination paths in routing tables.

4.1 The AODV (Adhoc on Demand Vector) Protocol.

AODV is a routing protocol of MANET based on demand. In AODV, the sensor node has a table of routing and defines fresh routes using the sequence number for all routing information.

Once the sensor node receives a packet of control in each request, the table of routing is refreshed using hop count or a sequence number. When a route to a destination is required, it's created in the discovery of route step and is kept in the route maintenance step.

4.2 The Architecture of the Proposed Algorithm

To improve and extend the life of the networks, we will use a priority approach to control the level of energy in the network nodes, on the one hand, and balance the load between nodes in terms of usage based on the energy metric. So, the principle is to give priority to high-reputation nodes in terms of energy level in the design, creation, and delivery of packages (Fig. 3).

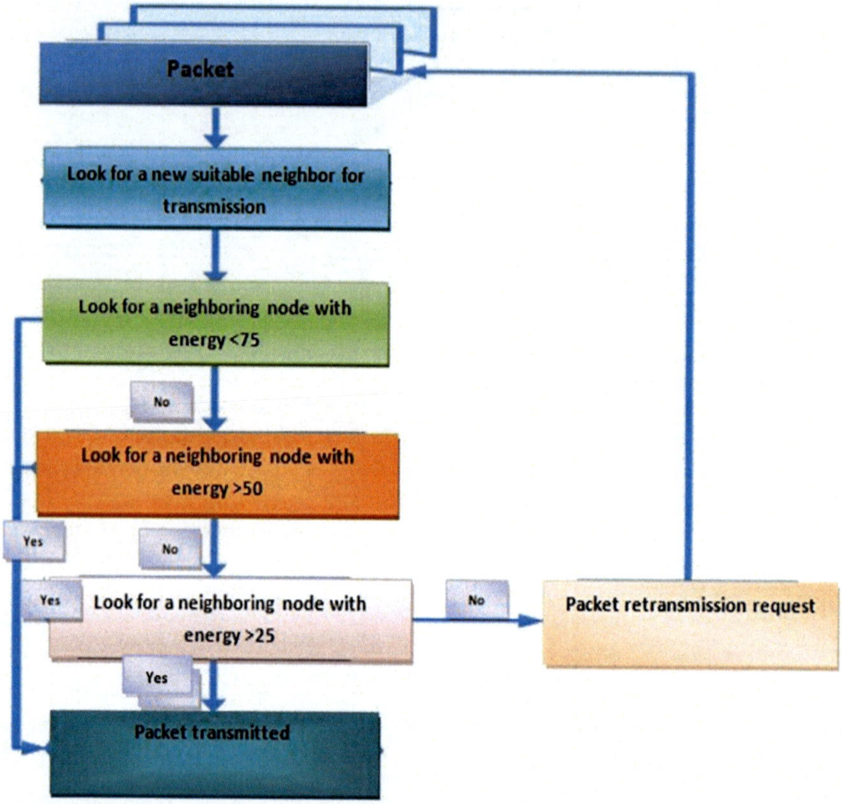

Fig. 3. Proposed algorithm model: it illustrates the flow graph of the proposed algorithm. The proposed idea makes it possible to identify the energy levels of each node and make transmission decisions for the routing of packets based on the level of this primordial resource. Each time it is necessary to calculate the energy levels of all the nodes and generate the power status report. Initially, all nodes have 100% energy.

5 Simulation and Results

We employed NS-2 simulator [15]. Figure 4 provides an outlook of NS-2, takes as input a TCL file. NS2 is based on two key languages: C++ and object-oriented tool control language (OTcl). While C++ describes the internal tool of simulation objects, OTcl configures the simulation by assembling and configuring objects as well as requiring discrete events (Fig. 4).

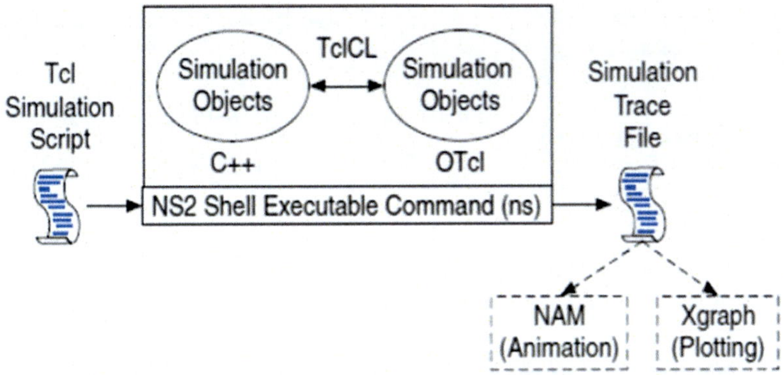

Fig. 4. NS-2.35 Simulation Overview, this tool takes as input a TCL file.

During the simulation, NS2 gives a trace file that will be implemented by various tools, such as Xgraph and NAM. We will make a simulation using NS-2 Scenario Generator [16]. Table1 displays the definition of the network parameters in the TCL file. The first parameter presents the simulation when sensor nodes send and get the packets via wireless channels. We use IEEE 802.15.4, which gives the media access control and the physical layer. Val (nn) is the number of sensor nodes, set to 20. Val (rp) is AODV protocol, it gives the routing protocol used in the simulation area of 500 m^2.

Table 1. Simulation résultats

Parameter	Value
Environment of simulation	500 * 500
Number of nodes	20
Time of simulation	10
Data Rate	1 Mb
Package size	512
Routing protocol	AODV
Energy	100 J

Figure 5 presents the NS-2 trace file we need to interpret the trace file result of our simulation. Field 1 is an event, it gives various symbols ('r', 'd', etc.). These symbols correspond to received and abandoned packages. Field 2 is the time when the event happens. Field 3 presents the source node to which the event happens. Field 4 is the destination node. Field 5 shows information about the packet type, which are UDP or TCP packets. Field 6 is the packet size. Field 7 presents information on certain flags. The 9 field is the flow ID, it is used to mention the flow color in the NAMdisplay. Field 10 is the source address. Field 11 is the destination node address. Field 12 is the sequence number of the packet in the network layer protocol, and the last field is the unique identifier of a packet.

```
r 2.798674924 _0_ AGT --- 271 ack 40 [13a 0 5 800] [energy 97.201036 ei 0.242 es 0.000 et 1.172 er
1.385] ------- [5:0 0:4 30 0] [11 0] 1 0
s 2.798674924 _0_ AGT --- 275 tcp 1040 [0 0 0 0] [energy 97.201036 ei 0.242 es 0.000 et 1.172 er 1.385]
------- [0:4 5:0 32 0] [23 0] 0 0
r 2.798674924 _0_ RTR --- 275 tcp 1040 [0 0 0 0] [energy 97.201036 ei 0.242 es 0.000 et 1.172 er 1.385] -
------ [0:4 5:0 32 0] [23 0] 0 0
```

Fig. 5. A part of AODVA.tr Trace File results extracted that presents the NS-2 trace file format we use to interpret the trace file result of simulation.

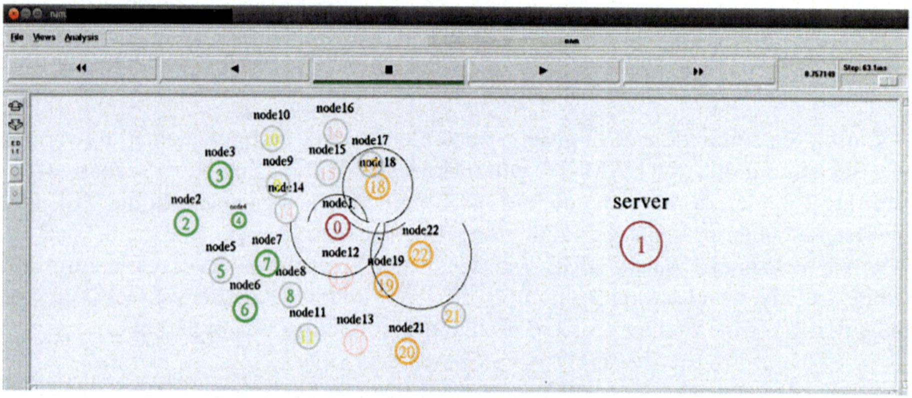

Fig. 6. The scenario of an alarm if the neighboring nodes have an energy E less than 70, The log file shows all the events during the different simulation moments.

Here we present simulation results of the priority-based approach and load balancing in comparison with the AODV and AODV protocol in the event of a denial-of-service (DOS) attack.

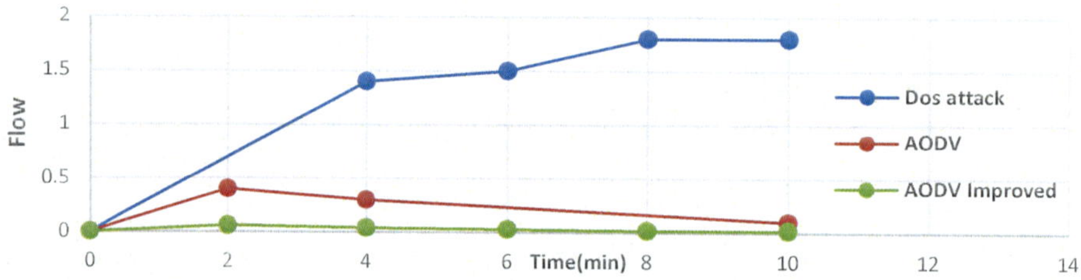

Fig. 7. Graph of Flow shows simulation results of the flow parameter, we can conclude that our improved model shows a flow rate less than AODV.

The simulation results presented in Figs. 7, 8, and 9 show that this enhanced AODV approach based on prioritizing load balancing using the AODV protocol minimizes delays and energy consumption and increases throughput compared to the AODV base protocol.

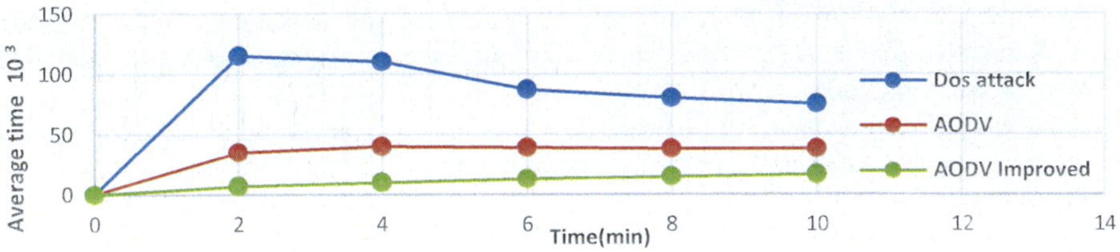

Fig. 8. This graph shows simulation results of the average time frames parameter, we can conclude that our improved model shows a reduced average time frame compared to AODV.

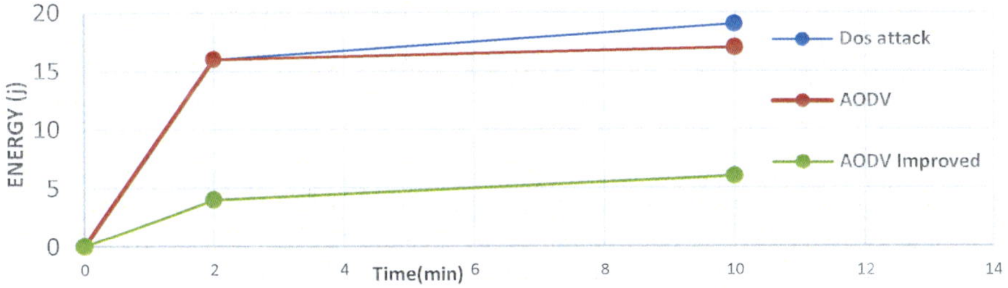

Fig. 9. Graph of Energy Consumption shows simulation results of the energy consumption parameter, we can conclude that our improved model consumes energy less than AODV.

6 Conclusion

Current wireless sensor network research concentrates more on protocols of communication than on energy consumption modeling. The traditional method of energy analysis is to minimize the energy consumption states of the sensor node and the whole network based on theoretical energy consumption data or theoretical solutions of system components.

By modeling the energy consumption of different node components in different phases of operation and different state, this work proposes a new energy management model, based on the improvement of the reactive routing protocol AODV, under the NS simulator, providing energy management and load balancing mechanism for the entire network, applying a technique of prioritizing nodes with more energy resources and event triggering. This model showed good results in energy consumption, average time, and flow.

As an outlook, we will try to implement our proposal in a real large-scale environment, so we will use the load balancing proposition in another type of routing protocol in a wireless sensor network.

References

1. Benahmed, K.: Surveillance distribuer pour la sécurité d'un réseau de capteurs sans fil, Thèse de Doctorateninformatique. Universitéd'Oran, Algérie (2011)
2. Wang, Y., Attebury, G., Ramamurthy, B.: A survey of security issues in wireless sensor networks. IEEE Commun. Surv. **8**(2), 2–23 (2006)

3. Akyildiz, I.F., et al.: Wireless sensor networks: a survey. Comput. Netw. **38**, 393–422 (2002)
4. Al-Karaki, N., Kamal, A.E.: Routing techniques in wireless sensor networks: a survey. IEEE Wirel. Commun. (2004)
5. Bian, F., Li, X., Govindan, R., Shenker, S.: Using hierarchical location names for scalable routing and rendezvous in wireless sensor networks. Int. J. Ad Hoc Ubiquitous Comput. **1**(4), 179–193 (2006)
6. Boukerche, A., Oliveira, H.A.B., Nakamura, E.F., Loureiro, A.A.F.: Secure localization algorithms for wireless sensor networks [security in mobile ad hoc]. IEEE Comm. Magazine **46**(4), 96–101 (2008)
7. Culler, D., Estrin, D., Srivastava, M.: Overview of sensor networks. Comput. Mag. IEEE Comput. Soc. **37**(8), 41–49 (2004)
8. Guoliang X., Lu C., Pless R., and Huang Q, "On Greedy Geographic Routing Algorithms in Sensing-Covered Networks," Mobi-Hoc'04, ACM Press, vol. 15 n. 6, pp. 31–42. 2004
9. Haeberlen, A., Flannery, E., Ladd, A.M., Rudys, A., Wallach, D.S., Kavraki, L.E.: Practical robust localization over large-scale 802.11 wireless networks. EURASIP J. Adv. Sig. Process. **14**(4), 415–433 (2008)
10. He, T., Stankovic, J.A., et al.: A spatiotemporal protocol for wireless sensor network. IEEE Trans. Parallel Distrib. Syst. **16**(10), 995–1006 (2005)
11. Intanagonwiwat, C., et al.: Directed diffusion: a scalable and robust communication paradigm for sensor networks. In: Proceedings of ACM MobiCom 2000, Boston, MA (2000)
12. Li, Z., Trappe, W., Zhang, Y., Nath, B.: Robust statistical methods for securing wireless localization in sensor networks. In: Fourth International Workshop Information Processing in Sensor Networks (IPSN 2005), vol. 4, no.1, pp. 91–98, April 2005
13. Lu, C., Blum, B.M., Abdelzaher, T., Stankovic, J., He, T.: RAP: real-time communication architecture for large-scale wireless sensor networks. In: Real-Time and Embedded Technology and Applications Symposium, vol. 38, no. 4, pp. 55–66, September 2002
14. Moulad, L.: 'Mise en oeuvre d'un mecanisme de sécurité pour RCSF basé sur la gestion d'énergie (CMT) (2017)
15. Devi, M., et al.: Routing with AODV Protocol for Mobile ADHOC Network. Int. J. Technol. Eng. Syst. **2**(1) (2011)
16. Fall, K., Varadhan, K.: The NS manual, 9 May 2010
17. Huang, J., Shahnasser, H.: A preprocessor Tcl script generator for NS-2 communication network simulation. San Francisco State University, USA, pp. 184- 187, 5 May 2011
18. Moulad, L.: 'Mise enœuvre d'un mécanisme de détectiond'intrusionhybridehiérarchique dans un RCSF basé sur la gestion de l'énergie, (RTIS) (2018)
19. Sahraoui, S.: mécanisme de sécurité pour l'intégration des RCSF à l'IoT, thesis (2016)

New Experimental Approach for MAC Layer Misbehavior of the IEEE 802.11 Protocol

Mohammed-Alamine El Houssaini[1](✉), Souad Ajjaj[2](✉), Souad El Houssaini[3], Abdellah Nabou[4], and Jamal El Kafi[3]

[1] ESEF, Chouaib Doukkali University, El Jadida, Morocco
elhoussaini.m@ucd.ac.ma
[2] ENSAM, Hassan II University, Casablanca, Morocco
souad.ajjaj-etu@etu.univh2c.ma
[3] Faculty of Sciences, Chouaib Doukkali University, El Jadida, Morocco
{elhoussaini.s,elkafi.j}@ucd.ac.ma
[4] Faculty of Sciences Semlalia, Cadi Ayyad University, Marrakech, Morocco
abdellah.nabou@uca.ma

Abstract. This paper is dedicated to the description of the procedure to be followed in the experimental validation of the fairness of the BEB (Binary Exponential Backoff) algorithm as well as the case of the presence of the bad behavior of this algorithm. We set up all the necessary tools for the experimentation, the generation and the analysis of communication traffic between the stations of our experimental bench. This work is conditioned by the availability of network cards that can support modifications to the protocols, including the BEB. The solution for this modification currently lies in the ATHEROS network cards (those which can support one of the three drivers: MADWIFI, ATH5K or ATH9K).

The proposed experimental network consists of at least three stations (laptops) including a receiver and a cheat with integrated wireless cards. Laptops operate under IEEE 802.11. We put all the devices inside the lab to make sure they are under the same channel conditions and the only difference between legitimate and misbehaved nodes is the Back-off timer.

JPerf, the commonly adopted network test tool, will be used to create traffic on the network, as well as generate User Datagram Protocol (UDP) streams that can fill the transmit queue on each device, putting all devices in full mode. Wireshark is employed to display the captured data and the Perl language to analyze the corresponding trace files.

Keywords: Mobile Ad Hoc network · MAC layer · Security · Experimental installation

1 Introduction

Mobile ad hoc networks MANET [1] continue to increase their presence in our daily lives. They are becoming a cornerstone in many fields mainly: military, scientific, logistics applications, etc. However, these networks operate mainly in open environments and

S. Motahhir and B. Bossoufi (Eds.): ICDTA 2022, LNNS 454, pp. 331–340, 2022.
https://doi.org/10.1007/978-3-031-01942-5_33

are therefore vulnerable to a large number of threats. The existing methods of securing networks based on prevention techniques, which are not easy to implement by IEEE. Particularly, the suggestion of a new MAC layer based on modification of the RTS, CTS and ACK frames exchanged from the CSMA/CA method. This consumes more bandwidth, the deployment of the security mechanism at the watchdog level in ad hoc mode, and the extension of the IEEE 802.11 standard by suggesting new algorithms for allocating waiting time (new Back-off algorithms) to access the transmission channel [2].

An emerging threat at this layer is the intentional misbehavior of the MAC media access control layer [3]. As wireless stations are generally under the control of individual users, a selfish or greedy user may try to gain more bandwidth at the expense of other legitimate users. In the extreme, when a station is constantly occupying the wireless medium, a DOS denial of service attack [4] is launched and the misbehaving station becomes an attacker. A greedy node intentionally changes the MAC layer [5] of the IEEE 802.11 protocol to gain more benefits than legitimate nodes. By this bad behavior of access to the channel, a greedy node can derive several benefits such as the increase of its speed and the reduction of its energy consumption.

In order to materialize the robustness of these security mechanisms, we propose an approach to be followed for the experimental validation of various contributions in this topic. We offer experimental platforms for such a practice in a real environment.

The rest of this paper is organized as follows: Sect. 2 describes the different equipment's of our experimental prototype as well as the preliminary installations. The third section exhibits the results of monitoring the network without the presence of MAC layer misbehavior. The following section details the required configurations to test the mode of the network under attack of the cheating type of the BEB.

2 Experimental Equipment and Installation

As shown in Table 1, the prototype consists of two transmitters and a monitoring receiver. The three stations are laptops equipped with wireless cards including the cheater is based on the ATHEROS AR9285 chipset. MADWIFI [6], ATH5K [7] or ATH9K [8] drivers (ATHEROS drivers for Wifi) provide full support for wireless cards using ATHEROS chipsets on a Linux platform. The numerous features of these drivers make it easier to implement cheating, as we will see later. While the same functionality may not be available in products from other manufacturers, the growing trend of moving more functions from firmware and hardware to software will make it easier to implement cheating on most chipsets.

The first step consists in installing and configuring an ad hoc network with the hardware and software configuration detailed in Table 1. However, the creation of ad hoc mode is disabled on Windows 10 [9]; we can force it in command mode as follows:

```
netsh wlan show drivers
```

We will check after the wireless network card supports the hosted network (Yes), if it is displayed (No), we need to update the Wifi driver.

We will then run the following command to create our ad hoc network:

```
netsh wlan set hostednetwork mode=allow ssid=mynetwork key=123456789
```

mynetwork is the name of the ad hoc network and 123456789 is the password (Windows 10 requires at least 8 characters for the password).

To start the ad hoc network, just type the following command:

```
netsh wlan start hostednetwork
```

Table 1. Network hardware and software components.

Station	Role	Brand	Operating system	Network card	Traffic generator	Traffic analyzer
A	Receiver	DELL LATITUDE E5540	WINDOWS 10	INTEL(R) DUAL BAND WIRELESS AC 7260	–	WIRESHARK 2.2.5
B	Transmitter	ACCENT NOTEBOOK	UBUNTU.12	ATHEROS AR9285 PCI-EXPRESS	JPERF 2.0.2	–
C	Transmitter	HP ELITEBOOK	WINDOWS 10	INTEL (R) CENTRINO (R) ADVANCED N 6205	JPERF 2.0.2	-

On station A, we have installed Wireshark version 2.2.5 [10] to analyze the traffic. On the transmitting stations, we installed JPERF 2.0.2 [11] to generate the traffic. For JPERF 2.0.2 to work, the Java language must be installed.

To install java on UBUNTU [12] we will need the following commands:

```
sudo apt-get install default-jre
sudo apt-get install default-jdk
```

Then we check the installation by:

```
java –version
```

3 Greedy Node-Less Network Monitoring

Figures 1, 2 and 3 show a snapshot of the wireless data collected and displayed on the screens of the three stations A, B and C mentioned in Table 1. Laptop A acts as the monitor.

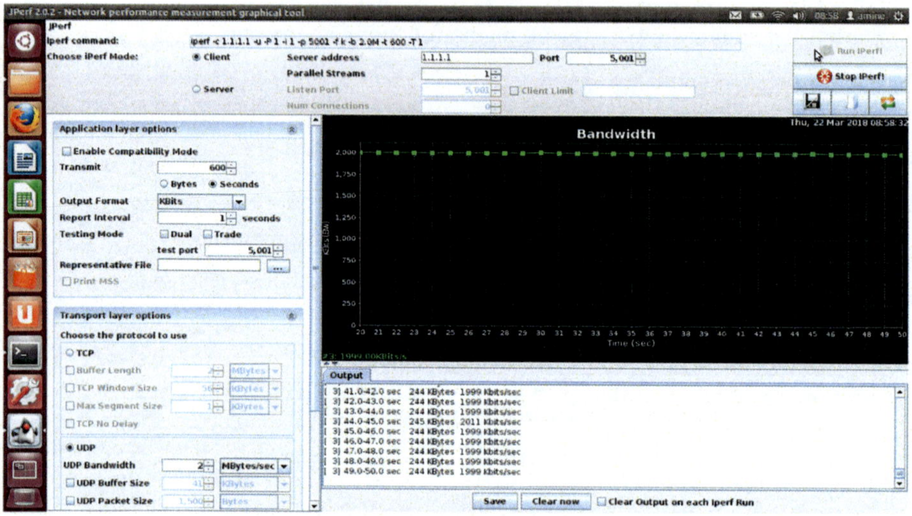

Fig. 1. Screen capture of the first transmitting station B (JPerf 2.0.2 window).

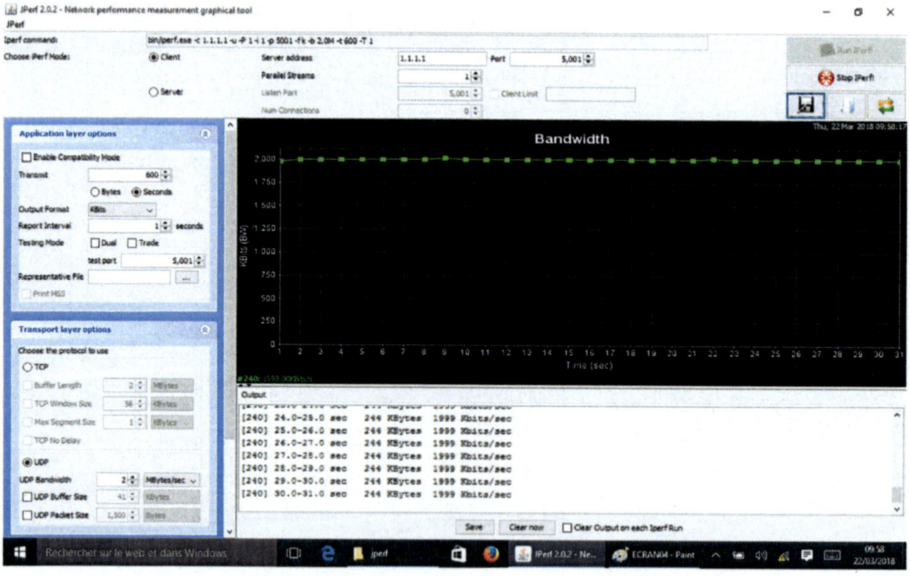

Fig. 2. Screenshot of the second transmitting station C (JPerf 2.0.2 window).

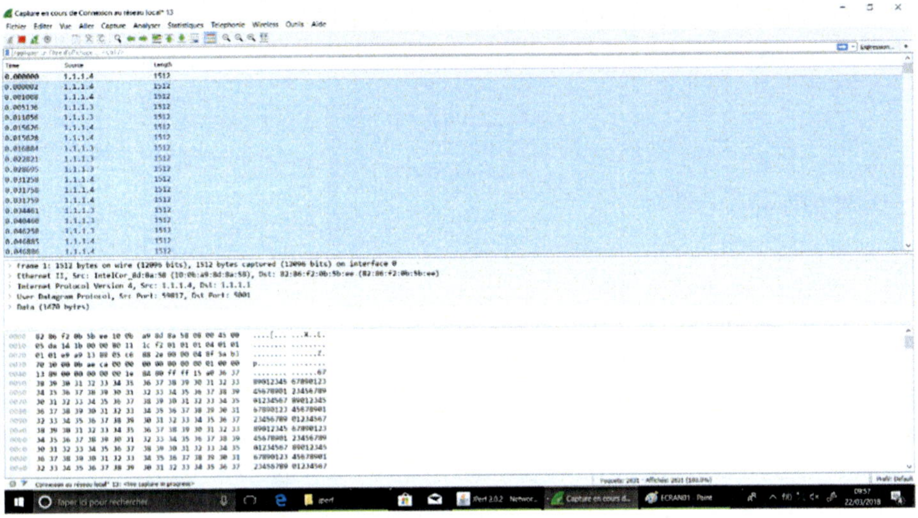

Fig. 3. Screenshot of receiving station A (Wireshark 2.2.5 window).

To observe the behavior of the transmitting stations, the monitoring mechanism must generally be installed on the receiving station. We use Wireshark to display the captured data and Perl [13] scripts to analyze the corresponding trace files. More precisely, each frame is associated with a timestamp indicating the moment when its reception is terminated. This analysis allowed us to extract our useful metrics (reception rate and inter-packet time).

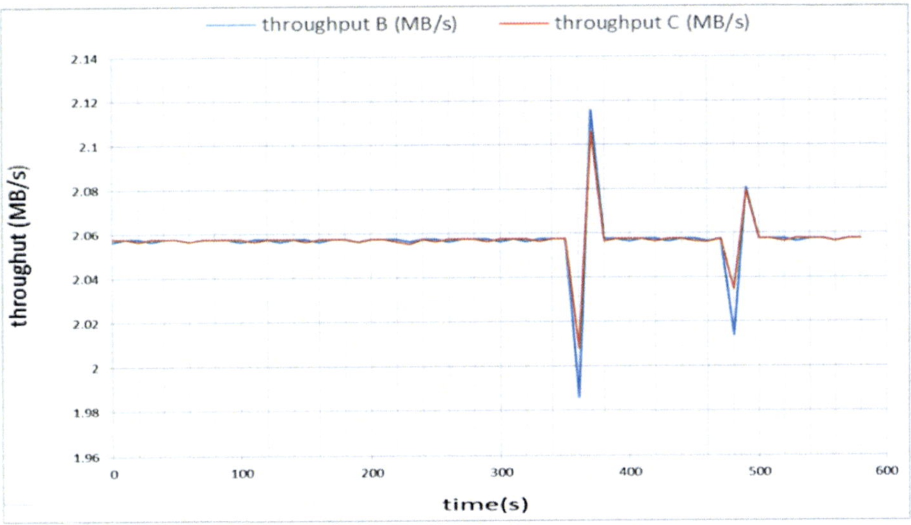

Fig. 4. Experimental flow of the two emitters B and C in the normal case without attack

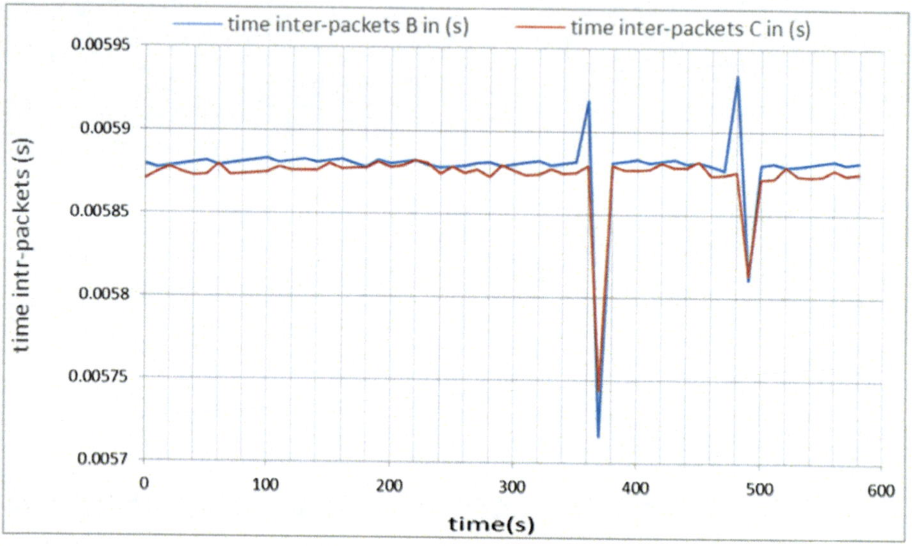

Fig. 5. Experimental inter-packet time of the two transmitters B and C in the normal case without attack

Figures 4 and 5 show the results of this data exploitation. According to these two figures, the two transmitting stations B and C are equal and oscillate around the same rate and the same inter-packet time. The oscillatory nature of the curves can be explained by the guarantee of the IEEE 802.11 protocol in terms of equity of access to the transmission channel.

4 Monitoring with Greedy Node

This step involves the implementation of the cheater attack in a transmitting station. In other words, we will need the MADWIFI wireless card driver.

We can note that this driver only works on a few specific ATHEROS brand cards on a Linux system (http://madwifi-project.org/).

We were able to find a laptop computer equipped with an ATHEROS wifi card (see specifications in Table 1).

To install MADWIFI on this machine, we have run a madwifi.sh script (Table 2).

Then we checked with the terminal the network interfaces that can be changed by the iwconfig command.

The iwpriv command is used to configure parameters specific to the Wi-Fi card.

Figure 6 shows the result of this check:

Wlan no private ioctls

The driver for this card does not provide any specific parameters to configure. So, the MADWIFI driver failed.

Table 2. Script for MADWIFI installation

madwifi.sh

```
#!/bin/sh
# This script will install the madwifi drivers for Atheros cards.
# The script must be run in a root terminal with 755 permission
script_name="madwifi.sh"
# Script must run as root
if [ $USER != "root" ]; then
echo "You need to run this script as root."
echo "Use 'sudo ./$script_name' then enter your password when prompted."
exit 1
fi
# Install essential tools
sudo apt-get -y install build-essential bin86
# Install the sharutils package
sudo apt-get -y install sharutils
# Change to the /usr/src folder
cd /usr/src
# Get the necessary drivers from www.stchman.com
sudo wget http://www.stchman.com/tools/madwifi/madwifi-0.9.3.3.tar.gz
# unpack the tarball
sudo tar -xzf /usr/src/madwifi-0.9.3.3.tar.gz
# Change to the folder that the tarball created
cd /usr/src/madwifi-0.9.3.3
# Make the drivers
sudo make clean
sudo make
sudo make install
# Make backup of /etc/modules file
sudo cp /etc/modules /etc/modules.bak
# Append to the end of the /etc/modules file by using >>
# First insert a carriage return in to the /etc/modules file
sudo echo -e '\n' >> /etc/modules
sudo echo "ath_pci" >> /etc/modules
# Enable WEP, WPA. WPA2 via network manager
# Install wpa supplicant
sudo apt-get -y install wpasupplicant
# Install Network Manager (this will allow you to see all available wireless networks)
sudo apt-get -y install network-manager-gnome network-manager
# Make backup of the /etc/network/interfaces file
sudo cp /etc/network/interfaces /etc/network/interfaces.bak
# Make new /etc/network/interfaces file with just "lo" entry in it.
# The single > will create a new file, >> appends to a file
sudo echo "# The loopback network interface" > /etc/network/interfaces
sudo echo "auto lo" >> /etc/network/interfaces
sudo echo "iface lo inet loopback" >> /etc/network/interfaces
# Create a file called /etc/default/wpasupplicant
# This file does not exist, but the > command will create it
sudo echo "ENABLED=0" > /etc/default/wpasupplicant
#sudo touch /etc/default/wpasupplicant
# Reboot the machine, wireless should now be enabled!!!!
sudo reboot
```

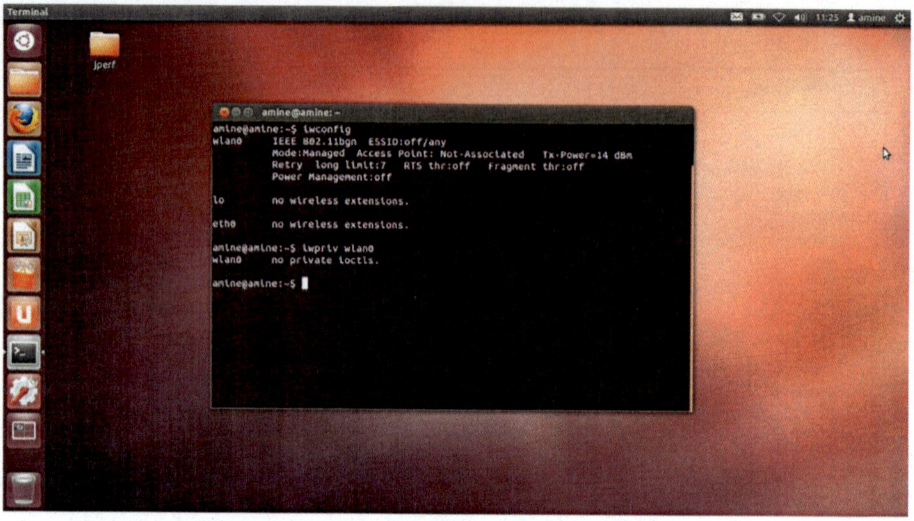

Fig. 6. Screen capture for the two commands iwconfig and iwpriv

Then we proceeded to install the ath9k driver since our Wi-Fi card supports this driver as indicated on the link: https://wireless.wiki.kernel.org/en/users/Drivers/ath9k.

To install it, you must first prepare the UBUNTU environment. The first step is to install the following packages in order to upgrade the KERNEL (Table 3).

Table 3. Installation of packages for KERNEL upgrade

Installation command lines
sudo apt-get install libncurses5-dev binutils-dev linux-source
sudo apt-get install fakeroot build-essential crash kernel-wedge kernel-package
mkdir ~/src
cd ~/src
wget http://www.kernel.org/pub/linux/kernel/v3.0/linux-3.2.17.tar.bz2
tar -xjvf linux-3.2.17.tar.bz2
cd linux-3.2.17/
cp -vi /boot/config-\`uname -r\` .config
make oldconfig
make localmodconfig

We then installed the "headers" with the following command:

```
sudo apt-get install linux-headers-$(uname -r)
```

Next, we proceeded to install the ath9k driver. This driver is only compatible with specific ATHEROS cards. The installation steps for this approach are described as follows:

Download the latest version of the "ath9k" package from the site https://backports.wiki.kernel.org/

Package installation (Table 4):

Table 4. The ath9k driver installation

Commands for ath9k driver installation
tar –xvf backports-4.2.6-1.tar.xz
cd backports-4.2.6-1/
sudo make clean
sudo make
sudo make install

Preparation of the MAC layer: to activate ath9k, we must first activate mac80211 via "make menuconfig" when compiling the KERNEL. Here are the options we need to activate ATH9K via "make menuconfig" (Table 5).

Table 5. Activation of ath9k driver

Commands to activate ath9k driver
Networking --->
Wireless --->
<M> Improved wireless configuration API
<M> Generic IEEE 802.11 Networking Stack (mac80211)
Activation of ATH9K:
Device Drivers --->
[*] Network device support --->
Wireless LAN --->
Atheros Wireless Cards ---->
<M> Atheros 802.11n wireless cards support

Finally, we verified the activation of the ATH9K driver with the iwconfig and iwpriv commands.

So, the ATH9K pilot was successful this time around. So, the test of the MAC layer misbehavior can be realized now.

5 Conclusion

In this paper, we have developed an approach to be followed in the experimental validation of the various MAC layer misbehaviors. We have implemented all the necessary steps for practice in a real environment. Indeed, we have experimentally generated and analyzed communication traffic between stations. This work is conditioned by the availability of network cards that can support modifications to the protocols, including the Back-off timer. The solution to this modification lies for the moment in the ATHEROS network cards (those which can support one of the three drivers: MADWIFI, ATH5K or ATH9K). The experimental implementation of the BEB's cheat detection mechanisms utilizing the approaches proposed in this research will be considered soon.

References

1. Al-Absi, M.A., Al-Absi, A.A., Sain, M., Lee, H.: Moving Ad Hoc networks—a comparative study. Sustainability **13**, 6187 (2021). https://doi.org/10.3390/su13116187
2. Ye, C., Li, Y., Reznik, A.: Performance analysis of exponential increase exponential decrease back-off algorithm. In: 2010 IEEE Global Telecommunications Conference GLOBECOM 2010, pp. 1–6 (2010). https://doi.org/10.1109/GLOCOM.2010.5683538
3. Hoang, T.-M., Bui, V.-K., Nguyen, T.-T.: The performance evaluation of an IEEE 802.11 network containing misbehavior nodes under different backoff algorithms. Secur. Commun. Netw. **2017**, 1–8 (2017). https://doi.org/10.1155/2017/2459780
4. Seth, A.D., Biswas, S., Dhar, A.K.: De-authentication attack detection using discrete event systems in 802.11 Wi-Fi networks. In: 2019 IEEE International Conference on Advanced Networks and Telecommunications Systems (ANTS), pp. 1–6 (2019). https://doi.org/10.1109/ANTS47819.2019.9118100
5. IEEE Standards Association: IEEE 802.11 Standard for Wireless LAN Medium Access Control (MAC) and Physical Layer (PHY) Specifications. IEEE Standards Association, pp. 818–840, March 2012
6. http://madwifi-project.org/. Accessed 28 Nov 2021
7. https://wireless.wiki.kernel.org/en/users/Drivers/ath5k. Accessed 28 Nov 2021
8. https://wireless.wiki.kernel.org/en/users/Drivers/ath9k. Accessed 28 Nov 2021
9. https://www.microsoft.com/fr-fr/software-download/windows10. Accessed 28 Nov 2021
10. https://www.wireshark.org/docs/relnotes/wireshark-2.2.5.html. Accessed 28 Nov 2021
11. https://sourceforge.net/projects/jperf/. Accessed 28 Nov 2021
12. https://www.ubuntu-fr.org/. Accessed 28 Nov 2021
13. https://www.perl.org/. Accessed 28 Nov 2021

Performance Evaluation of Pilot-Based Channel Estimation for Conventional, Cell-Free, and Small-Cell Massive MIMO Networks

Jamal Amadid(✉), Asma Khabba, Zakaria El Ouadi, and Abdelouhab Zeroual

Instrumentation, Signals and Physical Systems (I2SP) Group, Faculty of Sciences Semlalia, Cadi Ayyad University, Marrakesh, Morocco
{jamal.amadid,asma.khabba,zakaria.elouadi}@edu.uca.ac.ma, zeroual@uca.ac.ma

Abstract. Massive multiple-input multiple-output (M-MIMO) and net- work design are two contemporary technologies that have received a lot of attention. Moreover, an innovative network design known as cell-free (CF) has attracted considerable interest. Thus, addressing and assessing the channel estimation (CE) operation in a CF M-MIMO network is an appealing topic. Consequently, this paper focuses on the M-MIMO technology, in particular, the CE operation for three differents three-time division duplex-enabled M-MIMO networks, namely, the conventional M- MIMO network, the CF M-MIMO network, and the small-cell M-MIMO network. Furthermore, for each network design, we apply the minimum mean square error (MMSE) estimator to evaluate and determine the CE operation, including the expression of the large-scale fading model. On the other hand, we established the normalized mean square error formula for the MMSE estimator regarding each network configuration. Addition- ally, a pilot assignment scheme is proposed for CF M-MIMO network. Numerical results are displayed to confirm our theoretical expressions.

Keywords: 6G Wireless communication systems · Massive MIMO · Channel estimation · Conventional Massive MIMO network · Cell-Free · Small-Cell

1 Introduction

In recent years, the so-called wireless communication systems have undergone a significant improvement, thanks to the employment of many technologies that aim to bring many advantages to the current network [1]. One of these technologies is the Massive multiple-input multiple-output MIMO (M-MIMO) that offers many benefits to the current wireless communication systems (i.e., conventional M-MIMO network) [2, 3]. M-MIMO is the deployment of a large number of antennas (NoA) in each base station (BS) to serve a relatively large number of users equipment (UEs) (subject to the constraints that the NoA in each BS is very much larger than the number of UEs), where this technology enables each BS to benefit from the beam-forming technique [4]. Further, another technique that significantly boosts wireless performance is the cell-free

S. Motahhir and B. Bossoufi (Eds.): ICDTA 2022, LNNS 454, pp. 341–350, 2022.
https://doi.org/10.1007/978-3-031-01942-5_34

(CF) network [5] where the cell has no edge, i.e., the coverage area is not split as in the conventional M-MIMO. In addition, access points (APs) and UEs are free to locate anywhere on the network. As reported in [6, 7], combining the two technologies (CF and Massive multiple-input multiple-output (M-MIMO) have greatly enhanced system performance. In this work, we discuss the use of M-MIMO technology in two networks (CF and conventional networks) as this technology is a prospective wireless access technology for the fifth generation and beyond. Besides, high efficiency, reliability, and energy efficiency are delivered using only straightforward signal processing. Furthermore, we address the small cell as a particular case of a CF M-MIMO network in which each UE is served using only one AP (i.e., the AP that provides the best channel condition). There are two setups for processing the information received at each AP, (i.e., the processing can be either centralized [8] or localized). The processing in the centralized setup is done by a unit known as the central processing unit, which controls all the APs in the network. This unit also handles the channel estimate procedure, as well as precoding and decoding. On the other hand, the processing in the localized setup is conducted in each AP (i.e., the processing is done locally in each AP). However, as reported in [1], both setups provide similar performance. That is, the centralized or localized processing produces identical results in terms of channel estimation (CE), precoding, and decoding.

1.1 Related Works

Many studies in the literature have addressed the CF network under other descriptions, such as MIMO network and distributed antenna system. The conventional M-MIMO network is addressed in [2, 3], where the authors focus on the CE operation under a higher pilot contamination degree. Additionally, the authors propose a straightforward channel estimator to overcomes the unrealistic property on which the MMSE estimator is based. In addition, the authors drive the normalized mean square error (NMSE) expression regarding each estimator (i.e., the LS estimator, MMSE estimator and the proposed estimator). The authors conclude that the proposed estimator overcomes the issue and performs similarly with the MMSE, especially for a large number of BS antennas. The UL phase for CF M-MIMO considering spatially correlated channels has been examined in [9]. Additionally, the authors consider the Rician fading. Besides, the authors investigate the SE using three estimators. The authors conclude that spatial correlation is a beneficial factor for SE. Furthermore, the linear MMSE estimator achieves deteriorate performance since it has no phase knowledge.

1.2 Work Outlines

The remainder of this work is organized as follows:

In Sect. 3, we provide the signal model and the CE operation of the conventional M-MIMO network. In Sect. 4, we cover the signal model, the CE procedure and the proposed pilot assignment (PA) scheme for two networks (i.e., CF M-MIMO and small-cell M-MIMO networks). In Sect. 5, we provide numerical results to affirm our theoretical expression. Conclusion and future work and provided in Sect. 6.

1.3 Work Contributions

* We evaluate, compare and analyze the performance of three networks, namely, a conventional M-MIMO network, a CF network, and a small-cell network. Moreover, we address the CE using the MMSE estimator. In addition, for each network, we drive the Average MSE expression of the MMSE estimator.
* The recent studies are addressed the PA scheme sequentially. i.e., the PS occurs sequentially regarding each UE. As a result, in this work, we propose a PA scheme in which AP choices are performed initially regarding all UEs, followed by PAs. This is inspired by the fact that the former scenario necessitates extra complexity, which is defined by the time required to perform the APs choice as well as PA as it runs successively to every UE.

2 Conventional Massive MIMO Network

This section describes the system model and CE operation for the UL M-MIMO conventional network. Thus, L denotes the number of BSs (i.e., the number of cells), K indicates the number of UEs, and N symbolizes the NoA in each BS. In addition, M symbolizes the total NoA in all APs (i.e., $M = N \times L$).

2.1 System Model for the Conventional Massive MIMO Network

According to the time division duplex (TDD) framework, the coherence block (i.e., the time during which the channel consider as constant) is divided into three-phase, firstly a τ_p pilot is used for the CE, while the remaining phases are for the UL and DL data. In a practical network, the coherence block is limited, which leads to restricts the number of orthogonal pilots used in the network. Hence, for serving all UEs, a reuse pilot technique must be used [10]. In this work, we assume that the reuse pilot technique is used (i.e., the UEs will not be used orthogonal pilot) (i.e., $\tau_p \leq K$) [11]. Additionally, we consider that the k_{th} UE transmits the pilot $\phi_k \in \mathbb{C}^{\tau_p \times 1}$, which has the following property $\phi_k^H \phi_k = 1$. Besides, we assume that $\mathbf{g}_{ilk} \in \mathbb{C}^{N \times 1}$ (i.e., $\mathbf{g}_{ilk} \sim \mathcal{N}_{\mathbb{C}}(0, \mathbf{R}_{ilk})$, where $\mathbf{R}_{ilk}$ is the covariance matrix (CM) of the channel) stands for the channel vector that joins the k_{th} UE in the l_{th} to the i_{th} BS. As we examine the UL phase, the incoming channel vector at the i_{th} BS is expressed as

$$\mathbf{Y}_i = \sum_{l=1}^{L} \sum_{k=1}^{K} \sqrt{q}\mathbf{g}_{ilk}\phi_k^H + n_i \tag{1}$$

Where q denotes the UL power (i.e., the signal to noise ratio (SNR)), and $n_i \in \mathbb{C}^{N \times \tau_p}$ is a matrix describing the therminal noise, which follows $\mathcal{CN}(0, \sigma^2)$. Thus, the Eq. (1)) may be re-expressed without having to rely on the pilot as follows:

$$\mathbf{y}_{ik} = \mathbf{Y}_i\phi_k = \sqrt{q}\mathbf{g}_{iik}\phi_k^H\phi_k + \sum_{l=1, l\neq i}^{L} \sum_{j=1}^{K} \sqrt{q}\mathbf{g}_{ilk}\phi_j^H\phi_k + N_{ik} \tag{2}$$

Here, $N_{ik} = n_i\phi_k$ indicates the noise caused by the k_{th} UE in the i_{th} BS.

2.2 Channel Estimation for the Conventional Massive MIMO Network

In this segment, we address the CE operation performed in the i_{th} BS [12]. Additionally, we compute and evaluate the channel estimate using the MMSE estimator. Besides, as reported in [1], using the incoming channel at the i_{th} BS, the estimated channel using the MMSE estimator is formulated as

$$\hat{\mathbf{g}}_{iik} = \sqrt{q\tau_p}\mathbf{R}_{iik}\boldsymbol{\Upsilon}_{ik}\mathbf{y}_{ik} \tag{3}$$

Here, $\boldsymbol{\Upsilon}_{ik} = \mathbb{E}\{\mathbf{y}_{ik}(\mathbf{y}_{ik})^H\} = (\sum_{l=1}^{L} q\tau_p\mathbf{R}_{ilk} + \sigma^2\mathbf{I}_N)^{-1}$, where $\boldsymbol{\Upsilon}_{ik}$ denotes the received signal's correlation matrix or the sum of the CM of UEs in all cells that use the same pilot as the k_{th} UE. It is important to highlight that the error of estimation $\tilde{\mathbf{g}}_{iik} = \hat{\mathbf{g}}_{iik} - \mathbf{g}_{iik}$ and the vector estimate $\hat{\mathbf{g}}_{iik}$ are independent random variables and expressed as

$$\hat{\mathbf{g}}_{iik} \sim \mathcal{N}_{\mathbb{C}}(0, q\tau_p\mathbf{R}_{iik}\boldsymbol{\Upsilon}_{ik}\mathbf{R}_{iik}), \quad \tilde{\mathbf{g}}_{iik} \sim \mathcal{N}_{\mathbb{C}}(0, \boldsymbol{\Omega}_{iik}) \tag{4}$$

Here, $\boldsymbol{\Omega}_{iik} = \mathbf{R}_{iik} - q\tau_p\mathbf{R}_{iik}\boldsymbol{\Upsilon}_{ik}\mathbf{R}_{iik}$, which is the error correlation matrix. The NMSE is used to determine and evaluate the CE accuracy. Thus, using the MMSE estimator, the corresponding NMSE is defined as follows

$$\mathcal{NMSE}_{ik} = \frac{\mathbb{E}\{||\hat{\mathbf{g}}_{iik} - \mathbf{g}_{iik}||^2\}}{\mathbb{E}\{||\mathbf{g}_{iik}||^2\}} = \frac{tr(\boldsymbol{\Omega}_{iik})}{tr(\mathbf{R}_{iik})} = 1 - \frac{q\tau_p tr(\mathbf{R}_{iik}\boldsymbol{\Upsilon}_{ik}\mathbf{R}_{iik})}{tr(\mathbf{R}_{iik})} \tag{5}$$

For the uncorrelated channels scenario or in single BS antennas, the spatial correlation matrix $\mathbf{R}_{iik}$ is reduced to the average gain of channel β_{iik}, where the NMSE expression shown in Eq. (5) simplifies to

$$\begin{aligned} \mathcal{NMSE}_{ik} &= 1 - \frac{q\tau_p\beta_{iik}}{q\tau_p\beta_{iik} + \sum_{l\neq i}^{L} q\tau_p\beta_{ilk} + \sigma^2} \\ &= 1 - \frac{SNR_{iik}}{SNR_{iik} + \sum_{l\neq i}^{L} SNR_{ilk} + 1} \end{aligned} \tag{6}$$

Here, $SNR_{iik} = \frac{q\tau_p\beta_{iik}}{\sigma^2}$. The coefficient SNR_{iik} is known as the effective SNR, as it depends on the pilot sequence length. In addition, as the pilot sequence length increases, the effective SNR increases (since it depends linearly on τ_p), which highly increases the estimation quality.

3 Cell-Free Networks

This section describes the system model and CE operation for the UL CF M-MIMO network [13, 14]. Thus, L denotes the number of APs, K indicates the number of UEs, and N symbolizes the NoA in each AP. In addition, M symbolizes the total NoA in all APs (i.e., $M = N \times L$).

3.1 System Model for Cell-Free Network

This segment examines the CF M-MIMO system model that will be used throughout this work, where L APs are distributed uniformly on the coverage area with N antenna each [15]. In addition, these APs are intended to serve K single-antenna UEs, where only the APs that provide the best channel condition regarding each UE are allowed to serve it. We assume that $\tau_p \leq K$ (i.e., not all UEs use orthogonal pilot). In other words, τ_p UEs benefit from orthogonal pilots whereas, the remaining UEs use the previously assigned pilots [16], which leads to the pilot contamination problem. Besides, we consider that the k_{th} UE transmits the pilot $\phi_k \in \mathbb{C}^{\tau_p \times 1}$, which adhere the following property $\phi_k^H \phi_k = 1$. Furthermore, we assume that $\mathbf{g}_{ik} \in \mathbb{C}^{N \times 1}$ (i.e., $\mathbf{g}_{ik} \sim \mathcal{N}_{\mathbb{C}}(0, \mathbf{R}_{ik})$, where $\mathbf{R}_{ik}$ is the CM of the channel) is the channel vector that joins the k_{th} UE to the i_{th} AP. Thus, The incoming channel vector at the ith AP is represented as

$$\mathbf{Y}_i = \sum_{k=1}^{K} \sqrt{q}\mathbf{g}_{ik}\phi_k^H + n_i \tag{7}$$

Where, q stands for the UL power, and $n_i \in \mathbb{C}^{M \times \tau_p}$ denotes the noise matrix, which follows $\mathcal{CN}(0, \sigma^2)$. Thus, the Eq. (7) may be re-expressed without having to rely on the pilot as follows:

$$\mathbf{y}_{ik} = \mathbf{Y}_i\phi_k = \sqrt{q}\mathbf{g}_{ik}\phi_k^H\phi_k + \sum_{j \in \mathcal{P}_k \backslash \{k\}} \sqrt{q}\mathbf{g}_{ij}\phi_j^H\phi_k + N_{ik} \tag{8}$$

Here, $N_{ik} = n_i\phi_k$ indicates the noise caused by the k_{th} UE in the i_{th} AP and $\mathcal{P}_k$ indicates the subset/group of UEs that utilize identical pilot as the k_{th} UE.

3.2 Channel Estimation for Cell-Free Network

The CE procedure for CF M-MIMO is discussed and evaluated through this part, where the MMSE estimator is used to compute and assess the channel estimate. Furthermore, as reported in [1], and based on the received channel, the estimated channel using the MMSE estimator is represented as

$$\hat{\mathbf{g}}_{ik} = \sqrt{q\tau_p}\mathbf{R}_{ik}\mathbf{\Upsilon}_{is_k}\mathbf{y}_{ik} \tag{9}$$

Here, s_k indicates the index of the pilot allocated to the k_{th} (i.e., $s_k \in \{1, \ldots, \tau_p\}$) and $\mathbf{\Upsilon}_{ik}$ ($\mathbf{\Upsilon}_{is_k} = \mathbb{E}\{\mathbf{y}_{ik}\ (\mathbf{y}_{ik})^H\} = (\sum_{k \in \mathcal{P}_k} q\tau_p\mathbf{R}_{ik} + \sigma^2\mathbf{I}_N)^{-1}$)denotes the received signal's correlation matrix. It is important to highlight that the error of estimation $\tilde{\mathbf{g}}_{ik} = \hat{\mathbf{g}}_{ik} - \mathbf{g}_{ik}$ and the vector estimate $\hat{\mathbf{g}}_{ik}$ are independent random variables and expressed as

$$\hat{\mathbf{g}}_{ik} \sim \mathcal{N}_{\mathbb{C}}(0, q\tau_p\mathbf{R}_{ik}\mathbf{\Upsilon}_{ik}\mathbf{R}_{ik}), \quad \tilde{\mathbf{g}}_{ik} \sim \mathcal{N}_{\mathbb{C}}(0, \mathbf{\Omega}_{ik}) \tag{10}$$

Here, $\mathbf{\Omega}_{ik} = \mathbf{R}_{ik} - q\tau_p\mathbf{R}_{ik}\mathbf{\Upsilon}_{ik}\mathbf{R}_{ik}$, which is the error correlation matrix. To analyze and evaluate the CE operation (the CE is obtained using the MMSE estimator) accuracy in CF M-MIMO, we employ the NMSE. In this situation, the NMSE is defined as

$$\mathcal{NMSE}_{ik} = \frac{\mathbb{E}\{||\hat{\mathbf{g}}_{ik} - \mathbf{g}_{ik}||^2\}}{\mathbb{E}\{||\mathbf{g}_{ik}||^2\}} = \frac{tr(\mathbf{\Omega}_{ik})}{tr(\mathbf{R}_{ik})} = 1 - \frac{q\tau_p tr(\mathbf{R}_{ik}\mathbf{\Upsilon}_{ik}\mathbf{R}_{ik})}{tr(\mathbf{R}_{ik})} \tag{11}$$

One can remark that the spatial correlation matrix $\mathbf{R}_{ik}$ is reduced to the average gain of channel β_{ik} in the uncorrelated channels scenario or single-antennas AP. Thereby, the NMSE expression given in Eq. (5) simplifies to

$$\begin{aligned} \mathcal{NMSE}_{ik} &= 1 - \frac{q\tau_p\beta_{ik}}{q\tau_p\beta_{ik}+\sum_{j\in\mathcal{P}_k\backslash\{k\}} q\tau_p\beta_{ij}+\sigma^2} \\ &= 1 - \frac{SNR_{ik}}{SNR_{ik}+\sum_{j\in\mathcal{P}_k\backslash\{k\}} SNR_{ij}+1} \end{aligned} \tag{12}$$

Here, $SNR_{ik} = \frac{q\tau_p\beta_{ik}}{\sigma^2}$. As the SNR_{ik} is dependent on the length of the pilot sequence thus, it is known as the *effective* 0.1 cm *SNR*. In this regard, when a system uses a large number of orthogonal pilots (i.e., $\tau_p \geq K$, this can highly improve the system performance, as the pilot sequence length increases, as the *SNR* increases. It is worth noting that the CF network is simply a generic situation in which each UE is served with the APs close to it (i.e., the APs most capable of delivering a satisfactory service to each UE). Besides, this study also addresses a network known as the small-cell network, which is a type of CF network in which the UE is served only with one AP.

3.3 Proposed Pilot Assignment Scheme for Cell-Free Network

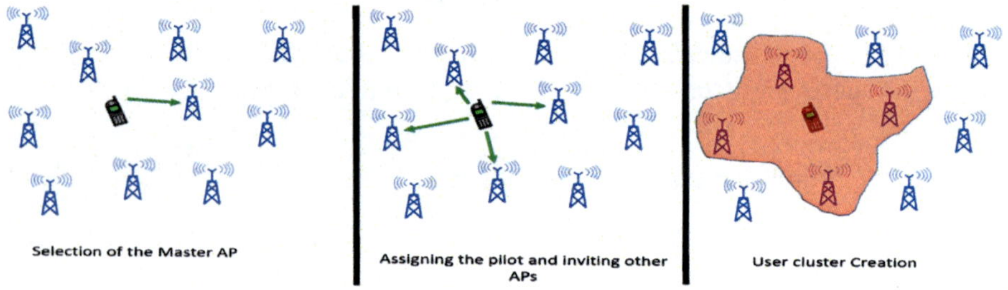

Fig. 1. The three stages of the proposed PA Scheme for CF network

After the APs choice stage, PAs are done regarding all UEs, and every UE passes through the these steps:

Once a UE requests network access, a PS must be assigned to the UE, with every UE goes through three stages, as illustrated in the Fig. (1).

1. First, each UE should choose a master AP that is close to them, as the APs closest to each UE are more capable of providing good service. As a result, the following equation is supplied to illustrate the point at which the UE accepts the i_{th} AP as its Master AP.

$$i = arg\ max_i\ \beta_i \tag{13}$$

Furthermore, every UE contacts its serving AP utilizing the normal random access mechanism [17, 18].

2. Following that, the pilot allocated to this UE is reallocated to some UEs on the network so that for the UE to encounter the lowest amount of interference; that is, a lower PC degree. Consequently, this point is described using the equation below

$$\tau = arg\ \ min_p\ \ tr(\boldsymbol{\Upsilon}_{is_k}) \tag{14}$$

In this case, just the K UEs are considered, where τ stands for the pilot allocated to the UE, so that the Master AP gets fewest PC degree. Then, each AP in the vicinity decides if it will serve the UE or not. Thus, in the CF network, two aspects for serving the UE are possible: whether the access point is not serving another UE on the pilot τ, or whether the newly connected UE has a stronger channel information than the one it is serving.

3. Each UE creates its cluster. Additionally, it is important to highlight that each UE must be served at least by one AP so that all UEs in the network be served. On the other hand, when a Master AP wants to serve a UE; thereby, it should immediately contact the subset/group of neighboring APs that it wishes to serve the said UE with the pilot τ.

4 Numerical Results

This section seeks to present numerical results to back up our prior theoretical assumptions. In addition, we provide the large-scale fading coefficients as a function of three parameters, i.e., the UE's distance from the serving APs/BSs, the height at which the APs/BSs are located, and a parameter that depends on the environment, (i.e., path-loss exponent) where it is setting to 3.6 during this work. Thus, the large-scale fading (i.e., β) is generated through the following model

$$\beta = -30.5 - 10\,\eta \log 10\left(\sqrt{(d^2 + h^2)}\right) \tag{15}$$

Here, η signifies the path-loss exponent, d the distance between the UE and the AP/BS, and h the height at which the APS/BSs are located.

In Fig. (2), The NMSE is displayed against the transmit power, q, where the MMSE estimator's performance is compared for three network types. According to Fig. (2), the MMSE estimator provides the best estimation using the small-cell network whereas, it is the worst for the conventional M-MIMO network. Besides, for small transmit power values, the CF M-MIMO provides a smaller gap in respect to the small-cell network, i.e., CF and small-cell networks perform similarly for low transmit power values, even if the UE in small-cell is provided with the best channel condition. However, as transmit power increases, the small-cell network achieves lower NMSE values than the other networks (e.g., when $q = 100$, the NMSE of small-free network reaches 1.7×10^{-5}, the NMSE of CF network reaches 7.8×10^{-4}, and the NMSE of conventional M-MIMO network reaches 0.044).

In Fig. (3), The NMSE is displayed against the length of the coverage area. Additionally, the MMSE estimator's performance is compared for three network types. According to Fig. (3), the MMSE estimator provides the best estimation using the small-cell network (i.e., smaller NMSE values) whereas, it is the worst for the conventional M-MIMO

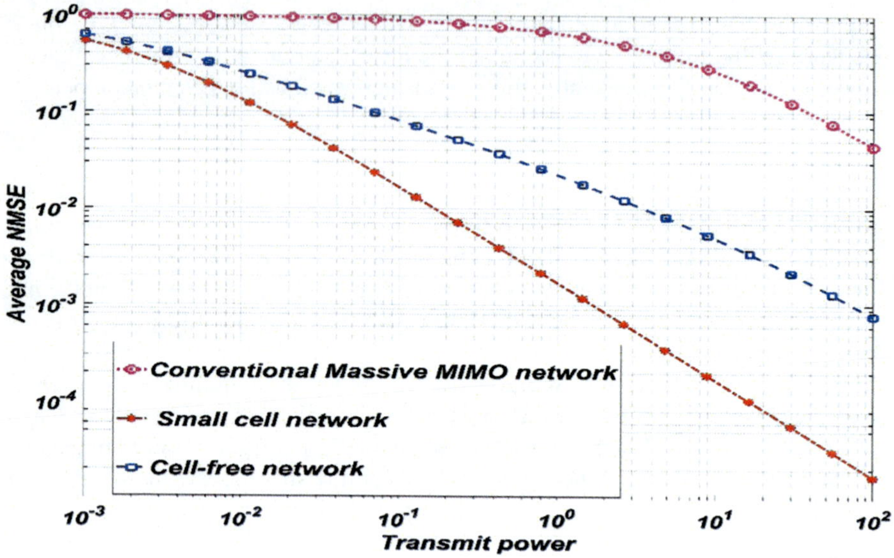

Fig. 2. Average NMSE versus the transmit power

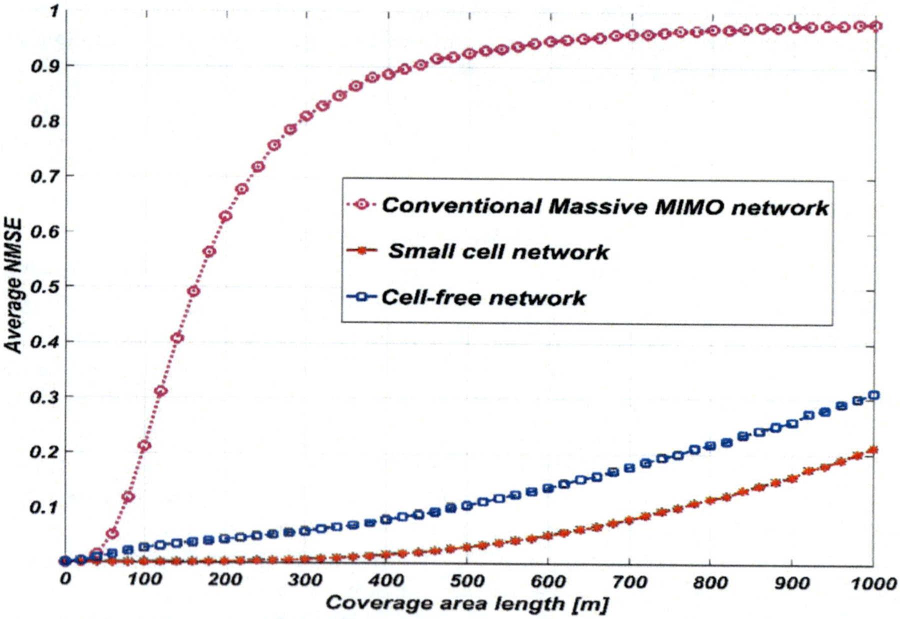

Fig. 3. Average NMSE versus the coverage area length.

network. Besides, for small coverage area length (<100 m), all studied networks provide approximatively the same NMSE values. When the coverage area surpasses 100 m, a smaller gap between the CF M-MIMO and the small-cell network is created whereas, the conventional M-MIMO network provides the worst performance. The small cell achieves the highest performance over a wide coverage area length since the UE in the small cell is provided with the best channel condition (e.g., when the coverage area length is 800m, the NMSE of a small-free network reaches 0.11, the NMSE of CF network reaches 0.21, and the NMSE of conventional M-MIMO network reaches 0.97).

5 Conclusion and Future Works

This paper has addressed the impact of M-MIMO technology in different net- works architecture. In effect, this work has addressed the CE operation for three TDD-enabled M-MIMO networks, that is, the conventional M-MIMO network, the CF M-MIMO network, and the small-cell M-MIMO network. Additionally, we have used the MMSE estimator to assess and examine the CE operation in each network design. We have concluded that the M-MIMO technology with small-cell M-MIMO provides higher performance compared to the others. Furthermore, the performance provided through the M-MIMO conventional network is the worst for all studied situations. Moreover, for each network architecture, we have developed the NMSE expression for the MMSE estimator.

References

1. Demir, Ö.T., Björnson, E., Sanguinetti, L.: Foundations of usercentric cell-free massive MIMO. arXiv preprint arXiv:2108.02541 (August 2021)
2. Amadid, J., Boulouird, M., Hassani, M.M.: Channel estimation for massive MIMO TDD systems and pilot contamination with uniformly distributed users. In: Bennani, S., Lakhrissi, Y., Khaissidi, G., Mansouri, A., Khamlichi, Y. (eds.) WITS 2020. LNEE, vol. 745, pp. 1037–1047. Springer, Singapore (2022). https://doi.org/10.1007/978-981-33-6893-4_94
3. Amadid, J., Boulouird, M., Hassani, M.M.R.: Channel estimation with pilot contamination in mutli-cell massive MIMO systems. In: 2020 IEEE 2nd International Conference on Electronics, Control, Optimization and Computer Science (ICECOCS), pp. 1–4. IEEE, December 2020
4. Molisch, A.F., et al.: Hybrid beamforming for massive MIMO: a survey. IEEE Commun. Mag. **55**(9), 134–141 (2017)
5. Marzetta Ngo, H.Q., Ashikhmin, A., Yang, H., Larsson, E.G., Marzetta, T.L.: Cell-free massive MIMO: uniformly great service for everyone. In: 2015 IEEE 16th International Workshop on signal Processing Advances in Wireless Communications (SPAWC), pp. 201–205. IEEE, June 2015
6. Ngo, H.Q., Ashikhmin, A., Yang, H., Larsson, E.G., Marzetta, T.L.: Cell-free massive MIMO versus small cells. IEEE Trans. Wirel. Commun. **16**(3):1834–1850 (2017)
7. Bjornson, E., Sanguinetti, L.: Scalable cell-free massive MIMO systems. IEEE Trans. Commun. **68**(7), 4247–4261 (2020)
8. Björnson, E., Sanguinetti, L.: Making celll-free massive MIMO competitive with MMSE processing and centralized implementation. IEEE Trans. Wirel. Commun. **19**(1), 77–90 (2019)

9. Wang, Z., Zhang, J., Björnson, E., Ai, B.: Uplink performance of cell-free massive MIMO over spatially correlated Rician fading channels. IEEE Commun. Lett. **25**(4), 1348–1352 (2020)
10. Amadid, J., Boulouird, M., Riadi, A., Hassani, M.M.: On channel estimation of uplink TDD massive MIMO systems through different pilot structures. In: Ben Ahmed, M., Teodorescu, H.N.L., Mazri, T., Subashini, P., Boudhir, A.A. (eds.) Networking, Intelligent Systems and Security. Smart Innovation, Systems and Technologies, vol. 237, pp. 161–176. Springer, Singapore (2022). https://doi.org/10.1007/978-981-16-3637-0_12
11. Boulouird, M., Amadid, J., Riadi, A., Hassani, M.M.: Channel estimation in massive MIMO systems for spatially correlated channels with pilot contamination. In: Ben Ahmed, M., Teodorescu, H.N.L., Mazri, T., Subashini, P., Boudhir, A.A. (eds.) Networking, Intelligent Systems and Security. Smart Innovation, Systems and Technologies, vol. 237, pp. 147–160. Springer, Singapore (2022). https://doi.org/10.1007/978-981-16-3637-0_11
12. Amadid, J., Boulouird, M., Riadi, A.: Channel estimation in massive MIMO-based wireless network using spatially correlated channel-based three-dimensional array. Telecommun. Syst. 1–18 (2022)
13. Amadid, J., et al.: On channel estimation for Rician fading with the phase-shift in cell-free massive MIMO system. Wirel. Pers. Commun. (2021)
14. Amadid, J., Belhabib, A., Zeroual, A.: On channel estimation in cell-free massive MIMO for spatially correlated channels with correlated shadowing under Rician fading. Int. J. Commun. Syst. e5011 (2021)
15. Amadid, J., Belhabib, A., Khabba, A., Zeroual, A., Hassani, M.M.R.: On channel estimation and spectral efficiency for cell-free massive MIMO with multi-antenna access points considering spatially correlated channels. Trans. Emerg. Telecommun. Technol. e4438 (2022)
16. Amadid, J., Belhabib, A., Boulouird, M., Hassan, M.M.R., Zeroual, A.: On channel estimation and power control for multi-cell multi-path massive MIMO TDD system (2022)
17. Rao, C.R., Rao, C.R., Statistiker, M., Rao, C.R., Rao, C.R.: Linear Statistical Inference and its Applications, vol. 2. Wiley, New York (1973)
18. Fisher, R.A.: On the mathematical foundations of theoretical statistics. Philosophical transactions of the royal society of London. Ser. A Contain. Pap. Math. Phys. Character **222**(594–604), 309–368 (1922)

Compact 5G Filtenna Operating at 26.17 GHz

Zakaria El Ouadi(✉), Asma Khabba, Jamal Amadid, Abdessalam El Yassini, Saida Ibnyaich, and Abdelouhab Zeroual

Instrumentation, Signals and Physical Systems (I2SP) Team, Faculty of Sciences Semlalia, Cadi Ayyad University, Marrakesh, Morocco

{zakaria.elouadi,asma.khabba,jamal.amadid}@edu.uca.ac.ma, {s.ibnyaich,zeroual}@uca.ac.ma

Abstract. A 5G Filtenna (filtering-Antenna) was described in this paper, it can operate at the n258 band (24.25–27.5 GHz). The structure was designed on an FR4 Epoxy substrate which has an electrical size of $0.52\lambda_0 \times 0.52\lambda_0 \times 0.087\lambda_0$, noting that, λ_0 is the free-space wavelength at the resonant frequency 26.17 GHz. The software used for the simulation is ANSYS HFSS, based on the finite element approach.

The simulation results show an omnidirectional radiation pattern, the reflection coefficient is less than 10 dB $|S_{11}| < 10$ dB, which means that our system captures at least 90% of the input power and just 10% of the power is lost due to reflection. The achieved selectivity reinforces the performance of our Filtenna as well as the (Voltage Standing Wave Ratio) VSWR result which is less than 2.

Keywords: Antenna · 5G · Filtenna · n258 band

1 Introduction

In these last years the field of telecommunications has recognized a mutation that has facilitated the life of people in different sectors: industrial, military, civil, ..., this mutation is thanks to the development of electronic devices that have allowed access to many services, for example, cell phones are not limited anymore to the simple telephone conversations but also they have allowed the transport of the multimedia (images, videos), the access to the television, the radio and the connection to networks such as Internet [1]. Figure 1 represents the evolution of the mobile generation.

Nowadays, we are working on the fifth-generation 5G, which is a technology characterized by its speed, greater reliability, very low latency and it can support a large number of devices. The 5G technology will also allow the development of uses completely different than 4G, such as telemedicine, autonomous vehicles, or factory automation [2].

According to the Federal Communications Commission (FCC), the 5G spectrum is divided into three important bands, the low-band which is devoted to frequencies less than (1 GHz), the sub-6 GHz band, or as we called it the mid-band dedicated to frequencies less than 6 GHz, and the third band is the high-band which includes the Millimetre-Wave band devoted to frequencies higher than 24 GHz [3].

S. Motahhir and B. Bossoufi (Eds.): ICDTA 2022, LNNS 454, pp. 351–358, 2022.
https://doi.org/10.1007/978-3-031-01942-5_35

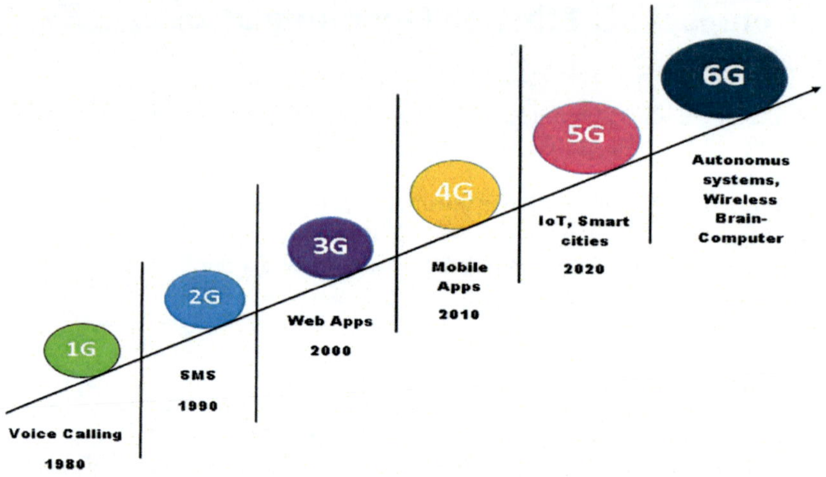

Fig. 1. Evolution of the mobile phone generation

In release 15, the 3rd Generation Partnership Project (3GPP) has divided the 5G spectrum into small operating bands [4]. The n258 band (24.25–27.5 GHz) is very interesting in Europe and particularly in France since it is allocated to the Ministry of the Armed Forces and meteorology, also it is reserved for space and terrestrial research using satellites: the Space Research Service (SRS) (space-to-Earth direction) and the Earth Exploration Satellite Service (EESS) (space-to-Earth direction) [5].

The Filtenna is a kind of multifunctional device which contains both filter and antenna integrated. At first to get an idea about the design we take some inspiration from the literature by looking at the different filters ready to be integrated with a microstrip antenna [6, 7]. A microstrip antenna is made up of a dielectric (substrate) that separates two parallel conducting layers having a permittivity of 2.2 to 12 [8, 9].

Filtenna, which is the combination of a filter and an antenna, can have a variety of structures [10–13].

In this work we had the idea to design a microstrip Filtenna that can operate at the n258 band (24.25–27.5 GHz), the structure is designed on an FR4 epoxy substrate having a dielectric constant $\epsilon_r = 4.4$ and a thickness (h) equal to 1 mm. The material used for the top (patch and resonators) and bottom (ground plane) layers is copper due to its perfect conductivity.

2 Filtenna Design

As shown in Fig. 2b, the top side of the Filtenna consists of two components, the filter is in the form of square open-loop resonators which are coupled with the transmission line, and this line feeds a patch antenna containing a rectangular slot of length L_{cut} and width W_{cut}. The underside of the Filtenna contains a partial layer of copper which represents the ground plane as shown in Fig. 2c. Figure 2a, it is a basic antenna that we have calculated its dimensions based on mathematical equations below.

$$W = \frac{c}{2f_0\sqrt{\frac{\epsilon_r+1}{2}}} \quad (1)$$

Knowing that $f_0 = 26.17$ GHz is the resonant frequency and $\epsilon_r = 4.4$ is the dielectric constant.

The microstrip synthesis gives the effective relative dielectric constant for $W > h$ as:

$$\epsilon_{reff}(W) = \frac{\epsilon_r + 1}{2} + \frac{\epsilon_r - 1}{2}\left[1 + 12\frac{h}{W}\right]^{\frac{-1}{2}} \quad (2)$$

Hammerstad presented the empirical equation that immediately determines the length extension:

$$\Delta L = 0.412h\frac{(\epsilon_{reff} + 0.3)(W/h + 0.264)}{(\epsilon_{reff} - 0.258)(W/h + 0.813)} \quad (3)$$

The effective length is expressed as:

$$L_{eff} = \frac{c}{2f_{res}\sqrt{\epsilon_{reff}}} \quad (4)$$

Subtracting the extension of the length from the effective length yields the patch's length L:

$$L = L_{eff} - 2\Delta L \quad (5)$$

As a result, the patch dimension values obtained are: $W_p = 2.21$ mm & $L_p = 3.48$ mm.

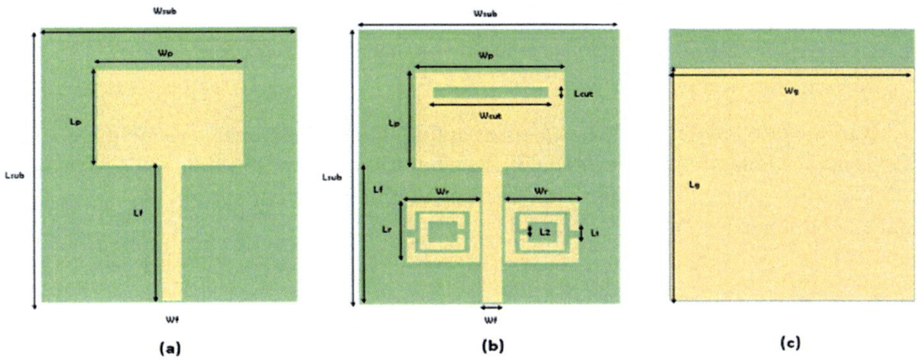

Fig. 2. The proposed Filtenna's evolution: (a) Basic Antenna (b) Proposed Filtenna (c) Bottom view

Table 1 contains all of Filtenna's dimensions.

Table 1. The proposed Filtenna's dimension.

Parameters	L_{Sub}	W_{Sub}	L_p	W_p	L_{cut}	W_{cut}	L_1	L_2	L_f	W_f	L_r	W_r	L_g	W_g
Dimensions (mm)	6	6	2.21	3.48	0.5	1	0.1	0.2	2.9	0.4	1.5	1.5	5.2	6

3 Simulated Results for the Proposed Filtenna

Figure 3 illustrates the simulation results of the reflection coefficient for the basic antenna and proposed Filtenna.

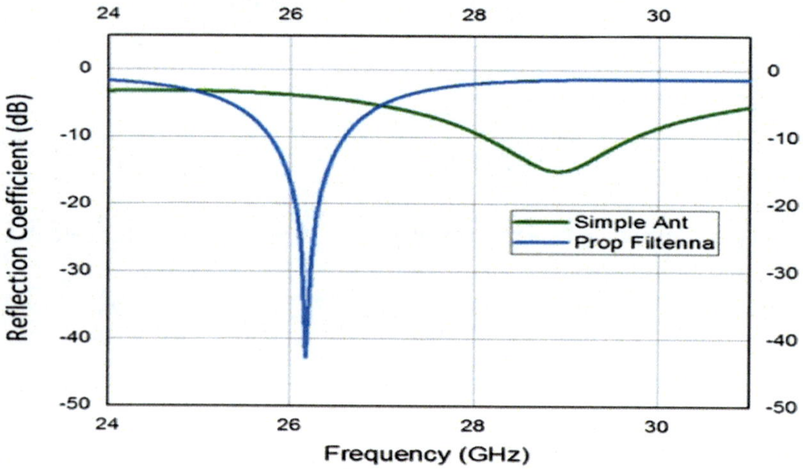

Fig. 3. Reflection coefficient as a function of frequency

From the previous figure, we can observe that the basic antenna's reflection coefficient is approximately equal to −15.03 dB, whereas the proposed Filtenna's reflection coefficient is around −41.92 dB. This last value implies that our system has a very low number of reflected waves.

3.1 Input Impedance and VSWR

Figure 4 depicts the real and imaginary parts of the proposed Filtenna's input impedance, which can be represented as $Z_e = 50.40 - 0.66j$.

At the resonance frequency, we can observe that we have an excellent impedance matching because the real component of the input impedance is near to (50) and the imaginary part is close to zero.

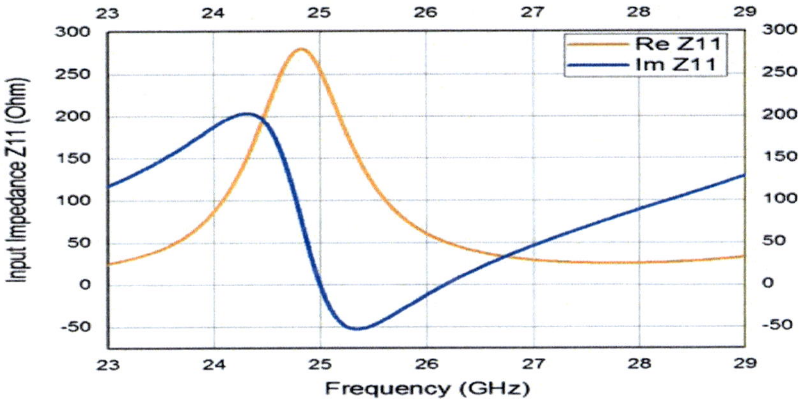

Fig. 4. Input Impedance Z_e as a function of frequency

The Voltage Standing Wave Ratio (VSWR) is less than 2 at the resonant frequency 26.17 GHz, as seen in Fig. 5.

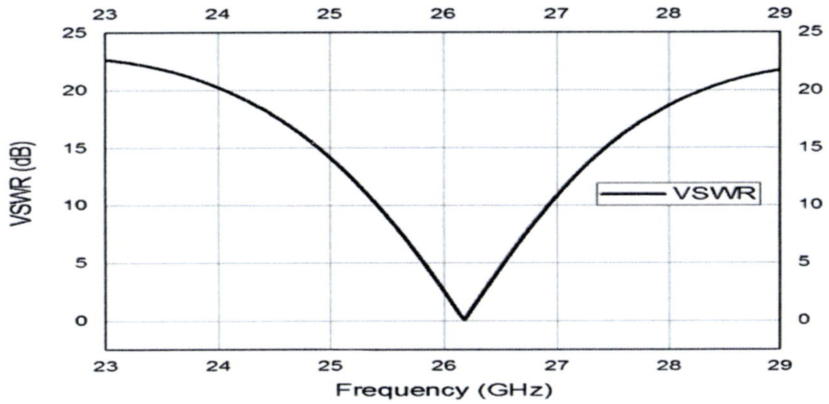

Fig. 5. The VSWR as a function of frequency

The VSWR is equal to 0.18, this involves that we are closer to the progressive regime than the stationary regime, which is suitable in this field.

3.2 Radiation Pattern, Gain and Directivity

Figures 6, 7 and 8 show the simulated 2D radiation pattern, simulated gain, and directivity for the proposed Filtenna designed at 26.17 GHz, respectively.

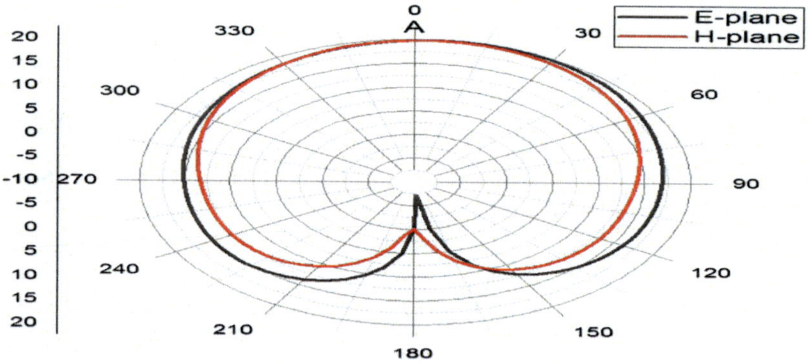

Fig. 6. The proposed Filtenna's 2D radiation pattern at 26.17 GHz,

In both, the electrical and magnetic planes, we remark that the radiation pattern is almost omnidirectional.

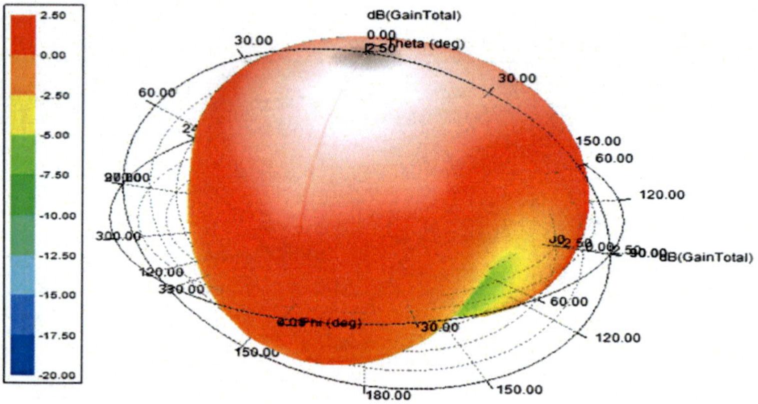

Fig. 7. The proposed Filtenna's Gain at 26.17 GHz

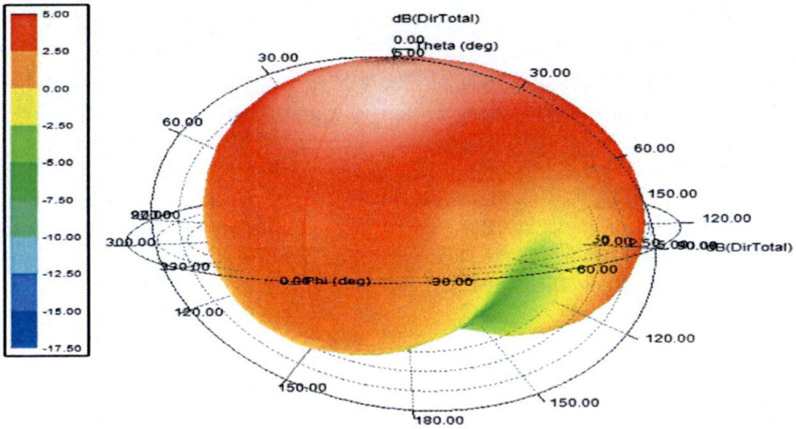

Fig. 8. The proposed Filtenna's Directivity at 26.17 GHz

According to the two previous figures we observe that the proposed Filtenna has a gain of 2.5 dB and a directivity equal to 5 dB.

Table 2 compares our Filtenna to other Filtennas that have been previously published in the literature:

Table 2. Comparative study.

References	Gain (dB)	Reflection coefficient (dB)	Size (mm^2)
Reference [11]	2.5	-32.5	29.26 × 20.5
Reference [12]	–	-32.82	34.24 × 26.5
Reference [13]	4.95	-29.81	111 × 47
Proposed Filtenna	2.5	-41.92	6 × 6

We can comment that our Filtenna had a very good reflection coefficient and a good gain.

4 Conclusion

A new Filtenna has been proposed in this paper, the reflection coefficient, input impedance, and VSWR results are excellent, we could observe the effect of the slot made on the rectangular patch of the basic antenna which allowed to reduce the resonant frequency towards the desired value, we could also remark the effect of the integrated filter on the selectivity of the antenna and the S_{11} value. The other obtained results from the radiation pattern, gain, and directivity are also acceptable and good.

In addition to the good obtained results, we also have the small dimensions of the Filtenna which means a low cost and ease of integration in various systems operating in the n258 band of the 5G spectrum.

References

1. Wang, M., Zhu, T., Zhang, T., Zhang, J., Shui, Y., Zhou, W.: Security and privacy in 6G networks: new areas and new challenges. Digit. Commun. Netw. **6**(3), 281–291 (2020)
2. Pisarov, J., Mester, G.: The impact of 5G technology on life in 21st century. IPSI BgD Trans. Adv. Res. (TAR) **16**(2), 11–14 (2020)
3. Lee, J., et al.: Spectrum for 5G: global status, challenges, and enabling technologies. IEEE Commun. Mag. **56**(3), 12–18 (2018)
4. Zhao, K., et al.: Spherical coverage characterization of 5G millimeter wave user equipment with 3G pp specifications. IEEE Access **7**, 4442–4452 (2018)
5. Kang, G., Liu, H., Li, K.: Analysis on the new progress of spectrum planning of IMT-2020(5G). J. Phys. Conf. Ser. **1437**(1), 012016 (2020)
6. Nouri, M., Jafarieh, A., Behroozi, H., Mallat, N.K., Jamaluddin, M.H., Aghdam, S.A.: Compact 5G millimeter-wave dual-band filter with application in filtenna. Microw. Opt. Technol. Lett. **63**(2), 620–625 (2021)
7. Coumar, S.O.: Miniaturized DGS based multi-band pass filters for satellite applications. J. Ambient Intell. Humaniz. Comput. **13**, 241–249 (2021)
8. Thatere, A., Khade, S., Lande, S.V.: A modified rectangular mid-band microstrip slot antenna for WLAN, WiFi and 5G applications. J. Univ. Shanghai Sci. Technol. **23**(1), 1–6 (2021)
9. Thummaluru, S.R., Ameen, M., Chaudhary, R.K.: Four-port mimo cognitive radio system for midband 5G applications. IEEE Trans. Antennas Propag. **67**(8), 5634–5645 (2019)
10. Lu, R., Yu, C., Hong, W.: Siw cavity-fed filtenna arrays for 5G millimeter wave applications. In: 2020 International Symposium on Antennas and Propagation (ISAP), pp. 441–442. IEEE (2021)
11. Hosain, M.M., Kumari, S., Tiwary, A.K.: Compact filtenna for WLAN applications. J. Microwaves Optoelectron. Electromagn. Appl. **18**(1), 70–79 (2019)
12. Rodrigues, L., Varum, T., Matos, J.N.: Reconfigurable filtenna for 5G/LEO constellations mobile terminals. In: 2020 14th European Conference on Antennas and Propagation (EuCAP), pp. 1–4. IEEE (2020)
13. Syrytsin, I., Shen, M., Pedersen, G.F.: Antenna integrated with a microstrip filter for 5G mm-wave applications. In: 2018 International Conference on Electromagnetics in Advanced Applications (ICEAA), pp. 438–441. IEEE (2018)

Web-Based Techniques and E-learning

E-learning Recommendation Systems: A Literature Review

Hicham Aberbach[1(✉)], Adil Jeghal[2], Abdelouahed Sabri[1], and Hamid Tairi[1]

[1] Faculty of Science Dhar-Mahraz, LISAC, University of Sidi Mohamed Ben Abdellah, P.B 1796, Atlas-Fez, Morocco
aberbachhicham@gmail.com

[2] National School of Applied Sciences, LISAC, University of Sidi Mohamed Ben Abdellah Fez, Fes, Morocco
adil.jeghal@usmba.ac.ma

Abstract. Lately, the problem of choosing and searching data is becoming more and more difficult to solve, especially with the significant increase in the mass of information that exists on the internet. Computer techniques exist to facilitate research and allow the relevant extraction of information, one of these techniques is the recommended process that guides the user during his exploration, seeking for him the information that seems relevant. Different systems are capable of providing personalized recommendations to guide the user to interesting and useful resources within a large data space. Their goal is to reduce information overload through a process of collecting, filtering, and proactively recommending information. Recommendation systems introduce the notions inherent in the recommendation, based, among other things, on information retrieval, filtering, machine learning, and collaborative approaches. It also deals with the evaluation of such systems and presents different application cases. This article aims to provide a review of recommendation systems used in an e-learning environment. In the first part, we will present the concept of recommendation systems, their operating principle, its different approaches with their advantages and disadvantages, the second one, concerned this type of system in an educational environment and especially e-learning with a precision of its different approaches used and adopted in the same field.

Keywords: Recommendation system · E-learning · Learning approach · Machine learning

1 Introduction

Recently, Recommendation systems are designed to help users of a platform identify what they might like, whether it is training content articles or even people. They facilitate navigation and prevent users from being lost to too much information. These systems have now become essential for news reading, online music, video on demand, online shopping, information retrieval, recruitment, job search, online learning or self-study, etc. The mass of data exchanged on the Internet is an unprecedented asset for every- one's

S. Motahhir and B. Bossoufi (Eds.): ICDTA 2022, LNNS 454, pp. 361–370, 2022.
https://doi.org/10.1007/978-3-031-01942-5_36

access to information. At the same time, it is also difficult because it requires important filtering processes to ensure that the returned data is relevant to the user. In this context, consideration of the context in which a request is submitted is given particular attention by the Information Seeking community. Helping users discover and choose resources in an information space of this magnitude is an important challenge that remains still relevant. Recommendation systems are a solution to this Information overload problem. The main purpose of these systems is to provide the user with recommendations that reflect his personal preferences. Also, the Recommender system is a specific type of intelligent system, which exploits historical user ratings on items and/or auxiliary information to make recommendations on items to the users. It plays a critical role in a wide range of online shopping, e-commercial services social networking applications, and online learning [1].

2 Working Principle of Recommendation System

The primary role of recommendation systems is to identify the subgroup of users to which a user belongs to propose results that may be of interest to the user. The identification of sub-groups of users to which a user belongs is generally done according to the history of use of the service by that user. The recommendation system may, however, be based on known characteristics of the user (age, socio-professional category, sex, occupation.) or a combination of these characteristics and history. It then remains for the recommendation system to find the other users sharing the most common points with this user, analyze the most ordered items, shared or acclaimed by these users to offer a personalized selection of recommended items. Recommendation systems generally rely on a clustering algorithm to distinguish between different sub-groups of users, which can be trained in unsupervised learning [2].

3 Recommendation Approaches

3.1 Content-Based Filtering Approach

Content-based recommendation systems have several things in common, such as the description of what to recommend, specifics about user profiles, and a set of methods and techniques for comparing profiles to recommendation items and drawing the most appropriate. Article recommendation is done based on two things, the description of the article and the set of interests of a given user. In this way, recommendations are made for web pages, articles to read online, learning capsules, etc. in this type of approach the task is to determine which items in the catalog best match the user's preferences. Such an approach does not require a large user community or system usage action in the past. This type of filtering technique filtering built on the customer preferences and description about the item. these algorithms aim to suggest items or products that are alike to those items that the user enjoyed in past or is looking at in the present-day [3], Content-based filtering recommends each learner participating in a learning journey which activities and items are appropriate for them based on the number of similarities that have been chosen by the same user before. This is the same principle on which the content-based filtering algorithms are based to recommend the items that best match the user with of course comparison for each item previously assessed by the same learner [4] (Fig. 1).

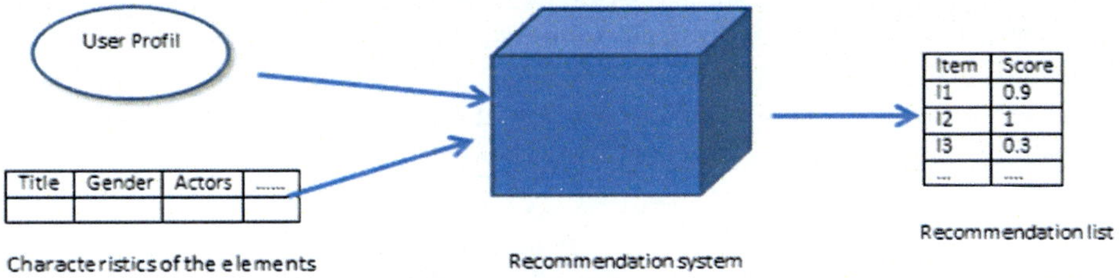

Fig. 1. Recommendation system based on content-based filtering this figure illustrates the principle of this type of recommendation.

Advantages:

Based on our analysis of these types of approaches, we were able to specify this set of advantages for this type of recommendation:

- Is the key to having the most relevant recommendations for each;
- Matching user preferences and element characteristics work for many types of data (textual, numeric, etc.) as keyword lists are generally used;
- Data on other users are useless;
- There is no problem with a cold start when a new item is added to the catalog or low density since it is a matter of matching user preferences and item characteristics;
- It is possible to make recommendations to users with "unique" tastes.

Disadvantages

- Elements represented by the same set of keywords cannot be distinguished;
- Users who have viewed a very large number of elements pose a problem (too much information in the user profile to coincide with the characteristics of the elements);
- History is unavailable once a user starts using the system;
- A difficulty in the development of user profiles and the evolution of user interests must be managed and taken into consideration;
- For the system to produce accurate recommendations the user must provide feedback on the returned suggestions but this is time-consuming for the user.

3.2 Collaborative Filtering Approach

Systems based on collaborative filtering produce recommendations by calculating the similarity between a user's preferences and those of other users. Such systems do not attempt to analyze or understand the content of the items to be recommended. The method consists of making automatic predictions about a user's interests by collecting opinions from many users, In social network environments, collaborative filtering presents a useful means for the selection of the best-published content, it uses algorithms for filtering data from proposals and user reviews to be able to draw personalized recommendations for people with common preferences. Specifies that the type of collaborative filtering is fixed in three categories: the first is based on the principle of neighbors or users are selected by making a comparison with the similarity of active users, the second is element by element which consists in using a matrix to detect it the resemblance of element pairs, and finally the third, it is based on the classification and makes its recommendation through how similar users liked this classification or category. For this type of recommendation we present this figure to detail the process (Fig. 2):

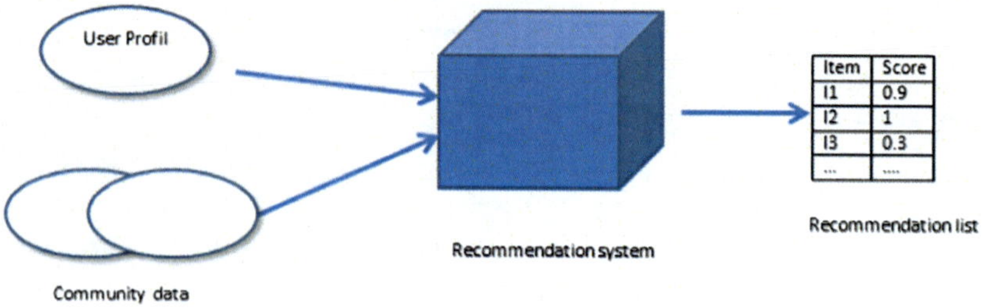

Fig. 2. Process of collaborative filtering approach

Advantages

- Use other users' scores to assess the usefulness of the elements;
- Find groups of users with the same interests as the current user;
- By increasing the number of users, the score rises and subsequently, we will obtain good results.

Disadvantages

- The recommendation system encounters the low density of the Users X Elements matrix;
- In addition, there is also the problem of cold start (cold-start problem): when a new user uses the system, his preferences are not known and once an item is created and added to the catalog, nobody has assigned him a score;
- With this type of system, we can find a large number of elements and users, computation grows linearly;
- It does not make sense to recommend to a user all films with a certain author who liked them before, it means a non-diversity of this system.

3.3 Hybrid Filtering Approach

This approach is a composition of several filtering methods: this type is used to solve the problems that we may encounter when using the two filtering approaches above, such as the cold start problem, over-customization. The goal of using this type of approach is to improve the accuracy and efficiency of recommender operations [5]. This technique presents the advantages of the approaches which compose it while limiting their drawbacks, this technique presents the advantages of the approaches which compose it while limiting their disadvantages.

4 Recommendation System in e-learning

E-learning is also one of the areas most affected by recommendation systems, as we will see below the different definitions of this concept: recommendation system oriented for E-learning, according to [6] Drachsler et al. (2007), recommendation systems in computer-based human learning environments must take into account characteristics that are specific to the learning context. These characteristics are learning goal, prerequisite knowledge, learner characteristics, and preferences, learner grouping, teaching resources, pathways, and learning strategies. The recommendation systems present an effective means that can be considered as a guide allowing in an e-learning environment to make recommendations of educational resources compatible with each type of learner, links, and training paths to follow. Although the development of the agents responsible for these recommendations should be based on data associated with users and their profiles such as interests, reviews, etc [7]. Recommendations systems have been developed as a solution to address information overload and to provide personalized services to appropriate users at the right time [8]. Recommendations systems have two important components: the first is the users to whom the system provides services, and the second is the online products, services or courses to be recommended to users, like movies, books, or learning journeys with certificates like we find in Massive Online Open Course 'MOOCs' [9]. The point that constitutes a challenge for designing a recommendation system or one of its components oriented for e-learning is to define the learning user, the objective of his context as well as the learning domain, for this it is necessary to research and discover the skills and knowledge levels of learners to arrive at a recommendation system that resolves information overload problems for learners [10].

E-learning is essentially the computer and network-enabled transfer of knowledge and skill, another new recommendation system proposed that predicts the best combination of subjects in which students are more interested using the learning management system such as Moodle to collect the data from a student regarding their course choices [11].In other sense, a new principle of recommendation of tasks and online learning operations makes it possible to suggest to a learner a set of tasks according to the process already carried out before the learner, the conception of this principle is made in two main parts: one "learning" module which contains the tasks carried out and already passed, and a "counseling" module which takes into account the past module to recommend tasks. Several techniques allow the realization of this process such as the exploration of rules of the association, collaborative filtering, or data clustering [12]. Recommendation systems in computer-based human learning environments aim to develop a recommended strategy based on the characteristics specific to the learning context. The goal is to be able to support learners through their learning process to achieve their learning goals [13]. The challenges and difficulties arising from information research-oriented learning are caused by the overload of this information, where the crucial role of personalized intelligent agents and recommendation systems emerges as solutions to solve these problems of learning overload and help learners find learning devices and materials that are useful, relevant and meet their learning needs[14]. In an online learning environment, we can define a recommendation system as a system agent that allows recommendations to be made intelligently of operations and work to a certain learner through the activities of previous learners. These activities can be different from doing an exercise, viewing posts posted to a forum or conference system, performing an online simulation, or an online learning resource [15]. Recommendation systems designed to facilitate online learning are a solution that reduces the seriousness of the problem of the overload and mass of information that we currently detect either on social networks or on the various means of online learning.

5 E-learning Recommendation System Approaches

Recommendation systems direct the natural operations regarding the choice of classmates and colleagues, training leaders, and any other resources that make the learning process successful and prevent deficiencies. These online recommendations are made based on personalized techniques based on the preferences, characteristics, and similar views of the learners, all this for a better proposal of learning activities [16]. Most of the work on e-learning recommendation techniques has focused on the recommendation of learning objects (materials/resources) or learning activities [17].

In what follows, we will take a look at all of the techniques and recommendation approaches used in online learning environments:

5.1 Collaborative Filtering Approach in e-learning

We note that the majority of recommendation systems use collaborative filtering, learning online using e-learning platforms becomes indispensable in the teaching process, this type of recommendation is used by [18] to set up a new method for recommending relevant courses to learners which are based on social filtering and collaborative filtering for defining the best way in which the learner must learn, and recommend courses which better much the learner's profile and social content [19]. develop an educational collaborative filtering recommender agent, with an integrated learning style finder. The agent produces two types of recommendations: suggestions and shortcuts for learning materials and learning tools, helping the learner to better navigate through educational resources [20]. Proposed a recommendation technique combining collaborative filtering and ontology to recommend personalized learning materials to online learners, provides an algorithm for personalizing learning content with the necessary educational standards by combining collaborative filtering algorithms with social choice theory [21]. The presented algorithm gives a ranking of "difficulty" assigned to a particular learner's questions by combining the rankings of similar learners, as measured by different aspects of their performance on common past questions, such as marks, the number of attempts, and the time spent in solving questions, this method aims to improve the adaptation of all the problems and exams to the learners individually by directing them to the areas where the lack of training persists [22].

5.2 Content-Based Filtering Approach in e-learning

There is a set of searches that have opted for content-based filtering. The content-based recommendation makes it possible to have suggestions regarding learning objects and activities that belong to the same family of learning elements that have been appreciated previously. These recommendations are derived from the individual characteristics of each learner [23]. The presentation of the key elements used in the content recommendations translates into the forms of the attributes, the prediction made to the learner by combining the evaluation of the elements and the elements of the lookalikes. Subsequently, the learner will have the opportunity to have recommendations without considering other learners. Two categories can be drawn from the techniques available and based on content [24, 25].

Achieve learning personalization by developing a new recommendation system based on a content-based filtering approach and learner's negative rating and recommending the interesting messages to the learners for better performance [26].

Case-based Reasoning Techniques

Systems using case-based reasoning (CBR: case-based reasoning) recommend items with a strong correlation with items valued by the learner in the past. It is useful to inform the learner of the intended learning objectives. Case-based reasoning techniques are independent of the domain. They do not require content analysis, and the quality of the recommendation increases over time when learners assess more items [27].

Techniques Based on Attributes

Attribute-based techniques recommend items based on their match to the learner's profile. Attributes can be weighted according to their importance to the learner. At- tributes can be weighted according to their importance to the learner. Attribute-based techniques are sensitive to changes in learner profiles [27]. They will have control over the personalization of the sponsorship system with all the conditions for modifying attributes and profiles.

5.3 Hybrid Filtering Approach in e-learning

In recent years, many researchers have suggested that recommendation systems need to combine more than one technique to better analyze learner behavior and select the sequencing of the recommendation list [28]. Indeed, several recommendation systems adopting hybrid filtering have been implemented in the field of e-learning. A learning management system was proposed by Tang and McCalla (2005) [29] to store and share digital learning resources. This system uses a hybrid recommendation process based on data clustering and collaborative filtering approaches to classify learners according to interests and tastes using an explicit evaluation, Salehi (2013) proposed an e-recommendation system. Learning that recommends pedagogical resources based on implicit and explicit evaluation subsequently leads to a combination of collaborative filtering and data mining techniques [30]. Bourkoukou et al. (2016) proposed a hybrid recommendation model based on collaborative filtering and sequential feature tools. The purpose of the designed model is the selection and sequencing of the most appropriate learning objects [31]. The proposed system accompanies the learners by offering them the most appropriate learning objects by considering the objects previously visited by them and those consulted by different other learners with the same profiles. The system developed is based on collaborative filtering techniques and association rules [32].

6 Discussion

Recommendations systems have emerged in education through learning management systems (LMS), to improve online learning and teaching processes. However, with the growth in the number of educational resources and their diversity, the problem of information overload is becoming more and more critical. Therefore, offering learners personalized pedagogical recommendation tools is a necessity, the personalization of learning in an LMS is characterized by the functionality which gives the possibility of the system to respond in a personalized way to the different needs, to the specific characteristics of a learner to support him in his progress. These characteristics include skill levels, prerequisites, prior knowledge, interests, preferences, and learning styles. This personalization model contributes to improving the overall quality of learning with a suggestion of educational content necessary to achieve the educational objectives. Unlike marketing-oriented systems, learning recommendation systems must involve learners by challenging them. Indeed, rather than providing them with an environment with many constraints, these systems are designed to leave a lot of room for exploration and to confront learners with unexpected content, thus allowing learning by discovery. All these differences between e-commerce or other domains and educational contexts call for an adaptation of recommendation systems so that they become powerful learning tools.

7 Evaluation

Evaluating recommender systems in a general context is a complex process. It is even more for in an educational context, which must assess in addition to general parameters, the impact of recommendations during learning. General evaluation standards for a recommender system include forecast accuracy, diversity, the fraction of coverage, serendipity, confidence coefficient, robustness, scalability, etc.

8 Conclusion

Recommendation systems aim to facilitate research and access to information by offering items adapted to the requirements and conditions of users, they now represent an increasingly key role in knowing how to offer products or services to consumers, to users. Through this paper we have outlined the concept of recommendation systems, their operating principle, their different approaches with their advantages and disadvantages, on the one hand, on the other hand, we have treated this type of system in an educational environment and especially in e-learning with a precision of its different approaches used and adopted in the same context. and finally, all participants in online courses follow the same learning path and therefore have access to the same training content, despite their information needs being different this is where we can identify the important role of recommendation systems that have changed teaching and learning models by offering content adapted to the needs of learners based on their profile or content already followed.

References

1. Wei, J., He, J., Chen, K., Zhou, Y., Tang, Z.: Collaborative filtering and deep learning based recommendation system for cold start items. Expert Syst. Appl. **69**(1), 29–39 (2017)
2. Elsa Negre,https://interstices.info/les-systemes-de-recommendation-categorisation,2018
3. Berkani, L., Nouali, O., Chikh, A.: Recommandation personnalisée des ressources dans une communauté de pratique de e-learning. INFORSID (2013). computer science
4. Ricci, F., Rokach, L., Shapira, B., Kantor, P.B.: Recommender Systems Handbook, 1st edn. Springer-Verlag, Berlin, Heidelberg (2010)
5. Domavicius, G., Tuzhilin, A.: Toward the next generation of recommender systems: a survey of the state-of-the-art and possible extensions. IEEE Trans. Knowl. Data Eng. **17**, 734–49 (2005)
6. Klašnja-Milićević, A., Vesin, B., Ivanović, M., Budimac, Z., Jain, L.C.: Recommender systems in e-learning environments. In: E-Learning Systems. Intelligent Systems Reference Library, vol. 112. Springer, Cham (2017). https://doi.org/10.1007/978-3-319-41163-7_6
7. Aher, S.B., Lobo, L.: Course recommender system in e-learning. Int. J. Comput. Sci. Commun. (2012)
8. Chee, S.H.S., Han, J., Wang, K.: RecTree: an efficient collaborative filtering method. In: Kambayashi, Y., Winiwarter, W., Arikawa, M. (eds.) Data Warehousing and Knowledge Discovery. DaWaK 2001. LNCS, vol. 2114, pp. 141–151. Springer, Berlin, Heidelberg (2001) https://doi.org/10.1007/3-540-44801-2_15
9. Drachsler, H., Verbert, K., Santos, O.C., Manouselis, N.: Panorama of recommender systems to support learning. In: Ricci, F., Rokach, L., Shapira, B. (eds.) Recommender Systems Handbook. Springer, Boston, MA (2015). https://doi.org/10.1007/978-1-4899-7637-6_12
10. Tarus, J.K., Niu, Z., Mustafa, G.: Knowledge-based recommendation: a review of ontology-based recommender systems for e-learning. Artif. Intell. Rev. **50**(1), 21–48 (2017). https://doi.org/10.1007/s10462-017-9539-5
11. Zaiane, O.R.: Building a recommender agent for e-learning. In: Conference on Computers in Education (2002)
12. Lu, J.: Personalized e-learning material recommender system. In: Proceedings of the international conference on information technology for application, pp. 374–379. Londres, Angleterre (2004)
13. Khribi, M.K., Jemni, M. and Nasraoui, O. Automatic recommendations for e-learning personalization based on web usage mining techniques and information retrieval. J. Educ. Technol. Soc. **12**(4), 30–42 (2009)
14. Madani, Y., Erritali, M., Bengourram, J., Sailhan, F.: Social collaborative filtering approach for recommending courses in an e-learning platform. Procedia Comput. Sci. (2019)
15. Dascalu, M.I., Bodea, C.N., Moldoveanu, A., Mohora, A., Lytras, M., Pablos, O.P.: A recommender agent based on learning styles for better virtual collaborative learning experiences. J. Comput. Hum. Behav. **45**, 243–253 (2015)
16. Tarus, J.K., Niu, Z., Khadidja, B.: E-learning recommender system based on collaborative filtering and ontology. Sch. Inf. Sci. (2017)
17. Segal, A., Katzir, Z., Gal, K., Shani, G., Shapira, B.: A collaborative filtering approach to personalization in e-learning. Educ. Data Min. (2014)

18. Billsus, D., Pazzani, M.J.: Learning collaborative information filters. In: Proceedings of International Conference on Machine Learning (ICML 1998), Madison, Wisconsin, USA, pp. 46–54, 24–27 Juillet 1998
19. Schmitt, S., Bergmann, R.: Applying case-based reasoning technology for product selection and customization in electronic commerce environments. In: The Twelfth International Bled Electronic Commerce Conference, Bled, Slovenia, 7–9 Juin 1999
20. Wilson, D., Smyth, B., Sullivan, D.O.: Sparsity reduction in collaborative recommendation: a case-based approach. Int. J. Pattern Recognity. Artif. Intell. **17**(5), 863–884 (2003)
21. Kandakatla, D.M.: A content-based filtering and negative rating recommender system for e-learning management system. In: Proceedings of the International Conference on Intelligent Computing and Sustainable System (2018)
22. Drachsler, H., Hummel, H.G.K., Koper, R.: Personal recommender systems for learners in lifelong learning networks: the requirements, techniques, and model. Int. J. Learn. Technol. **3**(4), 404–423 (2008)
23. Aher, S.B., Lobo, L.M.R.J.: Combination of machine learning algorithms for a recommendation of courses in E-Learning System based on historical data. Knowl. Based Syst. **51**, 1–14 (2013)
24. Tang, T., McCalla, G.: Smart recommendation for an evolving e-learning system: architecture and experiment. Int. J. E-Learn. **4**(1), 105–129 (2005)
25. Salehi, M.: Application of implicit and explicit attribute-based collaborative filtering and bide for learning resource recommendation. Data Knowl. Eng. **87**, 130–145 (2013)
26. Bourkoukou, O., El Bachari, E., El Adnani, M.: A recommender model in e-learning environment. Arabian J. Sci. Eng. **42**, 607–617 (2017)
27. Imran, H., Belghis-Zadeh, M., Chang, T.W., Graf, S.: Plors: a personalized learning object recommender system. Vietnam J. Comput. Sci. **3**, 3–13 (2016)
28. Chughtai, M.W., Selamat, A., Ghani, I., Jung, J.J.: Retracted: e-learning recommender systems based on goal-based hybrid filtering. Int. J. Distrib. Sens. Netw. (2014)
29. Das, D., Sahoo, L., Datta, S.: A survey on recommendation system. Int. J. Comput. Appl. (0975–8887) **160**(7) (2017)
30. Thorat, P.B., Goudar, R.M., Barve, S.: Survey on collaborative filtering, content-based filtering and hybrid recommendation system. Int. J. Distrusted Sens. Netw. (2014)
31. TechTarget Contributo, August 2017. https://whatis.techtarget.com/definition/collaborative-filtering
32. Buder, J., Schwind, C.: Learning with personalized recommender systems: a psychological view. Comput. Hum. Behav. **28**(1), 207–216 (2012)

Factors Influencing Students' Satisfaction in Learning with Virtual Platforms: Case of Higher Education in Morocco

Ghita Faouzi[1](✉), Naila Amrous[2], and Nour-Eddine El Faddouli[1]

[1] RIME Team, MASI Laboratory, Mohammadia School of Engineers (EMI), Mohammed V University in Rabat, Rabat, Morocco
ghita.faouzi123@gmail.com, faddouli@emi.ac.ma

[2] Laboratory of Mediation-Information-Knowledge-Society (MIKS), School of Information Science, Rabat, Morocco

Abstract. E-learning provides many technological and pedagogical advantages thanks to the merger result of education sciences, telecommunication, and the internet. It is a solution adopted at the moment of the changes imposed by the Coronavirus pandemic to achieve distance learning. This paper sheds light on the impact of online education and the students' satisfaction identifying problems encountered and solutions adopted, which represents a solid basis for better strengthening distance education through either the resolution of the problem or the proposal of remedies. Findings show that Moroccan students' in engineering schools are not satisfied due to many issues related to connectivity and technical problems, but they proposed many suggestions that can help to build a hybrid educational model.

Keywords: E-learning · Coronavirus pandemic · Satisfaction · Hybrid model

1 Introduction

At the time of the coronavirus pandemic, higher education is undergoing a radical transformation right now. The COVID-19 disease quickly led to the closing of universities and colleges around the world in the hope that social distancing could help to reduce total fatalities and confirmed cases of this virus. Following the logic of the exception, the ministry of education in Morocco announced the close of all educational institutions from March 16, 2020, and adopted e-learning as an educational tool instead of face-to-face education; it has become more important than ever to ensure pedagogical continuity.

There are many problems that students and teachers have faced during the pandemic, sometimes they are technical and other times they are non-technical related to resources and economic issues. Our research paper emphasizes the problem of how to resolve the issues that are slowing us down in the development of academic life by focusing on some indicators cited by students which are mainly extracted from reports and papers

S. Motahhir and B. Bossoufi (Eds.): ICDTA 2022, LNNS 454, pp. 371–379, 2022.
https://doi.org/10.1007/978-3-031-01942-5_37

measuring students satisfaction that potentially prove the engagement of students in the educational life. It is carried out among students of engineering schools in Morocco. Its objective is to study students' satisfaction with the educational offer during the COVID-19 era. Those indicators are grouped using K-means to identify the clusters; each cluster refers to the issues related to connectivity, motivation, or technical problems, in the hope of proposing an adaptive model through the recommendations for each group of students. This study explores higher education students' experience, feedback, and perceptions of online education and recent changes, particularly those in engineering schools.

The majority of students are not satisfied for many reasons related to the online environment that can be measured by various criteria like motivation, interaction, and concentration. In [1] the authors affirm also that system quality and social influence positively impact e-learner satisfaction, on the other hand, learner computer anxiety has a negative impact on student satisfaction. As a solution, they propose in their research to focus on providing suitable infrastructures for professors and students, motivation, positive attitudes, good governance, and effective pedagogical engineering to give to students the best environment to relate online classes to their practices to conceptualize their learning [2]. In the Maghreb, from Algeria, the authors treat the subject of the perceptions of Algerian students regarding the abrupt transition to online learning amid the COVID-19 pandemic by variance ANOVA and SPSS software. The result is that students have a negative perception of distance learning [3]. In Tunisia, the state of higher education is discussed from a different perspective, because it treats the subject from the vision of teachers who affirmed that distance learning was difficult due to the lack of interactivity, psychological, pedagogical, and technical problems. The teachers suggested solutions to involve students in the study and for them, it is recommended that the ministries of education, higher education, and scientific research as well as the ministry of telecommunications invest in this new learning paradigm by improving infrastructure through the installation of optical fibers, the implementation of 5G, the intensification and generalization of digital equipment in schools and universities [4]. In Morocco, the article of Soumaya Elmendili and Sara Saaidi has the goal of taking stock of the universities in Morocco, especially in Mohammed V University in Rabat, during the crisis, through the documentary method and collection of data from different resources and highlight the difficulties which faced distance learning [5].

Several papers have been published by researchers around the world presenting studies on the various impacts of the COVID-19 pandemic on education, particularly methods used to decrease the impact of this crisis and measure students' satisfaction which becomes an important target on higher education. The main method used in this context is the Hierarchical linear modeling (HLM) analysis to determine the factors affecting satisfaction on online education, which are: the adaptation of faculty members to online education and capacities of the university [6]. In China, the Technology Satisfaction Model was used to highlight factors that impacted students' satisfaction: Computer self-efficiency and the perceived ease of use and usefulness of the platforms [7]. Students' satisfaction level is measured after modifying some educational practices based on students' recommendations [8]. In Saudi Arabia students are highly satisfied with online teaching, the data collected are analyzed using SPSS, Average Mean Score and validated using the Standard Deviation [9]. Other approaches are adopted by authors also such as the PACT method (Polarity Approach for Continuity and Transformation) are used in the article [10] that consist of seeing, mapping, and leveraging polarities that will have the greatest positive impact on teams. This method helps to understand the issues and opportunities then build the polarity map that contains desired results that can be achieved. In [11] the authors proposed as solution the implementation based on many modules: "Legal foundations of higher education and the digital economy", "Higher school pedagogy in the digital economy era", "Psychology of higher education in the digital economy era", "Basics of working in LMS" that involves updating the knowledge of teachers to work with the electronic information educational environment, "Working with podcasts" provides training in the basic principles of using podcasts: autonomy, interactivity, mobility, "Smart-learning", "Video processing and video recording", the module "creating a site" involves studying the algorithm for creating a teachers' site on free and paid platforms, "Tools for webinars", "Basics of working with online platforms", the module "Economics of educational programs" studies the main provisions of marketing and economics in the implementation of educational programs, both basic and additional and "Using social networks as a learning tool" which is familiar for students. Other techniques and methods of machine learning are used to predict students' academic performance such as ANCOVA, Logistic Regression, Support Vector Regression, and Log-linear Regression, to enhance education efficacy [12]. The contribution of our method: K-means and decision tree is to regroup the major problems of educational life for Moroccan engineering students' and try to propose a model starting from these issues.

2 Research Methods

This research was conducted in Moroccan higher education institutions from March 2020 to March 2021, participants were mainly students of engineering schools. It was based on a qualitative study utilizing the deductive study of a content analysis which is a research method for making valid inferences from data to a context to provide new insights of facts and a practical guide to action [13], and data collection via a questionnaire produced by Google forms that contain questions related to the platforms used, satisfaction with distance learning and a mastery level of digital tools. The surveys' goal is to know the situation of educational life at the time of the crisis, expressed by students of higher education, the majority of them are in the following schools in Morocco: Preparatory Year for Graduate Studies in Agriculture, Mohammadia School of Engineers, School of Information Science, National School of Applied Sciences, National Institute of Posts and Telecommunications, National School of Computing and Systems Analysis, National Institute of Statistics and Applied Economics, Agronomic and Veterinary Institute Hassan II, Higher Normal School of Technical Education, National Higher School of Electricity and Mechanics, National Higher School of Mines of Rabat, Higher School of Textile and Clothing Industries, Hassania School of Public Works, Mohammed VI International Academy of Civil Aviation, Preparatory Classes for High Schools. 190 of them answered the questionnaire.

3 Results

Various systems are used to assess the effectiveness of the education system. The most common criteria to an assessment are the level of satisfaction with the measures taken by the supervisory ministry, the level of satisfaction in handling TIC, the teachers' engagement, the level of satisfaction with the work required, the level of satisfaction with the home study environment, the network quality, internet access and the quality of platforms and videoconferences [14].

3.1 Demographic Data

One hundred and ninety students from the engineering schools and preparatory classes for high schools are responding to this survey, including 43.7% of them from Agronomic and Veterinary Institute Hassan II, 34.2% from Mohammadia School of Engineers, 4% from Hassania School of Public Works, 3.7% from Preparatory Classes for High Schools, 2.6% from Higher Normal School of Technical Education, 2.6% from National Institute of Statistics and Applied Economics and the rest from the other engineering schools. Students are required to answer many questions to identify their issues with the e-learning at the time of coronavirus pandemic, which is grouped by K-means clustering method that intended to partition data set into k number of clusters. It works iteratively to assign each data point to one of the k groups; three groups in our study are obtained: Motivation, technical problems, connectivity (Table 1).

Table 1. Problems impacted distance learning.

Impact (%)	Causes
Motivation (39.4%)	E-learning can influence the behavior of students, The majority of them claim their decreased motivation and absence of concentration due to the lack of interaction between the teacher and students or students among themselves (Teamwork) and the absence of practical work (TP)
Technical problems (21.2%)	Related to Laptop performance and the use of educational tools
Connectivity (39.4%)	Summarized in connection cuts in some places

To analyze the three clusters, some questions were asked. The first one is related to the first cluster (motivation), the question is: The level of satisfaction with the home study environment? (See Fig. 1), the second question concerning technical problems: How would you rate your use of digital educational tools? (See Fig. 2) And finally, connectivity: What type of internet access do you have? (See Fig. 3). For the first question, 76.8% of students are not satisfied with the environment of study at home, 17.9% are satisfied, while 5.3% are very satisfied.

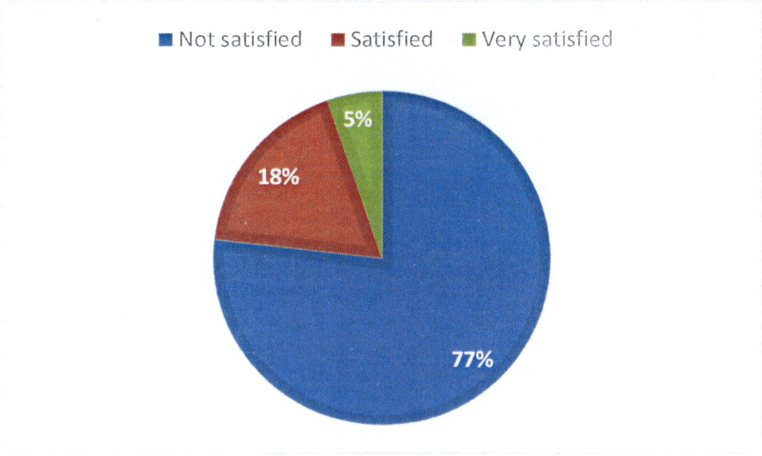

Fig. 1. Level of students' satisfaction about the learning environment at home.

According to students' answers, among the problems faced by e-learning, there is an issue related to the mastery of digital tools that requires training programmers. 52.6% of students are weak in using digital tools, while 39.5% have important competence to implement digital tools and 7.9% have very important skills.

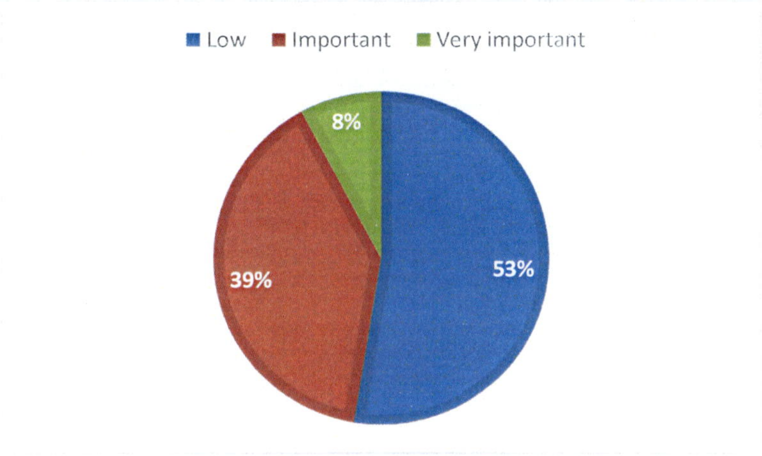

Fig. 2. Mastery of digital tools.

When asked what students use during the lockdown to access the internet, 42.1% of them have a broadband connection, 43.2% of them connect via a smartphone, and the rest via a modem.

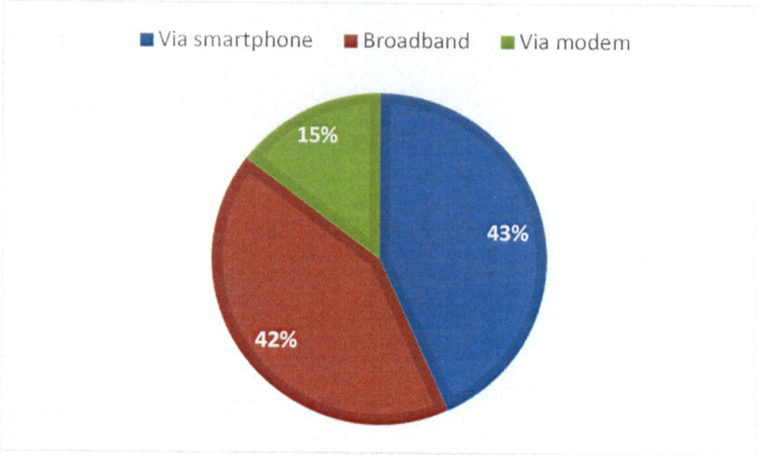

Fig. 3. Students' connection mode.

The problems above-mentioned impact the students' productivity, because 54.3% of them replied that they can't turn in the required work, while 45.7% can do it at the time (See Fig. 4).

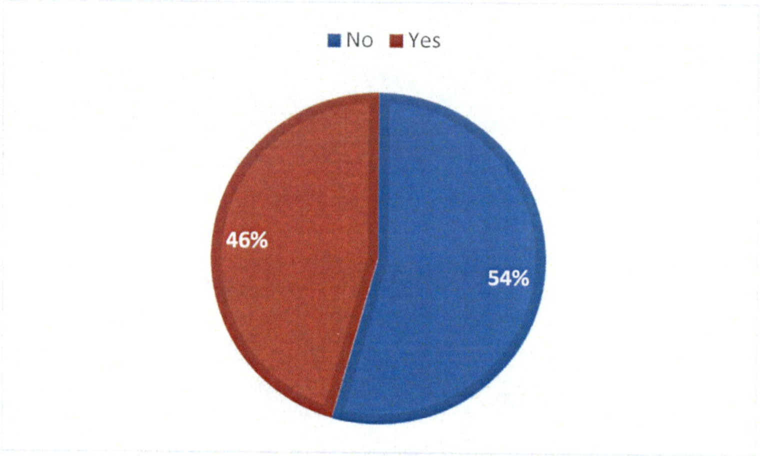

Fig. 4. Returned work requested by professors.

This study explores the perceptions of Moroccan engineering students', the result showed that the situation of online teaching during the lockdown can affect the choice of the teaching model, 84.2% of students choose the face-to-face model and 11.6% chooses the hybrid model while the rest choose online learning (See Fig. 5).

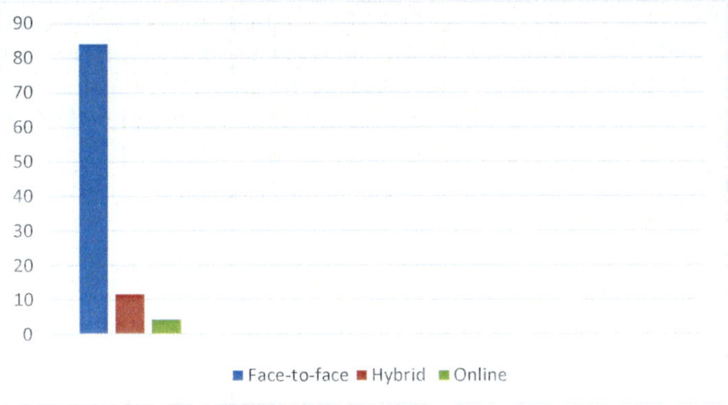

Fig. 5. Teaching models are chosen by students.

4 Discussion

The COVID-19 pandemic is posing challenges to our educational system. Ensuring effective learning to students of higher education in the present scenario is so challenging for teachers and students at the same time. It seriously affects the global educational system in all its aspects, forcing the world to enter another new-style society. Measuring the educational system becomes a necessity to find out the weakness that is presented as a signal indicating the quality of distance learning, which is measured through many indicators: the interest in the courses and training, the pedagogical quality of courses, the availability of teachers, the guidance assistance and the support of the professional project [15].

The measurement instruments contingent on many variables such as system quality, instructor quality, social Influence, learner computer anxiety, e-learning system use, and e-learner satisfaction [16] are based on many warning indicators which differ between face-to-face education and distance learning.

Warning signs when using face-to-face teaching:

- Group work is delayed
- Dissatisfaction of students with the quality of offered classroom material
- Increase in incidents of conflicts between students
- Special needs students do not attend

Warning signs when using distance learning:

- Dissatisfaction of students with faculty engagement
- Failure of students to submit assignments
- Students can't identify all the names of teachers
- Students fail to attend the synchronous online classrooms

According to the findings, it appears that the majority of students were dissatisfied with distance education. Hence, the need to highlight the main factors that impacted students' satisfaction. Engin Karadag proposes Hierarchical linear modeling to determine those factors. Our model based on K-means can serve to a group the similar factors and adapt suggestions according to each group of students. Through our analysis and referring to the paper: Model for utilizing distance learning post-COVID-19 using PACT a cross-sectional qualitative study, we are allowed to put another measurement instrument: the students' level satisfaction, mastery of digital tools as well as the returned work to study students' feedback about the online teaching. Students claim many problems which disrupt the normal functioning of the online study, but on the other hand, they recommended also many beneficial solutions to remedy these problems. Primary, it is proposed to develop the teaching quality through the diversification of teaching models. Variety is therefore likely to be a pedagogical yardstick to evaluate the educational system. For example, we can adapt face-to-face model in practical works to foster teamwork and interactivity between the students and teachers and adopted distance learning for the theoretical class. We can also give priority to first-year students and those who are not in the field of computing science. To discuss the results in the light of other research cited in the introduction, we find that students also had a negative perception about distance learning, but they propose some recommendations related to the investment in the new learning paradigm to improve online teaching and updating teachers' knowledge to work with the electronic environment and this more or less reduces the problems encountered by teachers and students undoubtedly. The analysis presents certain peculiarities related to students' performance and satisfaction that offer the possibility to create a hybrid education model as an alternative that is coupled to the new reality. It can be based on the platforms created by IT companies to ensure the follow-up of and the evolutionary maintenance of this model to overcome students' difficulties. The integration of artificial intelligence tools (decision tree, for example) for academic monitoring and the recommendation of activities are proposed. It can be used to identify students' categories grouped by the k-means to treat each group.

5 Conclusion

In conclusion, online education in Morocco and all over the world is spreading more and more. Therefore, it becomes requisite to analyze all its aspects, especially in the period of COVID-19, to provide a suitable system of education that extends to the period post-COVID-19. Many challenges faced by e-learning impact students' educational life. This work proposed the use of k-means to cluster students' issues into different groups to propose an adaptive model for each group of students.

In this context, the strategic watch can have an important role in monitoring the students' problems through creating monitoring offices at universities, to communicate directly with students and capitalize the knowledge created in all this period to reserve this institutional heritage and build the students' learning.

References

1. Abdelaziz, O., Chafik, K., Omar, B.: Measuring e-learning systems success: data from students of higher education institutions in Morocco. Data Brief **35**, 3–9 (2020)
2. Abdelmajid, J.: Issues in e-learning during Covid-19 in Morocco: a focus on EFL master students' voices. Int. J. Soc. Sci. Hum. Res. **4**, 466 (2021)
3. Blizak, D., Blizak, S., Bouchenak, O., Yahiaoui, K.: Students' perceptions regarding the abrupt transition to online learning during the COVID-19 pandemic: case of faculty of chemistry and hydrocarbons at the university of Boumerdes. J. Chem. Educ. C (2020)
4. Khaoula, O.G., Henia, B. A.: L'enseignement universitaire à distance en Tunisie: promesses et obstacles à son adoption. ISTE Ltd, 15–16 (2020)
5. Soumaya, E., Sara, S.: Les pratiques de l'enseignement à distance dans l'université marocaine à l'ère du coronavirus: Cas de l'université Mohammed V de Rabat. J. Qual. Educ. **10**(16), 75–76 (2020)
6. Engin, K., Ahmet, S., Hatice, E.K.: Multi-level analysis for distance education capacity, faculty members' adaptation, and indicators for student's satisfaction in higher education during COVID-19 pandemic. Int. J. Educ. Technol. High. Educ. **18**(57), 8–13 (2021)
7. Jiang, H., Islam, A.Y.M.A., Gu, X., Spector, J.M.: Online learning satisfaction in higher education during the COVID-19 pandemic: a regional comparison between Eastern and Western Chinese universities. Educ. Inf. Technol. **26**(6), 6747–6769 (2021). https://doi.org/10.1007/s10639-021-10519-x
8. Fayyaz, A.F., Muhammad, N.: Evaluation and Improvement of Students' satisfaction in online learning during COVID-19. Open Praxis **12**(4), 496–499 (2020)
9. Norah, M.A., Shabir, H.K.: Students' satisfaction with online learning experiences during the COVID-19 pandemic. Int. J. Emerg. Technol. Learn. **15**, 254–255 (2020)
10. Samar, A.A., et al.: Model for utilizing distance learning post COVID-19 using (PACT)™ a cross sectional qualitative study. BMC Med. Educ. **20**(400), 3–5 (2020)
11. Voronin, D.M., Saienko, V.G., Tolchieva, H.V.: Digital transformation of pedagogical education at the university. In: International Scientific Conference "Digitalization of Education: History, Trends and Prospects", vol. 473, pp.760. Springer Nature (2020)
12. Inssaf, E.G., Zakaria, B., Rim, M., Aimad, Q.: Forecasting students' academic performance using different regression algorithms, In: International Conference on Digital Technologies and Applications, Fez (2021)
13. Satu, E., Helvi, K.: The qualitative content analysis Process. JAN Res. Methodol. **62**(1), 108 (2007)
14. Aziz, H.: Les conditions de l'enseignement à distance pendant le confinement du au COVID19. HAL Open Science, pp. 11–14 (2020)
15. Jean-Françoit, G.: Quelle satisfaction pour les étudiants à distance et en présentiel dans les universités françaises ?. HAL Open Science, 4 (2020)
16. Abdelaziz, O., Chafik K., Omar B.: op.cit, p. 3

A Framework for Adaptive E-learning Systems Based on Self-regulated Learning Strategies

Abdelghani Babori(✉), Noureddine Falih, and Hicham Ouchitachen

LIMATI Laboratory Polydisciplinary Faculty, Sultan Moulay Slimane University Beni Mellal, Beni Mellal, Morocco

a.babori@uhp.ac.ma, nourfald@yahoo.fr, h.ouchitachen@gmail.com

Abstract. The majority of researches propose theoretical frameworks of adaptive e-learning suggesting generic models for studying adaptive e-learning systems without concretely providing elements of each model (learner, domain, and adaptation models) that could constitute the basis of the proposed framework. Previous research on adaptive e-learning systems stated that the generic framework of these systems is composed of three components: learner model, domain model, and adaptation model. Since the self-regulated learning model is the most used in online environments because the learner is on his own with or without the intervention of an online tutor, it seems interesting to us to consider this concept as an important characteristic of the learner model. Thus, this article suggests a framework based on self-regulated learning (SRL) to build a more comprehensive tool that serves to analyze adaptive e-learning systems. The research presents the elements (in dynamic interaction) of the conceptual framework for adaptative e-learning as well as a model for detecting self-regulated learning strategies using log files.

Keywords: Adaptive E-learning · Distance learning · E-learning 4.0 · Artificial intelligence

1 Introduction

Universities around the globe are increasingly concerned about the integration of IT (information technology) in higher education [1]. Thus, all these universities should be equipped with the software and hardware infrastructure for teaching and learning digital content. In this regard, E-learning as a kind of distance learning utilizing ICT is becoming increasingly popular in universities all over the globe [2, 3]. A plethora of E-learning definitions have been presented. The one offered by the European Union: "the use of new multimedia technologies and the Internet to improve the quality of learning by simplifying access to resources and services, as well as exchanges and cooperation," seem to us adequate.

From an educational standpoint, distance learning has generated and guided various E-learning researches. However, these studies reveal a scarcity of researches concentrating on web 4.0 in E-learning environments.

S. Motahhir and B. Bossoufi (Eds.): ICDTA 2022, LNNS 454, pp. 380–387, 2022.
https://doi.org/10.1007/978-3-031-01942-5_38

The findings of a recent literature review about E-learning 4.0 reveal that the major topics covered in research about using artificial intelligence in e-learning contexts focused on profiling and prediction, evaluation and assessment, personalization, and intelligent tutoring systems. The literature review which highlights a paucity of research dealing with the study of the profiles of learners in online learning platforms (LMS) during the coronavirus crisis, leads to the conclusion that the analysis of adaptive e-learning based on the full potential of artificial intelligence, Big data, machine learning, and deep learning deserve special attention in future research.

More particularly, given that adaptive e-learning systems are considered among the gaps in the literature review identified in the research trends about learning in a distance context [4, 5], it would be of great interest to investigate this topic. More specifically, this article proposes a framework for adaptive e-learning systems based on self-regulated learning (SRL).

This paper is organized as follows: after the introduction, Sect. 2 deals with related work about frameworks for adaptive e-learning systems. Section 3 describes our proposed framework giving more details about its elements and their interactions. Section 4 concludes the paper by presenting implications for future work.

2 A Generic Framework for Adaptive E-learning Systems

Previous research on adaptive e-learning systems stated that the generic framework of these systems is composed of three components: learner model, domain model, and adaptation model. We describe below the major specification of these elements.

2.1 Learner Model

To be "intelligent", a system must be able to adapt to the learner in front of the machine. This can only be achieved by knowing the learner's model. The knowledge set will help to define the characteristics of the learners so that it will be easier to adapt the content to their learning profile, monitor their progress, provide appropriate administrative and cognitive support, etc.

Behaz and Djoudi [6] have organized the intrinsic knowledge of the learner model into four categories in addition to its identity (name, age, etc.):

a) Capacity: the state of knowledge (beginner, intermediate, advanced). The evolution of the learner's knowledge level during a learning session is induced by an evaluation method.
b) Objectives: tasks or activities to be completed. This category defines the educational objectives. These objectives will influence the behavior of the system. Either the learner is concerned by the information aspect of the subject, in this case; a free exploration of the document without any guidance is provided by the system. In other words, either the learner is concerned by the course element; he needs to be guided and evaluated, in which case the system must be able to provide this assistance
c) Preferences: they allow the learner to specify the preferred media types. Thus, the learner can define a classification on the physical media for the presentation content (text, image, video).

d) The student's past: keep track of the state of his history. For example, the memorization of navigation and elements read in the documents, information on the current session of the learner.

2.2 Domain Model

The domain model has two main tasks. The first is the analysis of the domain content in terms of concepts, principles and procedures. The second is the organization of the content formulated in terms of their interrelationships: for example, the relationships of prerequisites, composition, etc. [7].

Most of the adaptative e-learning systems distinguish several types of concepts that represent different kinds of knowledge elements and different kinds of links between concepts. The types of links that are used are classical conceptual links such as "is-a" and structural links such as "part-of". Adaptative e-learning systems generally use prerequisite links between concepts that represent the fact that one concept must be learned before another. Brusilovsky [7] classified the elements of the domain model into three categories:

a) One level model: the simplest case of a domain model is that its concepts are not related to each other.
b) Two-level model: a domain model is a level two model if the concepts are linked together. Thus, they form a semantic network.
c) Three-level model: in level three models, each concept is represented as a set of attributes. This model is also called the frame-based domain model.

2.3 Adaptation Model

An adaptation model could be used to optimize the structure of learning material and how a student learns it in a limited amount of time. It may also serve as a foundation for the creation of personalized learning paths and the provision of relevant tips and feedback to students as needed. It primarily considers the learner model and the domain model by matching appropriate instructional content, or sequences of learning objects, to individual learners' requirements and characteristics.

There are three types of adaptable content: adaptive content, display, and navigation [8, 9]. Adaptive content and presentation approach deal with a variety of tasks, such as content insertion and modification, as well as interface zooming and layout changes. Selective learning pathways, curricular sequencing, link creation, direct guidance, link concealing, and link sorting are all examples of adaptive navigation.

3 Our Proposed Framework for Adaptive E-learning Systems

3.1 Self-regulated Learning

The majority of researches propose theoretical frameworks of adaptive e-learning suggesting generic models for studying adaptive learning systems [10] without concretely providing elements of each model (learner, domain, adaptation) that could constitute the angles of attack of the proposed framework. A minority of research specifically state frameworks focusing on the learner model with an entry by the learning style considered as a key factor of learning [9].

Since the self-regulated learning model is the most used in online environments because the learner is on his own with or without the intervention of an online tutor [10], it seems interesting to us to consider this concept as an important characteristic of the learner model. Thus, the proposed framework incorporates the concept of self-regulated learning strategies (SRL) to build a more comprehensive tool that serves to analyze the adaptive e-learning systems. We detail below the elements of this concept.

SRL represents a learning modality in which students take control of their learning [10, 11], Cohen & Magen-Nagar [11] define SRL as a form of learning that is largely influenced by students' thoughts, emotional strategies, and behaviors that are directed toward the achievement of their goals. This model makes it possible to identify the different cognitive and metacognitive strategies used by learners in the E-learning context and how resources are managed.

Cognitive strategies refer to strategies for processing information, such as *memorization, elaboration, and organization.* **Metacognitive strategies** refer to the situations in which cognitive strategies can be useful and how they should be adopted in various tasks. Thus, in addition to cognitive strategies, learners need to develop metacognitive strategies about these cognitive strategies to control the application of the strategies and evaluate their effects (plan, regulate, and monitor) [10]. *Planning* is the prior development of the way to approach a task as well as the choice of adequate strategies. *Regulation* corresponds to the implementation of cognitive processing when the initial processing is not satisfactory by increasing efforts and/or managing the time available [11]. *Monitoring* is considered as the conscious follow-up of the activity and its evaluation. Thus, the learner verifies that what he is doing is in line with the learning objectives already set. Resource management describes the strategies for managing time and the study environment.

3.2 Elements of the Framework in Dynamic Interaction

The framework consists of three major components: the learner model, the domain model, and the adaptation model. As previously stated, many adaptive e-learning systems mentioned these components. The proposed framework incorporates a new and original characteristic to the learner model, which is self-regulated learning strategies. Figure 1 presents the elements, in dynamic interaction, of the conceptual framework for adaptative e-learning. The framework is elaborated in four steps. The acquisition of relevant information about learners can be done by automatic methods or by direct questioning of the learner. Thus, the mode of acquisition can take three forms (see Table 1).

Table 1. Acquisition of information modalities.

Explicit acquisition	Implicit acquisition	Mixed acquisition
Ask explicitly the learner's preferences at the beginning of the activity by asking learners, for example, to answer a questionnaire	Observe the learner during his use of the system. The model is inferred from the navigation in the platform	Combine implicit and explicit acquisition

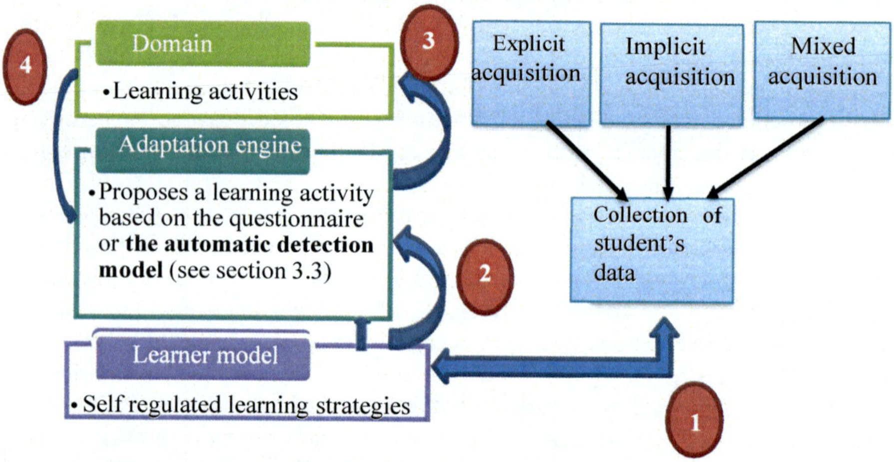

Fig. 1. The proposed framework for adaptive e-learning

a) The first step is to obtain the learner information (characteristics) by using explicit (questionnaire), implicit (log files) or mixed acquisition methods. More specifically, to extract explicitly learner's information, using a questionnaire of the "Motivated Strategies for Learning questionnaire" [10, 11] is recommended to identify self-regulated learning strategies. To extract implicitly learner's information, we will use the self-regulated learning strategies based on machine learning techniques (Sect. 3.3)
b) The data gathered from the learner will be sent to the adaptation engine.
c) By analyzing data, the intervention system learning activities could be chosen depending on the self-regulated learning strategy adopted by a student to study in the e-learning environment. For example, the system will provide learning activities that prompt students to take notes or memorize concepts (for students who prefer memorization strategies) or activities that will make connections to prior knowledge or experience (for students who prefer elaboration strategies) or activities that will

use diagrams or charts to organize knowledge (for students who prefer organization strategies).

d) Based on the data analyzed by the adaptation engine, a learning activity (a part of the domain model) is proposed to the learner. During the learning process, the learning model is examined and if there is a change in the self-regulated learning captured by the questionnaire (if it exists) or detected automatically by the machine learning techniques, the data will be updated in the system.

3.3 Self-regulated Learning Strategies Detected Based on Machine Learning Techniques

Based on the model of Rasheed and Wahid [12] which suggests the detection of students' learning style based on the power of artificial intelligence, we propose the detection of self-regulated strategies following these steps:

a) The information gathered from the questionnaire will be used to see if our model could appropriately predict the self-regulated learning strategies.
b) The online learning system keeps track of the actions from log files. The indications from log files are subsequently used by the model to predict SRL.
c) After that, the data is cleaned and processed, and machine learning techniques will be used to produce predictions. Several techniques are proposed such as decision trees, Support Vector Machines, K-Nearest Neighbour, Naïve Bayes, Linear Discriminant Analysis, Random Forest, and Logistic Regression [12].
d) Finally, to evaluate the accuracy of the proposed predictive model detecting self-regulated learning strategies, the questionnaire findings and the anticipated value are compared to see if the suggested self-regulated learning strategy is consistent. The model is consistent if they are equal; otherwise, the model is not consistent so we have to choose another machine learning algorithm.

The figure below summarizes the steps to follow to detect self-regulated learning strategies.

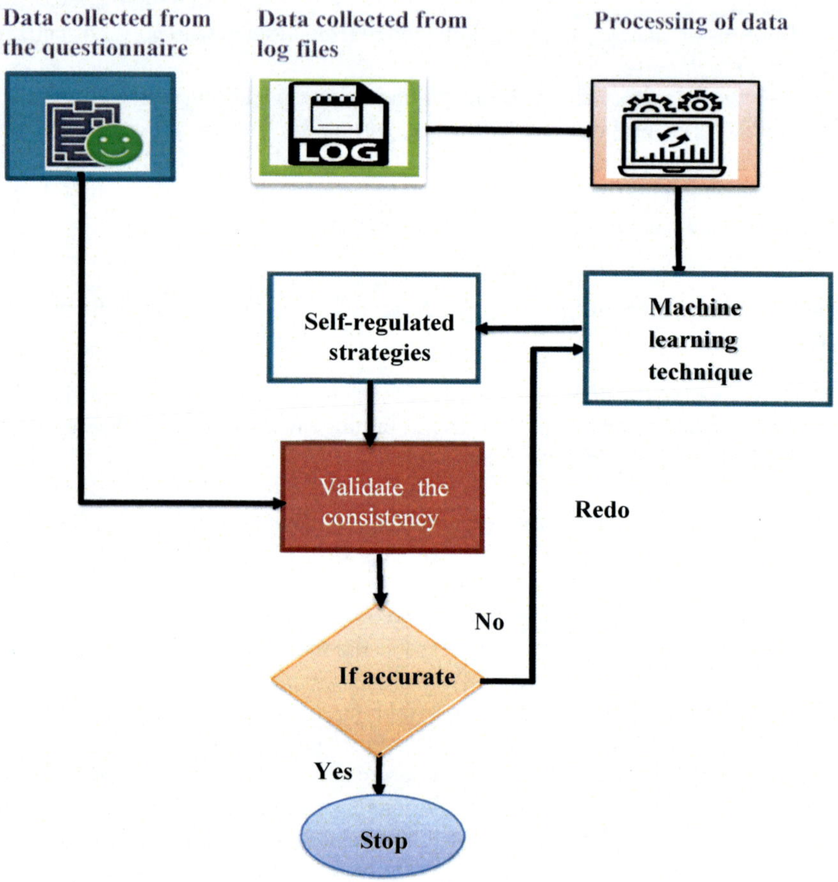

Fig. 2. The proposed model for detecting self-regulated learning strategies

4 Conclusion and Implications

This research work aims to propose a theoretical framework for adaptive learning systems. The literature review highlights a paucity of research the study of the profiles of learners in online learning platforms (LMS) during the coronavirus crisis, leads us to conclude that the analysis of adaptive e-learning based on the full potential of artificial intelligence, Big data, machine, and deep learning deserves special attention in future research. More specifically, we note that adaptive e-learning systems are considered among the gaps in the literature review. Indeed, the majority of studies propose theoretical frameworks of adaptive e-learning suggesting generic models for studying adaptive learning systems without concretely providing elements of each model (learner, domain, adaptation). This article proposes a specific approach to self-regulated learning within the framework for e-learning systems to build a more comprehensive tool that serves to analyze the adaptive e-learning systems. It also suggests a model for detecting self-regulated learning strategies using log files.

The framework presented is easy to use and can easily be applied to all disciplines (mathematics, physics, computer sciences, and so on). However, an implementation of the machine learning algorithms and online experimentation of this system will be conducted to validate the approach to get results about learning outcomes and learner satisfaction when using the implemented system for adaptive e-learning resources.

References

1. Babori, A.: Trends in MOOCs research: analysis of educational technology journals. Int. J. Emerg. Technol. Learn. **15**(17), 47–68 (2020)
2. Babori, A., Fihri, H.F., Zaid, A.: Research on MOOCs: current trends and taking into account of content. In: The 2nd International Conference on Networking, Information Systems & Security, pp. 1–9. ACM, Rabat (2019)
3. Babori, A., Ghoulam, K., Falih, N., Ouchitachen, H.: E-learning 4.0 for higher education: literature review, trends and perspectives. In: The International Conference on Digital Age & Technological Advances for sustainable development. IEEE, Marrakech, Morocco (2021)
4. Babori, A., Falih, N., Ouchitachen, H.: Systematic Literature Review on E-learning 4.0. In: Meeting-Days. The Annual IOSIS Meeting-Days, Mohammedia, Morocco (2021)
5. Bozkurt, A., Karadeniz, A., Baneres, D., Guerrero-Roldán, A.E., Rodríguez, M.E.:Artificial intelligence and reflections from educational landscape: a review of AI studies in half a century. Sustainability **13**(2), 1–16 (2021)
6. Behaz, A., Djoudi, M.: Création dynamique de documents hypermédias adaptatif. In:10ème Colloque International sur le Document Electronique, CIDE'10, Nancy (2007)
7. Tadlaoui, M., Chikh, A., Bouamrane, K.: ALEM: Un modèle de référence pour les applications web adaptatif educatif. In LEAFA 2010: The First International Conference on e-Learning For All, Hammamet, Tunisia (2010)
8. Dhupia, B., Alameen, A.: Adaptive e-learning system: conceptual framework for personalized study environment. In: International Conference on Advanced Informatics for Computing Research, pp. 334–342. Springer, Singapore (2019)
9. Alshammari, M. Anane, R., Hendle, R.J.: An e-learning investigation into learning style adaptivity. In: Hawaii International Conference on System Sciences 2019, pp. 11–20 (2019)
10. Lee, D., Watson, S.L., Watson, W.R.: Systematic literature review on self-regulated learning in massive open online courses. Australasian J. Educ. Technol. **35**(1), 28–41 (2019)
11. Cohen, L., Magen-Nagar, N.: Self-regulated learning and a sense of achievement in MOOC among high school science and technology students. Am. J. Dist. Educ. **30**(2), 68–79(2016)
12. Rasheed, F., Wahid, A.: Learning style detection in E-learning systems using machine learning techniques. Expert Systems with Applications 174, 202

"AskBot": The AI Chatbot that Enhances the Learning Process

Khadija El Azhari[1(✉)], Imane Hilal[2], Najima Daoudi[2], and Rachida Ajhoun[1]

[1] Smart Systems Laboratory, E.N.S.I.A.S, Mohammed V University, Rabat, Morocco
khadja_elazhari@um5.ac.ma

[2] LyRica Labs, School of Information Sciences, Rabat, Morocco
{ihilal,ndaoudi}@esi.ac.ma

Abstract. The application of artificial intelligence (AI) in the field of education has increased significantly over the last decade. In fact, several works have been carried out to implement intelligent systems to help learners in their learning processes. Educational chatbot is one of the AI solutions that help tutors to capitalize their knowledge and transfer it to learners. Its main feature is to interact effectively with human requests using natural language processing (NLP) tools, in order to understand the human intention and respond to it through text or voice. As a result, chatbot improves efficiency and productivity in the field of education.

In this paper, we present our chatbot's architecture "AskBot" that supports students in their learning process by giving them appropriate answers to their questions. Our proposed architecture guarantees a continuous and autonomous update of AskBot's knowledge base (KB). To demonstrate the feasibility of our approach, we implemented the kernel components of the proposed architecture.

Keywords: Artificial intelligence · Smart learning · Chatbot · Natural language processing

1 Introduction

In recent decades, AI technologies have been implemented in a wide range of domains: medicine, Finance, Commerce, Industry, E-learning, etc. In fact, they have succeeded in developing these domains by implementing intelligent solutions that can automate tasks and perform processes [1]. AI chatbot are among the AI technologies widely used. it's a computer program that simulates a conversation with a person, either written or spoken. When a user asks a question or formulates a command, the chatbot will answer or perform the requested action [2]. Chatbots have been implemented in several domains including education to automate tasks, respond to users' questions and perform tasks.

The E-learning domain suffers from various problems that can be presented as follows:

1. Professors haven't enough time and energy to give answers to all repetitive and massive students' questions [3].

S. Motahhir and B. Bossoufi (Eds.): ICDTA 2022, LNNS 454, pp. 388–396, 2022.
https://doi.org/10.1007/978-3-031-01942-5_39

2. Students' feel lost in several resources (books, online links, encyclopedia, etc.) and they waste a lot of time looking for the information needed [3].
3. Students can't ask all their questions because they fear to be judged [4].

In order to solve these issues, chatbots are becoming widely used in the E-learning domain [5] because they can offer an increasing number of options to help students improving their learning process. In fact, they can help students find appropriate answers by responding in real time to the repetitive and massive questions asked by students.

The chatbot focuses on understanding student's intents and try to respond efficiently and effectively to their questions within the limits of its knowledge base [5].

The purpose of this article is to propose an architecture of an educational chatbot that can offload the teachers by answering in real time the massive and repetitive students' questions and guarantee a continuous and autonomous update of its knowledge base.

The remainder of this paper is structured as follows: The next section explains the main concepts that are used in this paper. The third section presents literature review, the fourth section proposes our chatbot architecture and its implementation and the last section provides the conclusion and future works.

2 Chatbot

Chatbot is an AI technology that can understand natural language and simulate conversation with humans by text or voice. The chatbot works by relying on a KB which is a collection of questions and their answers [6] (see Fig. 1). The KB stores and centralizes all available knowledge. It enables the chatbot to understand user requests and interact effectively by giving the appropriate responses [6].

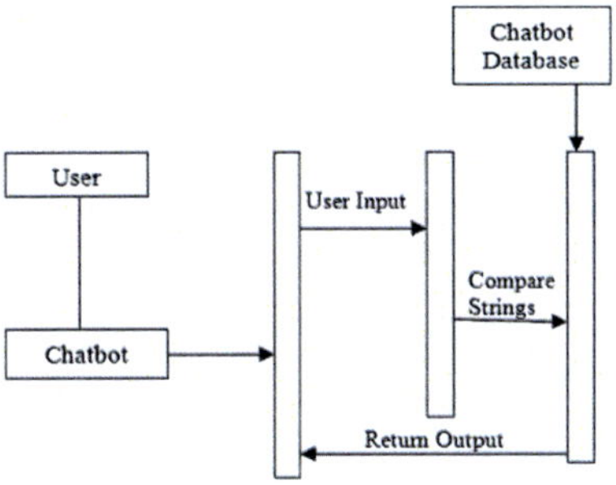

Fig. 1. Sequence diagram representing design of the Chatbot [6].

3 Literature Review

In this section, we will present some research that focuses on the use of chatbots in the educational field, evaluate them and compare their approaches.

Jizhou Huang and all in [7] focus on the use of the data available in an online discussion forum to construct the chatbot knowledge base which allows collecting a very large KB automatically which saves time and helps chatbot to enhance its ability to interact with users. However, the Chatbot's KB is limited on the previously extracted discussions which make the KB not up to date. Shafquat Hussain and Athula Ginige in [8] propose an approach to extend the chatbot's KB beyond its local KB to external knowledge source Wikipedia which enhance the analysis ability of the chatbot and allows it interacting with users correctly but using Wikipedia as the only external source makes the chatbot's KB limited. Aliane Loureiro Krassmann and all in [9] present a process to extract the chatbot's KB from text corpora which improves the chatbot KB and reduces the time needed to construct it. Nevertheless, the extraction process of the Chatbot's KB is semi-automated because it requires the evaluation of the expert in several steps of the process, which consumes time and makes the chatbot fixed and limited on the content already extracted. Bhavika R.Ranoliya and all in [10], and Menal Dahiya in [6] propose an educational chatbot to answer students' questions about the university but the chatbot's KB is limited to the answers stored manually. Daniel Carlander-Reuterfelt, and all in [11] propose a chatbot named "JAICOB" to facilitate learning in the field of data science through machine learning algorithms and cognitive computing, but the construction of the chatbot's KB is made manually which is time consuming. Also, the KB is limited to the content already extracted. Matthew Verleger and James Pembridge in [12] propose an implementation of an educational chatbot « EduBot», it was developed for an introductory computer programming course. EduBot sought an answer from a Question & Answer (Q&A) database. If an answer could not be found, then, the bot informs the user that no response is found, and that the professor had been notified. The professor can respond to the question and update the Q&A database. A questionnaire was made to evaluate the chatbot. The result shows that 33.3% of the students don't find the answers of their questions and 28% of the students find that chatbot responses were not helpful and just 16.7% of the responses received were clearly addressed questions. The chatbot users become disengaged when they don't find what they ask for, and they should wait for the instructor's response. The proposed process of updating the KB is done manually, which is time consuming, and it negatively impacts the satisfaction of students and makes them disengaged and unmotivated to use the chatbot. Sameera A. Abdul-Kader and John Woods in [13] and Pavel Smutny and Petra Schreiberova in [14] affirm that AI chatbots are in general limited and they need improvements by designing more comprehensive KB. Bing Liu and Sahisnu Mazumder in [15] argue that an intelligent chatbot shouldn't be limited to its offline local KB, thus, the chatbot should learn continuously on the job by using real conversation with users.

The papers presented in this review positively impact the development of educational chatbot by (i) automating the extraction process of a huge size knowledge which can't be done manually [7–9], (ii) extending the chatbot KB to external source [8] which enhance the interaction ability of the chatbot.

However, there are some limitations in the chatbot knowledge base, which are presented as follows:

- The chatbot's KB is fixed on the knowledge already stored [6–11], which can negatively impact chatbot's answers, by giving information out of date to users.

- The chatbot KB is limited to one external source [9], which is not enough to satisfy users' requests.
- Manually extracted KB is time consuming, and difficult to adapt to all users' requests [6, 10–12].

Our approach focuses on these limitations and proposes the architecture that automatically updates the chatbot KB and makes it enriched and autonomous based on its users' knowledge.

4 Our Approach

4.1 AskBot's Architecture

In this section, we will present our architecture which focuses on extending and updating the chatbot' KB automatically. AskBot's architecture is based on several components to analyze users' requests and give them the appropriate answer (see Fig. 2).

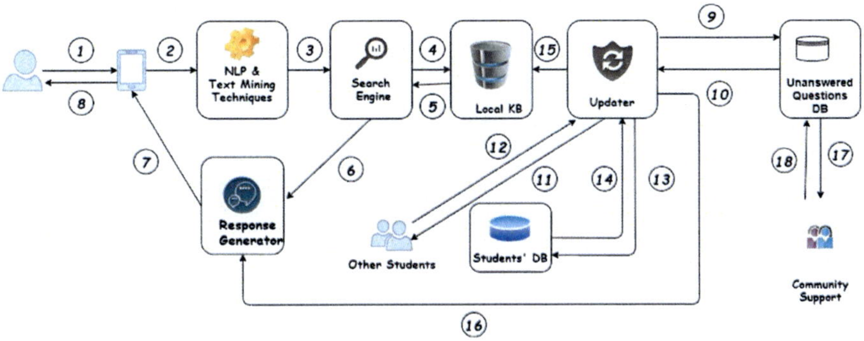

Fig. 2. The AskBot's architecture.

The architecture of AskBot is based on nine main components, namely (1) Graphical user interface (G.U.I): It provides a user-friendly interface for the use of the chatbot. (2) Natural Language Processing (N.L.P) and Text Mining (TM) component: It processes user message by applying NLP and TM Techniques to retrieve user's question, user's intent and entities and send them to the Search Engine component. (3) Search Engine: Its main task is to receive user's intent/entities and search for the appropriate answers in the local KB. (4) FAQ Local KB: It's the brain behind the chatbot. It contains a series of questions and their answers about the domain of Data Science, in fact, we have focused on the common questions that can be asked by students, and we added them to AskBot's KB. (5) Unanswered Question: It is a database that contains all questions that have no answer in the local KB. (6) Student's DB: It contains information about users, their preferences, and their engagement's score. (7) Updater component: It automatically updates the local KB, by asking other users to give responses to unanswered questions. The update process can be described as follows: Updater sends questions to users. Then, it evaluates each response received based on various criteria that will be explained in the next section. When the response is confirmed as right, the updater component updates the local KB automatically by adding the question and its response. (8) Response Generator: It uses the information extracted from the local KB and transforms it to a suitable response, then sends it to user. (9) Community Support: It contains professors that can check the Unanswered Question database, give the right answers, and update the local KB.

4.2 Communication Flow

As presented above, AskBot's architecture contains several components to analyze effectively and efficiently the information flow passed between users and chatbot.

In this section, we will present the approach adopted to analyze the large amount of information exchanged between different components of the chatbot (see Fig. 3).

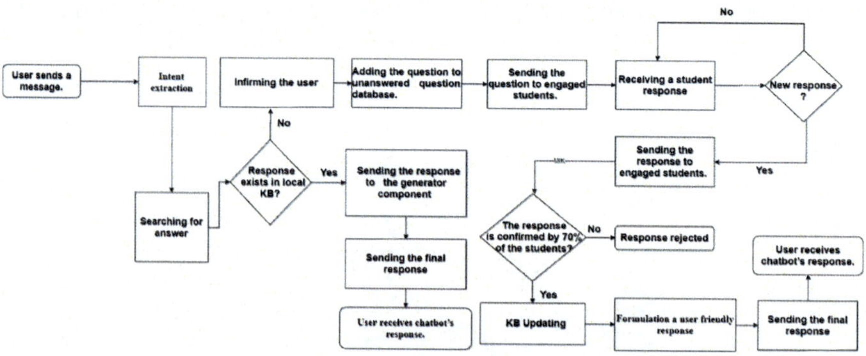

Fig. 3. The flowchart diagram of AskBot's architecture.

The architecture of AskBot allows it to extend its KB automatically and rapidly using users' knowledge. Also, students have an engagement score corresponding to their engagement in developing the chatbot KB, which encourages them to give the right answers, make them more engaged with the chatbot and save time. Hence, the rating system helps the updater to evaluate the responses given by the students, and deal with the wrong knowledge.

5 Implementation Process

5.1 AI Chatbot Framework

To demonstrate the feasibility of our architecture, we implemented the first iteration of our proposed chatbot. Several frameworks exist and integrate several AI components to help people in the chatbot's development process. We made a benchmark between the top AI chatbot frameworks, to detect the framework that can boost us in our implementation (see Table 1).

Table 1. Benchmark of top AI chatbot frameworks.

Framework	SDK	Free plan
Microsoft bot	JS, Python, C#	Available
Wit.ai	Node.js, Python, Ruby	Available
Dialogflow	Node.js, java, python	Available
IBM Watson	Python, Node.js,.NET	Available but limited
Pandorabots	JS, Python, C#	Available and open source
RASA	Python	Available

After analyzing the result of the benchmark presented above in Table 1, we decided to use Google Dialogflow technology, because of the following reasons:

- It supports the Python language which contains rich libraries that we can use to integrate machine learning algorithms.
- It supports voice which makes our chatbot friendlier and inclusive.
- Its free edition proposes several tools and features which help us in the process of integration with different communication channels.
- It can be integrated with several platforms which facilitate the use of the chatbot by students.

5.2 Technical Details

In this section, we will present the technical details behind the implementation process of AskBot. Three main technologies were used to implement the chatbot:

- **Google Dialogflow technology:** As mentioned above several AI components are integrated in Dialogflow which helped us in the implementation process of various components, namely (1) G.U.I because Dialogflow provides a user friendly interface for chatbot, (2) NLP and TM component because it's already integrated in Dialogflow and its main task is to understand user's intent and extract entities, (3) Search Engine component which is a powerful component of Dialogflow that can search for information in the FAQ database, (4) Response Generator component that can formulate a friendly response and send it to user, (5) FAQ Local KB is a FAQ Database in Dialogflow that can be fed manually or automatically. The reason behind using Dialogflow is to test the first iteration of the chatbot rapidly and focus our energy on the updater component which is the central component of our proposed architecture.
- **Flask Framework** is used to develop different functions that can manage the continuous and autonomous update of AskBot's KB. The connection between Dialogflow and Flask Framework is done using the fulfillment method integrated in Dialogflow.
- **MySQL Database service** is used to create students' Database and Unanswered Database. The databases can communicate with the updater through a Restful API developed by using Postman platform.

5.3 Conversation with AskBot

In this section, we will present two examples of conversation with our chatbot.

Example 1: When a user asks a question about the difference between Gini Impurity and Entropy in a Decision Tree, AskBot responds to this question based on its local knowledge base (see Fig. 4).

Example 2: When a user asks a question about the definition of cross-validation, and since AskBot did not find the appropriate answer in its local knowledge base, it informs the user that the answer does not exist and that it will try to find it quickly (see Fig. 5).

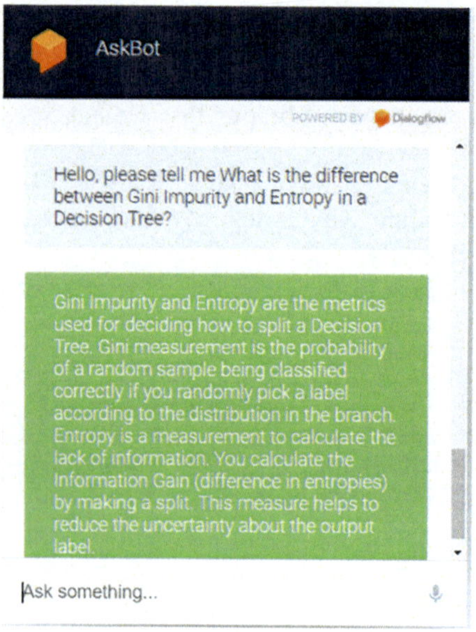

Fig. 4. AskBot responds to the user's question using its FAQ local KB.

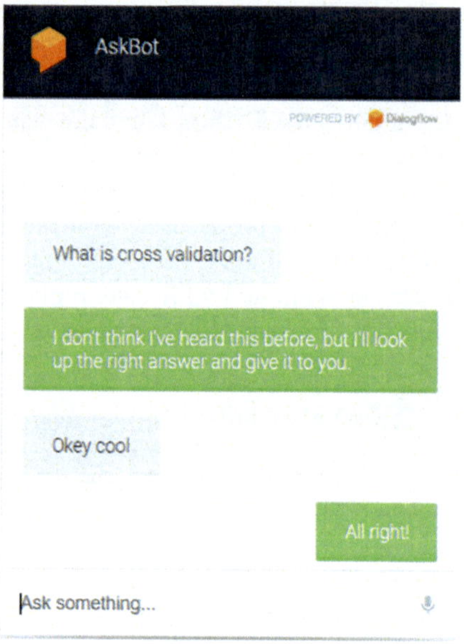

Fig. 5. AskBot doesn't find the response in the KB.

5.4 The Updater Component

As detailed above in the architecture section, if no response is found, updater searches for appropriate answers by sending questions to users (see Fig. 6).

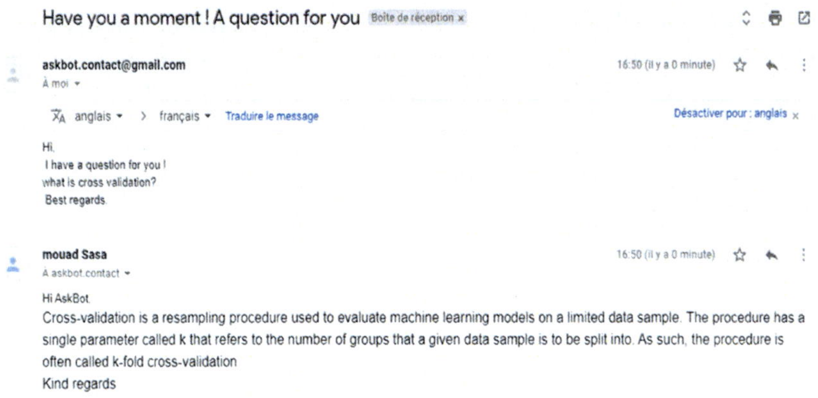

Fig. 6. Example of asking questions to other students

After analyzing and evaluating students' answers, the appropriate answer is sent to the student who asked the question (see Fig. 7).

Fig. 7. Example of sending the appropriate response to the student.

6 Conclusion and Future Works

Several works are being conducted to develop educational chatbots to improve the quality of learning, make learners more engaged and encourage them to ask questions without fear of being judged. However, previous studies used either a manual or semi-automatic process to construct the chatbot KB. In this paper, we proposed AskBot's architecture that aims to feed its knowledge base in real time with students' responses, which will encourage them to use the chatbot and develop its knowledge base. We proposed an architecture to ensure continuous updating of the local knowledge base through the evaluation process detailed above to improve the relevance and quality of the knowledge base. In fact, before adding new knowledge, the response should be evaluated and reviewed automatically by the update component, and then it can be added to the local knowledge base. In addition to that, community support members can also participate in this evaluation process by viewing the new answers, evaluating them, and providing answers to unanswered questions. The proposed architecture enhances the chatbot's learning capability, extends its knowledge base and makes it larger and more extensive by learning from users' knowledge in real time, which will save time and reduce costs.

As a future work, we plan to enrich AskBot's KB by adding advanced NLP and Machine Learning algorithms, improve students' engagement with the bot and enhance the interactivity between AskBot and students.

References

1. Azan Basallo, Y., Estrada Senti, V., Martinez Sanchez, N.: Artificial intelligence techniques for information security risk assessment. IEEE Lat. Am. Trans. **16**(3), 897–901 (2018)
2. Forbes. https://www.forbes.com/sites%20/gilpress/2017/01/23/top-10-hot-artificial-intelligence-ai-technologies/. Accessed 25 Oct 2021
3. Xian, L.: Artificial intelligence and modern sports education technology. In: 2010 International Conference on Artificial Intelligence and Education (ICAIE), pp. 772–776. IEEE, Hangzhou, China (2010)
4. Chen, L., Chen, P., Lin, Z.: Artificial intelligence in education: a review. IEEE Access **8**, 75264–75278 (2020)
5. Molnár, G., Szűts, Z.: The Role of Chatbots in Formal Education. In: 2018 IEEE 16th International Symposium on Intelligent Systems and Informatics (SISY), pp. 000197–000202. IEEE, Subotica, Serbia (2018)
6. Dahiya, M.: A tool of conversation: chatbot. Int. J. Comput. Sci. Eng. (IJCSE) **5**(5), 158–161 (2017)
7. Huang, J., Zhou, M., Yang, D.: Extracting chatbot knowledge from online discussion forums. In: Proceedings of the 20th International Joint Conference on Artificial Intelligence (IJCAI), pp. 423–428 (2007)
8. Hussain, S., Ginige, A.: Extending a conventional chatbot knowledge base to external knowledge source and introducing user based sessions for diabetes education. In: 2018 32nd International Conference on Advanced Information Networking and Applications Workshops (WAINA), pp. 698–703. IEEE, Krakow, Poland (2018)
9. Loureiro Krassmann, A., Marcos Flach, J., Raquel Cestari da Silva Grando, A., Margarida Rockenbach Tarouco, L., Bercht, M.: A process for extracting knowledge base for chatbots from text corpora. In: 2019 IEEE Global Engineering Education Conference (EDUCON), pp. 322–329. IEEE, Dubai, United Arab Emirates (2019)
10. R. Ranoliya, B., Raghuwansh, N., Singh, S.: Chatbot for university related FAQs. In: 2017 International Conference on Advances in Computing, Communications, and Informatics (ICACCI), pp. 1525–1530. IEEE, Udupi, India (2017)
11. Carlander-Reuterfelt, D., Carrera, A., Iglesias, C., Araque, O., Sanchez Rada, J., Munoz, S.: JAICOB: a data science chatbot. IEEE Access **8**, 180672–180680 (2020)
12. Verleger, M., Pembridge, J.: A pilot study integrating an AI-driven chatbot in an introductory programming course. In: 2018 IEEE Frontiers in Education Conference (FIE), IEEE, San Jose, CA, USA (2018)
13. A., S., John, D.: Survey on Chatbot Design Techniques in Speech Conversation Systems. Int. J. Adv. Comput. Sci. Appl. (IJACSA) **6**, 72–80 (2015)
14. Smutny, P., Schreiberova, P.: Chatbots for learning: a review of educational chatbots for the Facebook Messenger. Comput. Educ. **151**, 1–11 (2020)
15. Computer Science. https://arxiv.org/abs/2009.10750v1. Accessed 16 Oct 2021

OSIBR: Ontology Focused Semantic Intelligence Approach for Book Recommendation

Ramanathan Nachiappan[2] and Gerard Deepak[1](✉)

[1] Department of Computer Science and Engineering, National Institute of Technology, Tiruchirappalli, India
gerard.deepak.christuni@gmail.com
[2] SRM Institute of Science and Technology, Ramapuram, Chennai, India

Abstract. With the dawn of digital books on the web, finding reliable book suggestions has become a complicated question, and several significant machine learning algorithms were developed to address such requests. Also, there exists a problem of the prevailing recommendation system not presenting a well fit to the semantically designed nature of the world wide web. Moreover, due to the vast digitization of books, a plethora of the old book discovery strategic approaches is intended for physical selection only. Therefore, research interest is increasing in the innovative approach to search, browse, and use the recommendation engine. In this paper, an ontology focused semantic intelligence approach for the book recommendation is proposed using Jaccard similarity, NPMI, and Kullback Leibler divergence with the optimization done by the Krill Herd algorithm. The proposed system OSIBR reaches an overall Precision and Accuracy of 94.16% and 95.55% respectively and with the highest Normalized Discounted Cumulative Gain (nDCG) of 0.96.

Keywords: Book recommendation · Decision tree · Ontology · Random forest semantic similarity

1 Introduction

With increasing internet users within the last decade, a tremendous amount of data has been produced. For users to choose or personalize and request the right data have become a burdensome job. This difficulty has been addressed with the Informal Retrieval System. They have partially solved the problem, but the more significant issue of personalization remains unsolved [1]. Hence, for this ultimate purpose recommender systems were introduced. Personalization or selections are based on the idea or interest of the user or a group, and it is a filtering system so-called recommender system. This paper has put forward a fresh approach using ontology infused with machine learning and semantic measure for the book recommendation. With the increased availability of books online and offline and the era of e-books services like kindle and digital libraries, old methods and strategies to find suitable books are complex and time-consuming.

On the other hand, finding a concept rich and technical related book is relatively easy compared to non-technical books with genres like mystery, thriller, romance, western,

S. Motahhir and B. Bossoufi (Eds.): ICDTA 2022, LNNS 454, pp. 397–406, 2022.
https://doi.org/10.1007/978-3-031-01942-5_40

fantasy, historical, fictional, bildungsroman. The key challenge in creating a workflow for recommender systems during runtime without human involvement. Therefore, it isn't easy to classify and recommend books. So, a best-in-class recommendation approach has been established in this paper with existing metadata, preface and description of the books.

Motivation: The present rise of digitization of semantic web or web 3.0. So, there is always a need for distinct recommendation strategies which are not available or insufficient, and there is no specific good approach for the book recommendation. Therefore, a good recommendation approach for books that is semantically compliant has been proposed. The put forward outline is a pace in the path to specify how remarkable are the semantic similarity essential to produce the adjoining recommendation. In book recommender systems, there is constantly a need for the user-oriented way, which is also knowledge-centric and to analyze metadata of the books.

Contribution: A framework for recommending books that combines metadata in addition to dynamically modeled ontology, that contemplates user obligation on top of semantically stirred approach using KL divergence, SemantoSim, Jaccard similarity and NPMI. Classification is completed by Bagging of Decision Tree and Random Forest algorithms. Experimentation demonstrates that using a framework that is Ontology-driven attained a very high Accuracy when related to traditional recommender systems. This method also uses Structural Topic Modelling (STM) of entity population. The OSIBR structure achieves a total Precision, Accuracy, and Recall of 94.16%, 95.5%, and 96.84% correspondingly.

Organization: The following Section of the paper shows the flow of the process involved. The 2nd Section briefly reports the related research done in Book Recommender System. 3rd Section describes the strategic system architecture titled OSIBR. 4th Section comprises the implementation method, its results, and the evaluation. Lastly, the 5th Section consists of a conclusion.

2 Related Works

This segment confers the recommender systems past studies towards recommending books to the user. Darekar et al. [1], have put forward an algorithm that uses traditional collaborative filtering with content grounded filtering in addition to demographic filtering. Tsuji et al. [2], has introduced a machine learning constructed method. It hybridizes Support Vector Machine (SVM) with Random Forest along with adaboost. Mclean et al. [3], have projected an approach that uses a vector space model for computing Semantic Similarity and Doc2vec model. Vishal et al. [4], have put forward an approach that infuses STM with NPMI and recurrent neural networks to organize the query-linked documents. Sabitha et al. [5], has proposed a method that uses filtration of data grounded by the user's requirements. This approach uses Cosine similarity, KNN, and Pearson similarity to recommend books. Krishnan et al. [6], has projected a text summarization method with infused ontology and semantic measures. Intended for the question-and-answer classification, LSTM is used in this approach. Giri et al. [7], has proposed an algorithm semantically preserving the associations among the ontological entities. Ru et al. [8], has put forward a method that emphasizes dependency and relative phrases in the knowledge base. To filter incorrect labels using semantic similarity with term embedding and Jaccard similarity. In [9–20], Several models supporting the proposed literature have been discussed.

3 Proposed System Architecture

The architecture of the put forward OSIBR framework for book recommendation is presented in Fig. 1 and Fig. 2. The put forward system entails 3 phases that produce an end outcome as a convenient and relevant Book. The initial segment pacts with user information query pre-processing. The subsequent stage emphasizes structural topic modelling and ontology construction for the book recommendation from book repositories. Finally, the tertiary phase includes classifying data with the pre-processed user facts, besides raised ontology and semantic similarity which is commutated using it and finally recommended to the user.

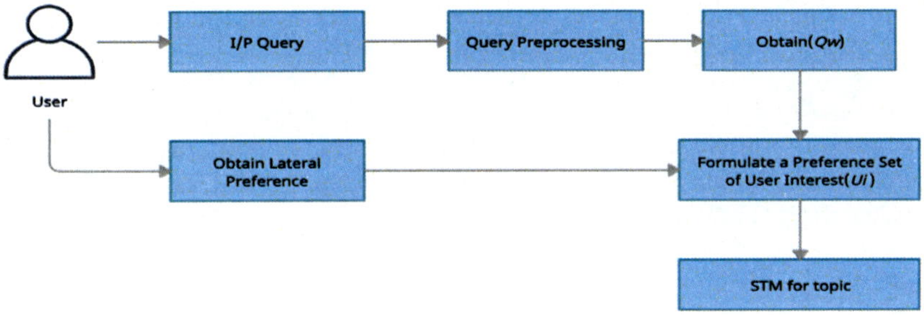

Fig. 1. Proposed system architecture design for the OSIBR model

The first phase entails pre-processing the input and output queries, depicted in Fig. 1. The user query is taken as the primary source of input. User feedback is implicitly taken much later, and the other lateral preferences of the user are considered. So, initially, the user query is subjected to pre-processing, which involves Tokenization, Lemmatization, Stop Word Removal and Word Sense Disambiguation, Pos Tagging, named entity recognition to obtain the individual query words (Qw). To extract separate terms from the input and output queries, Tokenization is used. At the same time, Lemmatization includes the conversion of words from inflexion form to base form. In the same way, removing words like if, and, are, is called stop word removal as they tend to surge the noise besides with no effect on gaining increased Accuracy.

The second phase is where the already modelled Ontology of subcategories is incorporated into the framework. Ontology is dynamically generated or manually modelled from the user and library contributed metadata. Then, this subcategory ontology is matched with the obtained query words in order to formulate a preference set of user interests. Also, the User's lateral preference is considered to develop the initial set of user interests (Ui). So, this set contains all the user preferences based on the query, current user needs and lateral preferences to Structural Topic Modelling. STM is mainly done to uncover hidden relationships among several domains and subdomains in order to ensure that the proposed approach is a full-term cover search.

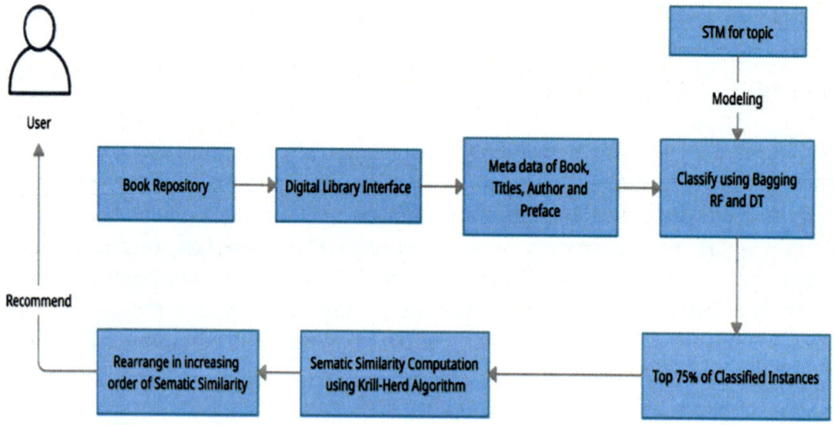

Fig. 2. Proposed system architecture design for the OSIBR model

STM is specialized in using metadata to optimize the latent subjects in the document. First, the metadata, preface, titles, and authors are extracted from the book depository like google books, Indian book depository and Bloomsbury India, Harvard Depository, Project Gutenberg, E-Book Foundation. Then, the book depository is crawled using a customized web crawler to get those features. Then, STM is used to ingest and manipulate text data and calculate covariate effects on latent topics. The goal line of the Structural Topic Model is to let investigators learn topics and their associations with the documented metadata. where the outputs of the models can be used for supposition testing of these associations.

The third phase entails the data taken from the book repository based on the initial terms from the queries, Ui, which is subjected from STM, used to classify by Bagging. Classification of data is done using Random Forest and Decision Tree through Bagging. Bagging is an ensemble learning technique that can be used in both classification algorithms and regression algorithms. We bag Random Forest as it provides competitive accuracy and captures non-linear and non-monotonic relationships. The Random Forest algorithm makes a minor nip to bagging and results in a robust classifier. A tree-identical flowchart is known as a Decision Tree. Decision trees can handle a variety of continuous phase variations.

In order to recommend the instances, the initial user interest, as well as the top75% of classified instances are taken and semantic similarity is computed using SemantoSim measure with the threshold of 0.75, NPMI with the threshold of 0.5, Jaccard and Kullback Leibler divergence with step deviation of 0.25 is considered. Then, the top 75% of the classified instances are pushed to the semantic similarity phase and the Krill Herd algorithm. Krill herd addresses optimization with the help of meta-heuristic swarm intelligence, modelling-based techniques. These semantic similarity algorithms with the current threshold are used as objective functions. Objective functions of semantic similarity models and Kl divergence are used as multi-objective functions. The Krill herd algorithm is deployed mainly to choose the optimized and best solution from the initially computed solutions and recommendations.

The Jaccard index is a tool to find similarities between two sets. It is characterized as the size of the crossing point separated by the size of the association of two sets. The Jaccard similarity coefficient is exemplified in Eqs. (1), where J is Jaccard distance. Q and W are the two sets given. It returns float esteem between zero and one, increasing it by a hundred will provide the level of likeness. The KL divergence measures the difference in the distribution of one probability to reference probability. Equations (2) exemplified the KL divergence, and Q and W are the probability distributions. Pointwise Mutual Information is a degree of association among two explicit actions, namely Q and W, an alternative of converging on an average of all events. It uses singular events and can be used to normalize the info between negative one and positive one, where -1 designates never together, 0 designates independent, and 1 designates occurring together Eqs. (3) NPMI where x (q, w) is the self-information.

$$\mathrm{J(Q,W)} = \frac{\mathrm{Q} \cap \mathrm{W}}{\mathrm{Q} \cup W} \tag{1}$$

$$D(Q||W) = \sum\nolimits_{x \in X} Q(x) log \frac{Q(x)}{W(x)} \tag{2}$$

$$Npmi(q; w) = \frac{pmi(Q; W)}{x(Q, W)} \tag{3}$$

From the recorded classes, a set of alike groups are framed. Furthermore, the semantic similarity of the book classes existing in the set and the uppermost 50% of the classified book categories are considered for a threshold value >0.75. Then, the books are rearranged in growing order of semantic similarity, and it is finally recommended to the User.

4 Implementation and Result Analysis

The implementation of the put forward semantically infused book recommendation's dataset was completed using Google Collab notebook for python, with i5, 4.0 Ghz processing power by intel and 32 GB Ram. Sklearn machine learning library was used to devise Bagging, an Ensemble technique with the collection framework of python. Data is crawled using a customized web crawler in order to get features. The key study goal is to devise a method called OSIBR. The data set used in this projected model is taken from Kaggle, which comprises 271,379 books in whole. The Algorithm of the proposed OSIBR for recommending books is made known in Algorithm 1.

ALGORITHM 1: ALGORITHM FOR PROPOSED OSIBR

Input: Queries composed from the user in addition to the dataset
Output: Recommendation of Books built on user query

Query pre-processing and obtain the individual query words (*Qw*)
Ontology of subcategories is incorporated into the framework
Formulate a preference set of User Interests
Structural Topic Modelling is done for the extracted object of the query words
Dataset is classified using Random Forest Classifier, and Decision Tree assisted with the Extracted Query Word as the user input.
Semantical modelling with NPMI, Jaccard Similarity, KL divergence
Optimize results using Krill Herd Algorithm [9]
    Population P of AB krill, V f, Dmax, generation counter X, and N max is Initialized, and fitness is calculated
    for X< MaxGeneration **do**
        Class the population as claimed to corresponding fitness.
        while j=1: AB **do**
            Accomplish the subsequent motion calculation
            Motion induced via further individuals
            Implement the genetic operatives.
            Searching motion Physical diffusion
            Calculate fitness intended for separate krill corresponding to its new pos
        end
        X=X+1
    end
Rearrange in increasing order of semantic similarity and recommended to the User.

The performance of the put forward OSIBR, an ontology-focused semantic intelligence approach for the book recommendation is computed by Recall, F-Measure, Precision, Accuracy, FDR, and nDCG as the potential metrics. FDR depicts the number of false positives that's been furnished. nDCG which is normalized Discounted Cumulative Gain quantifies the diversity of results. The recommendation is that the higher the nDCG, the higher the variety and the better and unbiased the results. Standard formulations for Precision, Recall, Accuracy, F-Measure, FDR, and nDCG were used. The effectuation of the put forward approach is estimated by comparing it with HMBR [1], BRMLM [2] and SBDBR [3]. Three present methods are measured as the baseline, made known in Table 1. The foremost use HMBR [1] and the model's performance has Precision of 81.63%, Recall of 83.17%, Accuracy of 82.40%, F-measure of 82.39%, False Discovery of 0.19 and nDCG of 0.81. The following method uses BRMLM [2] to accomplish the identical task. The model's performance has a Precision of 84.44%, Recall of 86.32%, Accuracy of 85.38%, F-measure of 82.39%, False Discovery of 0.16 and nDCG of 0.82. The third method is SBDBR [3], and the model's performance related to the problem has a Precision of 90.65%, Recall of 92.47%, Accuracy of 91.56%, F-measure of 91.55%, False Discovery of 0.10 and nDCG of 0.95. The fourth approach is the projected OSIBR model. Its performance related to the problem has Precision of

94.16%, Recall of 96.84%, Accuracy of 95.50%, F-measure of 95.78%, False Discovery of 0.06 and nDCG of 0.97.

Table 1. Assessment of performance of the proposed OSIBR using added Approaches

Search Technique	Average Precision %	Average Recall %	Accuracy %	F-Measure	FDR	nDCG
HMBR [1]	81.63	83.17	82.4	82.39	0.19	0.81
BRMLM [2]	84.44	86.32	85.38	85.36	0.16	0.82
SBDBR [3]	90.65	92.47	91.56	91.55	0.10	0.95
Proposed OSIBR	94.16	96.84	95.5	95.48	0.06	0.97

The HMBR [1] yields the lowest Recall, Precision, F-Measure, Accuracy and highest FDR. The lowest nDCG value stays as it incorporates traditional collaborative filtering with content-based and demographic filtering. However, CBF requires ratings for the particular entity, so there can be a biased rating, then the entire recommendation can be affected. Apart from rating, collaborative filtering, content and demographic filtering alone are used without learning or inferential models. BRMLM [2] is a machine learning-based tactic that uses statistics. It hybridizes Random Forest with Support Vector Machine, and user preferences are also considered along with Adaboost. Since BRMLM uses a series of learning algorithms alone and a very high complexity based Adaboost and SVM, the approach is complex. The lack of semantic similarity or inferential model makes the FDR of 0.16 comparatively high with a low nDCG of 0.82.

Fig. 3. Precision Percentage Vs No. of Recommendations of the proposed model and other baseline models

SBDBR [3] is a semantic approach that uses a vector space model for the computation of Semantic Similarity and Doc2Vec model. The topic model has been used, which ensures that a very high nDCG of 0.95 is achieved. Furthermore, collaborative filtering is used in SBDBR confirming that the rating biased problem may occur and problems arise in the recommendation. However, infusing semantic similarity, topic model, and content-based filtering is highly commendable. That's why it has comparatively lower Recall, Precision, Accuracy when related to OSIBR. Moreover, the semantic similarity is quite old, and the topic model can also be improved, given that collaborative filtering alone is the drawback of this approach.

This Fig. 3 presents the analysis of the proposed OSIBR relation with HMBR, BRMLM and SBDBR systems. The chart tells us that the OSIBR has higher Precision. This is because this approach uses a semantic approach infused with ontology to recommend books. Thereby supplying the need for auxiliary data and background data as the result of ontology bringing diversification of results. Thus, the put forward system dynamically models ontology from or after the data, and it is a lot better approach than the prevailing models. It used the proposed (Average) – for 16192 Queries whose ground truth has been collected. OSIBR tends to improve this by using an ontology of subcategories. The use of ontology makes sure that knowledge description models are incorporated in the approach. Also, Static Modeling of Ontologies by users is achieved, and the dynamically versioned Ontology is more used for combined knowledge. Furthermore, the use of lateral preferences makes sure that the user interest computation and STM is better than the LDA. As a result, newer topics are uncovered, thereby ensuring nDCG value is high. Finally, Random Forest and Decision Tree take care of the machine learning capability in the algorithm. However, learning alone is insufficient. Therefore, it is coupled with semantic similarity using four measures: Jaccard similarity, SemantoSim, NPMI, and Kullback Leibler divergence. By amalgamating all these algorithms with both learning and inferential, the put forward OSIBR is much better than other existing models.

5 Conclusion

The put forth model assist the User in deciding the books they would need using the recommender system. We displayed the best approach that customs Random Forest and Decision Tree with Bagging and also formulating Ontologies. We presented that learning semantic features can outstrip traditional collaborative filtering for a recommender system in book search. Learning semantic features would boost content-grounded recommenders. So, we have introduced an ontology infused semantic intelligence approach for Book Recommendation to solve the specified problem. An overall 95.48% F-Measure has been reached with very little FDR of 0.06 and with a high nDCG of 0.97, which brands the projected OSIBR as the superior tactic mainly meant for recommending the books, which can learn and recommend with a minimal fault rate besides higher Accuracy.

References

1. Darekar, R., Dayma, K., Parabh, R., Kurhade, S.: A hybrid model for book recommendation. In: 2018 Second International Conference on Inventive Communication and Computational Technologies (ICICCT), pp. 120–124, April 2018
2. Tsuji, K., Yoshikane, F., Sato, S., Itsumura, H.: Book recommendation using machine learning methods based on library loan records and bibliographic information. In: 2014 IIAI 3rd International Conference on Advanced Applied Informatics, pp. 76–79. IEEE, August 2014
3. Mclean, N., Davis, J.: Utilizing semantically rich big data to enhance book recommendation engines. In: 2016 IEEE 18th International Conference on High Performance Computing and Communications; IEEE 14th International Conference on Smart City; IEEE 2nd International Conference on Data Science and Systems (HPCC/SmartCity/DSS), pp. 1434–1441. IEEE, December 2016
4. Vishal, K., Deepak, G., Santhanavijayan, A.: An approach for retrieval of text documents by hybridizing structural topic modelling and pointwise mutual information. In: Innovations in Electrical and Electronic Engineering, pp. 969–977. Springer, Singapore (2021)
5. Sabitha, S., Choudhury, T.: Proposed approach for book recommendation based on user k-NN. In: Bhatia, S., Mishra, K., Tiwari, S., Singh, V. (eds.) Advances in Computer and Computational Sciences. AISC, vol. 554, pp. 543–558. Springer, Singapore (2018). Doi: https://doi.org/10.1007/978-981-10-3773-3_53
6. Krishnan, N., Deepak, G.: KnowSum: knowledge inclusive approach for text summarization using semantic alignment. In: 2021 7th International Conference on Web Research (ICWR), pp. 227–231. IEEE, May 2021
7. Giri, G.L., Deepak, G., Manjula, S.H., Venugopal, K.R.: OntoYield: a semantic approach for context-based ontology recommendation based on structure preservation. In: Proceedings of International Conference on Computational Intelligence and Data Engineering, pp. 265–275. Springer, Singapore (2018)
8. Ru, C., Tang, J., Li, S., Xie, S., Wang, T.: Using semantic similarity to reduce wrong labels in distant supervision for relation extraction. Inf. Process. Manage. **54**(4), 593–608 (2018)
9. Kandhway, P., Bhandari, A.K., Singh, A.: A novel reformed histogram equalization based medical image contrast enhancement using krill herd optimization. Biomed. Signal Process. Control **56**, 101677 (2020)
10. Deepak, G., Gulzar, Z., Leema, A.A.: An intelligent system for modeling and evaluation of domain ontologies for Crystallography as a prospective domain with a focus on their retrieval. Computers & Electrical Engineering, 107604 (2021)
11. Ojha, R., Deepak, G.: Metadata driven semantically aware medical query expansion. In: Villazon-Terrazas, B., et al. (eds.) KGSWC 2021. CCIS, vol. 1459, pp. 223–233. Springer, Cham (2021)
12. Yethindra, D.N., Deepak, G.: A semantic approach for fashion recommendation using logistic regression and ontologies. In: 2021 International Conference on Innovative Computing, Intelligent Communication and Smart Electrical Systems (ICSES), pp. 1–6 (2021)
13. Surya, D., Deepak, G., Santhanavijayan, A.: KSTAR: a knowledge-based approach for socially relevant term aggregation for web page recommendation. In: Motahhir, S., Bossoufi, B. (eds.) ICDTA 2021. Lecture Notes in Networks and Systems, vol. 211, pp. 555–564. Springer, Cham (2021). Doi: https://doi.org/10.1007/978-3-030-73882-2_50
14. Krishnan, N., Deepak, G.: Towards a novel framework for trust driven web URL recommendation incorporating semantic alignment and recurrent neural network. In: 2021 7th International Conference on Web Research (ICWR), pp. 232–237 (2021)
15. Rithish, H., Deepak, G., Santhanavijayan, A.: Automated assessment of question quality on online community forums. In: Motahhir, S., Bossoufi, B. (eds) ICDTA 2021. Lecture Notes

in Networks and Systems, vol. 211, pp. 791–800. Springer, Cham (2021). Doi: https://doi.org/10.1007/978-3-030-73882-2_72
16. Deepak, G., Kasaraneni, D.: OntoCommerce: an ontology focused semantic framework for personalized product recommendation for User targeted e-commerce. Int. J. Comput. Aided Eng. Technol. **11**(4–5), 449–466 (2019)
17. Roopak, N., Deepak, G.: KnowGen: a knowledge generation approach for tag recommendation using ontology and honey bee algorithm. In: Musleh Al-Sartawi, A.M.A., Razzaque, A., Kamal, M.M. (eds.) EAMMIS 2021. LNNS, vol. 239, pp. 345–357. Springer, Cham (2021). https://doi.org/10.1007/978-3-030-77246-8_33
18. Deepak, G., Santhanavijayan, A.: UQSCM-RFD: A query–knowledge interfacing approach for diversified query recommendation in semantic search based on river flow dynamics and dynamic user interaction. Neural Computing and Applications, 1–25 (2021)
19. Varghese, L., Deepak, G., Santhanavijayan, A.: A fuzzy ontology driven integrated IoT approach for home automation. In: Motahhir, S., Bossoufi, B. (eds.) ICDTA 2021. Lecture Notes in Networks and Systems, vol 211, pp. 271–277. Springer, Cham (2021). https://doi.org/10.1007/978-3-030-73882-2_25
20. Arulmozhivarman, M., Deepak, G.: OWLW: ontology focused user centric architecture for web service recommendation based on LSTM and whale optimization. In: Musleh Al-Sartawi, A.M., Razzaque, A., Kamal, M.M. (eds.) EAMMIS 2021. Lecture Notes in Networks and Systems, vol. 239, pp. 334–344. Springer, Cham (2021). Doi: https://doi.org/10.1007/978-3-030-77246-8_32

IntelliLegalRec: An RDF Based Metadata Driven Semantically Compliant Recommendation System for Socio-legal Judicial Documents

Akhil S. Krishnan[2] and Gerard Deepak[1](✉)

[1] Department of Computer Science and Engineering, National Institute of Technology, Tiruchirappalli, India
gerard.deepak.christuni@gmail.com
[2] Department of Computer Science and Engineering, SRM Institute of Science and Technology, Ramapuram, Chennai, India

Abstract. Pertinent socio-legal judicial documents are a necessity as backing for similar judicial cases. However, incomplete and irrelevant particulars while reviewing potentially related judicial documents pertaining to a case may prove unfavourable to its direction. This paper proposes a Semantically inclined recommendation system IntelliLegalRec for the recommendation of socio-legal judicial documents, complemented by the generation of metadata and the Resource Description Framework. The approach is user query centric and uses the Latent Dirichlet Allocation model for topic modelling. Furthermore, two stages of semantic similarity computation with the help of semantic measures like Concept Similarity, Normalized Pointwise Mutual Information and Normalized Google Distance, is carried out and further optimized using the Golden Eagle Optimization algorithm. As a result, IntelliLegalRec achieves best in class accuracy of 94.60%, making it an efficient and semantically compliant system for the recommendation of judicial documents.

Keywords: Golden Eagle Optimization · Metadata · NGD · NPMI · RDF · Topic modelling

1 Introduction

With the advent of the internet, the content available on the World Wide Web has skyrocketed. Easier accessibility of the internet and a decrease in data prices have resulted in the internet being accessible by a larger demographic than ever before. This surfeit of data has resulted in a spot of bother for the end-user, who generally ends up with a handful of data that may or may not suit their needs. Thus arises the need to recommend accurate and relevant data to the user's needs. A recommender system is a system of filtration that recommends necessary and relevant data to a user depending upon the input queries and the data furnished from the user side.

S. Motahhir and B. Bossoufi (Eds.): ICDTA 2022, LNNS 454, pp. 407–416, 2022.
https://doi.org/10.1007/978-3-031-01942-5_41

Relevant Judicial case documents are of utmost importance in finding support for decisions that may be favourable pertaining to another case. Attorneys spend an intemperate amount of time finding the proper reference documents for their merit briefs. This task generally involves screening numerous document sources from the present and the past, resulting in wasted time. Such onerous processes may pose problems such as obtaining incomplete or irrelevant data. A single missed point resulting from a missing document may be detrimental to how the case proceeds and may impair the litigation process. Thus, a judicial case recommender system is essential for providing pertinent and accurate documents about a particular case, which may boost the attorneys' ability to rebut a point of law.

Motivation: With the World Wide Web's current structure, namely the Web 3.0 or the semantic web, there emerges a need for semantically infused recommendation strategies. Recommendations involving highly specialized domains like the judicial domain are challenging using traditional means, mainly because a lot of domain knowledge is needed. Moreover, the judicial aspects differ from country to country in the socio-legal judicial domain. Hence, it is generally difficult to generalize such systems, as specific domain knowledge from different countries is essential for support. Thus arises a need for the semantically inclined recommendation of socio-legal judicial documents.

Contribution: The proposed model is metadata-driven and incorporates generated Resource Description Framework (RDF). The topic modelling is carried out by Latent Dirichlet Allocation (LDA) and is subject to classification using logistic regression. The model uses the CAIL 2018-Small dataset, classified using the Support Vector Classifier (SVC) and an Indian judicial case repository facilitating RDF generation crawled from the web. A two-stage semantic similarity computation with class label and instance-level similarity computation is done using concept similarity, Normalized Pointwise Mutual Information (NPMI) and Normalized Google Distance (NGD). The results are optimized using the metaheuristic Golden Eagle Optimization algorithm [1]. The aforementioned features have contributed to the proposed approach exhibiting higher accuracy and a very low False Discovery Rate (FDR).

Organization: The latter part of the paper is structured as follows. Section 2 encompasses the study of relevant research carried out in relation to the field. The architecture of the proposed approach is elucidated in Sect. 3. The means of implementation and the evaluation of performance exhibited by the proposed approach and baselining with state-of-the-art models are presented in Sect. 4. Finally, the conclusion drawn from the proposed approach is presented in Sect. 5.

2 Related Works

Yang et al. [2] put forward a model for the recommendation of judicial cases based on the similarity in texts of judgment. They devise a novel approach by modelling a TF-Word2Vec similarity model with the deployment of a neural network to carry out the same. Guo et al. [3] have put forth a system for recommendation based on content, which uses semantic methods to retrieve important information from judicial documents from a dataset of 200,000 cases, using TF-IDF and Latent Dirichlet Allocation for topic modelling. Mones et al. [4] discuss a network science approach for prediction in judicial systems; the patterns of citation structures are used to bolster existing citations and establish a link between current and further references for the benefits of court cases, making use of the link prediction model. Thomas et al. [5] have discussed the system "Quick check" for the judicial recommendation, which derives contentions based on the user's case and recommends suitable opinions. Zihuan et al. [6] have put forward a deep learning-based analysis of various facts based on a case and further recommendation of them. The recommendation is based on the similarity of the word vector to that of the training cases. Tieke He et al. [7] present a model for deducing the "sanction amount" of judicial cases based on the similarity of documents. Word embeddings are used to evaluate the document distance. The cases from a legal dataset are grouped into different classes based on each class's sanction, and they surmise the class with which the case is associated. Tian et al. [8] propose a recommendation system for university books in libraries based on a recommendation algorithm with a hybrid approach. Safaryan et al. [9] discuss a semantically inclined system of recommendation for research papers. The system is implemented using a pipeline of three steps: pre-processing, w2w vectorization involving vector mapping, and finally, the computation of cosine similarity of the text. Dai et al. [10] discuss a recommendation system of references essential to a research paper. They design the model based on Latent Dirichlet Allocation and factorization matrix, under the scope of a semantically compliant topic learning and detection of groups methods. In [11–22], several models in support of the proposed literature have been discussed.

3 Proposed System Architecture

Figure 1. illustrates the architecture diagram for the proposed judicial document recommendation system. In phase 1, the web usage data obtained from the user is analyzed for judicial cases. The judicial case information and the input query obtained are subject to pre-processing. The pre-processing of the query data entails Tokenization, Stop Word removal, Lemmatization, and POS Tagging – (which associates a word from the text to a part of speech). On pre-processing, the query words are extracted and is subject to topic modelling. The model chosen for topic modelling is LDA which ensures the generation and modelling of new topics synonymous with the query words. For the query words generated, query word enrichment is carried out, and for each of these, metadata is generated.

Metadata is also further generated from the World Wide Web and other tools like Open Calais and "Gleaning Resource Descriptions from Dialects of Languages" (GRDDL). Open Calais is an efficient tool that generates metadata at a blazing fast rate and goes beyond the primary identification of entities by returning various facts which otherwise seem to be hidden in the text of the document you provide. GRDDL is a format of markup to obtain the gleaning resource descriptions of a language. The metadata obtained from the World Wide Web is a relevant source for query terms enriched or modelled by the LDA.

The metadata is extensive; hence a robust classifier like logistic regression is necessary for classification of the same. The logistic regression model estimates the chances of an event happening.

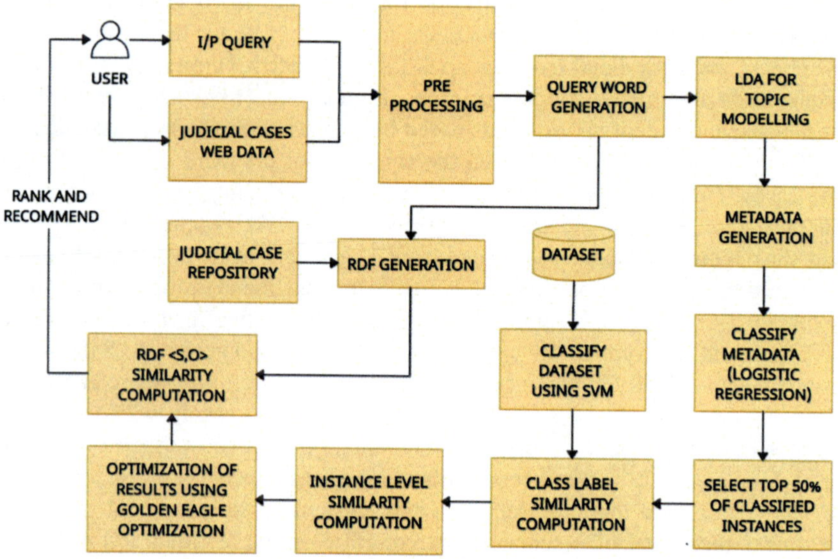

Fig. 1. Proposed system architecture design for the IntelliLegalRec model

For the classification of metadata, the initial query words are chosen as features. The extensive nature of the metadata means only the top 50% is used to classify the instances. The approach also uses the CAIL 2018-Small dataset. This dataset is classified using the SVC. The SVC is used because the dataset need not be extensively classified. Since the dataset is smaller in scale, SVC being a binary linear classifier is light weighted and reduces computational load, making the approach computationally inexpensive. As the approach is based on recommendation and not classification, classification is infused into the framework.

In phase 2, the computation of semantic similarity takes place. This process is carried out in two stages. The class label similarity computation is carried out first instead of following just the class and then the instances, using concept similarity. Further, all the matching classes are shortlisted, and instance-level semantic similarity computation is computed using two approaches: NPMI and the NGD.

Pointwise mutual information is the means of measuring the association between two events, x and y. Normalized pointwise mutual information (NPMI) normalizes data between -1 and 1, where -1 refers to never occurring together, 0 for independent and 1 for occurring together. Equation (1) illustrates the formula for NPMI where h (x, y) is the joint self-information.

$$\text{NPMI}(x; y) = \frac{\text{pmi}(x; y)}{h(x, y)} \tag{1}$$

The Normalized Google Distance (NGD) is a measure of semantic similarity, procured from the figure of relevant results shown by Google's search engine. The keywords closest to the exact meaning are close, whereas the distinct ones are farther. The equation of Normalized Google Distance is illustrated in Eq. (2), where 'a' and 'b' are search terms, 'N' is the number of the web pages that come up on google search, 'f(a)' and 'f(b)' are the number of results found for different search terms and 'f(a,b)' is where both the terms return search results.

$$\mathrm{NGD}(a, b) = \frac{max\{\log f(a), \log f(b)\} - \log f(a, b)}{\log N - min\{\log f(a), \log f(b)\}} \tag{2}$$

The threshold for NPMI and NGD are considered 0.5 and 0.75, respectively. Based on the matching labels and the instances, a hash set is formulated of all instances under each matching class label. The semantic similarity computed using NGD and NPMI is subject to the Golden Eagle Optimization algorithm [1]. The Golden Eagle Optimization algorithm is nature-inspired and is based on the intellect of the Golden eagle to adjust its speed and trajectory based on a spiral in the process of hunting for prey. The eagle tends to revolve around the prey in the initial stages and attack in the final stages. The algorithm is the mathematical modelling of the effective use of these two traits for catching prey quickly. The objective functions used for optimization are the NPMI values and the NGD.

From the query words and a repository comprising cases of several types and natures from India, RDF is generated. This repository is constructed using customized web crawlers and the Beautiful Soup API. Resource Description Framework or RDF is the standard for exchanging data of significant interconnection on the World Wide Web The RDF is generated using the Onto Collab tool. It generates the OWL format, and it is then developed into RDF.

Finally, the instances matching the RDF are rearranged based on ascending semantic similarity and recommended to the user. This process continues till there are no further user clicks from the user's end. If there are user clicks, the process is re-iterated. Several matching sub-topics and highly diverse terms will be yielded to the user, and on the click, the exact document will be furnished.

4 Implementation and Performance Evaluation

The proposed IntelliLegalRec was implemented using Jupyter notebook for python, with an i7 processor of 4.0 GHz,16 GB Ram and 60 GB of hard disk storage. The gensim module is imported to use the LDA model for topic modelling. The CAIL 2018 small dataset with 196,000 cases from the Supreme people's court of China is used as the dataset in this proposition. "Beautiful Soup", an HTML parser, is used for crawling the judicial data repository. The case details for the repository are mainly extracted from the national portal of India, which provides judicial case data associated with India over the years. The algorithm for the system presented IntelliLegalRec is illustrated in Algorithm 1.

Algorithm 1: Algorithm for the Proposed IntelliLegalRec

Input: User query, web usage data, CAIL dataset, Case Repository.
Output: Judicial documents recommended in ascending semantic similarity.

```
Pre-process user query cum web usage data and obtain the query words.
Perform Topic modelling using LDA for the query words obtained.
Generate metadata from query words, WWW and other tools.
classify metadata using Logistic Regression and select top 50% instances.
Classify dataset using SVC.
Compute class label semantic similarity with concept similarity and instance
level similarity using NPMI and NGD.
Optimize Semantic similarity computed using Golden Eagle Optimiza-
tion [1]
    Initialize eagle population, fitness function, cruise tendency (t_p) and
    attack tendency(t_a)
    for each recurrence do
        Update the t_p and t_a, Calculate the distance of crowding
        for each eagle k do
            Select catch present in archive and Compute attack vector B⃗
            while (len.vec! = 0) do
                Compute the cruise vector K⃗ , step vector S⃗ and fitness
                function, update pos
                if pos.new is non governed to members of archive then
                    if archive.ext is not full then
                        Add solution which are current to archive.ext
                    else
                        Calculate dist of sparsity and restore the old soln
                    end
                end
            end
        end
    end
Generate RDF using query terms and Indian judicial case repository and
compute semantic similarity
Recommend the judicial documents to the user pertaining to the ascend-
ing order of semantic similarity
```

Good performance metrics are essential for any recommendation system to give accurate and satisfying results for the user. The performance metrics used for evaluating the proposed model are Precision, Accuracy, Recall, F-Measure and False Discovery rate, which is an essential measure, especially in the case of recommendation systems, as a lower FDR value equates to a better recommendation system. All standard formulae are used for the evaluation of the different performance metrics.

Table 1. Performance comparison of the proposed IntelliLegalRec with other baseline methods

Search technique	Average precision %	Average recall %	Average Accuracy %	F-Measure	FDR
CSCR [2]	82.12	85.12	83.62	83.59	0.17
CBRF [3]	88.14	92.28	90.21	90.16	0.11
NB + TF-IDF	74.12	76.08	75.10	75.08	0.25
FCM + SVM + W2Vec	84.02	86.47	85.42	85.22	0.15
Proposed IntelliLegalRec	93.18	96.02	94.60	94.57	0.06

The proposed IntelliLegalRec's performance is evaluated and furnished in Table 1. and is compared with other baseline approaches. It is observed that the proposed approach achieves superior performance compared to the baseline approaches and benchmark approaches. The performance comparison is made in contrast to models like CSCR [2], CBRF [3], Naïve bias with TF-IDF, Fuzzy C- means Clustering including SVM and Word2Vec. All the baseline models were assessed in a similar environment as the proposed model with similar user queries and usage data from the web.

The CSCR [2] uses word embedding frequency, TF-Word2Vec and traditional cosine similarity. Since the conventional cosine similarity is used, the relevance of results tends to be low in this approach, resulting in lower performance metrics. The CBRF [3] uses traditional cosine similarity and TF-IDF. However, it performs better than the CSCR [2] because LDA is used for topic modelling. The topic modelling ensures background knowledge is used. In the approach involving Naïve Bayes and TF-IDF, the Naïve Bayes is insufficient on its own and TF-IDF is used to obtain more relevant documents. However, since there is no regulating mechanism for semantics or machine learning relevance on a high-level classifier, very low-performance metric values are reported with the highest FDR value.

When Fuzzy C-means Clustering, SVM and Word2Vec is combined the computational complexity is lower. However, classification and clustering with Word2Vec give better results but are very computationally expensive. Finally, it can be observed quite clearly that the proposed approach of IntelliLegalRec furnishes the highest performance overall, mainly because it facilitates the generation of metadata. When the metadata is generated, relevant data content to the domain of choice is incorporated into the framework. Hence, it transforms into a dense knowledge-driven approach where the density of knowledge incorporated becomes very high, which is again subjected to classification using logistic regression, a powerful classifier. The role of RDF generation ensures the highly cohesive relationship of subject and object being instilled into the system. Further, the combination of NGD and NPMI for instance-level similarity and usage of concept similarity for computation of class label similarity, ensures three different semantic similarity measure methods are used with different thresholds. Most importantly, the Golden eagle optimizer to generate the best results ensures that the recommendations are highly relevant. As a result, the proposed approach brings out better performances in comparison to the baseline approaches.

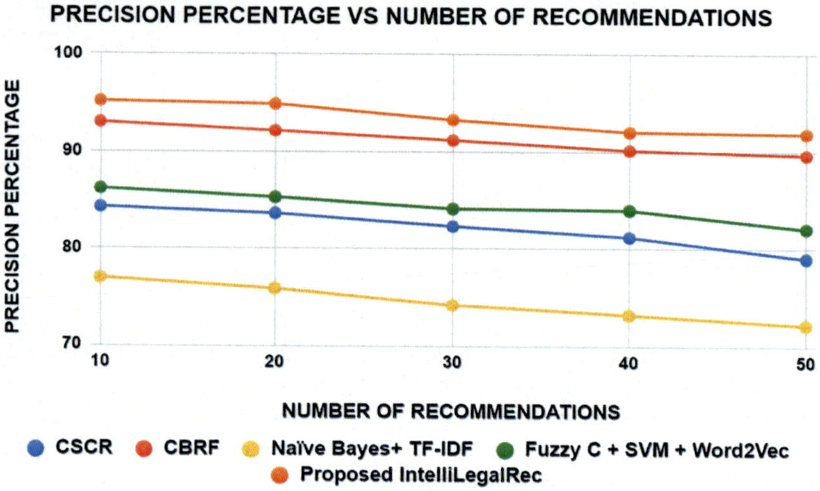

Fig. 2. Precision Percentage vs the number of recommendations of the proposed IntelliLegalRec and other baseline models

Figure 2 Presents the precision percentage for the Proposed IntelliLegalRec system compared with the other baseline models over a range of 10 to 50 recommendations. It is observable that the proposed IntelliLegalRec has higher Precision for the number of recommendations when compared to the other baseline models. This is owing to the generation of metadata and RDF, two stages of semantic similarity computation and, most importantly, the optimization of results under the Golden Eagle Optimization algorithm.

5 Conclusions

A socio-legal judicial document recommendation system, IntelliLegalRec is proposed, which is based on RDF generation and is metadata-driven. The proposed approach incorporates user queries and web utilization statistics of the user for topic modelling and metadata generation. Two stages of semantic similarity computation are carried out using NGD, NPMI and Concept similarity measures that the Golden Eagle Optimizer further optimizes. RDF is generated from query terms and judicial cases repository of Indian cases. Due to metadata and RDF generation, the proposed approach with increased efficiency furnishes higher performance metrics over other state-of-the-art models pertaining to the same domain. Especially a high accuracy of 94.60% and a low FDR value of 0.06 makes the IntelliLegalRec model one of the best-in-class approaches in the recommendation of socio-legal judicial documents.

References

1. Mohammadi-Balani, A., Nayeri, M.D., Azar, A., Taghizadeh-Yazdi, M.: Golden eagle optimizer: a nature-inspired metaheuristic algorithm. Comput. Ind. Eng. **152**, 107050 (2021)
2. Yang, F., Chen, J., Huang, Y., Li, C.: Court similar case recommendation model based on word embedding and word frequency. In 2020 12th International Conference on Advanced Computational Intelligence (ICACI), pp. 165–170, August 2020
3. Guo, Z., He, T., Qin, Z., Xie, Z., Liu, J.: A content-based recommendation framework for judicial cases. In: International Conference of Pioneering Computer Scientists, Engineers and Educators, pp. 76–88, September 2019
4. Mones, E., Sapieżyński, P., Thordal, S., Olsen, H.P., Lehmann, S.: Emergence of network effects and predictability in the judicial system. Sci. Rep. **11**(1), 1–10 (2021). Yoo, S., & Jeong, O. (2015)
5. Thomas, M., et al.: Quick check: a legal research recommendation system. In: NLLP@ KDD, pp. 57–60 (2020)
6. Xu, Z., He, T., Lian, H., Wan, J., Wang, H.: (2019, September). Case facts analysis method based on deep learning. In International Conference on Web Information Systems and Applications (pp. 92–97). Springer, Cham
7. He, T., Lian, H., Qin, Z., Zou, Z., & Luo, B. (2018, September). Word embedding based document similarity for the inferring of penalty. In International Conference on Web Information Systems and Applications (pp. 240–251). Springer, Cham
8. Tian, Y., Zheng, B., Wang, Y., Zhang, Y., Wu, Q.: College library personalized recommendation system based on hybrid recommendation algorithm. Procedia CIRP **83**, 490–494 (2019)
9. Safaryan, A., Filchenkov, P., Yan, W., Kutuzov, A., & Nikishina, I. (2020, October). Semantic Recommendation System for Bilingual Corpus of Academic Papers. In International Conference on Analysis of Images, Social Networks and Texts (pp. 22–36). Springer, Cham
10. Dai, T., Zhu, L., Cai, X., Pan, S., Yuan, S.: Explore semantic topics and author communities for citation recommendation in bipartite bibliographic network. J. Ambient. Intell. Humaniz. Comput. **9**(4), 957–975 (2017). https://doi.org/10.1007/s12652-017-0497-1
11. Deepak, Gerard, Gulzar, Z., Leema, A.A.: (2021). An intelligent system for modeling and evaluation of domain ontologies for Crystallography as a prospective domain with a focus on their retrieval. Computers & Electrical Engineering, 107604
12. Hybridized, K.C.N.: OntoKnowNHS: Ontology Driven Knowledge Centric Novel Hybridised Semantic Scheme for Image Recommendation Using Knowledge Graph. Knowledge Graphs and Semantic Web, 138
13. Ojha, R., Deepak, G.: Metadata driven semantically aware medical query expansion. In: Villazon-Terrazas, B., et al. (eds.) KGSWC 2021. CCIS, vol. 1459, pp. 223–233. Springer, Cham (2021)
14. Yethindra, D.N., Deepak, G.: A semantic approach for fashion recommendation using logistic regression and ontologies. In: 2021 International Conference on Innovative Computing, Intelligent Communication and Smart Electrical Systems (ICSES), pp. 1–6. IEEE, September 2021
15. Adithya, V., Deepak, G.: HBlogRec: a hybridized cognitive knowledge scheme for blog recommendation infusing XGBoosting and semantic intelligence. In: 2021 IEEE International Conference on Electronics, Computing and Communication Technologies (CONECCT), pp. 1–6, July 2021
16. Surya, D., Deepak, G., & Santhanavijayan, A. (2021, January). KSTAR: A knowledge-based approach for socially relevant term aggregation for web page recommendation. In International Conference on Digital Technologies and Applications (pp. 555–564). Springer, Cham

17. Krishnan, N., & Deepak, G. (2021, May). Towards a Novel Framework for Trust Driven Web URL Recommendation Incorporating Semantic Alignment and Recurrent Neural Network. In 2021 7th International Conference on Web Research (ICWR) (pp. 232–237). IEEE
18. Rithish, H., Deepak, G., & Santhanavijayan, A. (2021, January). Automated Assessment of Question Quality on Online Community Forums. In International Conference on Digital Technologies and Applications (pp. 791–800). Springer, Cham
19. Deepak, G., Kasaraneni, D.: OntoCommerce: an ontology focused semantic framework for personalized product recommendation for user targeted e-commerce. International Journal of Computer Aided Engineering and Technology **11**(4–5), 449–466 (2019)
20. Roopak, N., Deepak, G.: KnowGen: a knowledge generation approach for tag recommendation using ontology and honey bee algorithm. In: Musleh Al-Sartawi, A.M., Razzaque, A., Kamal, M.M. (eds.) EAMMIS 2021. LNNS, vol. 239, pp. 345–357. Springer, Cham (2021). https://doi.org/10.1007/978-3-030-77246-8_33
21. Deepak, G., & Santhanavijayan, A. (2021). UQSCM-RFD: A query–knowledge interfacing approach for diversified query recommendation in semantic search based on river flow dynamics and dynamic user interaction. Neural Computing and Applications, 1–25
22. Tiwari, S., Al-Aswadi, F.N., Gaurav, D.: Recent trends in knowledge graphs: theory and practice. Soft. Comput. **25**(13), 8337–8355 (2021). https://doi.org/10.1007/s00500-021-05756-8

HSIL: Hybrid Semantic Infused Learning Approach for Course Recommendation

Dev Agrawal[1] and Gerard Deepak[2](✉)

[1] Department of Computer Science and Engineering, Jaypee Institute of Information Technology, Noida, India
[2] National Institute of Technology, Tiruchirappalli, India
gerard.deepak.christuni@gmail.com

Abstract. Course recommendation is essential to enable learners to choose the best course among many available courses. So, as a result, a semantically driven course recommendation system is required as the data over the world wide web is increasing, and the web is also tending towards Semantic Web 3.0. In this approach, the initial term pool is created based on the existing user learner profile and the current clicks of the user to determine the exact needs of the user for the term enrichment is done by incorporating ontologies of courses and metadata generation approach is proposed where the metadata is initially classified using XGBoost algorithm where the top 25% is used. Then, the core dataset is also classified using the XGBoost algorithm based on the enriched user terms. And then, the semantic similarity is computed using concept similarity under the genetic algorithm and is recommended to the user. Overall, the accuracy of 96.3% is achieved.

Keywords: Criminology · Knowledge base · LSTM · Ontology · Transformers

1 Introduction

In this world full of competition, wasting even a second to think about what to do can hamper growth badly. Moving with the same pace of life, with the correct selection of choices in hand, has become the mandatory task for all. With the rapid growth of e-learning resources, universities also provide many sources for learning courses and taking certificates. Still, if a learner opts for the irrelevant course, it's useless for him and a waste of time. People are confused about which course they should go for and waste less time just searching for it. Also, with the increase in competition among the instructors, information is getting divided with the decrease in the quality of education. So how to screen information has become a crucial issue to be resolved. And for this, we need an intelligent course recommendation system that can help any person opt for the right course. It analyses data automatically just by taking answers to a few questions from users. These systems help to become confident with the decisions and make things very simple.

Users must be careful in this social environment, where people are confused but love to advise others. Therefore, we should rely on solid data and statistics to make

S. Motahhir and B. Bossoufi (Eds.): ICDTA 2022, LNNS 454, pp. 417–426, 2022.
https://doi.org/10.1007/978-3-031-01942-5_42

decisions. To solve this problem and personalize information, course recommendation systems work with the help of artificial intelligence, which is used to recognize related traits of the users. As per the interest of users, it provides them with meaningful recommendations. This paper presents a model Hybrid Semantic Infused Learning (HSIL) approach for course recommendation. The proposed system architecture of the ontology focuses on metadata-driven classified using XGBoost classification. The semantic similarity is computed using concept similarity under genetic algorithms between the top 25% of the classified instances of the metadata and the classified instances of the dataset.

Motivation: The ultimate aim of all the recommendation systems is to give Users recommendations that will be positively judged and accepted by their users. This paper aims to design and develop a Hybrid Semantic Infused Learning (HSIL) approach for course recommendation. The proposed architecture focused on metadata-driven keeping in mind the exponential increase in the data over the world wide web.

Contribution: Initial term pools are generated based on the existing user learner profile and the current clicks of the user. Two-step classification is performed by classifying the dataset, and the metadata generated using the XGBoost algorithm. Then, ontologies of courses are applied for generating data and term enrichment, and semantic similarity computation is done using concept similarity by moderating it under the genetic algorithm for optimization. Using this accuracy of 96.30% and f measure of 96.29 is achieved for this dataset.

Organization: The latter content of this paper is structured as follows. Related Works are contained in Sect. 2. Section 3 briefly explains the Proposed System Architecture. Section 4 consists of the Implementation, and the Results and Performance Evaluation are discussed in Sect. 5. Finally, in Sect. 6, the paper is concluded.

2 Related Works

Zameer Gulzar et al. [1] includes static ontologies and the learner profile, which are given more importance, and feeding is done for the instances. It performs very well when its ontologies are perfectly separated into the approach, while when ontologies become chaotic, the approach fails to yield the required performance. Jing Zhang et al. [2] proposed a recommendation system based on machine learning criteria. Reinforcement learning ensures that the HRL is mainly dependent on the dataset. Hualong ma et al. [3] is a Semantic Similarity analysis-based course recommendation system that is semantically compliance while only the TFIDF word to vector along with cosine similarity is incorporated into the approach. Mohammed E. Ibrahim et al. [4] is an Ontology-Based Personalized Course Recommendation Framework where dynamic ontology mapping makes it quite envious as well as quite efficient. The approach uses collaborative filtering with content-based filtering.

N D Lynn et al. [5] introduced the hybrid recommendation method and concluded it to be the finest from other recommendation methods, such as content-based approaches, collaborative filtering for students to opt for the specific course. A hybrid method combines all the strengths of all other methods, making it a robust method for good recommendations. Amer Al-Badarenah et al. [6] presents a collaborative recommender system that exploits courses that other similar learners had taken and recommends courses based on likes and dislikes among learners to select elective courses offered by the university. To discover patterns between courses the proposed system has used the technique of rules mining algorithm. Hazra Imran et al. [7] established a structure to join learning management systems with a recommender system approach, making these systems give recommendations based on fortunate learning experiences of related learners. The approach uses a neighbourhood algorithm and association rule mining which help learners to go through the course and improve their learning performance and contentment.

Prem Melville et al. [8] shows that the predictions of a recommender system can be improved significantly by incorporating content information into collaborative filtering. They depicted that how content-boosted collaborative filtering is better than collaborative filtering, a pure content-based predictor, and a naive hybrid of the two. Bhaskar Mondal et al. [10] applied a K-means clustering algorithm on the dataset to create groups having similar learners. Based on their learning experience and past performances. After the creation of clusters, a frequent pattern mining algorithm is applied to them. Recommendation of courses will be done based on frequent pattern mining. Sanjog Ray et al. [11] compared two collaborative filtering approaches, first Users based collaborative filtering and other is items based collaborative filtering. They conclude that there was not much difference in accuracy, but the second one performs better for $x = 10\%$, while the first performs slightly good from higher values of x. In [12–25] the several ontological and knowledge-driven frameworks in support of the proposed literature have been justified.

3 Proposed Architecture

Figure 1 detects the proposed system architecture of the ontology focused on a metadata-driven course recommendation system. Here, the learner profile and the current clicks of the learner are taken as source repositories for obtaining the initial term pool. The initial term pool consists of the unique terms based on the current navigation of the user as well as the terms from the learner profile. First, the terms from the learner profile are subjectively pre-processed, which includes tokenization (it is the process where the vast body of the text is split into smaller lines or words), lemmatization (it is the process of making groups of the same words having different forms. In search engines, it allows end-users to question any kind of a base word and get appropriate outcomes), stop words removal (we don't want words like "a", "the", "in", "an" to waste our memory and precious time of processing, so the query engines are programmed to ignore the searches for them) and main entry recognition. After this, the initial term pool is formulated.

It includes the term enrichment by using ontologies of courses. Course ontology is statically modeled using protege and also from several other websites like Coursera, udemy are where course description and course title are taken, and ontology is automatically generated by using OntofCollab as a tool. Ontology of courses and the initial term pool is matched by using concept similarity for term enrichment. Further, the metadata is generated by using the ontology of courses. The procedure of drawing out extra essential

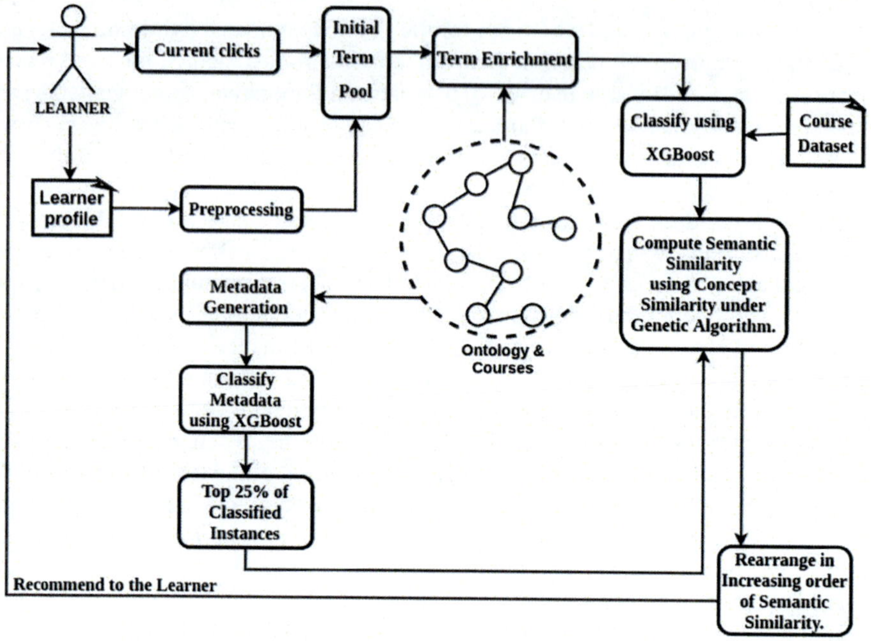

Fig. 1. HSIL system architecture

details from the Web resource with reference to ontology is called metadata extraction, and this is required for our experiments. Both semi-automatic and automatic methods and devices have been expanded. Various tools exist, including Cream-7, S-Cream, Semagix Freedom toolkit, and SemTag. Ontologies population has been carried out classically by using Semagix Freedom, which was part of IBM's WebFountain project. Other tools used to extract the metadata are OpenCalais, PyRdfa, GRDDL, RDF Distiller.

The generation of metadata is extensively very large, so there is a need to classify metadata. So for classification, the term pool will be the input. Based on the item in the initial term pool, the metadata is classified using XGBoost classification. XGBoost (eXtreme Gradient Boosted trees) can be meant as a distinctive approach that incorporates ensemble learning that uses a gradient boosting framework. It is a regularized boosting algorithm that prevents the model from overfitting. It uses a feature called tree pruning, so unlike normal decision trees where it is just branching once, it stops seeing a benefit of doing but here it will go very deep by default and then try to prune that tree backward so that generally results in deeper trees but more highly optimized trees making this algorithm successful and effective. It is a supervised learning methodology that optimizes specific loss functions based on function approximation. The binary classification with log loss objective function is given by Eq. (1).

$$yln(p) + (1-y)ln(1-p) \; where \, p = 1/(1+e^{-x}) \tag{1}$$

where y is the real label in $\{0, 1\}$ and p is the probability score? Users need to find first and second-order derivatives with respect to x to reduce the log loss objective function. Then elected the top 25% of the classified instances because we focused on highly relevant terms as the number of entities is extensive. Dataset is scrapped from the Coursera website. It is made categorical by performing annotations on it. After that, the core dataset is also classified using XGBoost. The semantic similarity is computed using concept similarity under genetic algorithms between the top 25% of the classified instances of the metadata and the classified instances of the dataset. Further, All the matched entities are rearranged in the increasing order of semantic similarity, i.e., concept similarity. Semantic similarity between two concepts is calculated using WordNet. The shortest path between the two concepts is calculated using the Lealock-Chodorow algorithm in WordNet. Then the concept name similarity is calculated using the shortest path. In Eq. 2 [9], $len(C_1, C_2)$ represents the shortest path between the two concepts, $2Depth$ represents the depth of WordNet.

$$sim_{semantic}(C_1, C_2) = len(C_1, C_2)/2Depth \quad (2)$$

Genetic Algorithms are the abstraction of real biological evolution which are used to solve complex problems quickly. It focuses on optimization. Evolutionary algorithms are the superset of genetic algorithms which are used in calculations. Genetics and natural selection concepts are used here to compute the solutions to the problems. These algorithms are based on the actions of chromosomes and their genetic shape. Every single chromosome provides solutions. It provides outstanding parallel capabilities and doesn't need derivative information. The answer provided by genetic algorithms improves over time.

4 Implementation

The proposed HSIL was designed and successfully implemented, python 3.9.0 was used to implement HSIL architecture. Google Colab Notebook was the IDE with an i5 processor, and a processing speed of 1.2–3.6 GHz has been utilized to design and implement the architecture. Google Colab offers two different models of GPU NVIDIA K80 / T4 for free usage and 100 GB of free space for data, and 16 GB of RAM to provide high-performance computing tasks. The XGBoost algorithm is implemented by using standard XGBoost libraries of python. Pre-processing was done using a natural language processing framework. The data set used in this architecture is scrapped from the Coursera website. It contains mainly six columns, i.e., cousrse_title, course_organization, courseCertificate-type, course_rating, course_difficulty, course students enrolled, and 890-course data is made categorical by performing annotations on it. Ontologies were modeled using protege by crawling several repositories, which consist of course as well as details of courses like standard websites udemy, Coursera using Beautiful Soup. Ontology is modeled using OntoCollab. Manual modeling of ontologies was also done. Indexes from the world wide web were also taken, relevant to e-learning and other learning domains. A large ontology was created, having 14916 concepts and sub-concepts and around 18421 individuals. The model ontology was implicitly axiomatized, and SWRL reasoning was also performed to maintain the ontologies' hierarchy.

5 Performance Evaluation and Result

This Experiment is done on 8192 queries whose ground truth has been collected. The performance of the proposed HSIL approach has been computed using precision, recall, accuracy, f-measure, FDR (False Discovery Rate), and NDCG (Normalized Discounted Cumulative Gain) as the potential matrix. Precision, Recall, Accuracy, F-measure quantify the relevance of the results furnished by the proposed approach. FDR computes the number of false positives; thereby lower the FDR, the better the system's performance. Finally, NDCG measures the diversity of the results evaluated by the proposed approach. Proposed model is a baselined with PCRS, HRL, SSA, OPCR for computing the proposed HSIL. These base papers were also implemented in the same environment as the proposed HSIL, and the experiment was one of 8192 queries whose ground truth has been collected.

Table 1. Comparison of Performance of the proposed HSIL with other approaches

Search Technique	Average Precision %	Average Recall %	Accuracy %	F-Measure %	FDR	nDCG
PCRS [1]	87.63	92.45	90.04	89.97	0.13	90.54
HRL [2]	88.44	91.15	89.79	89.77	0.12	85.63
SSA [3]	90.65	92.84	91.74	91.73	0.1	88.45
OPCR [4]	92.16	94.18	93.17	93.15	0.08	92.26
Proposed HSIL	**95.18**	**97.43**	**96.30**	**96.29**	**0.05**	**95.63**

From Table 1, it is highly indicative that PCRS yields Precision, Recall, Accuracy, F-Measure, FDR, nDCG as 87.63, 92.45, 90.04, 89.97, 0.13, 90.54, respectively. HRL yields Precision, Recall, Accuracy, F-Measure, FDR, nDCG as 88.44, 91.15, 89.79, 89.77, 0.12, 85.63, respectively. SSA yields Precision, Recall, Accuracy, F-Measure, FDR, NDCG as 90.65, 92.84, 91.74, 91.73, 0.1, 88.45, respectively. OPCR yields Precision, Recall, Accuracy, F-Measure, FDR, nDCG as 92.16, 94.18, 93.17, 93.15, 0.08, 92.36, respectively, for 8192 queries. The proposed HSIL has the highest Precision, Recall, Accuracy, F-measure percentage of 95.18, 97.43, 96.30, 96.29, 0.05, 95.63, respectively. The proposed HSIL has the lowest FDR compared to PCRS, HRL, SSAL, OPCR of 0.05. The nDCG of the proposed model is highest when compared with the baseline models.

PCRS [1] includes static ontologies, as well as the learner profile, which are given more importance. Feeding is done for the individual instances obtained by the learner. Moreover, euclidean distance alone is used, which makes the relevance computation of PCRS much low, and the use of static ontologies makes sure that ontologies have to be readily separated to the system, which is a major disadvantage, lower the quality of ontologies the system is unable to maintain the standard of the recommendation results. Thus, PCRS performs very well when their ontologies are perfectly separated into the approach, while when ontologies become chaotic, the approach fails to yield the required performance. Thereby there is a need to tailoring the PCRS system and improving on the course recommendation.

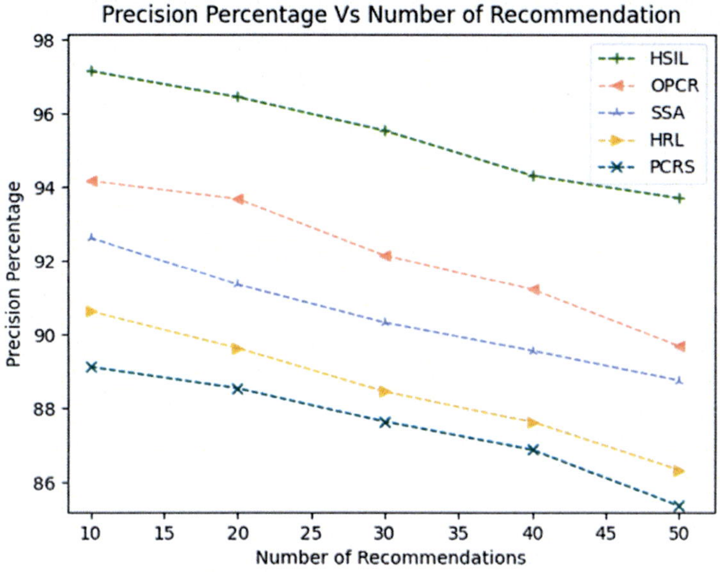

Fig. 2. Precision percentage vs no. of recommendations

HRL [2] includes hierarchical reinforcement learning. It is a course recommendation system that is mainly based on machine learning criteria. It fails to address a highly dense semantic environment as the HRL algorithm would make it highly complex when the dataset is highly dense and hierarchical in nature. So, courses users, as well as user behaviour, are taken into consideration which is a good feature criterion. While reinforcement learning ensures that the HRL is mainly dependent on the dataset, and if the dataset is not learned or feature engineering is not viable, the performance will be poor, resulting in either underfitting or overfitting problems. Moreover, it is highly difficult to learn all the relations, making the system computationally highly complex. As a result, the HRL must be improvised by making it semantically compliance as well as semantic incorporated machine learning would be much better with the lighter computational complexity. SSA [3] is a Semantic Similarity analysis-based course recommendation system that is semantically compliance while only the TFIDF word to vector along with cosine similarity is incorporated into the approach. The use of cosine similarity makes the semantic similarity analysis quite conventional and traditional. Moreover, TFIDF is a doc to vec to make sure that the results yielded a higher rate of irrelevance. As a result, there is a scope of improvement in SSA as well.

OPCR [4] is an Ontology-Based Personalized Course Recommendation Framework where dynamic ontology mapping makes it quite envious as well as quite efficient. But there is keyword similarity alone. The approach uses collaborative filtering with content-based filtering. The amalgamation of both collaborative and content-based filtering makes sure that when the approach is viable to collaborative filtering there is a need for rating and when the rating is given for each of the courses then it is highly difficult to choose a trade-off between the rating by the user or whether the relevance based on the ontology or term similarity. So as a result, this approach can definitely be improved by making it devoid of collaborative filtering.

The Fig. 2 graph depicts the comparison of Precision percentage vs No. of Recommendation for the proposed HSIL w.r.t the PCRS [1], HRL [2], SSA [3], OPCR [4], respectively. The proposed HSIL is based on metadata generation. So when the metadata is generated, a highly dense instance relevant to the dataset and the user's query term or the selection term is generated. The use of XGBoosting for classification makes the approach quite empowering. Also, XGBoosting is comparatively less complex, and ontologies of courses when they are used incorporated, definitely a wide arena of auxiliary knowledge with respect to the courses incorporated learner profile content, the user clicks to make sure that the personalization is incorporated and concept similarity when it is computed under genetic algorithm the relevance of result is quite high because of the large density of metadata as well as ontologies there is need of high degree of relevance, optimization algorithm like genetic algorithm is used under concept similarity so that the relevance of result is maintained and thereby the proposed approach is much better than the existing approaches.

6 Conclusion

The proposed HSIL was designed and successfully implemented, python 3.9.0 was used to implement HSIL architecture. Google Colab Notebook was the IDE with an i5 processor, and a processing speed of 1.2–3.6 GHz has been utilized to design and implement the architecture. Google Colab offers two different models of GPU NVIDIA K80/T4 for free usage and 100 GB of free space for data, and 16 GB of RAM to provide high-performance computing tasks. The XGBoost algorithm is implemented by using standard XGBoost libraries of python. Pre-processing was done using a natural language processing framework. The data set used in this architecture is scrapped from the Coursera website. It contains mainly six columns, i.e., cousrse_title, course_organization, courseCertificate-type, course_rating, course_difficulty, course students enrolled, and 890-course data is made categorical by performing annotations on it. Ontologies were modeled using protege by crawling several repositories, which consist of course as well as details of courses like standard websites udemy, Coursera using Beautiful Soup. Ontology is modeled using OntoCollab. Manual modeling of ontologies was also done. Indexes from the world wide web were also taken, relevant to e-learning and other learning domains. A large ontology was created, having 14916 concepts and sub-concepts and around 18421 individuals. The model ontology was implicitly axiomatized, and SWRL reasoning was also performed to maintain the ontologies' hierarchy.

References

1. Gulzar, Z., Anny Leema, A., Deepak, G.: PCRS: personalized course recommender system based on hybrid approach. In: 6th International Conference on Smart Computing and Communications, ICSCC 2017, 7–8 December 2017, Kurukshetra, India (2017)
2. Zhang, J., Hao, B., Chen, B., Li, C., Chen, H., Sun, J.: Hierarchical reinforcement learning for course recommendation in MOOCs. In: The Thirty-Third AAAI Conference on Artificial Intelligence (AAAI-19)
3. Ma, H., Wang, X., Hou, J., Lu, Y.: Course recommendation based on semantic similarity analysis. In: 2017 IEEE 3rd International Conference on Control Science and Systems Engineering
4. Ibrahim, M.E., Yang, Y., Ndzi, D., Yang, G., Almaliki, M.: Ontology-based Personalised Course Recommendation Framework IEEE Access. https://doi.org/10.1109/ACCESS.2018.2889635
5. Lynn, N.D., Emanuel, A.W.R.: A review on Recommender Systems for course selection in higher education. In: The 5th Annual Applied Science and Engineering Conference (AASEC 2020). https://doi.org/10.1088/1757-899X/1098/3/032039
6. Al-Badarenah, A., Alsakran, J.: An automated recommender system for course selection. (IJACSA) Int. J. Adv. Comput. Sci. Appl. **7**(3) (2016)
7. Imran, H., Hoang, Q., Chang, T.-W., Kinshuk, Graf, S.: A framework to provide personalization in learning management systems through a recommender system approach". ACIIDS 2014, Part I. LNAI, vol. 8397, pp. 271–280. Springer, Cham (2014). https://doi.org/10.1007/978-3-319-05476-6_28
8. Melville, P., Mooney, R.J., Nagarajan, R.: Content-boosted collaborative filtering for improved recommendations. In: Eighteenth National Conference on Artificial Intelligence (AAAI-2002) (2002)
9. Chongchong Zhao, and Aonan Cai., "The Similarity Calculation of Concept Name", 2016 2nd International Conference on Cloud Computing and Internet of Things (CCIOT)
10. Mondal, B., Patra, O., Mishra, S., Patra, P.: A course recommendation system based on grades. In: Conference: 2020 International Conference on Computer Science, Engineering and Applications (ICCSEA). https://doi.org/10.1109/ICCSEA49143.2020.9132845
11. Ray, S., Sharma, A.: A Collaborative Filtering Based Approach for Recommending Elective Courses
12. Deepak, G., Kasaraneni, D.: OntoCommerce: an ontology focused semantic framework for personalised product recommendation for user targeted e-commerce. Int. J. Comput. Aided Eng. Technol. **11**(4–5), 449–466 (2019)
13. Deepak, G., Rooban, S., Santhanavijayan, A.: A knowledge centric hybridized approach for crime classification incorporating deep bi-LSTM neural network. Multimed. Tools Appl. **80**(18), 28061–28085 (2021). https://doi.org/10.1007/s11042-021-11050-4
14. Deepak, G., Santhanavijayan, A.: UQSCM-RFD: a query–knowledge interfacing approach for diversified query recommendation in semantic search based on river flow dynamics and dynamic user interaction. Neural Computing and Applications, 1–25 (2021)
15. Krishnan, N., Deepak, G.: Towards a novel framework for trust driven web URL recommendation incorporating semantic alignment and recurrent neural network. In: 2021 7th International Conference on Web Research (ICWR), pp. 232–237. IEEE, May 2021
16. Roopak, N., Deepak, G.: OntoKnowNHS: ontology driven knowledge centric novel hybridised semantic scheme for image recommendation using knowledge graph. In: Iberoamerican Knowledge Graphs and Semantic Web Conference, pp. 138–152. Springer, Cham (2021). https://doi.org/10.1007/978-3-030-91305-2_11

17. Ojha, R., Deepak, G.: Metadata driven semantically aware medical query expansion. In: Iberoamerican Knowledge Graphs and Semantic Web Conference, pp. 223–233. Springer, Cham (2021). https://doi.org/10.1007/978-3-030-91305-2_17
18. Surya, D., Deepak, G.: USWSBS: user-centric sensor and web service search for IoT application using bagging and sunflower optimization. In: International Conference on Emerging Trends and Technologies on Intelligent Systems, pp. 349–359. Springer, Singapore (2021). https://doi.org/10.1007/978-981-16-3097-2_29
19. Arulmozhivarman, M., Deepak, G.: OWLW: ontology focused user centric architecture for web service recommendation based on LSTM and whale optimization. In: European, Asian, Middle Eastern, North African Conference on Management & Information Systems, pp. 334–344. Springer, Cham (2021). https://doi.org/10.1007/978-3-030-77246-8_32
20. Adithya, V., Deepak, G., Santhanavijayan, A.: HCODF: hybrid cognitive ontology driven framework for socially relevant news validation. In: Motahhir, S., Bossoufi, B. (eds.) ICDTA 2021. LNNS, vol. 211, pp. 731–739. Springer, Cham (2021). https://doi.org/10.1007/978-3-030-73882-2_66
21. Rithish, H., Deepak, G., Santhanavijayan, A.: Automated assessment of question quality on online community forums. In: Motahhir, S., Bossoufi, B. (eds.) ICDTA 2021. LNNS, vol. 211, pp. 791–800. Springer, Cham (2021). https://doi.org/10.1007/978-3-030-73882-2_72
22. Aditya, S., Muhil Aditya, P., Deepak, G., Santhanavijayan, A.: IIMDR: intelligence integration model for document retrieval. In: International Conference on Digital Technologies and Applications pp. 707–717. Springer, Cham (2021). https://doi.org/10.1007/978-3-030-73882-2_64
23. Tiwari, S., Al-Aswadi, F.N., Gaurav, D.: Recent trends in knowledge graphs: theory and practice. Soft Comput. **25**(13), 8337–8355 (2021)
24. Abhishek, K., Pratihar, V., Shandilya, S.K., Tiwari, S., Ranjan, V.K., Tripathi, S.: An intelligent approach for mining knowledge graphs of online news. Int. J. Comput. Appl., 1–9 (2021)
25. Tiwari, S., Gaurav, D., Srivastava, A., Rai, C., Abhishek, K.: A preliminary study of knowledge graphs and their construction. In: Emerging Technologies in Data Mining and Information Security: Proceedings of IEMIS 2020, vol. 3, pp. 11–20. Springer, Singapore (2021). https://doi.org/10.1007/978-981-15-9774-9_2

Image and Information Processing

MASSTagger: Metadata Aware Semantic Strategy for Automatic Image Tagging

Shivam Sawarn[1] and Gerard Deepak[1,2(✉)]

[1] Department of Computer Science and Engineering, DIT University, Dehradun, India
gerard.deepak.christuni@gmail.com
[2] National Institute of Technology, Tiruchirappalli, India

Abstract. As the social media usage is expanding and people are using more and more platforms to share their stories through pictures, the use of image tags is becoming more relevant in the social media world. On social media. It is often seen that the option of tagging the images that are uploaded. Since image tagging is optional to the users, many images are posted untagged. Now the problem which arises is without image tags these images are hard to find. To overcome this issue many automatic tagging tools have been proposed. But image tagging requires precision and accuracy so that the images are tagged automatically based on the content in them. The proposed methodology uses Structural Topic Modelling and semantic similarity to tag the images and classify them into their respective categories.

Keywords: Automatic · Component · Image tagging · Semantic similarity · Structural topic modelling

1 Introduction

The current scenario of social media and the number of images posted daily on the internet is so large that almost 300 million photos are uploaded on the inter- net every day and not all of them are annotated. The tags are used to classify the images uploaded into different categories and divide them into classes based on the content of the image and the context also. This encouraged the researchers to look for an automatic and optimal solution to ease the whole process. Automatic Image tagging based on various parameters have been implemented in past by many researchers to make the whole process more efficient. The main problem is the number of images coming on the internet every day. To tackle this problem Zhou et al. [1] presented a framework which was based on Hybrid Probabilistic model and consists of TIAM (Tag Image Association Matrix) and a collaborative filtering method based non-negative matrix factorization to handle the data sparsity in the matrix. Generally, the images which are tagged by the user contain noise and are biased. Qian et al. [2] presented an approach to re-tag the images based on the diverse semantic features of the images. The image tagging process has been manual for a very long time. Big social media sites and apps provide the feature of tagging your images manually according to the image content. To efficiently incorporate the

S. Motahhir and B. Bossoufi (Eds.): ICDTA 2022, LNNS 454, pp. 429–438, 2022.
https://doi.org/10.1007/978-3-031-01942-5_43

application of automatic image tagging many scholars have tried to propose an efficient framework. Zhou, N et al. An approach towards the Spider Mon- key algorithm was put forth by Sharma H et al. [3] which explained how the Spider Monkey algorithm works and how it is inspired by nature. An approach for continual prediction using the LSTM model was introduced by Gers F et al. [4]. It focused on how the LSTM model can reset itself whenever required. They called it the "forget gate". In the works of Sundermever et al. [5] a study was performed for language modelling. The language modelling was done for English and French language respectively and with the use of LSTM neural networks they were able to improve the whole network by 8%. Ma X et al. [6] presented an approach of bidirectional LSTM and the use of CNN. The ensembling of multiple techniques improved the results to 97%. A strategic algorithm of Se- mantoSim was put forth for the computation of semantic similarity. Wang S et al. [7] came up with a novel approach working with LSTM neural networks for making the model learn natural language processing. Another enhanced LSTM neural network was proposed was brought forward by Chen Q et al. [8] in which they parsed syntactic information due to which they obtained satisfactory results. Swami V et al. [9] The approach takes the content as well as the semantic compensations of the image to determine the suitable tag for a given list of im- ages. This framework aims at classifying the existing tags based on the content and the semantic similarity between them. The use of Spider Monkey Optimisation algorithm makes the whole model more robust and helps in achieving high accuracy and better results. [10–22] are the methods used to force the implied approach.

The remaining paper is organized as follows: Sect. 2 comprises of the methodology of the proposed framework and the technologies involved. Section 3 explains about the implementation of the whole algorithm based on the methodology. Section 4 comprises of Results and the Analysis based on the results. Section 5 concludes the paper.

2 Methodology and Proposed Architecture

2.1 Structural Topic Modelling

Structural Topic modelling is a type of Statistical Modelling which helps in dis- covering the different and abstract topics in various documents or in a collection of documents. Basically, it is a method for finding a specific set of words in a document or in a collection of documents that gives the idea about the con- tent of the document. In the proposed framework, this method is used to find a pattern of recurring words in a set of tags so that the initial information about the images can be extracted using the set of the tags in the dataset. Using the information and the metadata, tags can be classified using various algorithms. In this framework, LSTM classification model has been used to classify the tags based on the computed semantic similarity. Alternative document categorization and clustering approaches are less versatile and rely on subject matter experts than topic modelling. When using labels and techniques used in previous studies, it may, for example, reveal any number of topics concealed in a collection of incident reports that have been, or will be, assigned to the same group. STM is similar to LDA, but it improves the assignment of words to latent topics in a corpus by using metadata about documents (such as the author's name or the date the document was created).

2.2 LSTM

LSTM stands for Long Short-Term Memory is an artificial recurrent neural network used in deep learning. LSTM does not work like standard forward feed like frameworks. It has feedback connections due to which it can process sequences of data like textual data, video. In this framework, the metadata generated from the extracted tags of the images will be classified according to the enriched initial tags. The LSTM model uses enriched initial tags which are derived from the Topic Modelling and they use that enriched initial tags to classify the tags and then calculate the semantic similarity between the tags using NPMI and Normalized Google Distance. The LSTM is used along with the sigmoid activation which helps the network to select which information is required and which information is garbage and based on that it updates the cell and selects the output which is to be shown to the user. LSTM has user gates that can control the flow of information in this framework.

$$q_p = \sigma\left(W_q x_p + U_q h_{(p-1)} + b_q\right) \tag{1}$$

A supervised RNN with LSTM units can be trained on a collection of training sequences using an optimization algorithm, such as gradient descent, combined with back propagation through time to compute the gradients required during the optimization process, in order to adjust each weight of the LSTM network in proportion to the derivative of the error (at the output layer of the LSTM network) with back propagation through time as shown in Eq. 1.

2.3 Monkey Optimization

The foraging behaviour of spider monkeys influenced SMO. The problem's solutions are defined by spider monkey food sources. The superiority of a food source will be determined based on the measured fitness value. There are four control parameters in the SMO algorithm: Local Leader Limit, Global Leader Limit, Maximum Group (MG), and Perturbation Rate (pr). SMO is a meta- heuristic strategy influenced by spider monkeys' intelligent foraging behaviour. Spider monkeys' foraging activity is founded on a fission-fusion social system. The characteristics of this algorithm are determined by the social organization of a group in which a female leader decides whether to divide or merge the group. The national leader of the party is referred to as the global leader, while the leaders of the small groups are referred to as local leaders. The phenomenon of food scarcity is characterized by no change in the solution, according to the SMO algorithm. Since SMO is focused on swarm intelligence, each small group should have a set number of monkeys.

2.4 Normalized Google Distance

The Normalized Google Distance (NGD) is a semantic similarity metric derived from the number of hits returned by Google for a collection of keywords. Key- words with similar or identical meanings in natural language are "near" in terms of Normalized Google Distance, while words with dissimilar meanings are "far" away. The probabilities of Google search terms, calculated by dividing the frequencies of page counts returned by Google by the number of pages indexed by Google (multiplied by the average number of search terms in those pages) are close to the real relative frequencies of those search terms as used in society.

In Eq. 2, the Normalised Google distance is represented by relationships based on this assumption. Based on this assumption, the normalised Google distance represents the presumed true semantic relations governing the search terms to a reasonable degree. The World Wide Web and Google are included in the NGD.

The model starts with extracting the initial tags of the image and generating metadata using those tags. After that pre-processing is done using various methods. Processes like stemming, lemmatization, tokenization, stop words removal. After that using Named Entity Recognition the tags are categorized. After that structured topic modelling is done on the processed data and a few words are extracted from the tag which are useful information and can be used to classify the tags. In the next phase, the initial tags are enriched and they are used along with metadata to classify them using the Long Short Term Memory algorithm. Using enriched initial tags, the semantic similarity between the tags is calculated using monkey optimization. Using monkey optimization, this whole process is optimized. The semantic similarity is calculated using NPMI and Normalised Google Distance. After calculating the semantic similarity, the tags are arranged until the increasing order of the semantic similarity.

3 Implementation

The framework has been tested on Linux Mint operating system having 16 GB RAM and an i7 processor. The IDE used for the development of the algorithm was Spyder under Anaconda Environment. The model works by extracting the initial tags and generating metadata from them which is then classified using the LSTM classification model. On the other side, Structural topic modelling is done to extract the recurrent words that are in the document. After that, semantic similarity is calculated using NPMI and Normalized Google Distance which finally is combined with the top 2% of the classified metadata extracted from the tags and then they are arranged in increasing order with respect to the calculated semantic similarity. The algorithm for the proposed model is mentioned in Algorithm 1.

Algorithm 1 Proposed Algorithm for MASSTagger

Input - 7896 Queries whose ground truth has been collected. Output- Top 2% classified tags based on the metadata and thesemantic similarity between the tags.
Begin
Step 1 - Pre-processing of the tags are done after extracting the initial tags.
Stop word removal, lemmatization, stemming methodsare used to pre-process the data.
Step 2 - Metadata is generated after extracting the initial tags.
Step 3 - The initial tags are enriched after the structural topic modellingis done on the pre-processed tags.
Step 4 - The metadata is classified using the LSTM model and the semantic similarity between them is calculated.
Step 5 - The semantic similarity is calculated using NPMI and Normalized google distance.
Step 6 - Top 2% tags are obtained and then rearranged in increasing order.
Step 7 - The whole algorithm is optimized using spider monkey an algorithm whosepseudo-code is mentioned:
Part I: Set parameters and initialize
Randomly generate the initial population of NP positions of the monkeys (X), i=1,2,..................... NP, and j=1,2,..NV
Define low and high boundaries of Xi Assign values to parameters: a, b, c and d**Part II: Algorithm loops**
For I = 1: Imax (max. Iteration)
Climb process loop
For counter 1= 1: Nc (number of cycles)Randomly generate vector AX(1)
Generate the pseudo-gradient of the objective function f (Xi) Generate new monkeys' position Yi using a and f
Update Xi with Yi if feasible

End

For Nc

Watch-jump process
Generate new Yi(I) from (Xi-b, Xi+b)
If -f -f
Update Xi(I) with Yi(I) if feasible (i.e., within limits)
End If
Apply climb process loop again
Rank the positions and find the current best solution so far
Somersault process
Estimate Pivot P(I) from Xii, c and dCalculate Yi(I) around the pivot
Update Xi(I) with Yi(I) if feasible and repeat somersault until feasibleRank the positions and find the current best solution
If (the new monkey's position violates its boundary limits)Use Search-based function to handle constraints
Retinas the highest mountaintop (7/(X)) and its position X
End

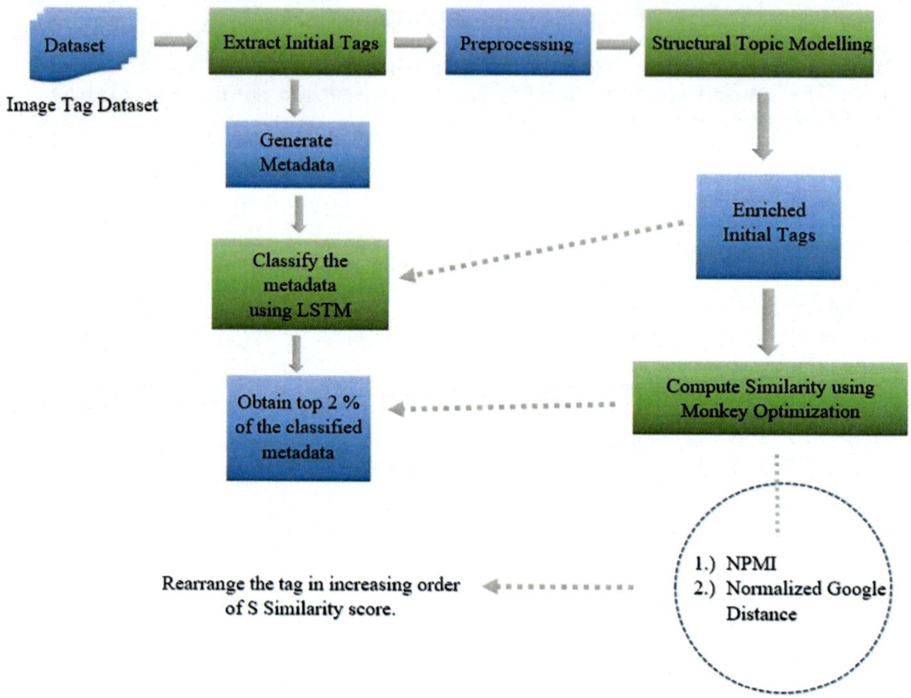

Fig. 1. System architecture of MASSTagger

4 Results and Performance Analysis

From Table 1, it can be inferred that as compared to ATWCA and ITRMR which are the base frameworks for this model, the proposed model performs better. The reason is the proposed framework takes in account the metadata of the image tags. Due to which it is easy to identify and classify them accordingly. As compared to the method used in [13], the proposed framework uses Spider Monkey Optimization Algorithm after calculating the semantic similarity which makes the whole process more robust. Standard formulation of Average precision, average recall, accuracy, F-measure and False Discovery Rate curve have been used to test and analyse the performance of the implied approach.

Table 1. Comparison of performance of the proposed HBLOGREC with other approaches

Search technique	Average precision %	Average recall %	Accuracy %	F-Measure	FDR	nDCG
ATWCA [1]	74.85	78.36	75.18	38.28	0.21	0.79
ITRMR [3]	76.62	79.36	77.81	38.98	0.21	0.78
HMAIA [2]	82.28	86.21	83.21	42.09	0.14	0.81

(continued)

Table 1. (*continued*)

Search technique	Average precision %	Average recall %	Accuracy %	F-Measure	FDR	nDCG
Image Tagging using Naive Bayes and Ontologies	80.12	83.26	81.87	40.82	0.17	0.91
Image Tagging using fuzzy c-means clustering and CNN	85.21	88.39	86.81	43.33	0.12	0.83
Proposed MASSTagger	94.27	97.21	95.86	47.85	0.05	0.93

The algorithm was tested for 7896 Queries whose ground truth has been collected. The results indicate that the proposed framework outperforms all the compared frameworks. The accuracy of the proposed framework is 95.86% which indicates that the model works exceptionally well on the tested dataset and has a great potential to get implemented on the real-world dataset. The main rea- son behind the high accuracy and the precision of the model is the use of two different criteria to classify the tags and using monkey optimization algorithm to calculate the semantic similarity between the tags indicating the relationship between Recall and No. of Recommendations shows that the general trend is the Recall will decrease as the number of recommendations increase, but even with that trend the proposed framework shows a high Recall indicating the model's better performance as compared to other modes.

Moreover, all the baseline models give predictions for a lesser number of data points while a considerable amount of data has been taken into the course of action for carrying out experimentation on the proposed system. Furthermore, this novel strategy of ensembling all the strategies mentioned prior together contributes toward enhancing model accuracy and overall performance. Each algorithm complements the other and hence, aids to the proposed strategy being liable enough to outperform all the existing baseline techniques. It can be observed from Fig. 2. that the proposed framework outperforms all the compared models. The Graph explains the relationship between the No. of recommendations and Recall and according to the plot, the proposed framework yields very high recall even after the no. of recommendations are increasing. The other compared framework and was unable to achieve a similar trend.

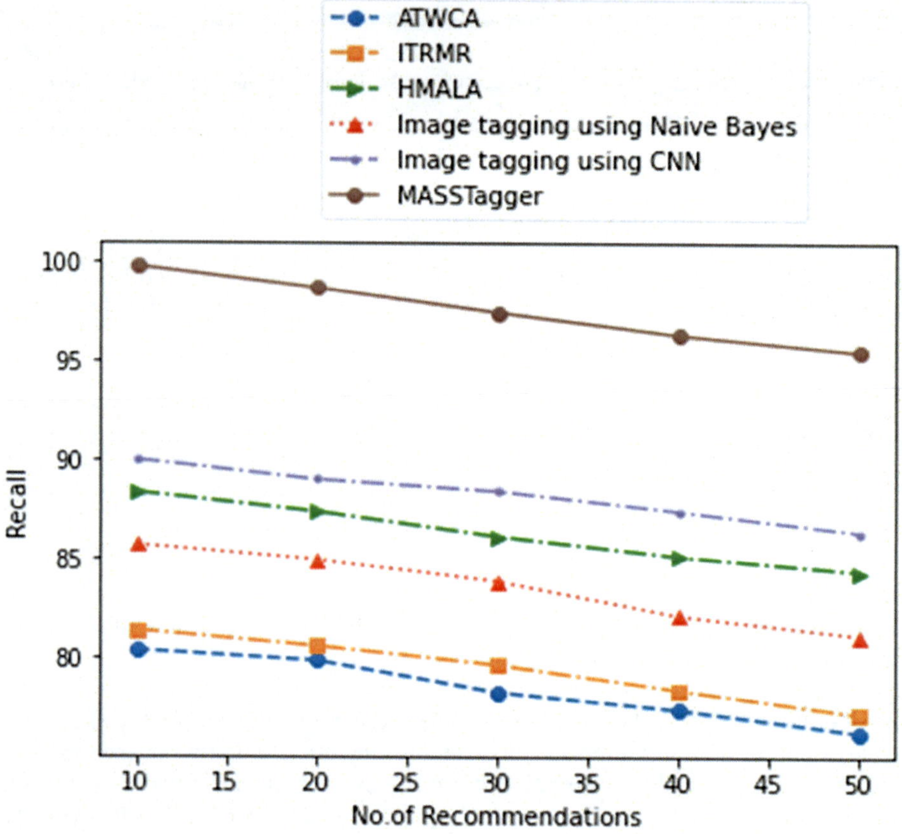

Fig. 2. Performance comparison based on recall

5 Conclusions

An ingenious automatic mass image tagger method is proposed. The method includes the use of various techniques working in a single framework contributing to the better performance of the model. The use of metadata extracted from the initial tags gives the model an extra criterion to classify the tags. Using the extracted metadata and by computing the semantic similarity between the different and multiple tags, the model classifies them using the LSTM classification model. The model uses NPMI and Normalized Google Distance to measure the semantic similarity between the tags. The algorithm uses the spider monkey algorithm to optimize the whole framework and provide better results. The ensembling of multiple algorithms makes the model more robust and helps reduce the statistical noise. The average precision of 94.27% and average Recall of 97.21 indicates that the model outperforms the compared frameworks. The average accuracy achieved by the model is 95.86 and the average F-Measure is47.85. All these estimations help in arriving at an approximation of MASSTagger being the best-in-class model for image tagging based on metadata and semantic similarity. The further work which can be done on this system could probably be to focus on raising the relevance tally on the whole. Moreover, distinct techniques of semantic similarity or priority estimation can be used to achieve even better values for evaluation metrics. An ensemble of different machine learning classifiers can be tried out to achieve better results.

References

1. Zhou, N., Cheung, W.K., Qiu, G., Xue, X.: A hybrid probabilistic model for unified collaborative and content-based image tagging. IEEE Trans. Pattern Anal. Mach. Intell. **33**(7), 1281–1294 (2010)
2. Qian, X., Hua, X.S., Tang, Y.Y., Mei, T.: Social image tagging with diverse semantics. IEEE Trans. Cybern. **44**(12), 2493–2508 (2014)
3. Sharma, H., Hazrati, G., Bansal, J.C.: Spider monkey optimization algorithm. In: Bansal, J., Singh, P., Pal, N. (eds.) Evolutionary and Swarm Intelligence Algorithms. SCI, vol. 779, pp. 43–59. Springer, Cham (2019). Doi: https://doi.org/10.1007/978-3-319-91341-4_4
4. Gers, F.A., Schmidhuber, J., Cummins, F.: Learning to forget: Continual prediction with LSTM (1999)
5. Sundermeyer, M., Schlu¨ter, R., Ney, H.: LSTM neural networks for language modelling. In: Thirteenth Annual Conference of the International Speech Communication Association (2012)
6. Ma, X., Hovy, E.: End-to-end sequence labelling via bi-directional lstm-cnns-crf (2016). arXiv preprint arXiv:1603.01354
7. Wang, S., Jiang, J.: Learning natural language inference with LSTM (2015). arXiv preprint arXiv:1512.08849
8. Chen, Q., Zhu, X., Ling, Z., Wei, S., Jiang, H., Inkpen, D.: Enhanced lstm for natural language inference (2016)
9. Swami, V., Kumar, S., Jain, S.: An improved spider monkey optimization algorithm. In: Soft Computing: Theories and Applications, pp. 73–81. Springer, Singapore (2018)
10. Deepak, G., Gulzar, Z., Leema, A.A.: An intelligent system for modelling and evaluation of domain ontologies for Crystallography as a prospective domain with a focus on their retrieval. Computers & Electrical Engineering, 107604 (2021)
11. Roopak, N., Deepak, G.: OntoKnowNHS: ontology driven knowledge centric novel hybridised semantic scheme for image recommendation using knowledge graph. In: Iberoamerican Knowledge Graphs and Semantic Web Conference, pp. 138–152. n. In IECON 2016-42nd Annual Conference of the IEEE Industrial Electronics Society (pp. 5657–5662). IEEE, November 2021
12. Ojha, R., Deepak, G.: Metadata driven semantically aware medical query expansion. In: Iberoamerican Knowledge Graphs and Semantic Web Conference, pp. 223–233. Springer, Cham (2021). Doi: https://doi.org/10.1007/978-3-030-91305-2_17
13. Deepak, Gerard, Gulzar, Z., Leema, A.A.: An intelligent system for modelling and evaluation of domain ontologies for Crystallography as a prospective domain with a focus on their retrieval (2021)
14. Yethindra, D.N., Deepak, G.: A Semantic Approach for Fashion Recommendation Using Logistic Regression and Ontologies. In 2021 International Conference on Innovative Computing, Intelligent Communication and Smart Electrical Systems (ICSES), pp. 1–6. IEEE, September 2021
15. Adithya, V., Deepak, G.: HBlogRec: A Hybridized Cognitive Knowledge Scheme for Blog Recommendation infusing XGBoosting and Semantic Intelligence, July 2021
16. Krishnan, N., Deepak, G.: Towards a Novel Framework for Trust Driven Web URL Recommendation Incorporating Semantic Alignment and Recur- rent Neural Network, May 2021

17. Sawarn, S., Deepak, G.: An Approach for Document Clustering Using Semantic Similarity and Whale Optimization. In: Musleh Al-Sartawi, A.M., Razzaque, A., Kamal, M.M. (eds.) EAMMIS 2021. Lecture Notes in Networks and Systems, vol 239, pp. 322–333. Springer, Cham (2021). https://doi.org/10.1007/978-3-030-77246-8_31
18. Krishnan, N., Deepak, G.: KnowSum: knowledge inclusive approach for text summarization using semantic alignment. In: 2021 7th International Conference on Web Research (ICWR), pp. 227–231 (2021)
19. Pushpa, C.N., Deepak, G., Kumar, A., T. J. and V. K.R.: OntoDisco: Improving web service discovery by hybridization of ontology focused concept clustering and interface semantics. In: 2020 IEEE International Conference on Electronics, Computing and Communication Technologies (CONECCT), pp. 1–5 (2020)
20. Adithya, V., Deepak, G.: OntoReq: an ontology focused collective knowledge approach for requirement traceability modelling. In: Musleh Al-Sartawi, A.M., Razzaque, A., Kamal, M.M. (eds) Artificial Intelligence Systems and the Internet of Things in the Digital Era. EAMMIS 2021. Lecture Notes in Networks and Systems, vol 239. Springer, Cham (2021). https://doi.org/10.1007/978-3-030-77246-8_34
21. Surya, D., Deepak, G., Santhanavijayan, A.: KSTAR: a knowledge based approach for socially relevant term aggregation for web page recommendation. In: Motahhir S., Bossoufi B. (eds) ICDTA 2021. Lecture Notes in Networks and Systems, vol 211. Springer, Cham (2021). https://doi.org/10.1007/978-3-030-73882-2_50
22. Vaish, K., Deepak, G., Santhanavijayan, A.: DSEORA: integration of deep learning and metaheuristics for web page recommendation based on search engine optimization ranking. In: Shetty, N.R., Patnaik, L.M., Nagaraj, H.C., Hamsavath, P.N., Nalini, N. (eds.) Emerging Research in Computing, Information, Communication and Applications. LNEE, vol. 790, pp. 873–883. Springer, Singapore (2022). https://doi.org/10.1007/978-981-16-1342-5_69

Optimization of Meixner Moments by the Firefly Algorithm for Image Analysis

Ahmed Bencherqui[1](✉), Mohamed Amine Tahiri[2], Hicham Karmouni[1,2], Achraf Daoui[1], Mohammed Alfidi[1], Mohamed Ouazzani Jamil[3], Hassan Qjidaa[2], and Mhamed Sayyouri[1]

[1] Engineering, Systems and Applications Laboratory, National School of Applied Sciences, Sidi Mohamed Ben Abdellah University, Fez, Morocco
ahmed.bencherqui@usmba.ac.ma

[2] CED-ST, STIC, Laboratory of Electronic Signals and Systems of Information LESSI, Dhar El Mahrez Faculty of Science, Sidi Mohamed Ben Abdellah-Fez University, Fez, Morocco

[3] Systems and Sustainable Environment Laboratory (SED), Faculty of Engineering Sciences (FSI) Private, University of Fez (UPF), Fez, Morocco

Abstract. Discrete orthogonal polynomials are widely used for image processing, these polynomials are characterized by parameters. However, it was very important to optimize these parameters to obtain a good result, in this context; this study suggests a new approach to optimize Meixner polynomial (*MPs*) parameters (β, u) based on the Firefly Algorithm (FA), and we use them to reconstruct 2D images. Using the FA algorithm determines the optimal parameters of the *MPs*, which improves the quality of the reconstruction. The simulation results, on the one hand, suggest that the proposed method based on the FA algorithm offers a high image reconstruction quality. On other hand, the comparison carried out with other algorithms indicates the advantage of the proposed method.

Keywords: Firefly algorithm · Meixner polynomial · Meixner moments · 2D images reconstruction

1 Introduction

The orthogonal moment has been widely used to characterize images in a variety of image processing applications, including classification [1], reconstruction [2], compression [3], video encoding [4], indexing of images [5], and watermarking [6]. Orthogonal moments are more powerful when representing images with low information redundancy and good noise resistance than non-orthogonal moments. The orthogonal moments are divided into two types, continuous and discrete. For the continuous, we find, for example, Zernike [7] Legendre [8], Jacobi [9], Gaussian Hermite [10], Fourier-Mellin [11], and Bessel-Fourier [12], these moments lead to errors of approximation and discretization, which decreases the quality of the reconstructed image. These errors encountered when using continuous moments are reduced by several researchers who have introduced discrete orthogonal moments, such as Charlier [13], Tchebichef [14], Meixner [15], Hahn [16],

S. Motahhir and B. Bossoufi (Eds.): ICDTA 2022, LNNS 454, pp. 439–448, 2022.
https://doi.org/10.1007/978-3-031-01942-5_44

Krawtchouk [17], Racah [18] and Dual-Hahn [19]. The efficiency of discrete orthogonal Meixner moments (*MMs*) has been confirmed by their extensive use in image processing applications [20]. At the beginning of this article, we developed the recursive relation of the *MPs* to the order n, and we suggested the usage of the modified Gram - Schmidt Ortho-normalization (MGSOP) [21], for solving the problem of numerical instability of *MPs*.

The orthogonal *MPs* are characterized by coefficients β and u, and the choice of these coefficients remains arbitrary since the optimal selection of the values of the *MPs* coefficients (β, u) allows to obtain a high-quality reconstruction. Therefore, we propose to use the Firefly algorithm (FA) [22], to optimize these parameters (β, u). The concept of the FA is mainly founded on the reduction of the objective function (MSE) to acquire an optimal parameter that, allows the reconstruction of 2D images with excellent quality.

The organization of the rest of this paper is as follows: The second section is devoted to presenting the solution of the problem which concerns the numerical instability during *MPs* computation, as well as the theoretical calculation of the two-dimensional Meixner moments. In the following section, we explain the adopted optimization technique which is mainly based on the use of the FA algorithm to determine the *MPs* parameters (β, u), and we used it for image reconstruction. Finally, simulation results are presented to verify the robustness of the FA-optimization method in 2D image reconstruction.

2 Meixner Discrete Orthogonal Moments

In this paragraph, we will explain the mathematical computation of discrete orthogonal *MPs*, including the recursive calculation, and we introduce the discrete orthogonal Meixner moment and its two-dimensional reconstruction.

2.1 Discrete Orthogonal Meixner Polynomials

The nth *MPs* are determined as a hypergeometric function of a discrete variable x specified in the interval $[0; N-1]$, with the two reals β and u defined as $0 < u < 1$ and $\beta > 0$, conforming to the following equation:

$$M_n^{(\beta,\mu)}(x) = (\beta)_n \, {}_2F_1\left(-n, -x, \beta; 1 - \frac{1}{\mu}\right) \tag{1}$$

The hypergeometric function ${}_2F_1(.)$ is defined by the following relation:

$${}_2F_1\left(\begin{matrix} x_1, x_2 \\ y_1 \end{matrix} | z\right) = \sum_{k=0}^{\infty} \frac{(x_1)_k (x_2)_k}{(y_1)_k} \cdot \frac{z^k}{k!} \tag{2}$$

It is worth mentioning that in the following we adopt the normalized *MPs*, which satisfies the recurrence relation for order n Eq. (3), to solve the problem of numerical instability of *MPs* values. The initial values of *MPs* Eqs. (4 and 5) and the recursive calculation of the *MPs* is summarized in Algorithm 1.

$$\tilde{M}_n^{(\beta,\mu)}(x) = A\tilde{M}_{n-1}^{(\beta,\mu)}(x) + B\tilde{M}_{n-2}^{(\beta,\mu)}(x) \tag{3}$$

where:

$$\tilde{M}_0^{(\beta,\mu)}(x) = \sqrt{\frac{\beta + x - 1}{x}}\tilde{M}_0^{(\beta,\mu)}(x-1) \text{ With } \tilde{M}_0^{(\beta,\mu)}(0) = \sqrt{(1-\mu)^\beta} \tag{4}$$

$$\tilde{M}_1^{(\beta,\mu)}(x) = \left(\beta + x - \frac{x}{\mu}\right)\sqrt{\frac{\mu}{\beta}}\tilde{M}_0^{(\beta,\mu)}(x) \tag{5}$$

Algorithm 1: The recursive computation of MPs to the order n

Inputs: ***N:*** **The maximum value of the variable** ***x***, $N_{\max}$ **:** ***MPs*** **order,** β and μ **:** ***MPs*** **parameters**

Output: ***MPs*** **matrix**

$\tilde{M}_0^{(\beta,\mu)}(0) = \sqrt{(1-\mu)^\beta}$

for $x \leftarrow 1$ to $N-1$ **do**

  Calculate $\tilde{M}_0^{(\beta,\mu)}(x)$ **by Eq. (4)**

end for

for $x \leftarrow 0$ to $N-1$ **do**

  Calculate $\tilde{M}_1^{(\beta,\mu)}(x)$ **by Eq. (5)**

end for

for $n \leftarrow 2$ *to* $N_{\max} - 1$ **do**

  for $x \leftarrow 0$ to $N-1$ **do**

  Calculate $\tilde{M}_n^{(\beta,\mu)}(x)$ **by Eq. (3)**

  end for

end for

It should be noted that the use of Algorithm 1 for the calculation of the values of *MPs* guarantees only numerical stability. However, the problem of numerical error propagation remains, and the orthogonality property changes. In the light of what has been mentioned so far, to address these issues, we will use the modified Gram-Schmidt orthogonalization method (MGSOP) [21], which makes it possible to considerably reduce the numerical errors and consequently to maintain the orthogonality property for the high orders.

We tested the efficiency of the new algorithm for great order using some 2D plots. Figure 1 shows the values of the *MPs* calculated using Algorithm 1 and MGSOP for **a)** $\beta = 1400$ and $u = 0.5$, **b)** $\beta = 1000$ and $u = 0.5$, **c)** $\beta = 550$ and $u = 0.5$. From this figure, we can observe that: i) the *MPs* values are well calculated and orthogonal, which indicates the recursive calculation and MGSOP overcomes the problem of the numerical error propagation of *MP*s for large orders. ii) Changing the *MPs* parameters directly affects the signal, which leads to a change in the region of interest.

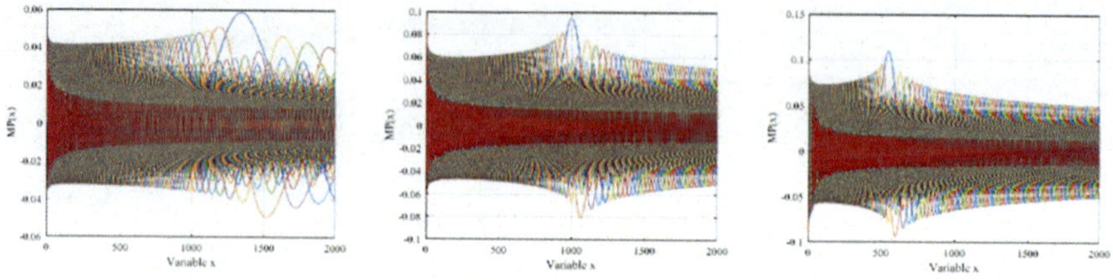

a) β=1400 and u=0.5 b) β=1000 and u=0.5 c) β =550 and u=0.5

Fig. 1. The 2D MPs plot for the different values of β and u.

2.2 Meixner's Discrete Orthogonal Moments

Mathematically, Meixner's discrete orthogonal moments (*MMs*) for an image $f = \{f(x, y)\}_{x,y=0}^{x,y=N-1}$ of size $N \times M$ are defined by the matrix form:

$$MM_{nm} = MP_n^T \times f \times MP_m \tag{6}$$

$$\text{With} \begin{cases} MP_n = \left\{\tilde{M}_n^{(\mu,\beta)}(x)\right\}_{n,x=0}^{n,x=N-1} \\ MP_m = \left\{\tilde{M}_m^{(\mu,\beta)}(y)\right\}_{m,y=0}^{m,y=M-1} \end{cases}$$

To reduce the computation time of the reconstructed 2D image, we adopt the following matrix form:

$$\hat{f} = MP_n \times MM_{nm} \times MP_m^T \tag{7}$$

It should be noted that the difference and the performance between the reconstructed 2D image and the original image would be defined by the following evaluation criteria Mean-Square Error (MSE) Eq. (8) and Signal-to-Noise Ratio (PSNR) Eq. (9):

$$MSE = \frac{1}{N \times M} \sum_{x=0}^{N-1} \sum_{y=0}^{M-1} \left(f(x, y) - \hat{f}(x, y)\right)^2 \tag{8}$$

$$PSNR = 10 \log_{10} \left(\frac{\sum_{x=1}^{N} \sum_{y=1}^{M} [f(x, y)]^2}{\sum_{x=1}^{N} \sum_{y=1}^{M} \left[f(x, y) - \hat{f}(x, y)\right]^2} \right) \tag{9}$$

Following the presentation of the *MPs* and their stable computation, it should be highlighted that, theoretically, the optimal selection of the parameters β and u is critical to fully profit from this polynomial. In this regard, we will offer a new optimization approach based on FA in the next section.

3 Optimal Reconstruction of MMs by the Firefly Algorithm (FA) for 2D Images

To reconstruct a 2D image, we use the discrete orthogonal Meixner moments which are computed by the *MPs* at order n, this polynomial is influenced by the local variables (β, u). Therefore, to get a high reconstruction quality of a 2D image, these parameters must be optimized, for this reason, we offer as a solution, to use the Firefly algorithm (FA) [22], which is a bio-inspired optimization algorithm and belongs to the family of metaheuristics, discovered in 2007 by Dr. Xin-She Yan of Cambridge University. The principle of this algorithm is based on the behavior of a swarm of fireflies in nature, and the idea of attraction between fireflies. And it also has a lot in common with other metaheuristics based on collective intelligence, namely the PSO (Particle Swarm Optimization) and Bee Colony Optimization (ABC) algorithm, but according to recent bibliographies, the FA algorithm has been able to show that it is more efficient than other methods at solving optimization issues.

The following steps constitute the development technique of our FA algorithm for image reconstruction.

Step 1: Random initialization of the fireflies.

The initial values used in our algorithm are.:

- Numbers of fireflies, $N = 20$
- Attractiveness coefficient, $\beta 0 = 1$
- The absorption coefficient, $\gamma = 1$
- The maximum number of iterations is, $T = 100$

Step 2: Calculate each Firefly's intensity using the following formula:

$$I(r) = I_0 e^{-\gamma r} \tag{10}$$

Step 3: Compare the light intensity of each firefly xi, with the light intensity of other fireflies, xj.

Step 4: Move each firefly i to the brightest fireflies, and updated the position of each firefly by the following equation:

$$x_i = x_i + \beta_0 e^{-\gamma r}(x_j - x_i) + \alpha(rand - 1/2) \tag{11}$$

Step 5: Evaluate the objective function (MSE) for any firefly xi, and update its light intensity.

Step 6: Sort the fireflies and choose the best one.

Step 7: Steps 2 through 6 should be repeated until the shutdown conditions are fulfilled.

Figure 2 summarizes the steps in implementing the FA algorithm for the image reconstruction.

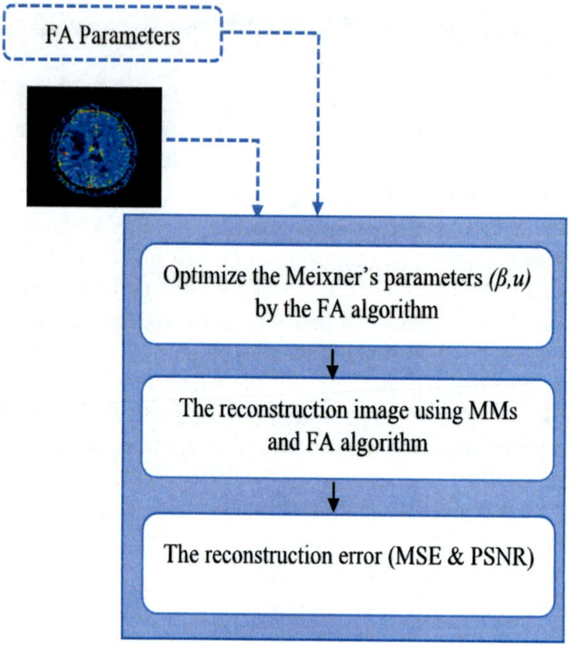

Fig. 2. The steps of the reconstruction method using the FA algorithm

4 Simulation Results

The robustness of the proposed reconstruction method on medical images is evaluated in this paragraph. The first test compares our method to the conventional method which uses a fixed value of the polynomial parameters, while the second test is devoted to proving the superiority of the proposed technique for different orders of reconstruction (10, 100, 500, and 1024). In the last test, FA was compared to optimization algorithms namely: Bee Colony Algorithm (ABC), Ant Colony Optimization (ACO), Differential Evolution (DE), Learning Teaching Based Optimization (TLBO), and Particle Swarm Optimization (PSO).

Firstly, we compare the performance of the proposed method based on *MMs* and the FA to the classical method based on *MMs* only with $\beta = 500$ and $u = 0.5$. Figure 4 illustrates the MSE and PSNR curves for the "Abdominal Aortic" 2D medical image reconstruction using both methods. From this figure, we observe that MSE decreases as the order of reconstruction increases, approaching zero.

The quality of the reconstruction increases with increasing Meixner moment order, and when we reach a maximum reconstruction order value, the reconstructed image and the original image become very similar. On the other hand, as the reconstruction order increases, the PSNR increases until it reaches the optimal value so that the reconstructed image becomes closely similar to the original image.

Furthermore, it is appreciated that the FA algorithm allows reaching high values of PSNR with a minimal MSE compared to the conventional method, which justifies the robustness of the optimization method based on the FA algorithm for 2D image reconstruction.

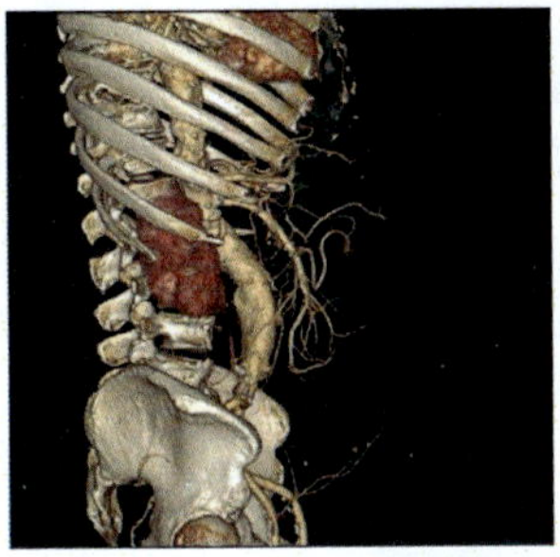

Fig. 3. Abdominal aortic 2D image 500 × 500.

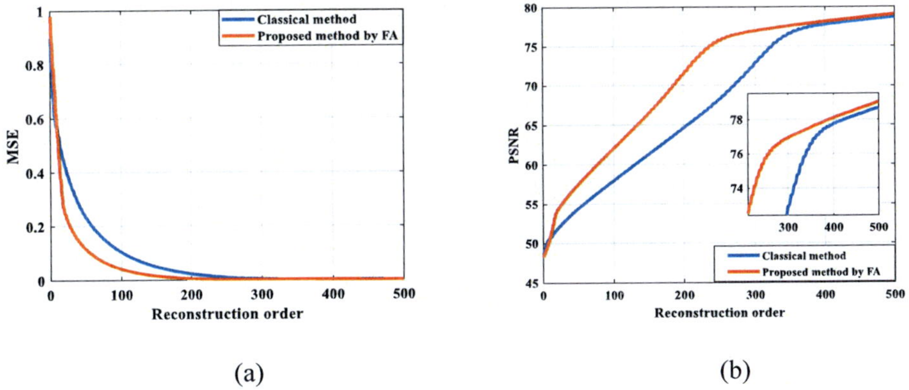

Fig. 4. Curves of: (a), MSE, and (b): PSNR for the proposed method and the classical method.

To also test the ability of the FA optimization method to reconstruct medical images, we have chosen to use true color images: "LEG" of sizes 1024 × 1024, which is extracted from the database [23]. We have reconstructed this image using the *MMs* optimized by the FA algorithm for orders ranging from 10 up to 1024.

Table 1 presents a set of images reconstructed by the *MMs* using the FA algorithm for the orders $n = 10$, $n = 100$, $n = 500$, and $n = 1024$. This table also presents the values of the local parameters (β, u) of the *MPs* calculated using the FA algorithm for each order and the MSE and PSNR reconstruction errors. From the reconstruction results obtained in Table 1, we observe that for each order, the FA algorithm determines the optimal values of the parameters (β, u) which make it possible to reconstruct the image with a very low MSE.

Indeed, we see that the analogy of the reconstructed image with the original image increases with the growth of orders of *MMs* (for order 10 we have MSE = 0.264299, for order 500 MSE = 0.00249634, and the order 1024 MSE = 0.000547) which justifies the superiority of our optimization method for the reconstruction of medical images.

Table 1. Reconstructed image from the original image "LEG" by the proposed method for different orders

Reconstruction order	MMs via FA	(β, u)	Reconstruction errors
(10, 10)	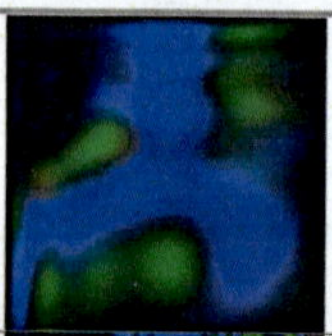	$\beta = 16.572$ $u = 0.9388$	$MSE = 0.264299$ $PSNR = 53.9098$
(100, 100)	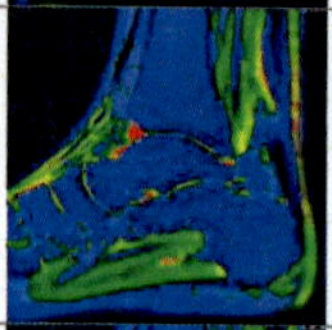	$\beta = 69.51721$ $u = 0.68591$	$MSE = 0.03107$ $PSNR = 63.2065$
(500, 500)	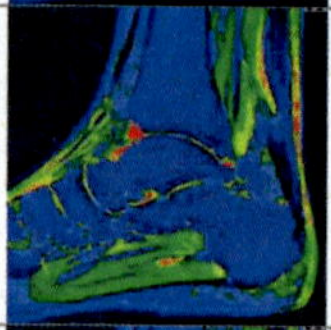	$\beta = 96.7046$ $u = 0.63474$	$MSE = 0.00249$ $PSNR = 74.157$
(1024, 1024)	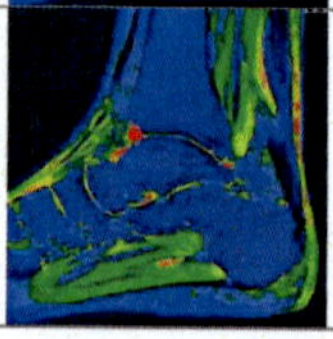	$\beta = 100$ $u = 0.6471$	$MSE = 0.000547$ $PSNR = 80.7463$

On the other hand, we have again validated the high efficiency of the proposed image reconstruction method, and this time we compare the PSNR reconstruction errors obtained during the reconstruction using the FA algorithm with the other errors obtained by the optimization algorithms: ACO, PSO, ABC, DE, and TLBO.

Error-values (PSNR) of the "Diabetic Retinopathy" image reconstructed by FA-optimized *MMs* and other optimization algorithms are displayed in Fig. 5. Through this figure, we can observe that the reconstruction errors based on our optimization method are much more optimal than those obtained by the other algorithms.

Note that the PSNR values for FA, ACO, PSO, ABC, DE, and TLBO are 79.64085, 71.6998, 72.4476, 71.6347, 72.17002, and 72.6202 for the 2D image, respectively. These results show that our method gives excellent results (high PSNR), the thing that allows us to obtain a high-performance reconstruction quality.

5 Conclusion

In this paper, we explained the discrete orthogonal *MPs* and their moments, then we solved the problem of selecting the local Meixner coefficients (β, u) using the FA algorithm, and we used it to reconstruct 2D medical images. And we confirmed the efficiency of our proposed method for 2D image reconstruction, which is shown in the simulation results.

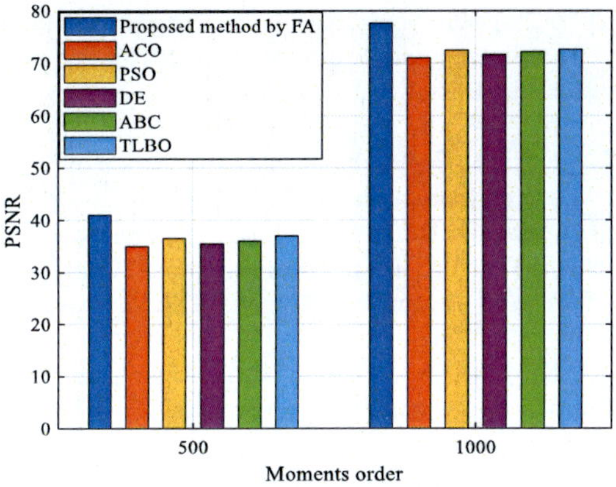

(a)

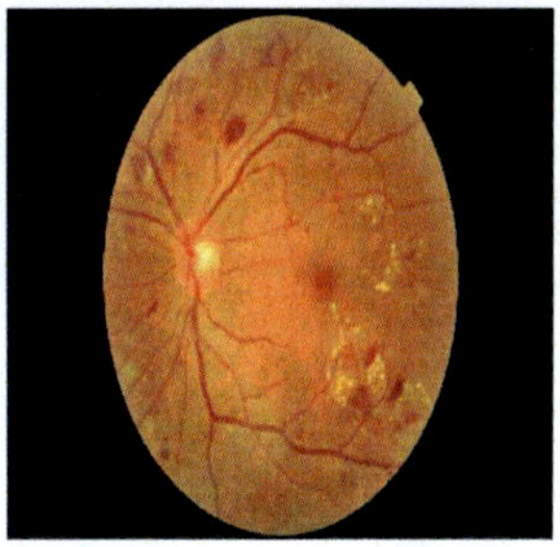

(b)

Fig. 5. a) The PSNR obtained by the proposed method compared with other algorithm for different order $n = 500$ and $n = 1000$. b) Diabetic Retinopathy 1000×1000.

References

1. Karmouni, H., et al.: Fast computation of 3D discrete invariant moments based on 3D Cuboid for 3D image classification. Circuits Syst. Signal Process. **40**(8), 3782–3812 (2021). https://doi.org/10.1007/s00034-020-01646-w
2. Karmouni, H., Jahid, T., Sayyouri, M., El Alami, R., Qjidaa, H.; Fast 3D image reconstruction by cuboids and 3D Charlier's moments. J. Real-Time Image Process, 1–17 (2019)
3. Ernawan, F., Kabir, N., Zamli, K.Z.: An efficient image compression technique using Tchebichef bit allocation. Optik (Stuttg). **148**, 106–119 (2017)
4. Oliveira, P.A., Cintra, R.J., Bayer, F.M., Kulasekera, S., Madanayake, A.: Low-complexity image and video coding based on an approximate discrete tchebichef transform. IEEE Trans. Circuits Syst. Video Technol. **27**(5), 1066–1076 (2016)
5. Mandal, M.K., Aboulnasr, T., Panchanathan, S.: Image indexing using moments and wavelets. IEEE Trans. Consum Electron. **42**(3), 557–565 (1996)
6. Yamni, M., Daoui, A., Karmouni, H., Sayyouri, M., Qjidaa, H., Flusser, J.: Fractional Charlier moments for image reconstruction and image watermarking. Signal Process. **171**, 107509 (2020)
7. Zhao, Z., Kuang, X., Zhu, Y., Liang, Y., Xuan, Y.: Combined kernel for fast GPU computation of Zernike moments. J. Real-Time Image Proc. **18**(3), 431–444 (2020). https://doi.org/10.1007/s11554-020-00979-8
8. El Ogri, O., Karmouni, H., Sayyouri, M., Qjidaa, H.: 3D image recognition using new set of fractional-order Legendre moments and deep neural networks. Signal Process. Image Commun. **98**, 116410 (2021)
9. Bencherqui, A., Karmouni, H., Daoui, A., Alfidi, M., Qjidaa, H., Sayyouri, M.: Optimization of jacobi moments parameters using artificial bee colony algorithm for 3D image analysis. In: Fourth International Conference on Intelligent Computing in Data Sciences (ICDS), pp. 1–7. IEEE, October 2020
10. Hosny, K.M., Elaziz, M.A.: Face recognition using exact Gaussian-hermit moments. In: Hassaballah, M., Hosny, K.M. (eds.) Recent Advances in Computer Vision. SCI, vol. 804, pp. 169–187. Springer, Cham (2019). https://doi.org/10.1007/978-3-030-03000-1_7

11. Zhang, H., Li, Z., Liu, Y.: Fractional orthogonal Fourier-Mellin moments for pattern recognition. In: Chinese Conference on Pattern Recognition, pp. 766–778 (2016)
12. Xiao, B., Ma, J.-F., Wang, X.: Image analysis by Bessel-Fourier moments. Pattern Recognit. **43**(8), 2620–2629 (2010)
13. Daoui, A., Yamni, M., Karmouni, H., Sayyouri, M., Qjidaa, H.: Stable computation of higher order Charlier moments for signal and image reconstruction. Inf. Sci. **521**, 251–276 (2020)
14. Sayyouri, M., Hmimid, A., Karmouni, H., Qjidaa, H., Rezzouk, A.: Image classification using separable invariant moments of Krawtchouk-Tchebichef. In: 12th International Conference of Computer Systems and Applications (AICCSA), Marrakech, Morocco (2015). https://doi.org/10.1109/AICCSA.2015.7507142
15. Karmouni, H., Yamni, M., El ogri, O., Sayyouri, M., Qjidaa, H.: Fast computation of 3D Meixner's invariant moments using 3D image cuboid representation for 3D image classification. Multimed. Tools Appl. **79**(39–40), 29121–29144 (2020)
16. Tahiri, M.A., Karmouni, H., Tahiri, A., Sayyouri, M., Qjidaa, H.: Partial 3D image reconstruction by cuboids using stable computation of hahn polynomials. In: Bennani, S., Lakhrissi, Y., Khaissidi, G., Mansouri, A., Khamlichi, Y. (eds.) WITS 2020. LNEE, vol. 745, pp. 831–842. Springer, Singapore (2022). https://doi.org/10.1007/978-981-33-6893-4_75
17. Tahiri, M.A., Karmouni, H., Azzayani, A., Sayyouri, M., Qjidaa, H.: Fast 3D image reconstruction by separable moments based on hahn and krawtchouk polynomials. In: Fourth International Conference on Intelligent Computing in Data Sciences (ICDS), pp. 1–7. IEEE, October 2020
18. Zhu, H., Shu, H., Liang, J., Luo, L., Coatrieux, J.-L.: Image analysis by discrete orthogonal Racah moments. Signal Process **87**(4), 687–708 (2007)
19. Zhu, H., Shu, H., Zhou, J., Luo, L., Coatrieux, J.L.: Image analysis by discrete orthogonal dual Hahn moments. Pattern Recogn. Lett. **28**(13), 1688–1704 (2007)
20. Jahid, T., Karmouni, H., Sayyouri, M., Hmimid, A., Qjidaa, H.: Fast algorithm of 3D discrete image orthogonal moments computation based on 3D cuboid. J. Math. Imaging Vision **61**(4), 534–554 (2019)
21. Daoui, A., Yamni, M., Ogri, O.E., Karmouni, H., Sayyouri, M., Qjidaa, H.: New algorithm for large-sized 2D and 3D image reconstruction using higher-order hahn moments. Circuits Syst. Signal Process. **39**(9), 4552–4577 (2020). https://doi.org/10.1007/s00034-020-01384-z
22. Senthilnath, J., Omkar, S.N., Mani, V.: Clustering using firefly algorithm: performance study. Swarm Evol. Comput. **1**(3), 164–171 (2011)
23. Radiopaedia Homepage. https://radiopaedia.org/. Accessed 15 Nov 2021

An Approach to Sri Lankan Sign Language Recognition Using Deep Learning with MediaPipe

Randika Jeewantha Herath(✉) and Piumi Ishanka

Department of Computing and Information Systems, Sabaragamuwa University of Sri Lanka, Belihuloya, Sri Lanka
hmygrjherath@std.appsc.sab.ac.lk, ishankauap@appsc.sab.ac.lk

Abstract. Currently, there are millions of people around the world with speech and hearing impairments, and according to the latest census, there are nearly 70,000 people who use Sri Lankan Sign Language. Sign language is a visual language, and it is the main medium of communication in their daily conversations. But they face obstacles when communicating with people who do not know sign language. There are communication barriers in different contexts, such as in work environments, knowledge exchange, and message sharing. Therefore, technology should play a major role in helping people with these hearing and speech impairments to improve their quality of life. This research aims to suggest models that use Google MediaPipe Hand Pose landmarks to identify Sri Lankan Sign Language. Moreover, this article compares vision-based approaches with convolutional neural network (CNN) and recurrent neural network (RNN). We also considered activation functions (such as ReLU, Linear, and Softmax), loss functions (mean squared error (MSE) and Categorical_crossentropy), and optimizations (Adam and Stochastic Gradient Descent (SGD)). The result showed that most algorithms built with Long Short-Term Memory (LSTM), CNN, and CNN-LSTM achieved an accuracy greater than 95%, both with the training dataset and the test dataset. In particular, models with MSE as the loss function and Adam as the optimizer showed higher accuracy.

Keywords: Computer vision · Gesture recognition · Deep learning · Sri Lankan sign language

1 Introduction

People who are hearing and/or speech impaired are those who are not able to hear and/or speak (or have some difficulty in hearing and/or speaking) an oral language. Thus, they tend to communicate using sign languages. There are nearly 430 million people among the world population who are reported as hearing impaired, and this is 5% of the total population [1]. Sign languages are any formal languages employing a system of hand gestures, facial expressions, body postures, and finger-spelling. Sri Lankan Sign Language is based on British Sign Language (BSL). However, Sri Lankan

S. Motahhir and B. Bossoufi (Eds.): ICDTA 2022, LNNS 454, pp. 449–459, 2022.
https://doi.org/10.1007/978-3-031-01942-5_45

Sign Language has many differences compared to BSL and currently consists of more than 2000 sign-based words [2]. Its alphabet consists of 56 characters [3].

Hearing- and/or speech-impaired people face various obstacles and difficulties in dealing with other people [4]. In Sri Lanka, most information is shared verbally in government offices and private institutions. Therefore, such information is inaccessible to hearing- or speech-impaired people. In this paper, we are trying to find a solution for this problem using deep learning technologies.

In the context of artificial intelligence, sign language recognition is an extensive field of research that encompasses all of the computer vision features that are intended to discover the ability of a machine to perceive human gestures. Several neural networks have been used for experiments, including the CNN and RNN versions. However, the development of such systems can be extremely challenging due to an enormous number of sign languages, the lack of large, annotated data sets, the resolution of frames, the discrimination ability of sensors, and the speed of pose movements [5]. Since the glove-based methodology seems to be somewhat uncomfortable to put into practical use, we are concerned with sign recognition using deep learning algorithms in vision-based approaches. This study was carried out using both approaches – pixel-based and landmark-based – while considering different algorithms such as CNN, LSTM, and CNN-LSTM and frameworks such as MediaPipe, Keras, and OpenCV to recognize Sri Lankan Sign Language. As for the method, we obtained relevant data sets and pre-processed them for model training, where the data can be predicted and evaluated using models. In the pixel-based approach, classification and extraction are done using spectral information from the image, whereas for the landmark-based approach, we used the Google MediaPipe Hands pose library.

Other researchers have used RNN versions only with MediaPipe and their accuracy is lower than the results of the proposed study. We used both CNN and RNN, and we could get a maximum accuracy of 98%. Also, vision-based and landmark-based approaches are compared in this study. The experiments were conducted for different activation functions (ReLU, Linear and Softmax), loss functions (MSE and Categorical_crossentropy), and optimizations (Adam and SGD).

2 Related Work

Over the past few years, many scientists and technologists have been researching in the field of sign gesture recognition. Existing research is characterized by variations in the performance of the recognition approaches, techniques, datasets, and models in each of those studies. Considering the approach they used, these can be broadly divided into glove-based or sensor-based methods and appearance-based or vision-based methods [6, 7]. Tables 1 and 2 summarize some of the research studies conducted in glove-based and vision-based categories.

In the context of sign language gesture recognition with MediaPipe, a variety of projects have been undertaken, and those are categorized into vision-based approaches. The most commonly used algorithms are RNN (LSTM, Bidirectional long-short term memory (BLSTM), and Gated recurrent units (GRU)), PyTorch Lightning, and Hidden Markov Models (HMMs) [32]. The authors who did the research on the recognition of Sri Lankan sign gestures are not currently focused on MediaPipe.

Table 1. Sample glove-based studies

Published year	Dataset details	Technique(s)/Algorithm(s)/Hardware/Software Details
2015 [8]	American	Arduino Uno, Advance Virtual RISC, Sensors, Haptic Glove, LabVIEW
2016 [9]	Japanese	Microsoft Kinect v2 sensor, HMMs, GMMs
2016 [10]	Sri Lankan	SEMG, IMU Sensors, Wearable Armband
2018 [11]	Spanish	Arduino Uno, Advance Virtual RISC, Sensors, Accelerometer, LabVIEW
2018 [12]	Indian	Arduino Uno, Advance Virtual RISC, Sensors, Bluetooth Module, Audio Module
2021 [13]	Thai	Leap Motion Sensor, LSTM

Table 2. Sample vision-based studies

Published year	Dataset details	Technique(s)/Algorithm(s)/Hardware/Software Details
2015 [14]	Arabic	SIFT, LDA, SVM, kNN
2018 [15]	Spanish	LeNet CNN, HMM, Gaussian Mixtures, Statistical Machine Translation
2019 [16]	SIBI	Lucas-Kanade, LSTM
2019 [17]	Pakistan	CNN, MLP, SVM, SURF Key Descriptors
2019 [18]	SIBI	DTSBN
2019 [19]	Thai	YOLO, CNN
2019 [20]	Chinese	CBAM-ResNet, Res2Net, C3D, C3D + SVM, MFnet, MFnet + CBAM, MFnet + CBAM + LSTM, CBAM-ResNet + optical flow
2019 [21]	Indian	Alexnet, Googlenet
2020 [22]	Japanese	bidirectional RNN
2020 [23]	Indian	kNN, Hue Saturation Value, Rule-based Classifier
2020 [24]	American	CNN VGG-16
2020 [25]	American	Random Forest, SVM, Multilayer Perceptron, Multi Scale Density Features
2021 [26]	Chinese	Tensor Train Decomposition, VGG16, LSTM
2021 [27]	Indonesian	Multi-Class PCA
2021 [28]	German, Chinese	Reinforcement Learning, Semantic Boundary Detection
2021 [29]	Arabic	Skeletal CNN
2021 [30]	–	CapsNet, SKNet
2021 [31]	Russian	HamNoSys, OpenPose, Leap Motion SDK

3 Method

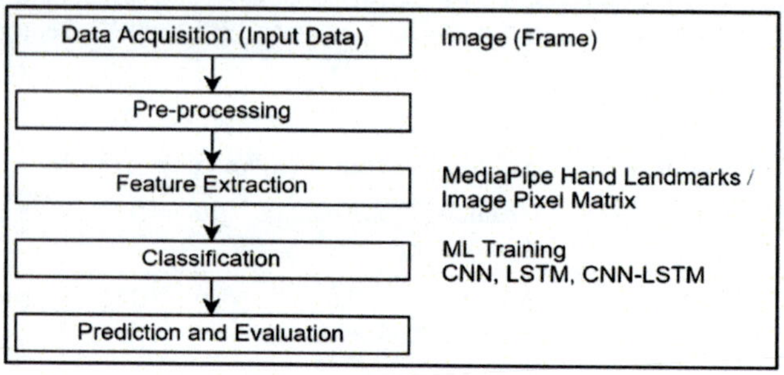

Fig. 1. Proposed method

Vision-based sign language recognition studies use two main methods for extracting features: extracting features directly from images (frames) or extracting the hand key points and linking their locations [32]. This study was carried out in both approaches—pixel-based and landmark-based—while considering different algorithms such as CNN, LSTM, and CNN-LSTM and frameworks such as MediaPipe, Keras, and OpenCV. Furthermore, both approaches are based on the pipeline, as shown in Fig. 1.

3.1 Pixel-Based Approach

The pixel-based approach is an effective and powerful traditional approach used to extract low-level features. The classification and extraction are done at the pixel level using the spectral information (pixel values within the locality) in the image.

3.2 Landmark-Based Approach

Nowadays, the location of hand landmarks (well-known, specific, and used with unique points such as the wrist and the tip of the thumb) is an important and trending source of information for the classification and extraction of hand gestures. Google MediaPipe Hands is one of the solutions for high-fidelity hand and finger tracking. This module can employ machine learning (ML) to infer twenty-one (21) 3D landmarks (x, y, and z enumerated values) of hands in a single frame and visualize these landmarks by linking their locations.

Each approach can be divided into four major steps: obtaining a dataset, preparing data (pre-processing) for model training, model training, and using models to predict and evaluate. Finally, the proposed study provides an insight into the context of those methods of sign language recognition. These steps are discussed in Sects. 3.3, 3.4, 3.5, and 3.6.

3.3 Input Data

We used Sri Lankan Sign Language datasets, and the sample data are shown in Fig. 2.

The collection included 5,538 images consisting of 19 classes (19 hand signs), which were 'One', 'Two', 'Three', 'Four', 'Five', 'Six', 'Seven', 'Eight', 'Nine', 'Ayubowan' or 'Welcome', 'Bad', 'Good', 'His', 'House', 'I Love You', 'What', 'Who', 'Why', and 'You'. The gestures of the images are Sri Lankan sign-language gestures that include different features such as size, angle, persons, and resolution.

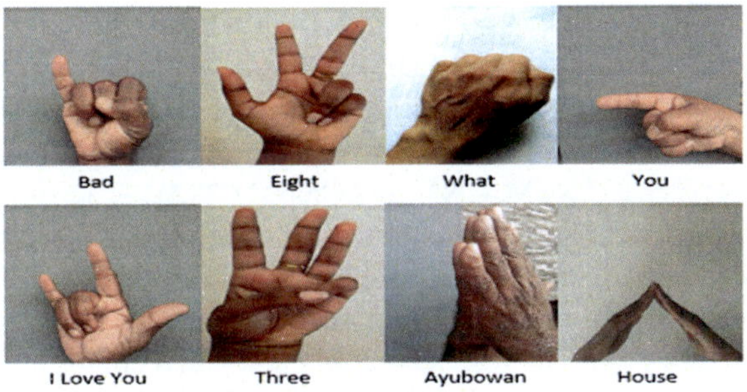

Fig. 2. Example signs in Sri Lankan sign language

3.4 Pre-processing

This covers resizing, orienting, colour adjustments, and more. Since there are two approaches, we have divided the pre-processing steps into two parts. In the pixel-based approach, each image is resized to 50 × 50 pixels. Then converted it to grayscale colour, extracted features from the loaded image and save them to a data frame. As the next step, shuffled the data in the data frame and normalized data into a range of 0–1. Converted the labels into a categorical representation and saved the data and targets as NumPy arrays. In the landmark-based approach, each image is resized to 100 × 100 pixels. Converted it to grayscale colour, extract features from the loaded image using Google MediaPipe hand pose (with "2 maximum number of hands" and "0.5 minimum detection confidence" configurations) and saved them to a data frame. Then saved data into a CSV file. Shuffled the data in the data frame. Converted the labels into a categorical representation and then saved the data and targets as NumPy arrays. MediaPipe provides a list of 21 hand landmarks, each consisting of x, y, and z. x and y are normalized to [0.0, 1.0]. Therefore, we extracted 63 landmarks (21 * 3) for each frame and saved them in a normalized and nonnormalized manner. The sample data or landmarks are as shown in Fig. 3.

class_type	wristX	wristY	wristZ	thumbCm	thumbCm	thumbCm	thumbMc	thumbMc	thumbMc	thumbIpX	thumbIpY	thumbIpZ	thum
Ayubowa	0.73955	0.528037	3.19E-06	0.733891	0.486507	-0.01299	0.733427	0.454607	-0.00531	0.731571	0.431515	0.008025	0.73
Ayubowa	0.439749	0.888397	-0.00017	0.328464	0.782833	0.20324	0.294671	0.671045	0.250114	0.278618	0.57999	0.273704	0.24
Ayubowa	0.437104	0.74793	-0.00012	0.346707	0.68829	0.173835	0.318319	0.605304	0.245483	0.311308	0.536191	0.285835	0.29
Ayubowa	0.496851	0.800801	-0.00018	0.500852	0.691894	0.232823	0.519741	0.598889	0.314056	0.548634	0.533466	0.343075	0.57
Ayubowa	0.580345	0.773884	-5.03E-05	0.595085	0.695367	0.034374	0.584756	0.617462	0.01146	0.576654	0.559919	-0.01815	0.56
Ayubowa	0.509096	0.440132	-0.0001	0.550414	0.437393	-0.1095	0.57156	0.434432	-0.18528	0.561168	0.434539	-0.23246	0.53
Ayubowa	0.484852	0.985775	-0.00029	0.632654	0.875867	0.037605	0.701271	0.764951	0.057219	0.722277	0.660401	0.078606	0.71
Ayubowa	0.577826	0.585335	4.72E-05	0.554632	0.52467	-0.02717	0.524942	0.486614	-0.09863	0.505535	0.448015	-0.15426	0.49
Ayubowa	0.524704	0.461718	-4.81E-05	0.470617	0.433514	-0.0722	0.427775	0.413722	-0.17233	0.426451	0.394438	-0.2437	0.44
Ayubowa	0.409017	0.549107	2.72E-05	0.417538	0.554092	-0.0456	0.416235	0.56007	-0.07253	0.408078	0.564443	-0.08915	0.40
Ayubowa	0.376455	0.594861	2.46E-05	0.381128	0.603363	-0.03526	0.377743	0.609866	-0.05394	0.370371	0.615959	-0.06413	0.3
[illegible]	[illegible]	[illegible]	[illegible]	[illegible]	[illegible]	[illegible]	[illegible]	[illegible]	[illegible]	[illegible]	[illegible]	[illegible]	[illegible]

Fig. 3. Saved hand landmarks to CSV

3.5 Model Training

Considering the data and features extracted for both approaches, referring to previous studies, primarily RNN and CNN can be used in both landmark-based approach and pixel-based approach. However, we decided to use different architectural variations on both RNN and CNN (such as LSTM, CNN, and CNN-LSTM) because CNN is commonly used in spatial data, such as images for image classification and object detection, whereas LSTM is commonly used in sequence data, such as time-series data for NLP and sentence classification as well as for making predictions based on it. So, CNN-LSTM is used for sequence-prediction problems with spatial inputs, which entails employing CNN to extract features from spatial input data and LSTMs to perform sequence classification and prediction. We also considered activation functions (ReLU, Linear, and Softmax), loss functions (MSE and Categorical_crossentropy), and optimizations (Adam and SGD). The experiment was carried out with different combinations of the above-mentioned activation functions, loss functions, and optimizations.

3.6 Predict and Evaluate

The whole dataset is divided into two sets as train and test datasets. An 80:20 aspect ratio was chosen to separate train and test datasets. Each algorithm (model) is evaluated against the test data.

4 Results

The experiments were carried out for several models, with different combinations of loss functions, optimizations, and activation functions as described in the model training section (Table 3). Tables 4, 5, and 6 show results based on the following parameters or configurations.

4.1 Pixel-Based Approach

We determined the number of epochs required to train by training several different models, as they need to be fair to all models. We identified more than 25 epochs; some models are over-fitting, and in contrast, less than 20 models are under-fitting. So here, we used 25 epochs to model training.

Table 3. Model-training configurations

Configurations	First layers	Last layer	Loss function	Optimization
	Activation function	Activation function		
Cfg. 1	Relu	Linear	MSE	Adam
Cfg. 2	Relu	Linear	Categorical_crossentropy	Adam
Cfg. 3	Relu	Linear	Categorical_crossentropy	SGD
Cfg. 4	Relu	Linear	MSE	SGD
Cfg. 5	Relu	Softmax	MSE	Adam

Table 4. Pixel-based approach models accuracy and loss comparison

Network	*Cfg. 1*	*Cfg. 2*	*Cfg. 3*	*Cfg. 4*	*Cfg. 5*
	Accuracy				
CNN	0.9945	0.0645	0.0845	0.5173	0.9891
LSTM	0.9900	0.0764	0.0745	0.4373	0.9827
CNN_LSTM	0.9964	0.9845	0.0218	0.1036	0.9909
	Loss				
CNN	0.0020	5.6999	8.8943	0.0381	9.0342e−04
LSTM	0.0011	8.0151	9.0261	0.0411	0.0013
CNN_LSTM	4.1789e−04	0.0558	7.2678	0.0460	6.5344e−04

4.2 Landmark-Based Approach

As mentioned earlier, we determined the number of epochs required to train by training several different models so that they would be fair to all models. We have identified more than 100 epochs, the accuracy of some models varies by a considerable margin, and on the contrary, less than 50 models are under-fitting. So here, 100 epochs were used for model training.

Table 5. Landmark-based approach models accuracy and loss comparison [normalized data]

Network	*Cfg. 1*	*Cfg. 2*	*Cfg. 3*	*Cfg. 4*	*Cfg. 5*
	Accuracy				
CNN	0.9705	0.9837	0.9613	0.1200	0.9715
LSTM	0.9797	0.0153	0.0631	0.1099	0.9797

(*continued*)

Table 5. (*continued*)

Network	*Cfg. 1*	*Cfg. 2*	*Cfg. 3*	*Cfg. 4*	*Cfg. 5*
CNN_LSTM	0.9848	0.0879	0.0172	0.1061	0.9758
	Loss				
CNN	0.0024	0.0640	0.1129	0.0473	0.0024
LSTM	0.0017	5.9521	7.0342	0.0463	0.0018
CNN_LSTM	0.0012	7.3264	9.1987	0.0456	0.0018

Table 6. Landmark-based approach models accuracy and loss comparison [non-normalized]

Network	*Cfg. 1*	*Cfg. 2*	*Cfg. 3*	*Cfg. 4*	*Cfg. 5*
	Accuracy				
CNN	0.0234	0.9817	0.0926	0.9349	0.0743
LSTM	0.9746	0.0631	0.1058	0.4669	0.9634
CNN_LSTM	0.9808	0.0192	0.0808	0.4515	0.9778
	Loss				
CNN	0.0977	0.1053	7.9525	0.0049	0.0926
LSTM	0.0020	9.3254	11.5926	0.0394	0.0024
CNN_LSTM	0.0017	6.6915	8.5475	0.0350	0.0017

5 Discussion and Conclusion

The result showed that most algorithms built with LSTM, CNN, and CNN_LSTM have an accuracy greater than 95% for both the training dataset and the test dataset, especially in the loss function, MSE, and the optimization, Adam. The pixel-based approach showed the highest accuracy: 99.00%, 99.45%, and 99.64%, respectively. Models with MediaPipe at lower resolution frames and similar postures have lower accuracy than models without MediaPipe. Some images with minimum detection confidence of less than 0.5 have lower accuracy. Furthermore, sequential data, such as MediaPipe data, is much faster at extracting features and training models compared to square pixels (image element or matrix). Moreover, CNN can take a longer time to train than RNN. Experiments show that CNN_LSTM is better than CNN and LSTM, whether it's pixel-based or landmark-based. This experiment revealed that it is better to use CNN_LSTM with images without MediaPipe to recognize the Sri Lankan sign language in low-resolution images. However, we recommend LSTM, CNN, and CNN_LSTM models with MediaPipe since those models are very fast in training and the accuracy gap between them is very little. Most experiments that have been done are pixel-based, but we proposed the landmark-based approach with MediaPipe and CNN_LSTM, which gives an accuracy of 0.9848. Therefore, further experiments with more pre-processing frames can increase the accuracy of the models. This experiment was performed on both the pixel-based and landmark-based approaches of the vision-based approach in a static setting. However, this experiment can be extended in dynamic settings, such as recognizing moving gestures, forming sentences, and returning the whole captured sentence.

Acknowledgement. I express my great appreciation to the Department of Computing and Information Systems, Faculty of Applied Sciences, Sabaragamuwa University of Sri Lanka for providing opportunities, guidance, advice, tremendous encouragement, endless experimental recommendations, a lot of precious time, and endless assistance for the success of this study.

References

1. Deafness and hearing loss. http://www.who.int/mediacentre/factsheets/fs300/en/. Accessed 01 Oct 2021
2. Fernando, P., Wimalaratne, P.: Sign language translation approach to SinhaleseLanguage. GSTF J. Comput. (JoC) **5**(1), 1–9 (2016). https://doi.org/10.7603/s40601-016-0009-8
3. Nirosha, W., Choolangika, S., Vinodya, S.: Sign language translator for deaf and speech impaired people using convolutional neural network. In: 12th International Research Conference on General Sir John Kotelawala Defence (2019)
4. Malinda, P., Gayan, M.: Computer interpreter for translating written Sinhala to Sinhala sign language. In: OUSL J. **12**(1), 70–90 (2017). https://doi.org/10.4038/ouslj.v12i1.7377
5. Bach, K., Duong, P., Ha, P., Anh, B., Son, N.: Vietnamese sign language detection using Mediapipe. In: 10th International Conference on Software and Computer Applications (ICSCA 2021) (2021). https://doi.org/10.1145/3457784.3457810
6. Ashok, S., Kiran, R.: Vision based Indian sign language character recognition. J. Theor. Appl. Inf. Technol. **67**(3) (2014)
7. Kishore, P., Panakala, K.: A video based Indian Sign Language Recognition System (INSLR) using wavelet transform and fuzzy logic. Int. J. Eng. Technol. **4**(5) (2012). https://doi.org/10.7763/IJET.2012.V4.427
8. Deepak, S., Deepak, V., Poras, K.: LabVIEW based sign language trainer cum portable display unit for the speech impaired. In: 2015 Annual IEEE India Conference (INDICON) (2015). https://doi.org/10.1109/INDICON.2015.7443381
9. Sako, S., Hatano, M., Kitamura, T.: Real-time Japanese sign language recognition based on three phonological elements of sign. In: Stephanidis, C. (ed.) HCI 2016. CCIS, vol. 618, pp. 130–136. Springer, Cham (2016). https://doi.org/10.1007/978-3-319-40542-1_21
10. Madushanka, A., Senevirathne, R., Wijesekara, L., Arunatilake, S., Sandaruwan, K.: Framework for Sinhala sign language recognition and translation using a wearable armband. In: Sixteenth International Conference on Advances in ICT for Emerging Regions (2016). https://doi.org/10.1109/ICTER.2016.7829898
11. Griselda, S., Jorge, C., Mario, B., Apolonio, P.: Recognition and classification of sign language for Spanish. Computacion y Sistemas**22**(1), 271–277 (2018). https://doi.org/10.13053/cys-22-1-2780
12. Hardik, R., Vishal, D., Dhroov, B., Hema, N.: Automated sign language interpreter. In: Eleventh International Conference on Contemporary Computing (2018). https://doi.org/10.1109/IC3.2018.8530658
13. Ponlawat, C., Kosin, C.: Backhand-view-based continuous-signed-letter recognition using a rewound video sequence and the previous signed-letter information. IEEE Access **9**, 40187–40197 (2021). https://doi.org/10.1109/ACCESS.2021.3063203

14. Tharwat, A., Gaber, T., Hassanien, A.E., Shahin, M.K., Refaat, B.: SIFT-based arabic sign language recognition system. In: Abraham, A., Krömer, P., Snasel, V. (eds.) Afro-European Conference for Industrial Advancement. AISC, vol. 334, pp. 359–370. Springer, Cham (2015). https://doi.org/10.1007/978-3-319-13572-4_30
15. Zuzanna, P., Carlos-D, M.: Sign language gesture classification using neural networks. In: IBER SPEECH (2018). https://doi.org/10.21437/IberSPEECH.2018-27
16. Aulia, P., Erdefi, R., Dadan, H.: Human skeleton feature extraction from 2-dimensional video of Indonesian language sign system (SIBI [Sistem Isyarat Bahasa Indonesia]) gestures. In: 5th International Conference on Computing and Artificial Intelligence, pp. 100–105 (2019). https://doi.org/10.1145/3330482.3330484
17. Sobia F., Yasar, A.: CNN and traditional classifiers performance for sign language recognition. In: 3rd International Conference on Machine Learning and Soft Computing, pp. 192–196 (2019). https://doi.org/10.1145/3310986.3311011
18. Darmana, A., Erdefi, R.: Generating of Sign System for Bahasa Indonesia (SIBI) root word gestures using deep temporal sigmoid belief network. In: ICCAI 2019: Proceedings of the 2019 5th International Conference on Computing and Artificial Intelligence, pp. 221–225 (2019). https://doi.org/10.1145/3330482.3330494
19. Pisit, N., Patcharee, M., Tatpong, K.: Thai finger spelling localization and classification under complex background using a YOLO-based deep learning. In: 11th International Conference on Computer Modeling and Simulation, pp. 230–233 (2019). https://doi.org/10.1145/3307363.3307403
20. Huang, C., Wang, F., Zhang, R.: Sign language recognition based on CBAM-ResNet. In: International Conference on Artificial Intelligence and Advanced Manufacturing, no. 48, pp. 1–6 (2019). https://doi.org/10.1145/3358331.3358379
21. Kirtee, P., Sreemathy, R., Akshay, V.: Recognition of Indian sign language alphabets for hearing and speech impaired people using deep learning. SSRN Electron. J. (2019). https://doi.org/10.2139/ssrn.3430055
22. Heike, B., Felix, L., Kazuhiro, N., Yuji, N.: Learning three-dimensional skeleton data from sign language video. ACM Trans. Intell. Syst. Technol. **11**(3) (2020). https://doi.org/10.1145/3377552
23. Neha, N., Saloni, M., Sandhya, A., Ekansh, T.: A dynamic gesture recognition system for mute person. In: Goyal, D., Gupta, A.K., Piuri, V., Ganzha, M., Paprzycki, M. (eds.) Proceedings of the Second International Conference on Information Management and Machine Intelligence. LNNS, vol. 166, pp. 33–39. Springer, Singapore (2020). https://doi.org/10.1007/978-981-15-9689-6_4
24. Amrita, T., Pujan, B., Sarmila, U., Shirish, S., Subarna, S.: Real time sign language recognition and speech generation. J. Innov. Image Process. **2**(2), 65–76 (2020). https://doi.org/10.36548/jiip.2020.2.001
25. Taniya, S., Soumi, P., Subhadip, B., Ayatullah, M.: Hand sign recognition from depth images with multi-scale density features for deaf mute persons. International Conference on Computational Intelligence and Data Science (2019). https://doi.org/10.1016/j.procs.2020.03.243
26. Biao, X., Shiliang, H., Zhongfu, Y.: Application of tensor train decomposition in S2VT model for sign language recognition. IEEE Access **9**, 35646–35653 (2021). https://doi.org/10.1109/ACCESS.2021.3059660
27. Astri, N., Fairuz, A.: Sign language recognition using principal component analysis and support vector machine. Int. J. Appl. Inf. Technol. **4**(1) (2021). https://doi.org/10.25124/ijait.v4i01.3015
28. Chengcheng, W., Jian, Z., Wengang, Z., Houqiang, Li.: Semantic boundary detection with reinforcement learning for continuous sign language recognition. In: IEEE Trans. Circuits Syst. Video Technol. **31**, 1138–1149 (2021). https://doi.org/10.1109/TCSVT.2020.2999384

29. Mohamed, B., et al.: Arabic Sign language recognition system using 2D hands and body skeleton data. IEEE Access **9**, 59612–59627 (2021). https://doi.org/10.1109/ACCESS.2021.3069714
30. Lu, J., Nguyen, M., Yan, W.Q.: Sign language recognition from digital videos using deep learning methods. In: Nguyen, M., Yan, W.Q., Ho, H. (eds.) ISGV 2021. CCIS, vol. 1386, pp. 108–118. Springer, Cham (2021). https://doi.org/10.1007/978-3-030-72073-5_9
31. Prikhodko, A., Grif, M., Bakaev, M.: Sign language recognition based on notations and neural networks. In: Alexandrov, D.A., Boukhanovsky, A.V., Chugunov, A.V., Kabanov, Y., Koltsova, O., Musabirov, I. (eds.) DTGS 2020. CCIS, vol. 1242, pp. 463–478. Springer, Cham (2020). https://doi.org/10.1007/978-3-030-65218-0_34
32. Anusorn, C., Kritsana, S., Thidalak, Y.: Thai sign language recognition: an application of deep neural network. In: 2021 Joint International Conference on Digital Arts, Media and Technology with ECTI Northern Section Conference on Electrical, Electronics, Computer and Telecommunication Engineering (2021). https://doi.org/10.1109/ECTIDAMTNCON51128.2021.9425711

Transfer Learning and Pressure Effect for Handwriting to Early Detection of Parkinson's Disease

Yassir El Ghzizal[1](✉), Nabil Aharrane[2], Ghizlane Khaissidi[1], and Mostafa Mrabti[1]

[1] Laboratory LIPI ENS, USMBA, Fez, Morocco
elghzizalyassir@gmail.com
[2] Laboratory LIIAN FSDM, USMBA, Fez, Morocco

Abstract. Parkinson's disease (PD) is a degenerative and progressive disorder that affects the nervous system. The diagnosis of this disease remains a very important area of research since there is no subjective method of diagnosis used clinically. In this work, to address this problem, we involved a deep learning model using transfer learning from a pre-trained CNN called MobilNetV2 to classify patients with Parkinson's disease. This model was retrained with a database of patients' handwriting spirals. To remedy the problem of the poor size of our database, we used augmentation techniques consisting of some pre-processing layers (mirror effect, rotation, saturation...) to have enough images to train the model. These images include the pressure exerted on the screen when drawing spirals by representing each level of pressure with a specified color. The obtained results have proved very promising.

Keywords: Deep learning · Transfer learning · CNN · Handwriting · Parkinson's disease

1 Introduction

Parkinson's disease (PD) is a neurodegenerative disease characterized by the destruction of a specific population of brain neurons called the dopamine neurons of the brain's dark matter. These neurons are involved in the control of human movement. The first criterion for diagnosis is the presence of at least two of the three major symptoms of the disease [1]: Slowness of movement, rigidity, and trembling at rest.

For neurological diseases, the early detection of PD allows a quick reaction to stop its progression and prescribe the right treatment. This makes the medical decision support systems an important tool to augment objectivity and help effective early diagnosis [2, 3]. In general, these systems make their decision based on some biomarkers [4]. Hence the quality of a decision depends on the relevance of these indicators. In the literature, several methods have been proposed to improve the accuracy of PD prediction and to quantify the severity of the disease and its progression. These methods focus on motor performance testing, olfactory testing, speech processing [5, 6], imaging techniques, and patient movement and walking modeling [7, 8].

S. Motahhir and B. Bossoufi (Eds.): ICDTA 2022, LNNS 454, pp. 460–469, 2022.
https://doi.org/10.1007/978-3-031-01942-5_46

In addition, patients writing and drawing skills are mainly influenced when these motor symptoms are signaled. According to Mclennan et al. [9], 30% of people with Parkinson's disease have an alteration in handwriting and 5% have a micrography even before the appearance of symptoms. For these reasons, many studies focus on disgraphy to try the early diagnosis of PD.

In this work, as a first step, we create images representing both the static and dynamic spirals of the patient drawing trajectory from the raw database containing the (x, y) coordinates and the pressure measure during drawing. The pressure is represented in the image as a color level assigned to each (x, y) point. The obtained images serve as input to train a convolutional neural network (CNN) to distinguish between parkinsonian and healthy samples. To highlight the effect of pressure, we compare the obtained results with the ones without adding the pressure information. To improve the performance of our system, we replace our trained CNN with the pre-trained neural network called MobileNetV2 taking the advantage of the already accumulated knowledge on the large database ImageNet. This operation called transfer learning consists of adapting the output of the pre-trained model to the problem output and performing a fine-tuning step to make the model more specific to the problem under study. It should be noted that a database augmentation technique was performed to have sufficient images to train the CNN and the pre-trained MobileNetV2.

This paper is organized as follows: Sect. 2 reports some related works to PD detection. Section 3 describes in detail the proposed systems. The obtained results on the enhanced spiral tests with a digitized graphic tablet database are discussed in Sect. 4. Finally, Sect. 5 is concluded with a general conclusion and perspectives of this work.

2 Related Works

This section essentially describes the research works already performing their experiments on the database of enhanced spiral tests with a digitized graphic tablet [10]. This choice is justified by the fact that this latter was also subject of our experiments. Indeed, several approaches were recently proposed to deal with the PD prediction problem taking advantage of the technology advances and the modern machine and deep learning algorithms development.

Khatamino et al. [11] proposed a convolutional neural network (CNN) architecture as follows: two convolutional, two Maxpooling, and two fully connected (FC) layers; of which the second FC has two neurons since it performs a binary classification. They applied this structure on the raw database as well; they evaluated the performance of their deep learning model by K- Fold Cross-validation and Leave-one-out cross-validation (LOOCV) techniques. Gil-Martín et al. [12] applied the same simplified CNN version of AlexNet of Khatamino et al., but they used the spectrum as CNN inputs instead of working with raw data since the tremors become more amplified in the spectral or frequency domain. They obtained good results, with a 96.5% recognition rate compared to the results of work carried out on the same basis. C. Gallicchio et al. [13] also chose to apply deep learning based on deep Echo State Networks (ESNs) for the identification of subjects with PD. they compared the results of their model with those obtained by the application of "shallow ESNs", and showed that ESNs perform better with an accuracy of 89.33%.

3 Proposed System

In this work, to handle the problem of the early detection of Parkinson's diseases, we proposed a classification system capitalizing on the effect of the patient pressure during spirals drawing. The pressure information is represented in images using color levels. The obtained images served as input to a classification model to distinguish between parkinsonian and healthy patients (Fig. 1).

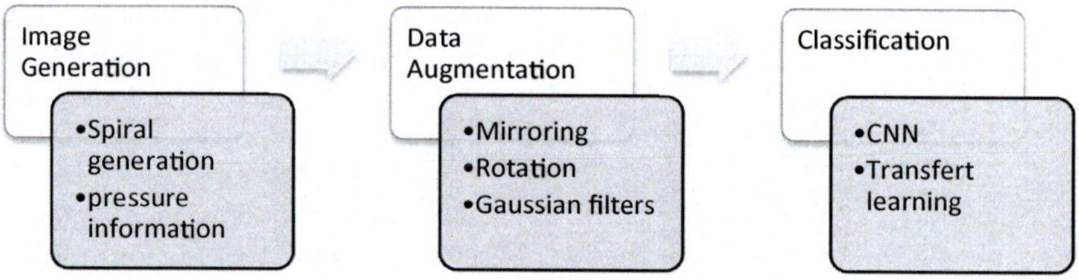

Fig. 1. Steps required to build the proposed system

The followed steps of the proposed system are described in detail in the following sections.

3.1 Images Generation

Database. In this work, we chose to work on a digital database acquired in the Department of Neurology of Cerrahpasa Medical Faculty, Istanbul University, using the Wacom Cintiq 12WX graphics tablet. The database consists of 62 people with Parkinson's disease as the patient group and 15 healthy people as the control group. All the samples were invited to accomplish spirals drawing tests. We were interested in the two tests: the Static Spiral Test (SST) and Dynamic Spiral Test (DST). Static Spiral Test (SST): An Archimedes spiral of three coiled turns appears and patients were asked to retrace the same spiral. Dynamic Spiral Test (DST): The same spiral for the static test but this time the spiral flashes to best identify the drawing abilities of the patient [10].

Among The information collected to generate images from the online digital trajectory data, we were interested in:

- $(X_i(t), Y_i(t))$ represent the abscissa and ordinate respectively of the ith sample of the spiral drawing in the SST and DST mode,
- $P_i(t)$: the pressure exerted on the tablet's screen in the ith sample while drawing.

The coordinates $(X_i(t), Y_i(t))$ of the spirals for both SST and DST modes were represented in a single image, as shown in Fig. 2, to process them simultaneously.

Pressure Information. To get the most out of the data recovered from the graphic tablet used for the database acquisition, we have inserted the effect of pressure to all the previously drawn images of the patient's spirals. Indeed, we gave each level exerted on the writing screen a specific color.

Fig. 2. Example of passage from a simple spiral to a spiral with the effect of pressure

3.2 Data Augmentation

Data augmentation is an essential part to provide additional images for classifier training. In this work, we used several augmentation techniques that allow us to keep the properties of the original handwriting images such as mirroring, rotation, and Gaussian filters. Figure 3 shows the resulting images after applying the said filters several times on the same image.

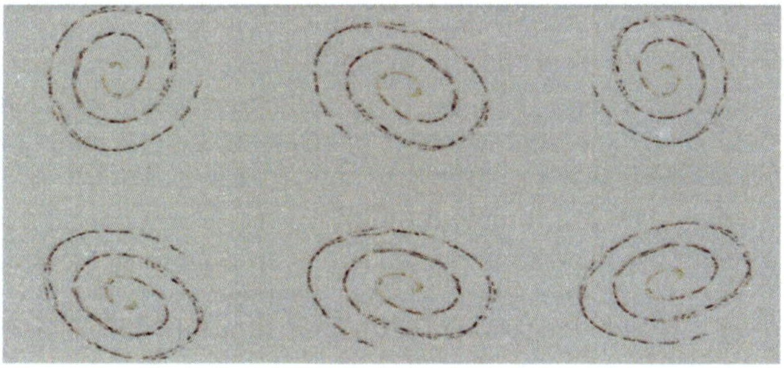

Fig. 3. Example of applying a few layers of pre-processing several times to the same image.

Subsequently, after applying augmentation techniques, the obtained color images were served as input to a CNN classifier to predict the corresponding class. We choose the CNN classifier to its ability to deal with images directly.

3.3 Classification

Image classification using learning algorithms can be done in two ways: either classically provide the extracted feature vectors as input, which requires the choice of a valid extraction algorithm. Or directly present the vectorized images or signals as input. In the concept of the second approach, convolutional neural networks (CNN) are designed to process data directly. Indeed, they are designed to automatically extract features from images without the need for manual extraction. In addition, several CNN architectures have been developed and pre-trained on large databases, including AlexNet, MobilNet, GoogLeNet, etc.

In this section, we present the structure and operation of a CNN with its classical layers and then describe the transfer learning technique using a pre-trained CNN named MobileNet V2 developed by google.

Convolutional Neural Network (CNN): is one of the most powerful algorithms in Deep Learning, which allows having for each input image a corresponding label while automatically extracting features.

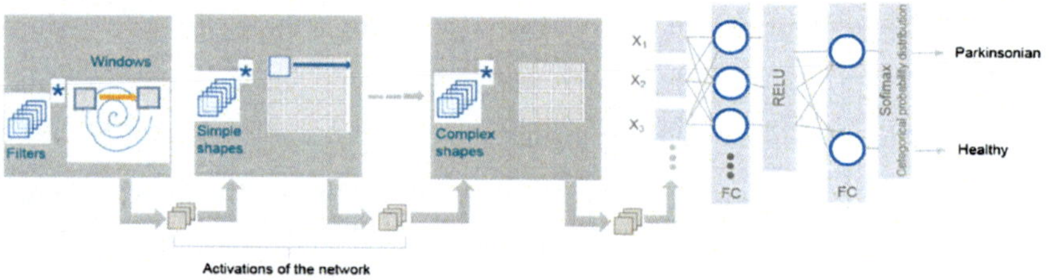

Fig. 4. The structure of a convolutional neural network

A CNN uses a model similar to the multilayer perceptron but designed to reduce the number of calculations. Figure 4 shows the structure of a CNN which is composed of a sequence of layers: an input layer that receives the image, a hidden layer composed of several convolutional layers, Maxpooling layers, fully connected layers, and normalization layers. The ultimate goal of this sequence of layers or filters is to create convolution maps in the form of images that will then be concatenated into a feature vector and finally an output layer, which is simply a multilayer MLP perceptron that receives as input the feature vector obtained from the hidden layer to perform classification.

Transfer Learning. To solve the problem of classification on a database with few examples, the researchers thought of leveraging the knowledge already acquired by pre-trained CNNs on important and general databases, taking advantage of the accumulated knowledge and integrating it into the new task being treated, i.e. one can exploit what has been learned in such a situation to better generalize another. This operation is called transfer learning.

Two popular techniques that are commonly used when applying transfer learning with pre-trained CNNs have been adopted, to customize a MobileNet V2 (developed at Google) pre-trained CNNs base model and to address our handwriting image classification task for the detection of Parkinson's disease patients, namely feature extraction and fine-tuning (Fig. 5).

Feature Extraction. In this part, the learned representations of the convolutional layers of a previously trained CNN will be used for the extraction of new significant features. This means that we don't need to train the whole MobileNet V2 system anymore, we just need to pass the new data to form a new classifier of the specific output for the new class numbers of the new task. In our case, two classes (Parkinsonian, healthy). In this technique, only the weights of the higher classifiers are updated, the convolution weights are maintained (frozen).

Fine-Tuning. This interesting step consists of selectively unfreezing and requalifying some of the upper layers of the convolutional base previously used for feature extraction. It allows us to jointly adjust its last layers to make them more relevant to the specific task.

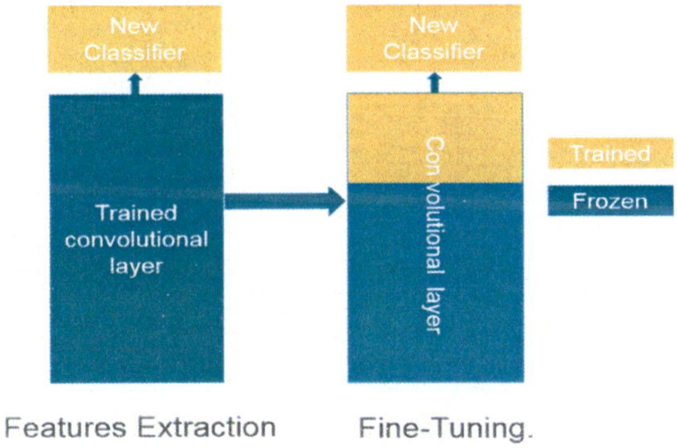

Fig. 5. Features extraction at the fine-tuning process in transfer learning

MobileNetV2 is a new efficient convolutional neural network system of 154 layers developed at google. It is an extended version of MobileNetV1, designed for mobile devices to advance automatic recognition applications such as classification, object detection, and semantic segmentation. It is previously trained on the ImageNet database; it is a very large database with a wide variety of categories, which contains 1.4 million images and 1000 classes. Like MobileNetV1, MobileNetV2 uses deep-separable convolution, introducing linear bottlenecks between layers to encode the intermediate inputs and outputs of the model and shortcuts between the bottlenecks that act as traditional residual connections, to ensure better and faster training.

4 Results and Analysis

This section presents the experimental results of our proposed system on the previously mentioned database (Sect. 3.1). The system consists of generating images from the time series of coordinates (X(t), Y(t)) representing a sample taken during the spiral drawing of the patient where each point (X(t),Y(t)) was colored depending on the pressure exerted at time t.

The resulting images were divided into training, validation, and testing set according to the ratios 60%, 20%, and 20% respectively. After applying some augmentation techniques to the training and validation sets, we train a CNN model with the architecture presented in Fig. 4.

To showcase the effect of representing the pressure in the images, we carry out a couple of experiments using images with and without representing the pressure to train the CNN model. Figure 6 represents the obtained accuracy on the training and validation sets for both kinds of images (with and without pressure information).

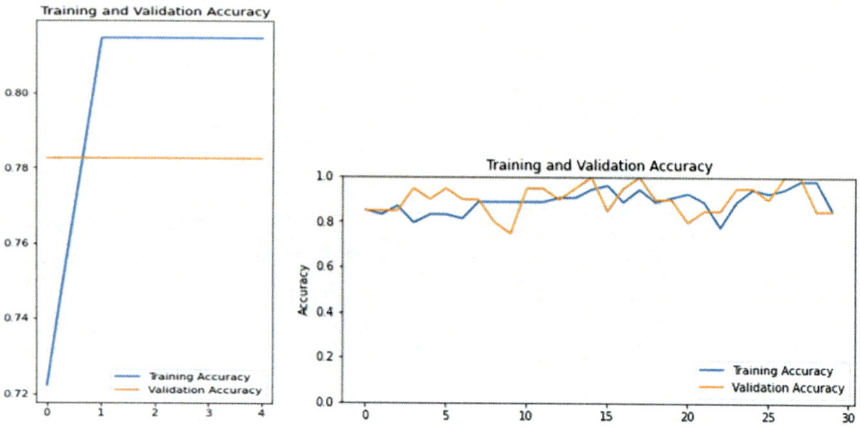

(a) (b)

Fig. 6. Results on the training and validation sets for images with and without pressure

As we can see in Fig. 6.a, and after one epoch, the learning phase ended with overfitting and, in the validation phase, the model classifies all the subjects in the parkinsonian class that has the largest number of samples. However, when introducing the pressure information in each point as a color level to the images, we overcome the overfitting problem as shown in Fig. 6.b. In the test phase, the model reaches a moderately promising accuracy rate of 67%, which can be explained by the small size of the dataset.

As proved in the literature, the use of pre-trained CNN models with previously formed layers, on a large database with a wide variety of categories, may improve the system performance and can help to overcome the problem of small datasets. Therefore, we conduct other experiments to substantiate the use of the pre-trained CNN (MobileNetV2) instead of the raw CNN in our system.

As explained in Sect. 3.3, the use of pre-trained CNN models is a two-step process. In the first step, we add to the pre-trained convolution layers a fully connected layer to fit the MobileNetV2 to our problem. The training operation is performed by freezing the weights of MobileNetV2, which means that only the weights of the last fully connected layer are updated. The second step is a fine-tuning step used to refine the weights of the 54 upper convolutional layers by training these layers in parallel with the fully connected layer so that the weights of these upper layers are more tuned to the weights learned by the fully connected layer in the first step.

The accuracy achieved on the training and validation sets in 160 epochs is shown in Fig. 7. The first 60 epochs were used to train the added fully connected layer while freezing the convolution layers and the last 100 epochs served for the fine-tuning step.

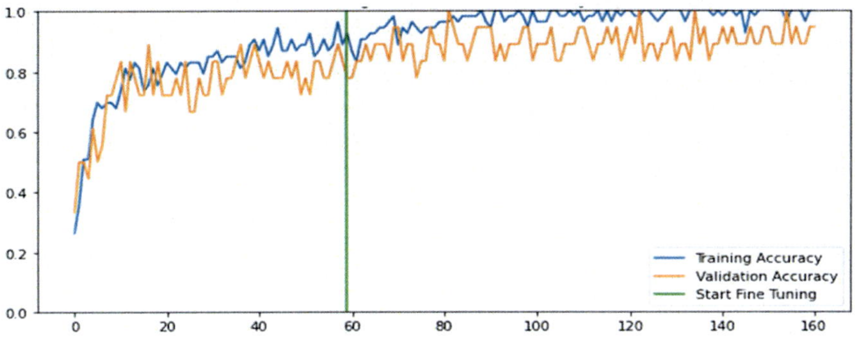

Fig. 7. Classification results after the fine-tuning process

One of the confronted drawbacks during the conception of our model was the difference existing between the complex colored images used to pre-train the MobileNet Model and our simple images. On the other hand, our model suffers from the smallness of our training dataset to best tune the model.

Despite these drawbacks, we can see a significant improvement in the prediction results on the test set with a classification rate of 94,44%. The following figure shows the classification results of some samples of the test set (Fig. 8).

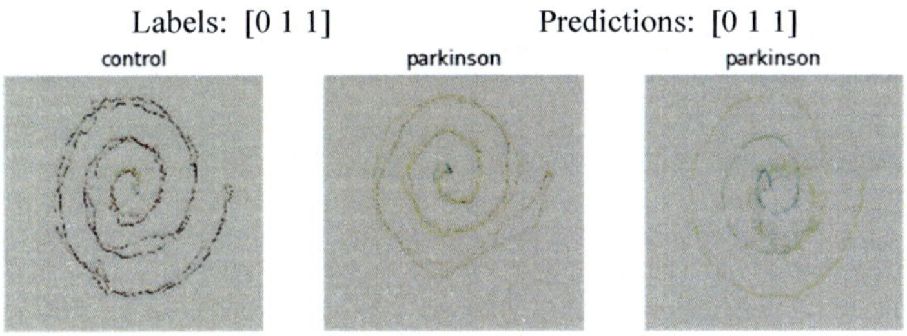

Fig. 8. Example of prediction on color test images with transfer learning.

The following table shows a comparison of our obtained accuracy with the other works mentioned previously in Sect. 2 (Table 1).

Table 1. Some previous works on the same database of enhanced spiral tests

Reference	Method	Accuracy
Khatamino et al. [11]	CNN from raw data	72.5%
Gil-Martín et al. [12]	CNN from spectrum	96.5%
Gallicchio et al. [13]	ESNs from raw data	89.33%
Our approach	Transfer learning with pressure effect	94.44%

5 Conclusion

In this paper, we address the problem of early automatic identification of Parkinson's disease using convolutional neural networks, specifically the transfer learning method using the pre-trained MobilNetV2 model. The latter is fed with images obtained from the signals extracted from the time series when drawing the spiral pattern. The main contributions of this paper are to complete the steps of the proposed system, including the addition of information about the pressure of the patients' handwriting translated by colors on the images, where each color is corresponding to a level of the pressure.

The experimental section compares the results obtained to justify, on the one hand, the choice of adding the pressure parameter and, on the other hand, why we opted for the transfer learning method. Indeed, the results proved to be very promising, as the CNNs succeeded in recording the essential information of the pressure. However, at the level of generalization for the Test-Set shows some inefficiency. For this reason, we thought of using the pre-trained MobileNetV2 model to benefit from its capabilities and adapt it to our task of identifying patients with Parkinson's disease. The results obtained are promising, especially the prediction results of the Test set, with a classification rate that reaches 94.44%.

References

1. Jankovic, J.: Parkinson's disease: clinical features and diagnosis. J. Neurol. Neurosurg. Psychiatry **79**, 368–376 (2018)
2. Ammenwerth, E., Nykanen, P., Rigby, M., de Keizer, N.: Clinical decision support systems: need for evidence, need for evaluation. Artif. Intell. Med **59**, 1–3 (2013)
3. Dreiseitl, S., Binder, M.: Do physicians value decision support? A look at the effect of decision support systems on physician opinion. Artif. Intell. Med. **33**, 25–30 (2005)
4. Mattison, H.A., Stewart, T., Zhang, J.: Applying bioinformatics to proteomics: is machine learning the answer to biomarker discovery for PD and MSA? Mov **27**, 1595–1597 (2012)
5. Lahmiri, S., Shmuel, A.: Detection of Parkinson's disease based on voice patterns ranking and optimized support vector machine. Biomed. Sig. Process. Control **49**, 427–433 (2018)
6. Gómez-García, J.A., Moro-Velázquez, L., Godino-Llorente, J.I.: On the design of automatic voice condition analysis systems. Part II: review of speaker recognition techniques and study on the effects of different variability factors. Biomed. Sig. Process. Control **48**, 128–143 (2019)
7. Viteckova, S., Kutilek, P., Svoboda, Z., Krupicka, R., Kauler, J., Szabo, Z.: Gait symmetry measures: a review of current and prospective methods. Biomed. Sig. Process. Control **42**, 89–100 (2018)
8. San-Segundo, R., Navarro-Hellín, H., Torres-Sánchez, R., Hodgins, J., De la Torre, F.: Increasing robustness in the detection of freezing of gait in Parkinson's disease. Electronics **8**, 119 (2019)
9. McLennan, J.E., Nakano, K., Tyler, H.R., Schwab, R.S.: Micrographia in Parkinson's disease. J. Neurol. Sci. **15**, 141–152 (1972)
10. Isenkul, M., Sakar, B., Kursun. O.: Improved spiral test using digitized graphics tablet for monitoring Parkinson's disease. In: ICEHTM, pp. 171–175 (2014)
11. Khatamino, P., Cantürk, I., Özyılmaz, L.: A deep learning-CNN based system for medical diagnosis: an application on Parkinson's disease handwriting drawings. In: Proceedings of the 2018 6th International Conference Control Engineering Information Technology, pp. 1–6 (2018)

12. Gil-Martín, M., Montero, J.M., San-Segundo, R.: Parkinson's disease detection from drawing movements using convolutional neural networks. Electronics **8**(8), 907 (2019)
13. Gallicchio, C., Micheli, A., Pedrelli, L.: Deep echo state networks for diagnosis of Parkinson's disease. arXiv:1802.06708 (2018)

Design a Controllable Band-Stop Microstrip Filter Based on Asymmetric Split Ring Resonator

Hanane Nasraoui[1(✉)], Youssef Mouzouna[2], Jamal El Aoufi[1], and Ahmed Mouhsen[3]

[1] Laboratory of Aeronautical Telecommunication, Mohammed VI, International Academy of Civil Aviation, Nouaceur, Morocco
hanane.nasraoui@gmail.com
[2] National High School for Electricity and Mechanics ENSEM, Hassan 2 University, Casablanca, Morocco
[3] Laboratory IMII, Faculty of Science and Technical, Hassan 1 University Settat, Settat, Morocco

Abstract. This work presents a design of a controllable bandstop microstrip filter based on an asymmetric split ring resonator (ASRR) . the proposed design is realized by a microstrip line loaded by ASRR. The structure of band-stop microstrip filters with ASRR is characterized by small size and simple structure. The ASRR offers the possibility of controlling the center frequency of the filter. These ASRR-based structures are thought to be of relevance for the creation of planar filters with possible use in future communication transceivers. The numerical and experimental results are in good agreement demonstrating the effectiveness of the proposed design.

Keywords: Band stop filter · Metamaterial · Microstrip · Asymmetric split ring resonator (ASRR)

1 Introduction

Artificial magnetic materials with a negative refractive index and a negative effective permeability on a defined frequency band establish a new electromagnetic concept. The unique properties of artificial magnetic materials enable the development of revolutionary applications of radio frequency (RF) circuits and engineering microwaves. The split ring resonators are key elements of metamaterials, many active and passive circuits contain split ring resonators how basic elements in their design as filters, diplexers, oscillators, amplifiers, and antennas [1]. In 1999 the first model of the split ring resonators was originally proposed by Pendry et al. [2], He introduced the first magnetic unit cell (Split Ring Resonator) and a network of metal structures in the form of Swiss roll. Since these years, several new resonators have been proposed to obtain improved magnetic properties. Changing the geometrical characteristics of unit cells allows both to create new structures and to adjust the magnetic properties as required. However,

S. Motahhir and B. Bossoufi (Eds.): ICDTA 2022, LNNS 454, pp. 470–479, 2022.
https://doi.org/10.1007/978-3-031-01942-5_47

each inclusion has advantages and disadvantages in terms of permeability function, dissipation factor, and scattering coefficient. The unique and extraordinary properties of split-ring resonators have encouraged researchers to use them in multiple applications. Researchers have proposed structures of filters based on metamaterials cells [3], and this is to achieve more compact filters, multi-band and reconfigurable. Early works on the filters using the resonators were through Garrcia et al. [4, 5]. They used the resonators SRR and CSRR to eliminate parasites band filters and improve their rejection [6–8]. In this work, we propose the design Band-stop filter based on a microstrip line charged by ASRR, we analyze the effect of the asymmetry of the gaps on the behavior of filter parameters. In addition, we adopt the idea of controlling the center frequency of the filter by ASRR. The filter was made on a Rogers FR4 with a relative dielectric permittivity of 4.4, a thickness of 1.6 mm, the filter size was $20 \times 38\ \text{mm}^2$.

2 Configuration and Design

The unit cell structures studied in this study are shown in Fig. 1. These are square-shaped single symmetric split-ring resonators (structure A), square-shaped single asymmetric split-ring resonators (B structure, structure C).

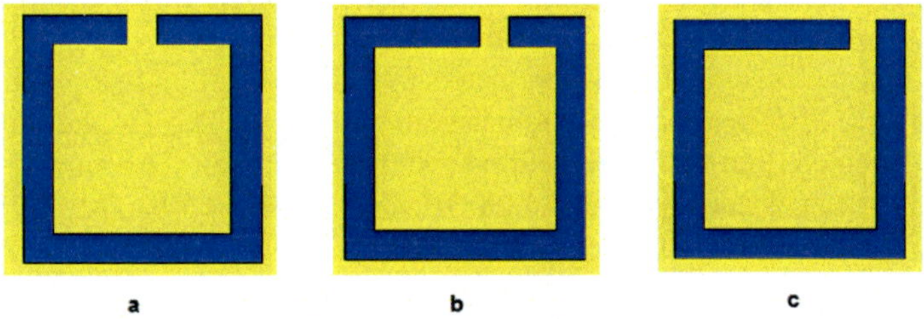

Fig. 1. The symmetric and asymmetric SRR structures that are used in this study

Currently, the microstrip line is used for the manufacture of microwave circuits [9, 10], mainly because it is going well has a Photolithography manufacturing process, [11] and also because it allows easy integration of active and passive surface mount components. The geometry of the proposed Band-stop microstrip filters presents the microstrip line structures loaded with split ring resonators symmetric and asymmetric (Fig. 1), the microstrip line is characterized by impedance 50 Ω, our structures have been printed on an FR-4 substrate with $\varepsilon = 4.4$ and at a height of about 1.6 mm. Dimensional view of Band-stop microstrip filters with split ring resonators symmetric and asymmetric are seen in Fig. 2.

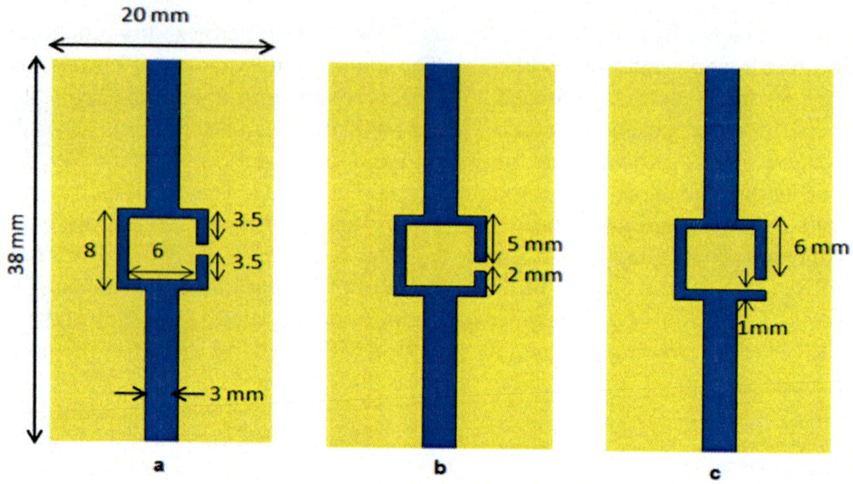

Fig. 2. The microstrip Band-stop filter structures with SRR symmetric and asymmetric

3 Simulation and Analysis

The results of S-parameters as a function of the frequency of microstrip Band-stop filters with split ring resonators symmetric and asymmetric are presented in Fig. 3.

Figure 3 reports the simulated S-parameters (S11&S21) of Band-stop microstrip filter structures with SRR symmetric (a) and asymmetric (b, c). The simulation shows the behavior of the notch filter with symmetrical split ring resonator (structure a) characterized by the operating frequency around 6.28 GHz and a-3 dB stop-band of 4.9 GHz from 3.8 GHz to 8.7 GHz. The S-parameters are presented in Fig. 3 of structure (b) shows a low reject level of −35 dB in the stop-band around the central frequency of 5.75 GHz, which is assumed for this filter; the stop-band of 5.2 GHz stretches from 3.8 GHz to 9 GHz. and the last bandstop microstrip filter with a split ring resonator asymmetric (ASRR) (structure c) characterized by a center frequency 4.83 GHz a-3 dB stop-band of 2.8 GHz from 3.2 GHz to 6 GHz, with a minimum rejection level of −39 dB around of the center frequency.

We observe the effect of the asymmetry of the gaps split-ring resonators on the behavior of notch filters parameters as shown in the simulation results of structures (Fig. 2). We can see the effect of the asymmetric split ring resonator, where replaces band-stop microstrip filter with symmetric split ring resonator by the new design a band-stop microstrip filter with an asymmetric split ring resonator, an improvement, reduction in the center frequency of new design with asymmetric split ring resonator compared to band-stop microstrip filter.

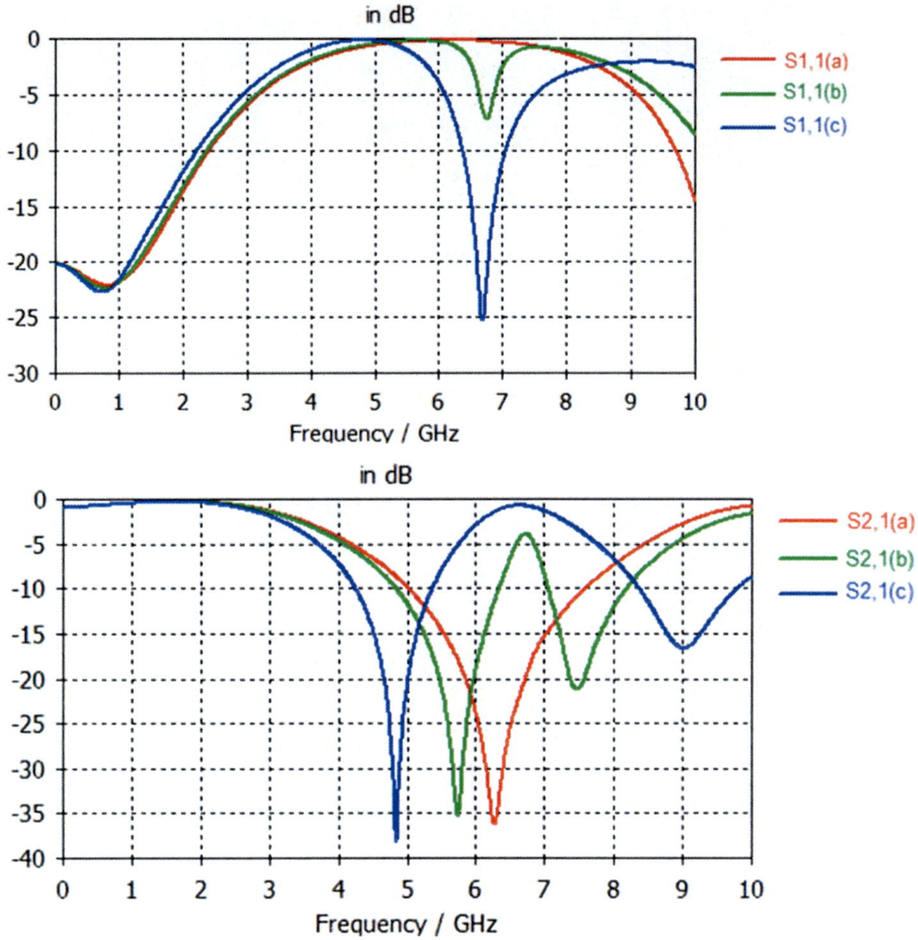

Fig. 3. S-parameters (S11&S21) of Band-stop microstrip filter structures with SRR symmetric (a) and asymmetric (b, c)

The surface current distribution of the band-stop microstrip filters loaded by ASRR is graphically presented in Figs. 4, 5 and 6, It is obvious that no power is communicated from Port 1 to Port 2 throughout the stop-band area of the stop-band microstrip filters loaded by ASRR.

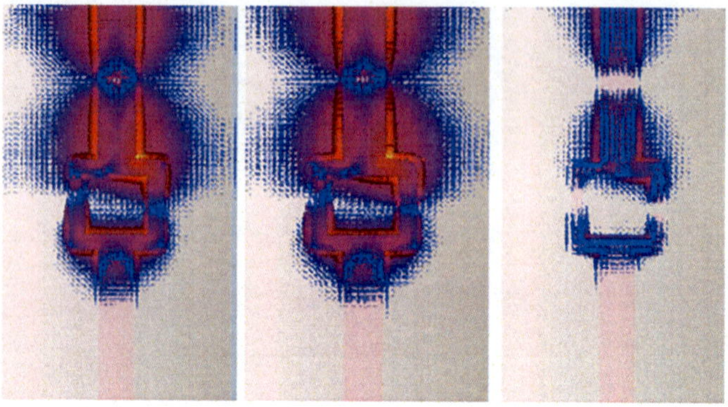

Fig. 4. The surface current distribution of the band-stop microstrip filter (a) at 6.28 GHz with a different phase.

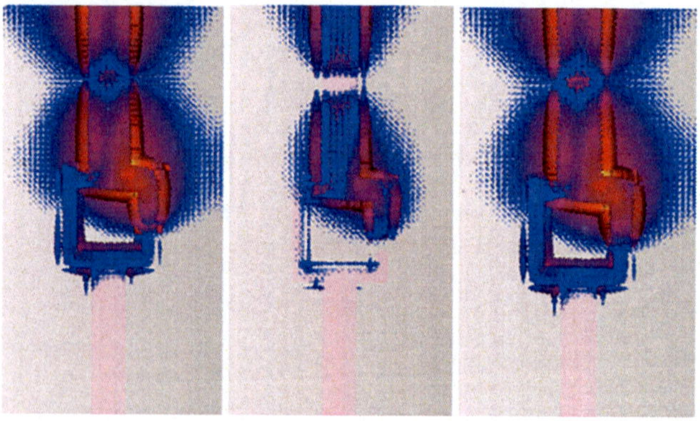

Fig. 5. The surface current distribution of the band-stop microstrip filter (b) at 5.75 GHz with a different phase.

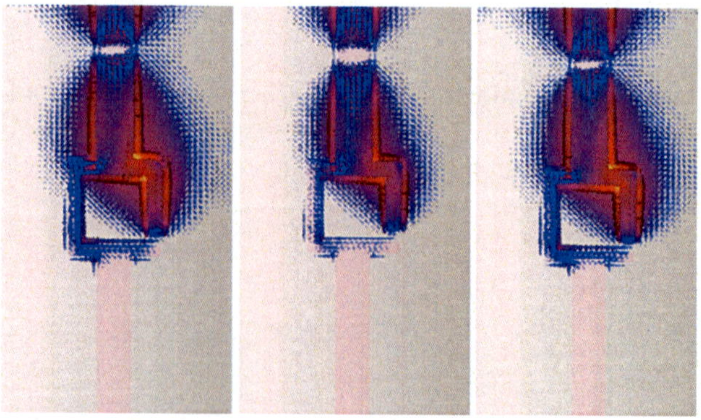

Fig. 6. The surface current distribution of the band-stop microstrip filter (c) at 4.83 GHz with a different phase.

4 Fabrication and Measurements

Figure 8 shows the photographs of the fabricated band-stop filters with dimensions described in Fig. 2, the proposed filters have been fabricated using an LPKF Protomat S100 mill/drill unit (see Fig. 7) and experimentally analyzed using Rohde & Schwarz ZVB20 analyzer (see Fig. 7) to verify the obtained simulation results.

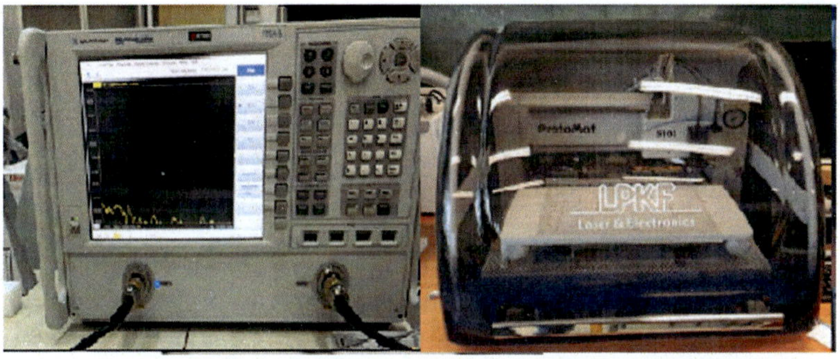

Fig. 7. Rohde & Schwarz ZVB20 analyzer and LPKF Protomat S100 mill/drill

The simulated and measured S parameters of the proposed printed filters are shown in Fig. 9, we see a good general accord between the S parameters measured and simulated.

Table 1 presents a comparison between the proposed band-stop filter and other band-stop filters recently published. We can conclude that the proposed filter offers high performance in terms in terms of miniaturization and stop-band characteristics.

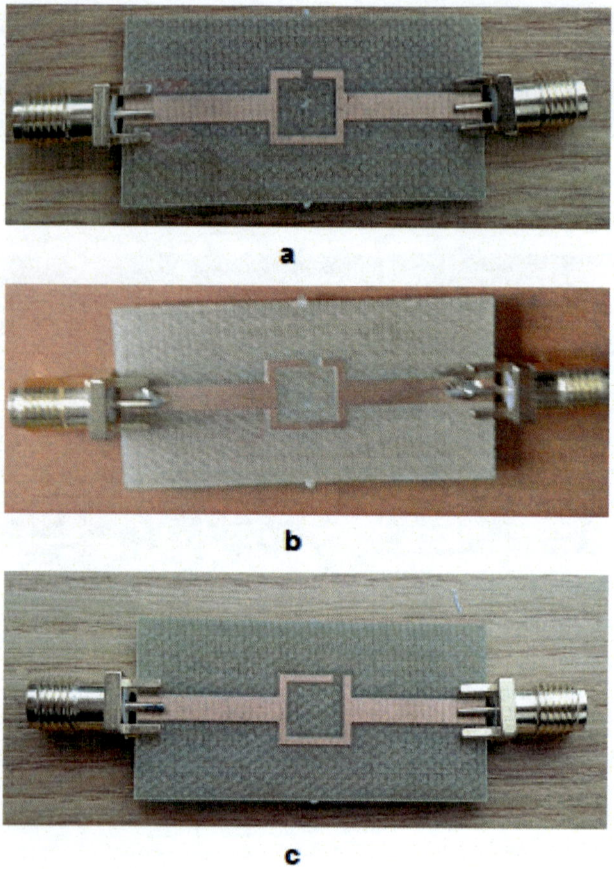

Fig. 8. Photograph of the fabricated printed band-stop microstrip filter structures with SRR symmetric (a) and asymmetric (b, c)

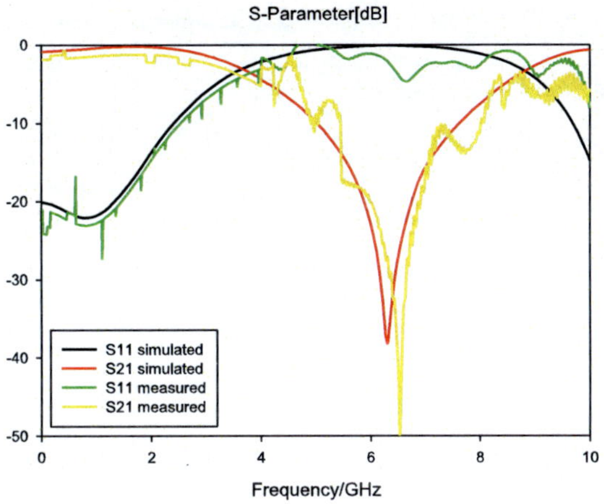

(a)

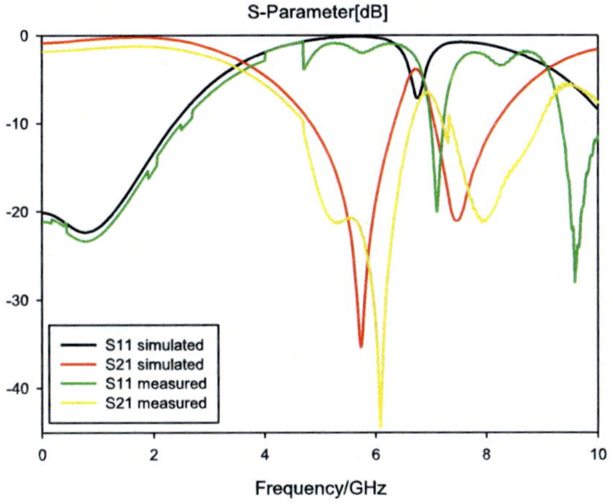

(b)

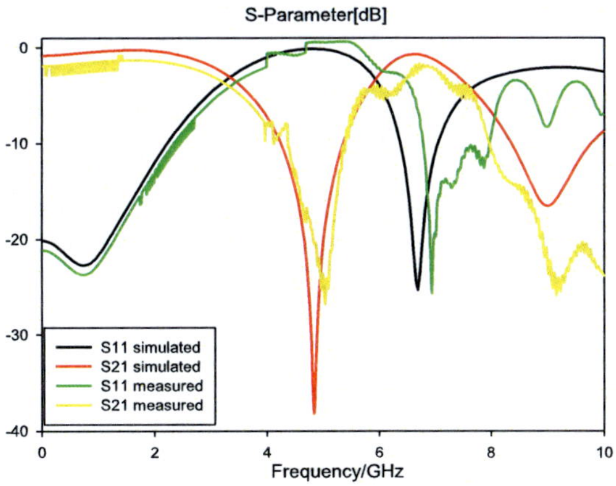

(c)

Fig. 9. Simulations and measurement results of Band-stop microstrip filter structures with symmetric SRR (a) and asymmetric (b, c)

Table 1. Comparison of this work with other passband filters.

Reference	Central frequency	Bandstop	Size
This work	4.83 GHz	2.8 GHz	$20 \times 30\ \text{mm}^2$
[12]	5.8 GHz	7.6 GHz	$35 \times 50\ \text{mm}^2$
[13]	3.15 GHz	0.5 GHz	$37 \times 40\ \text{mm}^2$
[14]	5.2 GHz	3.29 GHz	$100 \times 10\ \text{mm}^2$

5 Conclusions

This work has presented a study of Band-stop filter based on a microstrip line loaded by asymmetric split ring resonator (ASRR), We can see an improvement, reduction in the center frequency of new design with asymmetric split ring resonator compared to band-stop microstrip filter with asymmetric split-ring resonator, a new design methodology has been presented in this paper to exploit the property of the microstrip technology and asymmetric split ring resonator (ASRR) of other microwave applications.

References

1. Durán-Sindreu, M., Naqui, J., Paredes, F., Bonache, J., Martín, F.: Electrically small resonators for planar metamaterial, microwave circuit, and antenna design: a comparative analysis. Appl. Sci. **2**, 375–395 (2012)
2. Pendry, J.B., Holden, A.J., Robbins, D.J., Stewart, W.J.: Magnetism from conductors and enhanced nonlinear phenomena. IEEE Trans. Microwave Theory Tech. **47**(11), 2075–2084 (1999)
3. Disken, G., Pala, F., Demir, E., Korucu, H.D., Ekmekci, E.: Electromagnetic behavior of SRR loaded microstrip transmission lines: investigation for different SRR types and array topologies. In: Proceedings of the Progress in Electromagnetics Research Symposium (PIERS 2013), Stockholm, Sweden, pp. 1796–1799, 12–15 August 2013
4. Garrcia-Garrcia, J., et al.: Spurious passband supression in microstrip coupled line band pass filters using split ring resonators. IEEE Microwave Wirel. Compon. Lett. **14**(9), 416–418 (2004)
5. Garrcia-Garrcia, J., et al.: Microwave filters with improved stopband based on sub-wavelength resonators. IEEE Trans. Microwave Theory Tech. **53**(6), 1997–2006 (2005)
6. Casanueva, A., Leon, A., Gonzalez, O., Mediavilla, A.: A compact microstrip step-impedance low-pass filter (SILPF) using complementary split ring resonators (CSRRs). In: Institute of Electrical and Electronics Engineers (IEEE) MTT-S International Microwave Workshop on Wireless Sensing, Local Positioning, and RFID, Croatia, 24–25 September, pp. 1–4 (2009)
7. Hu, X., Zhang, Q., He, S.: Dual-band-rejection filter based on split ring resonator (SRR) and complimentary SRR. Microwave Opt. Technol. Lett. **51**(10), 2519–2522 (2009)
8. Zhang, X.Y., Huang, X., Zhang, H.-L., Hu, B.-J.: Novel dual-band bandstop filter with controllable stopband frequencies. Microwave Opt. Technol. Lett. **54**(5), 1203–1206 (2012)
9. Nasraoui, H., Capobianco, A., Mouhsen, A., El Aoufi, J., Mouzouan, Y., Taouzari, M.: Compact wideband band pass microstrip filter loaded by triangular split ring resonator. Microw. Opt. Technol. Lett. **59**(12), 3101–3107 (2017)

10. Lee, H.-J., Lee, J.-H., Jung, H.-I.: A symmetric metamaterial element-based RF biosensor for rapid and label-free detection. Appl. Phys. Lett. **99**(16) (2011). https://doi.org/10.1063/1.3653959
11. David m Pozar: Microwave Engineering. 4th edn, Wiley, New York (2011)
12. Gao, Z., Peng, W., Zhang, Y.: A new compact microstrip ultra-wideband (UWB) bandstop filter with good performance. Progr. Electromagn. Res. Lett. **93**, 9–12 (2020)
13. Fahmy, W.M., Farahat, A.E., Hussein, K.F.A., Ammar, A.-E.-H.A.: High Q-factor bandstop filter based on CPW resonator broadside-coupled to CPW through-line. Progr. Electromagn. Res. B **86**, 121–138 (2020)
14. Zheng, X., Jiang, T., Lu, H., Wang, Y.: Double-layer microstrip band stop filters etching periodic ring electromagnetic band gap structures. Electronics **9**(1216) (2020)

Government, Digital Society and Industry 4.0: Connective Action Against Covid-19 Fake News

Andi Luhur Prianto[1(✉)], Ihyani Malik[1], Nur Khaerah[1], Abdillah Abdillah[1(✉)], and Kittisak Jermsittiparsert[2]

[1] Universitas Muhammadiyah Makassar, Jl. Sultan Alauddin No. 259, Makassar, Sulawesi Selatan 90221, Indonesia
luhur@unismuh.ac.id, abdillah18001@mail.unpad.ac.id
[2] Dhurakij Pundit University, 110/1-4 Pracha Chuen Road, Thung Sondg Hong, Lak Si 10210, Bangkok, Thailand

Abstract. This article explores social media coverage of news related to the infodemic as a threat to panic and peace in society during the current global crisis due to Covid-19. The infodemic phenomenon has distorted scientific work and reduced public trust in state authorities. This research has identified and analyzed the government's response to fake news attacks on social media. The method used is qualitative-exploratory with data analysis used of Nvivo 12 Pro. The research results showed that the quality of people's digital literacy, public communication performance, and the effectiveness of government regulations are part of the challenges in mitigating the infodemic. The Indonesian people have been trying to mitigate the Covid-19 infodemic in Indonesia, both in groups and individually. The presence of the digital society and media movement to mitigate the Covid-19 infodemic that is rife in industry 4.0 within digital public spaces in Indonesia. The presence of fact-checking in the form of a report website to face hoaxes and disinformation is a form of the Indonesian digital 4.0 mitigation movement. In addition to the efforts of the Government of Indonesia, which has three strategic approaches, namely 1) Upstream, digital literacy, and capacity; (2) middle flow, direct action to remove harmful information through technology and collaboration with social media platforms; and (3) downstream, direct action to provide technical assistance in monitoring the spread of malicious information.

Keywords: Fake news · Government · Digital society · Connective action · Industry 4.0 · Covid-19

1 Introduction

Many studies have documented the benefits of using digital media in online activism [1, 2]. Social media platforms are widely used to carry out conversations including young people, urban activities - as well as socializing the government's policy to tackle fake news reports to the public [3]. The ability to disseminate information quickly, interactivity, and the ability to attract international attention, make social media an important feature and valuable resource [4]. Indonesia has different local wisdom that comes from

S. Motahhir and B. Bossoufi (Eds.): ICDTA 2022, LNNS 454, pp. 480–491, 2022.
https://doi.org/10.1007/978-3-031-01942-5_48

the development of science that accelerates the dissemination of information [5]. However, it is also the source of the threat of news uncertainty that invites chaos. According to Sufehmi [6] said that currently the term fake news has now become a global threat because it contributes to worsening the situation and does not help at all.

In addition to the challenges in trying to stop the spread of the new coronavirus or SARS-CoV-2, another obstacle facing society is the information epidemic about COVID-19. This kind of information overload can cause too much problem information, so its appearance can interfere with efforts to find solutions to the problem. Information epidemics can also be fatal, even leading to loss of life. For example, a phenomenon that often occurs in the community is misinformation about the antidote for COVID-19, making people feel safe with the drug, thus ignoring the health protocol recommendations. The consequences of this infodemic can be very fatal, even causing fatalities. For example, information about drugs is a hoax, so if you get hit, you can just give garlic, even though it's a hoax. There are various inciting narratives but hoaxes that cause panic in the community, which is already quite difficult because of this pandemic.

At the same time, research shows the weaknesses and bad threats of social media in the news. For example, Udalam [7] argues that people's digital activities often do not know what information is being collected, and when they do, it causes feelings of insecurity, anxiety, and caution in using social media. In addition, social media can allow powerful entities, such as organizations, to post important messages [7], and other groups can hijack hashtags to use for their purposes, which may conflict with shared goals [8], resulting in attempts to obtain information. trusted due to Covid-19, has become difficult to access.

This study uses an analysis of connective actions that can be understood through the concept of connective leadership [8–10] to mitigate fake news in Indonesia. Connective action online is different from collective action offline, where the leader is responsible for mobilizing and managing the interests of the general public, the connective leader functions as a filter for information. Their main task is to "connect people and information", "to give meaning and significance to the chaotic wealth generated by mass production", to act as "curator of raw material editing", to garner support and ensure that the information is valid. Internally (providing information to participants) and externally (providing information to the public and mass media) [9]. However, connective leadership has weaknesses that can be a threat to a movement. In the absence of centralized leadership and formal procedures for decision making, some individuals may subjectively assume a coordinating role, i.e. make decisions about movement developments, lobbies, events, and so on [8.5], which in turn can lead to power hierarchies and internal conflicts [8, 11].

Connective action logic can help explain social activism in events that communicate throughout their actions that they want to be leaderless or alienate parties and unions [12]. At the same time, they achieved an "impressive level of communication with the public", through messages and images that went viral on social networks. This connective action structure is loosely related and does not have the organizational hierarchy that can be found, informal groups. The connective structure also provides a platform for collective action [12, 13] (Fig. 1).

Fig. 1. Identification and analysis of research issues. Source: Processed by Nvivo 12 Pro, 2021

Several previous studies have found that there is a relationship between connective actions in explaining online activism and conceptualizing organizational structure. However, despite the significant emphasis on digital media, few studies explore how social media in the era of the industrial revolution 4.0 contributes to linking actions [10, 14]. Based on this background, this study examines how the connectivity of the digital community movement in the era of the industrial revolution 4.0 in Indonesia in mitigating fake news in Indonesia. At the same time, the government's strategy to deal with Covid-19 Fake News. The emergence of the hashtag #lawanCovid-19 #MasyarakatAnti Hoax is a form of digital community movement in Indonesia in the fight against the fake news of Covid-19 in Indonesia. Perseerita Fake Covid-19 Cases in Indonesia. The spread of fake news has caused a lot of anxiety and uncertainty, especially regarding the projection of the end of the Covid-19 crisis that hit the world in general, and in Indonesia in particular.

2 Research Method

This study aims to review the response of government institutions to attacks on fake news on social media in Indonesia. The methodological step is to identify, map and analyze the responses that have been carried out by the government. This study builds an analytical framework using qualitative and quantitative data. Data collection through exploration of content and news sources from social media and official government policies. This research was conducted with an interpretive-phenomenological approach, related to the digital activism movement in the spread and belief of hoaxes about Covid-19 that occurred in industry 4.0. The unit of analysis is the content of fake news spreaders on social media in 2020–2021 and the government's response through the Ministry of Communication and Information Affairs officially. The data analysis in this study uses the Nvivo-12 Pro tools, to help identify, map, and analyze the data.

3 Result and Discussion

3.1 Fake News About Covid-19 and the Digital Society Movement in the Industry 4.0 in Indonesia

An overview of fake news data traced from various sources. Especially from reports and identification of fake news about Covid-19 from the Directorate of Information Application Control, Directorate General of Applications, Ministry of Communication and Information Technology of the Republic of Indonesia (2021) [15]. It is also a digital society movement against the Covid-19 fake news in Indonesia. The spread of fake news is increasing in number in the realm of online media. In Indonesia's pluralistic society, this brings worry, tension, and anxiety. Mainly because of the difficulty of distinguishing true news from all existing news. By using the perspective of connective action as an effort to analyze, identify, select and evaluate the existing fake news. It is one of the efforts of the government and the digital society in mitigating the spread of fake news [16].

Social media activists and members of civil society made a declaration of the Indonesian Anti Hoax Society and disseminated information about the negative impact of hoaxes. This is a response to the widespread circulation of fake news or hoaxes on social media in Indonesia. It is hoped that many will be moved to join initiatives to combat hoaxes in the future, through the media, mass organizations, and other channels. The Indonesian Anti Hoax Community initially started from the resistance of Indonesian citizens who tried to combat hoaxes sporadically by establishing anti-hoax groups on social media.

Several groups were born because they were stifled by the rise of hoaxes, including the Anti-Slander, Incitement, and Hoax Forum (FAFHH), the Indonesian Hoax Extermination Fanpage & Group, the Indonesian Hoax Fanpage, and the Lifeboat Group. These four groups are all based on Facebook. The movement that was started to be rolled out was then put together in the Indonesian Anti Hoax Community Community forum to expand its reach to the online realm, with activities such as outreach and workshops on eradicating hoaxes. The Indonesian Anti Hoax Community is spread and active in various other regions in Indonesia. Apart from Jakarta, the launching and socialization of the Anti Hoax Society were also carried out simultaneously on the same day in Surabaya, Semarang, Solo, Wonosobo, and Bandung [17, 18].

The presence of anti-hoax community activists in Indonesia is expected to be able to assess the accuracy of news whether it is fake (hoax) or not, by comparing it with various other news sources. Until the Fake News countermeasures are better understood by the public. Hoax and disinformation cases as previously mentioned. As a result of advances in information and technology in this timeless. The spread of Palus news should be a valuable lesson and opportunity in improving the quality of media literacy in Indonesia. In the perspective of Mancur Olson [19], an individual will participate in a collective effort if he has more incentives than the resources he has to spend. Therefore, to be effective, collective action must be coordinated, have a leadership hierarchy and organizational strategy, have membership ties, and instill a collective identity or ideology. In contrast, in the connective action According to Bennet and Segerberg [12], individuals do not need to commit to a particular group. Participation can be done without the hassle of being a member. In cyberspace, political activism is fluid, flexible, and non-binding because it is carried out personally, but is connected by a shared concern for certain issues. So that in this study it can be said that there are two forms of movement to mitigate the Covid-19 infodemic in Indonesia in community activities, namely in groups and individuals. Like the rise of the media movement, the action is connected from groups to individuals and from individuals to groups [12].

Table 1. Social media fight infodemik lively hoax

No.	AccountSocial media	*#Hastag* Social media	Coverage
Indonesian government movement against infodemik (hoax and disinformation) Covid = 19			
2.	jabarprov.go.id I # Ayo5M + Vaccination @jabarprovgoid (Mar 5)	#Infodemic in West Java #JawaBarat #VaksinUntukKita #JabarSiapVaksinasi #CoronaVirus #Covid19 # Ayo5M + Vaksinasi On Twitter	How West Java residents deal with the infodemic (Covid-19 overstatement): (1) Check the source of the news; (don't just read the title; (3) find out who the author is; (4) check the date; (5) check the supporting evidence; (6) control ordinary thinking; (7) check with fact-checkers. At the same time, West Java applies 5M (wearing masks, keeping your distance, washing hands, staying away from crowds, and reducing mobility)

(continued)

Table 1. *(continued)*

No.	AccountSocial media	*#Hastag* Social media	Coverage
4.	Jabar Saber Hoax # AyoPakaiMasker@jabarsaberhoax (Aug 16)	#antihoax #antifitnah #hantamhoax #AyoPakaiMasker On Twitter	Initiation in hitting hoaxes in West Java by forming an online movement @jabarsaberhoax by Ridwan Kamil (Governor of West Java)
Indonesian community movement against infodemic (hoax and disinformation) Covid-19			
5.	IKRAM Malaysia@ikrammalaysia (Jul 25)	#Infodemic #Dissiplinilmu #BinaNegaraRahmah On Twitter	[Article] Defamation of the infodemic era: the importance of the principles of scientific discipline Next: https://bit.ly/era-infodemic
7.	NETIZEN.media @NETIZEN_media (Feb 7, 2019)	#melawanhoax #medsos #tipsjitu #bijak #coronavirus #COVID19 #infodemic #antihoax #antifitnah #hantamhoax #AyoPakaiMasker On Twitter	Fighting hoax is tips for playing social media wisely

Source: *Processed by researchers from Twitter and Instagram Social Media Platforms, 2021*

Table 1 above explains that in addition to the government's efforts in mitigating the infodemic, overcoming hoax news and disinformation in Indonesia regarding Covid-19. Shown in the hashtag #lawanCovid19 #InfodmikJawaBara #InfodmikBPOMRI #TNI-Infodenik. There are other movements from the Indonesian people with various forms of group and individual movements as an effort to mitigate the infodemic against hoaxes and disinformation about Covid-19. Shown in the hashtags #infodemic #melawanhoax #hantamhoax #antihoax and #antifitnah. Usually, the community's efforts are by holding online chats (webinars) and creating content tips for dealing with the infodemic. The path taken in overcoming the spread of hoaxes and disinformation in the Covid-19 infodemic case in Indonesia has various forms of movement in groups and individually by the Indonesian people. Nor by the government with the serious support of the people. One of the important efforts that can be done by the government is through a structural-systemic approach from existing regulations/policies [5, 20]. such as the Indonesian ITE Law to be able to curb the behavior of spreading Fake News. It is simpler to understand the movement of people in the digital space is fighting *Fake News* Covid-19, which is explained in Fig. 2 below:

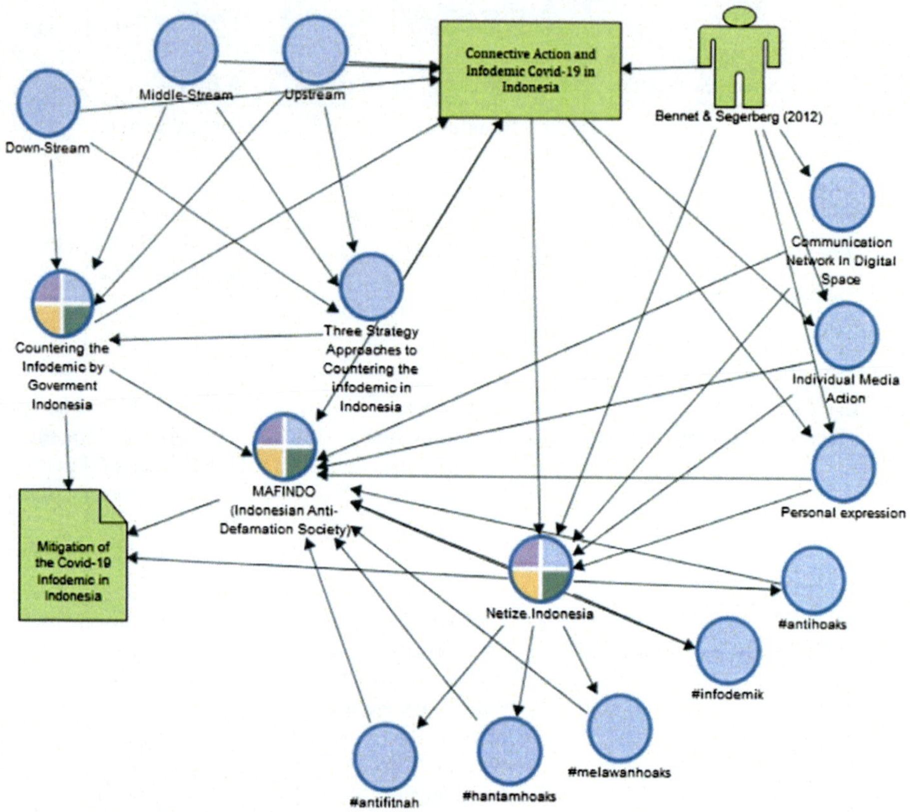

Fig. 2. Connective actions for infodemic mitigation efforts in Indonesia. Source: Processed by Researchers from Nvivo 12 Pro, 2021

In this regulation, the Government of Indonesia's efforts to control Fake News (Hoax and Disinformation) in Indonesia. The spread of fake news related to Covid-19 has a negative impact on society, such as giving rise to concern, panic, and distrust of information related to Covid-19 and its prevention efforts [21]. So that these regulations regarding Fake News (Hoax and Disinformation) are expected to regulate not only the fake newsmakers, who are given criminal sanctions. But also for the perpetrators who participated in sharing/transmitting (*share/forward*) the fake news. At the same time, as an effort to mitigate the Indonesian government's fake new infodemic that is developing in Indonesia. The following is a simple visualization of the infodemic mitigation efforts in Indonesia that build the relationship between the quality of digital literacy and the accuracy of government policies:

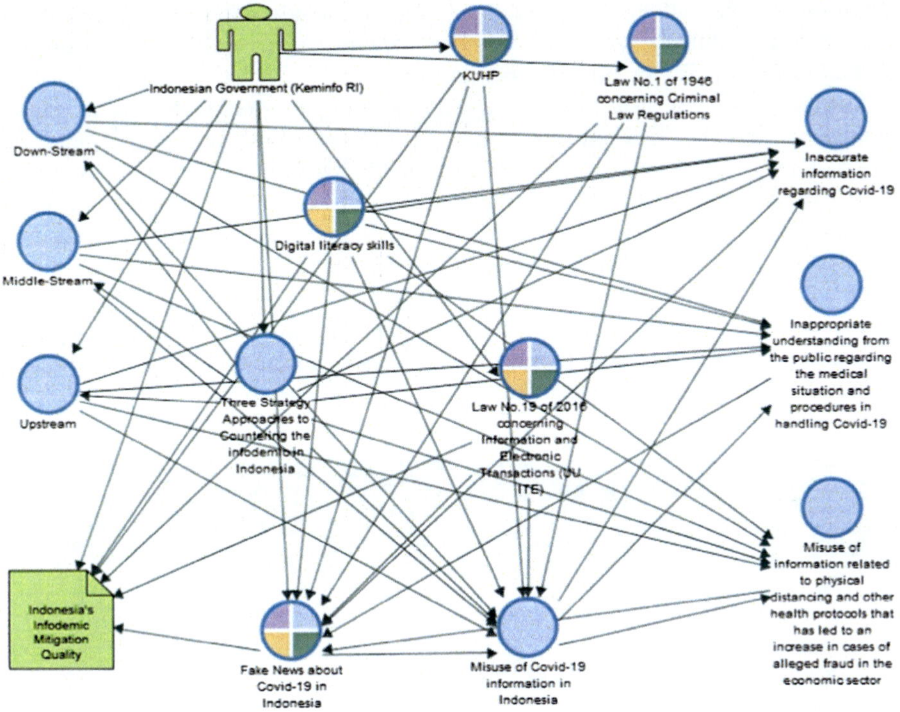

Fig. 3. Scenarios of infodemic mitigation efforts in Indonesia. Source: Processed by Nvivo 12 Pro, 2021

Based on the results of the analysis in Fig. 3 above, it is by the commitment of the Indonesian Government regarding the Infodemic mitigation. As the Government of the Republic of Indonesia voiced the importance of cross-sectoral collaboration in dealing with the *infodemic*. The Indonesian government focuses its attention on the rampant disinformation and misinformation that affects the handling of Covid-19 from a medical perspective [22]. According to Plate [22] Handling the Covid-19 pandemic is not just a medical problem. "The bigger problem lies in the problem of public perception and trust in information because there is too much information circulating and not a few of the information is misleading. This is called the *infodemic*. Emphasizing the importance of increasing literacy and adoption of digital technology to eradicate the *infodemic*. This initiative must be carried out through a cross-multistakeholder or stakeholder approach.

3.2 Approach Strategy in Threat of Fake News by the Government of Indonesian

Against Infodemic Anti-Hoax Society or called MAFINDO collaborates with the Government of Indonesia in combating hoaxes and *infodemic*. MAFINDO makes *fact-checking* in the form of a reporting website for the public, thereby helping the public to check the veracity of the information that is spread. Because Indonesia has experienced large-scale hoax incidents over the years. Creativity in handling cases using technology is often used by the Indonesian government [12, 15] Based on the Indonesian government applying the three approach method from downstream to upstream in fighting the spread of dangerous information circulating.

Looking at the assets of urban livelihoods, it is the accumulation of labor capital and financial capital [23]. Therefore, Indonesia needs to combat hoaxes and *infodemic* by recognizing three types of misuse of information, namely disinformation, misinformation, and misinformation. Director General Samuel explained, there are three issues of misuse of Covid-19 information that often arise in Indonesia, namely: (1) inaccurate understanding from the public regarding the medical situation and procedures in handling Covid-19, (2) inaccurate information related to Covid-19 19, and (3) misuse of information related to *physical distancing* and other health protocols which has led to an increase in cases of alleged fraud in the economic sector. The three-approach strategy mentioned above is shown in Fig. 4. below:

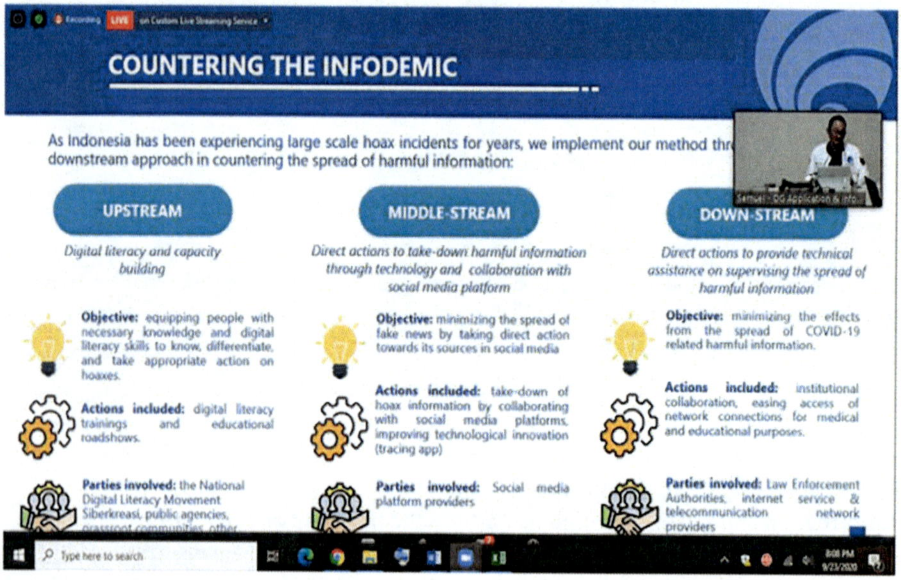

Fig. 4. Three strategies for countering the infodemic in Indonesia. Source: See Plate (2020), 2021

Figure 4 above fights the infodemic in Indonesia with three strategic approaches: (1) Upstream. digital literacy and capacity, the goal is to equip the public with the digital literacy knowledge and skills needed to identify, differentiate and take appropriate action against hoaxes. Actions include - digital literacy training and educational frog shows. Parties involved - cyber creation National digital literacy movement, public institutions, other grassroots communities; (2) Middle flow. direct action to remove harmful information through technology and collaboration with social media platforms. On the goal - minimizing the spread of fake news by taking direct action against its sources on social media. Actions taken include - suppression of hoax information by collaborating with social media platforms, increasing technological innovation (*tracing app*). Parties involved - social media platform providers; and (3) Downstream. direct action to provide technical assistance in monitoring the spread of malicious information. On the goal - minimizing the effects of the spread of covid-19 creating harmful information. actions include - institutional collaboration, easy access to network connections for medical and educational purposes. parties involved - law enforcement authorities, internet service providers, and telecommunications networks [15, 24, 25].

In sustainably dealing with hoaxes and infodemics, according to the Director-General of Aptika, the Government of Indonesia emphasizes efforts to educate and increase public digital literacy (capacity), in addition to efforts to handle content and also support for law enforcement [6, 22, 26] Broadly speaking, human activity is a place that is always become a destination and an attraction for the public to visit [27]. Thus, the community can be proficient and skilled in identifying misuse of information and taking follow-up actions if the information is wrong. As a concrete example, efforts to increase digital literacy are carried out through the National Digital Literacy Movement, Cybercreation, which collaborates with 108 communities, academics, and government institutions to carry out education and training related to literacy. digital. As a basis for conflict involving various groups of actors and forming a coalition framework [28, 29] To create community participation and the accuracy of government policies, it contributes to the improvement of critical social media posting practices, thus playing a role in mitigating the infodemic in Indonesia.

4 Conclusion

This study shows the quality of media literacy and the effectiveness of Indonesian government policies as an effort to mitigate the infodemic as a *Connective Action.* At the level of trust in information from online media and social media, which has an impact on decreasing trust in myths and false information related to COVID-19. Trust in news from online media, interpersonal communication, and social media, which is out of control has led to increased trust in the myths and false information of COVID-19. This in turn contributes to a less critical practice of social media posting, thereby exacerbating the infodemic in society. Public participation and the accuracy of government policies contribute to the improvement of critical social media posting practices, thereby playing a role in mitigating the infodemic in Indonesia.

The Indonesian people have been trying to mitigate the Covid-19 infodemic in Indonesia, both in groups and individually. The presence of the media movement with the hashtags #infodemic #melawanhoax #hantamhoax #antihoax and #antifitnah in the spirit of mitigating infodemic in public spaces in Indonesia. The presence of MAFINDO which makes *fact-checking* in the form of a reporting website for the public against hoaxes and disinformation is a form of the Indonesian people's mitigation movement. In addition to the efforts of the Government of Indonesia, which has three strategic approaches, namely 1) Upstream. digital literacy and capacity; (2) Middle flow. direct action to remove harmful information through technology and collaboration with social media platforms; and (3) Downstream. direct action to provide technical assistance in monitoring the spread of malicious information. Furthermore, the continuous handling of hoaxes and *infodemics* MAFINDO (Community Group Action) and the Government of Indonesia emphasize efforts to educate and increase public digital literacy (capacity), in addition to efforts to handle content and also support law enforcement.

References

1. Kumar, R., Thapa, D.: Social media as a catalyst for civil society movements in India: a study in Dehradun city. New Media Soc. **17**(8), 1299–1316 (2015). https://doi.org/10.1177/1461444814523725
2. Lee, F.L.F., Chan, J.M.: Digital media activities and mode of participation in a protest campaign: a study of the umbrella movement. Inf. Commun. Soc. **19**(1), 4–22 (2016). https://doi.org/10.1080/1369118X.2015.1093530
3. Howard, P.N., Duffy, A., Freelon, D., et al.: Opening closed regimes: what was the role of social media during the Arab Spring? (2011). https://ssrn.com/abstract=2595096, https://doi.org/10.2139/ssrn.2595096
4. Kharroub, T., Bas, O.: Social media and protests: an examination of Twitter images of the 2011 Egyptian revolution. New Media Soc. **18**(9), 1973–1992 (2016). https://doi.org/10.1177/1461444815571914
5. Prianto., A.L., Abdillah, A.: Dekolonisasi dan Indegenisasi Ilmu Pemerintahan. GOVERNABILITAS (Jurnal Ilmu Pemerintahan Semesta) **1**(2), 164–186 (2020). https://doi.org/10.47431/governabilitas.v1i2.86
6. Wibowo, A.: AncamanInfodemik Dapat Memperburuk Pandemi Covid-19 (2020). https://kominfo.go.id/content/detail/25895/ancaman-infodemik-dapat-memperburuk-pandemi-covid-19/0/virus_corona. Accessed Aug 2021
7. Uldam, J.: Social media visibility: challenges to activism. Media Cult. Soc. **40**(1), 41–58 (2018). https://doi.org/10.1177/0163443717704997
8. Azer, E., Harindranath, G., Zhen, Y.: Revisiting leadership in information and communication technology (ICT) enabled activism: a study of Egypt's grassroots human rights groups. New Media Soc. **21**(5), 1141–1169 (2019). https://doi.org/10.1177/1461444818821375
9. Della, R.D., Valeriani, A.: Remixing the spring! Connective leadership and read-write practices in the 2011 Arab uprisings. CyberOrient: Online J. Virtual Middle East **6**(1) (2012). http://www.cyberorient.net/article.do?articleId=7763
10. Poell, T., Abdulla, R., Rieder, B., et al.: Protest leadership in the age of social media. Inf. Commun. Soc. **19**(7), 994–1014 (2016). https://doi.org/10.1080/1369118X.2015.1088049
11. Toepfl, F.: From connective to collective action: internet elections as a digital tool to centralize and formalize protest in Russia. Inf. Commun. Soc. **21**(4), 531–547 (2018). https://doi.org/10.1080/1369118X.2017.1290127
12. Bennett, W.L., Segerberg, A.: The logic of connective action. Inf. Commun. Soc. **15**, 739–768 (2012). https://doi.org/10.1080/1369118X.2012.670661
13. Lim, M.: Framing Bouazizi: "White lies", hybrid network, and collective/connective action in the 2010–11 Tunisian uprising. Journalism **14**(7), 921–941 (2013). https://doi.org/10.1177/1464884913478359
14. Vaast, E., Safadi, H., Lapointe, L., et al.: Social media affordances for connective action: an examination of microblogging use during the Gulf of Mexico oil spill. MIS Q. **41**(4), 1179–1205 (2017). https://doi.org/10.25300/MISQ/2017/41.4.08
15. Kominfo: Laporan Isu Hoax Periode Maret 2021. In: Documen (Directorate General of Informatics Applications, Ministry of Communications and Informatics RI) (2021). https://eppid.kominfo.go.id/storage/uploads/2_31_Maret_2021_-_Isu_Hoax_Bulanan.pdf. Accessed July 2021

16. Malik, I., Prianto, A.L., Abdillah, R.Z., Amalia, A.A.: Urban resilience strategy in the climate change governance in Makassar City, Indonesia. J. Gov. Civil Soc. **5**(1), 31–50 (2021). https://doi.org/10.31000/jgcs.v5i1.3884
17. Kompas: Perlawanan "Netizen" Lahirkan Masyarakat Anti-"hoax" (2017). https://tekno.kompas.com/read/2017/01/08/10495047/perlawanan.netizen.lahirkan.masyarakat.anti-.hoax.?page=all. Accessed Aug 2021
18. Yunita: Ini Cara Mengatasi Berita "Hoac" di Dunia Maya (2017). https://kominfo.go.id/content/detail/8949/ini-cara-mengatasi-berita-hoax-di-dunia-maya/0/sorotan_media. Accessed Aug 2021
19. Olson Jr., M.: The Logic of Collective Action: Public Goods and the Theory of Groups. Harvard University Press, Cambridge (1965)
20. Abdillah, A., Deliarnoor, N.A., Yuningsih, N.Y., Fatmawati, F.: The position of auxiliary organ in government system of West Java Provincial Government. J. Contemp. Gov. Publ. Policy **1**(2), 67–81 (2020). https://doi.org/10.46507/jcgpp.v1i2.11
21. Hadna, A.H., Nasrulhaq, N.: Collaborating local government agencies to prevention adolescent reproductive health in Makassar, Indonesia. Otoritas: Jurnal Ilmu Pemerintahan **9**(1), 12–27 (2019). https://doi.org/10.26618/ojip.v9i1.2025
22. Plate, J.G.: Pemerintah RI NyatakanKomitmen Tangani Infodemik, May 2020. https://www.kominfo.go.id/content/detail/29609/siaran-pers-no-116hmkominfo092020-tentang-pemerintah-ri-nyatakan-komitmen-tangani-infodemic/0/siaran_pers. Accessed July 2021
23. Prianto, A.L., Amalia, A.A.: Combatting slums, suistaining poverty: dynamic urban governance in Makassar, Indonesia. Otoritas: Jurnal Ilmu Pemerintahan **9**(1), 28–41 (2019). https://doi.org/10.26618/ojip.v9i1.2008
24. Pond, P., Lewis, J.: Riots and Twitter: connective politics, social media and framing discourses in the digital public sphere. Inf. Commun. Soc. **22**(2), 213–231 (2019). https://doi.org/10.1080/1369118X.2017.1366539
25. Postill, J.: Book review: the logic of connective action: digital media and the personalization of contentious politics. Int. J. Press/Polit. **20**(4), 498–500 (2013). https://doi.org/10.1177/1940161215599940
26. Nurmandi, A., Purnomo, E.P., Prianto, A.L., Solahudin, J.: The current practice of citizen participation in Indonesia citizen and selected cases of local democracy in Indonesia. In: Citizen Participation in Selected Local Governance of Asia Pacific Countries, APSPA, Yogyakarta, pp. 21–48 (2015)
27. Faisah, N., Prianto, A.L.: Good environmental governance (Studi Kasus Pengelolaan Taman Macan Di Kota Makassar). Otoritas: Jurnal Ilmu Pemerintahan **5**(2), 174–188 (2015). https://doi.org/10.26618/ojip.v5i2.122
28. Prianto, A.L.: Model of advocacy coalitions in the formulation of spatial plan policy of Makassar in 2010–2030. Bisnis Birokrasi J. **20**(3), 163–170 (2013). https://doi.org/10.20476/jbb.v20i3.3205
29. Wijetunge, P.: Empowering 8: the information literacy model developed in Sri Lanka to underpin changing education paradigms of Sri Lanka (2005). http://archive.cmb.ac.lk:8080/research/bitstream/70130/4074/1/InformationLiteracy.pdf. Accessed Aug 2021

Inventory Management in a Refugee Crisis: Paving the Way Towards an Effective New Inventory Model

Youssef Malhouni(✉) and Charif Mabrouki

Faculty of Science and Technology of Settat, Hassan 1st University, Settat, Morocco
youssefmalhouni@gmail.com, Charif.mabrouki@uhp.ac.ma

Abstract. This paper aims to prepare a set of the most important assumptions to be considered when modeling the inventory management of a large humanitarian refugee organization. For this purpose, our paper attempts to link the results of conceptual works regarding refugee crisis management to the principles of inventory management in the context of the humanitarian supply chain. As a result, the paper paves the way for researchers to develop an efficient mathematical model to take under consideration the various parameters, such as the big demand fluctuations, the complexity of the situation of refugees, the choices of shipments, the budget, the optimal partitioning, and finally to the imperatives concerning the subsequent solution approaches.

Keywords: Refugee crisis · Inventory management · Refugee supply chain management · Humanitarian logistics · Disaster management

1 Introduction

Most refugee crisis creates a massive demand for emergency aid including food, medicine, shelter, water, and other resources. To meet these needs, effective emergency operations are needed to increase the supply of emergency aid (M.Safeer et al. 2014) [1]. In addition, managing emergency humanitarian logistics operations is more complex due to the risks and uncertainties that accompany any disaster (Ülkü, Bell, & Wilson, 2015) [2]. This poses challenges in the planning and mitigation phases in case a disaster is expected. Indeed, such a massive aid operation requires a complex network of stocks and distribution channels, for example, delivery times under current supplier agreements are variable and long, existing stocks may not be optimally located and stock shortages remain a fatal risk. This could spoil and create logistical bottlenecks, and sorting, storage, and distribution costs may even exceed the monetary value of the donations (R.Kian et al. 2021) [3]. Several authors have dealt with this topic in different ways for example Haavisto & Kovács (2014) [4] who studied the problems related to the transport of aid. On the other hand, authors G.Ferreira et al. (2018) [5] develop a model of continuous management of aid stocks, taking into account the uncertainties of both donations and demand. Recently, authors R.Garcia et al. (2021) [6] provide an example

S. Motahhir and B. Bossoufi (Eds.): ICDTA 2022, LNNS 454, pp. 492–500, 2022.
https://doi.org/10.1007/978-3-031-01942-5_49

of a system of drinking water supply in conditions of extreme drought and political and social conditions of a refugee camp.

Motivated by the increasing importance of the problems related to the storage of refugee aid, we propose a mathematical program's assumptions that will help optimize inventory management decisions based on the model of McCoy and Brandeau (2011) [7]. Thus, our contributions are the following: We propose new assumptions close to reality that aims to overcome the shortcomings of the model McCoy and Brandeau (2011) [7]. However, this work is based on only one model by McCoy and Brandeau (2011) [7]. This point constitutes both a weakness and a research opportunity. Indeed, since there is a lack of research addressing this problem, our work seeks to be the first one to propose a global model that is closer to the realities of field practice.

To this end, the remainder of the paper is structured as follows: Sect. 2 briefly summarizes the chronology of the evolution of the humanitarian supply chain management (HSCM) with an overview of analytical approaches developed for humanitarian inventory management, Sect. 3 describes the main challenges of inventory management related to refugee crises, Sect. 4 discusses the only model produced in this particular context (McCoy and Brandeau 2011) [7] before submitting some important assumptions to consider in future research.

2 Overview of HSC

In contrast to refugee logistics, HSCM research has undergone great evolution over time, which has earned it a certain maturity and confidence crowned by increasingly close cooperation with managers and practitioners in the field. This section traces briefly the evolution of this discipline before discussing the state of the art of inventory management.

2.1 Trend of Research on HSCM

To determine the chronological evolution of research in HSCM, we have opted for a particular approach. First of all, we have carefully analyzed the different literary journals established in this field, before carrying out a meticulous watch relating to the recent periods not covered by the above-mentioned publications. Then we have divided this evolution according to the advances common to each period. This method has allowed us to understand the path taken by this field of study and to determine the global orientation of the related research.

Thus, it appears that HSCM remains marked by an evolutionary, logical, and progressive approach starting with major definitions of boundaries and performance factors through the development of analytical and conceptual models before tackling the concept of inter-organizational cooperation to finally integrate industry 4.0 technologies, big data, and blockchain. On the other hand, the current year remains characterized mainly by the logistics of relief in a given area within the global crisis due to Covid-19.

2.2 Overview of Humanitarian Inventory Management

During the last decade, there has been a significant increase in the number of survey articles that focus on HL/DM. For example, Natarajarathinam et al. (2009) [8] focus on supply chain management in times of crisis, which can be caused by disasters or other causes such as a financial crisis. In contrast, other authors have focused on a single type of problem, while others address specific phases of the disaster management cycle by examining OR models for improving the performance of humanitarian operations. For example, Liberatore et al. (2009) [9] and Hoyos et al. (2015) [10] focus specifically on uncertainty modeling and present an analysis of stochastic models in the HL/DM literature. in addition, authors M.Tavana et al. (2018) [11] propose a multi-echelon humanitarian logistics network that takes into account the location of central warehouses, manages the inventory of perishable goods in the pre-disaster phase. Other authors S. Agarwal et al. (2020) [12] aim to explore the barriers to humanitarian supply chain management (HSCMB) and evaluate solutions to overcome these barriers to improve the implementation of humanitarian supply chain management (HSCM). S.Dermibas & M.Ertem et al. (2021) [13] study the benefits of operating equivalent warehouses (i.e., an appropriate level of inventory for each type of relief item) while reallocating a total available inventory for better coordination. In addition, it should be noted that most research focuses on pre-disaster studies, which aim to address a wide variety of disasters, as opposed to post-disaster research that focuses primarily on inventory decisions for a single item, which is one of the main limitations and also a real barrier to the use of these items in the field by practitioners.

3 Particularities of the Refugee Supply Chain Management

This section first briefly discusses the distinctive features and particularities of the refugee crisis before describing the importance of the inventory management strategy in this field through efforts of reflection, analysis, and synthesis that have led to the attached Figs. 1, 2, and 3.

3.1 Special Challenges of Refugee Logistics

To better address the logistics of refugees and the particularities related to them, it is necessary to elucidate the major challenges that refugees face. In this regard, we've highlighted in Fig. 1 the three levels that constrain their journey.

On the other hand, mass exoduses of refugees are most often provoked by armed conflicts where the affected populations are targeted because of their ethnicity, religion, or political opinions. This further complicates the supply chain due to chaotic conditions around key logistical facilities supporting the refugee camps, e.g. false dams, presence of armed militias, the existence of mines or explosive waste, the eventual desire of the government to seize humanitarian aid or to force its delivery to a part of the population that favors it.

Consequently, meeting the needs of international refugees and IDPs through logistics and SCM will have to include taking into account complex security and political contexts, continuous physical movement and mobility of the refugees and IDPs, and their constantly variable demand patterns, whether in sudden emergencies with high peaks or steady protract ongoing situations.

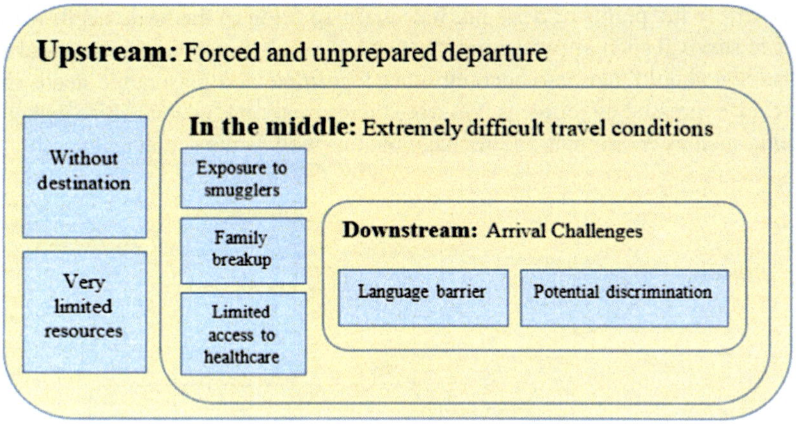

Fig. 1. Major characteristics of refugees

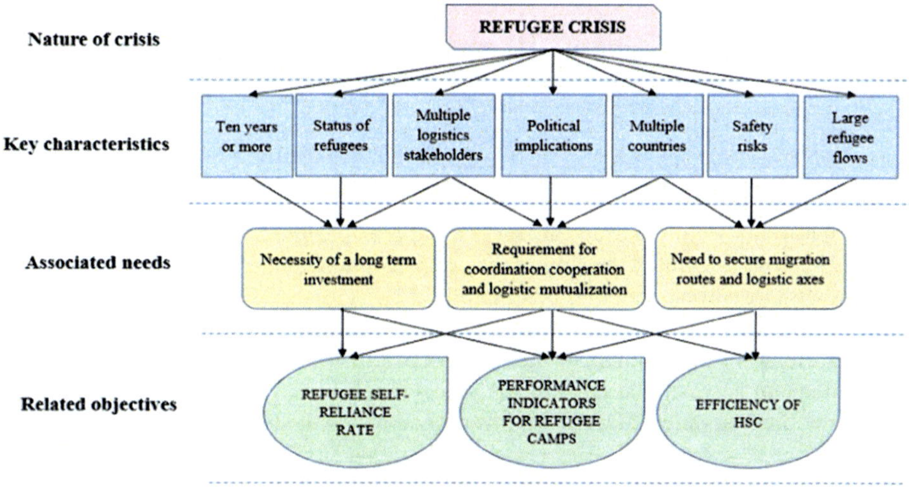

Fig. 2. Correlation between the most important challenges of a refugee crisis

Therefore, we drew Fig. 2 which outlines the most important factors of a refugee crisis and their correlation with the performance factors of the related supply chain. The figure is the result of our conception of the main axes that constitute the master plan of a high severity refugee crisis.

3.2 Major Criteria of the Inventory Management Strategy in the Refugee Crisis

From the above, in addition to the traditional concept of inventory management, it is now necessary to take a more general view of the support system of a refugee camp, through the appropriate choice of an "inventory management strategy". This support should not be limited to the acquisition of humanitarian aid, its storage, maintenance, and shipment to the points of demand. Indeed, in addition to the basic needs necessary for refugee survival such as protection, food, water, shelter, personal safety, and health, the logistician should take into account other legitimate and reasonable needs that are felt and even expressed by refugees. We were therefore prompted to analyze perspectives from fields such as economics and management as well as the social sciences.

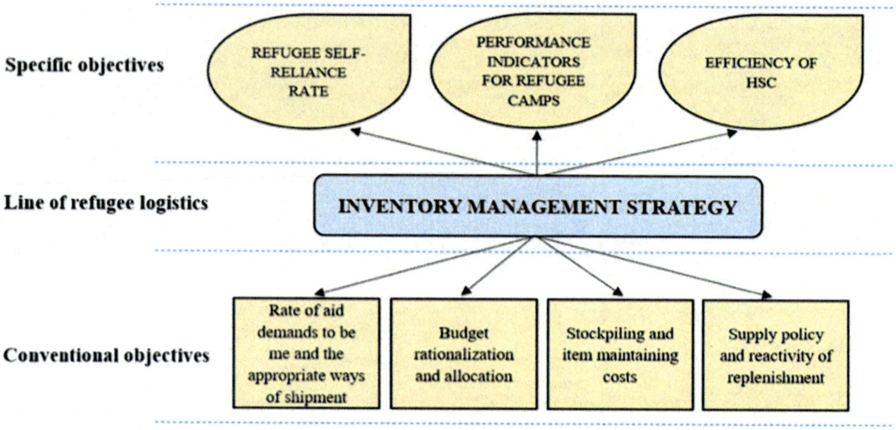

Fig. 3. Building blocks of inventory management strategy in a refugee crisis

Our questions were answered in several papers consulted, including Schön et al. (2018) [14], who illustrated the importance of self-reliance for refugee camp residents through long-term investment in infrastructure and services, and Betts et al. (2017) [15], who emphasized that the encouragement of refugee entrepreneurship within the camp, could contribute to a reduction in the dependence of refugees. Figure 3 clarifies the interrelationship between the level of long-term investment, refugee self-reliance rate, and other traditional dimensions of the inventory management strategy in this particular field.

4 Results and Discussion

To the best of our knowledge, the model developed by McCoy and Brandeau (2011) [7] is a fundamental paper that is exclusively addressing inventory management for refugee operations. It provides a valuable foundation for an organization's operational decision-making process geared towards efficient relief processes that combine limited resources with the needs announced by the administration of a given refugee camp. In this regard, we have identified four major limitations of this model that we have transcribed into imperative considerations for the design of an inventory management model dedicated to refugee crises.

This last section clarifies these limitations by starting with those related to the demand, then to the budget management, as well as to the shipping operation, and finally to the chosen mathematical solution.

4.1 Demand Assumptions

In practical and realistic terms, the demand of a refugee camp is based on a series of commodities whose quantity, importance, and urgency vary according to the situation on the ground.

Table 1. Limitations and proposals regarding the demand aspect

Demand criteria	Assumptions of McCoy and Bandeau (2011) [7]	Proposed assumptions
Number of items	01	Multi-items
Quantity of request	Independent and identically distributed across periods	Different and Interdependent
Importance and urgency	Unconsidered	Taken to account

Therefore, it is inconceivable in the real world to assume that demand is focused on a single product, nor to pretend that demand is fixed and independent of each other, since it varies according to the need identified in each period and remains interdependent. Indeed, the predominant factor is the number of refugees housed, which can change from one day to the next due to the voluntary departure of refugees or the arrival of a large number of refugees. Table 1 summarizes the contrast identified in the model of McCoy and Bandeau (2011) [7] as well as our proposed assumptions.

4.2 Shipment Assumptions

Once a demand is considered, the decision to fulfill it remains dependent on and directly related to the cost of shipping the items in question. Therefore, it is necessary to take into account the means chosen as well as the timeframe for shipping reliefs.

Furthermore, experience in the field shows that the choice of shipping method is closely linked to the degree of urgency of the demand to be met, as well as to the availability of infrastructure in the country of destination of humanitarian aid. Table 2 proposes to integrate into future models the means and time factors of shipment.

Table 2. Limitations and proposals regarding the shipment dimension

Shipment criteria	Assumptions of McCoy and Bandeau (2011)	Proposed assumptions
Means of shipment	Unconsidered	Depending on the urgency of the demand
Timeframe of shipment transit	Instantaneous	Depending on the chosen shipping mode according to the urgency of the demand and the budget ability

4.3 Budget Management

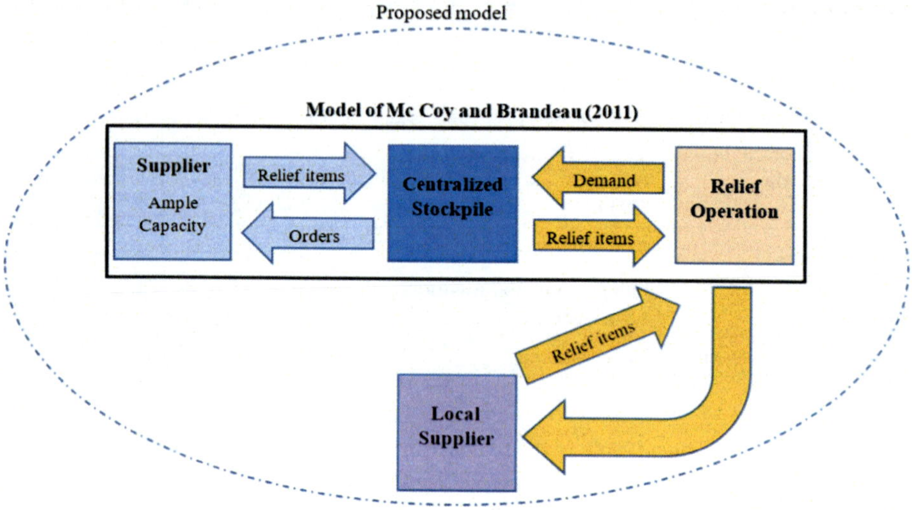

Fig. 4. Enhanced model proposal

Any organization or entity with a legal personality is characterized by the uniqueness of management and budget. in this context, it would be absurd to consider a procurement strategy that relies solely on shipments to rationalize expenses without accounting for purchases made from the local market by the refugee camp administration.

This suggests that the camp administration has an autonomous budget that is not derived from the central structure, which is not true. Figure 4 provides the necessary extension to McCoy and Brandeau's (2011) [7] model to fit the actual scenario experienced on the ground.

In the same spirit, reducing the cost of a product to the value of 1 would be unrealistic given the aforementioned objective of budget rationalization.

Furthermore, it would be smart to consider long-term investments such as school desks, educational aids and materials, and vocational training materials for refugees; this investment could be given a small percentage of the total budget for refugee support. Table 3 summarizes the proposed budget considerations for future models.

4.4 Solution Approach

The chosen heuristics by McCoy and Brandeau (2011) [7] refer to a huge deviation from the optimal solution. This percentage of error is due to the desire to find the best solution in the least amount of time. This was the objective of the simulator they delivered to the UNHCR at the end of their work.

Nevertheless, it is clear that this gap, reaching 21% from the 8th period onwards, is enormous given the financial stakes it generates. Table 4 proposes to choose a more efficient heuristic or meta-heuristic for solving the model.

Table 3. Limitations and proposals regarding the budget allocation

Budget criteria	Assumptions of McCoy and Bandeau (2011) [7]	Proposed assumptions
Locally procurement budget	Unconsidered	Taken into account
Cost of items	Normalized to 1	Variable according to the value of the article
Long term Investment	Unconsidered	Include a percentage of the total budget dedicated to camp development and refugee self-reliance

Table 4. Limitations and proposals regarding the mathematical solution

Solution criteria	Assumptions of McCoy and Bandeau (2011) [7]	Proposed Assumption
Percentage of error	Giant gap reaching 21% from the 8th Period	Choice of a more efficient heuristic or meta-heuristic

5 Conclusion

Ultimately, refugee logistics is a subfield of the humanitarian supply chain. The trend of the humanitarian supply chain, as well as the literature review concerning refugee logistics, shows that the latter is at its beginning. In this context, we believe that the assumptions we have proposed constitute a reorientation of both approaches and major contributions recorded to date, towards a more concrete and realistic model for optimizing inventory management in a refugee crisis context. Nevertheless, it turns out that some of the simplifications made in the case studies of refugee logistics stockpile management can marginalize or even lose sight of some major issues that are real bottlenecks for any large humanitarian organization operating in multiple theaters.

Starting from the field conditions, if the researchers were to work on the concretization of a model following the assumptions proposed during the work, the managers of humanitarian organizations, especially the largest ones such as the UNHCR, would be incentivized to take advantage of it and to reinforce the spirit of cooperation and recover "lost trust" between practitioners and academics.

References

1. Safeer, M., Anbuudayasankar, S.P., Balkumar, K., Ganesh, K.: Analyzing transportation and distribution in emergency humanitarian logistics. Proc. Eng. **97**, 2248–2258 (2014)
2. Ülkü, M.A., Bell, K.M., Wilson, S.G.: Modeling the impact of donor behavior on humanitarian aid operations. Ann. Oper. Res. **230**(1), 153–168 (2014). https://doi.org/10.1007/s10479-014-1623-5

3. Kian, R., et al.: Logistics planning of cash transfer to Syrian refugees in Turkey. Eur. J. Oper. Res. **296**(3), 1007–1024 (2022)
4. Haavisto, I., Kovács, G.: Perspectives on sustainability in humanitarian supply chains. Disaster Prevent. Manage. (2014)
5. Ferreira, G.O., Arruda, E.F., Marujo, L.G.: Inventory management of perishable items in long-term humanitarian operations using Markov Decision processes. Int. J. Disas. Risk Red. **31**, 460–469 (2018)
6. García, R., Naves, A., Anta, J., Ron, M., Molinero, J.: Drinking water provision and quality at the Sahrawi refugee camps in Tindouf (Algeria) from 2006 to 2016. Sci. Total Environ. **780**, 146504 (2021)
7. McCoy, J.H., Brandeau, M.L.: Efficient stockpiling and shipping policies for humanitarian relief: UNHCR inventory challenge. OR Spectrum **33**(3), 673–698 (2011)
8. Natarajarathinam, M., Capar, I., Narayanan, A.: Managing supply chains in times of crisis: a review of literature and insights. Int. J. Phys. Distrib. Logist. Manag. **39**(7), 535–573 (2009)
9. Liberatore, F., Scaparra, M.P.: Optimization and analysis of protection strategies for supply chains: Comparing regret and expected models. in Working Paper Series Kent Bus. School, Canterbury, U.K (2009)
10. Hoyos, M.C., Morales, R.S., Akhavan-Tabatabaei, R.: OR models with stochastic components in disaster operations management: a literature survey, Comput. Ind. Eng. **82** 183–19kent7 (2015)
11. Tavana, M., Abtahi, A.R., Di Caprio, D., Hashemi, R., Yousefi-Zenouz, R.: An integrated location-inventory-routing humanitarian supply chain network with pre-and post-disaster management considerations. Socioecon. Plann. Sci. **64**, 21–37 (2018)
12. Agarwal, S., Kant, R., Shankar, R.: Evaluating solutions to overcome humanitarian supply chain management barriers: a hybrid fuzzy SWARA–Fuzzy WASPS approach. Int. J. Disaster Risk Reduction **51**, 101838 (2020)
13. Demirbas, S., Ertem, M.A.: Determination of equivalent warehouses in humanitarian logistics by reallocation of multiple item type inventories. Int. J. Disaster Risk Reduction **66**, 102603 (2021)
14. Schön, A-M., Al-Saadi, S., Grubmueller, J., Schumann-Bölsche, .: Developing a camp performance indicator system and its application to Zaatari, Jordan J. Humanitarian Logist. Supply Chain Manage. **8**(3), 346–373 (2018).https://doi.org/10.1108/JHLSCM-10-2017-0047
15. Betts, M.G., et al.: Global forest loss disproportionately erodes biodiversity in intact landscapes. Nature **547**, 441–444 (2017). https://doi.org/10.1038/nature23285

Analyses and Improvement Iterative Decoding of Error Correction Codes for LDPC Codes Using Matlab-Mex

Hassan Touati[1(✉)], Rachid El Alami[1(✉)], Hajar El Ouakili[1], Abdennabi Morchid[1], and Youness Mehdaoui[2]

[1] Faculty of Sciences Dhar El Mehraz, LISAC Laboratory Fes, Sidi Mohammed Ben Abdellah University, Fez, Morocco
hassan.touati@usmaba.ac.ma, rachid.elalami@usmba.ac.ma

[2] LRPSI Laboratory-ISASTM Research Team, Polydisciplinary Faculty, USMS, Béni-Mellal, Morocco

Abstract. In this research, we have proposed methods for iterative decoding of the LDPC code based on the soft decision, belief propagation, Min-Sum, and its derivatives. We used regular R = 0.5 Matrices because the error correction coding guarantees the robustness of the data against white noise in digital communications. The split-row method facilitates the exploitation of column processing parallelism and greatly simplifies row processors. LDPC irregular tokens have received much attention from advanced standards, such as WIFI and DVB-S2 digital video broadcasting. In this sense, we have implemented the LDPC codes in Matlab-Mex. The results obtained from the simulation show that in the additive Gaussian white channels, the Optimized Min-Sum algorithm gets better performances and is closer to those in the propagation of belief.

Keywords: Low-density-parity-check-codes · MS · BP · Tanner graph · NMS · OMS

1 Introduction

Mobile networks, satellite communications, the Internet, and data transfer are modern communication systems essential in life. Unfortunately, all communication systems are noisy in practice. However, we can improve the physical characteristics of the channel and find a possible systematic solution in the form of error-correcting codes. Channel coding (digital) transforms the sequence of helpful information into a discrete coded sequence called a code word.

In 1948, Shannon [1, 2] demonstrated that when the transmission rate of the system is less than the capacity of the transmission channel, the errors caused by the noise can be reduced to an arbitrarily low-level using coding and appropriate decoding. From that point on, researchers began to study different methods of constructing error-correcting codes.

S. Motahhir and B. Bossoufi (Eds.): ICDTA 2022, LNNS 454, pp. 501–509, 2022.
https://doi.org/10.1007/978-3-031-01942-5_50

Gallagher initially invented the Low-Density parity check LDPC code in 1960 [3], a family of class linear Block code., The Mackay Neal Turbo code rediscovered them in 1995 due to growing interest in Iterative Decoding [4, 5].

The parity matrix characterizes LDPC Code. It is usually sparse, and the maximum weights for each row and column are restricted. All the parties are low-density, LDPC code is satisfied $H \times C^T = 0^T$ The columns of matrix H are called variable nodes, and the rows are called check nodes. Like other linear block codes, the vectors of a k-dimensional subspace submerged in an n-dimensional vector space are called code words. By the parameter N. R = K/N, the amount of information contained in a bit of a codeword denotes the length of the code, and the output quantifies that amount.

LDPC codes were initially determined using Gallagher's stricter restrictions. Indeed, the Parity matrix assumes that the must include precisely Wc non-zero values per row and Wr non-zero values per column; as a result, Wr parity check equations are used for each element of the code word, and each of these equations is made up of Wc bits. Wr and Wc use this set of LDPC codes as a parameter, and their performance is checked as follows by Eq. (1):

$$R \geq 1 - \frac{W_c}{W_r} \tag{1}$$

We have proposed methods for iterative decoding of LDPC code based on the soft decision (belief propagation and Min-SUM and its derivatives). Additional computing complexity is introduced during the real-time decoding process. In the final section, we will look at the "Min Sum" (MS) algorithm and its modified versions "optimized Min Sum" (OMS), the Normalized Min-Sum (NMS), and as well as Propagation algorithm (BP) to create efficient decoding for LDPC codes via the log-likelihood. Both regular and irregular LDPC codes can greatly enhance decoding performance. This research implemented LDPC codes in Matlab-Mex and used additive Gaussian white channels. The result of the simulation obtained by the two algorithms "BP" and "Min-Sum" compared parameters are determined after the decoding process; we discovered that BP is the best, but Min-Sum is flexible and less complicated. The last simulation showed that optimized Min-Sum was more efficient than Min-Sum and close to Belief Propagation algorithm.

The following section is the structure of this article: the next one deals with graphical representation. Section 3 presents the belief propagation algorithm. The Min-Sum algorithm in Sect. 4. the derivate of Min-Sum algorithm and Simulations are used to decrease the complexity and increase the algorithm's speed in 4.1 and 4.2. finally, Sect. 5 concludes the paper.

2 Graphical Representation (Tanner Graph)

The matrix H can also be represented by Tanner's graph [6, 7]. The parity matrix H can be represented using a bipartite graph called the Tanner graph: The nodes are divided into variable and parity nodes. Control nodes relate to parity equations, in Fig. 1 at the same time, Variable nodes represent codeword symbols. If the element hij of the parity check matrix H is not zero (hij = 1), a branch has a parity node Cj. The variable nodes will be represented by circles in Fig. 1, while squares represent the check nodes. Example: matrix 2. Represents the parity check matrix that generates an LDPC code with four bits of redundancy:

$$H = \begin{bmatrix} 1 & 1 & 000 & 1 & 0 & 101 \\ 0 & 1 & 100 & 1 & 0 & 010 \\ 0 & 0 & 110 & 0 & 1 & 101 \\ 0 & 0 & 011 & 1 & 0 & 010 \\ 1 & 0 & 001 & 0 & 1 & 010 \end{bmatrix} \tag{2}$$

The bipartite graphs representation of Matrix (2). Following in Fig. 1.

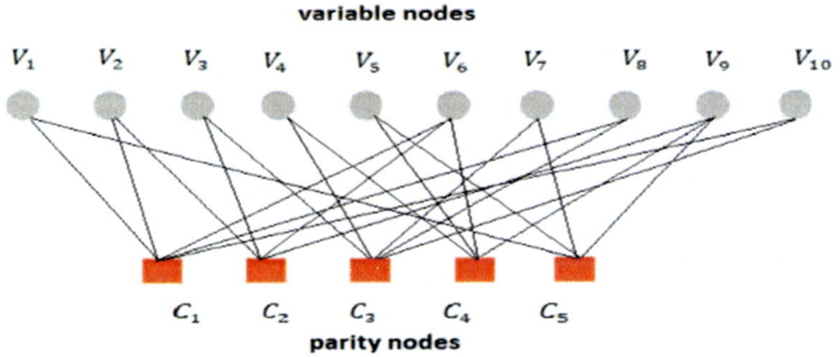

Fig. 1. Tanner graphical representation.

3 The Belief Propagation Algorithm (BP)

The BP algorithm is, also known as LLR (Log-Likelihood Ratios Sum-Product Algorithm), is a soft decision algorithm. Generally qualified as optimal since the BP decoding converges towards the maximum likelihood provided that the bipartite graph associated with the LDPC code does not contain short cycles. This algorithm achieves excellent overall performance in practical applications because it is miles near the channel's capacity (Shannon limit). It offers the best possible match between yield and performance Because the input bit probabilities were known in advance before performing the LDPC decoder. They are referred to as prior probability for the received bits. The posterior probabilities are the bit probabilities returned by the decoder. These probabilities are represented as LLR BP in the case of SPA decoding. The purpose of SPA decoding is to calculate the maximum posterior probability for each codeword bit, Pi = P {ci = 1|n}, which is the probability $i_{ème}$codeword bit is conditional on n that all parities check constraints are satisfied.

Binary phase-shifted modulation is supposed to be used to transmit a binary codeword (X1, X2, X3,......, Xn) (BPSK). The sign is then generated by sending the sequence via an additive white Gaussian noise channel (Y1, Y2, Y3,..., Yn).

We define:

The variable nodes involved in the test equation (i) are described. All the variable nodes V (i) without nodes j are designated by V (i)\j. The node j update variable's control nodes are set to the same value. C (j)\i: all control nodes C (j) without node i.

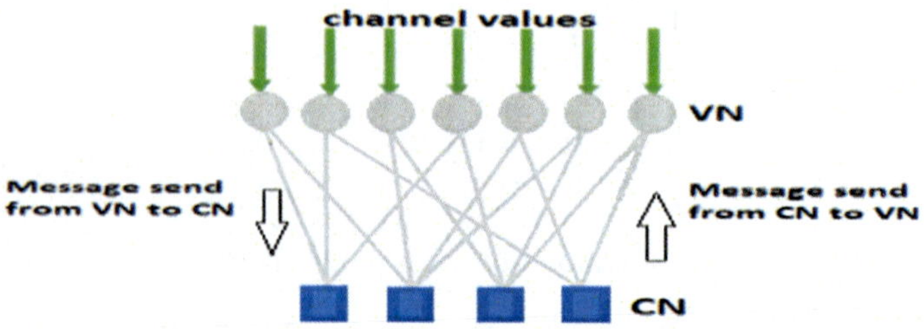

Fig. 2. The message flow decoding.

Figure 2 shows the directions of messages sent from variable nodes to parity nodes or from parity nodes to variable nodes.

Figure 3 shows the six components of the belief propagation algorithm [8] syndrome verification, initialization of the values of the variable nodes and check nodes, renewal of the variable nodes, renewal of the parity nodes, decision on bits, and verifying syndromes.

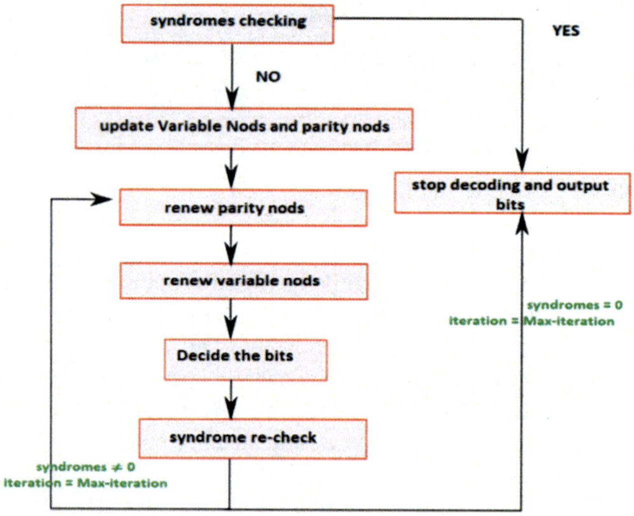

Fig. 3. The steps of the belief propagation algorithm.

Set the probability ratio for the sequence value received at each iteration of each in Eq. 3.

$$\lambda_j = \left(\frac{P(X_i = 1/y_j)}{P(X_i = 0/y_j)} \right) \tag{3}$$

The messages α and β are determined and traded between factor nods and control nods by the Tanner diagram's edges.

Step 1: This phase may speed up decoding since we can transform messages into binary (X1, X2, X3,............., Xn) without modulation by determining the received messages (Y1, Y2, Y3,............., Yn).

$$S(syndrome) = \sum_{j \in V(i)} X_j = 0 \tag{4}$$

in Eq. (4), if S equals 0, the received bits are error-free, we end decoding, and we have the codeword C: (c1, c2,...., Cn). If not, we will go on to the next phase.

Step 2: Row processing is used to calculate the Messages α_{ij} Eq. (5) uses the messages from the other variable nodes to update parity nodes except for information Vj, which is linked to the control node C(i).

$$\alpha_{ij} = \prod_{j' \in V(i) \backslash j} sign(\beta_{ij'}) \times \varphi\left(\sum_{j' \in V(i) \backslash j} \varphi(|\beta_{ij'}|)\right) \tag{5}$$

with the nonlinear function following in Eq. (6):

$$\varphi = -\log\left(\tanh\left|\frac{X}{2}\right|\right) \tag{6}$$

in Eq. (5), the parity update (sign) is the first product term, whereas the reliability update is the second product term (magnitude).

Step 3: Calculate βij in Eq. (7) except the control node Ci, messages employing channel information (λj), and inbound messages from all other control nodes connected to Variable Node Vj.

$$\beta_{ij} = \lambda_j + \sum_{i' \in C(j) \backslash i} \alpha_{i'j} \tag{7}$$

Step 4: Checking for syndromes with Eq. (8) and terminating early; when the column processing is complete, each bit of column j is updated by adding channel information (λj) and (α) messages from nearby control nodes.

$$Z_j = \lambda_j + \sum_{i' \in C(i)} \alpha_{i'j} \tag{8}$$

From the updated vector, an estimated vector $x_i = \{x_1, x_2, \ldots\ldots\ldots.., x_n\}$ and calculated as follows (9):

$$Z_j = \lambda_j + \sum_{i' \in C(i)} \alpha_{i'j} \tag{9}$$

$$x_i = \begin{cases} \text{One } si \quad Z_i < 0 \\ 0 \; si \quad Z_i \geq 0 \end{cases} \tag{10}$$

If $H \times x^T = 0$, so x is the codeword, the iterative cycle quits. Otherwise, the decoding repeats from step 2 until a valid codeword is obtained or until the maximum number of iterations is reached, ending the decoding process.

4 Min-Sum Algorithm

Using the MS algorithm, we can better understand how soft-decision algorithms are used to correct transmission errors in Fig. 4.

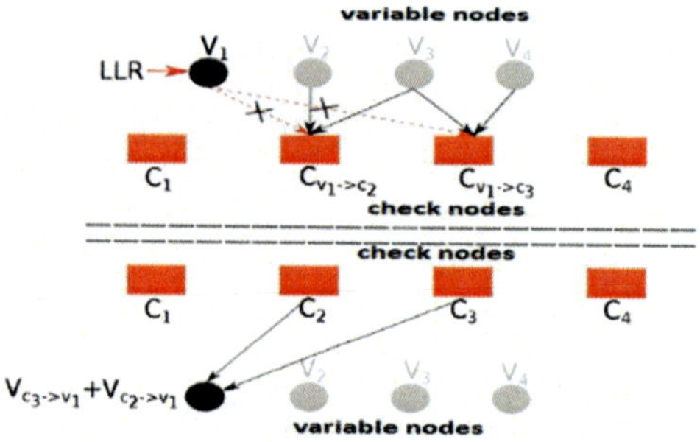

Fig. 4. Algorithm BP ou message passing.

Use the more reliable bits to correct the less reliable bits; if V1 is incorrect, use the correct bits V2 and V3 to correct it; if V1 and V2 are both incorrect, we must first use the other correct bits to correct V2, and then we can correct V1; if V1 and V2 are both incorrect, we must first use the other correct bits to correct V2, and then we can correct V1. With a minimum function approximating in Eq. (11), the calculation of tanh and $tanh^{-1}$, the decoding of the product of belief propagation may be simplified. The technique that uses this approximation is termed Min-Sum [9, 10].

$$\alpha_{ij} = \prod_{j' \in V(i) \setminus j} sign(\beta_{ij'}) \times \min\left(\left|\beta'_{ij}\right|\right) \tag{11}$$

4.1 Error Performance Analysis 1

For each curve, the iteration number is fixed at 100, and the size matrix is used (1000 × 2000) With efficiency R = 1/2.

We can observe in Fig. 5 that reaching a BER (bit error rate) of (10^{-2}) needs 2.3 dB of SNR for the Belief Propagation Algorithm (BP). However, the Min-Sum under the same conditions for TEB (10^{-2}) needs 2.46 dB of SNR. Therefore, we can see a 0.16 lag between the two algorithms (BP and MS).

Several techniques were implemented to keep Shannon's performances close while reducing material complexity. Different methods are used to acquire a form simplified by algorithm BP while closely measuring its performances. The most common methods are normalized Min-Sum (NMS) and optimized Min-Sum (OMS).

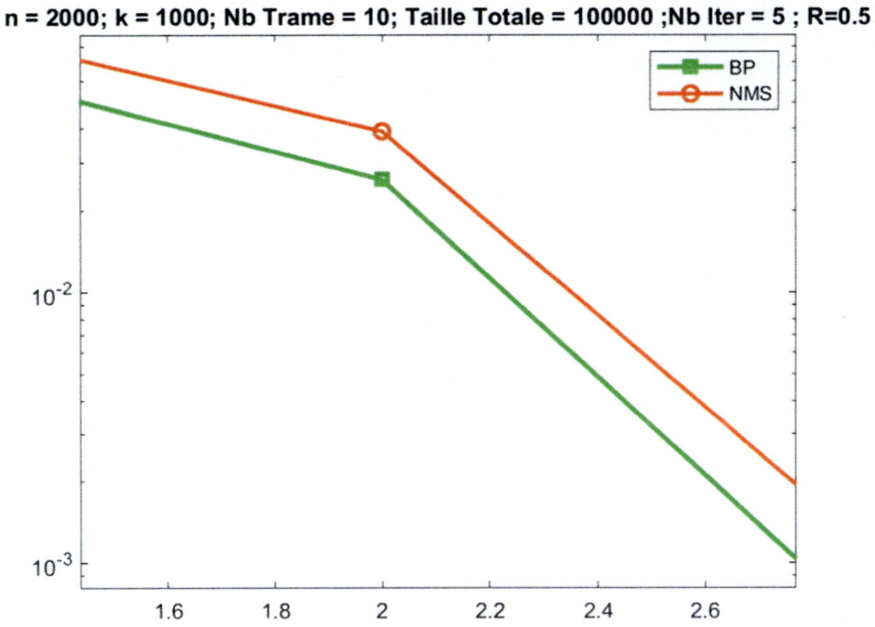

Fig. 5. Comparison between Min-Sum algorithm and belief propagation algorithms.

The MS decoding error performance loss can be improved by multiplying Eq. (9) by normalization factor s. The steps of the Min-Sum method are the same as the belief propagation for two derivatives, for example for step 2 for the Normalized Min-Sum (NMS):

$$\alpha_{ij} = s \times \prod_{j' \in V(i) \setminus j} sign(\beta_{ij'}) \times \min\left(\left|\beta'_{ij}\right|\right) \tag{12}$$

Nevertheless, we multiplied Eq. (9) by optimization factor b Min-Sum in step 2, we obtained Eq. 11. For 0 < b < = 1 in row processing when updating control nodes (step 2):

$$\alpha_{ij} = b \times \prod_{j' \in V(i) \setminus j} sign(\beta_{ij'}) \times \min\left(\left|\beta'_{ij}\right|\right) \tag{13}$$

Also, in column processing, when updating variable nodes (step 3). and obtained the following Eqs. (12):

$$\beta_{ij} = \lambda_{ij} + b \times \sum_{i' \in C(j) \setminus i} \alpha_{i'j} \tag{14}$$

4.2 Error Performance Analysis 2

We work with a matrix of size (504,1008). We employ LDPC codes with a coding rate of 1/2, a row weight of 12, and a column weight of 6. After BPSK modulation, the codes are retransmitted on the AWGN channel. The maximum number of iterations is set at 5. The Belief Propagation Decoding Algorithm [11], the Min-sum Algorithm [12], and the Belief Propagation Decoding Algorithm were compared. Normalized Min-sum (NMS) algorithm [13], Fig. 6 depicts the suggested Optimized Min-Sum (OMS) algorithms. Simulation findings reveal that the Optimized Min-sum method outperforms the Min-sum algorithm, is comparable to the Normalized Min-sum algorithm, and is closer to the Belief propagation approach. Figure 6 indicates that our approach can obtain a 0.2 dB decoding increase over Normalized for the BER value of 10^{-2}.

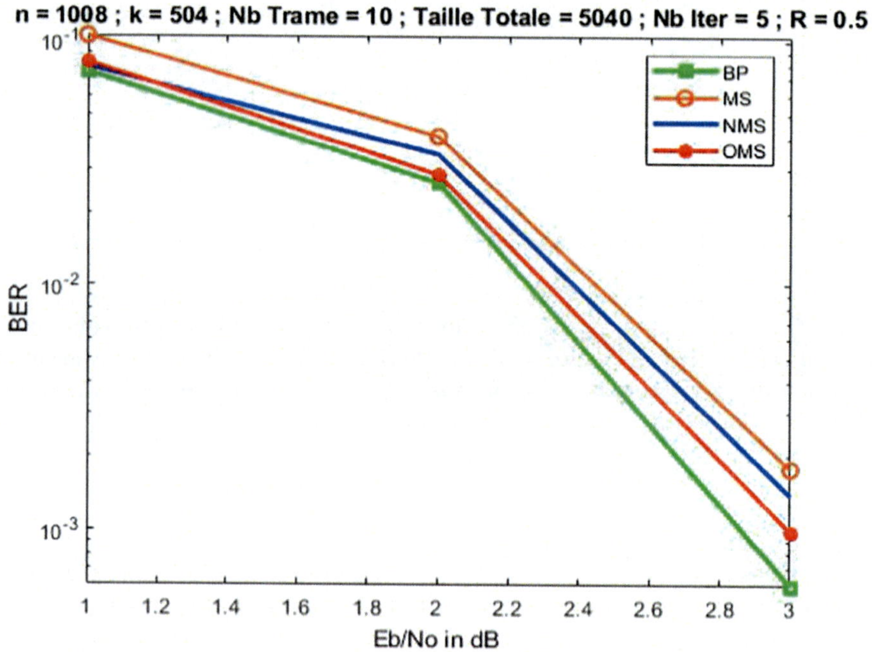

Fig. 6. Comparison performance between BP, MS, NMS, OMS algorithms.

From Fig. 6 we can also see that the improved Min-Sum technology has only −0.03 dB gain on the Belief propagation approach of rate code 1/2. The improved Min-Sum method is roughly equivalent to the belief propagation decoding technique.

5 Conclusion

We have introduced the optimization of LDPC codes using the optimized Min-Sum algorithm "OMS." We decreased the complexity of the BP belief propagation algorithm and found that the performance also increased more than the Min-Sum algorithm.

We observed at simulation diagram the normalized Min-Sum algorithm "NMS" very approached to the algorithm the belief propagation algorithm. The algorithm optimization factor must be determined experimentally before starting the decoding process. Thus, the proposed algorithms are competitive decoding techniques for regular LDPC block codes.

References

1. Benedetto, S., Montorsi, G.: Serial concatenation of interleaved codes: performance analysis, design, and iterative decoding. IEEE Trans. Inf. Theory **44**, 3 (1998)
2. Chung, S.Y., Forney, G.D., Richardson, T.J., Urbanke, R.: On the design of low-density parity-check codes within 0.0057 dB from the Shannon limit. IEEE Commun. Lett. **5**, 58–60 (2001)
3. Gallager, R.G.: Low-Density Parity-Check Codes. MIT Press, Cambridge (1963)
4. Author, F., Author, S.: Title of a proceedings paper. In: Editor, F., Editor, S. (eds.) Conference 2016. LNCS, vol. 9999, pp. 1–13. Springer, Heidelberg (2016)
5. Berrou, C., Glavieux, A., Thitimajshima, P.: Near Shannon limit error-correcting coding and decoding: Turbo-codes (1). In: Proceedings of the ICC 1993, vol. 2/3, pp. 1064–1071 (1993)
6. Forney, G.D.: Codes on graphs: normal realizations. IEEE Trans. Inf. Theory **47**(2), 520–548 (2001)
7. Tanner, R.M.: A recursive approach to low-complexity codes. IEEE Trans. Inf. Theory **IT-27**, 533–547 (1981)
8. Noor, I.U.S.: Construction and Performance Evaluation of QC-LDPC Codes
9. Kou, Y., Lin, S.: Low-Density Parity-Check Codes Based on Finite Geometries: A Rediscovery and New Results,' Department of Electrical and Computer Engineering, University of California
10. Fossorier, M.P.C.: Quasicyclic low-density parity-check codes from circulant permutation matrices. IEEE Trans. Inf. Theory **50**(8), 1788–1793 (2004)
11. Wiberg, N.: Codes and decoding on general graphs. Ph.D. thesis, Linkoping University, Sweden (1996)
12. Zhao, J., Zarkeshvari, F., Banihashemi, A.H.: On implementation of the min-sum algorithm and its modifications for decoding low-density Parity-check (LDPC) codes. IEEE Trans. Comms. **53**(4), 549–554
13. Wu, X., Song, Y., Cui, L., Jiang, M., Zhao, C.: Adaptive-normalized min-sum algorithm. In: 2010 2nd International Conference on Future Computer and Communication (ICFCC), vol. 2, pp. V2-661–V2-663, 21–24 May 2010

Advanced Technologies in Energy and Electrical Engineering

Towards the Implementation of a Digital Twin for Induction Motors

Adamou Amadou Adamou(✉) and Chakib Alaoui

INSA Euro-Méditerranéen, EUROMED University, Fes, Morocco
a.amadou@ueuromed.org, c.alaoui@insa.ueuromed.org

Abstract. Energy Efficiency (EE) is one of the main pillars of Industry 4.0. The road map of this new industrial revolution has imposed a certain degree of accuracy for conditions monitoring of industrial machines, where Induction Motors (IMs) represent the most important machines in terms of energy consumption (42.2% of the global energy demand). Achieving an accurate EE monitoring for IMs imposes a good Stray Load Loss (SLL) estimation by taking into consideration the In-situ aspect. In this paper, the five-dimension digital twin (DT) model is used to propose a new approach to integrate the EE for DT. The proposed method is the first step toward DT implementation for IM. Three state-of-the-art SLL models for IM are compared and the most accurate expression to integrate into the EE model for the proposed DT is then selected. The selected model takes into consideration all SLL sources in IM represented in the Electric Equivalent Circuit (EEC) by the stator and the rotor leakage inductances and the magnetizing inductance.

Keywords: Industry 4.0 · Digital Twin · Energy efficiency · Induction motor · Stray load losses

1 Introduction

One of the most important challenges in today's environment and Industry 4.0 is energy efficiency with low cost, where the energy management systems (EMS) allow to control and reduce the energy consumption of industrial systems without affecting their operational performance. According to the International Energy Agency (IEA) [1], 53% of electrical energy consumption goes to Electric Drives (ED) or Electric Motor-Driven Systems (EMDS). Among the EMDS, 77% of the electrical energy consumption is due to medium-sized motors, where 80% are Induction Motors of different types [2]. The use of a large quantity of IMs makes them a good candidate for energy efficiency improvements. The medium-sized IMs of 84% of energy efficiency consume 10,500 TWh per year, so if a mere one percent (1%) of EE is improved during the operation, 105 TWh could be saved [3]. The efficiency of IMs has emerged as having the biggest potential in energy saving [4]. Improving the efficiency of IMs cannot be achieved without accurate estimation of all losses of said motors, and in particular, the SLL. Beyond the design stage, in Industry 4.0, the operation condition of IM implies an accurate determination of efficiency, hence stray load losses.

S. Motahhir and B. Bossoufi (Eds.): ICDTA 2022, LNNS 454, pp. 513–523, 2022.
https://doi.org/10.1007/978-3-031-01942-5_51

The SLL has been the subject of various discussions, since 1912 with E. M. Olin [5] who suggested accounting for the stray load losses empirically relating no-load to load losses. Until today, the discussion on these losses is still on-going because it is confusing origins [6]. According to [5], the efficiency estimation methods of IM as the input-output method are precise but difficult to perform to obtain an accurate result. While the loss segregation method is easy to perform accurately for the conventional losses, the overall result, taking into account only these conventional losses (see Eq. 4) is less precise. The difference between the results of these two methods was quickly associated with the stray losses. The harmonics losses, also called high frequency stray no-load and stray load-losses, are related most closely to the loading operation of a machine, hence they are called stray load losses or load loss or unknown losses [5, 8]. Despite all disagreement, there is a consensus that the stray load losses are defined as the difference between the total power losses of the machine on load and the losses determined by the segregation method which are the conventional losses (see Eq. 1) [5–7].

Digital twin (DT) is one of the most capable enabling technologies for realizing smart manufacturing and Industry 4.0. They are characterized by the seamless integration between the cyber and physical spaces. The importance of DTs is increasingly recognized by research institutions and the industrial sector [10]. The condition monitoring using DT depends on the study goals such as design, production, manufacturing, and maintenance [11]. The stray loss models are suggested and the most accurate is then selected as a candidate to improve the DT implementation of IM. Among the energy-saving methods, Energy-based predictive maintenance is one of the intuitive approaches that we can use. This is clearly shown in [12], where the authors propose a new approach to predict and classified the Squirrel Cage Induction Motor failure based on only energy consumption data using machine learning techniques. The work is based on the energy consumption data (current, voltage, and time) which correlated with faulty SCIM motors vibration data regarding torque and speed.

2 Digital Twin of Induction Motor – Towards the Implementation of Stray Losses

2.1 Definition and Description

Several studies have defined the concept of a digital twin in different ways, although sometimes they are incomplete with a lack of information [11, 13]. In [14] DT is defined as "an integrated multi-physics, multi-scale, and probabilistic simulation of a complex product and uses the best available physical models, sensor updates, etc., to mirror the life of its corresponding twin". It is the most common definition of DT and it is similar to the definition dictated in [15]. In [10] the simulation aspect of the digital twin is defined as follows.

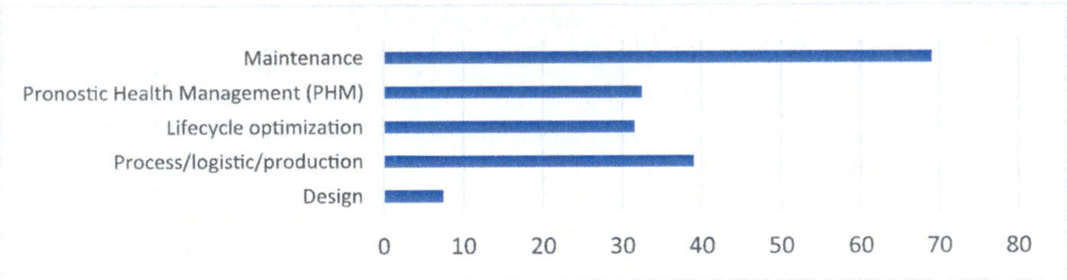

Fig. 1. Digital Twin applications [11]

The Digital Twin refers to a description of a component, product, or system by a set of well-aligned executable models with the following characteristics.

a. DT is the linked collection of the relevant digital artifacts including engineering data, operation data, and behavior descriptions via several simulation models. The simulation models making up the DT are specific for their intended use and apply suitable fidelity for the problem to be solved.
b. DT grows along with the real system along the whole life cycle and integrates the currently available knowledge about it.

DT presents many advantages to the modern industry, such as reduced production downtime; reduced breakdowns; cost savings; improved productivity; elimination of ambiguity in maintenance tasks; expended equipment life; improved customer service and reputation; reduced energy waste; and improved facility security [11]. These advantages may have different impacts on the various sectors, so the application domain of the DT is categorized into five groups (see Fig. 1) where maintenance is the dominant application of the DT.

2.2 The Proposed EE Model-Based Digital Twin

Choosing the right technologies and tools to implement DT is "la bête noire" of many researchers and professionals in engineering. Fortunately, the 5-dimension digital twin model allows for the implementation of this technology based on 5 keys [16, 17]. This model is constituted of Physical entities, Virtual model, DT data, Services in DT, and Connection. Based on this model, the proposed EE-based DT can be formulated as:

$$M_{DT} = f(HP, SP, DD, Ss, Com) \tag{1}$$

where *HP* is the hardware part, *SP* is the software part, *DD* is the data, *Ss* are the services and Com is the communication part.

Hardware Part: The physical motors installed in the industry and the smart sensors constitute the hard part of the physical system. At first, the EEC parameters are identified w.r.t the used laboratory motor using the algorithm established in [18]. These parameters are then integrated into the EE model to reduce the number of unknown parameters. The remaining parameters are then found using a minimization algorithm.

Software Part: It is the virtual environment (Digital Twin) that reproduces the defined behavior of the physical system. The EE model is implemented and run by the soft part as an RT simulation model. The Hard part information, the local and support services are the entities that supply the soft part as inputs, and the results of the RT simulation are shown by the HMI.

Services: The management of the DT ecosystem needs services to operate, these are local and external (support) services from the partners. The local services are carried out in the internal system by the company's collaborators, these can be the applications simulation and condition monitoring for EE, PHM, predictive maintenance, etc.

Data: This is the essential part of DT. It is the DT data server that is composed of heterogeneous, multi-sources, and multi-scale data from the previous three entities. The hard part provides static and dynamic data from the EEC to measure voltage, current, and temperature.

Communication Part: The information exchange in the DT system is managed by the communication part with help of the IoT technologies. It illustrates these iterations (Fig. 2).

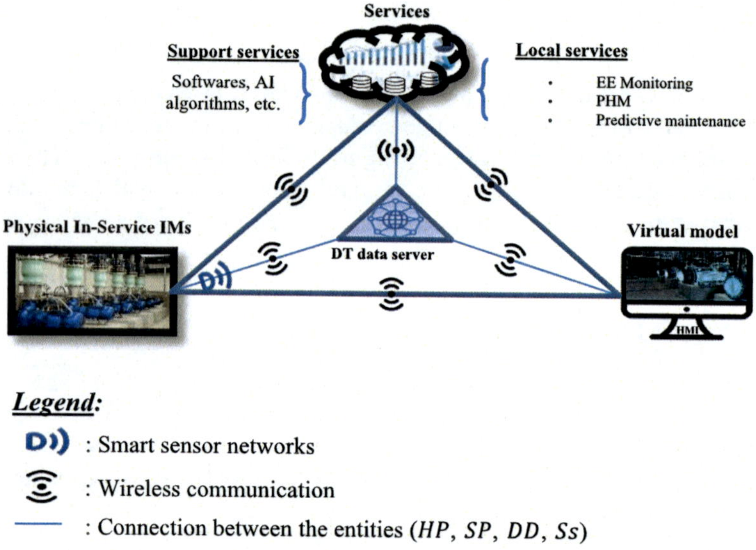

Fig. 2. The proposed DT system is based the on EE model.

3 Stray Losses Model of IM

The stray load losses are the only losses in induction motor where researchers are not unanimous on their characteristic aspects. In [19], Jimoh et al. detail 8 methods to evaluate the SLL, which are classified into 2 categories, full-load tests, and light-load tests.

Among methods proposed in the previous paper, the Full-Load Tests are inaccurate and non-economical. The Input-Output, Pump Back and Calorimetric methods are impracticable for a high efficient machine. However, the Light-Load Test methods are simple, economical, and indirect, based on too many assumptions, which have generated considerable controversy [19]. As mentioned previously, the origin of SLL can be divided into three sources: Leakage flux, space, and the time harmonics in the Air-gap flux density. According to the literature, there are 4 ways to evaluate the SLL which are, Measurement, Analytical/Numerical methods, Equivalent Circuit, and Empirical relations [3]. Among these methods, EEC, Analytical and Empirical relations are suitable for In-Situ EE monitoring. Equations (2) to (4) show these losses.

$$P_{Stray} = P_{Totalloss} - P_{Conventional} \tag{2}$$

$$P_{Totalloss} = P_{Electrical} - P_{Mechanical} \tag{3}$$

$$P_{Conventional} = P_{Js} + P_{Jr} + P_{Iron} + P_{Mec} \tag{4}$$

3.1 Model-Based on EEC

Several models have been proposed to evaluate the SLL based on the equivalent circuit, the first model with a modified EEC including SLL resistance was proposed by Lamine et al. in [25]. The approach used can be divided into two categories: Some evaluate the SLL based on the stator and rotor leakages inductances, and the magnetizing inductance in the equivalent circuit which represent the influence of slots on the magnetic field [3, 26], Others evaluate the SLL by adding resistance as an SLL resistance [9, 27] (Fig. 3).

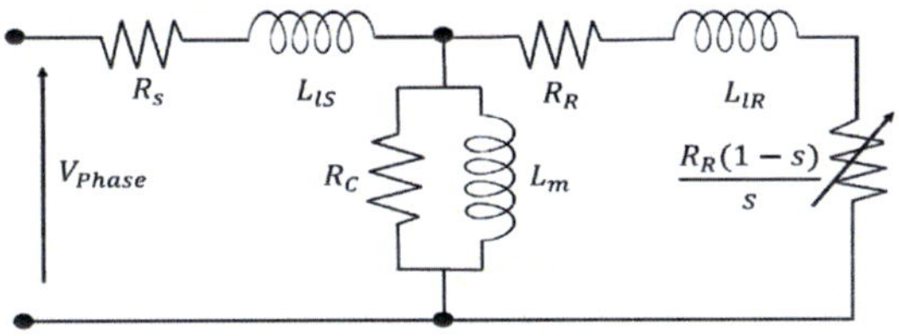

Fig. 3. T-model equivalent circuit of IM [3]

SLL Model Based on Leakage and Magnetizing Inductances. Rajendra et al. [3] proposed in their paper a new SL model for grid-connected IM which is judged accurate. This model can also be used for the design stage. The model is based on 11 sources (Table I in [3]) of SL which are represented by three inductances as stator leakage inductance (L_{ls}), rotor leakage inductance (L_{lr}), and the magnetizing Inductance (L_m). Find by using blocked rotor test and no-load test. The model is developed, as follows.

Model Formulation. Depending on the leakage flux, the SLL is directly proportional to the L_{lS} and L_{lr}, and it's also inversely proportional to L_m [26]. Based on these observations, eight (8) models were proposed in [26] and six (6) models in [3] to evaluate the SLL. Each study selects an accurate model based on certain criteria. In the first paper [26], standard deviation (S.D) of the errors and correlation coefficients (r) are calculated as KPIs to identify the most appropriate model (Eq. 5). In the 2nd paper [3], Akaike Information Criterion (AIC) and Bayesian Information Criterion (BIC) are used to select the suitable model (Eq. 6) for their relative capability of SLL representation. The model is based on the result reported by many papers that explained that the SLL is proportional to the output power (1% to 4% of the output power according to [28]), hence $SLL = K_{SL}P_{out}$.

$$K_{SL1} = \left(\sqrt{\frac{L_{lS}}{L_m}}\right)^m \left(\sqrt{\frac{L_{lr}}{L_m}}\right)^n .pf \tag{5}$$

$$K_{SL2} = \left(\frac{L_{lS}}{p.L_m}\right)^m \left(\frac{L_{lr}}{p.L_m}\right)^n \tag{6}$$

So the SLL models from the equations above are F-1: $K_{SL1}P_{out}$ and F-2: $K_{SL2}P_{out}$.

Accurate expressions of inductances have been expressed in [29] taking into consideration the leakage and the harmonics sources of SLL. The inductances expressions as reported in [3] are calculated using the procedure laid in IEC 60034-2-1 and the test data given by Toshiba Mitsubishi-Electric Industrial Systems Corporation (TMEIC).

SLL Model Based on Additional Resistances. These models involve additional resistances in the EEC to represent the SLL. In [27], Nasir proposed a new EEC including the Stray load and no-load losses by adding additional resistances called stray loss resistances, in parallel with the stator and rotor leakage reactances. In the usual EEC, the iron core resistances (stator and rotor) are not represented, the proposed circuit [27], includes these resistances to model the SLL in the stator and rotor circuit. These are expressed in (7) and (8).

$$R_{SSSL} = \frac{X_{lS}^2.R_{SiC}}{R_{SiC}^2 + X_{lS}^2} + \frac{X_{lS}^2.R_S}{R_S^2 + X_{lS}^2} \approx \frac{X_{lS}^2.R_S}{R_S^2 + X_{lS}^2} \tag{7}$$

$$R_{RSSL} = \frac{S^4.X_{lS}'^2}{R_{SiC}} + \frac{S^2.X_{lS}'^2}{R_r'} \approx \frac{S^2.X_{lS}'^2}{R_r'} \tag{8}$$

Hence the losses are calculated as: Stator: $P_{SSL} = 3.R_{SSSL}.I_S^2$; Rotor: $P_{SSL} = 3.R_{RSSL}.I_r'^2$.

However, in [9], Boglietti et al. proposed an EEC including the SLL due to the effects of the spatial harmonic in IM. Theses model is not convenient for our search because it considers only the additional losses due to the rotor cage currents induced by the air gap spatial harmonics. As performed in [21] by the sensitivity analysis, it is imperative to have the input power, the load torque, and the input current measured with high accuracy to guarantee a good estimation for the SLL.

4 Discussion

4.1 Comparison of Three States of the Art SLL Model

The stray loss model proposed in the previous papers by Rajendra et al. [26] and [3] include the leakage flux and the magnetizing reactance's, the difference is due on the pole pair which appear in the model (3), and the power factor which appear in the model (2). It's important to note that the exponent's m and n are calculated by regression analysis with the model and the SLL test data. At a steady-state, the efficiency can be expressed as $\eta = \frac{(ratedTorque)\times(ratedspeed)}{3UI\times pf}$, we know that the synchronous speed is inversely proportional to the pole pair ($n_s = \frac{120f_s}{p}$) which imply that the pole pair (p) is linked to the pf. Hence, we can conclude that the expressions (Eqs. (5) and (6)) above have the same parameter dependent (Table 1).

Table 1. Comparison of different SLL models.

SLL model	Include leakage flux	Include space harmonic	Include time-harmonic
$K_{SL1}P_{out}$	YES	YES	YES
$K_{SL2}P_{out}$	YES	YES	YES
$3.(R_{SSSL}.I_S^2 + R_{RSSL}.I_r'^2)$	YES	NO	NO

However, the model proposed by Nasir in [27] doesn't include the harmonics due to the Air Gap flux, the resistances which modeled the SLL represent only the leakage flux caused by the stator and rotor windings.

A comparison is proposed to select the best model between the models from model F-1 or F-2. Matlab is used to draw the SLL measured for 14 motors based on IEC-60034-2 standard, and the errors of the calculated SLL from the Eqs. (5 and 6) w.r.t the measured SLL (see Fig. 4). Standard deviation (SD) of the errors in calculated SLL from F-1 and F-2 w.r.t measured SLL are evaluated to be respectively 19.45446424 and 9.57042857. The most accurate model is the one with the lowest SD, hence the selected model is F-2.

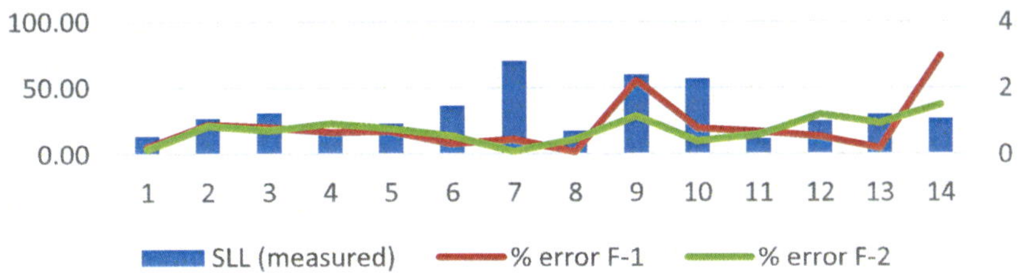

Fig. 4. SLL errors analysis with F-1 and F-2 (data provided from [3, 26])

4.2 Towards the EE Monitoring Based on DT

To achieve the proposed DT implementation, a first step of the process is illustrated (see Fig. 5). On a lab scale, the multi-physics model (including all the stator and rotor losses) outputs are used to run the 3D model which highlights the output magnitudes (Torque and speed). A virtual sensor is used to measure the output power and then combines with the input power to measure the EE of the machine. The way to measure virtually the internal temperature and the vibration from different measured points can be envisaged using the IM thermal model and the Finite Element Analysis (FEA) methods.

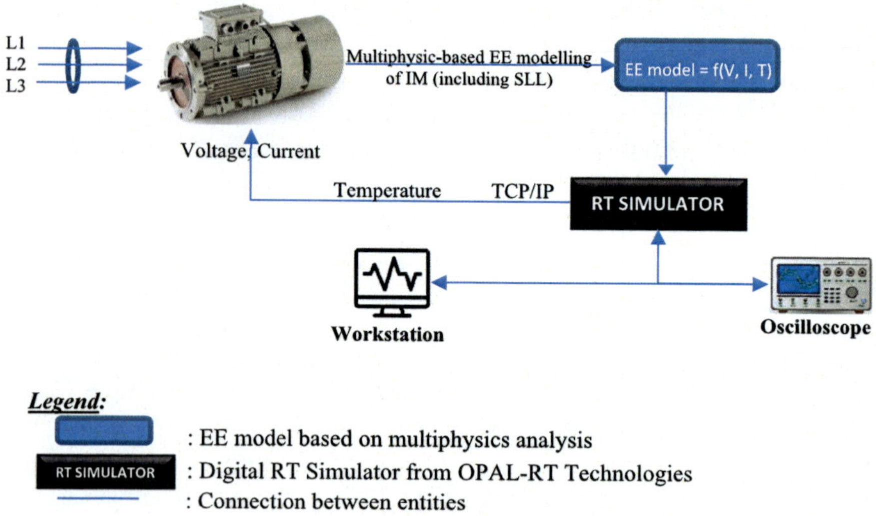

Fig. 5. The first step process towards the proposed DT model (lab test bench)

S. Bouzid et al. describe in their paper the development of RT DT of a wound rotor IM based on the finite element method. In this work [31], the RT outputs of electromagnetic quantities otherwise difficult to evaluate are computed using DT.

5 Conclusion

Stray load losses implementation in Induction Motor Digital Twin application is highlighted in this paper. Beyond the SLL measurement methods and models presented, the best model is selected for DT-based EE implementation.

This study is the first step towards an accurate EE model establishment for DT in Industry 4.0. Hence the next works can be directed towards the establishment of the so-called EE model, then, based on these models, the energy efficiency improvement methods during the operation can be defined. From the mentioned works, energy-based predictive maintenance is one of the intuitive methods to improve the EE during the operation. And a co-simulation with virtual Reality to the development of the proposed

DT can also improve the system as user-friendly as shown in [32]. In terms of the perspective, it is important to note that the historical EE data of the machine during operation allows to development of very good methods for energy consumption forecasting using machine learning (ML) methods. According to [33], among the ML algorithms such as Linear Regression (LR), Support Vector Machine (SVM), Decision Tree (DT), Artificial Neural Networks (ANN), and Recurrent Neural Networks (RNN) linear regression methods were compared, the linear regression shown the most accurate result for industrial energy consumption forecasting.

References

1. 4E_2017_Annual_Report_100518.pdf. https://www.iea-4e.org/wp-content/uploads/publications/2018/05/4E_2017_Annual_Report_100518.pdf
2. Waide, P., Brunner, C.U.: Energy-Efficiency Policy Opportunities for Electric Motor-Driven Systems. OCDE, Paris (2011). https://doi.org/10.1787/5kgg52gb9gjd-en
3. Kumar, R., Kumar, P., Kanekawa, T., Oishi, K.: Stray loss model for induction motors with using equivalent circuit parameters. IEEE Trans. Energy Convers. **35**, 1036–1045 (2020). https://doi.org/10.1109/TEC.2020.2964616
4. Mzungu, H.M., Sebitosi, A.B., Khan, M.A.: Comparison of standards for determining losses and efficiency of three-phase induction motors. In: 2007 IEEE Power Engineering Society Conference and Exposition in Africa - PowerAfrica, Johannesburg, South Africa, pp. 1–6. IEEE (2007). https://doi.org/10.1109/PESAFR.2007.4498051
5. Olin, E.M.: Determination of power efficiency of rotating electric machines: summation of losses versus input-output tests. Proc. Am. Inst. Electr. Eng. **31**, 1287–1310 (1912). https://doi.org/10.1109/PAIEE.1912.6660632
6. Wang, J., et al.: Stray load loss calculation for induction motor by combination of general airgap field modulation theory and 2D FEA. IEEE Trans. Energy Convers. **36**, 2524–2533 (2021). https://doi.org/10.1109/TEC.2021.3052365
7. International Standards for the Induction Motor Efficiency Evaluation A Critical Analysis of the Stray-Load Loss Determination.pdf
8. W. Novotny, D.: High-frequency losses in induction motors. University of Wisconsin, Department of Electrical and Computer Engineering, Wisconsin, USA (2021)
9. Boglietti, A., Cavagnino, A., Ferraris, L., Lazzari, M.: Induction motor equivalent circuit including the stray load losses in the machine power balance. IEEE Trans. Energy Convers. **23**, 796–803 (2008). https://doi.org/10.1109/TEC.2008.921467
10. Boschert, S., Rosen, R.: Digital twin—the simulation aspect. In: Hehenberger, P., Bradley, D. (eds.) Mechatronic Futures, pp. 59–74. Springer, Cham (2016). https://doi.org/10.1007/978-3-319-32156-1_5
11. Errandonea, I., Beltrán, S., Arrizabalaga, S.: Digital Twin for maintenance: a literature review. Comput. Ind. **123**, 103316 (2020). https://doi.org/10.1016/j.compind.2020.103316
12. Laayati, O., Bouzi, M., Chebak, A.: Smart energy management system: SCIM diagnosis and failure classification and prediction using energy consumption data. In: Motahhir, S., Bossoufi, B. (eds.) ICDTA 2021. LNNS, vol. 211, pp. 1377–1386. Springer, Cham (2021). https://doi.org/10.1007/978-3-030-73882-2_125
13. Grieves, M., Vickers, J.: Digital twin: mitigating unpredictable, undesirable emergent behavior in complex systems. In: Kahlen, F.-J., Flumerfelt, S., Alves, A. (eds.) Transdisciplinary Perspectives on Complex Systems, pp. 85–113. Springer, Cham (2017). https://doi.org/10.1007/978-3-319-38756-7_4

14. Tao, F., Cheng, J., Qi, Q., Zhang, M., Zhang, H., Sui, F.: Digital twin-driven product design, manufacturing and service with big data. Int. J. Adv. Manuf. Technol. **94**(9–12), 3563–3576 (2017). https://doi.org/10.1007/s00170-017-0233-1
15. Lermer, M., Reich, C.: Creation of digital twins by combining fuzzy rules with artificial neural networks. In: IECON 2019 - 45th Annual Conference of the IEEE Industrial Electronics Society, Lisbon, Portugal, pp. 5849–5854. IEEE (2019). https://doi.org/10.1109/IECON.2019.8926914
16. Mukherjee, V., Martinovski, T., Szucs, A., Westerlund, J., Belahcen, A.: Improved analytical model of induction machine for digital twin application. In: 2020 International Conference on Electrical Machines (ICEM), Gothenburg, Sweden, pp. 183–189. IEEE (2020). https://doi.org/10.1109/ICEM49940.2020.9270916
17. Qi, Q., et al.: Enabling technologies and tools for digital twin. J. Manuf. Syst. **58**, 3–21 (2021). https://doi.org/10.1016/j.jmsy.2019.10.001
18. Pedra, J., Córcoles, F.: Estimation of induction motor double-cage model parameters from manufacturer data. IEEE Trans. Energy Convers. **19**, 8 (2004)
19. Jimoh, A.A., Findlay, R.D., Poloujadoff, M.: Stray losses in induction machines: part I, definition, origin, and measurement. IEEE Trans. Power Appar. Syst. PAS **104**, 1500–1505 (1985). https://doi.org/10.1109/TPAS.1985.319165
20. Cao, W.: Comparison of IEEE 112 and new IEC standard 60034-2-1. IEEE Trans. Energy Convers. **24**, 802–808 (2009). https://doi.org/10.1109/TEC.2009.2025321
21. Boglietti, A., Cavagnino, A., Lazzari, M., Pastorelli, M.: International standards for the induction motor efficiency evaluation: a critical analysis of the stray-load loss determination. In: 38th IAS Annual Meeting on Conference Record of the Industry Applications Conference, 2003, vol. 2, pp. 841–848 (2003). https://doi.org/10.1109/IAS.2003.1257626
22. Boglietli, A., Cavagnino, A., Lazzari, M., Pastorelli, A.: Induction motor efficiency measurements in accordance to IEEE 112-B, IEC 34-2, and JEC 37 international standards. In: IEEE International Electric Machines and Drives Conference, IEMDC 2003, Madison, WI, USA, pp. 1599–1605. IEEE (2003). https://doi.org/10.1109/IEMDC.2003.1210664
23. Pillay, P., Al-Badri, M., Angers, P., Desai, C.: A new stray-load loss formula for small and medium-sized induction motors. IEEE Trans. Energy Convers. **31**, 1221–1227 (2016). https://doi.org/10.1109/TEC.2016.2539959
24. Lu, B., Habetler, T.G., Harley, R.G.: A nonintrusive and in-service motor-efficiency estimation method using air-gap torque with considerations of condition monitoring. IEEE Trans. Ind. Appl. **44**, 1666–1674 (2008). https://doi.org/10.1109/TIA.2008.2006297
25. Lamine, A., Levi, E.: Dynamic induction machine modeling considering the stray load losses. In: 39th International Universities Power Engineering Conference, UPEC 2004, vol. 1, pp. 582–586 (2004)
26. Kumar, R., Kumar, P.: Modelling of stray-load loss for medium power induction motors. In: IECON 2018 - 44th Annual Conference of the IEEE Industrial Electronics Society, pp. 571–576 (2018). https://doi.org/10.1109/IECON.2018.8591650
27. Nasir, B.A.: Modeling of stray losses in the equivalent circuit of induction machines. In: AIP Conference Proceedings, vol. 2307, p. 020006 (2020). https://doi.org/10.1063/5.0032902
28. Agamloh, E.B.: An evaluation of induction machine stray load loss from collated test results. IEEE Trans. Ind. Appl. **46**, 2311–2318 (2010). https://doi.org/10.1109/TIA.2010.2070474
29. Boldea, I., Nasar, S.A.: The Induction Machines Design Handbook, 2nd edn. CRC Press, Boca Raton (2020)
30. Use of magnetic circuits in leakage reactance calculations. In: Introduction to AC Machine Design, pp. 125–191. Wiley, Hoboken (2017). https://doi.org/10.1002/9781119352181.ch4
31. Bouzid, S., Viarouge, P., Cros, J.: Real-time digital twin of a wound rotor induction machine based on finite element method. Energies **13**, 5413 (2020). https://doi.org/10.3390/en13205413

32. Havard, V., Jeanne, B., Lacomblez, M., Baudry, D.: Digital twin and virtual reality: a co-simulation environment for design and assessment of industrial workstations. Prod. Manuf. Res. **7**, 472–489 (2019). https://doi.org/10.1080/21693277.2019.1660283
33. Bahij, M., Labbadi, M., Cherkaoui, M., Chatri, C., Lakrit, S.: A comparison study of machine learning methods for energy consumption forecasting in industry. In: Motahhir, S., Bossoufi, B. (eds.) ICDTA 2021. LNNS, vol. 211, pp. 165–175. Springer, Cham (2021). https://doi.org/10.1007/978-3-030-73882-2_16

Check for updates

Design of a New and Improved Non-linear Non-adaptive Backstepping Vector Controller for a Permanent Magnet Synchronous Motor

El-Houssine Bekkour(✉), Badre Bossoufi, Adil Essalmi, and Houda El Alami

LIMAS Laboratory, Faculty of Sciences Dhar El Mahraz, Sidi Mohamed Ben Abdellah University, Fez, Morocco
elhoussine.bekkour@usmba.ac.ma

Abstract. The design of an efficient control strategy, combining the absence of overshoot during excessive speed variation and the elimination of external load interference, as well as high overall system reliability, is the goal to achieve, even in the case of a strong coupling characteristic between phase currents and speed that indicates the non-linearity of the PMSM. This paper is devoted to the development of a new backstepping controller with known parameters based on the Lyapunov stability judgment. We implemented the system in MATLAB/Simulink environment. The results show that the proposed controller not only solves the impact of input saturation and obtains fast, stable, and accurate response, but also has a higher impact on position tracking performance despite the load torque disturbance and parameter variation over time. This further shows the strong robustness of the controller, the value added by the SPWM inverter used with the carrier frequency adjustment will minimize the output current distortion indicator THD ~ 0.18% which makes the current proportional to the voltage. Thus, this research gives a new solution for the optimal control of PMSM.

Keyword: Permanent magnet synchronous motors · Speed control · Non-adaptive Backstepping control · Robustness

1 Introduction

The industrial field has become a semi-imaginary development in recent years if we do not say these last hours, but it is not possible to talk about a developed industry without mentioning the main reason for the quality of the engines used by these companies. For a better industry, the best should be used, so it is emphasized that the light is on a permanently synchronous motor with magnets; due to the many advantages including speed synchronization and rotor containment of a magnetic float, making the system supported by a single stator source [1].

Regulating the speed of electric motors is an important requirement in many industrial applications such as robotics, computer numerical control, air conditioning engines, pumps, etc.... So that synchronous engines PMSM grow and replace the induction engine in applications such as electric cars and household appliances, but the main disadvantage

S. Motahhir and B. Bossoufi (Eds.): ICDTA 2022, LNNS 454, pp. 524–534, 2022.
https://doi.org/10.1007/978-3-031-01942-5_52

of PMSM is that its model is not linear, which complicates the design of the control module [2].

This imbalance has also led to many reports of advanced control strategies, some of which attempt to overcome many of the problems of conventional (IP) control. However, many of these control units lead to complex control laws that often lead to an impractical solution to the technical task, thus why the engine community and manufacturers are not enthusiastic about these proposals [3].

FOC also remains the backbone of PMSM industrial applications because of its simplicity, i.e., ease of operation, reliability, and good performance, so this control strategy consists of two proportional internal current loops (IP) and one external speed loop (IP). However, this method is still desirable, but not the best [4].

In this paper, I want to propose a controller that gives a new lease of life to the science of electro-automatic control, provides an effective solution to the non-linear system, and can achieve accurate servo control by decoupling the nonlinear terms of current and velocity in the mathematical model of PMSM. It is a backstepping control with a non-adaptive approach, which derives its stability strength from the Lyapunov theorem. The principle of backstepping control is to simplify the state space of the controlled system into many reduced-order subsystems by designating the virtual control function and the actual control law can be derived by selecting the appropriate property of the Lyapunov function. The backstepping control is a useful design method for nonlinear systems with uncertainties. Therefore, it has good characteristics such as rejection of external load variations and fast dynamic response and can achieve satisfactory results [5] (Fig. 1).

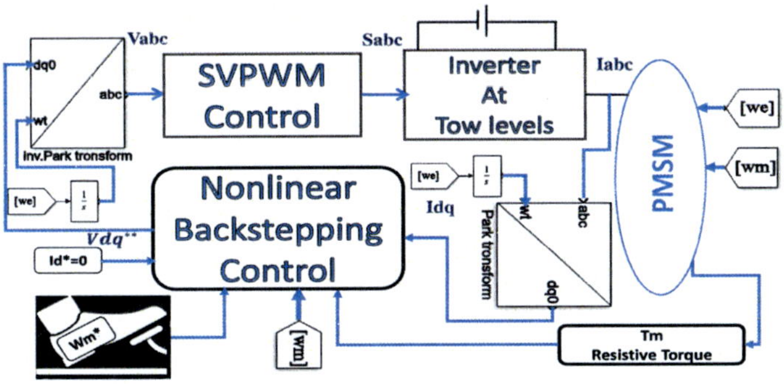

Fig. 1. The global system of the non-adaptive Backstepping control.

1.1 Mathematical Model of PMSM

The mathematical version of the permanent magnet synchronous machine is mainly based on the following assumptions: the stator windings are located sinusoidally along the air gap as long as the impact of mutual flux with the rotor is concerned. The rotor inductances no longer vary much with the variation of the rotor role concerning the stator slots. In addition, the stator windings are considered symmetrical, the winding resistances are constant, and the damper windings and capacitances are omitted. Therefore, the power losses remain constant, while magnetic hysteresis and saturation results are omitted [6].

The equations of the stator voltages:

$$\begin{pmatrix} V_d \\ V_q \end{pmatrix} = [R_S]\begin{pmatrix} I_d \\ I_q \end{pmatrix} + \frac{d}{dt}\begin{pmatrix} \varphi_d \\ \varphi_q \end{pmatrix} + w_e\begin{pmatrix} 0 & -1 \\ 1 & 0 \end{pmatrix}\begin{pmatrix} \varphi_d \\ \varphi_q \end{pmatrix} \tag{1}$$

e equations Flux at the stator:

$$\begin{pmatrix} \varphi_d \\ \varphi_q \end{pmatrix} = \begin{pmatrix} L_d & 0 \\ 0 & L_q \end{pmatrix}\begin{pmatrix} I_d \\ I_q \end{pmatrix} + \begin{pmatrix} \varphi_f \\ 0 \end{pmatrix} \tag{2}$$

φ_d is the d-axis flux linkage; φ_q is the q-axis flux linkage; φ_f is the permanent magnet flux linkage.

By replacing Eqs. (2) with (1), the voltage equation model of the PMSM can be expressed as Eq. (3) on the d-axis and q-axis, respectively. Equation (3) can be expressed in terms of applied voltage, resistance, inductance, and back emf [7]:

$$\begin{cases} V_d = I_d R_s + L_d \frac{dI_d}{dt} - P w_m L_d I_q \\ V_q = I_q R_s + L_q \frac{dI_q}{dt} + P w_m (L_d I_d + \varphi_f) \end{cases} \tag{3}$$

$w_e = P w_m$ is the electrical speed in $\frac{\text{rad}}{\text{s}}$; w_m is the mechanical speed in $\frac{\text{rad}}{\text{s}}$. [8] gave the electromagnetic torque:

$$T_e = \frac{3}{2}P\left((L_d - L_q)I_d I_q + \varphi_f I_q\right) \tag{4}$$

And the equation for the engine dynamics is:

$$T_e = T_m + B_m w_m + J_m \frac{dw_m}{dt} \tag{5}$$

T_m is the load torque; B_m is the damping coefficient; Jm is the moment of inertia.

By rearranging Eqs. (3) and (5), we obtain the final differential model of the PMSM [9]:

$$\begin{cases} \frac{dI_d}{dt} = -\frac{R_s}{L_d} I_d + P w_m \frac{L_q}{L_d} I_q + \frac{V_d}{L_d} \\ \frac{dI_q}{dt} = -\frac{R_s}{L_q} I_q - P w_m \frac{L_d}{L_q} I_d - P \frac{\varphi_f}{L_q} w_m + \frac{V_q}{L_q} \\ \frac{dw_m}{dt} = \frac{1}{J_m}\left\{\frac{3p}{2}\left(\varphi_f I_q + (L_d - L_q) I_d I_q\right) - T_m - B_m w_m\right\} \end{cases} \tag{6}$$

$$* SPMSM : L_q \neq L_d * \quad * SPMSM : L_q \neq L_d *$$

V_{dq} is the stator voltage respectively on the dq axis in V; I_{dq} is the stator currents respectively on the dq axis in A; L_{dq} is the inductance respectively on the dq axis in H; R_s stator resistance in Ω.

1.2 Inverter at Two Levels

1.2.1 Objective

To ensure the conversion of direct current into alternating current, it is necessary to work according to the design of a three-phase two-level inverter [10].

1.2.2 Modeling

The two-level voltage source structure inverter is shown in Fig. 2.:

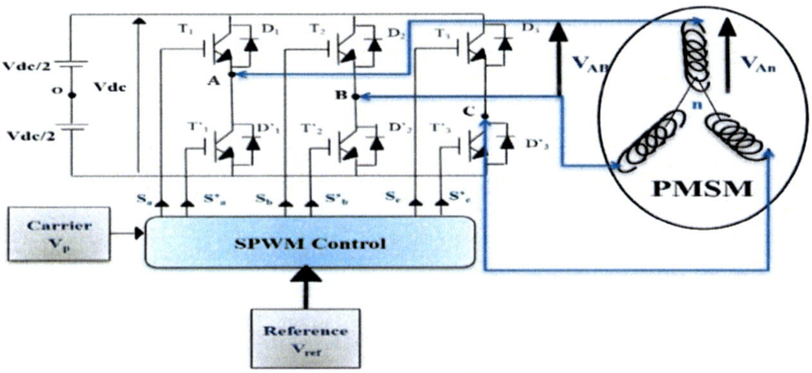

Fig. 2. Inverter at two levels.

It is a three-phase voltage inverter connected to a load (PMSM); this load is star-coupled; it contains three switching arms; each arm is composed of two cells; each one includes a diode and a transistor of different types (IGBT or THYRISTOR or MOSFET) according to our need. We can consider all these elements as ideal switches. Therefore, there is a combination of six switching elements to control the three phases of a motor. As the connection of these two switching elements is to one phase, the direction of the phase current is controlled depending on whether the upper switching element is turned on or the lower switching element is turned on. The upper and lower switching elements of the same phase do not turn on at the same time In Fig. 2, (n) is the neutral point of the motor, and (o) is the divisor of the dummy input [11].

First of all, let's start by defining the different voltage terms used:

(V_{AO}, V_{BO}, V_{CO}): inverter input voltages.

(V_{AB}, $V_{BC}V_{CA}$): line voltages.

(V_{An}, V_{Bn}, V_{Cn}): phase voltages.

Each arm has a logic function Si that determines the six conduction states (Fig. 3):

$$S_i = \begin{cases} 1 \ \text{ON} \\ 0 \ \text{OFF} \end{cases} (i = a, b, c) \tag{7}$$

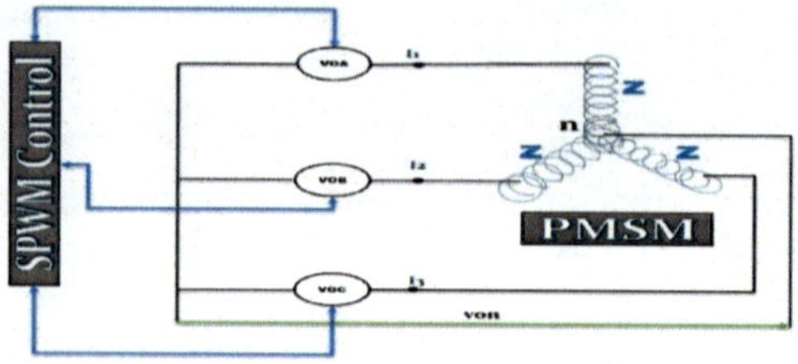

Fig. 3. Simplified representation three-phase equilibrate system-fed PMSM.

Voltages at the output of the inverter:

$$\begin{cases} V_{On} = V_{OA} + V_{An} = V_{OA} + ZI_1 \\ V_{On} = V_{OB} + V_{Bn} = V_{OB} + ZI_2 \\ V_{On} = V_{OC} + V_{Cn} = V_{OC} + ZI_3 \\ 3V_{On} = V_{OA} + V_{OB} + V_{OC} + Z(I_1 + I_2 + I_3) \\ Z(I_1 + I_2 + I_3) = 0 \end{cases} \quad (8)$$

After using the Schalk relationship, the relationship between the simple voltage and the input voltage is obtained in the form of a matrix:

$$\begin{pmatrix} V_{An} \\ V_{Bn} \\ V_{Cn} \end{pmatrix} = \frac{1}{3}\begin{pmatrix} 2 & -1 & -1 \\ -1 & 2 & -1 \\ -1 & -1 & 2 \end{pmatrix}\begin{pmatrix} V_{AO} \\ V_{BO} \\ V_{CO} \end{pmatrix} \quad (9)$$

Depending on the conduction of static switches or diodes in an inverter branch, the branch voltages can be equal to either $\frac{V_{DC}}{2}$ or $-\frac{V_{DC}}{2}$.

Then the input voltages applied to the machine can be obtained directly from the control signals S_a, S_b, S_c where each takes the value 1 or 0 as follows:

$$\begin{cases} V_{jO} = S_i V_{DC} - \frac{V_{DC}}{2} \\ S_i = 1 \text{ ON} \\ S_i = 0 \text{ OFF} \end{cases} \quad \begin{cases} j = A, B, C \\ i = a, b, c \end{cases} \quad (10)$$

Replacing (10) in (9) find the final expression that relates the phase-to-neutral voltages to the logic control (7) is (Fig. 4):

$$\begin{pmatrix} V_{An} \\ V_{Bn} \\ V_{Cn} \end{pmatrix} = \frac{V_{DC}}{3}\begin{pmatrix} 2 & -1 & -1 \\ -1 & 2 & -1 \\ -1 & -1 & 2 \end{pmatrix}\begin{pmatrix} S_a \\ S_b \\ S_c \end{pmatrix} \quad (11)$$

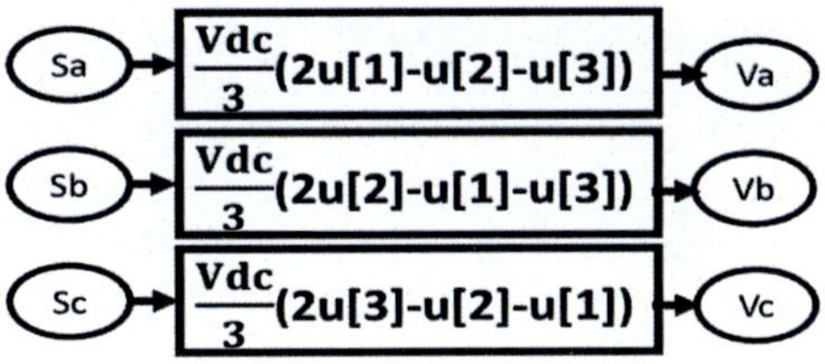

Fig. 4. Modeling a three-phase two-level inverter in MATLAB-Simulink.

1.3 SPWM Control

1.3.1 Objective

Minimizes total harmonic distortion to provide better voltage and current output quality when using the machine [12].

1.3.2 The Method of Work

This technique is based on the comparison between two signals reference and carrier, the intersection of these signals gives the switching times of the switches, and this technique is characterized by two main parameters:

$$\begin{cases} m_F = \frac{F_p}{F} \\ m_m = \frac{V_m}{V_P} \end{cases} \tag{12}$$

m_F is the modulation Index, m_m is the modulation rate, F_p is the modulation frequency, F is the reference frequency, V_m is the amplitude of the reference voltage, V_p is the amplitude of the carrier (Fig. 5).

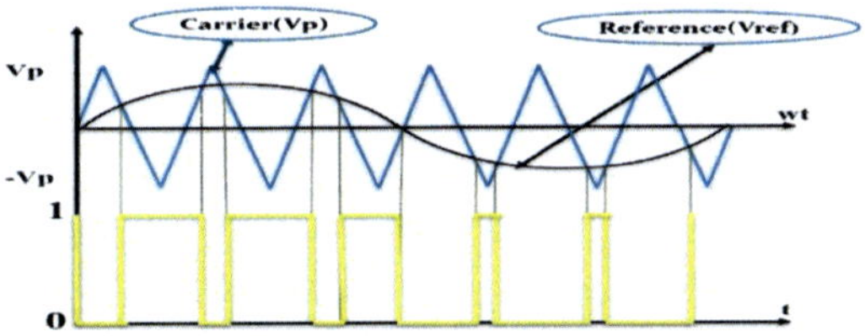

Fig. 5. Control principle by sinusoidal pulse width modulation Triangular.

The following algorithm can therefore summarize the principle of this strategy:

$$\begin{cases} \text{If } V_m = V_{ref} > V_p \;\; S_{a,b,c} = 1 \\ \quad \text{Else If} \;\; S_{a,b,c} = 0 \end{cases} \tag{13}$$

2 Nonlinear Backstepping Controller:

The equations of the IPMSM (d, q) model can be written as follows [13]:

$$\begin{cases} \frac{dI_d}{dt} = -\frac{R_s}{L}I_d + Pw_mI_q + \frac{V_d}{L} \text{ (a)} \\ \frac{dI_q}{dt} = -\frac{R_s}{L}I_q - Pw_mI_d - P\frac{\varphi_f}{L}w_m + \frac{V_q}{L} \text{ (b)} \\ \frac{dw_m}{dt} = \frac{1}{J_m}\left\{\frac{3p}{2}\varphi_f I_q - T_m - B_m w_m\right\} \text{ (c)} \\ T_e = \frac{3p}{2}\varphi_f I_q \text{ (d)} \end{cases} \tag{14}$$

The state vector : $[X] = [X_1X_2X_3]^T = \left[I_dI_qw_m\right]^T$, The Control vector : $U = \left[V_dV_q\right]^T$.

The objective of the backstepping control is to regulate the speed of the machine towards its reference value w* whatever the external variations. We supposed that the motor parameters are known and invariant, i.e., in non-adaptive mode [14].

Errors and their dynamics are defined by

$$\begin{cases} e_w = w^* - w_m(a);\ e_q = I_q^* - I_q(b); e_d = I_d^* - I_d(c) & (15) \\ \dot{e}_w = \dot{w}^* - \dot{w}_m(a); \dot{e}_q = \dot{I}_q^* - \dot{I}_q(b); \dot{e}_d = \dot{I}_d^* - \dot{I}_d(c) & (16) \end{cases}$$

2.1 Step1: Speed Backstepping Controller

To control the motor and achieve the desired speed, we adopt the following Lyapunov function [15]:

$$V_1 = \frac{1}{2}e_w^2 \tag{17}$$

Performing the derivative of (17) and replacing (14) (c), we obtain:

$$\dot{V}_1 = e_w\dot{e}_w = e_w\left[\dot{w}^* - \dot{w}_m\right] = e_w\left[\dot{w}^* - \frac{1}{J_m}\left\{\frac{3p\varphi_f}{2}I_q - T_m - B_m w_m\right\}\right] \tag{18}$$

We consider that

$$\frac{dw^*}{dt} = \frac{d^2w^*}{dt} = \frac{d^3w^*}{dt} = 0 \tag{19}$$

So that the velocity error is equal to 0, and the sum of the moments tends to 0, resulting in a smooth circular velocity, we will choose the two direct quadratic currents as the virtuous control, Therefore, to ensure that $\dot{V}_1 \leq 0$ i.e., the asymptotically stable system by the Lyapunov rule it is necessary that:

$$\begin{cases} I_d^* = 0 \\ I_q^* = \frac{2J_m}{3p0_f}\left(K_w e_w + \frac{T_m}{J_m} + \frac{B_m w_m}{J_m}\right) \end{cases} \tag{20}$$

So: $\dot{V}_1 = -K_w e_w^2$ with $K_w > 0$ to ensure that $\dot{V}_1 \leq 0$.

2.2 Step2: Curent Backstepping Controller

For the following, the assurance of stability, and convergence of the components Idq to the reference I* dq leads to choose the following Lyapunov function and derivation is [16]:

$$\begin{cases} V_2 = \frac{1}{2}\left(e_w^2 + e_q^2 + e_d^2\right) & (21) \\ \dot{V}_2 = e_w\dot{e}_w + e_q\dot{e}_q + e_d\dot{e}_d & (22) \end{cases}$$

Introduce the velocity dynamics in Eq. 23 we find:

$$\dot{V}_2 = e_w\left\{-\frac{1}{J_m}\left\{\frac{3p\varphi_f}{2}I_q - T_m - B_m w_m\right\}\right\} + e_q\dot{e}_q + e_d\dot{e}_d \tag{23}$$

By replacing Eq. (15) (b) and (20) by (23), we find:

$$\dot{V}_2 = -K_w e_w + e_q\left\{\frac{3p\varphi_f}{2J_m}e_w + \dot{e}_q\right\} + e_d\dot{e}_d \tag{24}$$

For Eq. (24) to be asymmetrically stable, we impose the following control laws on the system: with Ld = Lq = L:

$$\begin{cases} \mathbf{V_q}^* = \mathbf{L}\left(\mathbf{K_q e_q} + \frac{\mathbf{R_s}}{\mathbf{L}}\mathbf{I_q} + \mathbf{P w_m I_d} + \mathbf{P}\frac{\varphi_f}{\mathbf{L}}\mathbf{w_m} + \frac{2(\mathbf{K_w J_m - B_m})}{3\mathbf{p}\varphi_f}\dot{\mathbf{e}}_\mathbf{w} + \frac{3\mathbf{p}\varphi_f}{2\mathbf{J_m}}\mathbf{e_w}\right) \\ \mathbf{V_d}^* = \mathbf{L}\left(\mathbf{K_d e_d} + \frac{\mathbf{R_s}}{\mathbf{L}}\mathbf{I_d} - \mathbf{P w_m I_q}\right) \end{cases} \tag{25}$$

So: $\dot{V}_2 = -k_w e_w^2 - k_q e_q^2 - k_d e_d^2$ with $K_w, K_q, K_d > 0$ to ensure that $\dot{V}_2 \le 0$ (Fig. 6).

3 Simulation and Test Performance

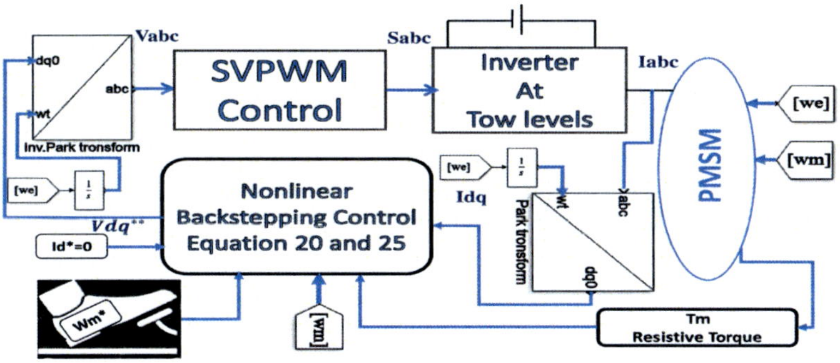

Fig. 6. The main scheme used to process the results of the backstepping command.

3.1 Simulation of the Backstepping Command

To test the performance of the backstepping control, we use the MATLAB-SIMULINK environment. We will analyze the performance of the real system, in terms of speed, stability, accuracy and robustness of the system [Table 1].

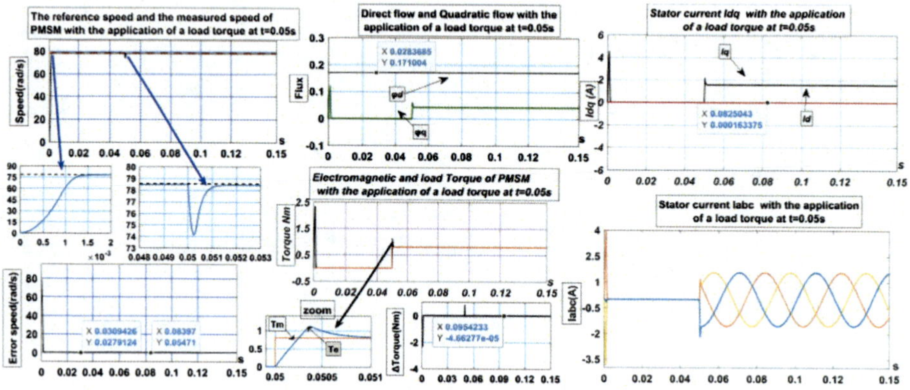

Fig. 7. Dynamic response of the PMSM.

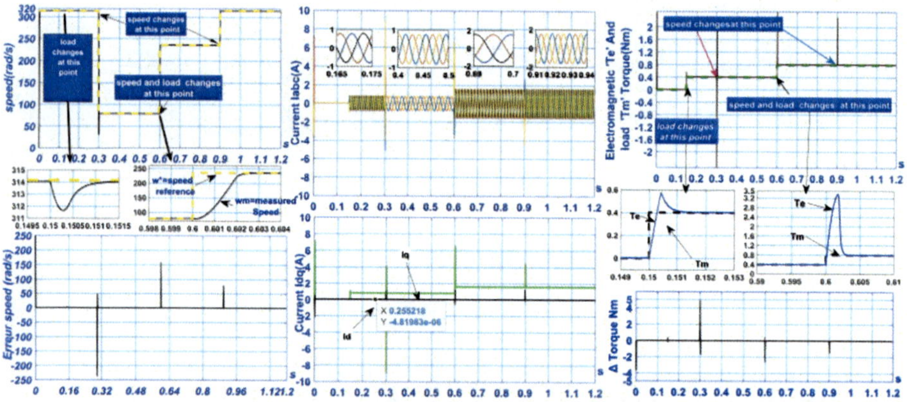

Fig. 8. Tracking of the trajectory and rejection of load torque variations.

- The results obtained in Figs. (7, 8, 9) show that the theoretical predictions have been verified, the linearity between the electromagnetic torque and the square current is evident and shows the responsibility of the quadratic stator current on the electromagnetic torque, before the application of the load, the velocity profile shows that all quantities stabilize after a response time of 1, 48110−3 s, without overshoot and with good damping, and that the tachometric response of the system is able to recover its equilibrium value in (1, 4910−3 s).

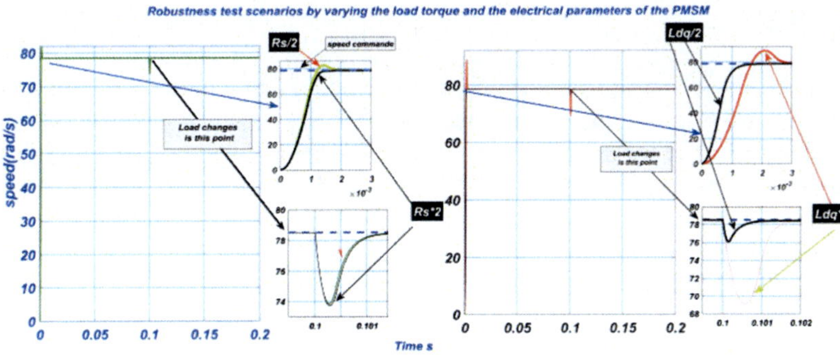

Fig. 9. Robustness test scenarios of a PMSM.

- Fig. (7), for a trajectory w* (=750 rpm at t = 0 s) and (Tm = 0.8 N.m 'nominal torque value' at t = 0.05 s) and a simulation time of 0.15 s, we see that the decoupling is maintained (the direct current is cancelled), and that when the load is applied the electromagnetic torque reaches its reference value to compensate for this solicitation with an almost instantaneous response.
- Concerning Fig. (7), for the same site trajectory previously and Fig. (8) for a trajectory w* (=3000 rpm at t = 0 s, = 750 rpm at t = 0.3 s, = 1500 rpm at t = 0.6 s, = 3000 rpm at t = 0. 9 s) and (Tm = 0. 4N. m at t = 0.15 s and increase at T = 0.8 Nm at t = 0. 6 s) with a simulation time of 1.2 s. We noted the good performance of the position tracking even during the existence of the disturbance, which shows the efficiency of our controller for the rejection of resistive variations in a non-existent time (10^{-3}s).
- Fig. (9), for a trajectory w* (=750 rpm at t = 0 s) and (Tm = 0.8 N.m a t = 0.1 s). We can notice that the variations in stator resistance Rs and inductance Ldq and resistive torque do not affect the dynamic characteristics of the speed. This is mainly due to the recursive nature of the latter, which allows the whole system by considering it as a cascaded subsystem and kept stable as it goes along, which shows the robustness of the nonlinear-non adaptive backstepping control.

Table 1. PMSM drive parameters for simulation:

Rated output power (kW)	Rated speed (rpm)	No of Poles
0.25	3000	4
Rated Torque (Nm)	Stator Resistance (Ω)	Stator d and q axis inductance (H)
0.8	0.9585	0.007
Inertia (Kg·m^2)	Friction Factor (Nm·s)	Flux Linkage in (wbr)
0.0006329	11e−5	0.171

4 Conclusion

The feedback control strategy based on Lyapunov theory can reduce the pulsation of the stator current and improve the control accuracy of the system. It enables reliable and stable operation of the three-phase PMSM system, achieving satisfactory torque and speed control effects and reducing the THD value of the three-phase stator current. The simulation results verify the feasibility and accuracy of the proposed strategy and show that it is suitable for engineering applications.

References

1. Youness, E.M., et al.: Implementation and validation of backstepping control for pmsg wind turbine using dspace controller board, energy rep., vol. 5, pp. 807–821, November 2019
2. Nadji, S., Benaicha, S.: robust integral backstepping control of pmsm for electric vehicle application. In: 2017 5th International Conference on Electrical Engineering-Boumerdes (ICEE-b), pp. 1–6 (2017)
3. Linares-flores, J., García-rodríguez, C., Sira-ramírez, H., Ramírez-cárdenas, O.D.: Robust backstepping tracking controller for low-speed pmsm positioning system: design, analysis, and implementation. IEEE Trans. Ind. Inform. **11**(5), 1130–1141 (2015)
4. Van den broek, H.W., Skudelny, H.C., Stanke, G.: Analysis and realization of a pulse width modulation based on voltage space vectors. IEEE Trans. Ind. Appl. **24**, 142–150 (1988)
5. Guo-jun, T., Zong-Bin, Y., Wei, J.: simulation and realization of the controlling method for three-level inverter. Power Electron. **41**, 86–88 (2007)
6. Bilhan, A.K., Akbal, E.: Modelling and simulation of two-level space vector PWM inverter using photovoltaic cells as dc source. IJEMME **2**, 311–317
7. Bhat, A.H., Agarwal, P.: Modelling and simulation of two-level space vector PWM inverter using photovoltaic cells as DC source. IJEMME **2**, 311–317
8. El-sousy, F.F.M.: Robust wavelet-neural-network sliding-mode control system for permanent-magnet synchronous motor drive. IET Electr. Pow. Appl. **5**(1), 113–132 (2011)
9. Liu, L., Zheng, X.H., Cui, L.L.: Backstepping control of speed sensorless permanent magnet synchronous motor. Trans. China Electrotech. Soc. **26**(9), 67–72 (2011)
10. Ke, S.-S., Et, J.S., Lin, S.: Ensorless speed tracking control with backstepping design scheme for permanent magnet synchronous motors. In: Proceedings of 2005 IEEE Conference on Control Applications, Toronto, Ontario, vol. 3, pp. 487–492, 28–31 August 2005
11. Bossoufi, B., Karim, M., Lagrioui, A., Taoussi, M.: FPGA-based implementation nonlinear backstepping control of a pmsm drive. IJPEDS Int. J. Power Electron. Drive Syst. **4**(1), 12–23 (2014)
12. Bossoufi, B., Karim, M., Lagrioui, A., Taoussi, M., El Ghamrasni, M.: Backstepping adaptive control of dfig-generators for variable-speed wind turbines. IJCT Int. J. Comput. Technol. **12**(7), 3719–3733 (2014)
13. Bossoufi, B., Karim, M., Lagrioui, A.: FPGA based implementation sliding mode control and nonlinear adaptive backstepping control of a permanent magnet synchronous machine drive. WSEAS Trans. Syst. Control **9**(1), 86–100 (2014)
14. Bossoufi, B., Karim, M., Ionita, S., Lagrioui, A.: Nonlinear non-adaptive backstepping with sliding-mode torque control approach for PMSM motor. J. Electr. Syst. JES **8**(2), 236–248 (2012)
15. Bossoufi, B., Karim, M., Ionita, S., Lagrioui, A.: Low-speed sensorless control of PMSM motor drive using a NONLINEAR approach BACKSTEPPING control: FPGA-based implementation. J. Theor. Appl. Inf. Technol. JATIT **36**(1), 154–166 (2012)

Model Reference Adaptive System Based DPC-SVM Control for Permanent Magnet Synchronous Generator

Btissam Majout[1](✉), Badre Bossoufi[1], Mohammed Karim[1], Najib El Ouanjli[2], Ikram Saady[1], Zakaria El Zair Laggoun[3], Mohammed El Mahfoud[1], and Mourad Yessef[1]

[1] Laboratory of Engineering Modeling and Systems Analysis, SMBA University, Fez, Morocco
btissammajout@yahoo.com

[2] Faculty of Sciences and Technology, LMIET, Hassan First University, Settat, Morocco

[3] Laboratory of Electrotechnique Constantine, Mentouri Brother's Constantine 1 University Constantine, Constantine, Algeria

Abstract. This paper presents a new optimization direct power control strategy based on the space vector modulation (DPC-SVM) applied to a PMSG-wind power system. The main purpose of this study is to overcome certain drawbacks of conventional direct power control on the wind power systems, such as variable switching frequency and high currents and power ripples. Hence, the system performance will be improved. In addition, to increase the control performance, a speed observer based on a Model Reference Adaptive System (MRAS) was used to overcome the problems caused by the use of the rotor sensor. To verify the effectiveness and the robustness of the proposed DPC-SVM with MRAS, the whole system was tested under a variable wind speed in the MATLAB/Simulink environment. Furthermore, a comparative analysis was performed between the proposed control DPC-SVM and conventional DPC to highlight the performance of the proposed approach.

Keywords: Wind turbine · DPC-SVM · MRAS observer · PMSG

1 Introduction

The exploitation and development of renewable energy have experienced significant progress during the last three decades. Among various kinds of renewable energy, wind power has become a viable solution for the production of energy [1]. To exploit this potential efficiently, the Wind Energy Conversion System (WECS) must be controlled with a robust controller that can make it able to handle any variation of the wind, quickly and with high robustness [2]. For these reasons, many techniques have been developed in the literature, such as direct and indirect field-oriented, sliding mode, and Backstepping controls. Despite their effectiveness, these controls require more calculation time and good skills to adjust their parameters [3, 4].

S. Motahhir and B. Bossoufi (Eds.): ICDTA 2022, LNNS 454, pp. 535–544, 2022.
https://doi.org/10.1007/978-3-031-01942-5_53

Unlike the techniques mentioned previously, the advantage of the direct power control is that it doesn't require the PWM modulation blocks and the internal regulation loops. Where the converter switching states are chosen from a switching table according to the digital value of the errors generated by the hysteresis controllers [5]. However, DPC suffers from high power and current ripples caused by the hysteresis controllers. To improve the classical DPC, this study suggests a robust model of direct power control based on Space Vector Modulation (SVM) to control the powers delivered by the generator PMSG. The latter enables the system to operate with a constant switching frequency, which can reduce the power and the currents ripple problem. Hence, the performance of the PMSG wind power system will be improved [5–8].

Furthermore, conventional controls obtain information about rotor speed and position from speed and position sensors which present a lot of problems such as complicated installation and maintenance, lower reliability, and higher cost [9]. To overcome the drawbacks associated with the use of rotor position sensors in a PMSG, this paper presents a position sensorless technique based on the model reference adaptive system (MRAS). Its principle is to estimate the rotor speed indirectly, using the generator variables (voltage and currents) [10–12].

The goal of this study is to combine the benefits of DPC-SVM control with the efficiency of the MRAS algorithm to reduce the problem of power and current ripples and develop a control law that does not rely on mechanical sensors to measure PMSG speed.

This work is structured as follows: Sect. 2 presents the dynamic model of a wind energy conversion system which includes both the WECS and PMSG models. Section 3 describes the principle of both proposed control DPC-SVM and MRAS algorithms. The simulation results are presented and discussed in Sect. 4 using a variable wind speed. Finally, Sect. 5 is devoted to a brief conclusion.

2 System Modeling

2.1 Description of the System

The wind energy conversion system, as illustrated in Fig. 1, consist of a three-blade wind turbine to convert wind energy into mechanical energy and a permanent magnet synchronous generator PMSG to convert mechanical energy into electrical energy which is then delivered to a grid via two back-to-back converters and an RL filter [2].

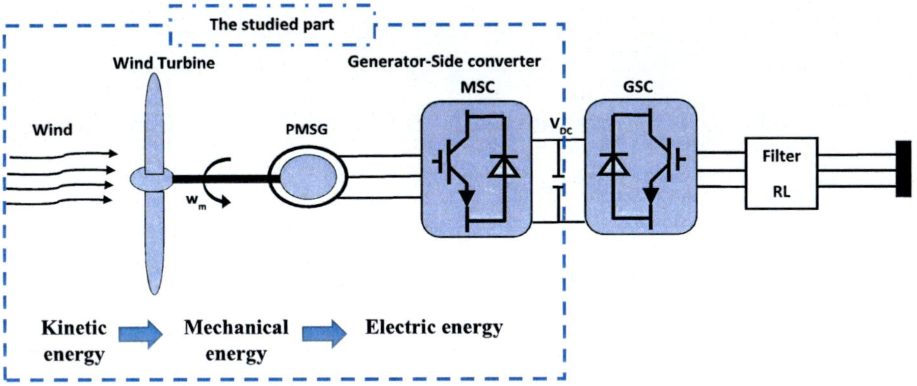

Fig. 1. Configuration of PMSG-based wind power system.

2.2 Wind Turbine Model

The wind power is converted into mechanical power according to the following model [2].

$$P_{wind} = \frac{1}{2}.\rho.S.v_w^3 \tag{1}$$

$$P_{turb} = C_p(\beta, \lambda).P_{wind} = \frac{1}{2}.\varphi.S.C_p(\beta, \lambda).v_w^3 \tag{2}$$

$$T_{turb} = \frac{P_{turb}}{\Omega_t} = \frac{1}{2}.\varphi.S.C_p(\beta, \lambda).v_w^3.\frac{1}{\Omega_t} \tag{3}$$

where C_p is a ratio between the mechanical power available at the turbine shaft (P_{turb}) and the power available in wind (P_{wind}). Its expression adapted for this wind turbine is as follows [3].

$$\begin{cases} C_p(\beta, \lambda) = [0.35 - 0.167(\beta - 2)]\sin\left(\frac{\pi(\lambda+0.1)}{14.34-0.3(\beta-2)}\right) - 0.00184(\lambda - 3)(\beta - 2) \\ \lambda = \frac{\Omega_t.R}{Vw} \\ C_p^{max}(\beta, \lambda) = \frac{16}{27} = 0.5930 \end{cases} \tag{4}$$

as indicated in Eq. 4, the value of coefficient Cp depends on the pitch angle (β) and tip speed ratio (λ). To ensure that the system operates around its optimum rotation where the energy extracted is maximum, the Cp coefficient must reach its optimum ($Cp = 0.36$) determined by Betz's limit law when $\beta = 2°$ and $\lambda = 0.707$.

2.3 Permanent Magnet Synchronous Generator Model

The PMSG model in the synchronous rotating reference frame (α, β) is as follows [5].

- Stator voltages:

$$\begin{cases} v_{s\alpha} = R_s.i_{s\alpha} + \frac{d\Psi_{s\alpha}}{dt} \\ v_{s\beta} = R_s.i_{s\beta} + \frac{d\Psi_{s\beta}}{dt} \end{cases} \tag{5}$$

- Stator fluxes: $$\begin{cases} \Psi_{s\alpha} = L_s.\, i_{s\alpha} + \emptyset_f.\cos\theta_e \\ \Psi_{s\beta} = L_s.\, i_{s\beta} + \emptyset_f.\sin\theta_e \end{cases} \tag{6}$$

The electromagnetic torque expression and the mechanical equation are given by:

$$\begin{cases} \frac{d\Omega_{\text{mec}}}{dt} = \frac{\text{T}_{\text{turb}}}{\text{J}_{\text{tot}}} - \frac{T_{em}}{\text{J}_{\text{tot}}} - \frac{\text{f}}{\text{J}_{\text{tot}}}.\Omega_{\text{mec}} \\ T_{em} = \frac{3}{2}.p.\left[\left(L_s.i_{s\alpha} + \emptyset_f.\cos\theta_e\right)i_{s\beta} - \left(L_s.i_{s\beta} + \emptyset_f.\sin\theta_e\right)i_{s\alpha}\right] \end{cases} \tag{7}$$

where $i_{s\alpha}$ and $i_{s\beta}$ are the αβ-axes stator currents components of the PMSG, respectively. R_s and L_s are the stator resistance and inductance of the PMSG, respectively. $\theta_e = p.\theta_{mec}$ is the electrical angle of the rotor (p is pole pair number) and $\emptyset_f$ is the permanent-magnet flux linkage. T_{em} is the electromagnetic torque and T_{turb} is the mechanical torque produced by the wind turbine. J_{tot} is the overall rotor inertia of the wind turbineand PMSG.f is the viscous friction coefficient.

Furthermore, the expression of the active and reactive power is as follows [5]:

$$\begin{cases} P_s = \frac{3}{2}\left[v_\alpha.i_\alpha + v_\beta.i_\beta\right] \\ Q_s = \frac{3}{2}\left[v_\beta.i_\alpha - V_\alpha.i_\beta\right] \end{cases} \tag{8}$$

3 Synthesis of Control

3.1 Principe of the Proposed Control

In general, the classical DPC directly controls both the active and reactive power of the PMSG by selecting the optimum voltage vector from a switching table [5]. As is shown in Fig. 2, this selection depends on the digital errors H_P and H_Q generated by the hysteresis controllers and the voltage vector $v_{\alpha\beta}$ location θ_e [4, 5].

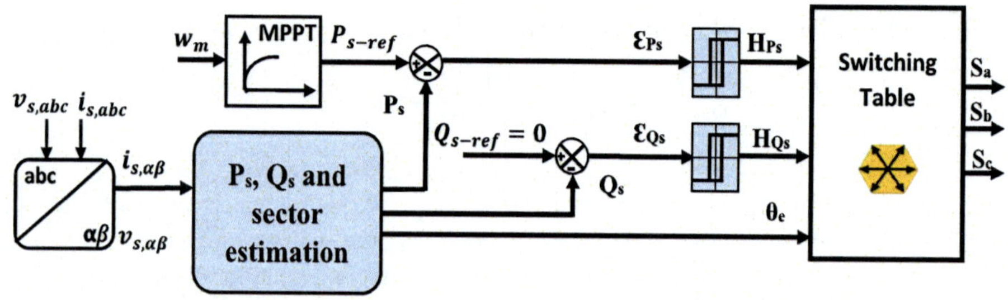

Fig. 2. General structure of the classical DPC applied to the generator-side converter.

Despite the various advantages of the classic DPC (lower parameter dependency, fast dynamic response, and high reliability) [4], it suffers from variable switching frequency and high powers ripple,which are proportional to the hysteresis bandwidth [7]. To overcome the classical DPC drawbacks, a direct power control based on the space vector modulation DPC-SVM combined with an MRAS observer was proposed.

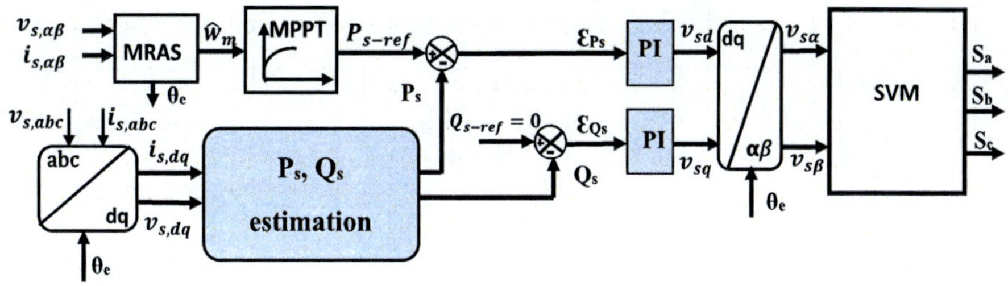

Fig. 3. General structure of the proposed DPC-SVM with MRAS observer applied to the generator-side converter.

As depicted in Fig. 3. The hysteresis controllers and the switching table used in the classical DPC are compensated by PI controllers and an SVM in the proposed DPC-SVM control.

Figure 3 describes the principle of the proposed control, Where the active and reactive power PI controllers generate the quadratic and direct stator voltages (v_{sq},v_{sd}), respectively. These two voltages must be then transformed to a fixed reference (α, β) and delivered to the SVM to generate the converter switching states (S_a, S_b, S_c) [7, 8]. The active power reference P_{s-ref} is generated by the MPPT strategy which uses the estimated speed $\hat{w}_m$ generated by the MRAS observer, whereas the reactive power reference Q_{s-ref} was set to zero to achieve unity power factor.

3.2 Basic SVM Algorithm

The SVM principle is based on the concept of approximating the reference voltage vector V_{ref} which is determined as a combination of two adjacent non null voltage vectors at the sector i [6]. Where the reference voltage vector can be considered constant for each switching period [6, 7] at high switching frequency. Therefore, its expression in the sector $i(i = 1 \ldots .6)$ can be presented as follows:

$$T_e.\overline{V}_{ref} = T_i\overline{V}_i + T_{i+1}\overline{V}_{i+1} + T_0\overline{V}_0 \tag{9}$$

$$T_0 = T_e - T_i - T_{i+1} \tag{10}$$

T_i and T_{i+1} are the applications times of the adjacent vectors $\overline{V}_i$ and $\overline{V}_{i+1}$, and T_0 is the time of application of the null voltage vectors $\overline{V}_0$ and $\overline{V}_7$. Where the application time value of each vector in sector $i(i = 1 \ldots .6)$ is given in Table 1.

Table 1. Application times of each sector.

Sector 1	Sector 2	Sector 3
$T_1 = \frac{(\sqrt{6}V_\beta - \sqrt{2}V_\propto)}{2V_{dc}}T_e$	$T_2 = \frac{(\sqrt{6}V_\beta + \sqrt{2}V_\propto)}{2V_{dc}}T_e$	$T_3 = \frac{(\sqrt{2}V_\beta)}{V_{dc}}T_e$
$T_2 = \frac{(\sqrt{2}V_\beta)}{V_{dc}}T_e$	$T_3 = \frac{(\sqrt{6}V_\beta + \sqrt{2}V_\propto)}{2V_{dc}}T_e$	$T_4 = \frac{-(\sqrt{6}V_\beta + \sqrt{2}V_\propto)}{2V_{dc}}T_e$
$T_0 = T_e - T_1 - T_2$	$T_0 = T_e - T_2 - T_3$	$T_0 = T_e - T_3 - T_4$
Sector 4	**Sector 5**	**Sector 6**
$T_4 = \frac{(-\sqrt{6}V_\beta + \sqrt{2}V_\propto)}{2V_{dc}}T_e$	$T_5 = \frac{-(\sqrt{6}V_\beta + \sqrt{2}V_\propto)}{2V_{dc}}T_e$	$T_6 = \frac{-(\sqrt{2}V_\beta)}{V_{dc}}T_e$
$T_5 = \frac{-(\sqrt{2}V_\beta)}{V_{dc}}T_e$	$T_6 = \frac{(\sqrt{6}V_\beta - \sqrt{2}V_\propto)}{2V_{dc}}T_e$	$T_1 = \frac{(\sqrt{6}V_\beta + \sqrt{2}V_\propto)}{2V_{dc}}T_e$
$T_0 = T_e - T_4 - T_5$	$T_0 = T_e - T_5 - T_6$	$T_0 = T_e - T_6 - T_1$

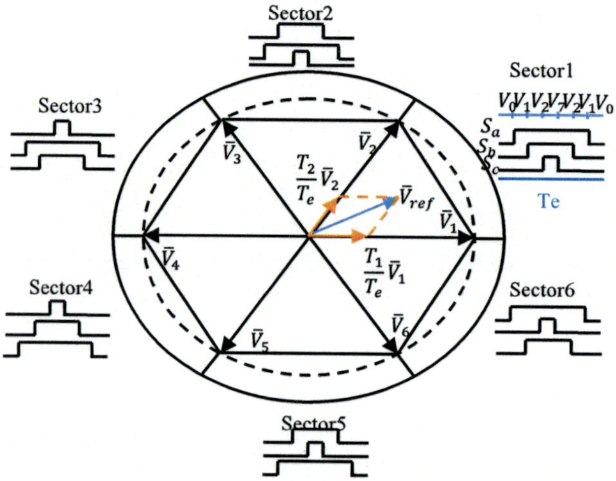

Fig. 4. Block diagram of the voltage vectors and space vectors.

Where for a two-level inverter, the vector to be applied for different positions of the reference voltage vector are indicated by Fig. 4.

3.3 Model Reference Adaptive Speed Observer

To overcome the drawbacks of the speed transducer and the position encoder used in the conventional controls. Different estimation methodologies have been proposed in the literature for PMSG. Among them, MRAS has gained significant attention in recent years, due to its simplicity, high reliability, and efficiency [11]. The MRAS structure, as illustrated in Fig. 5, consists of two basic models (a reference model and an adaptive mode). The error between these two models drives an adaptation mechanism that generates the estimated speed which is then used in the adaptive model through a closed-loop [12].

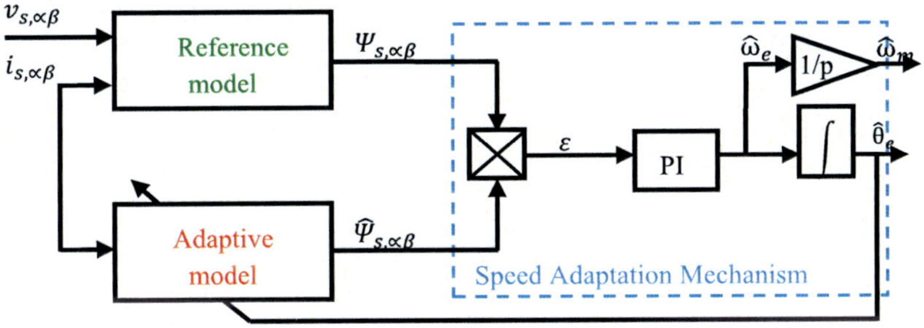

Fig. 5. MRAS observer structure.

The flux deduced from the generator voltage model is used as a reference model, and the flux presented by Eq. 12 is used as an adaptive model.

- the reference model is: $$\begin{cases} \Psi_{s\alpha} = \int (v_{s\alpha} - R_s . i_{s\alpha})dt \\ \Psi_{s\beta} = \int (v_{s\beta} - R_s . i_{s\beta})dt \end{cases} \quad (11)$$
- the adaptive model is: $$\begin{cases} \hat{\Psi}_{s\alpha} = L_s . i_{s\alpha} + \emptyset_f . \cos \hat{\theta}_e \\ \hat{\Psi}_{s\beta} = L_s . i_{s\beta} + \emptyset_f . \sin \hat{\theta}_e \end{cases} \quad (12)$$

The estimated rotor angle is calculated from the estimated rotor speed as follows:

$$\hat{\theta}_e = \int \hat{\omega}_e dt \quad (13)$$

Moreover, the error ε between the two flux vectors ($\Psi_{\alpha\beta}$, $\hat{\Psi}_{\alpha\beta}$) is calculated by the cross product of the reference and estimated fluxes, as illustrated on Fig. 5.

$$\varepsilon = \Psi_{s\alpha} . \hat{\Psi}_{s\beta} - \Psi_{s\beta} . \hat{\Psi}_{s\alpha} \quad (14)$$

this error ε was driven to zero using a PI controller. The estimated speed $\hat{\omega}_e$, which has the following equation, is the output of this controller:

$$\hat{\omega}_e = \left(k_p + \frac{k_i}{p}\right)\varepsilon \quad (15)$$

with k_p and k_i are the PI controller gains.

4 Results and Analysis

To evaluate the performance of the suggested technique a simulation test is performed in the MATLAB/Simulink environment using a wind power system based on a PMSG

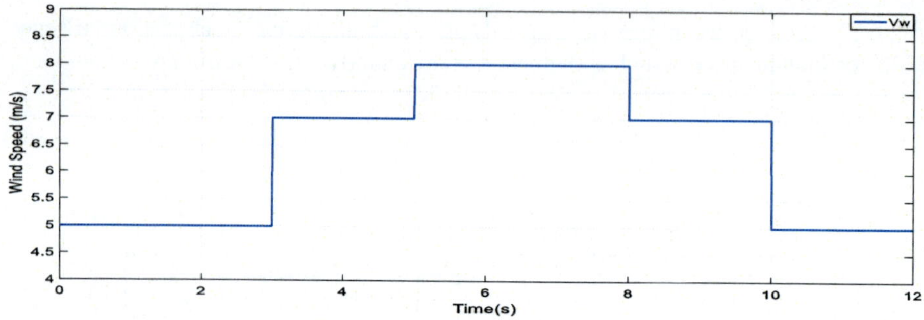

Fig. 6. Wind Speed profile

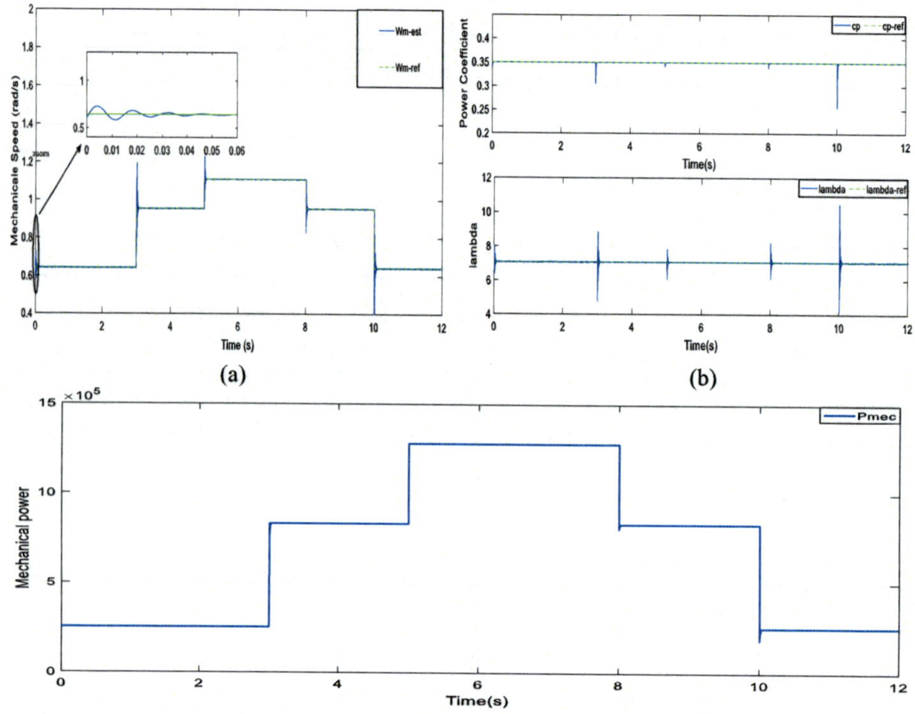

(c)

Fig. 7. Performance of the system using MRAS observer: **(a)** Mechanical speed w_{mec} of the PMSG, **(b)** Power coefficient C_p and lambda λ, **(c)** Mechanical power P_{mec}.

model of 1.5 MW. The simulation of the WECS control system was performed using a variable wind speed ranging from 5 to 8 m/s as shown in Fig. 6.

Figure (7a) presents the estimated mechanical speed $\hat{\omega}_{mec}$. of the PMSG determined by the proposed MRAS observer. As can be seen, the estimated mechanical speed varies in the same way as the wind speed, and follows its reference with a short response time and acceptable overshoot, indicating that the proposed technique is efficient.

Figure (7b) shows that the power coefficient C_p and the tip speed ratio λ (lambda) are equal to their optimal values reference 0.36 and 7.07, respectively. Moreover, the shape of the mechanical power is very similar to that of the wind Fig. (7c). These results ensure that the system is operating around its optimum rotational speed which proves the MPPT control effectiveness.

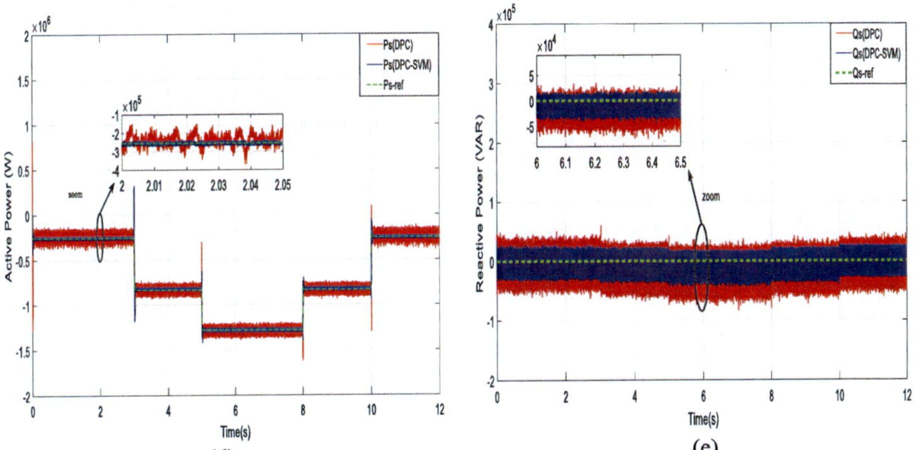

(d) (e)

Fig. 8. Stator powers: (d) active power Ps and (e) reactive power Qs.

It can be seen from Fig. 8 that the stator powers (Ps, Qs) track their reference values with high accuracy for both types of control. It can also be seen that DPC-SVM offers better performance than that of DPC in terms of oscillation and disturbance rates. While the reactive power Qs of both controllers remain zero, indicating that the operation is with a unitary power factor.

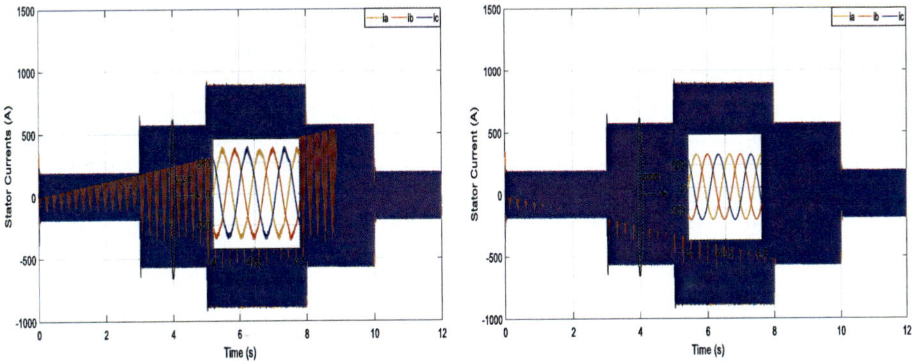

(f) (g)

Fig. 9. Stator current of: (f) DPC and (g) DPC-SVM.

Figure 9 shows that the stator current amplitude varies with the wind speed variation, with a better sinusoidal waveform and a lower ripple rate under DPC-SVM control than that of the DPC control.

5 Conclusion

In this work, the performance of the proposed DPC-SVM and MRAS observer was demonstrated through a simulation study conducted under variable wind speed in the MATLAB Simulink environment. It can be seen from the results obtained that the operation of the PMSG wind power system with DPC-SVM gives good performance than the classical direct power control DPC, whether for the reference tracking, power and current ripples and their waveform. Thus, it can be concluded that this control is a very effective solution to ensure the robustness of the system and to overcome the dynamic problems caused by both the DPC control structure and the speed sensors.

References

1. Bossoufi, B., et al.: Rooted tree optimization for the backstepping power control of a doubly fed induction generator wind turbine: dSPACE implementation. IEEE Access **9**, 26512–26522 (2021)
2. El Mourabit, Y., Derouich, A., Allouhi, A., El Ghzizal, A., El Ouanjli, N., Zamzoumyes, O.: Sustainable production of wind energy in the main Morocco's sites using permanent magnet synchronous generators. Int. Trans. Electr. Energ. Syst. e12390 (2020)
3. Majout, B., Abrahmi, D., Ihedrane, Y., El Bakkali, C., Mohammed, K., Bossoufi, B.: Improvement of sliding mode power control for WECS based on DFIG-generator. WSEAS Trans. Comput. Res. **8**(5), 26–33 (2020)
4. Freire, N., Estima, J., Cardoso, A.: A comparative analysis of PMSG drives based on vector control and direct control techniques for wind turbine applications, PRZEGLĄD ELEKTROTECHNICZNY (Electrical Review), ISSN 0033–2097, R. 88 NR 1a/2012
5. Hammoumi, D., El Bekkali, C., Karim, M., Taoussi, M., El Ouanjli, N., Bossoufi, B.: Direct controls for wind turbine with PMSG used on thereal wind profile of essaouira-Morocco City. Indonesian J. Electr. Eng.Comput. Sci. **16**, 1229–1239 (2019)
6. El Mourabit, Y., Derouich, A., El Ghzizal, A., El Ouanjli, N., Zamzoum, O.: DTC-SVM control for permanent magnet synchronous generator based variable speed wind turbine. Int. J. Power Electr. Drive Syst. **8**(4), 1732–1743 (2017)
7. Laggoun, Z.E.Z., Khalile, N., Benalla, H.: A comparative study between DPC-SVM and PDPC-SVM. In: 2019 International Conference on Advanced Electrical Engineering (ICAEE), pp. 1–5 (2019). https://doi.org/10.1109/ICAEE47123.2019.9014796
8. Elmahfoud, M., Bossoufi, B., El Ouanjli, N., Mahfoud, S., Taoussi, M.: Improved direct torque control of doubly fed induction motor using space vector modulation. Int. J. Intell. Eng. Syst. **14**(3), 177–188 (2021)
9. Sabir, O.: Robust DPC-SVM control strategy for shunt active power filter based on H∞ regulators. Int. J. Electr. Power Energy Syst. **117**, 105699 (2020)
10. Kumar, R., et al.: Speed sensorless model predictive current control of doubly-fed induction machine drive using model reference adaptive system. ISA Trans. (2018). https://doi.org/10.1016/j.isatra.2018.10.025
11. Mateusz, K.: A review on MRAS-type speed estimators for reliable and efficient induction motor drives. ISA Trans. **93**, 1–13 (2019)
12. Giribabu, D., Srivastava, S.P., Pathak, M.K.: Modified reference model for rotor flux-based MRAS speed observer using neural network controller. IETE J. Res. **65**(1), 80–95 (2019)

An Improved MPPT Based on Maximum Area Method for PV System Operating Under Fast Varying of Solar Irradiation

Abdelfettah El-Ghajghaj[1](✉), Najib El Ouanjli[2], Hicham Karmouni[1], Mohammed Ouazzani Jamil[3], Hassan Qjidaa[4], and Mhamed Sayyouri[1]

[1] Laboratory of Engineering, Systems, and Applications, National School of Applied Sciences, Sidi Mohamed Ben Abdellah University, Fez, Morocco
abdelfettah.elghajghaj@usmba.ac.ma

[2] LMIET, Faculty of Sciences and Technology, Hassan First University, Settat, Morocco

[3] Laboratoire Systèmes et Environnement Durables (SED), Faculté des sciences de l'ingénieur (FSI), Université Privée de Fès (UPF), Fez, Morocco

[4] Laboratory of Electronic Signal and Systems of Information (LESSI), Faculty of Sciences Dhar el Mehraz, Sidi Mohamed Ben Abdellah University, Fez, Morocco

Abstract. In this article, a new method of maximum power point tracking (MPPT) is proposed to improve energy production by a photovoltaic system; this method presents a modification of the perturb and observe (P&O) algorithm. This proposed method which is inspired by the classic P&O seeks to follow the maximum area of the curve (P, V), namely that the maximum area located at the zone of the maximum power point (MPP). The efficiency of the proposed algorithm is evaluated in the Matlab/Simulink environment and compares it with the classical and modified P&O algorithm. The comparative analysis shows that our methods offer a better follow-up of the MPP against the rapid variations of irradiation with less oscillation.

Keywords: PV · MPPT · P&O · Matlab/simulink

1 Introduction

The whole world is increasingly aware that renewable energies and energy efficiency are essential to fight climate change, it seems to be a good solution to reduce CO2 emissions. Then the growing demand for energy has allowed a remarkable development of renewable energies, such as photovoltaic (PV), wind, hydraulic [1–3].

Over the past decade, PV power has become a reliable energy source with a steadily decreasing production rate and maintenance cost due to the advancement of technology and the evolution of the PV market. Several applications of the energy obtained using photovoltaic generators are exploited such as the electrification of buildings, electric vehicles, water pumping and public lighting [4, 5].

The electrical characteristics of PV cells is a nonlinear, so the performance of PV systems can be improved by solutions using MPPT techniques. There are several MPPT

S. Motahhir and B. Bossoufi (Eds.): ICDTA 2022, LNNS 454, pp. 545–553, 2022.
https://doi.org/10.1007/978-3-031-01942-5_54

controls developed in the literature to increase the efficiency of PV systems like: perturb and observe (P&O) [6–8], modified P&O [9], incremental Conductance (IC) [10], fuzzy Logic Controller (FLC) [11].

In [6], the classical P&O technique is reliable, simple and gives good results in the tracking of MPP, besides it is very used because of its easy implementation on different digital and analog platforms. Despite these advantages of the P&O technique, a problem of maximum power tracking arises due to the increase of irradiation which influences the P-V curve and consequently the wrong decision of duty cycle. To overcome the irradiance increase problem, Killi et al. [9] propose a modified P&O technique that adds the current change information (ΔI). However, even though the irradiance increase problem has been solved, the maximum power tracking still exhibits remarkable oscillations around MPP.

In this article, a new P&O technique count on the area observation of the points of the (P-V) curve is proposed to reach the maximum area that lies near MPP. To the left of the maximum area, the variation in area (ΔA) is positive and to the right of maximum area (ΔA) is negative, then the duty cycle D is varied until reaching the maximum area that corresponds to a MPP point. The condition on the current variation (ΔI) which is approved on the modified P&O of [9] is also added to ensure better performance. To test the effectiveness and importance of the new method, a simulation in MATLAB/Simulink of a PV system is performed and then analyzed and compared with the classical and modified P&O. The results of simulation are clearly show the effectiveness of our solution in terms of the rapidity of pursuing the MPP and the remarkable elimination of oscillation in steady state.

The article is presented by the following sections: Sect. 2 is dedicated to the development of the PV model. In Sect. 3, an analytical study of the MPPT controls is presented. Section 4 is focused on the implementation of the method control proposed on Matlab/Simulink. Finally, a conclusion and proposal for future research works are adopted for Sect. 5.

2 System Modeling

To ensure the transformation of PV energy into continuous electrical suitable for the load, the PV installation connected to the load mainly consists of the following elements: 5

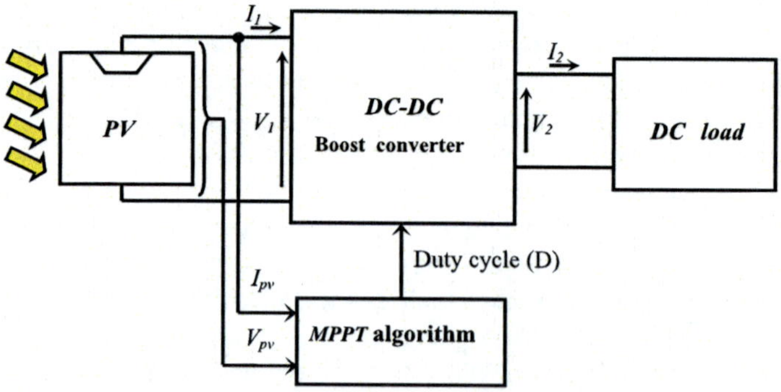

Fig. 1. The PV system with MPPT control.

PV panels connected in parallel, a boost transformer and an MPPT control, in order to bring the PV system worked in the MPP, as shown in Fig. 1.

2.1 PV Module

A PV cell alone is often not very usable, its small thickness makes it very fragile on the one hand, its very low voltage and current make it unusable directly in current electrical applications. The operating voltages of electrical equipment are standardized (12, 24, 48V, etc...) and a single PV cell cannot supply them directly. Thus, in order to increase the available power and energy produced by PV cells, manufacturers assemble PV cells in series and parallel to form a PV module.

The desired voltage is achieved by connecting several cells in series (N_s). To obtain the desired power, several of these series of cells can be connected as parallel branches (N_p).

2.2 Mathematical Model

The conversion of sunlight into electrical energy with a photovoltaic panel is done by photovoltaic effect. Figure 2 shows the single diode model of a single PV cell [12].

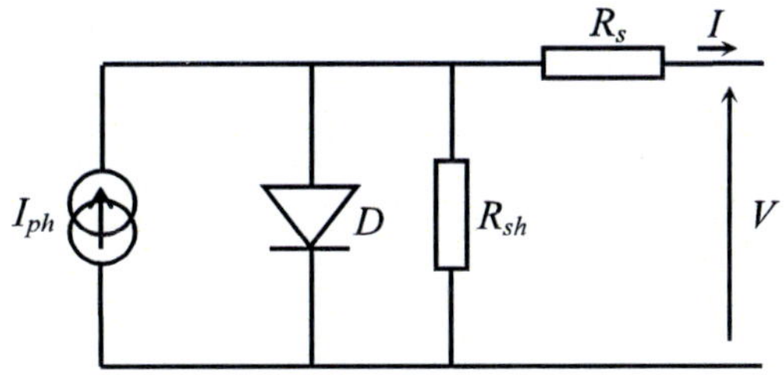

Fig. 2. Equivalent circuit of a PV cell

The single diode model is more suitable for power electronics because of its simplicity and efficiency.

As shown in Fig. 2, the single diode model of the photovoltaic panel incorporate a photonic current source Iph connected to a diode that emulates the PN junction, besides a series resistor Rs and a shunt resistor Rsh. So, the PV current can be expressed by the following equation [13]:

$$I = I_{ph} - I_0\left[\exp\left(\frac{V + R_s I}{aKTN_s}\right) - 1\right] - \left(\frac{V + R_s I}{R_{sh}}\right) \tag{1}$$

The 1Soltech 1STH-220-P panel is used in this study; Table 1 shows its specifications (Fig. 3).

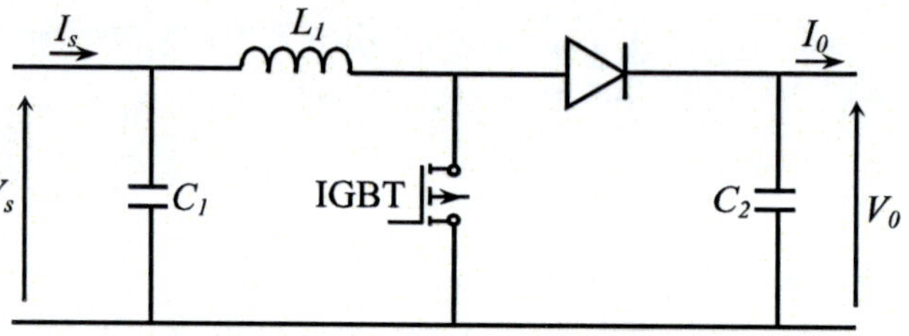

Fig. 3. The boost converter

2.3 DC/DC Boost Converter

The DC/DC Boost converter converts the input DC voltage to a higher output voltage. This converter includes the IGBT switch, the inductor (L), the diode which protects IGBT by blocking the return of current and the capacitor C2 which serves to smooth the output voltage [14].

The expression of the output current and voltage are given by the following equations:

$$\begin{Bmatrix} V_0 = V_S/(1-D) \\ I_0 = I_s(1-D) \end{Bmatrix} \quad (2)$$

with D is the duty cycle.

The parameters of this converter used are: $C_1 = 100 * 10^{-6}$, $L_1 = 5 * 10^{-6}$ and $C_2 = 3300 * 10^{-6}$.

Table 1. Characteristics of the 1Soltech 1STH-220-P PV Pannel.

Characteristics	Values
Maximum Power (W)	213.15
Open circuit voltage Voc (V)	36.3
Voltage at maximum power point Vmp (V)	29
Temperature coefficient of Voc (%/deg.C)	-0.36099
Cells per module (Ncell)	60
Short-circuit current Isc (A)	7.84
Current at maximum power point Imp (A)	7.35
Temperature coefficient of Isc (%/deg.C)	0.102
Light-generated current IL (A)	7.8649
Diode saturation current I0 (A)	2.9259 e-10
Diode ideality factor	0.98117
Shunt resistance Rsh (ohms)	313.3991
Series resistance Rs (ohms)	0.39383

3 Area-Based P&O Proposed for MPPT

The new technique proposed in this work is based on an idea that gives very important results especially on MPP tracking and also ripple minimization. Figure 4 shows that each point on the curve (*P*-*V*) defines an area *A* equal to *P***V*, this area is maximum A_{max} around MPP such as $A_{max} \approx P_m * V_m$. In Fig. 8. $A_1 < A_2$, $A_2 < A_3$, $A_3 > A_4$, $A_3 = A_{max}$. The proposed P&O algorithm is developed on the basis of the slope variation (*dA/dV*), we can visualize that the slope is positive to the left of the point which corresponds to A_{max}, and negative to the right of A_{max}. Depending on the sign of the slope, the duty cycle must be perturbed in order to follow the point corresponding to the maximum area to get closer to the MPP. So when there is an increase in irradiation $dA > 0$ and $dV > 0$, and to detect this change we use the same technique presented on the modified P&O [9] that adds the parameter *dI* which is positive on this change, as show in Fig. 5 (a&b).

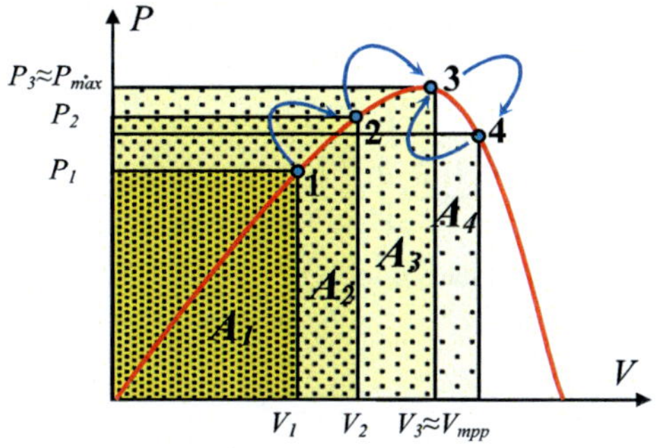

Fig. 4. Analysis with area-based P&O MPPT

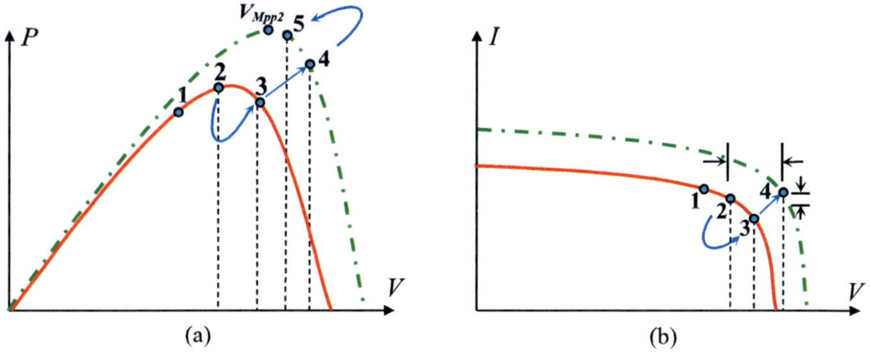

Fig. 5. (a) One time increase in irradiation, (b) Observation of change in current (dI)

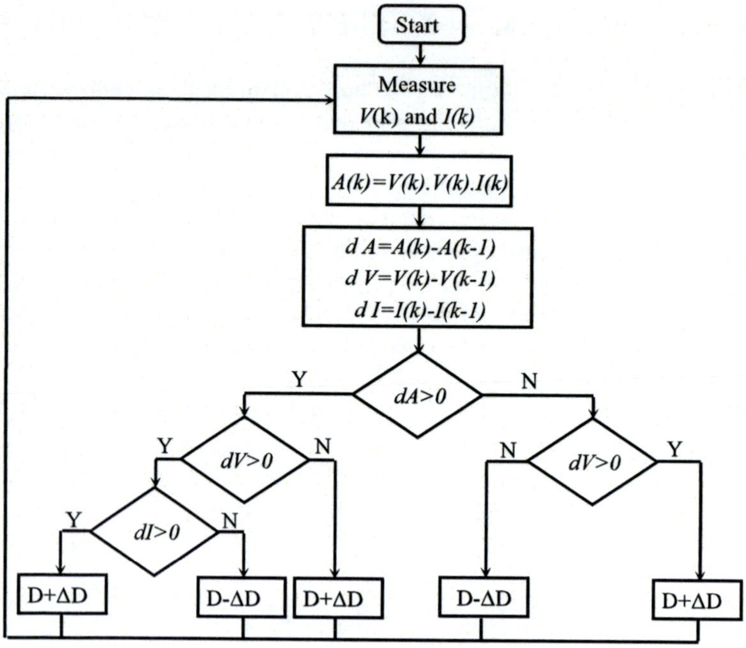

Fig. 6. Flowchart of the Area-based P&O MPPT

Figure 6 shows a flowchart of the Area-based P&O proposed MPPT.

4 Simulation Results

The Area-based P&O algorithm was tested on Matlab/Simulink under a temperature of 25 degrees and an irradiation change of 270 to 480 W/m^2 at time 1s, we choose the fixed perturbation step size $\Delta D = 0.000125$ and the load value at the converter output is 5 Ω.

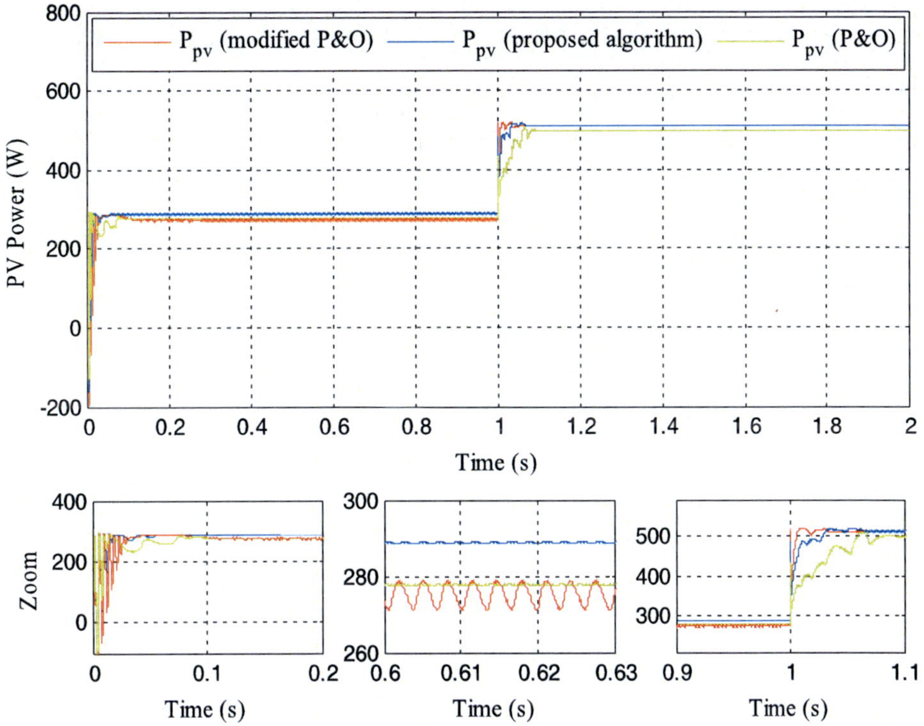

Fig. 7. Output power of the PV

The simulation result illustrated in Fig. 7 clearly shows the effectiveness of our proposed method compared to the classical [6] and modified [9] P&O on MPP tracking, which reach a maximum power in a short time. Moreover, the PV module power obtained by our solution is characterized by less oscillation.

The efficient tracking of MPP by our method allowed us to guarantee the power at the converter output with better performance when compared with other algorithms as shown in Fig. 8.

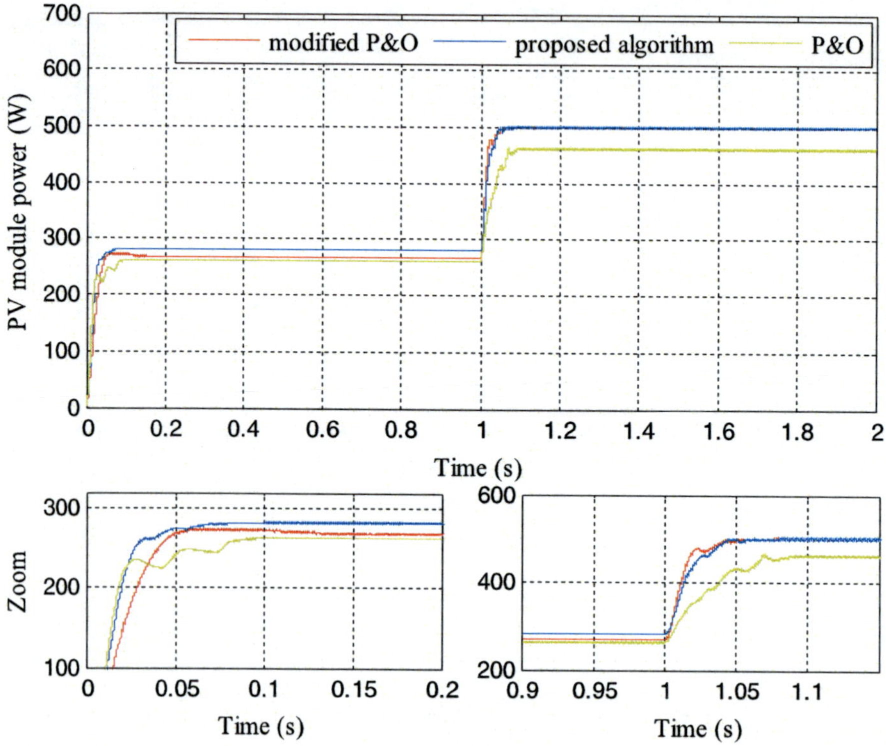

Fig. 8. Output power of the converter

5 Conclusion

This paper proposes a new P&O algorithm based on the observation of the area of the curve (P-V) applied to a photovoltaic system composed of five solar panels. This method consists in searching the maximum area closest to the MPP. After a modeling of the PV system and a development of the MPPT control algorithm, a simulation test is presented to validate the proposed technique on Matlab/Simulink software, and then compared with the results obtained by conventional and modified P&O. The comparative study analysis shows that the MPPT by our proposal tracks the power continuously with fewer ripples, less overshoot and fast tracking time compared to other algorithms. To advance technical and scientific research, future work will be devoted to the validation of the proposed algorithm on the FPGA board.

References

1. Motahhir, S., El Hammoumi, A., El Ghzizal, A.: The most used MPPT algorithms: review and the suitable low-cost embedded board for each algorithm. J. Clean. Prod. **246**, 118983 (2020)

2. El Mourabit, Y., Derouich, A., Allouhi, A., El Ghzizal, A., El Ouanjli, N., Zamzoum, O.: Sustainable production of wind energy in the main Morocco's sites using permanent magnet synchronous generators. Int. Trans. Electr. Energy Syst. **30**(6), e12390 (2020)
3. Tan, Y.T., Kirschen, D.S., Jenkins, N.: A model of PV generation suitable for stability analysis. IEEE Trans. Energy Convers. **19**(4), 748–755 (2004)
4. Mkahl, R., Nait-Sidi-Moh, A., Wack, M.: Modeling and simulation of standalone photovoltaic charging stations for electric vehicles. World Acad. Sci. Eng. Technol. Int. J. Electr. Comput. Energ. Electron. Commun. Eng. **9**(1), 72–80 (2015)
5. Errouha, M., Derouich, A., Motahhir, S., Zamzoum, O., El Ouanjli, N., El Ghzizal, A.: Optimization and control of water pumping PV systems using fuzzy logic controller. Energy Rep. **5**, 853–865 (2019)
6. Femia, N., Petrone, G., Spagnuolo, G., Vitelli, M.: Optimization of perturb and observe maximum power point tracking method. IEEE Trans. Power Electron. **20**(4), 963–973 (2005)
7. Pandey, A., Dasgupta, N., Mukerjee, A.K.: High-performance algorithms for drift avoidance and fast tracking in solar MPPT system. IEEE Trans. Energy Convers. **23**(2), 681–689 (2008)
8. Femia, N., Granozio, D., Petrone, G., Spagnuolo, G., Vitelli, M.: Predictive & adaptive MPPT perturb and observe method. IEEE Trans. Aerosp. Electron. Syst. **43**(3), 934–950 (2007)
9. Killi, M., Samanta, S.: Modified perturb and observe MPPT algorithm for drift avoidance in photovoltaic systems. IEEE Trans. Industr. Electron. **62**(9), 5549–5559 (2015)
10. Errouha, M., Derouich, A., El Ouanjli, N., Motahhir, S.: High-performance standalone photovoltaic water pumping system using induction motor. Int. J. Photoenergy **2020**, 1–13 (2020)
11. Al Nabulsi, A., Dhaouadi, R.: Efficiency optimization of a DSP-based standalone PV system using fuzzy logic and dual-MPPT control. IEEE Trans. Industr. Inf. **8**(3), 573–584 (2012)
12. da Rocha, M.V., Sampaio, L.P., da Silva, S.A.O.: Comparative analysis of MPPT algorithms based on Bat algorithm for PV systems under partial shading condition. Sustain. Energy Technol. Assess. **40**, 100761 (2020)
13. Ram, J.P., Babu, T.S., Rajasekar, N.: A comprehensive review on solar PV maximum power point tracking techniques. Renew. Sustain. Energy Rev. **67**, 826–847 (2017)
14. Chalh, A., El Hammoumi, A., Motahhir, S., Ghzizal, A.E.L., Derouich, A.: Development of an improved GMPPT based on scanning method for PV system operating under a dynamic partial shading conditions. Technol. Econom. Smart Grids Sustain. Energy **6**(1), 1–7 (2021). https://doi.org/10.1007/s40866-021-00118-7

Design and Analysis of a Switching Power Supply Controller Chip for a Flyback Converter

Meryem Lakdah[1(✉)], Karim El Khadiri[1], Zakia Lakhliai[1], Driss Chenouni[1], Mouhammed Ouazzani Jamil[3], Hassan Qjidaa[2], and Ahmed Tahiri[1]

[1] Laboratory of Computer Science and Interdisciplinary Physics (LIPI) Normal Superior School, Sidi Mohamed Ben Abdellah University Fez, Fez, Morocco
mameryemlakdah@gmail.com

[2] Department of Physics at the Faculty of Sciences, Sidi Mohamed Ben Abdellah University, Fez, Morocco

[3] Laboratoire Systèmes et Environnements Durables, Université Privée de Fès, Fez, Maroc

Abstract. Power supply faces many challenges, they must meet stability, accuracy, efficiency and cost one of the best power supplies that meets all of these constraints is the switch mode power supply is a technique that continuously switches between on and off states to manage power consumption and reduce wasted energy. A flyback converter is a BUCK-BOOST converter that can raise or lower the voltage. It has an electrical isolation between the input and the output by means of an inductor coupled or split to form a transformer. In the first phase the switch is closed, the magnetic flux stores energy and in the second phase the switch is open, the energy stored in the transformer is transferred to the load the magnetic flux drops, which causes the diode to be forward biased it fires to transfer energy to the output and allows current to flow from the transformer and charge the capacitor and the load. This paper describes the design and testing of a standalone power supply control device for a flyback converter. The flyback controller was designed, simulated and implemented in Cadence using TSMC's 180 nm SOI technology, with an input of 170 V and an output of 12.5 V, an efficiency of 91.7% and an optimized area of 0.0156 mm^2.

Keywords: Switch mode power supply · Flyback converter · SPS · Pulse Width Modulation (PWM)

1 Introduction

The flyback converter is a simple switching power supply topology that has many advantages, it does not require an output inductor - the transformer itself works as an inductor at the same time - so the power supply has reduced dimensions and an even higher efficiency than a direct converter. However, due to some practical limitations, the flyback topology is rarely used with output powers above a few hundred watts - a direct converter is used instead [1, 2].

Thanks to the recent advances in semiconductor technology that make it much easier to work with, it is possible to build high power flyback converters. The objective of this

S. Motahhir and B. Bossoufi (Eds.): ICDTA 2022, LNNS 454, pp. 554–562, 2022.
https://doi.org/10.1007/978-3-031-01942-5_55

paper is to exploit these new semiconductors, in particular MOSFETs and silicon carbide (SiC) diodes, to realize a simple, efficient and high-power flyback power supply [3].

In the simplest variant, the flyback converter uses a single-switch topology. The problem of this topology is that the transistor switch needs to have a high breakdown voltage, The Flyback converter, has a single switch topology which has a switching problem ie the switch transistor needs a high breakdown voltage. IGBT transistors are available with a very large nominal voltage which reaches 1200 V and they can switch high currents but, they are too slow for an efficient flyback converter so the solution is to use MOSFETs instead as MOSFETs are ideal in power electronics for fast switching and linear operation. However, the other constraint is that MOSFETs with such nominal voltages have high on-resistance, which means that they would dissipate a lot of power in a high-power application [5].

Fortunately, and thanks to the rapid evolution of electronics we can find suitable MOSFETs for our application are silicon carbide (SiC) MOSFETs. The use of SiC allows a reduction in the number of components, smaller passives, better cooling and an efficiency of more than 98%. Most importantly, these transistors combine a high breakdown voltage (generally) with low on-state resistance and very high switching speeds at the same time [4]. These new transistors are perfectly suitable for the flyback converter.

This paper is organized as follows. The second section describes the literature review and fundamental design theory of the SPS Flyback converter. The third section discusses the design and architecture of the SPS Flyback on the cadence platform. The fourth section discusses the results obtained in terms of efficiency and behavior. The last section concludes this paper and suggests future projects.

2 Flyback SPS Design

Flyback Fig. 1 below shows the external circuit of the SPS Flyback, including the view of the symbol of the chip which presents two coupled inductances playing the role of transformer. In this case we distinguish a phase of energy storage in the magnetic circuit and a phase of restitution of this energy. The dimensioning of the magnetic circuit defines the quantity of energy that can be stored and also the speed with which it can be stored and

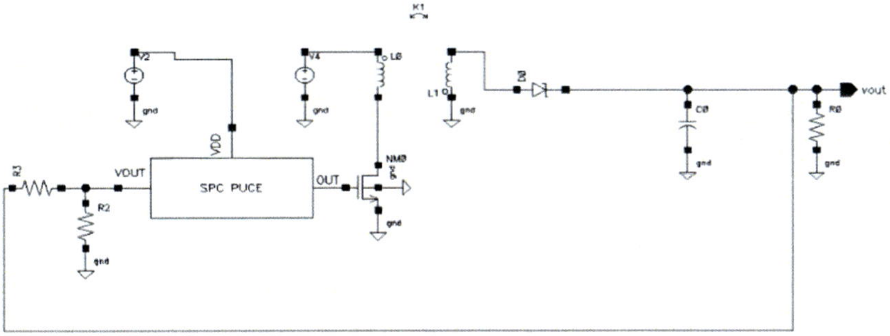

Fig. 1. Flyback SPS

retrieved. This is an important parameter that determines the power that the power supply can provide. The control signal of the switch is in the form of a square or rectangular signal of frequency F whose duration is in the high state (conduction time of the switch) is adjusted by the parameter α is called the duty cycle.

$$\alpha = \frac{\mathbf{T_{on}}}{\mathbf{T}} = \frac{\mathbf{T_{on}}}{\mathbf{T_{on}} + \mathbf{T_{off}}} \tag{1}$$

The NMOS transistor has the role of an interceptor and has two states: on or off. In the on state (transistor on and D diode off), the NMOS switch is closed and the primary of the transformer is connected directly to the input voltage source. This results in an increase of the magnetic flux in the transformer. The voltage across the secondary is negative, blocking the diode. It is the output capacitor that provides the energy required by the load and in the off state the switch is open. The energy stored in the transformer is transferred to the load thanks to D, which switches on to transfer the energy to the output. This NMOS is controlled by the controller chip which will turn it on or off. This application of constant voltage will linearly increase the number of hours the transformer will operate. The choice of a Schottky diode is always a compromise between the breakdown voltage and the threshold voltage. At the level of switching power losses, the higher the threshold voltage, the more these losses increase, and the faster the diode is to follow the switch variations. The optimal values of the load between and allow to have a current of 10 mA to 2.5 A other values of load resistances will change the obtained output. The output capacitor has the role of smoothing the output voltage.

3 SPS Controller Chip Design

The design of the chip is presented in Fig. 2 below the chip of the controller is constituted by the following component that we will treat in detail in the following axis: a reference circuit with band gap, a comparator, a NAND gate, an oscillator and a buffer. This chip will receive a voltage Vout of 12.5 V, through the external circuit, and through the voltage divider made by two resistors of and Vout it mottles 10 times to have a voltage

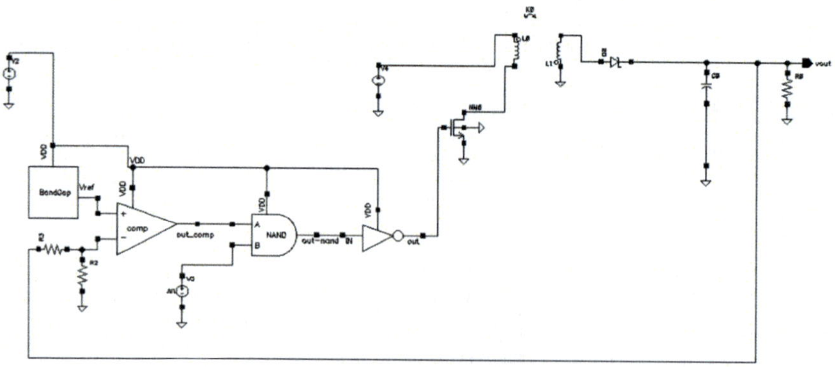

Fig. 2. The SPS chip

of almost 1.25 V, so that the voltage divider creates a voltage around the same range as the reference voltage 1.25 V to compare them through a comparator. This signal will control the NMOS. Then this voltage will travel through a buffer (driver that strengthens the signal and reduces capacitive effects) to establish a good square wave signal.

3.1 Bandgap Reference Circuit

The BGR circuit is based on the principle of the sum of a temperature dependent voltage called PTAT (Proportional to Absolute Temperature) and a voltage inversely proportional to temperature called CTAT (Complementary to Absolute Temperature) in order to compensate for the effect of temperature and to achieve a voltage at the output that is independent of any variation in the ambient temperature. The purpose of this Bandgap circuit or reference voltage generator is to generate a voltage of 1.25 V stable that is ideally independent of changes in temperature and variation of supply voltage. The Bandgap circuit in Fig. 3 is divided into three essential components, a start-up circuit that avoids the blocking state at the beginning of operation or the situation where a variation in the VDD (VDD = 1.8 V) power supply, a CTAT and PTAT circuit, and finally a combination of BJT bipolar junction transistors and resistors.

The values α1 and α2 should be chosen such that:

$$\mathbf{V}_{\mathrm{ref}} = \boldsymbol{\alpha}_1 \ \mathrm{PTAT} + \boldsymbol{\alpha}_2 \ \mathrm{CTAT} \tag{2}$$

$\frac{\partial V_{ref}}{\partial T}$ Must be zero to have an ideal signal.

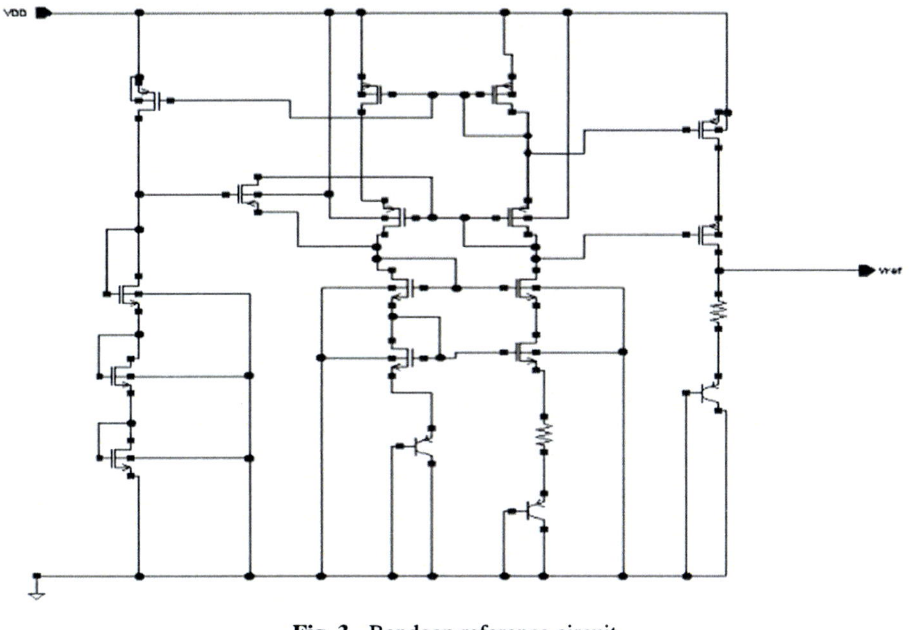

Fig. 3. Bandgap reference circuit

The following simulation (Fig. 4) shows the behavior of this circuit as a function of supply voltage variation VDD.

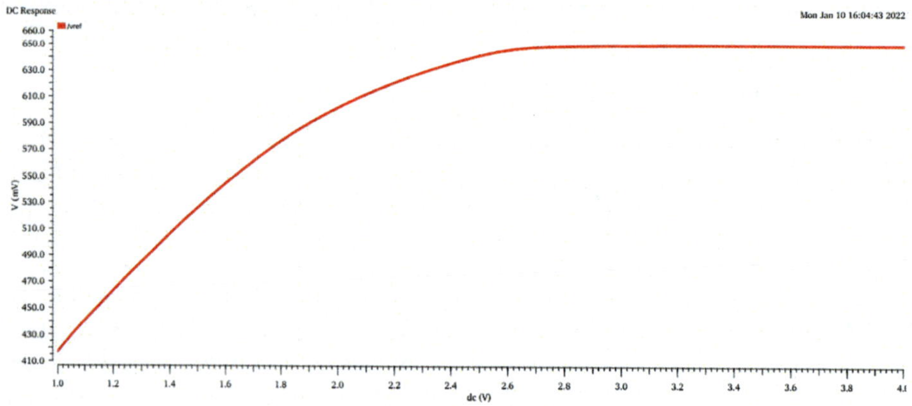

Fig. 4. Vref with Increasing VDD

3.2 Comparator

The comparator in Fig. 5 is a circuit that compares an analog signal (voltage) with another analog signal, hence the reference voltage, and delivers a binary signal as output Vp is the input voltage applied to the positive input terminal of the comparator and Vn is the reference voltage applied to the negative terminal of the comparator. If Vp, is the input of the comparator it will have a potential greater than the voltage Vn, while the output of the comparator will be a logical 1, and when the voltage Vp will have a potential lower than the voltage Vn, the output of the comparator will be at the logical level 0.

$$if\ \ Vp > Vn \Rightarrow Vout = 1 \text{ or } \ \text{VDD}$$

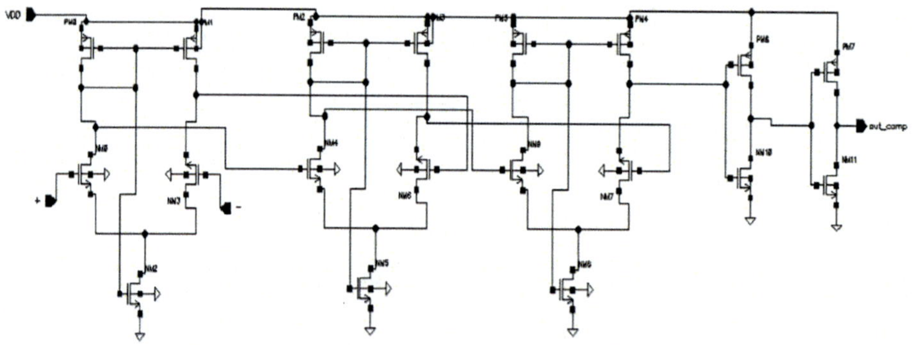

Fig. 5. Schematic comparator with three diff-amps and two inverters

$$if\ \ Vn > Vp \Rightarrow Vout\ = 0\ or\ VSS$$

This each amplifier has different characteristics than the others for ours it has a very high scanning speed and a very small resolution of 37 mV it is composed of 3 differential amplifiers and 2 inverters, all in parallel do not reach completely these values. These inverters better understand that the output of this comparator actually reaches the full logic levels. The output signal (Fig. 6) of this comparator is important because it compares a stable signal of Vref from Bandgap and another variable signal from Vout to have an output signal of two values VDD = 1.8 V or VSS = 0 able to control the NMOS.

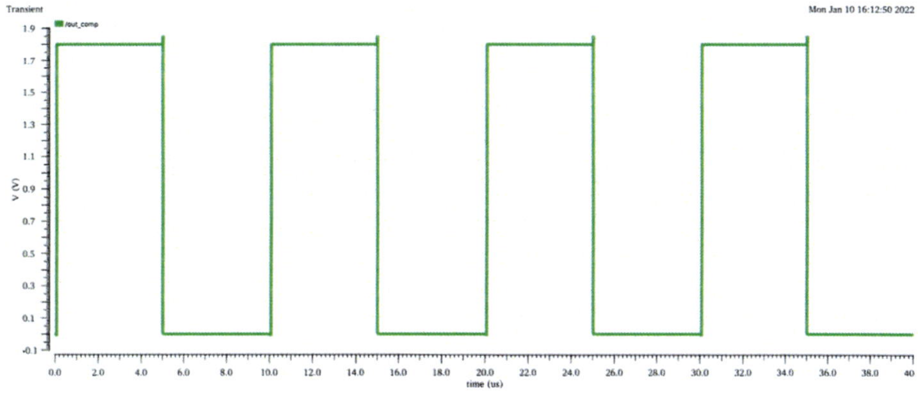

Fig. 6. Simulation of comparator

3.3 Buffer

The buffer in Fig. 7 is a circuit performs an electrical shaping, so if the signal is provided by a logic gate with a low load and the latter is important on the next stage, the signal may be strongly degraded the buffer serves to compensate for this. For this case the buffer is a circuit capable of driving the NMOS which has a large capacity of 1.2 pf Since this device is so large, we used a three-stage inverter The first inverter consists of a PMOS and an NMOS. To find the input capacitance of this first inverter, we can do the following calculation:

$$\begin{gathered} C_{in1} = C_{oxn} + C_{oxp} \\ C_{in1} = \frac{3}{2} * \frac{2.5f}{\mu^2} F * 0.18\mu * 2\mathrm{u} + \frac{3}{2} * \frac{2.5f}{\mu^2} F * 0.18\mu * 2\mu \\ C_{in1} = 2.7fF \end{gathered} \tag{3}$$

with this calculated, the input capacity of subsequent inverters can be found using similar steps. By choosing an A of four and a final stage multiplier of 4, Similar steps can be used to determine the input capacity of the following calculations.

$$\begin{gathered} C_{in2} = C_{in1} * D,\ C_{in3} = C_{in2} * D * \mathrm{F} \\ C_{in2} = (18.9fF) * 4,\ C_{in3} = (18.9fF) * 4^2 * 4 \\ C_{in2} = 75.6fF,\ C_{in3} = 1.209pF \end{gathered} \tag{4}$$

Therefore, the buffer will use three stages: 0.18u to drive the MOSFET switch.

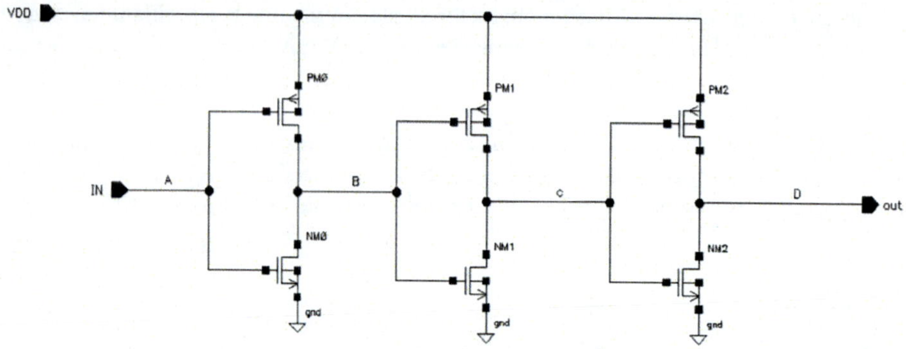

Fig. 7. Schematic of three stage buffer

Here the simulation (Fig. 8) that shows us how the buffer improves and correct the signal.

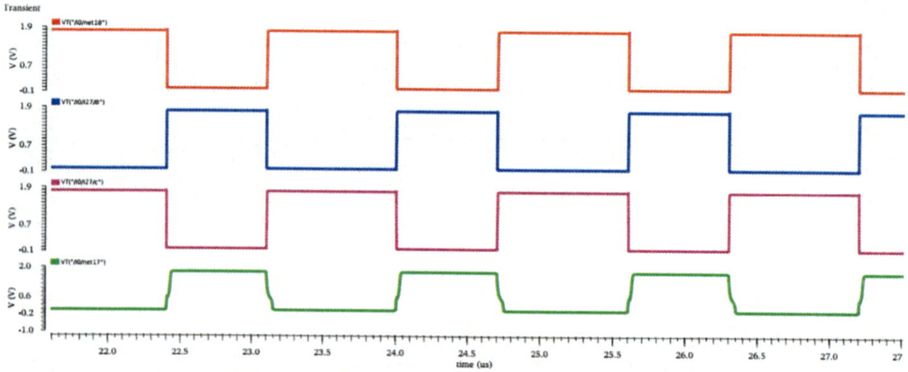

Fig. 8. Buffer output

4 Simulation Results

4.1 Layout Specifications

We made the drawing of the masks of each block trying to respect all the DRC rules (we only use metals 1 and 2) and LVS by minimizing the maximum possible surface finally here is the general layout (Fig. 9) of our circuit for who has an optimal ratio of 0.0156 mm^2.

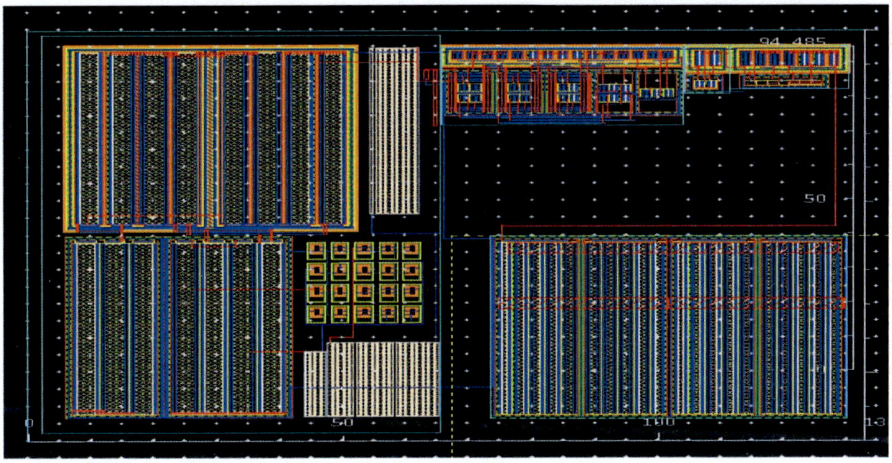

Fig. 9. Full layout measuring 0.0156 mm^2.

4.2 Simulation of the SPS Chip

Vout is the output (Fig. 10) voltage of the Flyback circuit under a load of 2.5 A and a VDD supply of 1.8 V the output signal has the value of 12.5 with a small ripple caused by the oscillator (Table 1).

Average Power and Efficiency Calculations

$$P_{avg} = I_{avg} * VDD = 200.3\ mA * 1.8 = 360.54\ mW$$

$$Efficiency(\%) = \frac{(V_{out} * I_{load})}{VDD * avg(I(VDD))}$$

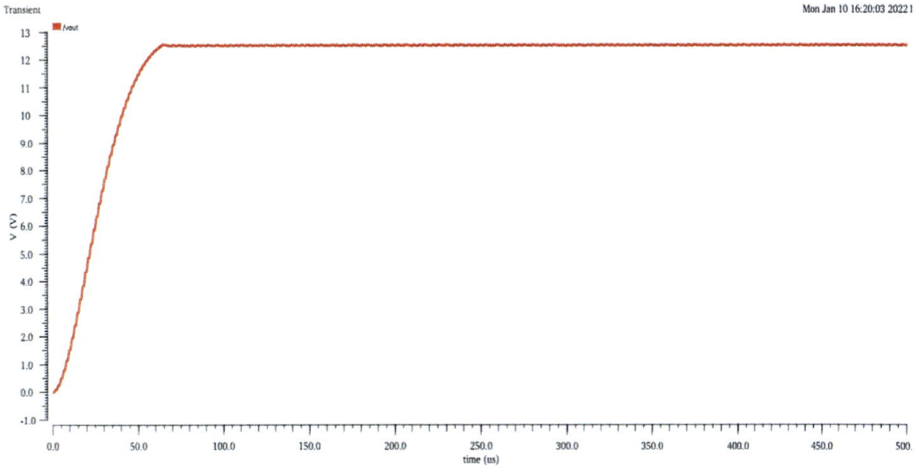

Fig. 10. Simulation of Vout

$$Efficiency(\%) = \frac{12.5V * 2.5\ A}{170V * 200.3\ mA} = 91.7\% \tag{6}$$

Table 1. Proposed flyback converter performance

Paratmeter	2017 [6]	2016 [7]	2020 [7]	This work
Technology	0.35 um HVCMOS	0.35 um BCD	0.18 um BCD	180 nm SOI
Die size (mm^2)	1.105	1.11	1.56	0.0156
Input voltage (V)	90–265	90–264	36–72	170
Output voltage (V)	12	12	12	12.5
Efficiency (%)	84.6	88	87.3	91.7

5 Conclusion

The objective of the presented work was to realize a DC/DC flyback converter powered by 1.8 V input of 170 V and output voltage of 12.5 v in conclusion, a new switching power supply controller chip for a Flyback converter was successfully designed in 180 nm silicon-on-insulator (SOI) technology. The assembly is based on integrated circuits and consists of two parts: the control part is the controller chip and the power part. The results of this study indicate good efficiency and low distortion. The chip has super characteristics suitable for various applications such as automotive application.

References

1. Connaughton, A., et al.: "Investigation of a soft-switching flyback converter with full secondary side-based control. IEEE Trans. Ind. Appl. **53**(6), 5587–5601 (2017)
2. Liu, P.H.: Design consideration of active clamp flyback converter with highly nonlinear junction capacitance. IEEE Appl. Power Electron. Conf. Expos. (APEC), 783–790 (2018)
3. Xue, L., Zhang, J.: High efficient secondary-resonant active clamp flyback converter. IEEE Trans. Ind. Electron. **65**(2), 767–770 (2018)
4. Chang, C., et al.: Design of a highly accuracy PSR CC/CV AC-DC converter based on a cable compensation scheme without external capacitor. IEEE Trans. Power Electron. **34**, 9552–9561 (2019)
5. Wi, S.-M., Kim, M.: Precise control strategy of dual-mode flyback DC-DC converter. IET Power Electron. **12**(2), 220–227 (2018)
6. Wu, Q., Zhu, Z.: A versatile OCP control scheme for discontinuous conduction moe flyback AC/DC converters. IEEE Trans. Ind. Electron. **64**(8), 6443–6452 (2017)
7. Chang, C., Xu, Y., Bian, B., Zhao, X.: A high-precision CV/CV AC-DC converter based on cable and inductance compensation schemes. IEEE Trans. Power Electron. **31**(9), 6372–6382 (2016)
8. Li, Y., Zhu, Z.: A capacitor-free control scheme for a primary-side controlled flyback converter. IEEE Trans. Power Electron **35**(9), 8484–9495 (2020)

Current Situation of Green Building Development in Morocco: Standards and Certification Systems

Redouan Assadiki(✉) and Fouzi Belmir

Laboratory of Materials, Processes, Catalysis, and Environment, School of Technology, Sidi Mohamed Ben Abdellah University, Fez, Morocco
redouan.assadiki@usmba.ac.ma

Abstract. The development of green building has emerged as a global trend to reduce the ecological footprint and well-being of societies, but in developing countries, the concept is still emerging. In such circumstances, the establishment of a national green building standard would be an important first step. Morocco has made significant progress in establishing national standards for environmental protection and assessment. However, this progress is not specifically applied to a sustainable policy in the construction sector. The development of green building in morocco is in its beginning phase, which currently lacks governmental and voluntary support to elaborate appropriate standards and regulations for green building. This article analyses the situation and progress of green standardization in the construction industry and the certification. In this context, Morocco's green buildings are certified with 51 projects evaluated by HQE, LEED, and BREEAM.

The findings of this paper indicate the national regulation is not specific to the construction industry except for the energy part of the thermal regulation. A framework for developing specific regulations for sustainability categories inspired by international systems was proposed that can help decision-makers adopt strategies to stimulate the development of green buildings in Morocco.

Keywords: Green building · Standards · Certification systems · Developing countries

1 Introduction

Buildings have a significant direct and indirect impact on the environment during any life cycle (design, construction, occupation, and demolition). In general, as mentioned by Bribian et al. (2011) [1] manufacturing activities represent 60% of raw non-renewable consumption. In 2017, Gursel et al. [2] found that cement is a carbon-intensive material, contributing to about 5–7% of global anthropogenic carbon emissions. Each year, construction accumulates 16% of the total iron and steel production [3], while the steel industry is responsible for about 6.7% of global carbon emissions [4]. This describes construction as a major user of non-renewable resources that generates waste, pollutes air, wastewater and decreases built-on land footprint. In developing countries, construction

S. Motahhir and B. Bossoufi (Eds.): ICDTA 2022, LNNS 454, pp. 563–572, 2022.
https://doi.org/10.1007/978-3-031-01942-5_56

activity is an important economic sector as it creates more employment and boosts the industrial sectors (cement, metal products, sanitary products, and mining). In Morocco, this sector contributed 6.1% to the national GDP in 2019 before the COVID-19 crisis. It has generated more than one million jobs as well as an outstanding housing credit of about 214 billion MAD (1€ = 9.1 MAD) [5], although the importance of the construction industry also highlights the need for more attention to sustainable building [6]. These facts have stimulated the creation of green building standards, certifications, and rating systems to mitigate the impact of buildings on the natural environment by sustainable design. In the context of developed countries such as the UK, the USA, Switzerland has argued that the regulatory framework is essential to promote a transition to green buildings [7][8]. Thus, developing countries such as Nigeria have presented a regulatory framework that can reinforce the Nigerian building code with sustainability criteria adapted from the LEED standard [9]. In the same continent, the study W.Goulding et al. (2015) [10] demonstrated a gap between green building practices (implementation phase) and legislative requirements in South Africa through a survey of companies, the stakeholders have a high degree of unfamiliarity with green building legislation and practices. The authors stated an urgent need to ensure both cultural and behavioral transformation. Given this, the government needs to establish a regulatory compliance monitoring strategy. This is supported by government incentive programs for green building. Hilde Nykamp (2020) [11] further recommends it that the policy-mix (policy instruments and stakeholders) should have a regulatory 'push' by the government and it should accompany the transition to green buildings in Norway through cumulative incremental changes.

The movement towards sustainable design increased with the launch in 1990 of the first green building rating system in the world by Building Research Establishment Environmental Assessment Method (BREEAM). In 2000, the U.S. Green Building Council (USGBC) followed the published criteria to improve the environmental performance of buildings through its LEED (Leadership in Energy and Environmental Design) and the French Haute Quality Environmental (HQE) 1996 system. LEED is one of the most famous GB rating systems and it is largely applied in more than 165 countries with 140000 certified projects worldwide [12]. Green building in developing countries, Asia and Latin American, is particularly remarkable which is a rapidly developing emerging market, while in Africa the development of certification and research in the field of buildings is very late [13]. Morocco, as well as other developing countries, need strategies to produce green buildings. However, existing studies focus mainly on developed countries, and few research has systematically addressed and quantified some key factors of green building in developing countries. The purpose of this study systematically investigates key aspects of the green building industry in Morocco, including regulation and certification.

2 Research Methodology

We have established the current situation of deployment of green building certification in Morocco. Therefore, we have based on the websites of international green building certification systems with information on certified projects (type of building, level certification) of 51 certified buildings since 2014. Concerning the inventory, as resources, we used the legislative texts from the websites of the Ministry of Housing, Moroccan Agency for Energy Efficiency, the Ministry of Energy Transition and Sustainable Development concerning eco-construction, and the standards catalog of the Moroccan Institute of Standardization.

Finally, to develop the future green building standard framework, we based on the categories suggested from building certification systems such as LEED in developed countries and the Egyptian GPRS system of a developing country assimilated to Morocco's climate and culture, the following performance categories were taken into account: material/waste, water efficiency, environmental quality, energy efficiency, and site.

3 Finding and Discussion

3.1 Regulatory Technical Standards

Morocco is engaged in responding to the challenges of the 21st century by making development sustainable. In this respect, the Ministry of Energy and Mines, Water, Environment, and the Ministry of Housing present in its portal an exhaustive review of the national legal arsenal in the form of laws, Dahirs, decrees, orders, and circulars. In parallel, the Moroccan Institute for Standardisation (IMANOR) is proposing a set of standards related to sustainable construction. Figure 1 summarizes the distribution of regulations in the field of construction.

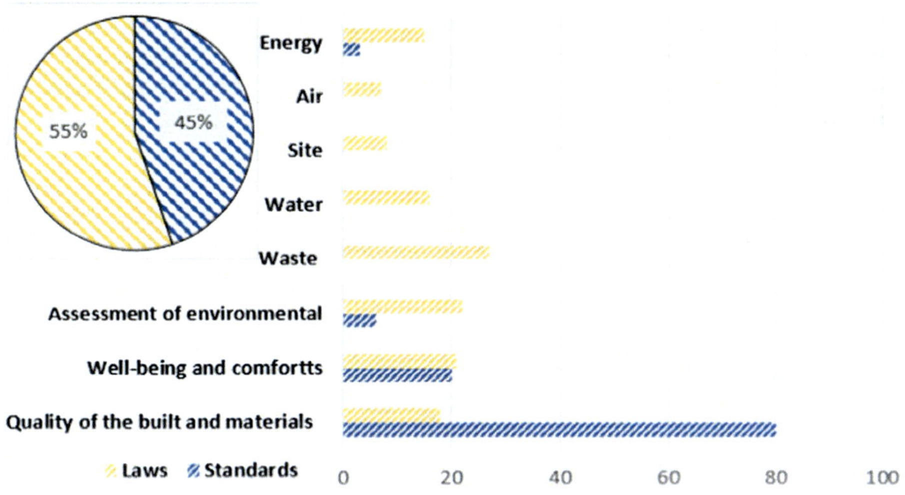

Fig. 1. The number of laws and standards by category in Morocco.

Morocco has made concrete progress in the supervision of the construction sector through a series of fundamental laws, in particular; the "Paraseismic Regulation" (RPS 2000), the thermal regulation in the building (RTCM), law 10–03 on architectural accessibility, etc. On the other hand, 150 standards were either reformulated or adapted from the bibliography of the International Organisation of Standardization or the European.

From all this, we can see that there is a balance between the laws and the standards adopted about sustainable construction. While some laws have existed since the 1980s, the standards have been created recently and updated annually by IMANOR. The technical parameters of the building materials and the quality of the building are the most predominant focus of the standards (80/117 standards), which characterize the structural, insulation, and finishing materials. More than 20% of the laws are oriented toward comfort (thermal/acoustic) and the well-being of the occupants. The laws relating to energy are still in their embryonic stage and do not cover energy systems such as (HVAC). In this instance, the designers in Morocco have recourse to the use of international standards (ASHRAE, EN, NF), which are not adapted to the national context.

In this field of energy efficiency, Morocco has developed a new energy strategy to reduce the energy bill, which resulted in the adoption of Law 47–09, which aims to achieve a saving of about 20% of energy consumption by 2030 [14]. In this perspective, the Moroccan Agency for Energy Efficiency (AMEE) has implemented a thermal regulation of buildings (RTCM) in 2014, which involves the integration of insulation and improve HVAC systems according to the standards in force, but since their introduction in 2015 is not applied. According to the study by M'LAHFI et al. (2020) [15], the main reasons preventing the application of the RTCM by building professionals in Morocco are related firstly to the "additional cost" that will affect the consumer and the real estate and secondly to the difficulty of application. M'LAHFI et al., suggest simplifying the procedure for obtaining a building permit, which is already rather slow and complex. Adding mechanisms for sanctions in case of non-application of the RTCM. After 2 years in 2016, the Moroccan government has presented the sectorial framework plan "Eco-construction and sustainable building", financed by the European Union. The general objective is to reach a situation by 2030 in which: all stakeholders involved in the design, construction, operation, and deconstruction of buildings have an understanding of sustainable buildings and are provided with the knowledge, resources, and incentives necessary [16]. However, the five main strategic axes of this plan are based only on the design phase of buildings. In parallel, actions on the development of support systems and capacity building for stakeholders. However, in the 21 proposed actions, the government and private stakeholders have not made any engagement or proposed schedules to realize these actions.

The national code lacks the technical and engineering efficiency to achieve sustainable green building construction. At present, there are no standards directly implementing green building assessment standards (heat island effect assessment, indoor air quality, site ecology, green roofs, vertical gardens, etc.). With this situation, we propose a future framework for developing a "green building assessment" based on the current legislation (see Fig. 2). Based on national regulations and building performance evaluation criteria by international systems, numerous regulations and standards can be elaborated in the future by the different ministerial departments during the project life cycle.

3.2 The State of Green Building Certification

Currently, the construction industry has been expanding rapidly, and knowledge about the position of green buildings is essential to propose the necessary measures taken for the deployment of sustainable construction. Of the responses to the question on the

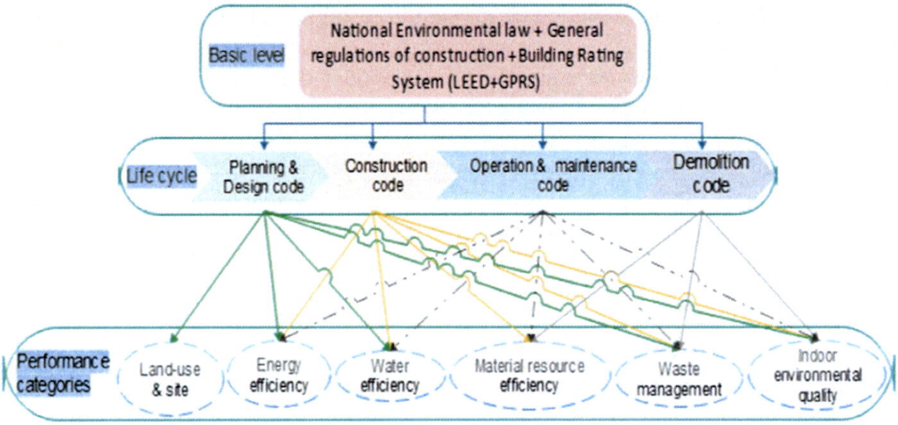

Fig. 2. Foundation of green building assessment in morocco.

deployment of green buildings, 44% of the participants felt that they have just started and have developed slowly, which shows the emergence of the notion of green building projects in the real estate market in Morocco[17]. To boost the situation, certification systems are an important part of achieving sustainability goals. This is why property investors are interested in certification schemes (HQE, BREEAM, and LEED); which are proving the benefits users have and the environmental performance of buildings and territories.

The new certifications established in the national construction industry signify new opportunities nationally. As shown in Fig. 3, 51 buildings were certified with the majority beyond "very good" for the HQE system as well as two green projects certified by the LEED system for all three levels (certified-silver-gold). Buildings that are at the top

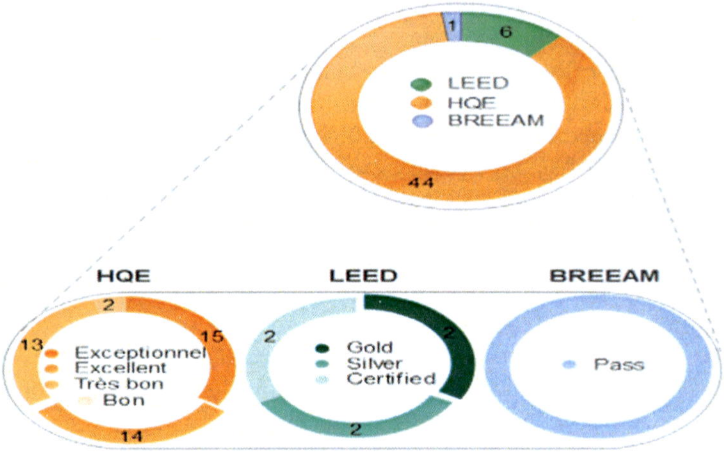

Fig. 3. The number of certified projects in Morocco

of the rating scale (gold, exceptional) may entail a higher additional cost than other projects.

In 2014, only two construction projects were labeled by the assessment systems. The number increased in the following years. By the end of 2020, 13 residential and tertiary building projects certified as green buildings. As can be seen in Fig. 4 and Fig. 5, certified buildings increased from 2015 to 2020 except for the year 2019. The number and area of green buildings in 2020 were approximately equal to the sum of the previous two years (2018–2019). At the end of 2020, the total area of green buildings was more than 200 thousand square meters.

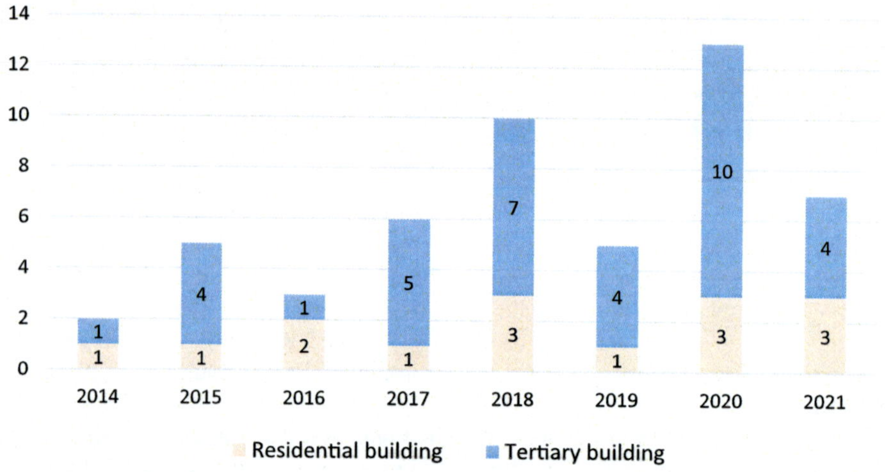

Fig. 4. Number of certified projects by building type.

The predominance of the total certified area of green residential buildings is evident in Fig. 5 for two years (2016–2018), by multi-family buildings (co-housing) or villas as well as temporary housing in the region Taghazout-Agadir. However, the number of these projects is a minority compared to the number of residential projects under construction. On the other hand, in the other years the predominance of tertiary buildings (offices, agencies, hotels…) invested by the private sector are developed by real estate developers in Morocco, banking agencies, and private educational institutions such as the International University of Rabat, which has been voluntary, implemented the certification process. Hence, public institutions need to do a lot of work to be an example and that was already launched in the national strategy of sustainable development between 2016 and 2030.

The proportions of certified cases in HQE and LEED are respectively 86.27 and 11.7%. Based on the survey responses, 43.50% used HQE, 20% of the respondents applied LEED being the vast majority worldwide. Finally, 10% of participants chose the British BREEAM system (of which 16 participants used all three systems)[17]. The French HQE system was still the most widely used in Morocco due to the accessibility, primarily because the French language of this system encourages professionals to be more interested in this system, whereas the other systems BREEAM and LEED

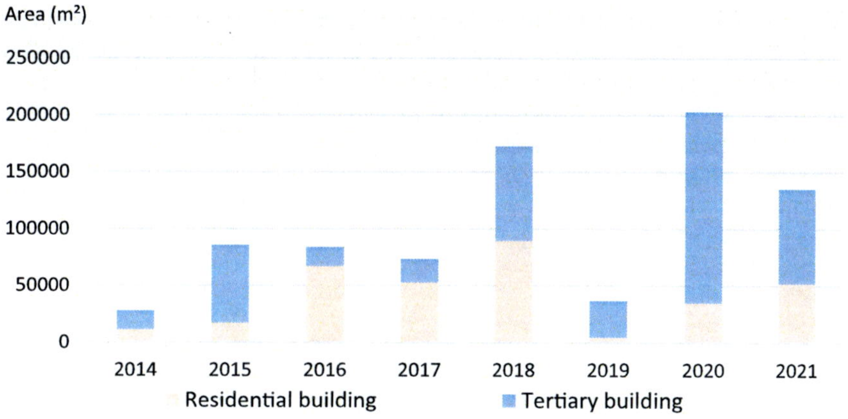

Fig. 5. The total area of green buildings.

use English for their technical manuals and certification process although they offer a translation.

On the other hand, accredited professional (AP) is vital to accompany and prepare the documentation for certified projects. As shown in Fig. 6, there are insufficient accredited professionals and assessors for all sustainability assessment tools in Morocco. The HQE system is more than thirty accredited referents in Morocco, a number that is considered acceptable, the BREEAM referents are fewer. Registration, accompaniment, and certification fees play a catalytic role in encouraging properties. The American LEED system is more expensive than BREEAM. Other systems are more affordable than these three systems, for example, the Egyptian Green Pyramid Rating System (GPRS). In the case of Morocco, with the limitations of funds in the public and private sectors, it is essential to use a system that is affordable and accessible to all.

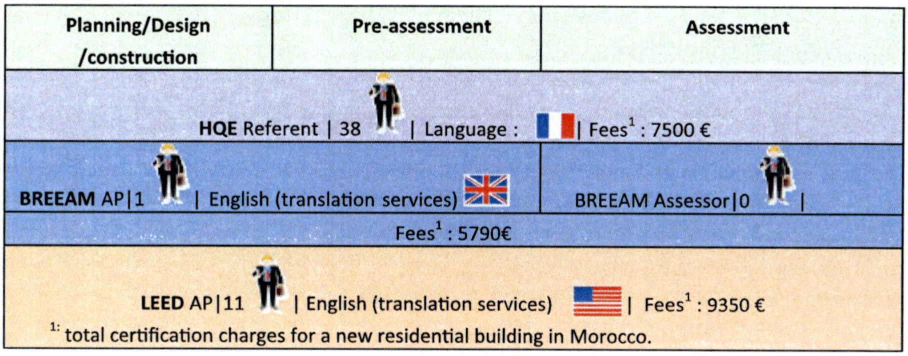

Fig. 6. Information on certification systems in Morocco.

In terms of building type, the majority of projects are office buildings (52.6%); followed by commercial buildings (21.1%), school buildings (13.2%), and residential buildings (10.5%). Most of the projects were certified in the design phase (81.6%), while 18.4% of the studied projects adopted the construction. Based on our analysis of the distribution of certified projects in Morocco, 12 cities have certified green buildings. As shown in Fig. 7, most of these projects are situated on the Casablanca-Rabat axis, which is in first place in terms of surface area and the number of projects, which means that the provinces with a larger scale of construction tend to have a greater number of green buildings. The province of Agadir ranked second in terms of surface area due to tourism projects under the national AZUR plan for the development of the tourism sector.

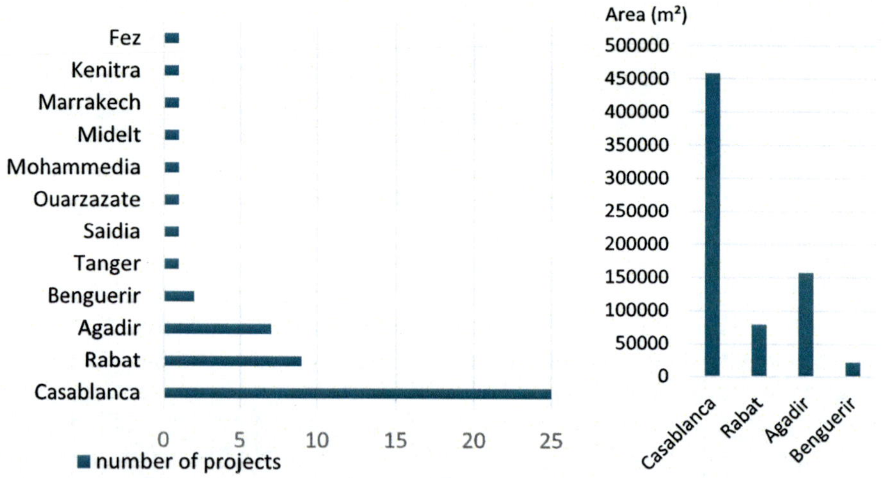

Fig. 7. Distribution of the number/area certified buildings in Morocco

The distribution of green buildings is irregular between the regions of the country. Regional GDP data is used to determine whether regions or cities with higher GDP tend to have more green buildings.

The regression analysis between GDP data and green buildings (see Fig. 8) shows that both the number and area of green buildings are significantly ($R^2 > 0.75$) related to regional GDP.

These relationships through linear regression analyses show that regions or cities with higher GDP (economic activities) tend to have more green buildings.

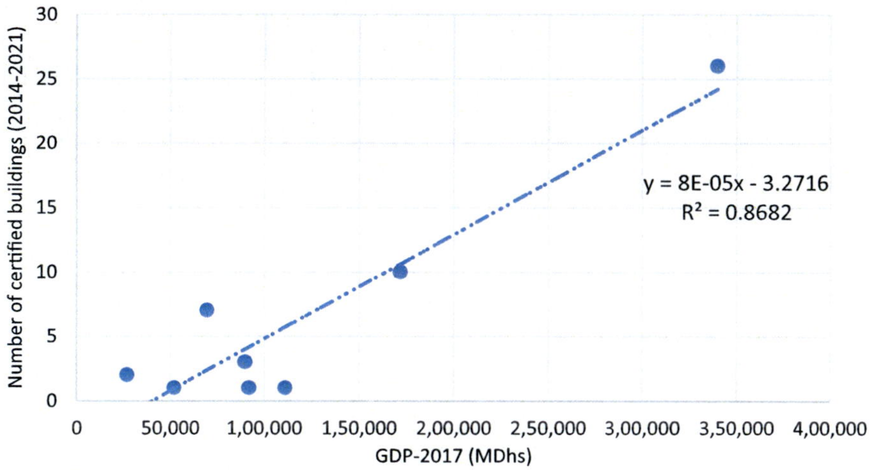

Fig. 8. Linear regression between GDP and the number of certified buildings.

4 Conclusions and Recommendations

Sustainability in the building sector contributes to achieving long-term development in environmental, social, and economic aspects. Sustainability rating systems provide a systematic approach to measuring sustainability through a set of weighted criteria. In particular, the French HQE system is more deployed in the Moroccan market than other systems with 51 projects certified between 2014 and 2021. Although these systems are voluntary and not mandatory, the fact remains that they are effective in promoting sustainable cities. In another, the standards required to include sustainability assessment categories that can guide decision-makers towards the proposed future legislative framework that takes into consideration energy, water, land use, and indoor environmental quality even though these criteria have been, to some measure, taken into account in Morocco's current regulations. These criteria need to be examined in the context of country-specific conditions during the different life cycles (design, construction, operation, and demolition), based on the experience acquired in the construction of already certified green buildings.

Future studies should put more focus on the challenges and opportunities of green building development in the Moroccan context. Finally, increased commitment and collaboration between stakeholders need to improve the current situation with the support of the government and university researchers.

Acknowledgments. We would like to thank Dr. Catherine BUHE, Gérard MERLIN, and Hervé BOILEAU (LOCIE UMR CNRS 5271) for their orientation, and helpful comments.

References

1. Darko, A., Chan, A.P.C., Yang, Y., Shan, M., He, B.J., Gou, Z.: Influences of barriers, drivers, and promotion strategies on green building technologies adoption in developing countries: the ghanaian case. J. Clean. Prod. **200**, 687–703 (2018). https://doi.org/10.1016/j.jclepro.2018.07.318
2. Gursel, A.P., Ostertag, C.: Comparative life-cycle impact assessment of concrete manufacturing in Singapore. Int. J. Life Cycle Assess. **22**(2), 237–255 (2016). https://doi.org/10.1007/s11367-016-1149-y
3. Yüksel, İ: A review of steel slag usage in construction industry for sustainable development. Environ. Dev. Sustain. **19**(2), 369–384 (2016). https://doi.org/10.1007/s10668-016-9759-x
4. Olmez, G.M., Dilek, F.B., Karanfil, T., Yetis, U.: The environmental impacts of iron and steel industry: a life cycle assessment study. J. Clean. Prod. **130**, 195–201 (2016). https://doi.org/10.1016/j.jclepro.2015.09.139
5. M. de l'Aménagement du T. National and de l'Habitat et de la P. de la V. de l'Urbanisme. Tableau de Bord du Secteur de l'Immobilier 2020 (2021)
6. Ministère de l'Energie des Mines de l'Eau et de l'Environnement du Maroc. Deuxième Rapport Biennal Actualisé (2019)
7. ONeill, K.J., Gibbs, D.C.: Towards a sustainable economy? socio-technical transitions in the green building sector. Local Environ. **19**(6), 572–590 (2014). https://doi.org/10.1080/13549839.2013.818954
8. Darko, A., Chan, A.P.C., Huo, X., Owusu-Manu, D.G.: A scientometric analysis and visualization of global green building research. Build. Environ. **149**, 501–511 (2019). https://doi.org/10.1016/j.buildenv.2018.12.059
9. Atanda, J., Olukoya, O.: Green building standards: opportunities for Nigeria. J. Clean. Prod. **227**, 366–377 (2019). https://doi.org/10.1016/j.jclepro.2019.04.189
10. Windapo, A., Goulding, J.: Understanding the gap between green building practice and legislation requirements in South Africa. Smart Sustain. Built Environ. **4**(1), 67–96 (2015). https://doi.org/10.1108/SASBE-01-2014-0002
11. Nykamp, H.: Policy mix for a transition to sustainability: green buildings in Norway. Sustain **12**(2), 15–17 (2020). https://doi.org/10.3390/su12020446
12. LEED project profiles | U.S. Green Building Council.2021
13. Oguntona, O.A., Aigbavboa, C.O., Thwala, W.D.: A scientometric analysis and visualization of green building research in Africa. J. Green Build. **16**(2), 83–86 (2021). https://doi.org/10.3992/jgb.16.2.83
14. Department of Housing and City Policy. Buildings | Energetic Efficiency
15. Basma M'lahfi, M.E.Q., Amegouz, D.: A new approach for the mandatory application of the thermal regulation of construction (RTCM) in the future moroccan buildings. SN Appl. Sci. **2**, 1–16 (2020). https://doi.org/10.1007/s42452-020-03367-w
16. A. nationale pour le D. des E. Renouvelables. Plan Cadre National pour les Mcpd Plan Sectoriel « Eco-Construction Et Bâtiment Durable » Maroc (2016)
17. Slaoui, S., Karibi, K., Mekaoui, N. T., El Harrouni, K.: Sustainable architecture and energy efficiency a university campus project in fez city, Morocco, vol. 6, no. Isbs 2017, pp. 65–79 (2018). https://doi.org/10.1007/978-3-319-63709-9_6

Optimal Design of Standalone Hybrid Renewable Energy Systems Using HOMER

Mouna Lamnadi[1](✉), Abdelkader Boulezhar[1], and Badre Bossoufi[2]

[1] Renewable Energy and Dynamic of Systems Laboratory, Faculty of Sciences Ain-Cock, University of Hassan II, Casablanca, Morocco
mouna.lamnadi@gmail.com

[2] LIMAS Laboratory, Faculty of Sciences Dhar El Mahraz, University of Sidi Mohamed Ben Abdellah, Fez, Morocco

Abstract. Hybrid systems can compensate the intermittent character of renewable energy sources and offer a higher overall energy efficiency. Their main interest is the possibility of energy autonomy that they permit since they do not depend on a single source. This paper discusses a new approach for designing a stand-alone hybrid system (photovoltaic, wind and battery) combining three objectives: power loss probability, life cycle cost and embodied energy for an optimal design. The simulation results are presented using Matlab/Simulink software and discussed with a comparison of developed methodology and an existing commercial tool (HOMER software).

Keywords: Hybrid system · Dynamic simulation · Optimization · HOMER

1 Introduction

In recent years, the world has been facing unprecedented energy and climate challenges for which strong alternatives are expected [1]. Meeting these challenges will have to go through a period of transition where the use of renewable energies, the improvement of energy efficiency, and the moderation of consumption in behavior will be the main vectors of change. The efforts to be undertaken in the field of renewable energies are considerable because their share in final energy is only 13.4% in 2015 while the target for 2020 is 42% for Morocco [2]. Different climatic, energy and economic factors such as global warming, the scarcity of fossil fuels, the liberalization of the energy market are supporting an increased awareness in the field of renewable energies on the part of users and political decision-makers [3]. Communities must play an important role in the development of new energies in their territory by implementing several actions such as the rehabilitation of heritage to reduce energy consumption, the use of local energies [4], and the integration of sustainable criteria in the public procurement" or even support for actors Since 2008, Morocco has based its energy model on the development of low-cost electricity with low greenhouse gas emissions, with a view to reducing dependence on fossil fuels. Today, the trend is to consume the energy produced locally [5]. In such a context, in order to reduce the percentage of use of traditional energies in overall

S. Motahhir and B. Bossoufi (Eds.): ICDTA 2022, LNNS 454, pp. 573–583, 2022.
https://doi.org/10.1007/978-3-031-01942-5_57

consumption, the difficulties linked to the use of clean energies should be removed. One of these difficulties is the optimization of the combination of several energy sources (such as the sun, the wind…) and the use of storage devices to compensate for the intermittent nature of these energies. Energy from renewable sources can be reinjected directly into the electricity grid, by providing either a storage device (batteries for example), or an intelligent management system in order to match production with the use of electricity. Energy. It is in this context that the main issue of this thesis is situated: designing and optimizing a multi-source system to meet the needs of a residential home in terms of electrical energy. To solve this problem [6].

Global energy consumption is growing with population and industrialization. Conventional energy resources are unable to meet these demands even though they have challenges such as greenhouse gas emissions and high life cycle costs [7]. Electrical energy generated from renewable sources, especially wind and solar, is being considered as an attractive generation alternative in today's world [8]. The combination of renewable energy resources with conventional fossil resources in addition to the storage is creating hybrid renewable energy systems (HRES) [9]. However their design is crucial [4], in this context based this study with a real simulation to create a stand-alone hybrid system.

The following sections compose this paper: Data Collection in Sect. 2. In Sect. 3, Hybrid system models. Section 4 Design and evaluation criteria for a hybrid system. Finally, Sect. 5 details the Optimization method, followed a Results & discussion. Finally, the Hybrid system simulation using HOMER software is discussed.

2 Data Collection

In this paragraph, the load power obtained for a residential house using the PV system software for a period (one year) also for renewable energy sources (solar energy, wind energy). The data should be collected carefully in order to obtain the appropriate sizing of any hybrid system.

The stand-alone hybrid system prposed (PV/W/Battery) to satisfy the charges of a residential house located in Province of Nouaceur (Casablanca), Morocco [5].

2.1 Renewable-Energy-Sources

In this work, we used hourly values used from the Metronome 6 software [2] (latitude: 33° North, longitude: 7° West, altitude: 1562 m) and sustainable energy sources are calculated by numerical integration for each year as shown in Fig. 1 (solar radiation), Fig. 2 (wind speed) and Fig. 3 (ambient temperature)

$$Energy(T) = \sum_{i=1}^{8760} \frac{1}{2}(Power(i) + Power(i+1)) * T \quad (1)$$

Where T (hour): the period of sampling.

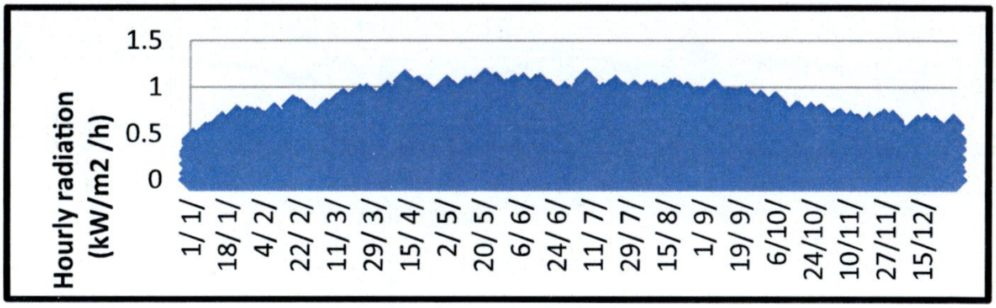

Fig. 1. Hourly radiation for one year

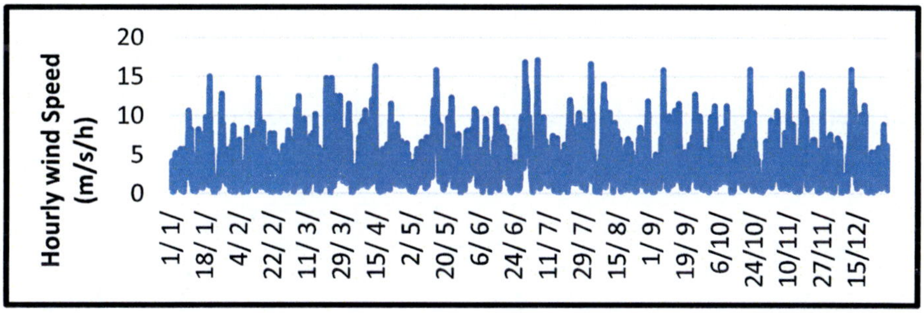

Fig. 2. Hourly wind speed for one year

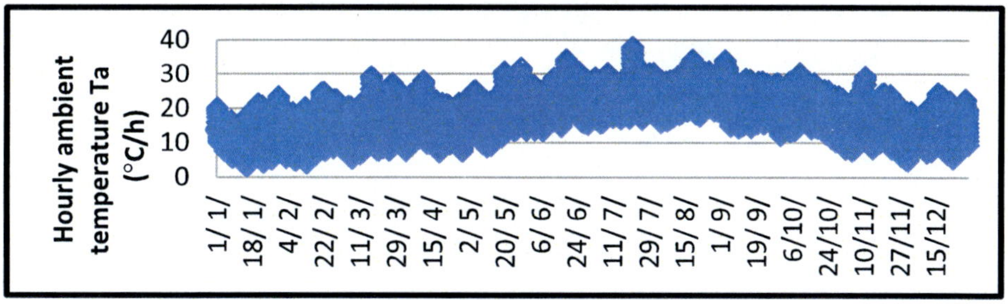

Fig. 3. Ambient temperature (year)

2.2 Energy Profile of the Charge

Regarding the electrical energy needs, the values were taken (one year) for a living house with four occupants, not including the heating, the cooking and the hot water production. The residence is used all year round and all loads are provided by 230 V/AC. The annual demand load profile with a one-hour acquisition period is shown below (see Fig. 4).

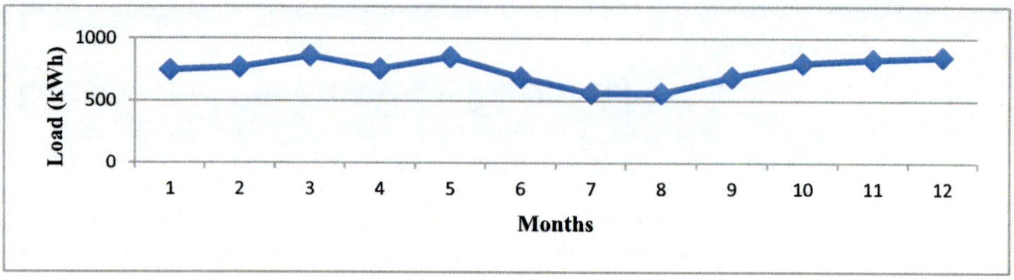

Fig. 4. Load profile for one year

3 Hybrid System Models

The modeling stage is necessary before proceeding to any optimal sizing step. The hybrid system is constructed of a mix of three energy sources: wind, solar and battery storage, the figure below shows the hybrid energy system for this project [10] (Fig. 5).

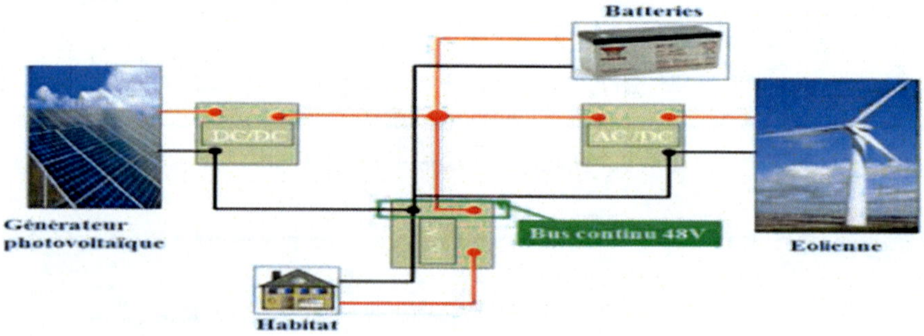

Fig. 5. Diagram of the hybrid system with storage.

4 Design and Evaluation Criteria for a Hybrid System

The goal is to satisfy energy needs with a lowest system life cycle cost and reduced environmental impact. This study focuses on identifying the most appropriate configuration for the hybrid system: PV energy (Apv), wind energy (Awt) and battery bank (Cn).

4.1 Reliability Requirements

The reliability of the system in all cases has been described by LPSP which is the probability of the state of charge (SOC) of the batteries, either less than or equal to SOC _min (minimum value) and the energy generated by the PV-Wind P_ren which must be lower than P_dem necessary for the load:

$$LPSP[\%] : \{P_{ren} \leq P_{demand} \; and \; SOC(t) \leq SOC_{min}\} \tag{2}$$

So, the first criteria is loss of power supply probability that we can defined by a mathematical expression: all energy losses to the load demand during the period (Δt) [8]:

$$LPSP(\Delta t, A_{PV}, A_{WG}, C_n) = \frac{\sum_{\Delta t=1}^{T} DE(t, A_{PV}, A_{WG}, C_n) * \Delta t}{\sum_{\Delta t=1}^{T} P_{load}(\Delta t) * \Delta t} \quad (3)$$

Or $DE(t, A_{PV}, A_{WG}, C_n)$ describes the power deficit. It considers when the total energy available between (t − Δt) and t and the energy stored in the batteries are not sufficient to meet the charge demand.

4.2 Cost Considerations

A second optimizing objective is to minimize the total life cycle cost (LCC) of the hybrid system: the cost accumulated (in $ or Dh) over its life cycle, from conception to recirculation. In this study, the life cycle cost model composed: equipment cost (C_o), installation cost (C_{ins}), replacement (PV_{rem}) and maintenance (PV_{mai}). It can be estimated as follows [9–12]:

$$PW = P_r * C_o \quad (4)$$

Or C_o is product cost and P_r represents the present value of an item that will be purchased (n) years later, which is given by:

$$P_r = \left(\frac{1+i}{1+d}\right)^n = x^n \quad (5)$$

With i is the inflation rate and d is the update rate. The life cycle cost equation is expressed as:

$$LCC = C_O + C_{ins} + PV_{rem} + PV_{mai} \quad (6)$$

Using C_{ins} for installation cost; PV_{mai} present of maintenance cost C_{mai} and is calculated based on cumulative present worth factor as follows:

$$PV_{mai} = (C_{mai}) * x * x * \frac{1-x^n}{1-x} \quad (7)$$

4.3 Environmental Impact

This criteria is defined by the embodied energy (the energy required for all activities in the manufacturing process, given by MJ or kWh). In this situation, the hybrid system needs no external energy for its operation, but the requires energy for the manufacturing of the all his components.

This energy is modeled and used for an optimization design to minimize the environmental effects [13, 14].

5 Optimization Method

5.1 Optimization Program Formulation

The optimization of the hybrid systems is considered as a multi-objective problem, its solution allows to find the best possible combination between the probability of dissatisfaction of the demand, the cost of the cycle and the intrinsic energy.

It can be represented by the following mathematical equation [13] (Fig. 6).

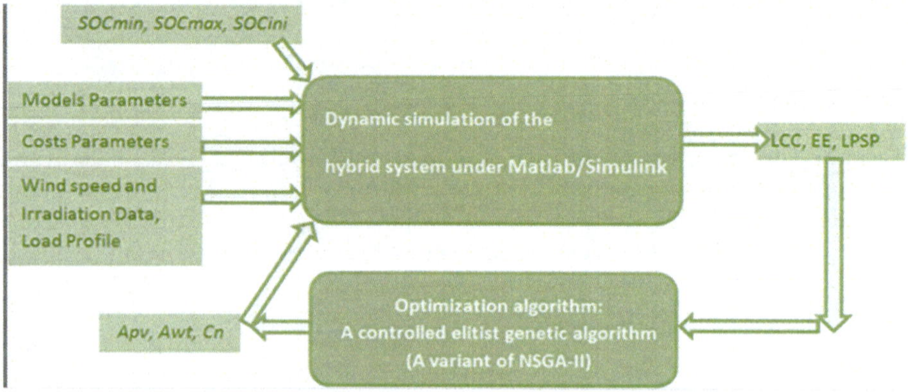

Fig. 6. Design of optimization program.

Objective functions:

- Minimum ≥ Hybrid system life cycle cost

$$LCC(A_{PV}, A_{WG}, C_n) = PV_{LCC}(A_{PV}) * WG_{LCC}(A_{WG}) * Bat_{LCC}(C_n) \quad (8)$$

- Minimum ≥ Hybrid system primary embodied energy cost

$$[EE(A_{PV}, A_{WG}, C_n) = PV_{EE}(A_{PV}) * WG_{EE}(A_{WG}) * Bat_{EE}(C_n)] \quad (9)$$

- Minimum $[LPSP \geq (\Delta t, A_{PV}, A_{WG}, C_n)]$

Constraints: possibility of an stand-alone house:

$$A_{PVmin} \leq A_{PV} \leq A_{PVmax} \quad (10)$$

$$A_{WGmin} \leq A_{WG} \leq A_{WGmax} \quad (11)$$

$$C_{n\,min} \leq C_n \leq C_{n\,max} \quad (12)$$

With (A_{PV}, A_{WG}, and C_n) are the decision variables. A_{PVmin}, A_{PVmax}, A_{WGmin}, A_{WGmax}, $C_{n\,min}$ and $C_{n\,max}$ are lower and upper decision variables limits specified by the user.

5.2 Calculation Method and Algorithm Design

Figure 7 shows the concept of optimization program. It is a repetitive procedure. It proceeds: at the current iteration (k = 0), we have a set of inputs (initial decision functions); the dynamic simulation used to estimate the output requirements (LCC en Dh or €; EE en MJ and LPSP en %).
At the same time, the sensitivity analysis involved in this optimization method evaluated the direction of change of the parameters with regarding if the stopping instructions are not respected, the optimization method is applied to calculate a new set parameter taking into account the results of the last simulation. This parameter set is used to determine the data in the next calculation.

All the necessary models for the simulation of the hybrid system implemented in the Matlab/Simulink software, such as: the wind and the PV model, the ideal battery bank for energy storage with his controller model, the energy calculator and the objective function estimator (LPSP, LCC and EE).

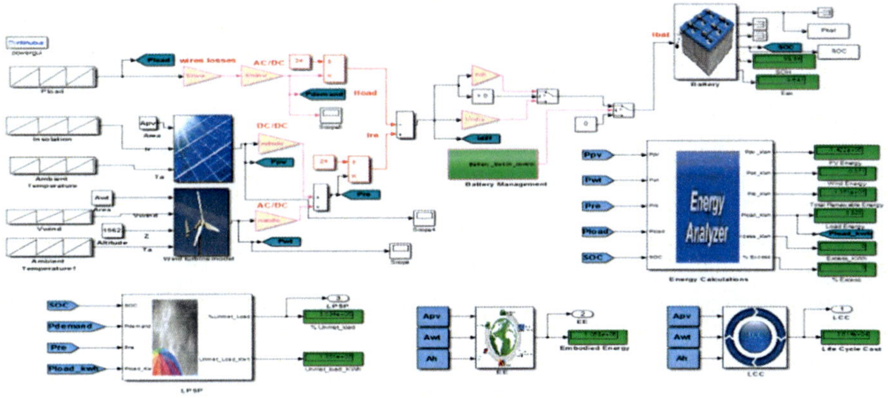

Fig. 7. Hybrid system using Matlab/Simulink software.

6 Results and Discussion

- The hybrid system can be configured according to:
- Configuration 1: PV modules with battery bank
- Configuration 2: wind turbine with battery bank
- Configuration 3: PV modules with wind turbine and battery bank

Input data:

- Installed area of the PV modules Apv = 14.8 m^2 (9 Sharp ND-240QCJ 240 W panels).
- Swept area of the wind turbine Awt = 3.49 m^2, an Aeromax Engineering (Lacota S, SC) wind turbine (900 W at 13 m/s) has a swept area close to 3.43 m^2 with a low LCC.

- The storage capacity of the battery bank gives Cn = 178.6 Ah, 4 Numax Gel SLG180-12.12 V/180 Ah VRLA batteries in series can be installed.
- The LPSP objective function must be lower than 5%.

Table 1. Evaluation of the three configurations.

Configuration	Demand (kWh)	PV (kWh)	Wind (kWh)	Excess (KWh)	Unmet load (kWh)	% Unmet load (LPSP)	SOH (%)
PV/Bat	2250	3312	0	1047	160.1	7.11	91.24
W/Bat	2250	0	742	96.06	1519	67.52	98.1
PV/W/Bat	2250	3312	742.4	1688	85.91	3.81	92.54

According to the results shown in Table 1, the third configuration (PV/Wind/Battery) can supply more than 95% of the electricity need for the concerned house over a year (the consumer doesn't need any external energy source) which means that the LPSP criteria is well respected (less than 5%).

In addition, the total life cycle cost of the proposed hybrid system is approximately 26 281$, distributed as follows: 38% for the PV system, 26% for the wind turbine, and 36% for the batteries bank.

The embodied energy required to obtain the required energy is estimated to be 106,080 MJ. The distribution between the system components is as follows: 49.3% for PV system components, wind system components estimated to 8.5% and 42.2% for the batteries, This distribution is due to the fact that the batteries consume primary energy because of regular replacement every 4.2 years and that the small wind turbines have a better energy efficiency than the PV panels.

For the first scenario the proposed system supplied 93% of the energy demand, i.e. the consumer needs an alternative energy source, and the same for the second configuration with an LPSP of 33%.

In Addition, about 27,019 ($) affordable life cycle cost is achieved and 97,604 MJ of embodied energy is necessary for configuration 1, and 23,295 ($) (LCC), 56,183 MJ (EE) for scenario number 2.

7 Hybrid System Simulation Using HOMER Software

HOMER software is a simulation tool used to evaluate and analyze different off-grid and grid-connected power hybrid system. In order to produce an optimized hybrid system configuration, it is important to verify each system components (number, size, etc.) in relation to the available energy resources [13]. The simulation scenarios (as can be seen in Fig. 8) have been based on the sensitivity analysis which consists of selecting wind speeds, daily solar radiation, scaled annual primary load averages as sensitivity variables.

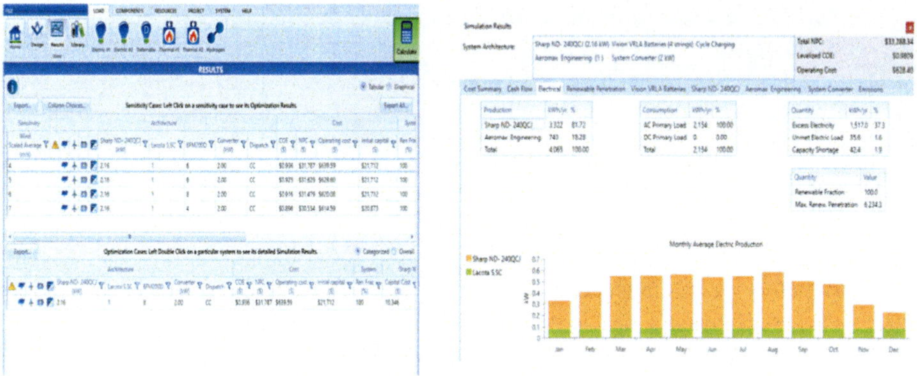

Fig. 8. Hybrid system evaluation using HOMER Pro software.

Table 2 compares the results obtained by HOMER Pro and by our simulator developed with Matlab/Simulink. For the comparison, we use for both, 200 Ah batteries with a depth of discharge of 60% ($SOCmin = 40\%$ and $SOCmin = 100\%$).

Table 2. Comparison of the results obtained using two types of methods.

Parameter	Results using HOMER	Results using Matlab/Simulink	Percentage of Différence
Electrical consumption (kWh)	2154	2250	4.27
Photovoltaic production (kWh)	3322	3312	0.30
Wind production (kWh)	743	742.4	0.08
Excess of production (kWh)	1517	1688	10.13
Unsatisfied load (kWh)	35.2	85.91	59.03
LPSP (%)	1.6	3.8	57.89
Life cycle cost (€)	32524	36138	10.00

Concerning the PV production, for the HOMER software, it is calculated by a mathematical equation. However for Matlab/Simulink software it based on the PV efficiency so we have a production increase. For the wind production, HOMER uses the data of the power graph as a function of the wind speed through a table of values (lookup table).

While in our Simulink model, we use a polynomial function but the results obtained are almost identical. For the other energy parameters (electricity consumption, excess of production and the load dissatisfaction rate), the differences observed are mainly due to the modeling of the cables losses, the load, the state of charge for batteries.

For the life cycle cost estimation, the difference is due to the component cost data for each simulation.

In conclusion, we note that the two simulation and dimensioning tools have practically the same orders of magnitude but with their own specificities.

8 Conclusion

In this study, a design and control methodology has been applied for a hybrid system (PV-wind-battery) for a domestic house, to find the best compromise between the three criteria: the loss of power supply probability, the life cycle cost and the embodied energy of the hybrid system. The optimization was achieved by a dynamic model of the system using Matlab/Simulink software. Then to validate the proposed model, the same simulation is realized with the Homer software in order to compare the results and to determine the reliability of the proposed model.

References

1. Bahramara, S., Parsa Moghaddam, M., Haghifam, M.R.: Optimal planning of hybrid renewable energy systems using HOMER: a review. Renew. Sustain. Energy Rev. **62**, 609–620 (2016)
2. Meteonorm: Irradiation data for every place on Earth. [En ligne]. Disponible sur http://www.meteonorm.com/. Consulté le: 04-août-2016
3. Mokhatara, C., Negrou, B., Bouferrouk, A., Yao, Y., Settou, N., Ramadan, M.: Integrated supply-demand energy management for optimal design of off-grid hybrid renewable energy systems for residential electrification in arid climates. Energy Conv. Manage. **221**, 113193 (2020)
4. Maheri, A.: Multi-objective design optimisation of standalone hybrid wind- PV- diesel systems under uncertainties. Renew. Energy **66**, 650–661 (2014)
5. Lamnadi, M., Trihi, M., Boulezhar, A., Bossoufi, B.: Optimal design of stand-alone hybrid power system using. Int. J. Energy Technol. Policy **15**(2–3), 280–300 (2019)
6. Abouzahr, I., Ramakumar, R.: Loss of power supply probability of stand-alone photovoltaic systems: a closed form solution approach. IEEE Trans. Energy Conv. **6**(1), 1–11 (1991)
7. Yang, H.X., Lu, L., Burnett, J.: Weather data and probability analysis of hybrid photovoltaic–wind power generation systems in Hong Kong. Renew. Energy **28**(11), 1813–1824 (2003)
8. McEvoy, Markvart, T., Castañer, L., Markvart, T., Castaner, L.: Practical Handbook of Photovoltaics: Fundamentals and Applications. Elsevier, Oxford (2003)
9. Salih, T., Wang, Y., Adam, M.A.A.: Renewable micro hybrid system of solar panel and wind turbine for telecommunication equipment in remote areas in Sudan. Energy Procedia **61**, 80–83 (2014)
10. Saady, I., et al.: Optimisation for a photovoltaic pumping system using indirect Field Oriented Control of Induction Motor. Electronics **10**(21), 2670 (2021)
11. Bouderbala, M., et al.: Experimental validation of predictive current control for DFIG: FPGA implementation. Electronics **10**(21), 2670 (2021). https://doi.org/10.3390/electronics10212670
12. Bossoufi, B., et al.: Rooted tree optimization for backstepping power control of DFIG wind turbine: dSPACE implementation. IEEE Access **9**(1), 26512–26522 (2021)

13. Bossoufi, B., Taoussi, M., Alami Aroussi, H., Bouderbala, M., Motahhir, S., Camara, M.B.: DSPACE-based implementation for observer backstepping power control of DFIG wind turbine. IET Electric Power Appl. **14**(12), 2395–2403 (2020)
14. Bossoufi, B., Karim, M., Lagrioui, A., Taoussi, M., Derouich, A.: Observer backstepping control of DFIG-generators for wind turbines variable-speed: FPGA-based implementation. Renew. Energy J. **81**, 903–917 (2015)

Exploration of Indoor Energy Harvesting

Sneha Patil[1(✉)], Mahesh Goudar[2], and Ravindra Kharadkar[1]

[1] G H Raisoni Institute of Engineering and Technology, Pune University, Pune, Maharashtra, India
patilsneha19@gmail.com, r.kharadkar@raisoni.net
[2] MIT Academy of Engineering, Pune University, Pune, Maharashtra, India
mdgoudar@mitaoe.ac.in

Abstract. An energy harvester is the best way to avoid having to use primary and secondary batteries. Thermal, vibration, solar, and wind energy are among the energy collection methods being investigated by researchers. Indoor light, which is a mix of artificial light and scattered sunlight, can be collected using solar cells. Despite the minimal power (in microwatts) generated, the energy can be used to power wireless sensor nodes. This research looks at a number of energy harvesting approaches and explores the need for indoor energy harvesting as well as the challenges that it presents..

Keywords: Energy harvester · Photovoltaic · Solar cell · Indoor energy · Neural Network

1 Introduction

In the subject of energy harvesting, solar energy is commonly recognised as one of the most widely employed renewable energy sources [1–4]. The use of this renewable energy source is mainly contingent on sunlight exposure. Due to the limits of the corresponding energy source, researchers have created a 'Hybrid Energy Harvesting System (HEHS)' by combining it with other sources. A photovoltaic (PV) array can also directly convert solar energy into direct current (DC) electricity. A vital aspect of this energy harvesting system is tracking the power from the low-voltage level, which comes from the solar array (SA) and the thermo-electric module (TEM) [5]. This fact corresponds to the 'Maximum Power Point (MPP)' of the PV array [6, 7]. In order to optimise the generation system's process, it's also necessary to solve the issue of non-linear electrical characteristics of PV arrays, especially for real-time systems. The current voltage property (I-V property) of a PV array, on the other hand, displays a nonlinear feature that fluctuates with the environment [8].

Many of the studies focused at solar energy harvesting in the open air and hybrid energy harvesting systems, but there isn't much information on inside energy harvesting. Artificial and natural light sources are used in indoor illumination. Irradiance, or natural light, is light that originates from the sun. Light that comes from artificial sources is referred to as "artificial light." [9].

S. Motahhir and B. Bossoufi (Eds.): ICDTA 2022, LNNS 454, pp. 584–591, 2022.
https://doi.org/10.1007/978-3-031-01942-5_58

During the day, light indoors is usually a blend of natural and artificial light, depending on the time of day. The size of the windows, their direction, and the degree of overcast, as well as the geographic location, season (date), and time of day, all influence the amount of light that enters a room [10]. The distance between an open aperture – such as a window – and the point of view, as well as the obscurity of the open aperture, determine the quantity of solar irradiance that enters a room. The distance to and reflectivity of large windows that span the entire width of a room are the primary causes of attenuation, or the steady decrease in the intensity of solar irradiation. The overall irradiance level is thus controlled by the architecture of the structure as well as interior elements such as wall, ceiling, and other surface reflectance.

2 Related Work

Jun Junget al. [11] used a novel design 'PIMWEH', inquiry analytical mathematical simulation, to determine the most difficult 'PIMWEH' form for application, which was used as a power basis of WSNs, in 2015. The investigative findings revealed that the 'optimal PIMWEH' produced 2.8 mW (RMS value) and did not shatter after 40 h. In addition, they have experimented with using a resolving circuit, a 'AC–DC converter,' and a 'electrical charger' to generate power for the sensor's operation. The exploratory findings revealed a clean DC voltage signal of 3.3 V with a yielded power of 1.0 mW. In addition, they achieved a charging energy of 0.845 J in 24 min. Following that, they formulated the efficacy of the 'PIMWEH' for performance evaluation. They achieved a general PIMWEH power conversion efficacy of 3.2% using the specified 'three-step' energy conversion technique.

In 2016, A. Nasir et al. proposed a multi-cell network deployment in which each cell's base station (BS) has direct contact with cell-edge users via a relay node, dubbed 'amplify-and-forward (AF)'[12]. Their main goal was to create a resource allocation strategy that optimised some characteristics such as BS transmitting power, established factors such as 'power-splitting' for the purpose of energy harvesting and information processing at the relays, and relay transmission power. They also advocated the use of a sequential convex approximation procedure and the formulation of iterative algorithms based on geometric programming and 'difference-of-convex-function' programming. The methods developed transformed a problem known as "non-convex problems" into a series of convex problems, each of which was expertly solved using the interior-point method. They've demonstrated the effectiveness of the algorithms they've created. They've also applied their findings to a specific scenario of 'decode-and-forward (DF)' relaying over variable time slots. The findings of the analysis confirmed that the created joint optimization approaches improved network performance over time, and they compared it to other circumstances that were optimized separately.

N. Rezaei-Hosseinabadiet al. developed the "wideband piezoelectric" architecture for wind energy harvesting in 2016, as well as an optimization method for gathering more energy that took into account the wind speed distribution model [13]. Furthermore, the produced device proved to be more useful in the operation of WSN in the smart monitoring process. The created gadget consists of a piezoelectric beam that vibrates due to interactions between'miniaturized permanent magnets (PMs)' that were implanted in a small turbine and a magnet at the 'tip' of the beam. In terms of the wind speed distribution model, the suggested model has determined the significance frequency of the beam. The model was compared to previous wideband topologies, which achieved wideband functionality by the grouping of PMs within the turbine. The 'wideband feature' was explored analytically with the help of a device prototype, and rotational speeds of 155.0, 193.5, 257.9, and 387.0 rad/s were attained. Furthermore, the produced device's supreme power density, supreme efficacy, and cut-in wind speed were 0.59 mW/cm^3, 24%, and 2.1 m/s, respectively.

A. Sultan [14] proposed a cognitive radio scenario in 2012, in which the subordinate user is in an energy harvester with a fixed capacity battery. The primary user operated in a time-slotted manner. From the start of each time slot, the subordinate user's goal is to increase throughput, which may have stayed inactive, or to send out spectrum sensing for the detection of major activity. The decision was made based on a subordinate belief about the primary activity as well as the amount of stored energy. A Markov decision procedure was used to formulate the problem. It was also shown how to create the best policy and compared it to a'myopic policy.' The investigation was carried out using the difference in throughput as a function of a number of system characteristics.

A.M. Yusopet al. [15] proposed the development of a new'maximum-power point-tracking (MPPT)'model based on an input shaping system controller in 2017. The proposed model was able to transform the first input response into a'shapeable MPPT algorithm' based on an exponential input function. This input function was chosen because it can stabilise the system, which happened at the end of the simulation time. The produced model was compared to other traditional models, and the results revealed that the system with the proposed model provided the highest output power over the system with the perturb and observe P&O process. The results also showed that it was capable of effectively tracking the maximum output power.

In 2015, Yin Li and Ronghua Shi [16] devised an intelligent solar energy harvesting system for providing reliable and long-term power. A solar panel, lithium battery, and control circuit were included in the constructed system. They employed hardware rather than software to keep the battery charged. The system's stability and reliability have both increased as a result of this. The system has implemented an MPPT circuit to maximise the use of solar energy while also ensuring the battery's long life through a proper charging approach. The designed method could be implemented with little power equipment, making it ideal for outdoor-oriented WSNs in the Internet of Things (IOT).In the year 2020, Chen et al. [17] created a day illumination technique based on the electrowetting principle. Initially, a model for the liquid prism was presented, which simplified the production process, enhanced reliability, and allowed for less maintenance. Finally, the performance evaluation was claimed in terms of cost, illumination time, and energy after creating liquid prism utilising the proposed approach.

The Table 1 outlines the various energy collecting approaches that have been described in this study thus far. The advantages and disadvantages of each strategy are discussed.

Table 1. Data synthesis of several energy gathering strategies

Author	Method	Features	Challenges
JunJunget al. [11]	PIMWEH method	Highly durable Real-world applications are possible	Needs improvement for more efficiency in performance Wind speed condition is very high
A. Nasir et al. [12]	SCA approach	The Karush-Kuhn-Tucker (KKT) criterion is met Increases throughput while lowering the sum-power	Problems may occur in beam forming design and relay selection Cannot applicable in all real time cases
N. Rezaei-Hosseinabadiet al. [13]	Turbine characterization topology	Increases the captured energy Can use in remote and inaccessible areas	Wind speed fluctuation changes the performance rate Adjustment is needed in frequency and power load
A. Sultan [14]	Markov decision process	Improves the performance of optimal policies Improves the immediate reward	Has more computational extensive Decreases the secondary throughput
A.M. Yusopet al. [15]	Exponential input function-based MPPT	Controls the output behavior of the entire system Consumes less energy	Response changes as per the condition
Yin Li and Ronghua Shi [16]	MPPT algorithm	Power loss is very low More reliable Extends the lifetime of lithium battery	Changes in circuit would reduce the efficiency Load balancing is difficult
Chen et al. [17]	Electro wetting technology	Low cost More stable and reliable	On-site performance is not known Fibre loss is more

(*continued*)

Table 1. (*continued*)

Author	Method	Features	Challenges
Yang et al. [18]	MPPT algorithm	Offers high cell efficiency	Not suitable for narrowband light signals
Teng et al. [19]	RTOS model	Highly efficient model	Standard testing method is not applied
Clément et al. [20]	TMM model	Predicted values are accurate More efficient	Because of changes in open circuit voltage and fill factor output power may differ
Khairul et al.[21]	Particle Storm Optimization	More precise Less energy consumption	Visual comfort and indoor lightening conditions are not considered
Nandha et al. [22]	Artificial Neural Network	Flexible	More focused on set point tracking only

The features and problems of various energy harvesting models in WSN are summarised in Table 1. More specifically, the PIMWEH approach [11] is more durable and applicable in real-world scenarios. However, it requires improvement in order to achieve greater efficiency in performance. The wind energy predictor (WEP) system improves performance while lowering prediction error [2]. However, the cause of climate change may alter the model's performance rate, and the diurnal cycle for wind energy is not viable in harsh environments. The Successive Convex Approximation (SCA) method meets the Karush-Kuhn-Tucker (KKT) criterion while also increasing throughput and lowering sum-power [12]. However, it is not applicable in all real-life scenarios. Wind speed fluctuations change the performance rate, necessitating more frequency and power load adjustments [13]. The defining topology of a turbine improves the gathered energy and allows it to be used in remote and inaccessible regions, but wind speed fluctuations change the performance rate, necessitating more frequency and power load adjustments. The performance of optimal policies and the immediate reward are both improved by the Markov decision process [14]. However, it requires more computational resources and reduces secondary throughput. The Novel Energy Efficient Clustering (NEEC) algorithm is more stable, and it improves efficiency while also increasing the amount of repair packets [6]. However, if the route is disrupted, the model may be affected, and nodes may require longer waiting time in the first round. The success ratio of data transmission is improved by using a threshold sensitive resurrection mechanism [7]. However, it reduces the number of nodes that survive, making real-time applications impractical. Although the modified earliest deadline first (MEDF) algorithm maintains timing and increases energy performance, interruptions may occur during scheduling, and the execution becomes useless if the super capacitor voltage falls below the operation [8]. Perturbation and observation method-MPPT requires fewer calculations and may estimate directly without using dc [9]. The experimental setup is tough with this model, and more improvement is required for effective findings. IDA-MPPT shows encouraging results and tracks the actual MPP [10]. However, the performance of TEM can be diminished in some instances, and adding extra loads can reduce the performance rate. Exponential input function-based MPPT controls the entire system's output behaviour while also consuming less energy [11]. It is, however, not entirely accurate. Although the MPPT method is more dependable, load balancing is difficult [12].

As a result, the solar energy harvester captures light from both inside and outside. Many historic works have examined the outside performance of solar modules and solar cells; however, the inner performance of solar cells and modules has received little attention in outdoor appliances. In principle, the indoor scenario is determined to be more challenging than outdoor settings due to the wide variety of illuminance sources. As a result, changes in the indoor light source's properties may have an impact on the solar harvester's and PV cell's power production.

Furthermore, light estimation in computer vision is regarded to be a more challenging task, especially in indoor contexts [23]. In these sorts of situations, there are numerous light sources of varied forms, intensities, sizes, and spectral properties. The multi-sensor model is the result of the majority of recent literatures evaluating lighting design by taking into consideration the lighting sensors located with the luminaries [24, 25]. However, this strategy makes installing light sensors in a lighting system more difficult and expensive [26, 27].

The proposed method in Fig. 1 might be one of the solutions to the issues connected with indoor light harvesting. Figure 1 depicts the suggested illuminance/irradiance prediction model for analysing solar module voltage-current behaviour. We offer an improved Neural Network (NN) model for predicting illuminance/irradiance as measured by photometry and radiometry. This project will make use of an improved ANN that will be fed various attributes as input. The expected illuminance and irradiance are calculated using photometry/radiometry.

A number of factors impact the performance of solar cells. The most obvious factor is light intensity, which everyone is aware of. Light intensity, temperature, light source, circuit connection, height, and angle are all factors that might impact a solar panel's performance. In light of this, the purpose of this abstract is to offer a novel prediction model for estimating artificial light setup parameters. This will be accomplished using a

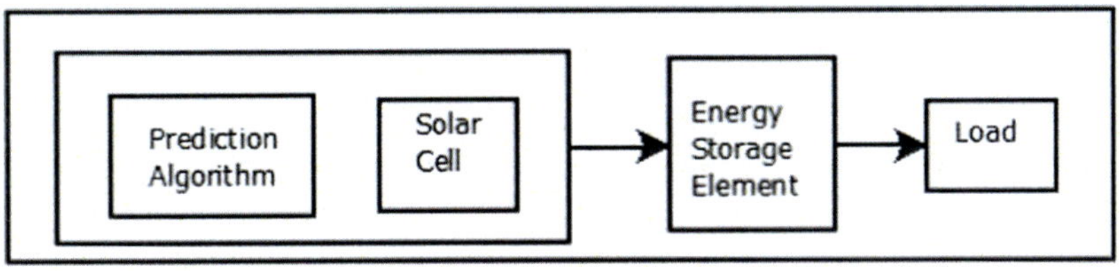

Fig. 1. Proposed system for indoor energy harvesting

Neural Network (NN) model that predicts some of the variables (e.g., control angle and height). Furthermore, a novel Improved Brain Storm Optimization (BSO) for training the model by tweaking the optimal weights is proposed to make the prediction model more accurate.

3 Conclusion

Indoor energy harvesting has the potential to overcome main and secondary batteries' drawbacks. Low power levels, solar cell selection, luminary positioning, fluctuating distributed sunshine properties, and so on are only a few of the challenges. Using lux sensors to monitor the illuminance of artificial light sources as well as the optimum solar cell, the suggested method would improve efficiency. In terms of prediction rate, accuracy, and efficiency, the suggested method may increase performance.

References

1. Yao, Y., Yu, Z., Wang, G.: Clustering routing algorithm of self-energized wireless sensor networks based on solar energy harvesting. J. China Univ. Posts Telecommun. **22**(4), 66–73 (2015)
2. Kosunalp, S.: An energy prediction algorithm for wind-powered wireless sensor networks with energy harvesting. Energy (2017)
3. Bozorgi, S.M., Rostami, A.S., Hosseinabadi, A.A.R., Balas, A.E.: A new clustering protocol for energy harvesting-wireless sensor networks. Comput. Electr. Eng. (2017)
4. Yang, H., Zhang, Y.: A task scheduling algorithm based on supercapacitor charge redistribution and energy harvesting for wireless sensor nodes. Journal of Energy Storage **6**, 186–194 (2016)
5. Yusop, A.M., Mohamed, R., Mohamed, A.: Inverse dynamic analysis type of MPPT control strategy in a thermoelectric-solar hybrid energy harvesting system. Renewable Energy **86**, 682–692 (2016)
6. Lyden, S., Haque, M.E.: Maximum Power Point Tracking techniques for photovoltaic systems: a comprehensive review and comparative analysis. Renew. Sustain. Energy Rev. **52**, 1504–1518 (2015)
7. Esram, T., Chapman, P.L.: Comparison of photovoltaic array maximum power point tracking techniques. IEEE Trans. Energy Convers. **22**(2), 439–449 (2007)
8. Joe-Air Jiang, J.A.: On a hybrid MPPT control scheme to improve energy harvesting performance of traditional two-stage inverters used in photovoltaic systems. Renewable Sustain. Energy Rev. **69**, 1113–1128 (2017)
9. Randall, J.F.: Designing Indoor Solar Products- Photovoltaic Technologies for AES. Wiley, Chichester (2005)
10. Ma, X., Bader, S., Oelmann, B.: Characterization of indoor light conditions by light source classification. IEEE Sensors J. **20** (2017)
11. Jung, H.J.: Design and optimization of piezoelectric impact-based micro wind energy harvester for wireless sensor network. Sensors Actuators Phys. **222**, 314–321 (2015)
12. Nasir, A.A., Ngo, D.T., Zhou, X., Kennedy, R.A., Durrani, S.: Joint resource optimization for multicell networks with wireless energy harvesting relays. IEEE Trans. Veh. Technol. **65**(8), 6168–6183 (2016)

13. Rezaei-Hosseinabadi, N., Tabesh, A., Dehghani, R.: A topology and design optimization method for wideband piezoelectric wind energy harvesters. IEEE Trans. Industr. Electron. **63**(4), 2165–2173 (2016)
14. Sultan, A.: Sensing and transmit energy optimization for an energy harvesting cognitive radio. IEEE Wireless Commun. Lett. **1**(5), 500–503 (2012)
15. Yusop, A.M., Mohamed A. Ayob, R., Mohamed, A.: Shapeable maximum-power point-tracking algorithm to improve the stability of the output behavior of a thermoelectric-solar hybrid energy-harvesting system. J. Assoc. Arab Univ. Basic Appl. Sci. **22**, 1–8 (2017)
16. Li, Y., Shi, R.: An intelligent solar energy-harvesting system for wireless sensor networks. EURASIP J. Wirel. Commun. Netw. **2015**(1), 1–12 (2015). https://doi.org/10.1186/s13638-015-0414-2
17. Chen, Q., Oh, S.J., Burhan, M.: Design and optimization of a novel electrowetting-driven solar-indoor lighting system. Appl. Energy **269**, Article 115128, 1 July 2020
18. Yang, C., RuiPu Xue, X., Li, X.Z., ZhenYu, W.: Power performance of solar energy harvesting system under typical indoor light sources. Renewable Energy **161**, 836–845 (2020)
19. Teng-Chun, W., Long, Y.-S., Hsu, S.-T., Wang, E.-Y.: Efficiency rating of various PV technologies under different indoor lighting conditions. Energy Procedia **130**, 66–71 (2017)
20. Reynaud, C.A., Clerc, R., Lechêne, P.B., Hébert, M., Arias, A.C.: Evaluation of indoor photovoltaic power production under directional and diffuse lighting conditions. Solar Energy Materials Solar Cells **200**, Article 110010, 15 September 2019
21. Wagiman, K.R., Abdullah, M.N., Hassan, M.Y., Mohammad Radzi, N.H.: A new optimal light sensor placement method of an indoor lighting control system for improving energy performance and visual comfort. J. Build. Eng. **30**, Article 101295, July 2020
22. Kandasamy, N.K., Karunagaran, G., Spanos, C., Tseng, K.J., Soong, B.-H.: Smart lighting system using ANN-IMC for personalized lighting control and daylight harvesting. Build. Environ. **139**, 170–180 (2018)
23. Miranda, R., Babilio, E., Singh, N., Santos, F., Fraternali, F.: Mechanics of smart origami sunscreens with energy harvesting ability. Mech. Res. Commun. **105**, Article 103503, April 2020
24. Li, Y., Grabham, N.J., Beeby, S.P., Tudor, M.J.: The effect of the type of illumination on the energy harvesting performance of solar cells. Sol. Energy **111**, 21–29 (2015)
25. Vlachokostas, A., Madamopoulos, N.: Daylight and thermal harvesting performance evaluation of a liquid filled prismatic façade using the Radiance five-phase method and Energy Plus. Build. Environ. **126**, 396–409 (2017)
26. Sarker, M.R., Julai, S., Mohd Sabri, M.F., Mohd Said, S., Tahir, M.: Review of piezoelectric energy harvesting system and application of optimization techniques to enhance the performance of the harvesting system. Sensors Actuators Phys. **300**, Article 111634, 1 December 2019
27. Berardi, U., Manca, M.: The energy saving and indoor comfort improvements with latent thermal energy storage in building retrofits in Canada. Energy Procedia **111**, 462–547 (2017)

Inner Differential Current Suppression in Modular Multilevel Converter

Mbarek Outazkrit[1](✉), Faicel El Aamri[2], Essaid Jaoide[1], Azeddine Mouhsen[1], and Abdelhadi Radouane[1]

[1] University of Hassan First, Settat, Morocco
outazkrit.mbarek@gmail.com
[2] University of Hassan II, Mohammadia, Morocco

Abstract. The applicability of the modular multilevel converter is expanding in diverse industrial applications. The presence of inner differential current can affect the performances and efficiency of this converter, increase the conduction losses, and add more ripples in capacitor voltages. This paper compares two control methods to suppress the AC component of the inner differential currents in each phase. The first uses a controller based on a DQ synchronous reference frame, while the second is based upon a moving average filter (MAF). The inverter structure of the MMC is modeled and simulated in the PLECS environment to evaluate the impact of inner differential current on capacitor voltage ripples and arm currents.

Keywords: Modular multilevel converter (MMC) · Inner differential current · Capacitor voltage ripples

1 Introduction

Modular multilevel converter (MMC) has attracted the attention of researchers in academic fields and global technology companies because of its advantages as good quality of output waveforms, low rating of power semiconductor devices, high efficiency, scalability, and modular structure [1]. In [2], R. Marquardt was the first researcher to introduce this converter in 2001. The first industrial application was completed in March 2010 [3].

Due to these features, this topology is widely used in different applications, such as variable speed drives [4, 5], transmission systems based on direct current (DC) [6], Static synchronous compensator STATCOM [7], and PV applications [8–10]. However, circulating current control, efficient and optimal capacitor voltage balance, and implementation of appropriate modulation techniques for each application remain significant concerns in MMC control. Each one of these concerns has a major impact on the output voltage, total harmonic distortion (THD), capacitor voltage fluctuations, and power losses.

In [11], it is reported that mitigation of internal differential currents can reduce the peak of arm currents and capacitor voltage ripples. The proportional-resonant (PR) regulator [12, 13], the repetitive controller [14], and proportional-integral (PI) regulator [15] are some methods offered for implementing the inner differential current controller.

S. Motahhir and B. Bossoufi (Eds.): ICDTA 2022, LNNS 454, pp. 592–602, 2022.
https://doi.org/10.1007/978-3-031-01942-5_59

The authors of [16] proposed two methods for eliminating harmonic components in the inner differential current.

The first approach operates directly on the inner current's AC harmonics, allowing the DC component to form naturally, while the second method directly regulates the current to its DC reference.

A great variety of modulation schemes are available for MMC operation and control [17]. Carrier-based pulse-width modulation (PWM) methods produce high switching frequency [18]. In other hand, non-carrier-based modulation techniques like staircase modulation have low switching frequency [19].

The paper is organized as follows. The model of the MMC is presented in Sect. 2. In Sect. 3, the inner differential current controllers are presented, and then, Sect. 4 provides simulation results. The conclusion is summarized in Sect. 5.

2 Model of the MMC

Figure 1 shows an inverter based on MMC topology connected to a grid. The upper and lower arm contains N identical submodules (SMs) in series with arm inductor L and resistor R. The last one represents the arm equivalent resistance and L is used to reduce DC fault current and the inner current.

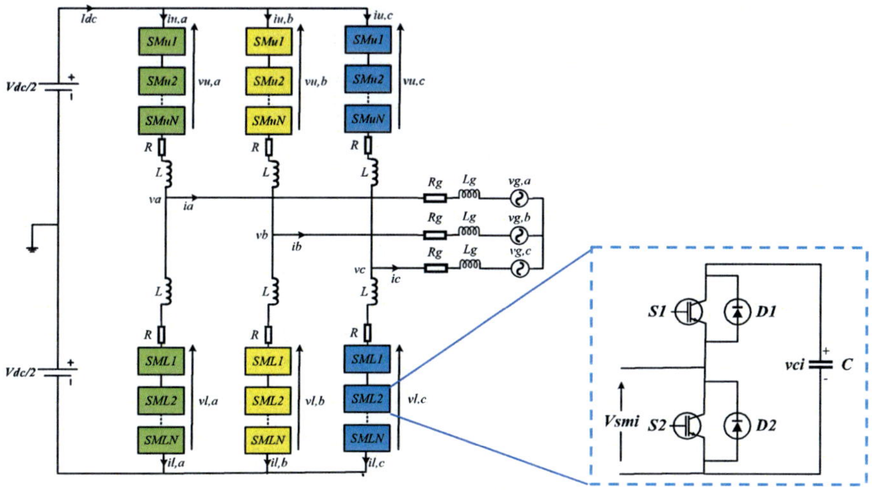

Fig. 1. General MMC three-phase structure and submodule-half bridge topology.

Half-bridge SM is the most used submodule topology. It consists of a capacitor C, two antiparallel diodes, and two IGBT switches (S1, S2), as depicted in Fig. 1. Table 1 summarizes the output voltage of each SM based on half-bridge topology. When IGBT1 is ON and IGBT2 is OFF, the SM is inserted. On the other hand, when IGBT1 is OFF and IGBT2 is ON, the SM is bypassed. The capacitor voltage is unchanged when the SM is bypassed. On the other hand, the SM capacitors charge and discharge with a positive arm current and a negative arm current, respectively.

Table 1. SM voltage for the Half-Bridge topology.

Submodule state	IGBT1	IGBT2	SM voltage
Inserted	ON	OFF	Vc
Bypassed	OFF	ON	0

From Fig. 1, the upper and lower arm voltages $v_{u,j}$ and $v_{l,j}$ can be expressed by:

$$v_{u,j} = \frac{Vdc}{2} - v_j - Ri_{u,j} - L\frac{di_{u,j}}{dt} \quad (1)$$

$$v_{l,j} = \frac{Vdc}{2} + v_j - Ri_{l,j} - L\frac{di_{l,j}}{dt} \quad (2)$$

where v_j represents the output voltage of phase j ($j = a, b, c$) and Vdc is the DC voltage. The corresponding arm currents are $i_{u,j}$ and $i_{l,j}$, which can be expressed:

$$i_{u,j} = i_{diff,j} + \frac{i_j}{2} \quad (3)$$

$$i_{l,j} = i_{diff,j} - \frac{i_j}{2} \quad (4)$$

where i_j is the phase current and $i_{diff,j}$ is the differential current, which flows through phase-leg is given as:

$$i_{diff,j} = \frac{i_{u,j} + i_{l,j}}{2} \quad (5)$$

The phase current i_j can be represented by arm currents as:

$$i_j = i_{u,j} - i_{l,j} \quad (6)$$

By subtracting Eq. (1) from Eq. (2), the output voltage v_j is yielded:

$$v_j = e_j - \frac{R}{2}i_j - \frac{L}{2}\frac{di_j}{dt} \quad (7)$$

where e_j is the electromotive force (emf), it can be written as:

$$e_j = \frac{v_{l,j} - v_{u,j}}{2} \quad (8)$$

From (7) end (8), It may be deduced that the output current i_j can be regulated by e_j. By adding Eqs. (1) and (2), the inner unbalance voltage $v_{diff,j}$ of phase j can be expressed as:

$$v_{diff,j} = \frac{V_{dc} - (v_{l,j} + v_{u,j})}{2} = Ri_{diff,j} + L\frac{di_{diff,j}}{dt} \quad (9)$$

According to (9) the voltage difference between the DC bus voltage and the summation of the upper $v_{u,j}$ and lower $v_{l,j}$ arm voltages gives rise to the inner current $i_{diff,j}$. The equivalent circuit of MMC considering inner current is represented in Fig. 2.

The only variable that can be used when controlling inner current is the arm voltage sum $v_{sum,j}$, which can be expressed:

$$v_{sum,j} = \frac{(v_{l,j} + v_{u,j})}{2} \tag{10}$$

3 Inner Differential Current Controllers

The output voltage and current are not affected by the circulation of $i_{diff,j}$ in each phase-leg. This current charges the floating capacitors in each arm with its DC component, which leads to the transfer of active power from the DC bus to the grid. However, AC components especially the second harmonic, with a negative sequence that increase the root mean square value of $i_{u,j}$ and $i_{l,j}$, conduction losses, and capacitor voltage ripples [20].

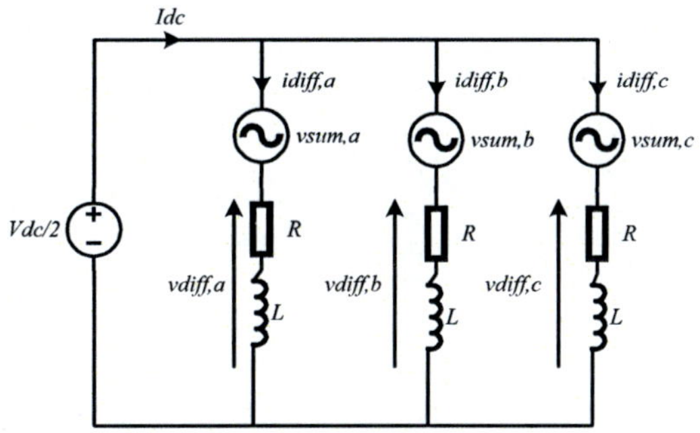

Fig. 2. Equivalent circuit of MMC for inner current

The expression for the inner current is as follows:

$$i_{diff,j} = \frac{I_{dc}}{3} + i_{AC,2f} \tag{11}$$

Idc represents DC current used to charge the capacitors. While $i_{AC,2f}$ is the second harmonic in the inner current.

The main control objective is to suppress the AC component, so that $i_{diff,j}$ becomes equal to $\frac{Idc}{3}$. This paper presents a comparison between two control methods proposed to suppress the inner current.

3.1 Method 1 Based on Dq Frame

The first inner current suppression method is based on a pair of PI regulators. The second-order harmonic of inner differential current is transformed to a DC component by using ABC to DQ transformation, as shown in Fig. 3. However, when the system is single-phase, this strategy is not practical.

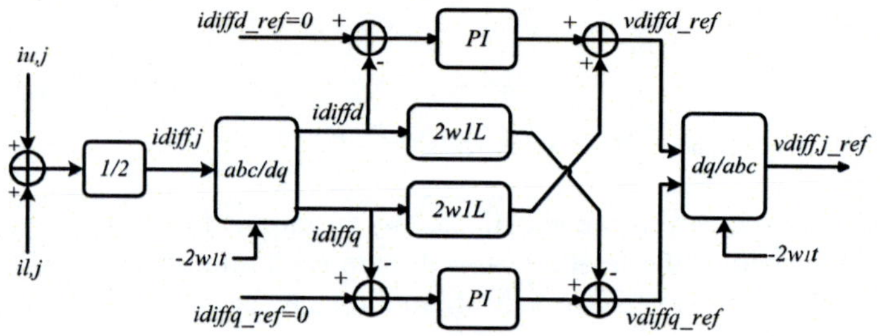

Fig. 3. Inner current controller based on dq frame

This method transforms the inner current from *abc* to *dq* coordinates using the Park transformation with twice the fundamental frequency. To eliminate the inner current, references i_{diffd_ref} and i_{diffq_ref} are set to zero.

3.2 Method 2 Based on a MAF

Unlike the first method, no transformation from *abc* to *dq* coordinates is needed; just a simple P controller and a high-pass filter (HPF) to extract the AC component of the inner current. A low-pass filter based on a moving average filter (MAF) technique can be used to implement the HPF. The control scheme of this method is shown in Fig. 4. It can be observed that the controller's reference is set to zero, and the DC component is not regulated because it will be set to the right value naturally.

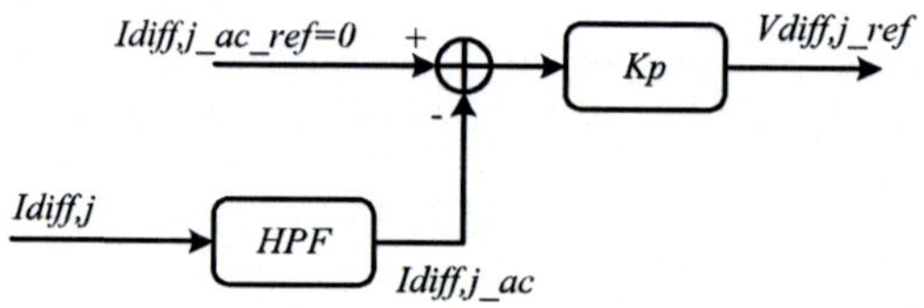

Fig. 4. Inner current controller based on P controller

The structure of HPF based on moving average filter (MAF) is presented in Fig. 5(a), and then the implantation of the MAF is shown in Fig. 5(b).

The frequency of MAF is equal to twice the grid frequency ($f_w = 2f_1$) to eliminate the second-order harmonic from the inner current.

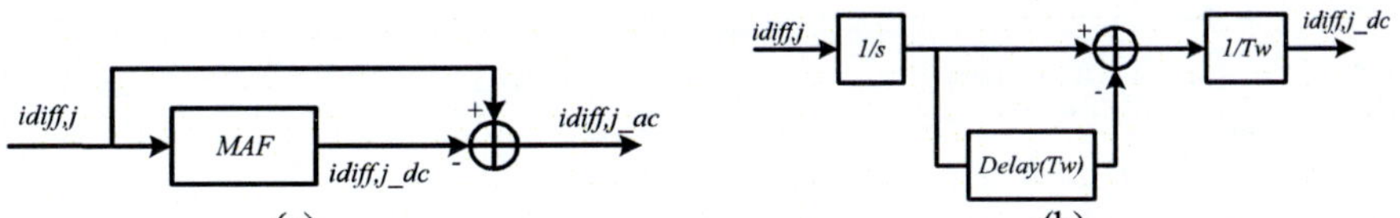

Fig. 5. (a) HPF based on MAF; (b) implementation of the MAF

Figure 6 shows the overall control scheme of the MMC. The outer power and output current controllers are conventional systems, which create the reference of inner electromotive force e_{j_ref}, described in (7) and (8). The inner current controller generates the reference of the inner unbalance voltage v_{diff,j_ref}.

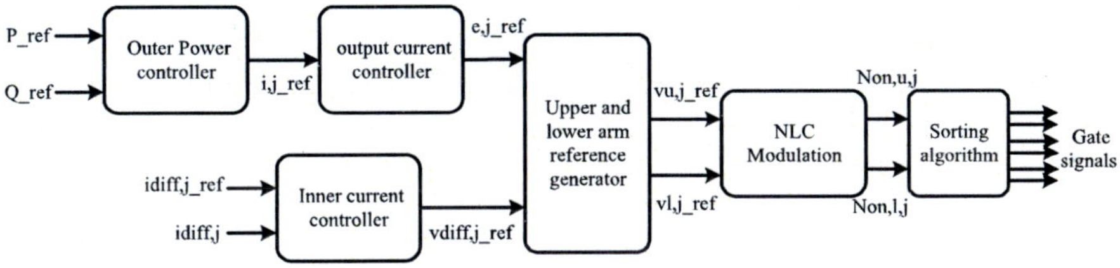

Fig. 6. Overall control of the MMC

The upper and lower arm voltage references v_{u,j_ref} and v_{l,j_ref}, which can be expressed:

$$v_{u,j\_ref} = \frac{Vdc}{2} - e_{j\_ref} - v_{diff,j\_ref} \tag{12}$$

$$v_{l,j\_ref} = \frac{Vdc}{2} + e_{j\_ref} - v_{diff,j\_ref} \tag{13}$$

Based on the arm voltage references, the modulation block calculates the number of SMs to be inserted in each arm (upper and lower for each phase). Then the sorting process generates the gate signals of all IGBTs and balances all capacitor voltages.

4 Simulation Results

A three-phase MMC connected to a grid with 20 SMs per arm is simulated using a PLECS environment to compare the proposed inner current suppression methods. Table 2 shows the simulation parameters.

The time for simulation is considered as 1 s, at t = 0.2 s the inner current controller is activated. Figure 7 shows inner current waveforms with and without inner current suppression controller (ICSC), before enabling the ICSC, idiff,a has a significant oscillation amplitude due to the presence of second-order harmonic.

However, when the controller is enabled at t = 0.2 s, the amplitude of the inner current is significantly reduced. To verify the ability of these controllers to suppress the second order-harmonic, Fig. 8 shows the harmonic spectrum of inner differential current.

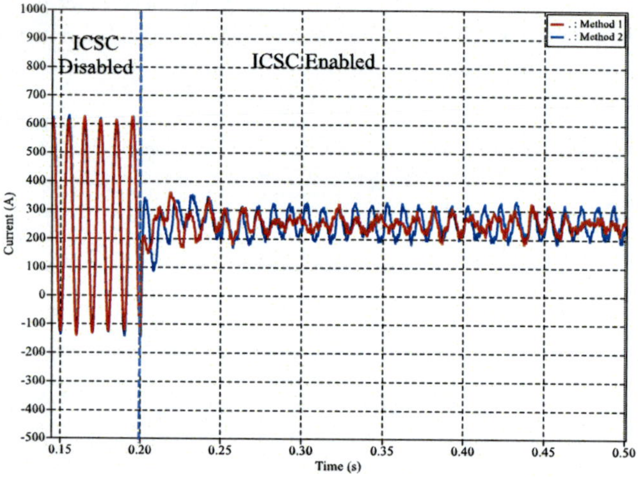

Fig. 7. Inner current idiff,a of phase a

Table 2. System parameters.

Parameter	Value
Power S	150 MVA
Grid voltage vg	100 kV
Grid inductance Lg	3.2 mH
Grid resistance Rg	0.1 Ω
Arm inductance L	51 mH
Arm resistance R	1.6 Ω
Number of SMs per arm N	20
DC bus voltage Vdc	200 kV
SM capacitance C	0.75 mF
Sampling frequency fs	15 kHz
Grid frequency f1	50 Hz

From Fig. 8(a), it can be seen that the inner current contains significant harmonics, especially the 2nd and 4th-order harmonics. As shown in Fig. 8(b), the magnitude of the second order-harmonic is mitigated from 498A to around 55.4A for method 2 and around 17.9A for method 1. In addition, the fourth harmonic is well reduced by 32% (12.5A to 8.5A) with method 2 and 71.2% (12.5A to 3.6A) with method 1.

From the arm current shown in Fig. 9, it can be seen that this current becomes less distorted after ICSC is enabled. The corresponding spectrum of arm current is depicted in Fig. 10. With the addition of a first harmonic, this current contains the same second and fourth-order harmonics as in the inner current spectrum shown in Fig. 8.

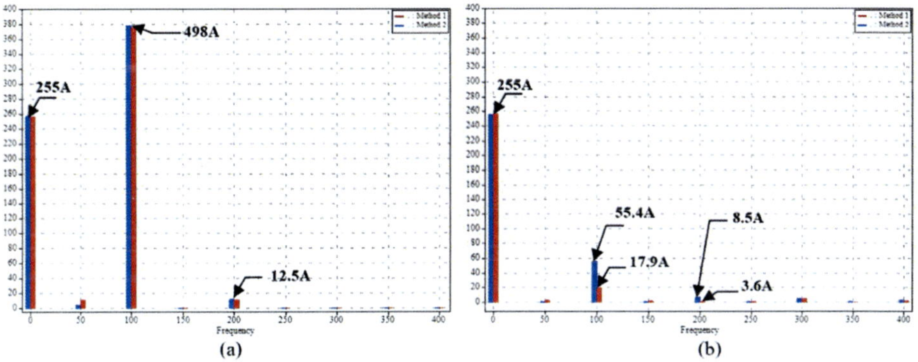

Fig. 8. Spectrum of inner current (a) without ICSC; (b) with ICSC

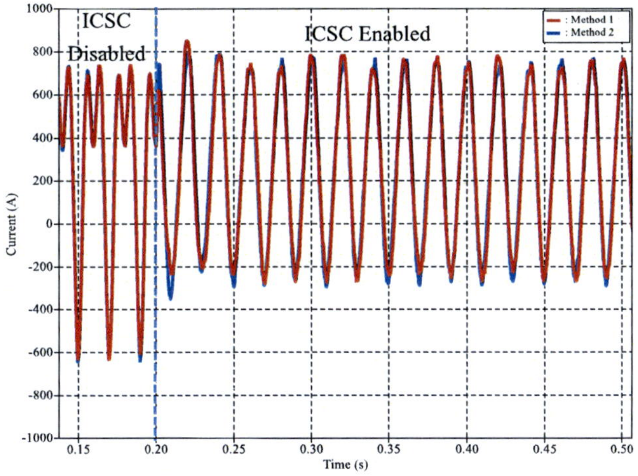

Fig. 9. Arm current of phase a

The first harmonic (f = 50 Hz) represents half of the output current its magnitude is not affected by the ICSC.

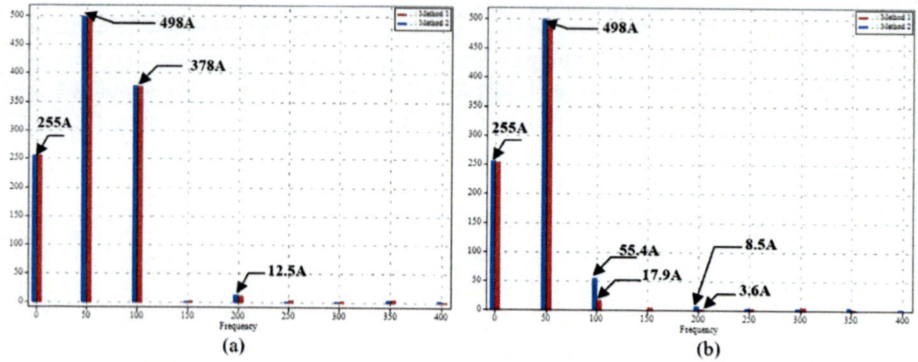

Fig. 10. Spectrum of arm current (a) without ICSC; (b) with ICSC

Figure 11 shows the impact of mitigating the inner current on the capacitor voltage ripples. It can be concluded that capacitor voltage peak-to-peak for method 1 is around 2946 V before t = 0.2 s and is reduced to 1472 V after the controller is enabled. Method 2 can reduce the capacitor voltage peak-to-peak to 1552 V, as shown in Fig. 11(a). The average capacitor voltages are regulated to 10 kV (Vdc/N). The SM capacitor voltages are well balanced due to the usage of the balancing algorithm.

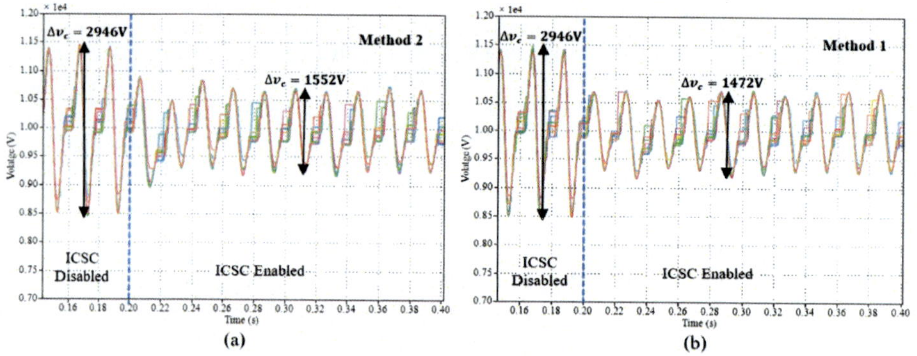

Fig. 11. SM capacitor voltages of upper arm (a) method 2; (b) method 1

5 Conclusion

This paper presents a comparison between two control methods proposed to suppress the inner differential current. The first method is based on a DQ synchronous reference frame with a pair of PI controllers and the second one uses a simple moving average filter (MAF) controller. PLECS simulation results for both controllers show that the first one has a superior capability to reduce the second and fourth-order harmonic over the MAF controller. Furthermore, both controllers result in lower capacitor voltage ripples and better arm current forms.

References

1. Wang, Y., Aksoz, A., Geury, T., Ozturk, S.B., Kivanc, O.C., Hegazy, O.: A review of modular multilevel converters for stationary applications. Appl Sci. **10**, 7719 (2020)
2. Marquardt, R.: Stromrichterschaltungen mit verteilten energiespeichern, German Patent DE20 122 923 U1, 2001
3. Teeuwsen, S.P.: Modeling the trans bay cable project as voltage-sourced converter with modular multilevel converter design. IEEE Power and Energy Society General Meeting, San Diego, USA (2011)
4. Hagiwara, M., Nishimura, K., Akagi, H.: A medium-voltage motor drive with a modular multilevel PWM Inverter. IEEE Trans Power Electron. **25**, 1786–1799 (2010)
5. Kumar, Y.S., Poddar, G.: Control of medium-voltage AC motor drive for wide speed range using modular multilevel converter. IEEE Trans. Ind. Electron. **64**, 2742–2749 (2017)
6. Saeedifard, M., Iravani, R.: Dynamic performance of a modular multilevel back-to-back HVDC system. IEEE Trans. Power Deliv. **25**, 2903–2912 (2010)
7. Sotoodeh, P., Miller, R.D.: Design and implementation of an 11-level inverter with FACTS capability for distributed energy systems. IEEE J. Emerg. Sel. Top. Power Electron. **2**, 87–96 (2014)
8. Khazaei, J., Pavlak, G., Lee, B., Elsenbaty, M.: A novel application of modular multi-level converters for partially shaded PV systems. In: 2019 IEEE Texas Power and Energy Conference, TPEC, TX, USA (2019)
9. Acharya, A.B., Ricco, M., Sera, D., Teoderscu, R., Norum, L.E.: Performance analysis of medium voltage grid integration of PV plant using modular multilevel converter. IEEE Trans. Energy Conv. 34, 1731–1740 (2019)
10. Rong, F., Gong, X., Huang, S.: A novel grid-connected PV system based on MMC to get the maximum power under partial shading conditions. IEEE Trans Power Electron. **32**, 4320–4333 (2017)
11. Song, Q., Liu, W., Li, X., Rao, H., Xu, S., Li, L.: A steady-state analysis method for a modular multilevel converter. IEEE Trans Power Electron. **28**, 3702–3713 (2013)
12. Li, S., Wang, X., Yao, Z., Li, T., Peng, Z.: Circulating current suppressing strategy for MMC-HVDC based on nonideal proportional resonant controllers under unbalanced grid conditions. IEEE Trans. Power Electron. **30**, 387–397 (2015)
13. Yang, S., Wang, P., Tang, Y., Zagrodnik, M., Hu, X., Tseng, K.J.: Circulating Current Suppression in Modular Multilevel Converters with Even-Harmonic Repetitive Control. IEEE Trans. Ind. Appl. **54**, 298–309 (2018)
14. Geng, S., Gan, Y., Li, Y., Hang, L., Li, G.: Novel circulating current suppression strategy for MMC based on quasi-PR controller. In: Applied Power Electronics Conference and Exposition, APEC. CA, USA (2016)
15. Tu, Q., Xu, Z., Xu, L.: Reduced switching-frequency modulation and circulating current suppression for modular multilevel converters. IEEE Trans. Power Deliv. **26**, 2009–2017 (2011)
16. Darus, R., Pou, J., Konstantinou, G., Ceballos, S., Agelidis, V.G.: Controllers for eliminating the ac components in the circulating current of modular multilevel converters. IET Power Electron. **9**, 1–8 (2016)
17. Samajdar, D., Bhattacharya, T., Dey, S.: A reduced switching frequency sorting algorithm for modular multilevel converter with circulating current suppression feature. IEEE Trans Power Electron. **34**, 10480–10491 (2019)
18. Antonio-Ferreira, A., Collados-Rodriguez, C., Gomis-Bellmunt, O.: Modulation techniques applied to medium voltage modular multilevel converters for renewable energy integration: a review. Electric Power Syst. Res. **155**, 21–39 (2018)

19. Darus, R., Pou, J., Konstantinou, G., Ceballos, S., Agelidis, V.G.: A modified voltage balancing sorting algorithm for the modular multilevel converter: Evaluation for staircase and phase-disposition PWM. In: IEEE Applied Power Electronics Conference and Exposition, APEC, TX, USA (2014)
20. Qiu, J., Hang, L., Liu, D., et al.: Novel method for circulating current suppression in MMCs based on multiple quasi-PR controller. J. Power Electron. **18**(6), 1659–1669 (2018)

Optimization of P&O Algorithm for Load Variation Scenario Through PV System Physical Model

Tariq Limouni[1,3](✉), Reda Yaagoubi[2], Khalid Bouziane[3], Khalid Guissi[1], and El Houssain Baali[1]

[1] Energy and Agro-Equipment Department, Institute of Agronomy and Veterinary Hassan II, 10112 Rabat, Morocco
t.limouni@iav.ac.ma

[2] School of Geomatics and Surveying Engineering, Hassan II Institute of Agriculture and Veterinary Medicine, 10112 Rabat, Morocco

[3] High School of Energy Engineering, College of Engineering and Architecture, LERMA, International University of Rabat, Sala el Jadida, 11000 Technopolis, Morocco

Abstract. To face global warming and fossils fuels depletion, the use of renewable energy is becoming mandatory to satisfy the energy demand that increases further each year. For instance, photovoltaic systems are widely used thanks to their high performance and small cost. The photovoltaic (PV) systems can be connected to the grid for night supply or standalone systems with storage devices like batteries. To study and analyze the power produced by PV systems in different weather conditions, several mathematical models have been used in the MATLAB environment. In this paper, the modeling of each element of the photovoltaic system has been done including a PV module, a Boost Converter, a load, and a MPPT to operate the PV system in high efficiency in different weather conditions and scenarios. Moreover, the maximum power point tracking (MPPT) was investigated in this work using the Perturb and Observe (P&O) approach. The newly optimized P&O has been compared to an optimized P&O in several cases. The results show that our P&O protocol is very close to the optimized P&O for the solar irradiance variation scenario but more suitable and efficient for the load variation scenario.

Keywords: Photovoltaic systems · MPPT · P&O algorithm · Modeling · MATLAB/Simulink

1 Introduction

Renewable energy like wind energy and solar energy has been largely deployed as clean energy sources to satisfy energy demand and reduce greenhouse emissions. The PV system that uses solar energy has been deployed recently for power supply thanks to its small operation and maintenance costs [1]. The issue with the PV system is that

S. Motahhir and B. Bossoufi (Eds.): ICDTA 2022, LNNS 454, pp. 603–612, 2022.
https://doi.org/10.1007/978-3-031-01942-5_60

current and voltage change with weather conditions [2].To investigate the behavior of the photovoltaic system in various conditions, several mathematic models have been developed and used to model PV cells. Various models have been used previously to assess the performance of PV systems. The single diode with a current source in parallel is the simplest equivalent circuit model of PV cells [3]. To increase the accuracy of this model, series resistance and a shunt resistance are added to the circuit [3]. A more advanced model named the Two diode model of PV cells has been developed consisting of two diodes instead of one [4]. To ensure the stability of the output voltage, the DC-DC converter was integrated with the PV system. Different DC-DC converter types exist including Buck converter and Boost converter [5]. DC-DC converter type depends on the required application. Furthermore, the Maximum Power Point Tracking (MPPT) algorithm is applied to track the point where the power is maximum of the PV system and increase its efficiency. Various algorithms have been proposed to track the MPP such as the Perturb and Observe, the constant voltage control method and the Incremental Conductance (IC) [6]. The P&O has been widely used thanks to its simplicity to track the MPP and has been optimized to be more efficient in different scenarios. Among these optimization works, there is [7] based on the modification of duty cycle only after 5 periods and increasing/decreasing of PV Power by a value higher than small positive value. The optimized P&O has been compared to the basic P&O and results show that the optimized one has a good efficiency compared to the conventional one with good response in a different operating scenario. Moreover, a proposed algorithm for MPPT was recently reported in [8] based on the limitation of the PV system voltage in an area where the maximum power point is inside. The results show that the proposed algorithm is converging faster with no oscillation compared to other MPPT methods like P&O with good efficiency.

In this work, the modeling of each element of the photovoltaic system has been done in MATLAB/Simulink to investigate the photovoltaic system operation under a different scenario. For PV cells, the mathematical model that has been deployed is the single diode model. Furthermore, the DC/DC converter used in this work is Boost Converter. The MPPT algorithm used is the P&O algorithm which has been optimized to increase its efficiency in a different operating scenario, and compared to an optimized P&O proposed by [7].

2 PV System Modeling

The photovoltaic system modeled in MATLAB/Simulink consists of a PV module as a power supply, a Boost converter to fit the power supply to the load, a PWM controller, a load, and an MPPT to get the maximum power point (Fig. 1). The specifications of the PV system used in this work are shown in Table 1.

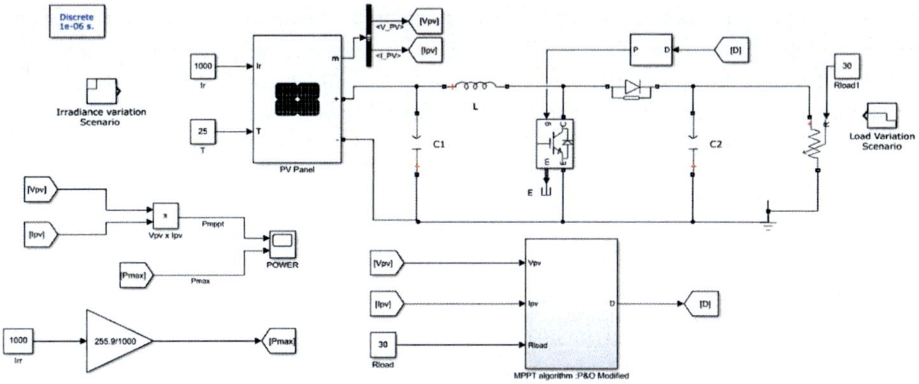

Fig. 1. Simulink model of PV system

Table 1. The PV system specifications

Sun model plus SW 255 Mono (STC)	Value
The maximum power P_{max}	255 W
The voltage at MPP V_{mpp}	31.4 V
The current at MPP I_{mpp}	8.15 A
The voltage at open circuit V_{oc}	37.8 V
The short circuit current I_{SC}	8.66 A
The configuration of cell	60 cells in series
The module dimensions	156 mm * 156 mm
The short circuit current temperature coefficient TCI_{SC}	0.04%/°C
PMW	*Value*
Switching frequency	1 kHz
Load	*Value*
Load resistance	30 Ω
Boost converter	*Value*
Capacitor	1000 μF
Inductor	13 mH

2.1 PV Module Modeling

The equivalent circuit of the PV cell is used to model the PV module in MATLAB/Simulink. As mentioned earlier, there are different equivalent circuits for the PV module. In our work, the PV cell single diode equivalent is employed as shown in Fig. 2. The single diode equivalent circuit is used to determine the current-voltage (I-V) characteristics under different irradiance and temperature conditions [1]. The circuit consists of two resistances R_S and R_{sh} and a single diode. R_s stands for the series resistance, while

R_{sh} is shunt resistance. The cross current associated with P-N junctions in semiconductor devices is represented by a diode [2, 5]. The other parameters indicated in Fig. 2 are the photocurrent I_{PH}, the diode current I_D, the current flowing across the shunt resistor I_{SH}, output current of PV cell I_{PV}, and output voltage of PV cell V_{PV}.

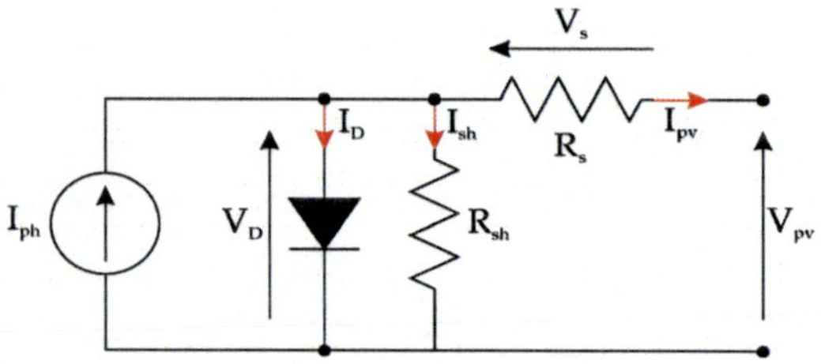

Fig. 2. The photovoltaic cell single diode equivalent circuit

By using Kirchhoff law in the PV cell equivalent circuit, the PV cell output current is represented by the following equation: [1]

$$I_{PV} = I_{PH} - I_D - I_{SH} \tag{1}$$

The photocurrent I_{PH} can be determined as follows [1]:

$$I_{PH} = \frac{I_r}{I_{r0}}\left[I_{sc} + K_I\left(T - T_{ref}\right)\right] \tag{2}$$

where I_r is solar irradiance (W/m^2), I_{r0} the solar radiation at standard test conditions STC (1000W/m^2), I_{sc} the short circuit current of a cell at STC (A), K_I the temperature coefficient short-circuit current, T temperature of a PV module in kelvin (K) and T_{ref} is the reference temperature of a PV module at STC (K).

The diode current I_D can be evaluated by using the equation below [1]:

$$I_D = I_s\left[\exp\left(\frac{q(V_{PV} + I_{PV}R_s)}{nKT}\right)\right] \tag{3}$$

where I_s is saturation current of the diode, q is charge of the electron (C) $(q = 1.6 \times 10^{-19}C)$, Boltzmann constant K (J/K) $(K = 1.38 * 10^{-23}J/K)$, n the diode ideality factor that depends on technology of the PV cell used.

The diode current saturation can be computed as follows [1]:

$$I_s = I_{rs}\left(\frac{T}{T_{ref}}\right)^3 \exp\left[\frac{qE_g}{nK}\left(\frac{1}{T_{ref}} - \frac{1}{T}\right)\right] \tag{4}$$

where *the* I_{rs} is reverse saturation current of the diode and E_g is band-gap energy of the semiconductor employed for P-N junction.

Reverse saturation current of the diode I_{rs} could be found as follows [1]:

$$I_{rs} = \frac{I_{sc}}{\exp\left[\frac{qV_{oc}}{N_sKnT}\right] - 1} \tag{5}$$

where *the* V_{oc} is open-circuit voltage of PV cell (V), and N_s is the number of cells (in series).

The current passing across the shunt resistance I_{SH} can be determined as follows [1]:

$$I_{SH} = \frac{V_{PV} + I_{PV} R_s}{R_{SH}} \tag{6}$$

2.2 DC-DC Converter Modeling

DC-DC converter ensure the stability of the output voltage and fit the power supply to the load. It transforms the DC input voltage to either higher or lower DC output voltage [9, 10]. The most used DC-DC converter for PV systems is the boost converter [11]. Its equivalent circuit is presented In Fig. 4, where I_{in} and V_{in} are the input current and voltage, I_{out} and V_{out} are output current and voltage. The circuit consists of a capacitor, a diode, an inductor, and a MOSFET transistor as a switch [12]. The relationship between output and input of Boost converter is [8, 13]:

$$V_{out} = V_{in} \frac{1}{1-d} \tag{7}$$

$$I_{out} = I_{in}(1-d) \tag{8}$$

where d is the duty cycle. By dividing Eq. (9) by Eq. (8), one can obtain [8]:

$$\frac{V_{out}}{I_{out}} = \frac{V_{in}}{I_{in}} \frac{1}{(1-d)^2} \tag{9}$$

Which is equivalent to [8]:

$$R_{out} = \frac{V_{PV}}{I_{PV}} \frac{1}{(1-d)^2} \tag{10}$$

where R_{out} is the load resistance R_{load}, V_{PV} and I_{PV} are the voltage and current in the output of PV. By extracting the duty cycle$d$, the equation can be expressed as follows [11]:

$$d = 1 - \sqrt{\frac{\frac{V_{PV}}{I_{PV}}}{R_{load}}} \tag{11}$$

To model a Boost converter in MATLAB/Simulink, the electronic components are used to build the Boost converter circuit. Figure 1 shows the converter circuit in MATLAB/Simulink by using an IGBT as a switch, controlled by a Pulse-width modulation (PWM) signal.

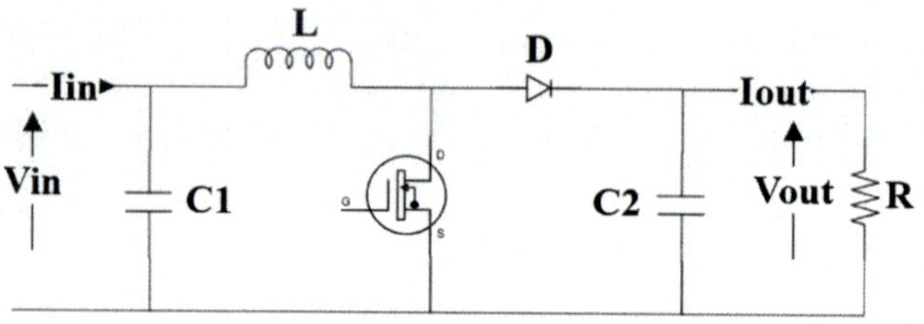

Fig. 3. The boost converter circuit.

2.3 MPPT Algorithm

As mentioned above, the PV output power Characteristics (P-V) are different due to irradiation and operating temperature variation. To maximize the efficiency of the PV module and the global system, the PV system must be operating at the MPP. For this reason, the MPPT algorithm is employed to provide the maximum power from the PV systems in different weather conditions [13]. The MPPT adjusts the Boost converter the duty cycle *d* to get an operation point where the power is maximum [14].

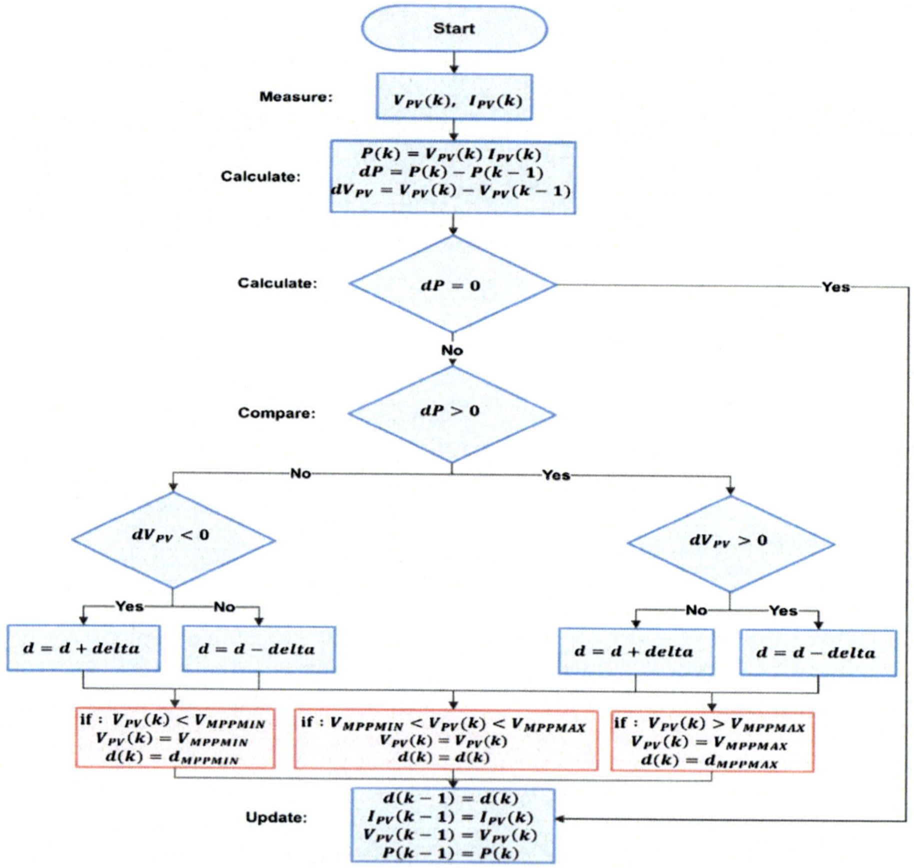

Fig. 4. P&O algorithm (Blue) and its optimization (Red)

Perturb and Observe Algorithm: P&O algorithm is the most used technique to track the point where the power is maximum due to its simple structure algorithm [3] (see Fig. 4). Initially, the voltage and current of the PV module are measured to calculate the PV power. The PV module voltage is then perturbed which can be translated to a decrease or increase in the PV power. If by increasing the voltage, the observed PV power increases, the operating power point is in the left (lower voltage values) of the MPP of the PV power curve, the PV voltage must therefore increase to converge to the maximum power point, [3]. Inversely, if the increase in the PV voltage decreases the PV power, the operating power point is on the right (higher voltage values) side of the P-V characteristic, and decreasing of PV voltage leads to the maximum power point [3]. The decreasing and increasing of the PV voltage can be translated by the variation of duty cycle$d$.

The Proposed P&O Algorithm: To optimize the P&O algorithm, the operating voltage is limited between a V_{MPPMIN} and a V_{MPPMAX} to get an operating power point as near as possible from the maximum power point for each weather conditions as shown in Fig. 5. If the operating voltage is inside of the area of MPP (between V_{MPPMIN} and a V_{MPPMAX}) the P&O algorithm will work normally. However, if the voltage is higher than V_{MPPMAX}, the voltage will be replicated by V_{MPPMAX} and duty cycle by d_{MPPMAX}. Inversely if the voltage is lower than V_{MPPMIN}, the operating voltage is replaced by V_{MPPMIN} and the duty cycle by d_{MPPMin}(see Fig. 4). It is worth mentioning that d_{MPPMAX} and d_{MPPMIN} can be calculated by using the Eq. (11) where V_{PV} is substituted either V_{MPPMAX} or V_{MPPMIN}.

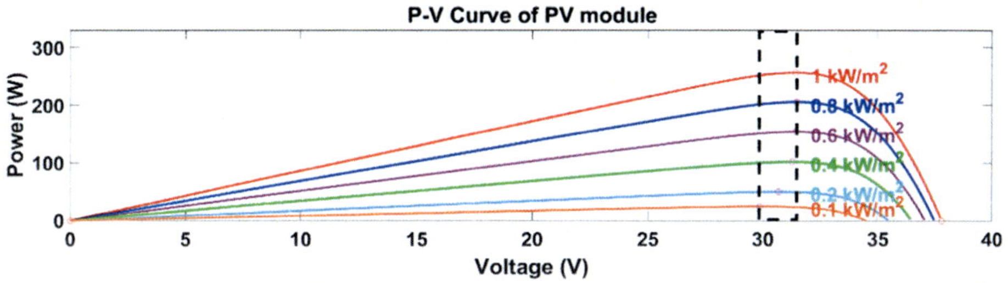

Fig. 5. Schematic representation of the investigated area encompassing the maximum power point.

3 Simulation Results

After modeling the PV system with all essential components, the simulation can be performed to study and analyze the proposed MPPT operation. to investigate its efficiency in different operating scenarios, our proposed P&O algorithm is compared with an optimized P&O algorithm used in [7] which based on P&O algorithm with some additions: The modification of duty cycle only after increasing/decreasing of PV Power by a value higher than a value (0.005 is the chosen value for this work).

The first scenario is based on varying the load and fixing the temperature at 25 °C and solar irradiance at 1000W/m^2. The load profile in this scenario is shown in Fig. 8. The second scenario is fixing the load (30Ω) and temperature (25 °C) and varying the solar irradiance as shown in Fig. 8.

For irradiance variation scenario, our proposed P&O algorithm has more oscillations than the other P&O algorithm as shown in Fig. 7, however, their root mean squared value of PV power given by their MPPT are so close and their efficiency is 98.18% for our proposed algorithm and 98.51% for the other approach as shown in Table 2. For the load variation scenario, Fig. 9 shows that our proposed P&O algorithm fits better the maximum power more than the optimized P&O. The root mean squared values of PV power given by our proposed P&O algorithm and the other algorithm are respectively 251.7 W and 209 W. The efficiency of our new P&O algorithm is found to be around 98.35% as compared to the other P&O method that has an efficiency not exceeding 81.67% (See Table 2).

To sum up, our new P&O algorithm is very efficient especially in load variation scenario with some oscillations as a drawback compared to the other P&O approach.

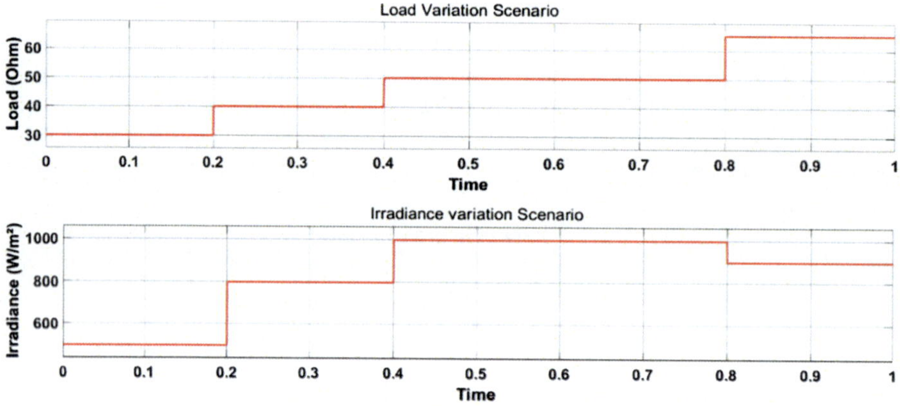

Fig. 6. Scenario profiles: Load variation scenario and irradiance variation scenario

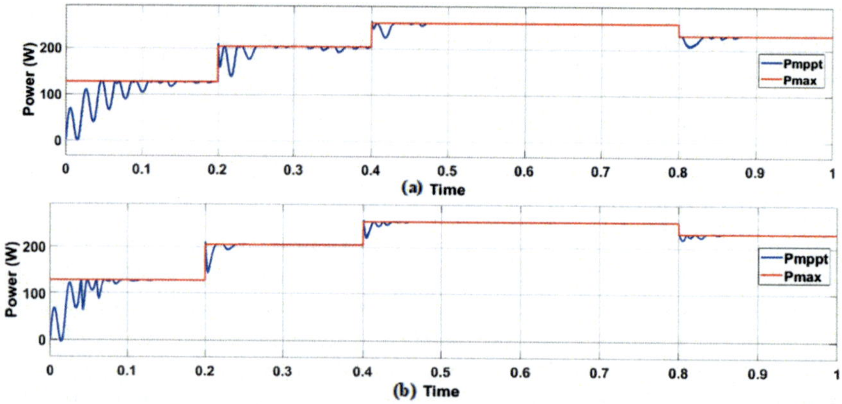

Fig. 7. Results of irradiation variation scenario for: (a) Proposed P&O algorithm, (b) optimized P&O algorithm

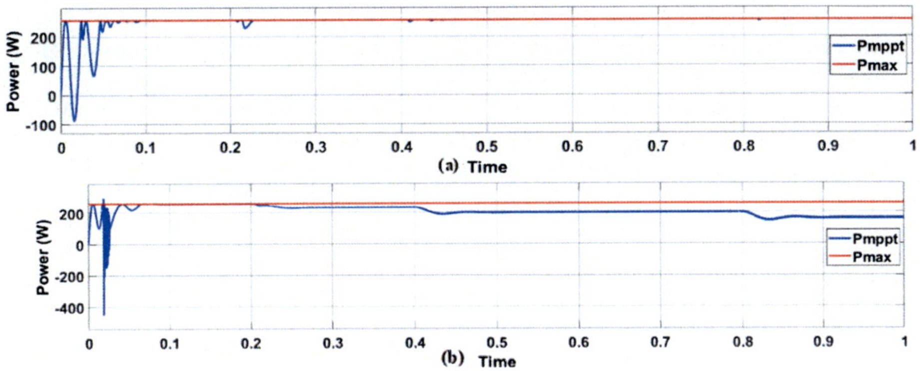

Fig. 8. Results of load variation scenario for: (a) Proposed P&O algorithm, (b) optimized P&O algorithm

Table 2. Efficiency and Root Mean Squared (RMS) of the Two P&O algorithms

	Irradiance variation scenario		Load variation scenario	
	Root Mean Squared of Pmppt	Efficiency	Root Mean Squared of Pmppt	Efficiency
Optimized P&O	217 W	98.59%	209 W	81.67%
Proposed P&O	216.1 W	98.18%	251.7 W	98.35%

4 Conclusion

In this work, the single diode model circuit has been used to model the PV cell in MATLAB/Simulink environment to extract the P-V and I-V characteristics under different environmental conditions. To track the maximum PV power under different operating conditions, the P&O algorithm has been used and optimized to be appropriate and suitable for different scenarios. The new P&O shows promising results in the load variation scenario with an efficiency of around 98.35% as compared to 81.67% obtained using of an optimized P&O algorithm. The two methods provide however similar results when varying the solar irradiation. For future works, the new P&O algorithm will be further optimized to get low oscillations in the output power, and it will be compared to different MPPT algorithms in different operation scenario.

References

1. Said, S., Massoud, A., Benammar, M., Ahmed, S.: A Matlab/Simulink-Based Photovoltaic Array Model Employing SimPowerSystems Toolbox. 12 (2012)
2. Yilmaz, U., Kircay, A., Borekci, S.: PV system fuzzy logic MPPT method and PI control as a charge controller. Renew. Sustain. Energy Rev. **81**, 994–1001 (2018)
3. Karim, S., Machkour, N., Zegrari, M., Aitelmahjoub, A., Sabiri, Z.: Modeling photovoltaic system using MATLAB/Simulink. SSRN J. (2018)

4. Jazayeri, M., Uysal, S., Jazayeri, K.: A simple MATLAB/Simulink simulation for PV modules based on one-diode model. In: 2013 High Capacity Optical Networks and Emerging/Enabling Technologies, pp. 44–50. IEEE, Magosa, Cyprus (2013)
5. Noman, A.M., Addoweesh, K.E., Mashaly, H.M.: A fuzzy logic control method for MPPT of PV systems. In: IECON 2012 - 38th Annual Conference on IEEE Industrial Electronics Society, pp. 874–880. IEEE, Montreal, QC, Canada (2012)
6. Mohanty, P., Bhuvaneswari, G., Balasubramanian, R., Dhaliwal, N.K.: MATLAB based modeling to study the performance of different MPPT techniques used for solar PV system under various operating conditions. Renew. Sustain. Energy Rev. **38**, 581–593 (2014)
7. Motahhir, S., El Ghzizal, A., Sebti, S., Derouich, A.: Proposal and implementation of a novel perturb and observe algorithm using embedded software. In: 2015 3rd International Renewable and Sustainable Energy Conference (IRSEC), pp. 1–5. IEEE, Marrakech (2015)
8. Chellakhi, A., El Beid, S., Abouelmahjoub, Y.: Implementation of a novel MPPT tactic for PV system applications on MATLAB/simulink and proteus-based arduino board environments. Int. J. Photoenergy, 1–19 (2021)
9. Boumaaraf, H., Talha, A., Bouhali, O.: Modelling and control of grid-connected photovoltaic systems. 14.3, 9 (2014)
10. Atallah, A.M., Abdelaziz, A.Y., Jumaah, R.S.: Implementation of perturb and observe MPPT of PV system with direct control method using buck and Buck- Boost converters. **1**, 15 (2014)
11. Dawidziuk, J.: Review and comparison of high efficiency high power boost DC/DC converters for photovoltaic applications. Bull. Polish Acad. Sci. Tech. Sci. **59**, 499–506 (2011)
12. Attia, H.A., Gonzalo, F.D.: Stand-alone PV system with MPPT function based on fuzzy logic control for remote building applications. IJPEDS **10**, 842 (2019)
13. Soori, P.K.: Modelling of a grid connected solar PV system using MATLAB/simulink. Int. J. Simul. Syst. Sci. Technol. (2016)
14. Abdul-Razzaq, I.K., Sakr, M.M.F., Rashid, Y.G.: Comparison of PV panels MPPT techniques applied to solar water pumping system. **12**, 11 (2021)

Effect of Number of PCM Tube Layers on the Dynamic and Thermal Behavior of a Cabinet Dryer

Dounia Chaatouf(✉), Mourad Salhi, Benyounes Raillani, Samir Amraqui, and Ahmed Mezrhab

Laboratory of Mechanics and Energetics, Faculty of Sciences, Mohammed First University, 60000 Oujda, Morocco
douniaextra1@gmail.com

Abstract. One of the biggest inherent problems with solar drying is the dependence on solar radiation. When the sun sets, the source of energy used for drying disappears, which increases the time required for drying as well as affects the quality of the products. To overcome this problem, a storage unit was introduced in an indirect cabinet dryer in the shape of copper tubes stuffed with gallium, a phase change material (PCM) that can store energy in the sun time and release it after sunset. This work contributes to the enhancement of the drying process by applying the CFD method, studying the effects of these tubes and the effect of multiple layers of them, as well as their inclination. To make this study more realistic, we used weather data from the city of Oujda (Eastern Morocco). The findings showed that as more layers of PCM tubes are used, more heat is stored at the time of sunshine, which significantly reduces the temperature of the chamber at day time, but when it comes to releasing this energy, the 2 layers showed the best performance. They release the energy in the same way as 3 layers, while for one layer, it barely releases anything. Regarding the inclination of these tubes, there is not much difference between vertical and inclined tubes, where the latter outperforms the vertical ones by about 1.6° at sun time.

Keywords: *CFD* · PCM tubes · Indirect solar dryer

1 Introduction

The sun-drying technique is one of the most commonly adopted methods of food preservation that utilizes solar energy for the drying process. Additionally, there are some disadvantages to this method, such as dependence on solar radiation. When the sun sets, the source of energy used for drying disappears, which increases the time required for drying as well as affects the quality of the products. In some cases, the products can rot before they dry. Thus, an energy storage unit is required to enhance the drying process. There has been a lot of research done on this topic; one example can be found in the literature is the work of El-Sebaii et al. [1].They studied a solar indirect dryer equipped with a Paraffin Heat Storage Unit as PCM to dry Thymus leaves. They found that with

S. Motahhir and B. Bossoufi (Eds.): ICDTA 2022, LNNS 454, pp. 613–620, 2022.
https://doi.org/10.1007/978-3-031-01942-5_61

the use of PCM, the drying duration was reduced by 50% in comparison with the system with no PCM, due to the continuity of the drying process after sunset. Ramirez et al. [2] investigate the performance of an indirect solar dryer with a PCM-based storage unit. The findings show that utilizing the PCM in the indirect solar dryer has decreased the drying time necessary for drying the products and, at the same time, increased the thermal performance from 24% to 28% of the system.

The CFD method is a widely used method for predicting thermal and dynamic behavior. Many CFD researchers who have worked on solar-powered systems together with PCM storage units have proven how the CFD model provides a strong correlation with the experimental results, thus an accurate result. (Allouche et al. [3] and Al-Abidi et al. [4]). In the majority of engineering processes, CFD is used to optimize the storage system before designing it in real life, in terms of thermal efficiency, dimensional design, the amount of PCM, and its position (Dutil et al. [5]). Therefore, in this study, we will focus on optimizing the storage system in terms of the number of layers of PCM tubes and their tilt to increase the efficiency of the solar dryer.

2 Mathematical Formulation and Boundary Conditions

The selected configuration consists of a drying chamber and a 1.5 m^2 solar air heater tilted at 34° (latitude of Oujda). The chamber is designed in two parts; the first one is where the products (fig fruits) are placed on trays, and the other one is a chamber mounted on the side which contains the PCM tubes as shown in Fig. 1.

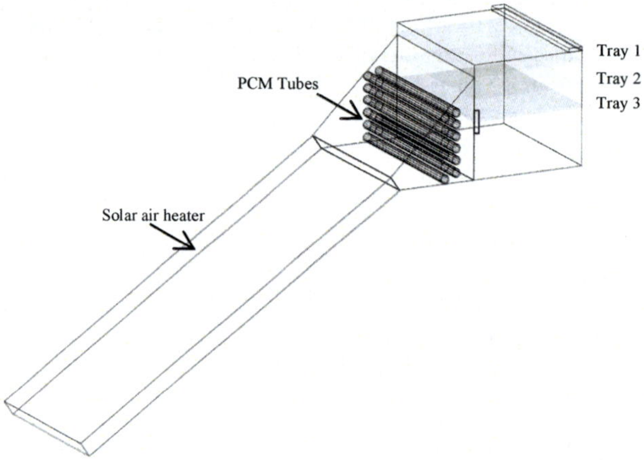

Fig. 1. The design scheme of the investigated solar dryer.

The Navier-Stokes equations are the basis of every fluid flow phenomenon. The air movements are ordinarily turbulent, to model it inside the dryer, we have chosen the K-eps standard model, which is defined by two transfer equations, one for the energy of kinetic turbulent k (Eq. 1) and the other for the dissipation rate (Eq. 2).

$$\rho(\frac{\partial k}{\partial t} + \frac{\partial (ku)}{\partial x} + \frac{\partial (kv)}{\partial y}) = (\mu + \frac{\mu_t}{\sigma_k})(\frac{\partial^2 k}{\partial x^2} + \frac{\partial^2 k}{\partial y^2}) + G_k - \rho\varepsilon \quad (1)$$

$$\rho(\frac{\partial \varepsilon}{\partial t} + \frac{\partial(\varepsilon u)}{\partial x} + \frac{\partial(\varepsilon v)}{\partial y}) = (\mu + \frac{\mu_t}{\sigma_\varepsilon})(\frac{\partial^2 \varepsilon}{\partial x^2} + \frac{\partial^2 \varepsilon}{\partial y^2}) + \frac{\varepsilon}{k}(c_1 G_k - c_2 \rho \varepsilon) \quad (2)$$

The equation of energy utilized to model the PCM is as follows:

$$\left[\frac{\partial h}{\partial t} + \frac{\partial H}{\partial t}\right] + \nabla.\left(\vec{V}h\right) - \nabla.\left(\frac{k}{\rho c_P}\nabla h\right) = S_h \quad (3)$$

Concerning the boundary conditions, the inlet velocity is fixed at 0.1 m/s with a variable ambient temperature. For the heat losses in the cabinet dryer walls, the heat coefficient is determined as follows:

$$h_0 = 2.8 + 3V_w \quad (4)$$

Vw is the wind velocity.

To model the trays and their resistance to airflow, the products in the trays are modeled as a porous medium with 50% porosity.

2.1 Climate Data

The weather data presented in this study are obtained from a metrological station located at Mohamed I University in Oujda. The high precision station gives us data such as wind speed and direction, pressure, humidity, and temperature, as well as irradiation values. The equations which were utilized in the UDF (User Defined Function) subroutine for solar radiation and ambient temperature are as described below:

$$T_{amb}(t) = 25 + 6\cos(\frac{\pi}{12}(t - 14)) \quad (5)$$

$$G_{sun}(t) = 965\sin(\frac{t-6}{14}\pi), \qquad 6 < t < 20 \quad (6)$$

2.2 Validation Results

The validation step is of great importance when it comes to the simulation, as it ensures the accuracy of the results. To do so, we have validated the melting process along the time of a pure Gallium in a rectangular box, which has been studied experimentally and numerically by Gau et al. [6] and Brent et al. [7] respectively. The progression of the melting front has been matched to these two studies, shown in Fig. 2, where it is evident that CFD gives a better accuracy close to the experimental results than numerical ones.

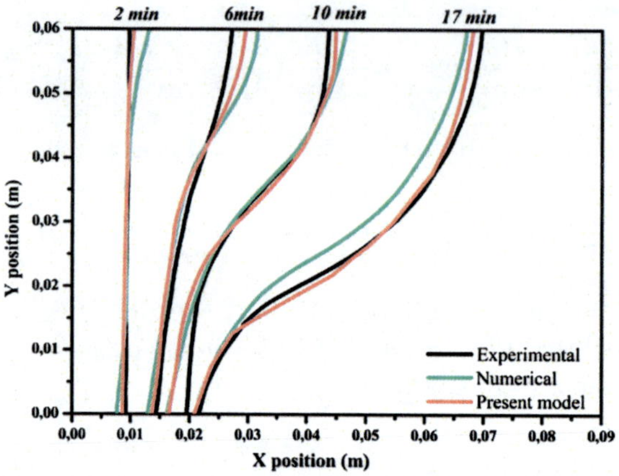

Fig. 2. The melting front progressions inside the cavity of pure gallium, experimentally [6], Numerically [7], and by using the present model.

3 Results

3.1 Effect of Number of PCM Tube Layers

The mesh independence test was conducted with five different grids and a cell number corresponding to 13068 cells was found to be enough to study geometry.

Gallium PCM tubes are placed at the entrance of the drying chamber to improve heat transfer and further absorb energy from the hot air. It is, therefore, necessary to know the amount of PCM needed, or, in another way, the number of layers that the air must pass through. To do so, we have presented the streamlines in the figure below, without tubes and with a different number of tube layers. Where the properties of both gallium and air are presented in Table 1. The first thing that can be noticed is that the adding tubes enhanced the airflow field by removing that big vortex created in the case without storage. And the more we add tubes, the better the distribution is, but 2 and 3 layers seem to have a similar effect. The tubes enhance the airflow distribution and make it more uniform inside the drying chamber, unlike in the case without PCM tubes (Fig. 3).

Table 1. Proprieties of Gallium and Air.

Parameters	Gallium	Air
Density (kg/m^3)	6093	1.177
Specific heat (J.kg$^{-1.}$ K^{-1})	381,5	1005
Thermal Conductivity (W. m^{-1}.K^{-1})	32	0.02462
Viscosity (kg. m.$^{-1}$s^{-1})	1.81×10^{-3}	$1.983.10^{-5}$
Melting temperature (K)	302,78	/
Latent heat (J. Kg^{-1})	80,160	/

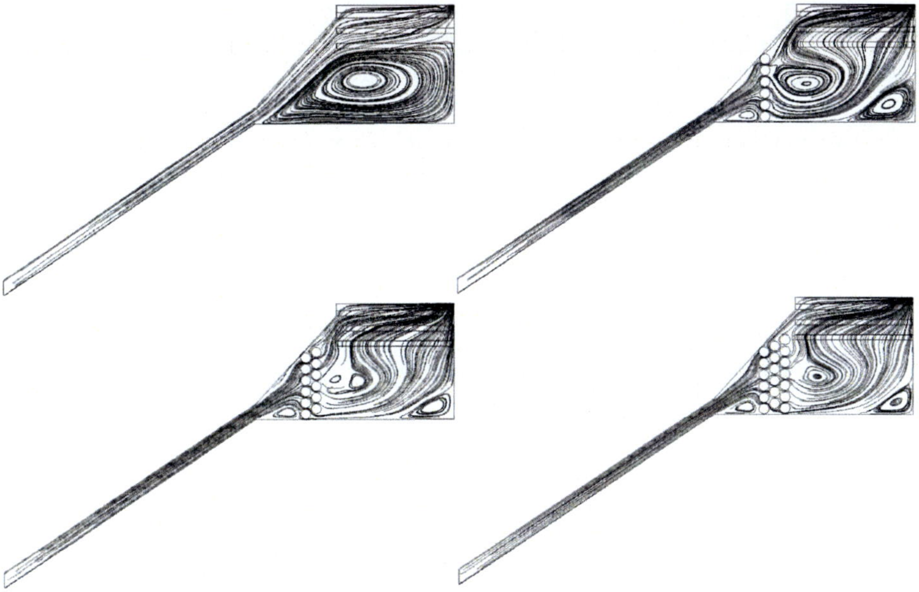

Fig. 3. Streamlines of the solar dryer for different PCM tube layers.

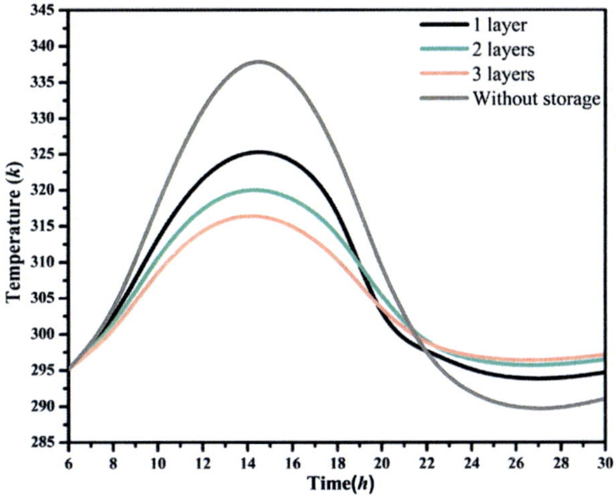

Fig. 4. Average temperature within the drying chamber throughout the day for various layers of PCM tubes.

Regarding the thermal effect of the tubes, Figs. 4, 5, and 6 show that PCM tubes absorb energy during sunny hours and release this energy during non-sunny hours. The number of layers of PCM tubes is very important because the amount of energy stored and released depends on these layers. The more layers of PCM tubes, the more energy is stored and discharged. However, in three layers of PCM tubes, the amount of stored energy is high, while the amount of released energy is the same as in two layers. This is actually due to the third layer, which has not been melted yet, as shown in Figs. 5 and 6. The hot air released by the first and second layers is stored by the third. It is clear from Fig. 6 that one layer of PCM tube does not release a large amount of energy compared to the two and three layers, where the amount of heat is high due to the small quantity of PCM.

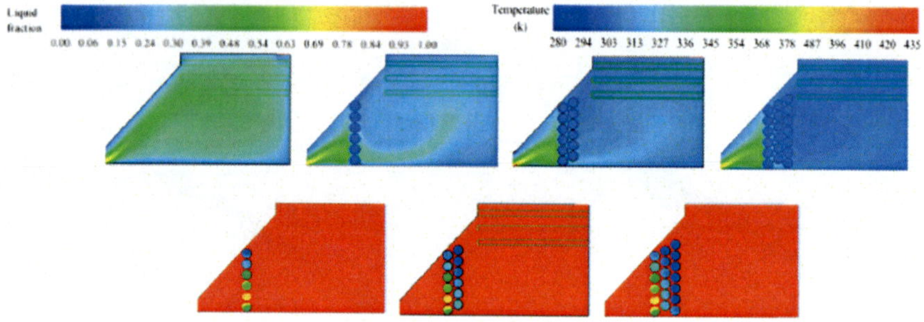

Fig. 5. Distribution of temperature and liquid fraction in the drying chamber at 2. am for different layers of PCM tubes

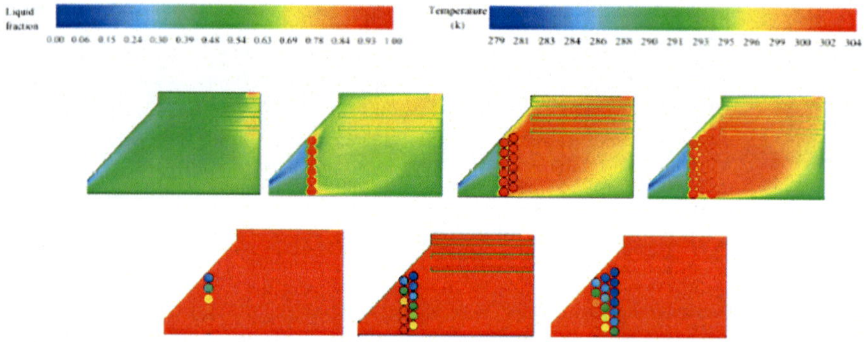

Fig. 6. Temperature and liquid fraction distribution within the drying chamber at midnight for various PCM tube layers

3.2 Effect of Inclination of PCM Tubes

In this section, we tried to change the position of the PCM tubes. The two-case studies are either vertical or inclined, as shown in the figure below, where we plotted the velocity streamlines. From this figure, it can be seen that although the vertical tube enhanced dramatically the airflow field compared to the case without storage by removing the vortex. As shown in the section above, there are still some small vortexes, and by making those tubes inclined, those small vortices disappear and the distribution becomes more uniform along the trays (Fig. 7).

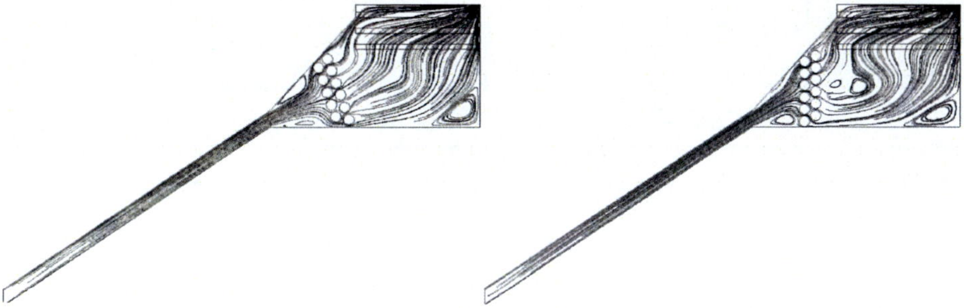

Fig. 7. Streamlines of the solar dryer for vertical and inclined tubes

Regarding the thermal behavior, we have plotted in Fig. 8 the mean temperature for the three trays in both cases throughout the day, and it seems that there is a small difference during sunlight hours, where the inclined tube showed better performance by outperforming the vertical tubes with a temperature difference reaching 1.6°. while in the non-sunny hours there was not much difference between the two cases, where the vertical tubes outperform the inclined tubes with a slight difference reaching 0.85°. Therefore, the amount of PCM has a much greater impact than its inclination (Fig. 9).

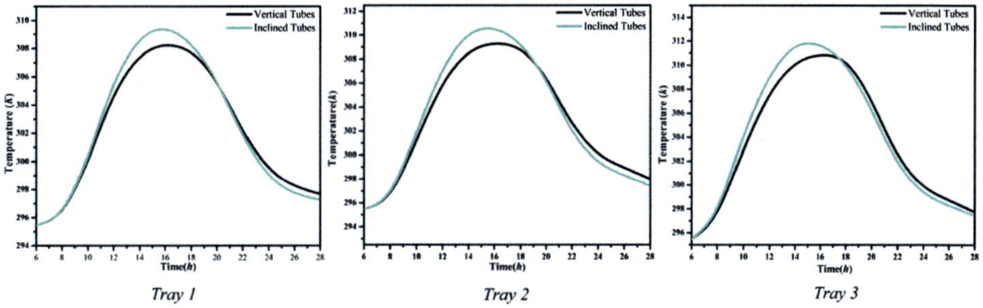

Fig. 8. The average temperature of the 3 trays along the day in case of vertical and inclined tubes

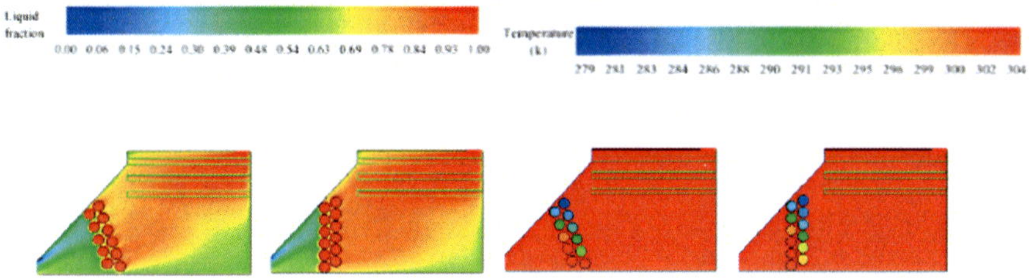

Fig. 9. Air temperature and liquid fraction distribution in case of Vertical and inclined PCM tubes at 10 pm

4 Conclusion

In the current work, a study on the impact of the number of PCM tube layers, as well as their inclination on the repartition of the air velocity and temperature inside the chamber, was presented, leading to the finding of a high temperature and a uniform velocity along the trays. The result showed that the two layers of PCM tubes stored and released a large amount of heat that reached. The addition of tubes improved the airflow field by removing the large vortex created in the case without storage, but by making the tubes inclined, the airflow distribution became more uniform by removing the small vortices. In addition, the inclination of the tubes improves the temperature along the trays by about 1.6 °C compared to the vertical case. Therefore, the amount of PCM has a much greater impact than its inclination.

References

1. El-Sebaii, A.A., Shalaby, S.M.: Experimental investigation of drying Thymus cut leaves in indirect solar dryer with phase change material. J. Solar Energy Eng. **139**(6) (2017). https://doi.org/10.1115/1.4037816
2. Ramirez, C., Palacio, M., Carmona, M.: Reduced model and comparative analysis of the thermal performance of indirect solar dryer with and without PCM. Energies **13**(20), 5508 (2020). https://doi.org/10.3390/en13205508
3. Allouche, Y., Varga, S., Bouden, C., Oliveira, A.C.: Validation of a CFD model for the simulation of heat transfer in a tubes-in-tank PCM storage unit. Renewable Energy **89**, 371–379 (2016)
4. Ambekar, A.S., Sivakumar, R., Anantharaman, N., Vivekenandan, M.: CFD simulation study of shell and tube heat exchangers with different baffle segment configurations. Appl. Therm. Eng. **108**, 999–1007 (2016)
5. Dutil, Y., Rousse, D.R., Salah, N.B., Lassue, S., Zalewski, L.: A review on phase-change materials: mathematical modeling and simulations. Renew. Sustain. Energy Rev. **15**(1), 112–130 (2011)
6. Gau, C., Viskanta, R.: Melting and solidification of a pure metal on a vertical wall. ASME. J. Heat Transfer **108**(1), 174–181 (1986). https://doi.org/10.1115/1.3246884
7. Brent, A.D., Voller, V.R., Reid, K.T.J.: Enthalpy-porosity technique for modeling convection-diffusion phase change: application to the melting of a pure metal. Numerical Heat Transfer, Part A Appl. **13**(3), 297–318 (1988)

Comparison of the Cleaning Periods Effect on the CSP Plants Under Summer Realistic Conditions

Benyounes Raillani(✉), Dounia Chaatouf, Samir Amraqui, and Ahmed Mezrhab

Laboratory of Mechanics and Energetics, Faculty of Sciences, Mohammed First University, 60000 Oujda, Morocco
benyounesraillani2@gmail.com

Abstract. In this work, the optical loss that occurs specifically due to the soiling effect will be emphasized. Specifically, we will study the optical performance of CSP technologies under the climatic conditions of the city of Oujda (34.64968°N; –1.90°E). A sample mirror with a size of 150 × 150 × 3 mm^3 used in CSP plants was installed in the climatic conditions of the city of Oujda to examine its optical efficiency, the daily variation of reflectivity is measured in two different cases, with regular cleaning (once a week) and without cleaning. The results showed that the energy production of the plant without cleaning can be reduced with a maximum rate of 80% during the summer period due to the high amount of aerosol particles, while for the case of regular cleaning, it showed that the losses are not as important as without cleaning, in addition, it can keep the rate of losses constant and at reasonable levels.

Keywords: CSP · Mirror · Soiling effect · Eastern Morocco

1 Introduction

Solar energy is a global and renewable source of energy, and the efficient use of this energy contributes to the stability of the world's need for energy, since it represents a promising alternative to the current problems related to the production of energy from conventional sources such as oil and gas, As the energy demand continues to increase every year, it is necessary to conduct extensive research in this area for development. Today, the systems used to convert sunlight into energy are diverse, and having highly efficient conversion systems could lead society to abandon or minimize their dependence on fossil fuel energy due to its many disadvantages, the most well-known of which is the release of greenhouse gases and pollutants into the global atmosphere, thus contributing to climate change.

The growth in energy demand has led to intensive studies and research in areas such as solar energy. As knowing, the photovoltaic and CSP power plants are the best-known systems in terms of conversion of this type of energy, therefore the studies focus on their development. The conversion process goes through many steps in each of these systems; the first is the collection of the energy received, it is in this step, that these technologies

S. Motahhir and B. Bossoufi (Eds.): ICDTA 2022, LNNS 454, pp. 621–629, 2022.
https://doi.org/10.1007/978-3-031-01942-5_62

have suffered from the weather condition that affects their optical efficiency. The main known effect is called soiling or the accumulation of dust on the mirror surface and is directly dependent on local conditions, defined by the number of aerosol particles suspended in the atmosphere and other meteorological parameters such as particle size, particle distribution, wind speed, and direction [1, 2] on which the transport of particles depends. Besides, it has been shown that the resulting adhesion is correlated with relative humidity (RH) as well [3].

The main difference between the effect of soiling on CSP and photovoltaic technologies is that, for the PV, the solar irradiation passes once through the soiling mask, whereas in the case of CSP, solar irradiation passes twice through the soiling layer, which can lead to greater diffusion and thus less effective solar irradiation. This difference in the soiling effect is identified in a CSP 5 to 10 times higher than that of photovoltaics [4]. As a result, many research papers contribute to the knowledge growth of the soiling phenomenon in mirror reflectance loss under specific outdoor conditions [5, 6]. The evaluation of the CSP mirrors in diverse locations has been investigated in Morocco with arid, semi-arid, and Saharan environments [7], in order to assess the risk of soiling and damage on CSP mirrors, such as the particle distribution, morphology, and dust composition.

The glass mirrors are not the only reflectors used in solar power plants, there are reflectors made of other materials such as aluminum with high reflectivity value that can play the same role [8, 9], each type should be investigated outdoor for more information about the antisoiling rate of the material [10, 11], The most commonly used mirrors in commercial CSP plants are made of glass, especially the so-called laminated or monolithic mirrors. This type has a high reflectivity value of about 96% and a life expectancy of up to 25 years, but they are still not very persuasive due to their high weight, which requires rigid structures [12].

The first step in understanding the effect of this problem is to study the variation in soiling for a mirror sample. To this end, an outdoor experiment was conducted to examine reflectivity during the summer period, when the concentration of aerosol particles is very high. Measurements of energy losses due to soiling are diverse, although this remains difficult to define due to the diversity of weather conditions, which require more and more investigations under different climatic conditions.

Following our foreword, the paper has been organized as follows: Sect. 2 presents the methodology and the steps of the experiment, as well as the presentation of the meteorological data; Sect. 3 presents the simulation results and the comparison of three different cases, in the last section, a conclusion summarizes the obtained results.

2 Methodology

Given the current situation, the optimal time required for cleaning is a major concern that we need to address, in order to research the optimal cleaning period and accurately identify the effect on the performance of the entire CSP system. The study of the optical performance of a mirror sample was conducted over two years. The period chosen for this work was the summer period due to the high amount of airborne aerosols according to [13]. The investigation method is applied to two different cases. First, the CSP installation scenarios consider the mirrors with a certain degree of soiling without any cleaning process, and second, with a cleaning period of one week. the degree of soiling is measured regularly for both scenarios. this section presents the details of this experiment as well as the simulation.

2.1 The Site of the Study

The location of exposition is situated in the eastern part of Morocco. Exactly the city of Oujda ((34.64968°N; –1.90°E) with an altitude of h = 617 m, at the Faculty of Letters and Human Sciences. This site was chosen because it is characterized by a high value of direct irradiation, which makes it suitable for the installation of CSP plants (Fig. 1).

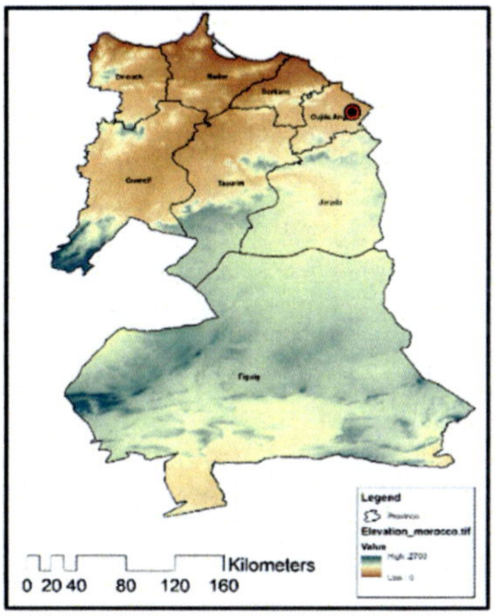

Fig. 1. Location of Oujda city on the map.

2.2 Measurement's Setup

The weather station equipment used for the measurements in this research is presented in Fig. 2. First, the mirror sample used is a FLABEG solar mirror with a dimension of 150 × 150 × 3 mm^3 and high initial reflectivity of 93%. The reflectivity variation was measured based on two pyrheliometers, the first pyrheliometer being mounted to be direct with the sun to measure direct solar radiation, while the irradiance reflected from the mirror was measured with the second pyrheliometer.

Other instruments are installed on the station to measure quantities such as pressure, relative humidity, wind direction, wind speed, temperature, and global irradiation. The irradiance instruments and mirror sample are mounted on the Sun Tracker to ensure that reflectivity data are collected throughout the day. In addition, the reflectivity must be measured under real outdoor conditions, as is the case in the CSP plant.

In order to calculate the soiling rate, the following equation is used:

$$R = {}^{I_r}/_{I_I} \tag{1}$$

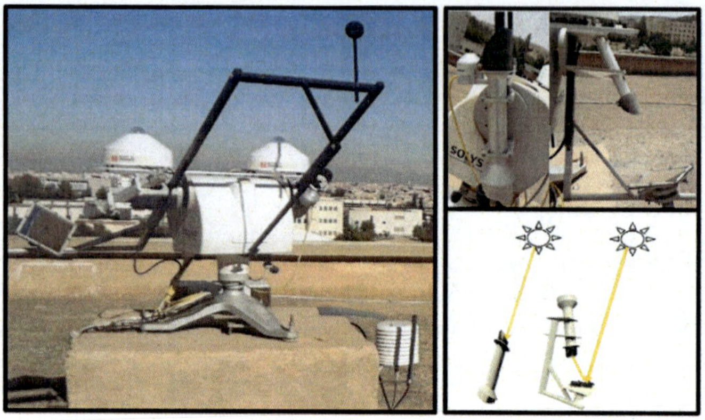

Fig. 2. The meteorological station.

where II, is the value of the irradiation coming directly from the sun and Ir is the value of direct irradiation reflected by the mirror.

The reflectivity results expressed by Eq. (1) are shown in Fig. 3. The monitoring periods start in June and end in August for both years.

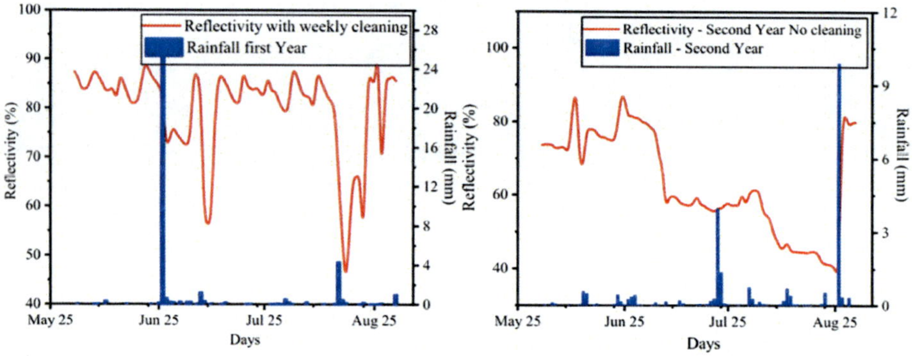

Fig. 3. Measured reflectivity on the two periods of experience.

2.3 The Plant Specification

The specific parameters chosen for the simulation are presented in Table 1. These parameters were chosen based on a preliminary optimization study in the region [14].

Table 1. The CSP plant parameters chosen for the simulation.

Description	Parameter	Value
Heliostat field	Solar multiple	1.4
Tower and receiver	HTF type	Salt (60% $NaN0_3$ 40% KNO_3)
	HTF hot temperature	574 C
	HTF cold temperature	290 C
Heliostat properties	Heliostat width	12.2 m
	Heliostat height	12.2 m
	Single heliostat area	144.375 m^2
Power cycle	Design turbine gross output	11 MWe
	Gross-to-net conversion factor	0.9
	Net output of the plant	10 MWe
	Type of the condenser	Evaporative

3 Results

As the optimal time for regular maintenance is considered to be one of the most important criteria, we have experimentally studied the soiling of a mirror sample during two periods, one with regular maintenance and one without. The reflection variations were recorded and integrated into SAM software so that the effect of the soiling can be taken into account in the prediction of the overall production of the plant. In this section, we present the results of the effect of the different maintenance periods on the CSP plant.

3.1 Energy Production

Figure 4 clearly shows that the losses of the CSP system due to dust accumulation are very high in both cases (with and without cleaning), although it can be observed that maintenance can improve the system performance. The losses due to dust accumulation without cleaning are higher than those due to weekly cleaning, which is very visible in the last month when the dust accumulation reaches relatively high values and approaches saturation, while regular maintenance can keep a small variation in losses close to stable values.

The highest energy loss was observed by the facility in August, when the maintenance processes are absent, thus granting a minimum reflectivity record during this period. The average rate of reflected irradiation during this period was 53%, while the corresponding losses of the facility are 82.33%.

In the second year, the simulation of the performance of the installation taking into account the fouling, has decreased the energy produced, and this decrease becomes very important again as the months go by, this is due to the accumulation of fouling throughout the period without any cleaning method even natural since this period is dry (Fig. 5).

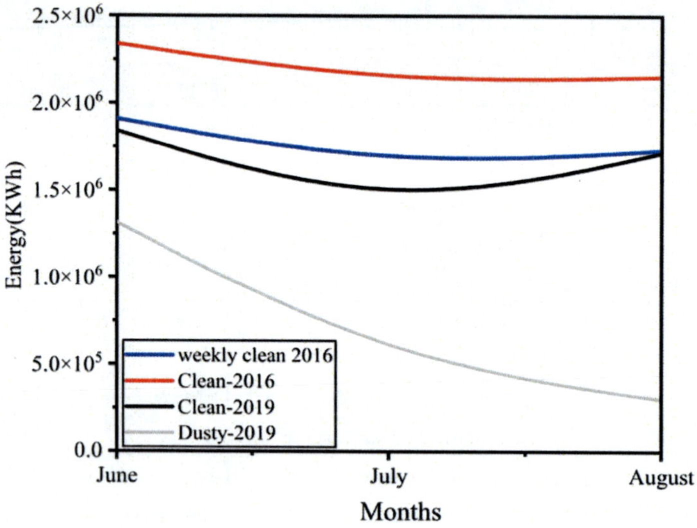

Fig. 4. Monthly energy production for Oujda city for the selected periods.

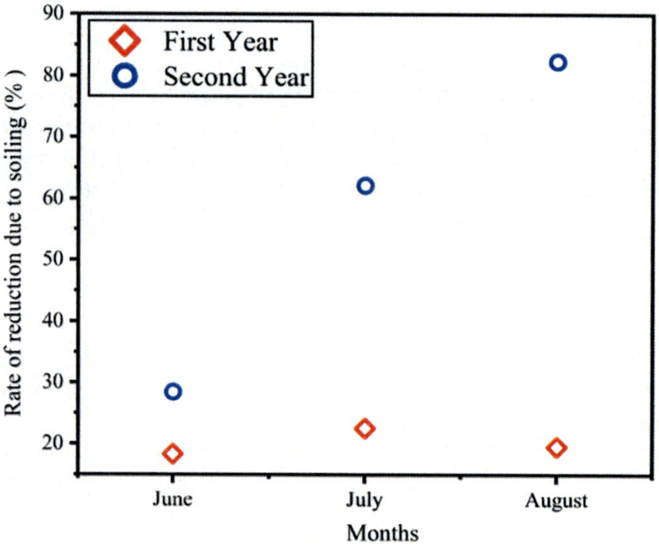

Fig. 5. Monthly energy reduction rate for selected periods.

3.2 Gross and Thermal Efficiency

Figure 6 shows the expected hourly average receiver thermal efficiency and power cycle efficiency for the studied CSP solar plant with both clean and dusty mirror cases. The gross and thermal efficiencies of the CSP system for the two years for which the mirrors are assumed to be clean (reflectivity constant and equal to 93%) showed high values.

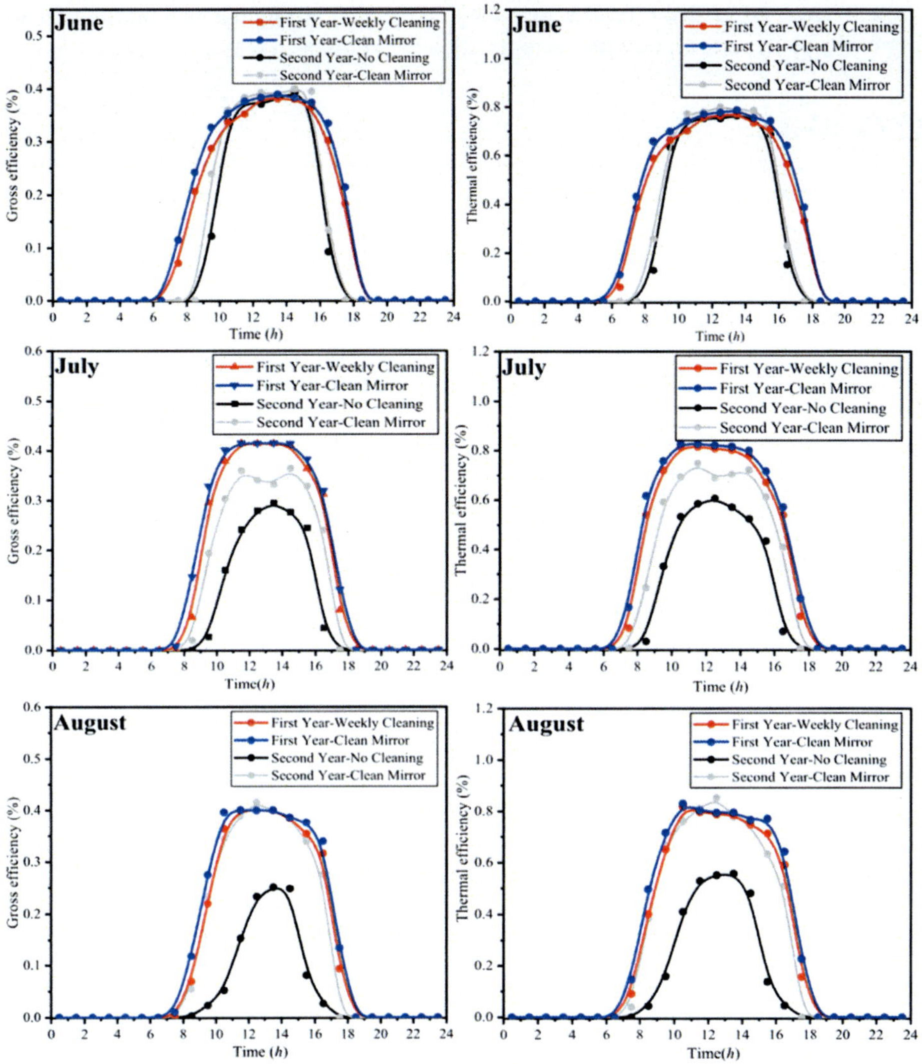

Fig. 6. Hourly gross and thermal efficiency for Oujda city for the selected periods.

The maximum was reached in August with values in the range of 38–41% and 80–85% for the gross and thermal efficiencies respectively.

While for the same two years, when the study considers the mirror as dirty, first with regular cleaning and then without any cleaning, the efficiency shows a significant reduction, the minimum is given by the case where there is no cleaning, and this is illustrated in August due to a large amount of dirt accumulated on the mirror throughout the period, while for regular cleaning, the rate of reduction in efficiency remained relatively the same for the three months.

It is clear from these results that cleaning makes a significant contribution to improving plant performance by keeping efficiency drops to a reasonable level.

4 Conclusion

The soiling mask that covers the mirror surface leads to a potential decrease in collector efficiency. This can be illustrated by the reduced power output of CSP systems. In the literature, a lot of research has been conducted on this technology by identifying the effects of weather conditions on their performance, In addition, special attention has been given to the effect of fouling due to its considerable effect on the performance, Therefore, in this work, we identify the effect of the maintenance period on the overall performance of the system, the period chosen is summer due to a large amount of dust particles and the fact that it is a dry period, so the natural cleaning is not too important, the results that have been found are as follows:

- The accumulation of dust on the mirror surface during the summer period can cause losses of up to 80% as a maximum value.
- Periodic cleaning can keep losses below 30%, which is good for the summer period when aerosol values are high.
- The cleaning keeps the efficiency drops by a reasonable level and remains relatively constant.

References

1. Miller, D.C., Kurtz, S.R.: Durability of Fresnel lenses: a review specific to the concentrating photovoltaic application. Sol. Energy Mater. Sol. Cells **95**, 2037–2068 (2011). https://doi.org/10.1016/j.solmat.2011.01.031
2. Bracco, G., Holst, B.: Surface Science Techniques (2013).https://doi.org/10.1007/978-3-642-34243-1
3. Zimon, A.D.: Adhesion of Dust and Powder. Second Edn. (1982)
4. Bellmann, P., Wolfertstetter, F., Conceição, R., Silva, H.G.: Comparative modeling of optical soiling losses for CSP and PV energy systems. Sol. Energy **197**, 229–237 (2020). https://doi.org/10.1016/j.solener.2019.12.045
5. Merrouni, A.A., Wolfertstetter, F., Mezrhab, A., Wilbert, S., Pitz-Paal, R.: Investigation of soiling effect on different solar mirror materials under Moroccan climate. Energy Procedia **69**, 1948–1957 (2015). https://doi.org/10.1016/j.egypro.2015.03.194
6. Karim, M., Delord, C., Raccurt, O., Naamane, S.: Exposure conditions effect on soiling of solar glass mirrors. In: AIP Conference Proceedings. 2033 (2018). https://doi.org/10.1063/1.5067234
7. Pescheux, A.C., le Baron, E., Raccurt, O.: Characterization of different Moroccan sands to explain their potential negative impacts on CSP solar mirrors. Sol. Energy **194**, 959–968 (2019). https://doi.org/10.1016/j.solener.2019.11.020
8. Meyen, S., Pernpeintner, J., Fend, T.: Optical characterisation of reflector material for concentrating solar power technology

9. Heimsath, A., Nitz, P., Heß, S.: Concentrating solar collectors for process heat up to 250°C and small-scale CSP – integrated optical design for improved perfoRmance. 1–8 (2016). https://doi.org/10.18086/eurosun.2010.09.14
10. Biryukov, S.A.: Degradation of optical properties of solar collectors due to the ambient dust deposition as a function of particle size. J. Aerosol Sci. **27**, S37–S38 (1996). https://doi.org/10.1016/0021-8502(96)00091-2
11. Bethea, R.M., Barriger, M.T., Williams, P.F., Chin, S.: Environmental effects on solar concentrator mirrors. Sol. Energy **27**, 497–511 (1981). https://doi.org/10.1016/0038-092X(81)90045-1
12. Jorgensen, G., Kennedy, C., King, D., Terwilliger, K.: Optical Durability Testing of Candidate Solar Mirrors. Nrel/Tp-520–28110. 1–58 (2000)
13. Raillani, B., Ouali, H.A.L., Amraqui, S., Moussaoui, M.A., Jeyar, M., Mezrhab, A.: The impact of Aerosol Optical Depth (AOD) on decreasing the optical efficiency of CSP solar power plants-a case study: Eastern Morocco. In: Proceedings - 2019 International Conference on Intelligent Systems and Advanced Computing Sciences, ISACS 2019, pp. 1–4 (2019). https://doi.org/10.1109/ISACS48493.2019.9068909
14. Ait Lahoussine Ouali, H., Raillani, B., Amraqui, S., Moussaoui, M.A., Mezrhab, A., Mezrhab, A.: Analysis and optimization of SM and TES hours of central receiver concentrated solar thermal with two-tank molten salt thermal storage. In: El Moussati, A., Kpalma, K., Ghaouth Belkasmi, M., Saber, M., Guégan, S. (eds.) SmartICT 2019. LNEE, vol. 684, pp. 666–673. Springer, Cham (2020). https://doi.org/10.1007/978-3-030-53187-4_73

Aluminum Heat Sink for Cooling PV Panel Technologies: Investigation of Flat and Lapping Fins

Abir Bria, Benyounes Raillani, Mourad Salhi, Dounia Chaatouf, Samir Amraqui(✉), and Ahmed Mezrhab

Laboratory of Mechanics and Energetics, Faculty of Sciences, Mohammed First University, 60000 Oujda, Morocco

samir.nova@gmail.com

Abstract. Nowadays, after the fourth industrial revolution, our economy has become very dependent on energy. The depletion of fossil resources is therefore a major concern. So, to be able to substitute our needs by renewable resources, we must pay more attention to the development of the latter. Among the most intrinsic problems that reduce the efficiency of photovoltaic (*PV*) panels is the operating temperature, due to the large amount of absorbed radiation that turns into heat, therefore, in this paper, an investigation of two types of fins for cooling the photovoltaic panel has been carried out. A comparison was made between the two systems as well as a study of the effect of lapping size on the cooling system. The proposed system is simple and suitable for integration. To be more realistic, a user-defined function *UDF* subroutine written in C+ + has been developed to integrate the unsteady evolution of the climatic conditions of a summer day in the city of Oujda (eastern Morocco). The computational fluid dynamics (*CFD*) investigation has revealed that the use of simple fins can reduce the average temperature of the panel by 18.16 °K, while when lapping fins are used, the temperature reduction was decreased by a value of 23.7 °K. Thus, it was found that the temperature of the PV panel decreases as the lapping height increases, according to the study of the three cases taken 0.003 m, 0.007 m, and 0.01 m.

Keywords: *CFD* · Cooling · Efficiency · Heat sink · Photovoltaic

1 Introduction

Nowadays, fossil fuels cover an important part of the planet's energy needs. But the limitation of these resources creates a problem of depletion, and even if this energy is quickly renewed, it cannot be used as much because of its high environmental impact, so the necessity of an alternative sources its priority for the human being. One of the best-known renewable energy technologies is photovoltaic solar energy, a subject that is currently in the spotlight because of its simplicity and easy implementation process, this technology is based on the direct conversion of radiation with the photovoltaic effect.

S. Motahhir and B. Bossoufi (Eds.): ICDTA 2022, LNNS 454, pp. 630–640, 2022.
https://doi.org/10.1007/978-3-031-01942-5_63

This was discovered in 1839 by Edmond Becquerel in order to convert solar radiation into electricity using photovoltaic cells.

The conventional crystalline silicon photovoltaic module can absorb 80–90% of the solar radiation, but only 12–18% of this solar radiation is converted into electricity, as shown in the study done by Ingersoll et al. [1], the rest being absorbed and transformed into heat. Therefore, its temperature increases and its efficiency decreases [2], Moreover, for each degree of increase in efficiency, the yield can generally be reduced by 0.5%, which was found by the Sala et al. [3] team in their investigations on the effect of temperature on the efficiency of photovoltaic cells. high temperatures significantly increase the stress of solar cells [4], as it has also another direct effect which appear on the voltage degradation, this is because the electronic excitation of thermal energy was greater than the electrical energy of the semiconductor material and therefore a drop in electrical voltage as the temperature increases [5], which makes the thermal regulation of the solar cells in these systems essential, to maintain the efficient operation of the photovoltaic cells and to avoid the irreversible degradation resulting from a high operating temperature in the long term and also to guarantee a long enough life span of these cells [6], Different techniques have been proposed to control the temperature of photovoltaic cells, and thus improve the efficiency of electricity production. These technologies are classified as active or passive cooling. For active cooling, a fan or a mechanical motor is usually required, this method has been studied theoretically by Mecheri et al. [7] in which studied the cooling by forced convection, using a mechanical pump to ensure the circulation of the heat transfer fluid and an external circuit to evacuate the heat of the system to the outside. In the same context, A temperature reduction of 30 °C is found by an experimental study conducted by Teo et al. [8] in which they used forced ventilation cooling through ducts with fins attached to its back. Unlike active cooling, passive cooling occurs without energy consumption, which makes it considered more effective in reducing the temperature of the *PV* module, many studies have investigated its efficiency on the operation of photovoltaic panel [9]. We found in another term the study given by the team Moharram et al. [10], in which they have developed another cooling technique for *PV* panels based on spraying water on the panels, and such a method has a dual function, which is cooling and cleaning the panel at the same time. This method has proven to be an effective method for cleaning and also for cooling *PV* panels. However, such a method consumes water which could be a problem in desert areas, where water is scarce and precious. Wu and Xiong [11] proposed a method that uses rainwater as a coolant and a gas expansion device to distribute the water on the roof of a house. The results of this method indicate that the efficiency of the *PV* module increased by about 8.3%.

In this study, a heat sink with a flat and lapped fins to cool the *PV* panel is investigated numerically, based on a 3*D* simulation model. The heat sink was attached to the bottom side of the *PV* panel. The study was performed using *ANSYS* Fluent software, while integrating real weather conditions for realistic approach.

2 Methodology

2.1 Physical Model

To study the thermal behavior of the *PV* panel, *ANSYS* Fluent is used, to have a more realistic study, a weather file of the city of Oujda (Eastern Morocco) has been used to integrate the evolution of the ambient temperature and solar radiation during the day.

Figure 1 shows the *PV* module with heat sink, where the photovoltaic panel is composed respectively of a glass layer, then a layer of ethylene vinyl acetate (*EVA*) which serves to protect the *PV* cells from ultra violet (*UV*) rays and moisture, followed by an anti-reflection coating (*ARC*) which contributes to the conversion efficiency of the *PV* model. Then, another layer of *EVA*, polycrystalline silicon (*pc-Si*) cell layer, and polyvinyl fluoride (*Tedlar*) which has almost the same purpose as *EVA*. An absorber plate equipped with fins was added under the *PV* panel as a cooling system, Aluminum was chosen as the heatsink material because of its high thermal conductivity [12]. All geometrical dimensions and thermophysical properties are presented in Table 1.

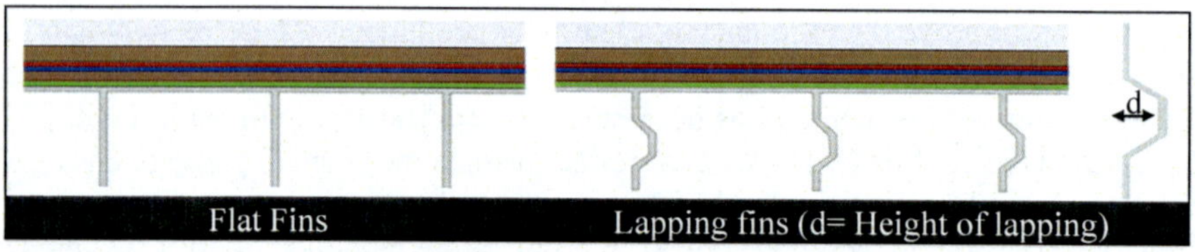

Fig. 1. The layers of the *PV* panel and the integrated absorber.

Table 1. Geometric dimensions and thermophysical properties of different layers of *PV* system.

Layers	Thickness (*m*)	Density (kg/m^3)	Thermal conductivity (*W/mK*)	Specific heat capacity (*J/kgK*)
Glass	0.004	2650	1.5	786
EVA	0.0005	960	0.35	2090
ARC	8.10^{-8}	4157	8.4	692
PV cells	0.0003	2330	100	700
Tedlar	0.0001	1200	0.2	1250
Heat sink	Plate: 0.003	2719	202.4	871

2.2 Mathematical Model and Boundary Conditions

Heat sink cooling is a natural convection-based cooling method used to increase the surface area and heat transfer rate. In this section, the mathematical model that governs heat transfer throughout the PV system is based on the Navier-Stokes equations, along with conservation of energy Eqs. (1–3).

$$\text{Continuity :} \qquad \vec{\nabla}.(\rho_f \vec{V_f}) = 0 \tag{1}$$

$$\text{Momentum :} \qquad \vec{\nabla}.(\rho_f \vec{V_f}\vec{V_f}) = -\vec{\nabla}P + \vec{\nabla}\tau_1 + \rho_f \vec{g} \tag{2}$$

$$\text{Energy :} \qquad \vec{\nabla}.(\rho_f C_{p.f} T_f \vec{V_f}) = \vec{\nabla}.(k_f \vec{\nabla} T_f) \tag{3}$$

All walls are exposed to convective losses, except for the outside of the glazing, which is exposed to convective and radiative losses according to the following equation.

$$-\lambda_v \frac{dT}{dn} = \varphi_{conv} + \varphi_{rad} \tag{4}$$

by replacing φ_{conv} and φ_{rad} by their terms, Eq. (4) will have the form of Eq. (5).

$$-\lambda_v \frac{dT}{dn} = h_0(T_v - T_{amb}) + \varepsilon_v \sigma (T_v^4 - T_{sky}^4) \tag{5}$$

with h_O is the convective exchange coefficient between the glazing and the ambient air [13], T_{sky} is the temperature of the sky, both of which are given by Eq. (6). and Eq. (7). respectively.

$$h_0 = 5.67 + 3.86V_{wind} \tag{6}$$

$$T_{sky} = 0.0552T_{amb}^{1.5} \tag{7}$$

The flow is considered as laminar, with a pressure-based solver, while for the pressure-velocity coupling, the SIMPLE (Semi implicit method for pressure linked equations) scheme is used. In terms of discretization scheme for the convection term, the "second order upwind" has been selected, and for the pressure the second order scheme has been used. The calculation is considered as convergent when the absolute residual decreases below 10^{-6}.

2.3 Climate Data

A typical hot day was taken into account, it was chosen in the month of August under the climatic conditions of the city of Oujda. This allows us to study the cooling of the PV panel with the integration of the heat sink in a more realistic way. The climatic data were obtained from our meteorological station installed at Mohammed I University Oujda, Morocco.

W. Chen et al. [14] have proposed approximate mathematical models that describe the variation of ambient air temperature and solar irradiance as a function of time presented in Eqs. (8 and 9)., these two approximative equations become in the form of the Eq. (10).and Eq. (11) representing our weather data of the chosen day.

$$T_{amb}(t) = \overline{T_{amb}} + T_{ar}\cos(\frac{\pi}{12}(t - 14)) \tag{8}$$

$$G_{sun}(t) = \hat{G}_{sun} \sin(\frac{t-a}{b-a}\pi) \quad , \quad a < t < b \tag{9}$$

$$T_{amb}(t) = 25 + 6\cos(\frac{\pi}{12}(t-14)) \tag{10}$$

$$G_{sun}(t) = 965 \sin(\frac{t-6}{14}\pi) \quad , \quad 6 < t < 20 \tag{11}$$

2.4 Validation Results

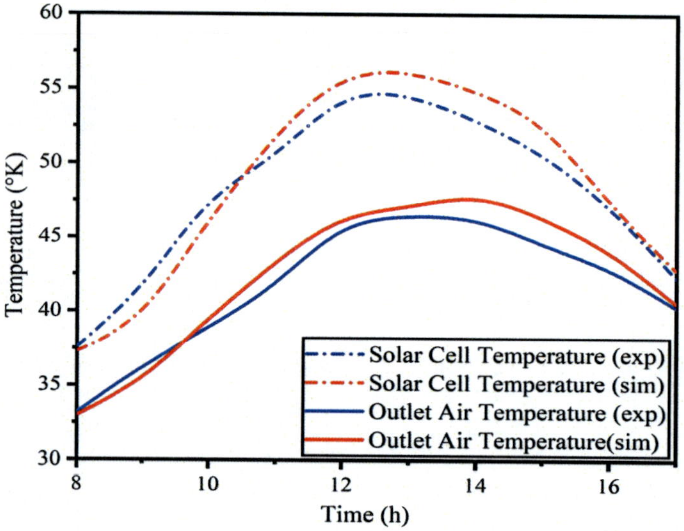

Fig. 2. Numerical validation of the cell temperature and the air outlet temperature.

The experimental results of Joshi et al. [15] for a photovoltaic/thermal (*PV/T*) air collector were used to validate the accuracy of the numerical results. The values of solar cell temperature and air outlet temperature obtained by the simulation were compared with the experimental results [15], and presented in Fig. 2. A good agreement between the experimental and simulation results is observed, having a maximum error of 5% and 4.2% for the solar cell temperature and the outlet air temperature, respectively.

3 Results and Discussions

In this section, the simulation results of *PV* panel with and without cooling system is presented, a heat sink with two forms of fins: flat fins and lapping fins are studied. Also, the effect of the lapping height on the temperature distribution and the average temperature of the *PV* are investigated.

A test on mesh independence was performed with five different grids, the test showed that a grid with a total number of nodes of 2305602 is good enough to study the geometry.

3.1 Effect of the Cooling System on the Average Temperature of the *PV* Panel

Figure 3 shows the average temperature of the *PV* panel with and without cooling system. The variation of solar flux and ambient temperature strongly influences the average temperature of the *PV* panel. At the beginning of the day, the panel records a low temperature due to the low solar radiation. Over times, the average temperature of the panel under different systems increases until reaching the maximum temperature at 12:00 with a value 349 °K, 331 °K and 325.6 °K, for the case of the photovoltaic panel without cooling system, the photovoltaic panel with flat finned heatsink, and the photovoltaic panel with lapped finned heatsink, respectively.

From a general point of view, it' s clear that the fin structure is advantageous, due to the increased dissipation capacity of its geometry. In comparison, the flat finned structure has intrinsic disadvantages: its parallel structure makes the airflow through the heatsink smoother, which is not beneficial for improving heat transfer. However, the lapping of the finned structure introduces more turbulence in the flow and generates more heat transfer. Therefore, the lapped fin structure gives better performance compared to the flat fins.

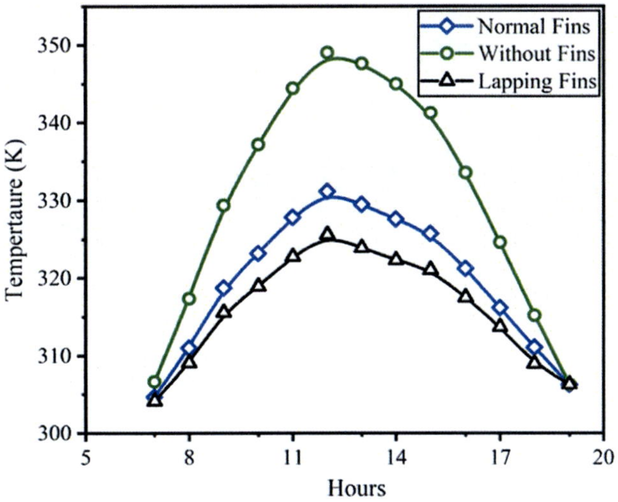

Fig. 3. Temperature of the photovoltaic panel with and without cooling systems in function of time.

Figure 4 shows the temperature contours of the *PV* panel with heat sink for the case of flat fins and lapped fins under a heat flux of 935 W/m^2. K which corresponds to 12:00. In this figure, it can be seen that for both types of fins, the temperature is high at the top surface of the *PV* panel and decreases towards the edges of the fins. This decrease is due to the intense heat transfer around the panel during cooling, caused by the enhanced airflow created by the fins. The maximum temperature of the *PV* panel for the flat fin type and the lapped fin type recorded are 332.5 °K and 327 °K, respectively. Thus, comparing the temperature contour and the average temperature of the two types we notice a decrease of 5.5 °K, which shows that the good performance is given by the lapped fin type.

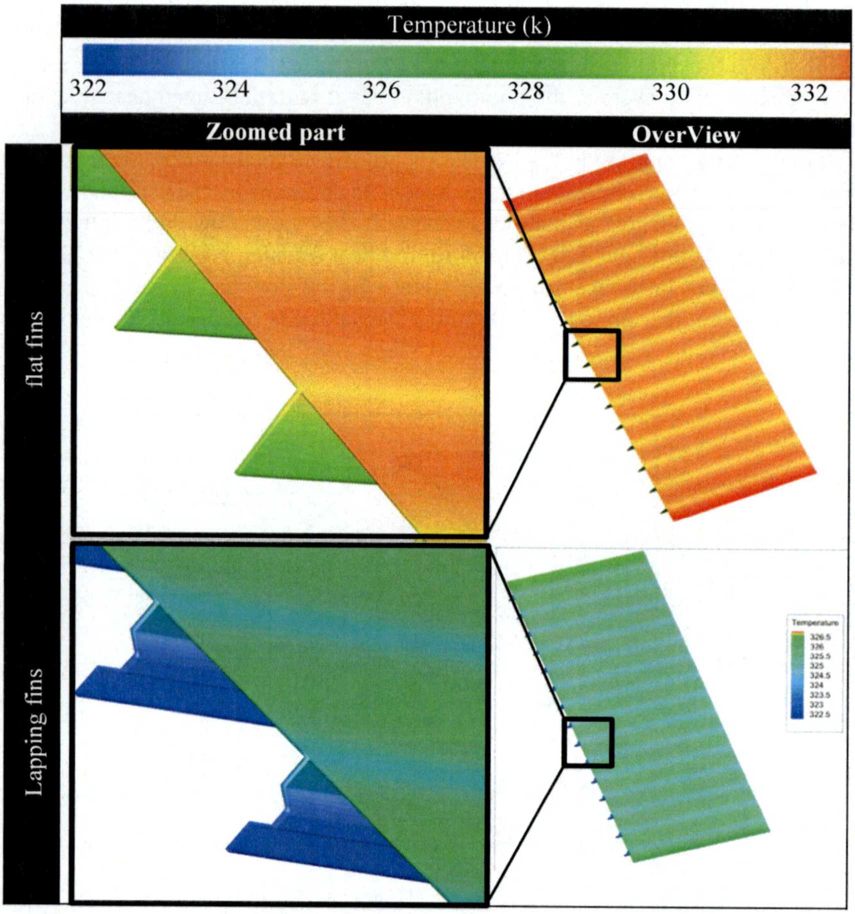

Fig. 4. Distribution of temperature for both, flat and the lapping fins.

3.2 Effect of the Lapping Height on the Average Temperature of the *PV* Panel

Figures 5 and 6 represent the temperature contour and average temperature of the *PV* panel with lapping, respectively. Three different lapping heights were presented: 0.003 m, 0.007 m and 0.01 m. In this figure, we can see that for the three cases of lapping height, the temperature is high at the top surface of the *PV* panel and decreases towards the fin edges with different intensity. The heat sink extracts more thermal energy from the panel by increasing the lapping height due to the improved airflow around the cooling system. The average panel temperature decreases with decreasing lapping height. For lapping heights taken 0.003 m, 0.007 m and 0.01 m, the recorded temperature is 329.92 °K, 327.88 °K and 325.6 °K, respectively.

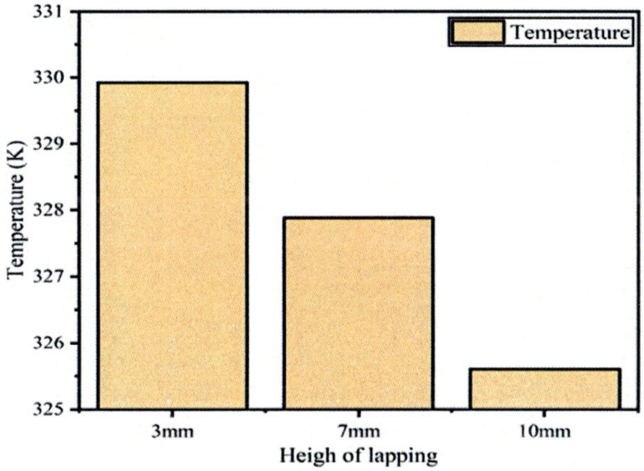

Fig. 5. *PV* panel temperature in function of the lapping heigh.

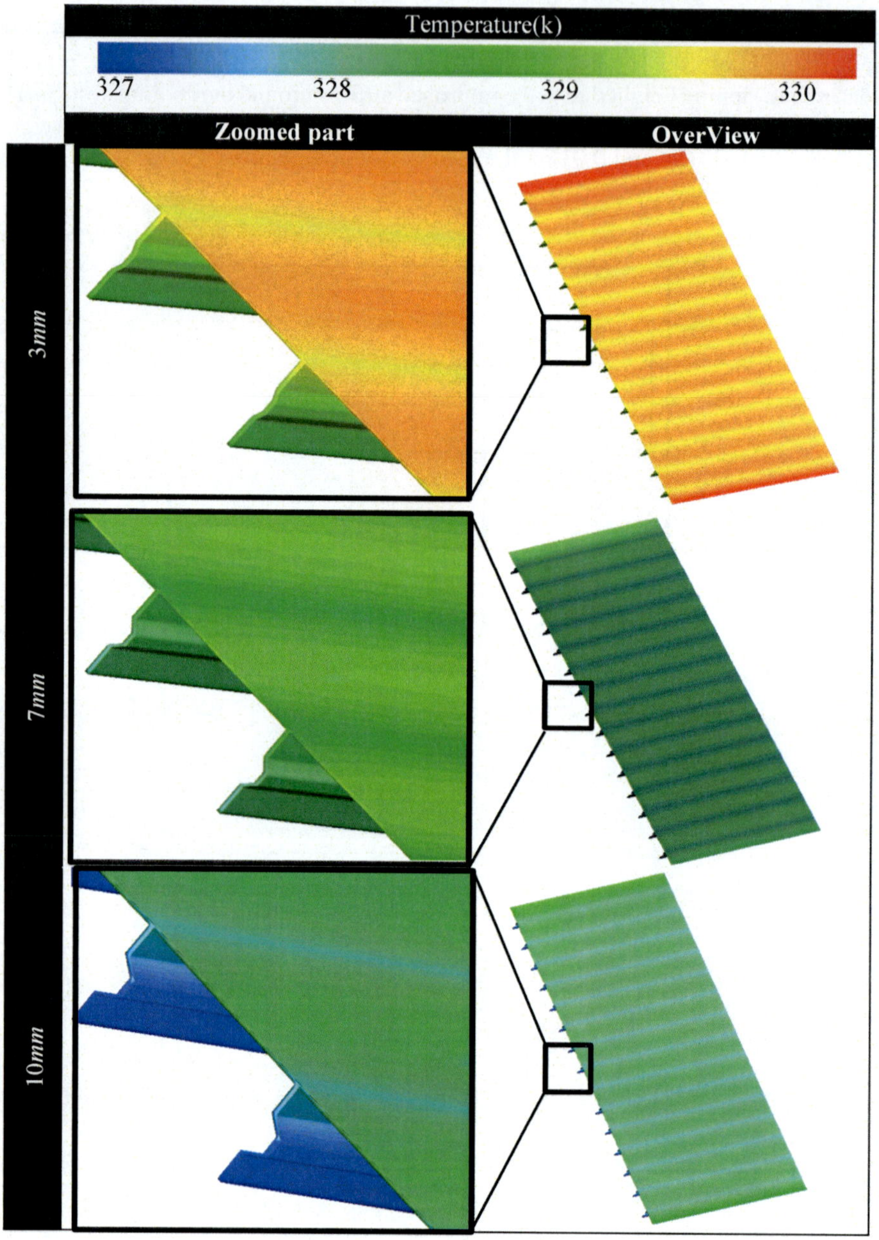

Fig. 6. Temperature distribution for different lapping heights.

4 Conclusion

As energy has a significant impact on economic development, especially after the fourth industrial revolution, where the life cycle of man is strongly related to energy, the development of solar energy technologies has become more interesting, so we have in this paper gives an investigation on photovoltaic cooling, in which the thermal behavior of the *PV* panel temperature with and without heat sink for both types of fins: flat fins and lapping fins have been studied. The results show that the average temperature of the *PV* panel decreases by 18.16 °K for the flat fins type and by 23.7 °K for the lapped fins type, this decrease is due to the improvement of the air flow around the fins. An investigation of the effect of lapping height was also carried out through three cases of height, we took 0.003 m 0.007 m and 0.01 m. The results show that the *PV* temperature decreases as the height increases.

References

1. Ingersoll, J.G.: Simplified calculation of solar cell temperatures in terrestrial photovoltaic arrays. J. Solar Energy Eng. Trans. ASME **108**, 95–101 (1986). https://doi.org/10.1115/1.3268087
2. Raillani, B., Ouali, H.A.L., Amraqui, S., Moussaoui, M.A., Mezrhab, A.: Analyze and study on photovoltaic parameters of mono-crystalline silicon solar cell. In: Proceedings - 2019 International Conference on Intelligent Systems and Advanced Computing Sciences, ISACS 2019, pp. 1–5 (2019). https://doi.org/10.1109/ISACS48493.2019.9068923
3. Gibbons, J., Szabo, P., Acosta, E., Bennett, N.: Investigating a radiative approach to solar cell cooling. In: 15th Photovoltaic Science Applications and Technology Conference, pp. 85–88 (2019)
4. Wohlgemuth, J.H., Cunningham, D.W., Monus, P., Miller, J., Nguyen, A.: Long term reliability of photovoltaic modules. In: Conference Record of the 2006 IEEE 4th World Conference on Photovoltaic Energy Conversion, WCPEC-4. 2, pp. 2050–2053 (2006). https://doi.org/10.1109/WCPEC.2006.279905
5. Cotfas, D.T., Cotfas, P.A.: Multiconcept methods to enhance photovoltaic system efficiency. Int. J. Photoenergy. 2019 (2019). https://doi.org/10.1155/2019/1905041
6. Brigand, S.: Les principes de l'énergie solaire photovoltaïque, p. 35 (2018)
7. Mecheri, B.: Refroidissement d'une armoire de Télécommunication avec Bouche Diphasique Thermosyphon. Université de technologie de belfort monbeliard, Thèse Doctorat (2011)
8. Teo, H.G., Lee, P.S., Hawlader, M.N.A.: An active cooling system for photovoltaicmodules. Appl. Energy **90**, 309–315 (2012). https://doi.org/10.1016/j.apenergy.2011.01.017
9. Siecker, J., Kusakana, K., Numbi, B.P.: A review of solar photovoltaic systems cooling technologies. Renew. Sustain. Energy Rev. **79**, 192–203 (2017). https://doi.org/10.1016/j.rser.2017.05.053
10. Moharram, K.A., Abd-Elhady, M.S., Kandil, H.A., El-Sherif, H.: Influence of cleaning using water and surfactants on the performance of photovoltaic panels. Energy Convers. Manage. **68**, 266–272 (2013). https://doi.org/10.1016/j.enconman.2013.01.022
11. Wu, S., Xiong, C.: Passive cooling technology for photovoltaic panels for domestic houses. Int. J. Low-Carbon Technol. **9**, 118–126 (2014). https://doi.org/10.1093/ijlct/ctu013
12. el Mays, A., et al.: Improving photovoltaic panel using finned plate of aluminum. Energy Procedia **119**, 812–817 (2017). https://doi.org/10.1016/j.egypro.2017.07.103

13. . Deceased, J.A.D., Beckman, W.A.: Solar engineering of thermal processes (1982). https://doi.org/10.1016/0142-694x(82)90016-3
14. Chen, W., Liu, W.: Numerical analysis of heat transfer in a passive solar composite wall with porous absorber. Appl. Therm. Eng. **28**, 1251–1258 (2008). https://doi.org/10.1016/j.applthermaleng.2007.10.017
15. Joshi, A.S., Tiwari, A., Tiwari, G.N., Dincer, I., Reddy, B.V.: Performance evaluation of a hybrid photovoltaic thermal (PV/T) (glass-to-glass) system. Int. J. Thermal Sci. **48**, 154–164 (2009). https://doi.org/10.1016/j.ijthermalsci.2008.05.001

Prediction of the Dust Accumulation Rate on a Parabolic Trough Mirror by CFD Method

Benyounes Raillani(✉), Dounia Chaatouf, Mourad Salhi, Samir Amraqui, and Ahmed Mezrhab

Laboratory of Mechanics and Energetics, Faculty of Sciences, Mohammed First University, 60000 Oujda, Morocco
benyounesraillani2@gmail.com

Abstract. In Industry 4.0, the development of intelligent power systems will contribute to providing people with better quality of electrical energy. The power generation system with solar power plants in Industry 4.0 is also the focus of development. The areas with high irradiation density have suffered from a high concentration of aerosol particles and thus, these technologies have experienced losses due to soiling. So, in this study, the behavior of dust particles with a cylindro-parabolic mirror has been carried out. For wind flow analysis, we have used the computational fluid dynamics (CFD) method, for the prediction in a 2D domaine, the simulation has been carried out based on the Shear Stress Transport k-w turbulence model and a discrete phase model for particle motion prediction. The results revealed that, unlike the large particle, the trajectory of the small ones creates vortices behind the heliostat, as the particle size increases, it can be noticed that the vortices start to dissipate. The maximum deposition rate due to gravity could be as high as 10% for 100 μm dust particles, which shows the important role played by gravity on the dust deposition rate for larger dust particles. The decrease in deposition after the size exceeds 100 μm is only because these particles do not reach the heliostat, they leave the area through the ground before reaching the mirror.

Keywords: *CFD* · Deposition rate · Soiling

1 Introduction

Since the beginning of life on Earth, human beings have ceased to exploit its natural wealth in all its forms to meet their needs, especially for heating/cooling and other several applications, and over time, the so-called "population explosion" emerged Mittal et al. [1], which refers to a significant increase in the Earth's population, the exploitation of natural wealth has reached record levels. In addition, the population explosion has been followed by a great industrial and commercial revolution, especially in recent decades. This has led to an increased burden of natural fossil fuel exploitation, and adverse effects on the environmental balance [1], given that fossil fuel exploitation is a crucial factor behind major climate change. With the increasing electricity demand, as well as

S. Motahhir and B. Bossoufi (Eds.): ICDTA 2022, LNNS 454, pp. 641–650, 2022.
https://doi.org/10.1007/978-3-031-01942-5_64

the depletion of fossil resources and their negative impact on the environment, which represents 29% of total greenhouse gas emissions [2, 3], it is necessary to search for alternative sources of energy. The most important option is the development of renewable energies as an environmental and economic alternative.

The world is currently focusing on harnessing one of the largest and most easily implemented renewable energy sources, namely solar energy, to generate electricity as an alternative to coal-fired power plants. For this purpose, many solar power plants are installed around the world, especially in regions with high solar radiation and suitable climatic and geological conditions. These plants are mainly based on Photovoltaic *(PV)* and Concentrated Solar Power *(CSP)* technologies, the latter consisting of concentrating sunlight through reflective mirrors, and then followed by a series of processes that result in the production of electricity [4].

Because of the high storage capacity offered by CSP technologies, whether by sensible heat or latent storage [5]. the global *CSP* capacity remains unchanged in the leading countries in this field, such as Spain and the United States until 2014, the global *CSP* capacity has been increasing since 2017 and reached 6.1 GW in 2019 by an increase of 11% (Fig. 1). In addition, in 2020, the CSP plants that have been under construction in the world include about 0.9 GW, 0.3 GW, 14 MW of trough, tower, and Fresnel systems, successively [6].

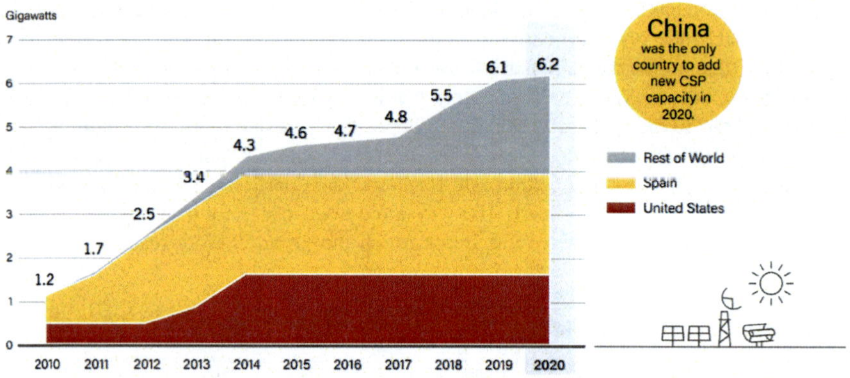

Fig. 1. Concentrating Solar Thermal Power Global Capacity, by Country and Region, 2009–2020.

Over time, some unfavorable factors for electricity generation in *CSP* systems began to emerge, the most important of which was the accumulation of dust on *CSP* mirror surfaces, especially in arid or semi-arid regions characterized by desert climatic conditions, where sandstorms are abandoned, this process reduces the efficiency of the plants, while increases the electricity cost. In this context, renewable energy researchers have conducted many studies, the ones that investigate the meteorological effect on the accumulation of dust particles,

In the current paper, we present a soiling modeling approach using *ANSYS* fluent software. This represents another way for further generalization of soiling phenomena, intending to take into account the disposition rate of different particle sizes, and promotes the development of a concentrated solar power system in Industry 4.0. All of which are structured as follows: Sect. 1, contains an integrated introduction with a literature review. Section 2, presents the geometric model and the governing equations of the *CFD* method. Section 3, includes a discussion of the results obtained from the study, and finally, Sect. 4, is the conclusion of the study.

2 Mathematical Formulation and Boundary Conditions

The chosen domain consists of a cylindro-parabolic heliostat mounted on a support with the heigh is H = 2.2 m, the mirror is inclined with 45°. All The dimensions of the domain are in Fig. 2.

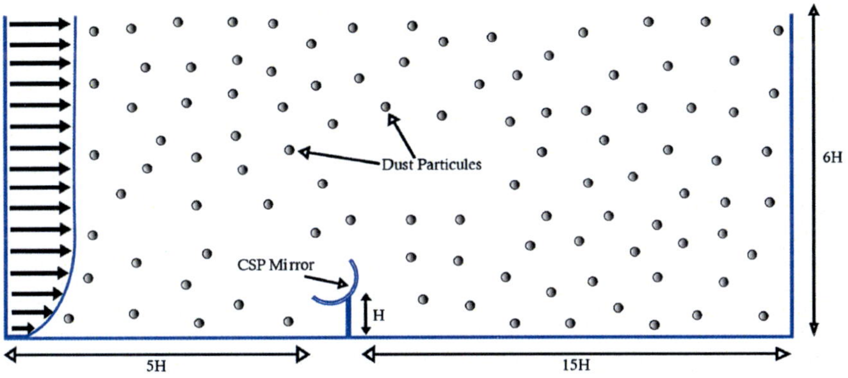

Fig. 2. Schematic design of the cilindro-parabolic mirror s domain.

The Reynolds averaged Navier-Stokes (RANS) approach has been adopted to predict the turbulent wind flow fields over the parabolic reflector in this study. The k-ω shear stress transport (SST) is used to predict the turbulent fields. Although this flow regime is continuous, it is governed by the Navier-Stokes conservation of mass and momentum equations, given by (1) and (2), respectively. While the k-ω transport equations are shown in (3) and (4) [7].

$$\frac{\partial \overrightarrow{u_i}}{\partial x_i} = 0 \tag{1}$$

$$\frac{\partial \overrightarrow{u_i}}{\partial t} + \overrightarrow{u_j}\frac{\partial \overrightarrow{u_i}}{\partial x_j} = -\frac{1}{\rho}\frac{\partial \overrightarrow{p}}{\partial x_i} + \frac{1}{\rho}\frac{\partial}{\partial x_j}\left(\mu\frac{\partial \overrightarrow{u_i}}{\partial x_j} - \rho\overrightarrow{u_i' u_j'}\right) \tag{2}$$

$$\frac{\partial}{\partial t}(\rho k) + \frac{\partial}{\partial x_i}(\rho k u_i) = \frac{\partial}{\partial x_j}\left(\Gamma_k\frac{\partial k}{\partial x_j}\right) + \underset{k}{\tilde{G}} - Y_k + S_k \tag{3}$$

$$\frac{\partial}{\partial t}(\rho \omega) + \frac{\partial}{\partial x_i}(\rho \omega u_i) = \frac{\partial}{\partial x_j}\left(\Gamma_\omega\frac{\partial \omega}{\partial x_j}\right) + \underset{\omega}{\tilde{G}} - Y_\omega + S_\omega \tag{4}$$

where, ρ is the density (kg/m^3), Yk and Yω are the turbulent kinetic energy dissipation term and the specific dissipation rate, respectively. Gω and Γω are the generation term and the diffusivity term of the specific dissipation rate. Gk and Γk are the generation term and diffusivity term of the turbulent kinetic energy. While, (u_i)- and p- are the time-averaged velocity and pressure, respectively [8].

The governing equation of dust deposition is presented in Eq. (5) as follow

$$\frac{\partial u_p}{\partial t} = \frac{1}{\tau}\frac{C_D \mathrm{Re}_p}{24}\left(u_g - u_p\right) + \frac{g\left(\rho_p - \rho_g\right)}{\rho_p}\xi\sqrt{\frac{\pi S_0}{\Delta t}} + \frac{2\rho \mathrm{K}_c \nu^{0.5}}{\rho_p d_p (S_{lk} S_{kl})} s_{ij}\left(u - u_p\right) \quad (5)$$

With τ the particle relaxation time, S the particle-air density ratio, C_C the Cunningham slip correction factor, and C_D is the drag coefficient, which is given by Eq. (6) [8].

$$C_D = \begin{cases} \frac{24}{\mathrm{Re_p}} \text{for } 1 < \mathrm{Re_p} \\ \frac{24}{\mathrm{Re_p}}\left(1 + 0.15\mathrm{Re_p^{0.687}}\right), \text{for } 1 < \mathrm{Re_p} < 400 \end{cases} \quad (6)$$

2.1 Boundary Conditions

The computational domain was modeled to obtain the kinetic energy and velocity distributions of the wind flow in the *2D* domain given in Fig. 3. The boundary conditions applied to the domain are given as follows. In the inlet, the turbulence kinetic energy and velocity profiles given by the experiment Tominaga et al. [9] were implemented as a function, using a User Defined Function (*UDF*) subroutine written in C++ language. the Out-flow condition was given at the outlet of the computational domain. The symmetry condition was applied to the top of the domain, the bottom was considered as a stationary wall without slip condition, the mirror and the support also took the same conditions.

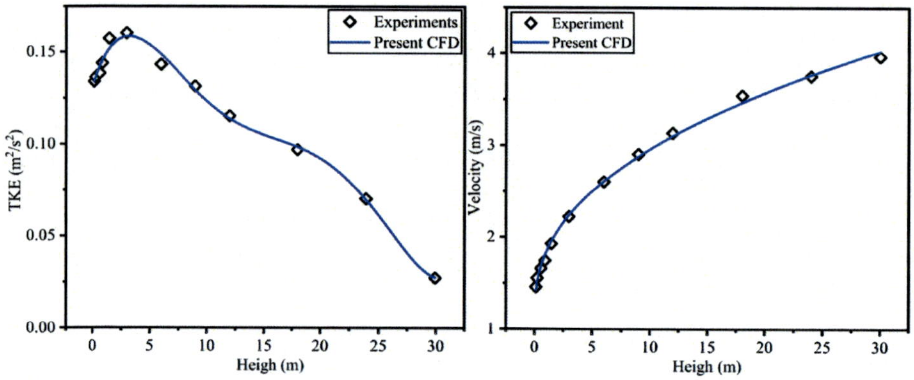

Fig. 3. The profiles of velocity and turbulence kinetic energy.

2.2 Mesh and Grid Independency

We examined three grids in this section to study their independence on the results and to determine the smallest grid size that would provide the most accurate results. The grids are as follows: fine (62,435), medium (49,729), and coarse (32,729). By comparison, it was concluded that the medium mesh would give the best results since it performs relatively the same as the fine mesh. The figure below shows the structured grid used in the study (Fig. 4).

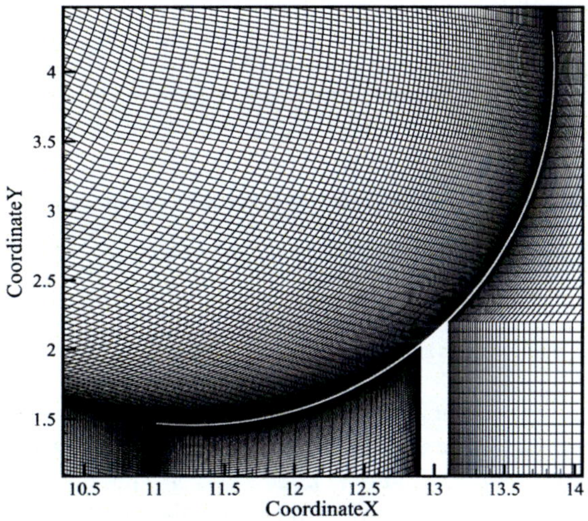

Fig. 4. Mesh grid structure.

2.3 Validation Results

For simulation studies, the validation step is crucial to guarantee the accuracy of the results. For this purpose, we used the experimental results of Abiola-Ogedengbe [10] to validate the average pressure coefficient. Our simulation results were compared and displayed in Fig. 5, from which it can be seen the high agreement between the experimental and our numerical results.

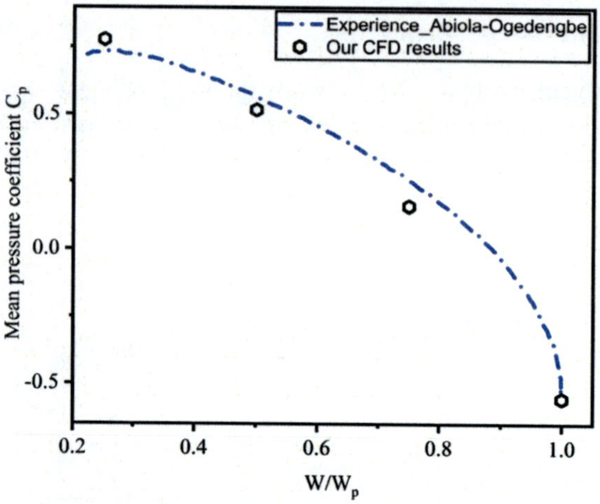

Fig. 5. The average pressure coefficient in function of the width of the panel.

3 Results

3.1 Airflow Fields Over the Heliostat

Turbulent kinetic energy and velocity distributions around the heliostat are important to investigate the motion and deposition of dust particles. Figure 6 shows those images of wind flow around parabolic cylindrical heliostat. Since the presence of the parabolic trough acts as a barrier to the wind flow, a rather complex velocity field can be observed. High turbulence intensity was shown in the field on the backside of the system, while on the front side the turbulence is relatively low, as is the velocity amplitude.

3.2 Effect of Gravity and Particle Trajectory

In this section, we have attempted to carry out a study of the effect of the gravitational force on the behavior of the dust particles, we have taken several sizes of particles, and then, studied their disposition on the mirror surface with and without gravity, the figure below shows the disposition rate of particles on the heliostat (Fig. 7).

To fully understand the behavior of different particles, the figure below illustrates the trajectory of different particles taking into account both gravity and no-gravity.

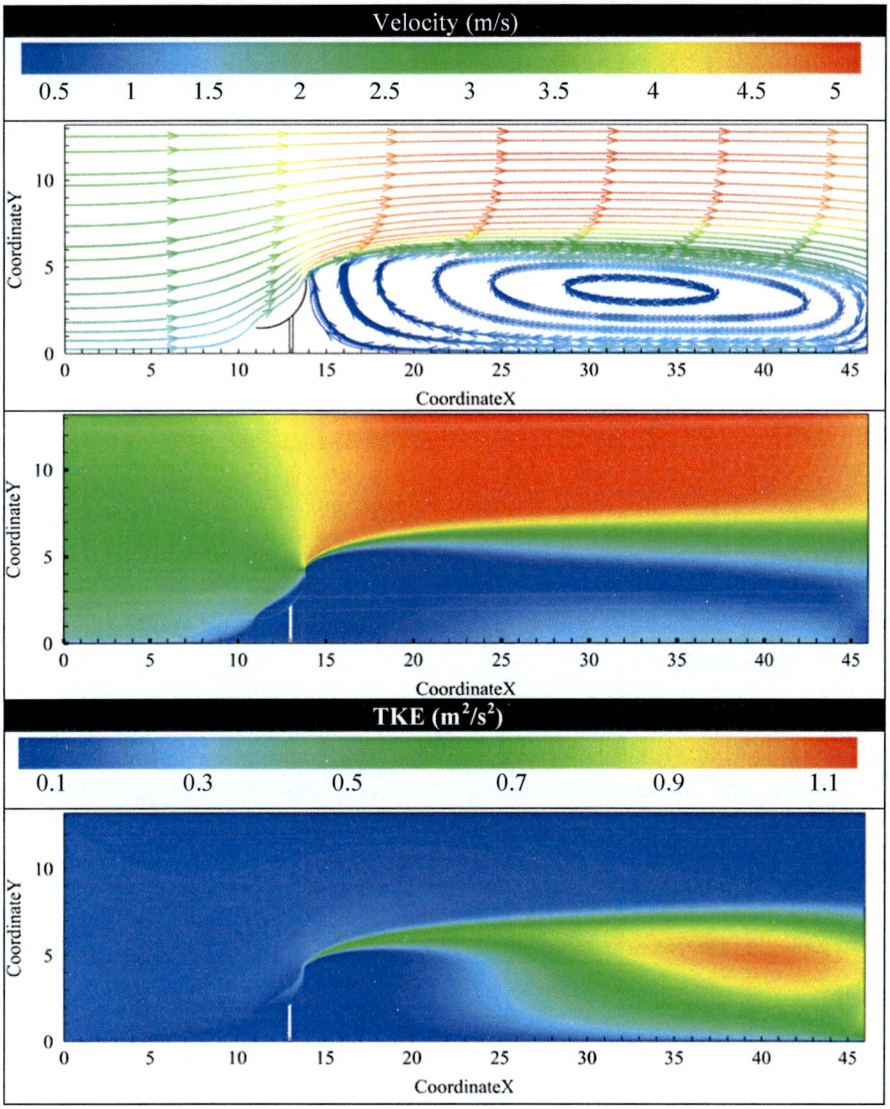

Fig. 6. Wind flows Domaine over the parabolic-through mirror.

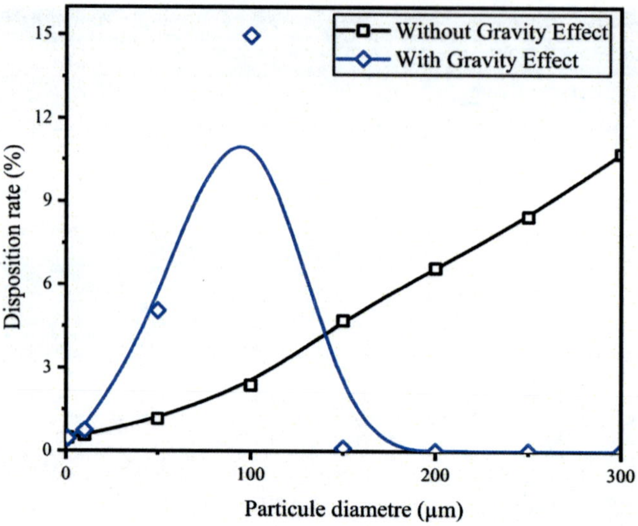

Fig. 7. The dust disposition rate in function of particle size.

Figure 8shows the trajectory of the particles that have been colored with the velocity magnitude along the X-axis.

As can be seen, small particles have relatively the same behavior in both cases, the trajectory of the particles creates vortexes behind the heliostat, when the size of the particles increases, we can notice that the vortexes start to dissipate.

From a particle diameter over 10 μm, the effect of gravity is very perceptible. As the particle diameter increases, the rate of dust deposition increases in the case without gravity, as it does in the case with gravity but with a different rate. When the particle size exceeds 100 μm, the gravitational force prevents a large number of particles from reaching the mirror surface, which decreases the deposition rate, unlike the case where the gravitational effect is not taken into account.

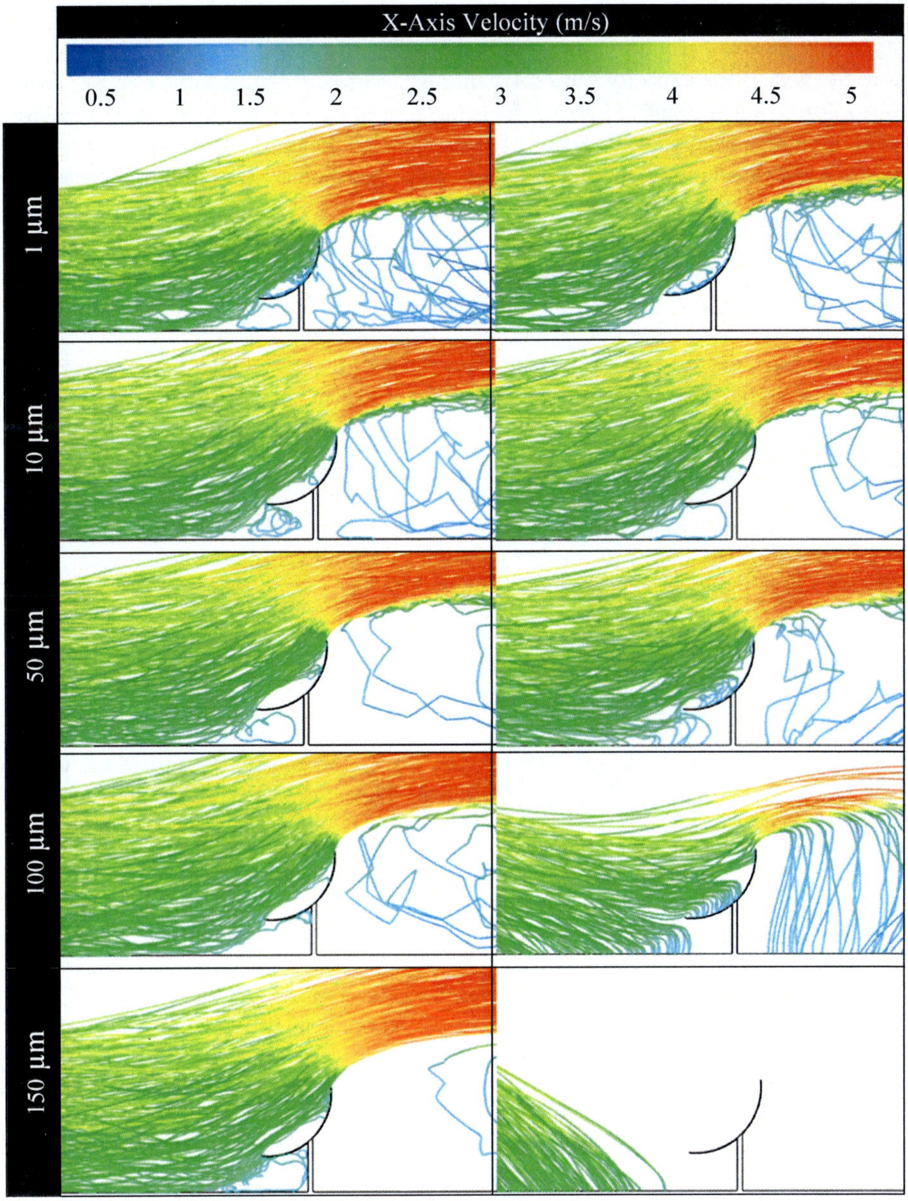

Fig. 8. The trajectory of different particle sizes with and without gravity effect.

4 Conclusion

Electricity generation capacity from solar energy has grown significantly in recent years, as it is a promising alternative to fossil fuels because of its good environmental behavior. The technologies used for conversion are diverse, CSP technologies have yielded high efficiency and thus have been encouraged to establish themselves in various locations, but this technology is suffered from the degradation of its performance due to the effect of soiling, So A deeper understanding of the dust deposition process and particle behavior on parabolic trough heliostats could contribute to the development of practical and effective solutions against soiling losses. Therefore, in this study, a CFD simulation was performed to determine the rate of particles deposited on the heliostat surface and to analyze the trajectory of particles of different sizes with and without the gravity effect, respectively. Unlike the large particle, the trajectory of the small ones creates vortices behind the heliostat. When the size of the particles increases, we can notice that the vortices start to dissipate. The maximum deposition rate due to gravity could be up to 10% for 100 μm dust particles, which shows the important role played by gravity on the dust deposition rate for the largest dust particles. The decrease in deposition after the size exceeds 100 μm is only since these particles do not reach the heliostat, they leave the area through the ground before reaching the mirror.

References

1. Mittal, R.: Impact of population explosion on environment. **1**, 5 (2013)
2. Weisser, D.: A guide to life-cycle greenhouse gas (GHG) emissions from electric supply technologies. Energy **32**, 1543–1559 (2007). https://doi.org/10.1016/j.energy.2007.01.008
3. Cao, G., Sun, K., Jiang, S., Lu, S., Wang, Y.: A Modular DC/DC photovoltic generation system for HVDC grid connection. Chin. J. Electr. Eng. **4**, 56–64 (2018). https://doi.org/10.23919/CJEE.2018.8409351
4. Qu, H., Zhao, J., Yu, X.: Simulation of parabolic trough solar power generating system for typical chinese sites. Zhongguo Dianji Gongcheng Xuebao/Proc. Chin. Soc. Electr. Eng. **28**, 87–93 (2008)
5. Li, J.: Scaling up concentrating solar thermal technology in China. Renew. Sustain. Energy Rev. **13**, 2051–2060 (2009). https://doi.org/10.1016/j.rser.2009.01.019
6. Members, R.E.N.: GSR2021_Full_Report (2021)
7. Lu, H., Zhao, W.: Effects of particle sizes and tilt angles on dust deposition characteristics of a ground-mounted solar photovoltaic system. Appl. Energy **220**, 514–526 (2018). https://doi.org/10.1016/j.apenergy.2018.03.095
8. Lu, H., Zhao, W.: CFD prediction of dust pollution and impact on an isolated ground-mounted solar photovoltaic system. Renew. Energy **131**, 829–840 (2019). https://doi.org/10.1016/j.renene.2018.07.112
9. Tominaga, Y., Akabayashi, S., Kitahara, T., Arinami, Y.: Air flow around isolated ga-ble-roof buildings with different roof pitches: wind tunnel experiments and CFD simula-tions. Build. Environ. **84**, 204–213 (2015). https://doi.org/10.1016/j.buildenv.2014.11.012
10. Abiola-Ogedengbe, A.: Experimental investigation of wind effect on solar panels. Electronic Thesis and Dissertation Repository. 1177 (2013). https://ir.lib.uwo.ca/etd/1177

Photovoltaic Water Pumping System: Modeling and Simulation of Characteristics for Direct Coupling

Abdelilah Hilali[1](✉), Yahya Mardoude[1], Mohammed El Mahfoud[2], Ali Essahlaoui[3], Majdoube Houssam[1], and Abderrafii Rahali[1]

[1] Electronics, Communication Systems and Energy Optimization Team, Faculty of Sciences, Moulay Ismail University, Meknes, Morocco
ab.hilali@edu.umi.ac.ma
[2] Laboratory of Engineering Modeling and Systems Analysis, SMBA University, Fez, Morocco
[3] Water Sciences and Environment Engineering Team, Faculty of Sciences, Moulay Ismail University, Meknes, Morocco

Abstract. Significant progress has been made recently in the field of photovoltaic (*PV*) generator conversion, with an emphasis on the use of multiple approaches. This has largely focused attention on irrigation fields, prompting several researchers to investigate ways to make photovoltaic installations more cost-effective and affordable. Additionally, several strategies for improving photovoltaic cell efficiency are considered, which are modeled at the level of increasing cell lifetime or extracting maximum power from the cells. With this in mind, the current study analyzes the various parameters that affect the manufacturing of *GPVs*, including their electrical characteristics. Additionally, this analysis is numerically based. A direct coupling is established between the *GPV*, which is composed of *REC 330NP* photovoltaic panels, and a solar water pumping system powered by a *SQF 0–6-2* motor pump. The numerical results of the proposed analysis are presented to emphasize the critical nature of a matching converter that maximizes the extraction of power generated by the *GPV*.

Keywords: GPV · PV modeling · Solar water pumping system · DC Motor

1 Introduction

The price of fossil energy is constantly increasing [1]. The fight against greenhouse gas emissions makes urgent control of energy consumption and the diversification of energy resources and its development. With its strategic location and favorable climate, Morocco provides ideal conditions for the use of renewable energies, particularly solar energy [2]. Numerous remote sites, such as solar water pumping systems, off-grid rural electrification, or direct injection of generated energy into the electricity grid, could easily accommodate the introduction of these energies.

S. Motahhir and B. Bossoufi (Eds.): ICDTA 2022, LNNS 454, pp. 651–660, 2022.
https://doi.org/10.1007/978-3-031-01942-5_65

In Morocco, the decrease in rainfall observed in recent years has made the quantity of available water resources insufficient to meet the water needs of agriculture [3]. Therefore, the use of pumped irrigation solutions has become an unavoidable necessity. Two methods are generally used to achieve this goal of water pumping system: the use of diesel or butane to pump, and the use of *PV* energy to power the pumping systems [4]. The first one is not only an environmentally friendly method, but also a solution that consumes a lot of money, which increases the production charge. The second requires a considerable initial installation cost. At the same time, several advantages are generated by the elimination of energy costs incurred after the installation phase.

PV generator (*GPV*) is a system that produces electricity from *PV* cells [5]. The process was completely silent, with no carbon dioxide emissions polluting the environment [1]. This energy is totally compatible with the current environmental policy of the government. Moreover, in his study, it found that the electrical power produced by the *GPV* varies from a fraction of a watt to a few megawatts, which allows the *GPV* to meet specific energy needs and to be integrated into any electrical system [6]. *PV* converters are characterized by very poor conversion efficiency. A further decrease in efficiency that depends mainly on the characteristics of the load and *GPV* in relation to the point of maximum power. To better exploit this type of energy, it is necessary to establish a knowledge model of *PV* panels based on the parameters that affect their performance.

Solar pumping systems are known for their conversion of solar radiation into electricity and are mainly used to power irrigation pumps, which allows them to use a renewable energy source and thus guarantee a clean and natural supply. These characteristics, among others, make solar pumping systems the most appropriate means for agriculture, especially in rural areas of Morocco, particularly in the city of Meknes. In their applications, these systems present constraints of different kinds. One of the most contested cases is the connection between a GPV and an irrigation pump, a very common problem in nature that has always attracted scientists in order to reduce the installation costs for a given power and thus to be able to mitigate or reduce the power loss between the power supplied by the GPV and the load. In this perspective, the present work aims to study the characteristics of a GPV according to the parameters that influence its performance and to analyze the direct coupling between a GPV and the water extraction pump without MPPT adapter.

2 Modeling of a PV Cell

It is easy to convert light energy into electricity using *PV* cells. In order to build a *PV* module, you need to start with *PV* cells. A typical cell is a few square centimeters in size and generates about 1 W of power. A module with a surface area of a few square meters is constructed by linking a series and parallel arrangement of cells on a panel. The following section discusses the modeling and simulation of a solar cell in further detail.

2.1 Electrical Model

The equivalent diagram of a real *PV* cell, includes a current generator (I_{ph}), a diode and two parasitic resistors (R_s and R_{sh}), which are represented by the equivalent diagram in Fig. 1 [6, 7]. The current generator (I_{ph}) it is an ideal current source. The diode (*D*) simulates a P-N junction by allowing the current to flow in one direction only. The series resistance R_s represents the loss of resistance in the solar cell: it is the resistance that the current meets on its path. The shunt resistance R_{sh} is the leakage resistance between the two zones *N* and *P* of the junction, it takes into account the inevitable leakage of current between the opposite positive and negative terminals of the solar cell.

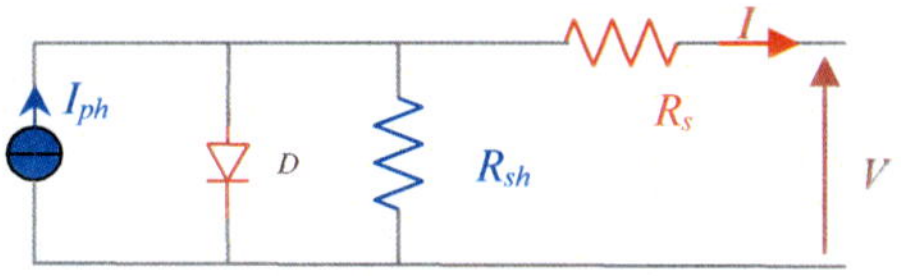

Fig. 1. Electrical model of a PV cell.

In order to increase *PV* efficiency, R_{sh} should be as high as possible in the cell because part of the current I_{ph} is derived from R_{sh} and cannot be transmitted into the load. According to the *PV* cell's current-voltage characteristic (*I, V*) at the considered irradiance and other factors. The operating power of the PV generator will be influenced by the load impedance. The *V* terminal is used to connect this load.

2.2 Mathematical Model

From the Fig. 1, the relationship between current and other parameters can be represented by the following equation:

$$I = I_{ph} - I_{obs} = I_{ph} - I_s\left[\exp^{\frac{V}{nV_{th}}} - 1\right] - \frac{V + I.R_s}{R_{sh}} \tag{1}$$

or,

$$V_{th} = \frac{K.T}{e}$$

with:
I_s *(A)*, corresponding to the saturation current;
V_{th} *(V)*, the thermodynamic potential;
K $(J.K^{-1})$, the Boltzmann constant;
T (K), the effective temperature of the cell;
e (C), the charge of the electron;
n, a factor of non-ideality of the junction;
I (A), that which comes from the cell as electricity;
V (V), the voltage applied across the cell,
I_{ph} *(A)*, the short-circuit current of the cell.

According to Eq. (1), the *PV* cell is a photoelectric current generator I_{ph} whose value is solely determined by the amount of light it receives. I_{ph} defines the hole-electron pairs that cross the *P-N* junction without recombination. The voltage at the terminals of this cell is reduced because of parasitic resistances such as the series resistance R_S, which is caused at the level of the metallic contacts and interconnections. It has a small effect on the short-circuit current and can be determined from the *I-V* curve by calculating its derivative around V_{oc}. To get the best performance from the *PV* cell, the R_S should be as low as possible. The other type of resistance is the parallel resistance R_{sh} which is due to impurities near the *P-N* junction. These impurities cause leakage currents. These impurities cause leakage currents. These cause leakage currents in the junction. Thus, the higher the R_{sh}, the more efficient the *PV* cell.

3 PV Module Simulation (*REC_330NP*)

PV module under consideration in this section is a monocrystalline solar panel. Table 1 shows the electrical parameters of this panel. According to the manufacturer, tests conducted under standard conditions ($E = 1000W.m^{-2}$, $AM = 1.5$, $Tc = 25$ °C) have revealed the following results:

Table 1. The *REC_330NP* Specifications.

R_{sh}	R_S	V_{mpp}	I_{mpp}	P_{mpp}	I_{sc}	V_{oc}
1 kΩ	73 mΩ	34.6 V	9.55 A	330 W	10.33 A	41 V

3.1 Characteristics Under Standard Conditions

To study this PV generator, we need its electrical model defined in Eq. (1), which takes the relationship between the various variables affecting *PV* production into account. Additionally, we can use simulation software. The Fig. 2 shows the electrical behavior of the *REC_330NP* module by simulation.

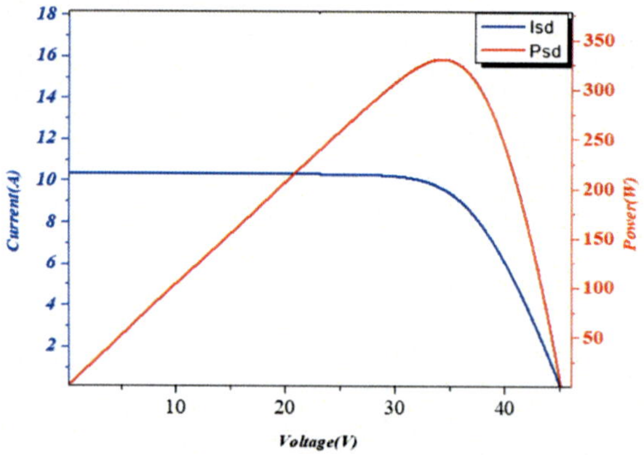

Fig. 2. Electrical characteristics (*I, V*) and (*P, V*) of the *REC_330NP* module

According to the simulation results, the maximum current ($I_{sc} = 10.33\ A$) occurs when the module terminals are short-circuited. It is known as short-circuit current, and it is highly dependent on the level of irradiance. For zero current, no-load voltage ($V_{oc} = 41\ V$) is used. This voltage is also referred to as the open-circuit voltage. The optimal operating point *MPP* ($V_{mpp} = 34.6\ V$, $I_{mpp} = 9.55\ A$) or the module's operating point at full power. Maximum delivered electrical power ($P = 330\ W_p$).

From the current-voltage characteristic, we can deduce the characteristic of the electrical power generated by the cell as a function of the voltage at its terminals. For this generated *PV*, there is a value of the electrical voltage where the power is maximum. This value is affected by a number of meteorological factors, including temperature and irradiation [8].

a) **Influence of irradiance**

Figure 3 depicts the effect of a variable irradiance, where the curves show the evolution of voltage and power as a function of current. In addition, the junction temperature is kept constant is *25 °C*.

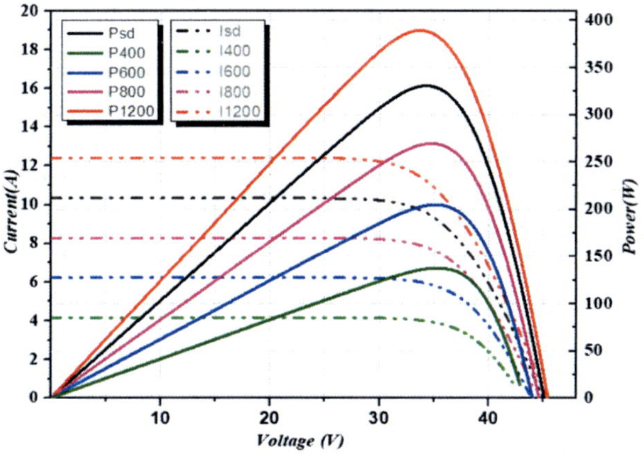

Fig. 3. PV characteristic under several radiation modes.

From the study of the curves in Fig. 3, which shows the characteristics of a *REC-330NP* panel, it appears that the maximum power increases from 130Wp for an irradiation of $400W.m^{-2}$ to almost double, a value of $380\ W_p$ for an irradiation of $1200\ W.m^{-2}$. In fact, this characteristic is due to the fact that in case of high irradiance, the current I_{sc} varies directly with the light radiation (*the relationship is proportional*), while the voltage remains relatively constant. It is important to emphasize that this behavior is interesting, especially for charging a battery.

b) **Influence of temperature**

Figure 4 illustrates the effect of varying the junction temperature, where the curves show the evolution of the voltage and power as a function of the voltage. The irradiance is kept constant at the value of *1000* $W.m^{-2}$.

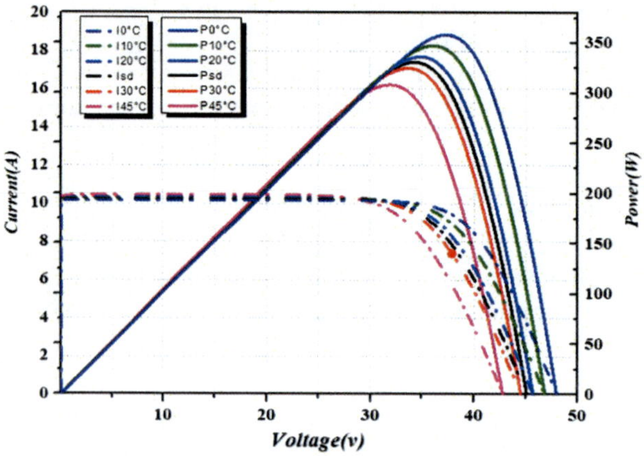

Fig. 4. PV characteristic under variable temperature

The curve in the above figure shows the drift of the characteristics of this crystalline silicon panel as a function of temperature. Figure 4, shows the current-voltage characteristics of this PV panel under constant irradiation and variable temperature. The temperature also has an important impact on the characteristics of the panel: we see that when the temperature increases, the voltage decreases, the power also decreases. We can see that the optimal power goes from *305* W_p to *360* W_p, so a gain of *55* W_p when the temperature goes from *0 °C* to *45 °C*. We must therefore take into account the temperature of use of the solar cell to apply the optimal power reduction factor at *25 °C*.

4 Direct Coupling to GPV

The *PV* system is a combination of components needed to reliably power an installation. However, a single *PV* panel is usually insufficient to power an application on a regular basis. Therefore, in a real installation, there are usually several *PV* panels combined in different modes to power the loads. This combination is a parallel/serial combination depending on the voltage and capacity, relating on the requirements of the application. The final characteristics (*I, V*) or (*P, V*) of this device under ideal conditions are obtained by the sum of the basic characteristics of constant current for n identical serial panels: $V_n = n.V_1$. For an array of p parallel panels, $I_n = p.I_1$. In a series array, the unit with the lowest current affects the whole array; if it is not taken into account, this characteristic will penalize the overall conversion very strongly. To prevent this situation, a diode must be connected in parallel. Commercial modules now include these parallel diodes to provide reverse voltage protection. In a dual fashion, an anti-reverse diode must protect the parallel channels from reverse currents [9].

The current overall electrical characteristic voltage of a *GPV* is therefore theoretically deduced from the combination of the characteristics of the $n_s.n_p$ elementary cells assumed to be identical which have the component of two affinities of ratio n_s parallel to voltages and of ratios n_p parallel to the axis of the currents. *GPV* is used to generate electricity. This energy will be charged with a source of storage, use or will be injected into an electrical network. In the case where the panels are used directly to supply a load, we can lose a fairly interesting part of this quantity. we cannot extract the entire production of the panels, this is the occupation of several researchers in order to increase its efficiency instead of adding panels. Take the case: solar pumping by a *DC* pump *SQF 0–6-2*, which has the characteristics $P = 1400\ W$, $I_n = 8.4A$ and a voltages 30-300VDC, and which will be supplied by a GPV consisting of five *REC_330NP* cells with total peak power of *1650 Wp* under standard conditions. To do this, there are two possibilities: direct coupling between *GPV* and the electric pump and another coupling through a *DC-DC* converter which is suitable for supply to the load and capable of extracting the maximum power produced by the *GPV*.

4.1 Simulation and Results

Take the characteristics of our *GPV* voltage $V_{mpp} = 172\ V$, $I_{mpp} = 9.6A$. Let's connect it directly to the pump terminals. The intersection of the characteristic curves (*V,I*) of the *GPV* and of the pump defines the point of operation of this installation.

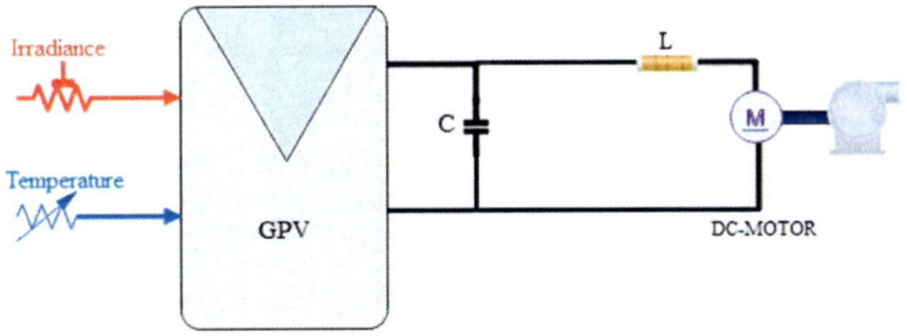

Fig. 5. Simulation scheme for a direct coupling between the GPV and the motor pump

Figure 6 illustrates the effect of irradiance variation on the behavior of the photovoltaic pump and PV generator in a time interval, where the irradiance corresponds to the different possible levels of variation and which models the natural behavior of the region. From the analysis of the pump (see Fig. 6), it is clear that the variation of irradiance leads to a large fluctuation in power and that the power transferred to the pump is much lower than that produced by the GPV. Moreover, under standard conditions where

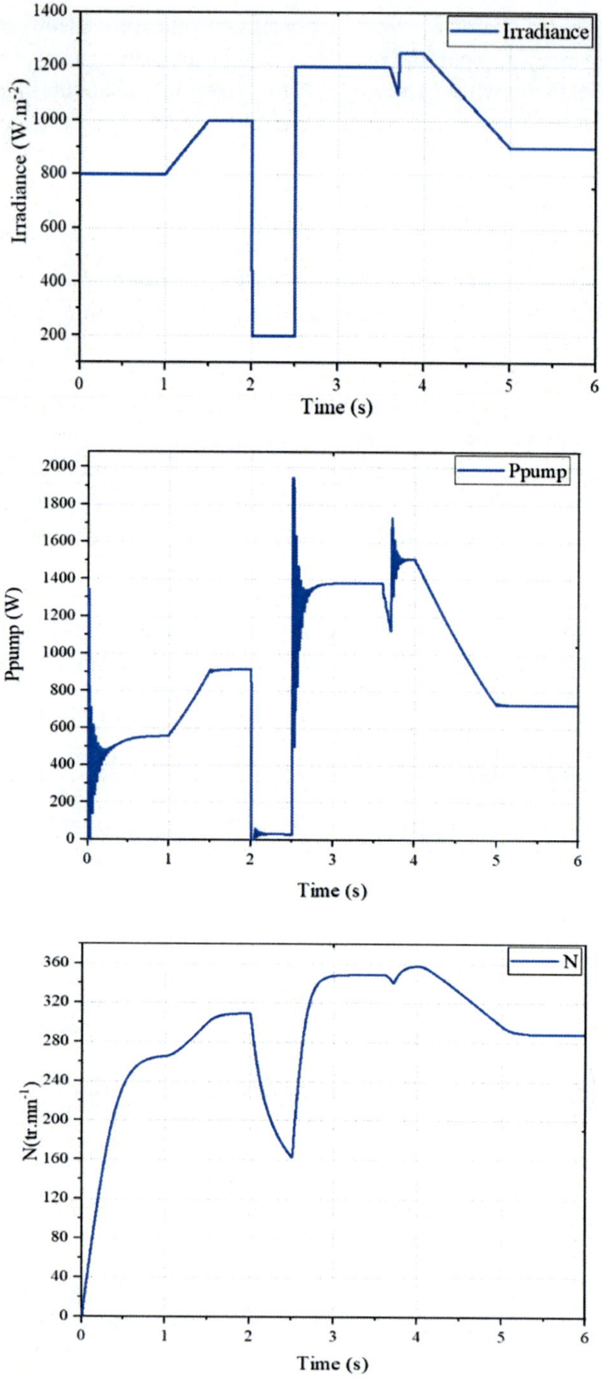

Fig. 6. Simulation results for a direct coupling between the GPV and the motor pump under standard conditions

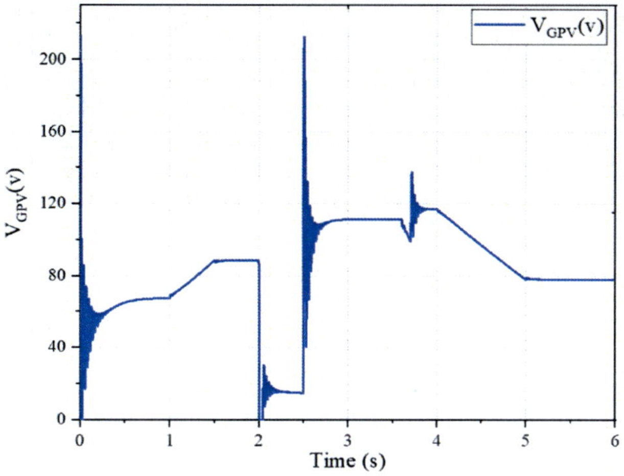

Fig. 6. continued

the irradiance has a value of 1000 $w.m^{-2}$, the power extracted by the pump is 917W, this value corresponds to only 55% of the power created (1650W) by the GPV.

The main disadvantage of the direct connection is that, on the one hand, a significant variation in the voltage of the motor pump is not desirable for the safety of the installation. On the other hand, the direct connection does not allow any type of limitation and/or readjustment of the voltage at the pump terminals, while the transfer of the maximum available power to the GPV terminals is not guaranteed. Moreover, the influence of this configuration lies in the dependence between the power supplied by the generator and the load. Finally, the optimization of the transmitted energy requires another configuration in which a matching device is inserted between the load and the GPV. It is also important to note that the matching device uses an efficient algorithm for MPPT to avoid a sudden system shutdown in case of low sunlight and to increase the system performance. Many methods have been proposed in the literature, such as perturb and observe (*P&O*), incremental conductance (*IC*), constant voltage, fuzzy logic, etc. [10]. It can be concluded that a part of the power produced by the panel will be lost although it will not be used.

5 Conclusion

The maximum power of an illuminated *GPV* is the essential quantity to evaluate its performances. There is a value that corresponds to the voltage at the terminals of this group which was optimal for this electrical power. This point is called the *MPP* (*Maximum Power Point*). In the case of direct coupling, a significant portion of the power produced by the *GPV's* does not reach the pump, reducing the installation efficiency. Indeed, the maximum power varies according to the climatic conditions. To address this issue, it is necessary to install a dynamic system that automatically adjusts the maximum power available from the panel each time.

References

1. Umar, M., Ji, X., Kirikkaleli, D., Alola, A.A.: The imperativeness of environmental quality in the United States transportation sector amidst biomass-fossil energy consumption and growth. J. Clean. Prod. **285**, 124863 (2021). https://doi.org/10.1016/j.jclepro.2020.124863
2. Allouhi, A.: Management of photovoltaic excess electricity generation via the power to hydrogen concept: a year-round dynamic assessment using Artificial Neural Networks. Int. J. Hydrog. Energy **45**(41), 21024–21039 (2020). https://doi.org/10.1016/j.ijhydene.2020.05.262
3. Ouatiki, H., et al.: Trend analysis of rainfall and drought over the Oum Er-Rbia River Basin in Morocco during 1970–2010. Arab. J. Geosci. **12**(4), 1–11 (2019). https://doi.org/10.1007/s12517-019-4300-9
4. Mergoul, K., Laarabi, B., Barhdadi, A.: solar water pumping applications in morocco: state of the art. In: 2018 6th International Renewable and Sustainable Energy Conference (IRSEC), pp. 1–6, December 2018. https://doi.org/10.1109/IRSEC.2018.8703015
5. Fakham, H., Lu, D., Francois, B.: Power control design of a battery charger in a hybrid active PV generator for load-following applications. IEEE Trans. Ind. Electron. **58**(1), 85–94 (2011). https://doi.org/10.1109/TIE.2010.2062475
6. Olayiwola, O.I., Barendse, P.S.: Photovoltaic cell/module equivalent electric circuit modeling using impedance spectroscopy. IEEE Trans. Ind. Appl. **56**(2), 1690–1701 (2020). https://doi.org/10.1109/TIA.2019.2958906
7. Slimani, M.E.A., Amirat, M., Kurucz, I., Bahria, S., Hamidat, A., Chaouch, W.B.: A detailed thermal-electrical model of three photovoltaic/thermal (PV/T) hybrid air collectors and photovoltaic (PV) module: comparative study under Algiers climatic conditions. Energy Convers. Manag. **133**, 458–476 (2017). https://doi.org/10.1016/j.enconman.2016.10.066
8. Patel, H., Agarwal, V.: MATLAB-based modeling to study the effects of partial shading on PV array characteristics. IEEE Trans. Energy Convers. **23**(1), 302–310 (2008). https://doi.org/10.1109/TEC.2007.914308
9. Bastidas-Rodriguez, J.D., Ramos-Paja, C.A., Saavedra-Montes, A.J.: Reconfiguration analysis of photovoltaic arrays based on parameters estimation. Simul. Model. Pract. Theory **35**, 50–68 (2013). https://doi.org/10.1016/j.simpat.2013.03.001
10. Motahhir, S., El Hammoumi, A., El Ghzizal, A.: The most used MPPT algorithms: review and the suitable low-cost embedded board for each algorithm. J. Clean. Prod. **246**, 118983 (2020). https://doi.org/10.1016/j.jclepro.2019.118983

Study of a Photovoltaic Pumping System Application for Irrigation the Palm in Beni Mellal City

Ikram Saady[1(✉)], Mohammed Karim[1], Badre Bossoufi[1], Saad Motahhir[2], Btissam Majout[1], Hassna Salime[1], and Nada Zine Laabidine[1]

[1] Laboratory of Engineering Modelling and Systems Analysis, Sidi Mohamed Ben Abdellah University Fez, Fez, Morocco
saady.ikram0@gmail.com
[2] ENSA, SMBA University, Fez, Morocco

Abstract. This paper presents a simple standalone photovoltaic pumping system that achieved satisfaction through the reliability installation, depending on the quantity of water needed and the solar energy deposit in the place concerned. This paper aims to design, size, and simulate the system that irrigates an area of 8823 m^2 of palm trees in Beni Mellal city located in the Middle Atlas of Morocco, which needs 60 m^3/day. The appropriate pump is selected according to the water required during the irrigation period, the total dynamic head, and the necessary hydraulic power, which is used to calculate the electric power to determine the required photovoltaic model and system elements. Finally, the PVWPS simulation is performed in the PVsyst software. Then, the photovoltaic pumping system scheme is validated using the MATLAB Simulink environment to discuss the obtained results.

Keywords: Solar photovoltaic module · Water pumping · SPVWPS (solar photovoltaic water pumping) · Sizing · PVSyst software · MATLAB/Simulink

1 Introduction

In the past years, many people in rural areas of developing countries faced a big problem of water deficit for irrigation and livestock due to expensive traditional resources such as fuel transportation and routine maintenance of diesel machines, which caused harm to the environment [1]. A photovoltaic solar water pumping system is the best solution, which consists of two ways depending on whether it works with or without a battery [2]. PV solar water pumping system without batteries is better than those use batteries, which increases the cost due to the high prices of batteries, especially in remote and off-grid areas. Solar pump performance (flow, pressure, efficiency) depends on the quantity of sunlight that varies during days and seasons [3].

Because of the benefits of the solar water pumping system nowadays, the Moroccan government has strongly supported and encouraged the diversification of energy supply. Solar agricultural irrigation has become an innovative and environment-friendly solar

S. Motahhir and B. Bossoufi (Eds.): ICDTA 2022, LNNS 454, pp. 661–671, 2022.
https://doi.org/10.1007/978-3-031-01942-5_66

water conservancy solution for the Moroccan agricultural economy [4], with a period of sunshine over the entire national territory that can last 3000 h (high plateaus and Sahara). The energy is received daily on a horizontal surface of 1m, about 5 kwh on most national territory, or nearly 1700 kWh/m^2/year in the north and 2263 KWh/m^2/year in the south of the country [5]. The critical parameters of a solar pump installation depend on different factors: Adequate quantities of solar radiation based on geographical location, assessment of water needs which vary according to yield and weather-related factors such as temperature, humidity, soil evaporation, wind speed, the season of the year under study, irrigation method [6] and head dynamic dimensions, which is determined by the confirmed fall, the height of the tank, and the maximum drawdown [7].

This paper contains four sections. In Sect. 2, the description of BNI MALLAL and case study site using the general system of PVWPS. Section 2 Validation by PVSYST software. The last Sect. 3, is critical to describe the solar pumping system when irradiance changes suddenly to improve the effective operation of PV water pumping. Two controls have been placed in the design, MPPT control with the perturb and observe (P&O) method between the boost converter and generator PV to pursue the maximum power from the PV generator, and indirect field-oriented control be-tween the inverter and AC motor to drive the pump. The simulation of the complete system is done by using MATLAB/Simulink.

2 The Description of the Site of BNI MALLAL

2.1 Geographic Location

The installation site is for the irrigation of a farm located in the Middle Atlas of Beni Mellal is characterized by geographical indications: 32° 20 32 N, 6° 22′ 32 W longitude, 509 m elevation, and 31° inclination.

2.2 Monthly Solar Radiation

The monthly average daily irradiation in Beni Mellal with the modules of the defined system (kWh/m^2) is presented in (Table 1), which has indicated the increase in solar energy in the hottest month of July up to 7.05 kWh/m^2/day and decreases during the coldest month of December up to 4.91 kWh/m^2/day.

Table 1. The monthly irradiation of the site.

Month	Jan	Feb	Mar	Apr	May	Jun
Irradiation (Kwh/m^2/d)	5.18	5.68	6.56	6.45	6.49	6.91
Month	Jul	Aug	Sep	Oct	Nov	Dec
Irradiation (Kwh/m^2/d)	7.05	6.98	6.49	6.15	5.19	4.91

3 The Description of the Site of BNI MALLAL

In general, the solar pumping system (Fig. 1) to be designed and chosen consists of a photovoltaic generator comprised of a set of photovoltaic modules coupled in series or/and parallel, a three-phase variable frequency inverter, a three-phase motor pump, and a tank [8, 9]. The water requirements of this system vary according to irrigation methods (drip) and types of crops. That is why we need the approved installation of the solar pump system to determine the size and control of the elements system.

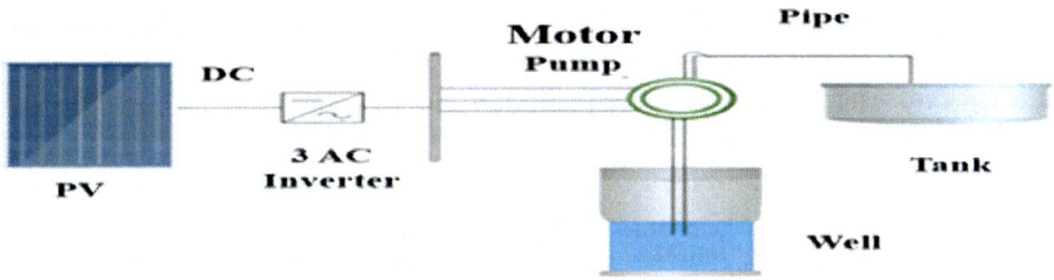

Fig. 1. Simplified synoptic diagram of PV pumping.

3.1 The Sizing of the Pump

Estimation of Water Requirements. The required hour flow is defined based on the water consumed during a specific service duration. On our farm, we need Q (m^3/day) = 60 to irrigate Palm. By dividing the daily requirement by the maximum number of sunshine, equal to 7 h.

$$Q(m^3/\mathrm{h}) = \frac{Q(m^3/day)}{h} \tag{1}$$

We get the hourly flow required for irrigation 60/7 = 8.57 m^3/h.

Total Manometric Height. The total dynamic height (TDH) is the sum of the distance between the ground and the point of water discharge in the tank H_r is 18.135 m, the distance between the land and the water surface equal to H_d is 128 m, and the pressure drop equivalent to 2.625 m.

$$H_{TDH} = H_r + H_d + \Delta H \tag{2}$$

The total dynamic height for irrigation is $H_{TDH} = 148\,\mathrm{m}$.

Hydraulic Energy. The average daily hydraulic power needed Eh is based on the water flow and total dynamic head (TDH) by the following formula.

$$E_h = \frac{\rho * Q * g * TDH}{3600} \tag{3}$$

The hydraulic energy E_h required is about $E_h = 24322, 26\ wh/Day$.

Electrical Power. The electrical power P_e required by the motor-pump to lift a certain amount of water to a certain given height during a day is calculated from the following equation:

$$P_e = \frac{E_h}{\eta_M \, \eta_{P} N_h} \tag{4}$$

where the efficiency of the motor is equal to 80%, and the pump efficiency 65%. The electrical power P_e need is about $P_e = 6.68$ KW.

The required daily flow rate and the total head are also responsible for selecting the correct type of pump to be installed. (Fig. 2) shows the total head in terms of flow rate, and the point (M) in the figure represents the pump's operating point, which allows us to choose a centrifugal pump for our installation.

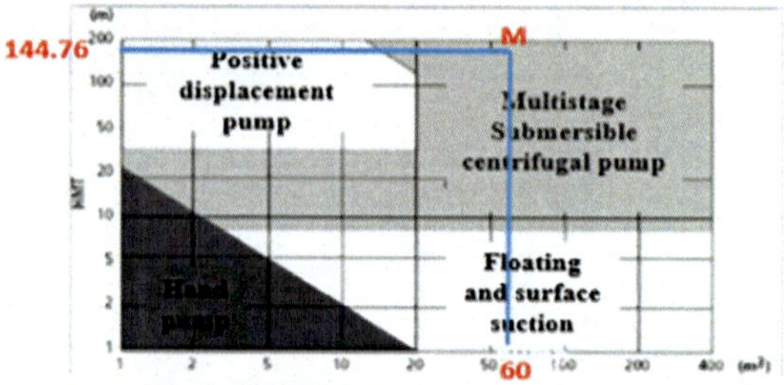

Fig. 2. The total head according to the flow rate.

The pump's electrical power in our farm is 6.63 kW, and since the power of the selected pump should be equal to or greater than 6.63 kW, we chose the submersible pump 100 R12/100 T, the characteristics of which are presented in (Table 2).

Table 2. Pump specifications 100R12/100.

Preference	100R12/100
Max flow rate (m^3/h)	9.5
Monomeric head (m)	152
Puissance (kw)	7.5

3.2 The Choice of DC/AC Inverter

The choice of an inverter depends on the electrical characteristics of the pump supplied (power, voltage) [10]. In our work, we chose variable frequency inverter drive from a Frecon is the type of PV200-4T-012G/015PB with characteristics presented in Table 3.

Table 3. The main characteristics of the inverter.

Inverter type	PV200-4T-012G/015PB
Manufacture	Frecon
Input Power at 25 °C	12 KW
Max inverter voltage Umax (input)	500 V
Optimal PV module voltage	230–500 V
Maxi inverter current	47.9 A
Yield	0.95

3.3 Design of a PV Array

The purpose of sizing is to determine the peak power of the solar generator, which depends on the sunshine data of the site and the electrical needs of the user

$$P_C = \frac{E_e}{E_{irr} * K} \tag{5}$$

E_{irr} The available solar energy during the lowest daily sunshine K $= (1 - pertes)$ is 20% losses attributable to temperature and dust.

The value of Pc is obtained $P_C = 11.9$ Kwc.

China sun energy CSUN 270-60M PV module is chosen, its characteristics are as follows Table 4:

Table 4. The parameter of PV module.

Parameter	Values	Unites
Max power Pmpp	270	Wc
Voltage Vmpp at Pmpp	30.7	V
Current Impp at Pmpp	8.77	A
Open circuit voltage (Voc)	37.9	V
Short circuit current (Isc)	9.07	A

The installation of the panel is determined according to three-step:

- The first step is a Total number of modules:

$$N_n = \frac{Pc}{Pn} = \frac{11.9 * 1000}{270} = 44 modul \tag{6}$$

where: P_c the peak power of the solar generator, P_n Maximum power of PV.

- The second step is the Number of modules connected in series:

$$Ns = E\left(\frac{V_{invmp}}{V_{oc} * 1.15}\right) = \frac{500}{37.9 * 1.15} = 11 \quad (7)$$

where: V_{inmp} the maximum inverter input voltage, V_{oc} the open-circuit voltage of the selected PV module, E is the integer part, and 1.25 is a modulus safety factor.

- The third step is the Number of parallel connections:

$$Np = E\left(\frac{I_{inmp}}{I_{sc} * 1.15}\right) = \frac{47.9}{9.07 * 1.15} = 4 \quad (8)$$

where: I_{invmp} the maximum inverter input current, I_{sc} Short circuit current of PV.

4 The Description of the Site of BNI MALLAL

After the sizing of the solar pumping system, the validation of the dimensioning of this system is done by the PVSYST software, which is an easy and comprehensive design process with an explanation of each component using diagrams [11].

4.1 Performance Analysis of Solar Pumping System

The complete one-year analysis of SPWPS shows (Table 5), which offers the value of the annual water used is 21900 m^3. At the same time, the pump system can supply about 21943 m^3, which gives 1%percent of water is missing throughout the year, which indicates that the water pumping system meets the design requirements. As we see, the array energy is 29934 KWh, and Energy at the pump side 15580 kWh, which gives the system's efficiency is 67.9%.

Table 5. Simulated results of the solar water pumping system.

	GlobEf	EArrMPP	E-Pmp	ETK-full	H-Pump	W-Pump	W-Used	W-Miss
Jan	167.1	1801	1336	256.5	147.6	1903	1860	0.000
Feb	168.3	1774	1176	313.0	147.8	1680	1680	0.000
Mar	198.4	2060	1302	392.6	147.8	1860	1860	0.000
Apr	201.1	2072	1285	457.4	147.4	1800	1800	0.000
May	197.2	2013	1327	432.4	147.1	1860	1860	0.000
Jun	202.6	2016	1288	477.5	147.1	1800	1800	0.000
Jul	205.7	2014	1327	444.8	147.2	1860	1860	0.000
Aug	203.5	1986	1292	422.8	147..3	1860	1860	0.000

(continued)

Table 5. (*continued*)

	GlobEf	EArrMPP	E-Pmp	ETK-full	H-Pump	W-Pump	W-Used	W-Miss
Sep	194.3	1937	1283	394.9	147.9	1800	1800	0.000
Oct	185.6	1882	1293	352.2	147.3	1792	1860	0.000
Nov	161.8	1705	1333	151.3	147.0	1868	1800	0.000
Dec	155.8	1673	1338	151.1	146.8	1860	1860	0.000
Year	2241.3	22934	15580	4246.3	147.4	21943	21900	0.000

GlobEff: Global Effective irradiation kWh/m, EArrMPP: Array virtual energy at MPP kWh, E-Pmp: Energy used by the pump KWh, ETKfull: Unused energy (tank full) kWh, H-Pump: Average total head at the pump meters, W-Pump: The amount of water pumped the, W-Used Water Used by the user m^3, W-Miss: Missing Water m^3.

4.2 Performance Ratio

The performance ratio (PR) is represented in (Fig. 3). It is the final yield of the PV system Yf over the reference yield Yr. The Design Performance Ratio (PR) is 63.1%, which is limited between 0.50 to 0.65. The PR result shows that the system has been installed and is working fine.

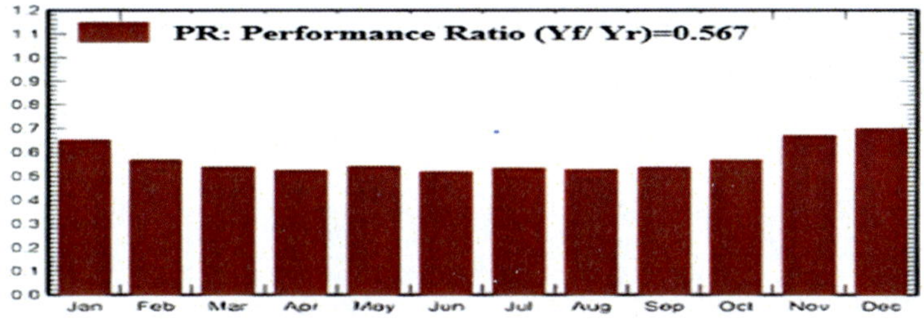

Fig. 3. Performance ratio.

4.3 Reference Incident Energy

The graph below (Fig. 4) shows the proposed Solar Photovoltaic water pumping system's energy production and loss factors distribution.

As we see, the adequate energy at the pump varies with the seasons of the year. It becomes maximum in the summer when the energy incident on the panels is higher. We also observe that the collection losses are more important than the system's (converter). They vary the same way throughout the year. Collection losses reach their maximum value of 0.78 KWh/KWp/day in the sunniest months (July and August), and the total value of converter losses surpasses 0.62 KWh/KWp/day in April.

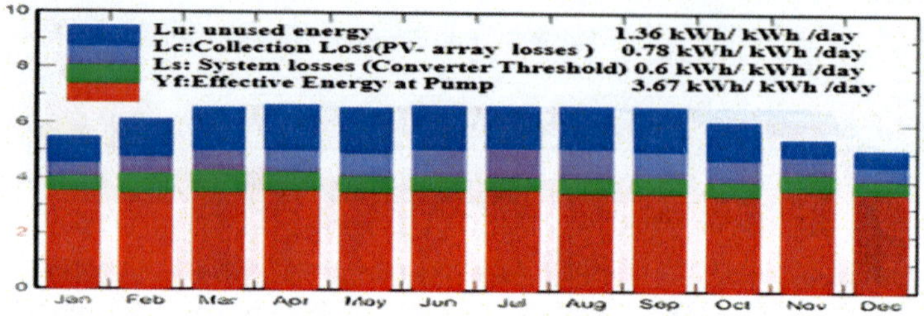

Fig. 4. Energy production.

5 Modeling of PV Pumping System and Simulation

Earlier studies helped us determine the parameters of the photovoltaic water pumping system (PVWPS). In this part, our system simulation is built using the MATLAB environment. We connected four modules in parallel and 11 modules in series to create an 11.9 kW array that can provide until 7.5 kW of the system to satisfy the system's needs and losses. The schematic circuit of the photovoltaic pumping system is shown in (Fig. 5), composed of a PV generator, a boost DC/DC converter, a voltage source inverter (VSI), a 3-phase induction motor, and a centrifugal pump. The PV generator is combined with the boost converter to control the inverter's input voltage. Then this inverter converts DC to a controllable three-phase AC supply according to the motor-pump set requirement. Our system has two controls. The first is the MPPT (P&O) control responsible for ensuring the maximum power when the irradiance changes by adjusting the boost converter's duty cycle. The second is An Indirect field-oriented control that controls the pumped flow. This technique is adequate because it ensures a decoupling between the torque and the flux with typical robustness.

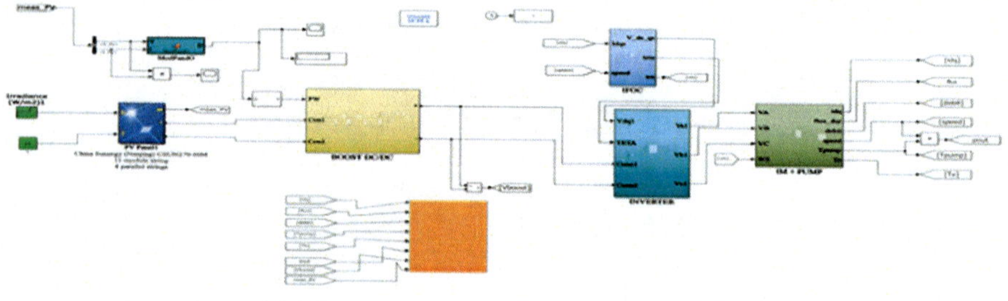

Fig. 5. The complete model of the solar pump.

The Proposed PVWPS was tested under different irradiation (500, 1000, 800) W/m^2 at (0, 1.5, 2.5) s with a constant temperature of 25 °C, which shows Fig. 6. (a).in order to see the outcome of the system radiation.

Figure 6 (b) the simulation illustrates the photovoltaic power. It started to increase until 6.8 KW and stabilize until 1.5 s under 500 W/m^3 irradiation, increased reach to 11.5 KW when the irradiation is 1000 W/m^3 and then dropped to 7.5 KW when the irradiance increased to 800 W/m^3.

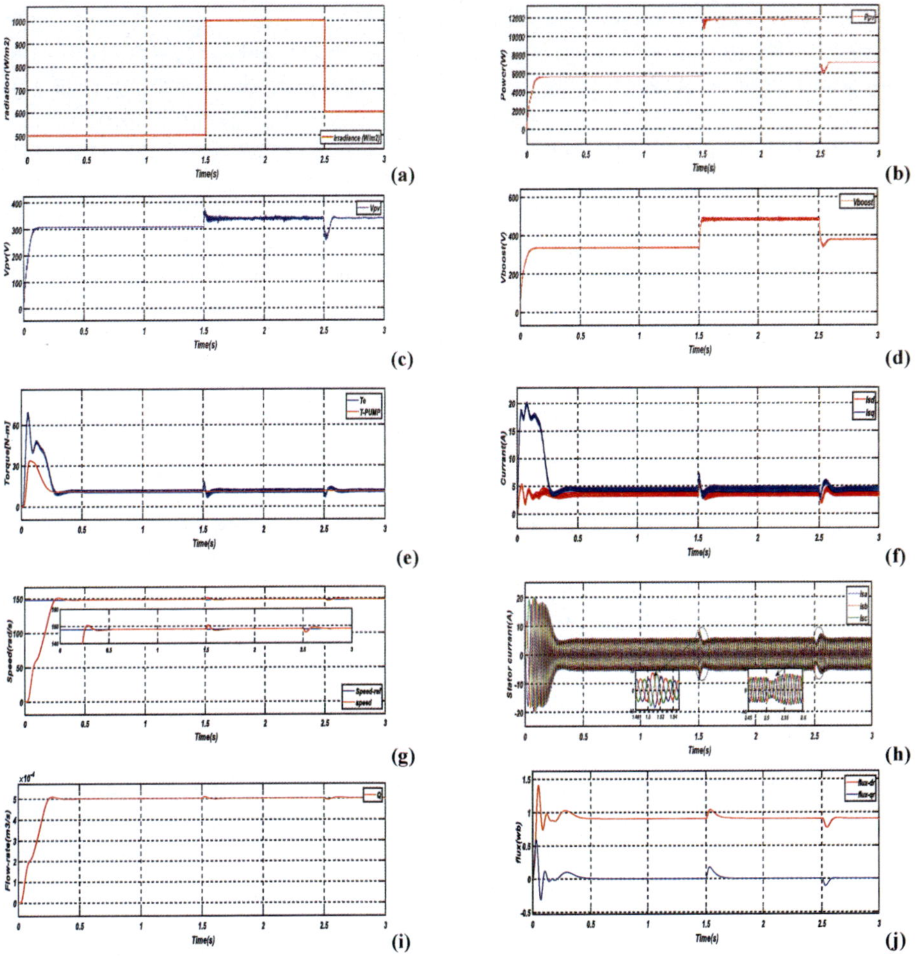

Fig. 6. (a) Solar radiation Profile. (b) The variation of Photovoltaic Power. (c) The output voltage of the PV panel. (d) The output voltage of the boost converter. (e) Electromagnetic Torque. (f) Currents wave-forms Isdq. (g) Mechanical Speed. (h) The Stator currents. (i) water flow evolution. (j) Rotor flux Φrdq.

Figure 6 (c). (d) shows the input and output of the boost converter that increases/decreases with increasing/decreasing irradiation. We can notice that the output voltage of the boost increases the PV voltage level up with less oscillation in a steady state due to the operating point of the MPP control, which is responsible for extracting the maximum power from the PV generator under sudden change of irradiation.

Figure 6 (e) illustrates the variation of electromagnetic torque. The results appear some distortion in torque at the beginning witch increase until 70 [Nm] and stabilize the calculated values with a slight rise at 1.5 s decrease at 2.5 s results from the irradiance change.

Figure 6 (f) demonstrates the variation of the direct current Isd and the quadratic current Isq. First, from the simulation, we can observe that Isd is proportional to the rotor flux. In contrast, the variation of Isq is commensurate with torque due to the use of IFOC control that is responsible for decoupling between the flux and torque. Second, we can see that the currents change under different irradiation levels (500, 1000, 800) W/m^3 at the time (0, 1.5, 2.5).

Figure 6 (g) shows the rotor speed of the IM. From the representative speed, we can see that on start-up, the speed increases sharply and then gradually returns to reach its reference value with a slight increase at 1.5 s decrease at 2.5 s, which is caused by the frequency of the inverter due to the change of the irradiance.

Figure 6 (h) shows the three-phase stator currents of the induction motor, which started with an increase of 20 A at a time of 0.2 s and then decreased until it reached a steady state. Then return to increase (8 amps at 1.5 s) and (9 amp at 2.5 s) due to the increase/decrease of motor speed due to radiation. The significant increase in current increases in core losses and thus the effect of photovoltaic solar pump performance degradation.

Figure 6 (i) ϕrd follows the set-point of 0.8 Wb and ϕrq remains zero a few moments after start-up and this throughout the operating cycle with a slight increase/decrease under variations in irradiation (500, 1000, 600) w/m^3 to (0,1,5,2,5).

The pump's flow rate in Fig. 6 (j) shows that the flow is proportional to the motor speed and achieves the actual calculated value.

6 Conclusion

In this work, the solar pumping system was designed and simulated for irrigating palm trees in a farm located in Beni Mellal.

The resulting design dedicated a good photovoltaic generator with 44 modules with a peak power of 270 each. Thus, the peak power of the PV generator is 11.44 KWp to provide 60 m^3 water per day.

The simulation is validated by PVSYST software, which works at an efficiency System of 67.9% and a performance ratio of 0.579. The requisite year water requirement 2900 m^3/year was computed with 1% of water missing throughout the year, which dedicates the energy performance of the SWPS is satisfied the water consumption needs.

Finally, the simulation under Matlab Simulink of a photovoltaic pumping system validated the calculation section. Electromagnetic torque, speed, and stator current show slight changes when the irradiance changes then return to follow their reference values. The simulation also dedicated that the water flow is proportionate to speed and gives good efficiency during the solar intensity increases.

References

1. Markson, I., Simeon, O., Iniobong, E., Abasi, O.: Sizing of solar water pumping system for irrigation of oil palm plantation in Abia State. Univers. J. Eng. Sci. **7**(1), 8–19 (2019)
2. Rakhi, S., Shivanshu, S., Sumit, T.: Design optimization of solar PV water pumping system. In: International Conference on Mechanical and Energy Technologies, no. 3, Vol. 21, pp. 1673–1679 (2020)

3. Hameed, H.: Design and implementation of photovoltaic water pumping system for use in Iraq. J. Eng. Sustain. Dev. **22**(06), 2520–0917 (2018)
4. Krishna, K., Kulkarni, P.S.: Design and simulation of solar photovoltaic water pumping system. In: International Conference on Energy Power and Environment (2021)
5. Abdennabi, B., Abdelouahed, A., Ghizlane, C., Aumeur, E.: Design and optimal choice of a 1.5 Kw photovoltaic pumping system for irrigation purposes. In: AIP Conference Proceedings
6. Asmae, H., Mohammed, K., Mohammed, B., Aziz, B.: A new approach for optimal sizing of photovoltaic pumping systems for irrigation. In: International Conference on Renewable Energies for Developing Countries (2020)
7. Naveen, K., Sudhakar, K.: Sizing and modelling of photovoltaic water pumping system design and optimal choice of a 1.5 kW photovoltaic pumping system for irrigation. In: AIP Conference Proceedings (2018)
8. Bossoufi, B., Karim, M., Taoussi, M., Aroussi, H.A., Bouderbala, M.: DSPACE-based implementation for observer backstepping power control of DFIG wind turbine. IET Electr. Power Appl. **14**(12), 2395–2403 (2020)
9. Ahmed, C., Mohamed, K., Massinissa, A.: Sizing a PV pumping system for an Algerian remote site. In: International Renewable Energy Congress (2015)
10. Santosh, R., Vikas, K.: Sizing and modelling of stand-alone photovoltaic water pumping system for irrigation. Energy Environ. **29**(4), 1–19 (2018)
11. Lamnadi, M., Trihi, M., Bossoufi, B., Boulezhar, A.: Comparative study of IC, P&O and FLC method of MPPT algorithm for grid connected PV module. J. Theor. Appl. Inf. Technol. JATIT **89**(1), 242–253 (2016)

An Improved Sinusoidal (SPWM) and Space Vector (SVPWM) Current Control for Multi-levels Three-Phase Photovoltaic Inverter Connected to a Non-linear Load

Abdelhak Lamreoua[1](✉), Anas Benslimane[2], Jamal Bouchnaif[1], and Mostafa El Ouariachi[1,2]

[1] Laboratory of Electrical Engineering and Maintenance (LEEM), BP: 473 Higher School of Technology, University of Mohammed I, Oujda, Morocco
a.lamreoua@gmail.com

[2] Laboratory Renewable Energy, Embedded System and Information Processing, National School of Applied Sciences, University of Mohammed I, Oujda, Morocco

Abstract. This article; is a contribution towards the improvement of the control of the three-phase two-level and multi-level photovoltaic inverter, with a new control strategy, by the sinusoidal (SPWM) and space vector (SVPWM) control for a non-linear load., so that the latter produces a voltage wave closer to that of the grid than that supplied by a conventional inverter. This contribution is based on the strategies of indirect control of the active and reactive powers injected by this inverter into the grid, to respond, in terms of cost and energy efficiency, to the problems encountered with the photovoltaic inverter. After comparing the results of the two systems, our findings suggest that the current THDi of the interlaced (multi-level) inverter is lower than that obtained with the two-level inverter, providing higher quality waveforms, reducing current and losses caused by high-frequency harmonics. The simulation results demonstrated the effectiveness of these techniques proposed in this work.

Keyword: Modeling · Three-phase inverter · Sinusoidal SPWM · Vector SVPWM · Total Harmonic Distortion (THD) · Multilevel inverter

1 Introduction

Currently, photovoltaic (PV) is the most promising clean source [1]. Since harmonic distortion sources have become an increasingly serious concern as more PV is integrated into the grid system, and can represent a high report of the total in small-scale systems [2, 3]. Many efforts have already been done to minimize the power quality issues are a particular problem for this converter [4–6]. Unfortunately, there are several disadvantages inherent to a conventional inverter, including larger size, a higher number of power switches, and higher cost, as well as the use of complex control methods [7, 8]. To avoid these problems of currents loss, we have proposed a new control and harmonic

S. Motahhir and B. Bossoufi (Eds.): ICDTA 2022, LNNS 454, pp. 672–683, 2022.
https://doi.org/10.1007/978-3-031-01942-5_67

compensation method for the three-phase multi-level inverter, based on space vector (SVPWM) and sinusoidal control (SPWM), like the one we applied on the inverter two levels [9, 10]; to maximizes exploitation of the converter hardware, limiting the effect of an inherent third harmonic and correct the unbalanced system [4, 11]. The proposed method control is based on the control of active and reactive power indirectly injected into the grid, and to respond, in terms of cost and energy efficiency, to the problems encountered with the photovoltaic inverter, to effectively filtered the harmonics of the grid current [12, 13]. These multilevel converters allow us to provide higher quality waveforms, reducing current and voltage ripple after reducing losses caused by high-frequency harmonics [14, 15]. This article presented here is about the simulation and modeling of a system *PV* connected to the power grid using an interlaced multi-level inverter with four cells in parallel, in a synchronously rotating d-q reference frame, to optimize the losses in the semiconductors as well as the improvement of yield and financial cost. The work presented here is about the proposed control and the modeling of the photovoltaic inverter presented in Sect. 1. the results obtained and the comparison of the order of the two and multilevel structures are presented in the second section.

2 Three-Phase Multi-level Inverter Modeling

2.1 Three-Phase Inverter Modeling

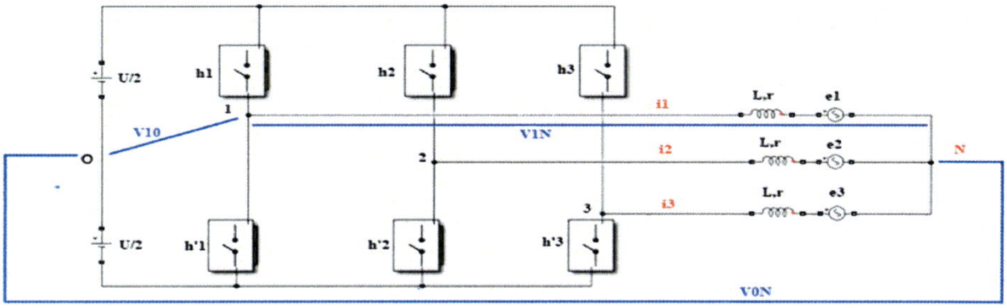

Fig. 1. Block diagram of the three-phase PV inverter

A three-phase inverter cannot be treated as three independent single-phase inverters and therefore special measures must be added to ensure the operation of the inverter.

The three-phase structure of Fig. 1 shows three connection functions (h1, h2 and h3), three phase-source voltages (v10, v20, v30), three phase voltages (v1N, v2N, v3N) and a qualified voltage v0N zero sequence voltage such as:

$$v_{k0} = h_k * \frac{U}{2} \tag{1}$$

$$v_{k0} = \frac{1}{3}\begin{vmatrix} 2 & -1 & -1 \\ -1 & 2 & -1 \\ -1 & -1 & 2 \end{vmatrix}\begin{vmatrix} h1 \\ h2 \\ h3 \end{vmatrix} * \frac{U}{2} \tag{2}$$

2.2 Three-Phase Multi-level Inverter Modeling (Interlaced Inverter Q = 4):

Figure 2 shows the proposed architecture of the system *PV* connected to the power grid using an interlaced inverter for one phase (q = 4). Each phase of the inverter is made up of $q = 4$ arms. The photovoltaic generator and the input capacitors of the inverter are represented by two voltage sources in series *Vdc*/2. Since the system is symmetrical, we will only be interested in one phase here. The $q = 4$ inductors $L1$ represent the filter inductances on the inverter side.

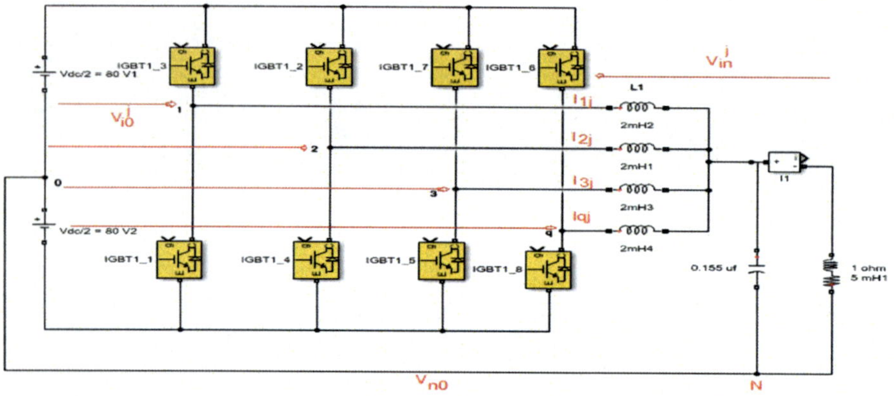

Fig. 2. Diagram of the *PV* system connected to the interlaced inverter grid (for one phase)

2.3 The Proposed Control Method

This command is similar to that applied to the conventional system with the difference that in part *MLI*, four carriers regularly phase-shifted by *Tsw*/4 are used. In this part, we directly control the output current of the inverter fed into the grid. The output current of the inverter is equal to the sum of the currents of the cells. Our command PWM is of type PWM intersective. The current loop generates the reference voltages $Ud*$ and $Uq*$ which are transformed in reverse Park to obtain three sinusoidal voltages in the reference *abc* noted $Vx*$ ($x = a, b, c$). Typically, the DQ control methods are associated with PI controllers because they have satisfactory behavior when regulating DC variables.

The matrix [L] transform the three-phase currents into the dq0 system as shown in (3)

$$\begin{bmatrix} i_d \\ i_q \\ i_0 \end{bmatrix} = [L] \begin{bmatrix} i_A \\ i_B \\ i_C \end{bmatrix} \quad with \quad [\mathrm{L}] = \sqrt{\frac{2}{3}} \begin{bmatrix} \sin\alpha & \sin(\alpha - \frac{2\pi}{3}) & \sin(\alpha + \frac{2\pi}{3}) \\ \cos\alpha & \cos(\alpha - \frac{2\pi}{3}) & \sin(\alpha + \frac{2\pi}{3}) \\ \frac{1}{\sqrt{2}} & \frac{1}{\sqrt{2}} & \frac{1}{\sqrt{2}} \end{bmatrix} \tag{3}$$

The reference currents I*d and I*q can be calculated from:

$$\begin{gathered} \mathrm{P{=}V_dI_d{+}V_qI_q} \overset{V_{q=0}}{\Longrightarrow} \mathrm{I^*{}_d{=}\frac{P^*}{V_d}} \\ \mathrm{Q{=}V_dI_q{-}V_qI_d} \overset{V_{q=0}}{\Longrightarrow} \mathrm{I^*{}_q{=}\frac{Q^*}{V_d}} \end{gathered} \tag{4}$$

The references voltages V*d and V*q can be obtained from the previous equation:

$$\begin{gathered} V_d^* = k_p(I_d^* - I_q) + K_I \int \left((I_d^* - I_q\right)dt - \omega L_f I_d + V_d \\ V_q^* = k_p(I_q^* - I_q) + K_I \int \left((I_q^* - I_q\right)dt - \omega L_f I_q + V_d \end{gathered} \tag{5}$$

The transfer function in the system DQ is given by Eq. (6) (Fig. 3):

$$\mathrm{G_{PI}^{(dq)}(s)} = \begin{bmatrix} \mathrm{K_p + \frac{K_i}{s}} & 0 \\ 0 & \mathrm{K_p + \frac{K_i}{s}} \end{bmatrix} \tag{6}$$

Phase-Locked Loop Algorithm. The PLL technique can be used to maintain the synchronism between the inverter and the grid (Fig. 4).

In this algorithm, the grid voltages are transformed to the stationary frame variables (Vα, Vβ) by the Clarke rotation matrix. After, the Vα and Vβ voltages are converted to the rotating (synchronous) frame variables (Vd, Vq) by the Park rotation matrix. The estimated phase angle (θ*) of the grid voltage is fed back to operate the ABC to DQ block so that the Park rotation can be performed (Fig. 5, Table 1).

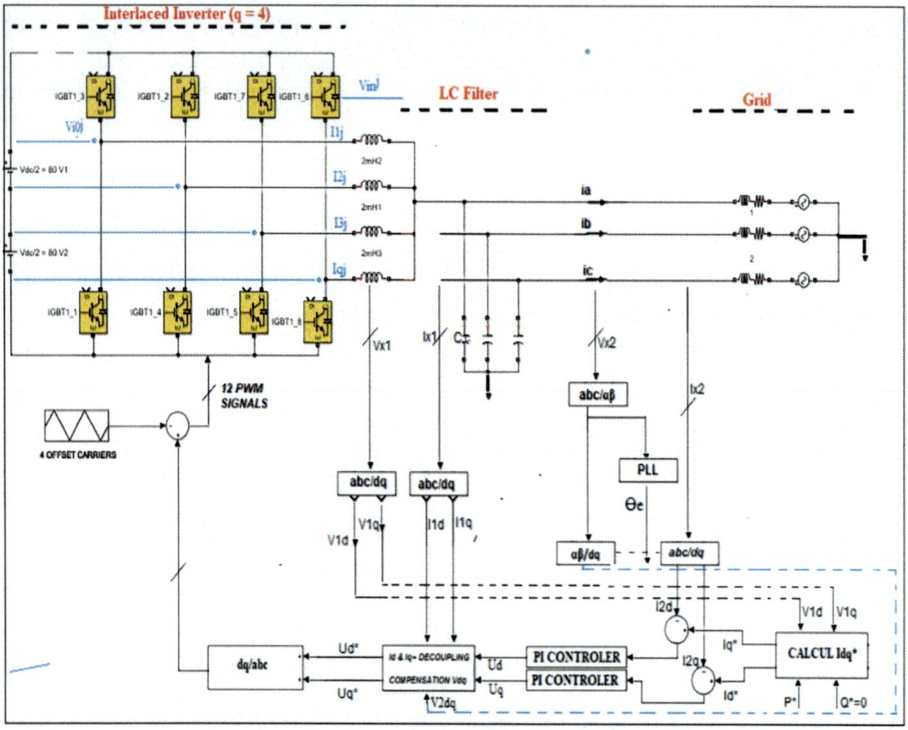

Fig. 3. Block diagram of the proposed control strategy

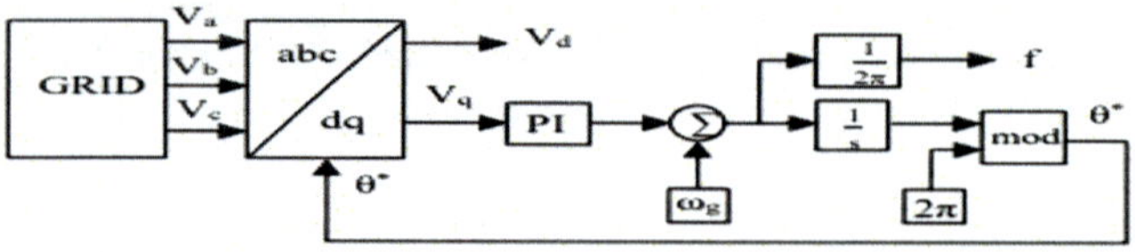

Fig. 4. Block diagram of the PLL algorithm

Table 1. Design parameters

Photovoltaic power	500 W
V_{dcref}	80 V
I_{min}	1 A
fg	50 Hz
f_{pwm}	10 kHz

3 Simulation Results

3.1 Three-Phase Two-Level Photovoltaic Inverter Connected to the Grid

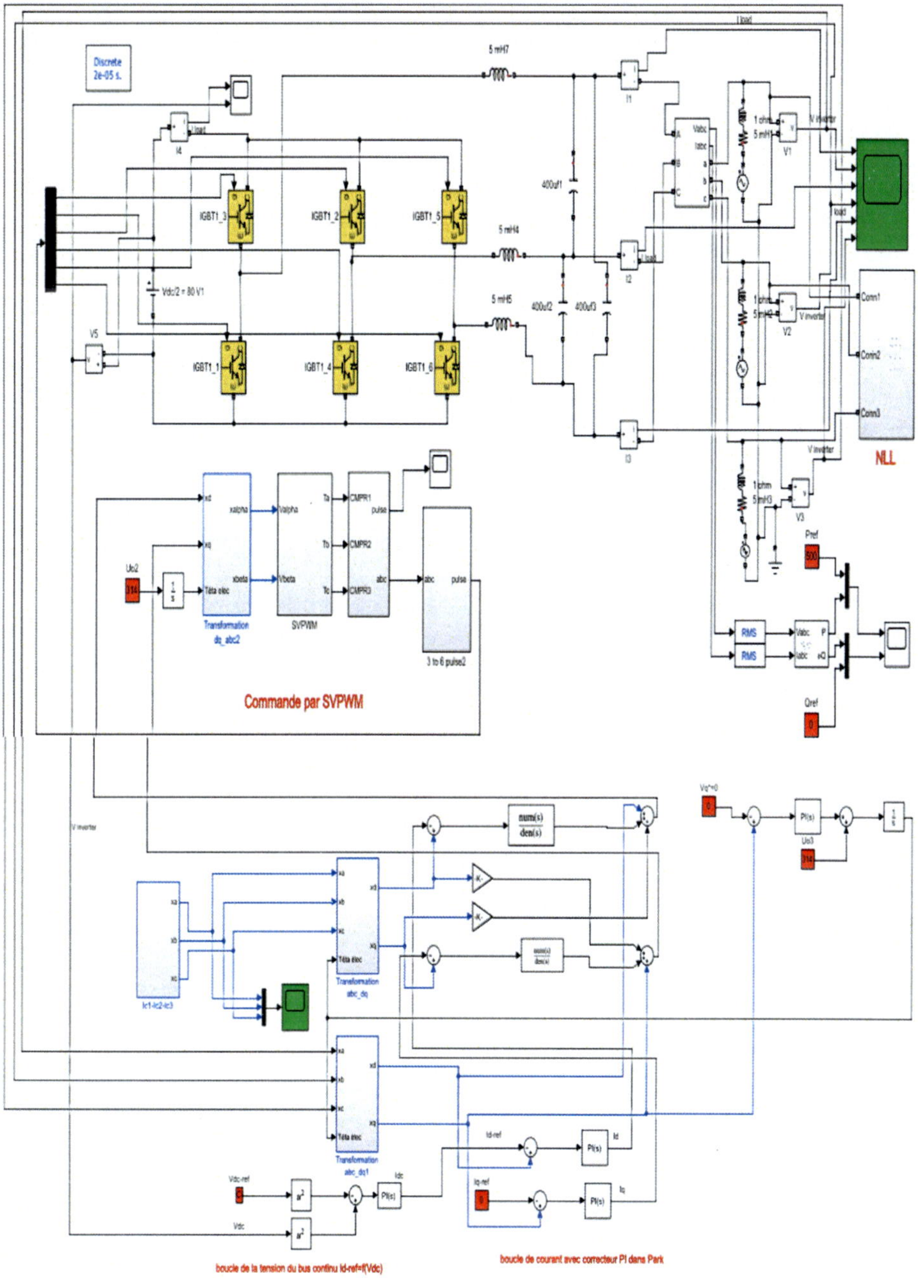

Fig. 5. Simulation model of three-phase photovoltaic inverter connected to the grid

Output current and voltage waveforms by sinusoidal control (SPWM)

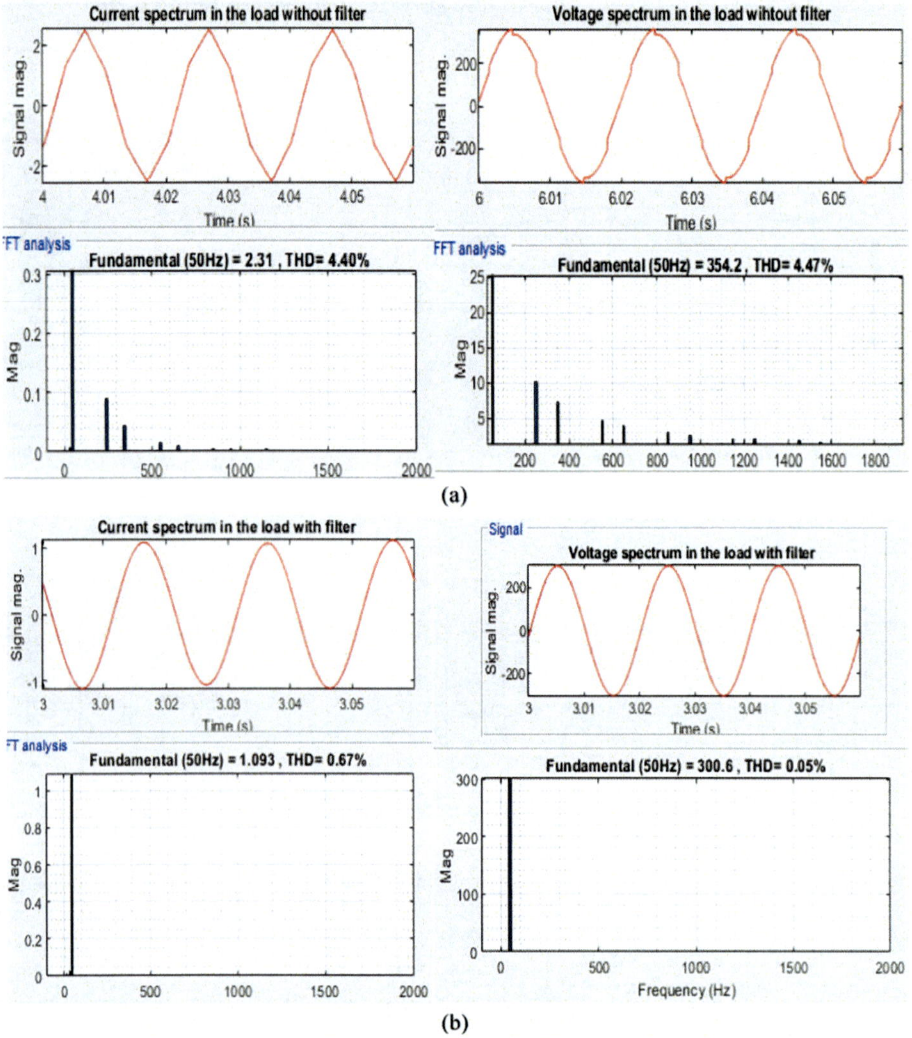

Fig. 6. Grid current and voltage FFT spectrum (a) without filtering (b) with filtering

Before Filtering. Figure 6(a) shows the harmonic spectra, without the harmonic compensation loop. the system current and voltage harmonic distortion rate is: THD value of the current and voltage equal to 4.40% and 4.47% respectively, with an output signal rich in harmonics of multiple order and are very far from the fundamental, cause power quality issues in the grid due to the connection to the non-linear load.

With Filtering. A system waveform using a harmonic compensation is shown in Fig. 6(b). The output harmonics and unbalanced current of phase in the PV inverter are compensated, when the active filter is connected to the system; Fig. 6(b) provides the results obtained from the simulation after the connection of dedicated compensation devices that can reduce the THD in the system from to THDi = 4.40% and THDv = 4.47% to THDi = 0.68% and THDv = 0.05% respectively < < 3%(international standard).

Output current and voltage waveforms by Space Vector control (SVPWM):

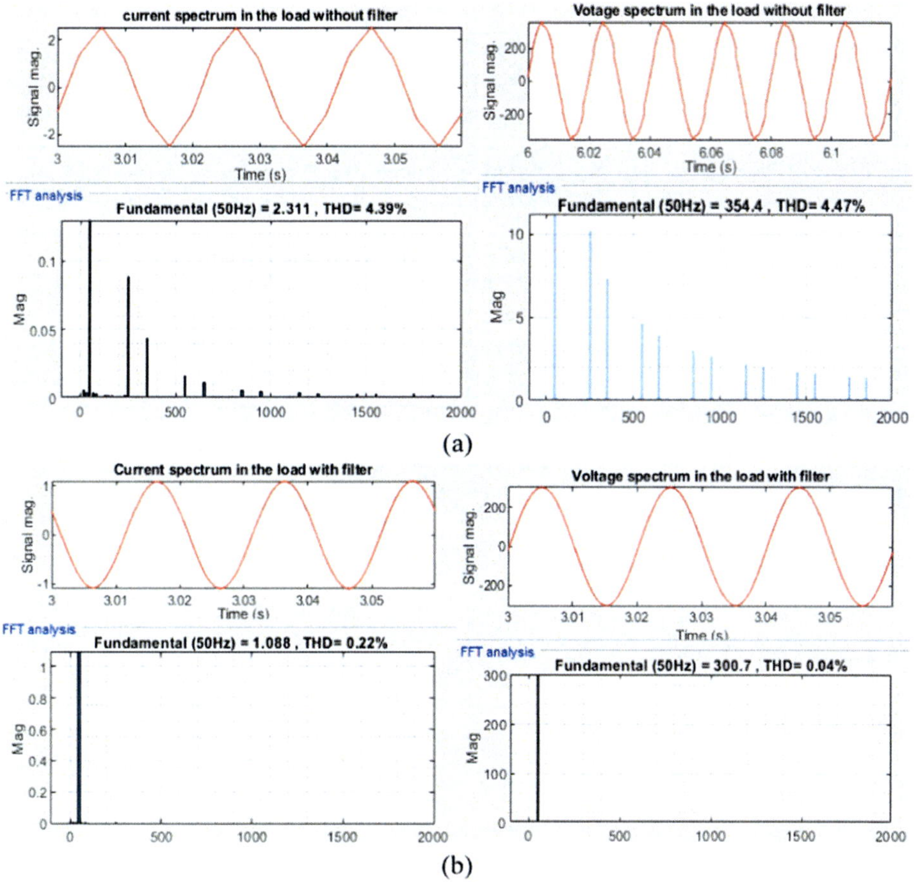

Fig. 7. Grid current and voltage FFT spectrum (a) without filtering (b) filtering

Before Filtering. When the load is connected to the grid, the THD in the system without any compensation is THDi = 4.39% and THDv = 4.47%. Figure 7(a) shows that the fundamentals obtained depend on those of reference with harmonics of large amplitudes and of frequencies close to that of the high carrier, which guarantees easy fundamental filtering.

With Filtering. Figure 7(b) shows that the output current quality for the PV inverter is improved after using the proposed harmonic compensator loop. The effects of the proposed control method can be observed by comparing the THD value of the system, which is lower than that obtained with the regulation before filtering, the THDi of current and voltage equal successively 0.22% and 0.04% < < 3%. In the end, the effects of the method proposed by SPWM and SVPWM control after compensation can be clearly identified.

3.2 Multi-level Three-Phase Inverter(Interlaced Inverter Q = 4):

Before the filtering of frequency harmonics in Fig. 8 (a), the one-shaped output signals are not purely sinusoidal because of the harmonics of currents and voltage. After applying to filter, the current and voltage curves in Fig. 8(b), the waveforms are almost sinusoidal and identical. Figure 8 shows the grid current and voltage FFT spectrums before and after filtering. We can notice the difference between FFT spectrums before and after

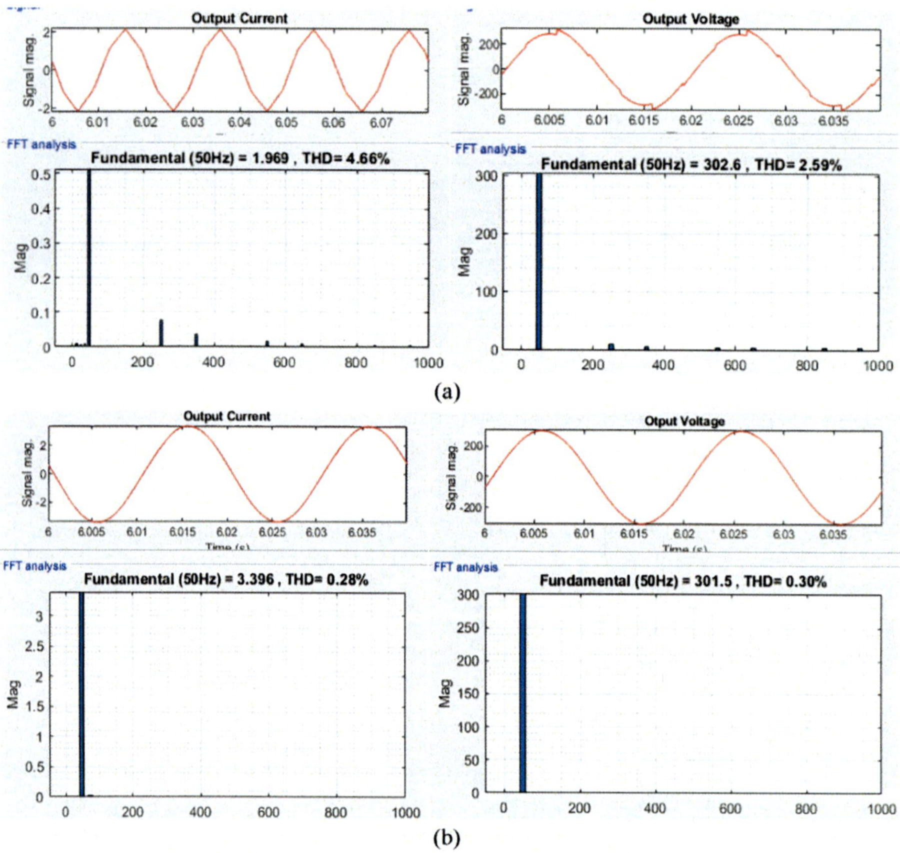

Fig. 8. Grid current and voltage FFT spectrums; (a) before filtering, (b) after filtering

filtering, such as, after filtering is perfectly sinusoidal while that before filtering is not. Figure 8(b). The Toral Harmonic Distortion of current and voltage are respectively THDi = 0.28% and THDv = 0.30% for the 50 first harmonics, unlike the indicators of *THD* obtained by the system before filtering (4.66% and 2.59%), the indicators *THD* of the system using an interlace inverter after filtering are much better. This shows that the *THD* indicators of the system using an interlaced inverter are much better than those obtained with the conventional inverter.

So by comparing the values and from the spectral analysis, it could be seen that the total THD of the multi-level three-phase photovoltaic inverter topology is equal to THDi = 0.28%(SPWM)- THDi = 0.23%(SVPWM) and THDv = 0.30%, is lower than in the two-level three-phase photovoltaic inverter typologies. Thus, we were able to observe a clear improvement in the transient regime and the waveforms in the case of the interlaced inverter system compared to the classic system. The *THD* of current and voltage are respectively 0.28% (SPWM)-0.23%(SVPWM) and 0.30% in the case of the system with interlaced inverter against 0.67%(SPWM)-0.22%(SVPWM) and 0.05% in the case of the two-level three-phase photovoltaic inverter (Fig. 9).

Active and reactive powers in the load.

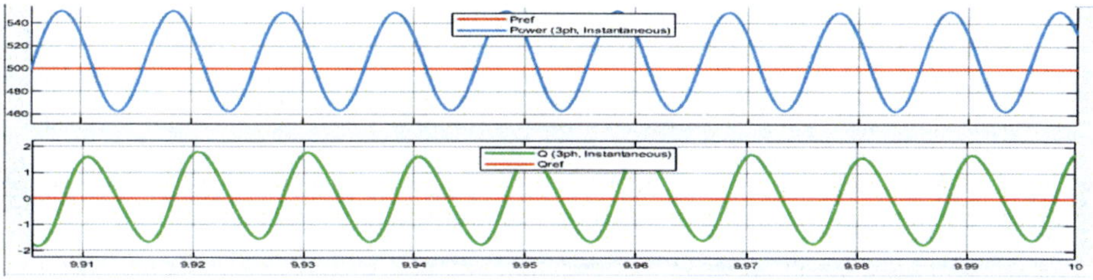

Fig. 9. Active and reactive power in the load and their references

The active and reactive power in the load in their references has an average value equal to the reference power value and the reactive power is maintained at zero, for the multi-level inverter typology after compensation.

4 Conclusion

This article; is a contribution towards the improvement of the control of the three-phase two-level and multi-level photovoltaic inverter, with a new control strategy, by the sinusoidal (SPWM) and space vector (SVPWM) control for a nonlinear load.

In the comparison of the results of the two systems, our findings suggest that the current THDi of the interlaced (multi-levels) inverter is lower than that obtained with the two-level inverter, providing higher quality waveforms, reducing current and losses caused by high-frequency harmonics. We notice too that the performance of the multi-level three-phase photovoltaic inverters is better than that obtained with the two-level three-phase photovoltaic inverter. We were to observe a clear improvement in the transient regime and the waveforms in the two cases. From the simulation results, we can demonstrate the efficiency of the multi-level three-phase photovoltaic inverters. The *THD*i of current are respectively 0.28% (SPWM) and 0.23% (SVPWM) in the case of the system with interlaced inverter against 0.67% (SPWM) and 0.22% (SVPWM) for the two-level photovoltaic inverter and obtain a high-quality grid (THDi < 5%). We result that this method responds, in terms of cost and energy efficiency, to the problems encountered with the photovoltaic inverter, and effectively compensates the harmonics current injected in the grid. To justify the suitability of the proposed inverter control, we started to validate the simulations performed via an experimental setup in the next step.

References

1. Naderipour, A., Guerrero, J.M.: An improved synchronous reference frame current control strategy for a photovoltaic grid-connected inverter under unbalanced and nonlinear load conditions. PLOS ONE (2017). https://doi.org/10.1371/journal.pone.0164856
2. Vadlamudi, S.D.V.R.: Decoupled DQ-PLL with positive sequence voltage normali-zation for wind turbine LVRT control, 25–27 October 2016, Sands Expo and Conven-tion Centre, Marina Bay Sands, Singapore (2016)
3. Bengourina, M.R.: Direct power control of a grid connected photovoltaic system, associated with an active power filter. Revue des Energies Renouvelables **20**(1), 99–109 (2017)
4. Purushotham, M.: Reinforced droop for active current sharing in parallel NPC inverter for Islanded AC microgrid application. MDPI Energies (2019)
5. Mahamat, C.,Petit, M., Costa, F.: Analyse et commandes des convertisseurs multi-niveaux pour un générateur photovoltaïque connecté au réseau électrique. Energie électrique. Université Paris Saclay (COmUE) (2018). Français. ffNNT: 2018SACLN024
6. Hasan, M.M., Abu-Siada, A., Islam, M.R.: Design and implementation of a novel three-phase cascaded half-bridge inverter. IET Power Electron. (2016). ISSN 1755-4535
7. Vadlamudi, S.D.V.R.: Decoupled DQ-PLL with positive sequence voltage normali-zation for wind turbine LVRT control. In: Sands Expo and Conven-tion Centre, Marina Bay Sands, Singapore, 25–27 October 2016
8. Lamreoua, A., Benslimane, A., El Ouariachi, M.: Modélisation et simulation des commandes directes et indirectes d'onduleurs photovoltaïque triphasé 2 niveaux connec-té au réseau. In: CIMAM 2018, Oujda, 07–08 Décembre 2018
9. Lamreoua, A., Benslimane, A., Bouchnaif, J., El Ouariachi, M.: An improved sinusoidal (PWM) and vector (SVPWM) Current control for a three-phase photovoltaic inverter connected to a non-linear load. In: Hajji, B., Mellit, A., Marco Tina, G., Rabhi, A., Launay, J., Naimi, S. (eds.) Proceedings of the 2nd International Conference on Electronic Engineering and Renewable Energy Systems. ICEERE 2020. LNEE, vol. 681, pp. 481–494. Springer, Singapore (2021). Online ISBN: 978-981-15-6259-4, https://doi.org/10.1007/978-981-15-6259-4_51
10. Babaei, E., Alilu, S., Laali, S.: A new general topology for cascaded multilevel inverters with reduced number of components based on developed H-bridge. IEEE Trans. Ind. Electron. **61**(8), 3932–3939 (2014)
11. Wu, T.-F., Chang, C.-H., Lin, L.-C., Yu, G.-R., Chang, Y.-R.: A D- Σ digital control for three-phase inverter to achieve active and reactive power injection. IEEE Trans. Ind. Electron. **61**(8), 3879–3890 (2014)
12. Chaouali, H., Othmani, H.: Improving the control strategy of a standalone photovoltaic pumping system by fuzzy logic technique. (IJACSA) Int. J. Adv. Comput. Sci. Appl. **8**(3) (2017)

13. Jain, P., Deshmukh, S.P.: Design of three-phase five-level cascaded H bridge inverter with boost converter. Int. J. Electron. 478–498 (2020)
14. Wu, T.F., Misra, M., Lin, L.C., Hsu, C.W.: An improved resonant frequency based systematic LCL filter design method for grid-connected inverter. IEEE Trans. Ind. Electron. (2017). https://doi.org/10.1109/TIE.2017.2682004
15. Noman, A.M., Addoweesh, K.E., Alabduljabbar, A.A., Alolah, A.I.: Cascaded H-Bridge MLI and three-phase cascaded VSI topologies for grid-connected photovoltaic systems with distributed MPPT. Int. J. Photoenergy **2019**, 22. Article ID 7642919. https://doi.org/10.1155/2019/7642919

Parametric Identification by Improved Levenberg-Marquardt Method of Solar Cell's Double Diode Model

Fayrouz Dkhichi(✉)

Network, Computer Science, Telecommunications and Multimedia Laboratory, Electrical Engineering Department, Superior School of Technology, Hassan II University of Casablanca, Casablanca, Morocco

dkhichi.fayrouz@gmail.com, fayrouz.dkhichi@etu.univh2c.ma

Abstract. The study of a photovoltaic module response is ensured by analyzing the behavior of its solar cells. To do this, the identification of the structural parameters of the solar cells is primordial. In literature, we distinguish several electrical models of solar cells which we consider the single and the double diode models. In this paper, we use an improved method "Levenberg-Marquardt combined with simulated annealing" to determine the five and the seven electrical parameters of the considered models. The performance of the proposed method is evaluated by comparing its identification accuracy with other methods discussed in the literature. A comparative study shows the better accuracy provided by the double diode model compared to the single diode model.

Keywords: Parametric identification · Single diode model · Double diode model · Improved Levenberg-Marquardt method

1 Introduction

The modeling of the intrinsic structure of the solar cells is crucial. In this context, the macroscopic, electronic, and physical phenomena are presented by electrical components. We distinguish in literature different electrical equivalent models, according to the intrinsic phenomena which are considered in each one. The models that we are dealing with in this study are the single and the double diode models, which are more reliable than the others. The single diode model presents five intrinsic parameters while the double diode contains seven parameters. The knowledge of the values of these structural parameters allows us to analyze the behavior of the cell and to know the intrinsic aspects and phenomena that influence the most photovoltaic module response. In this work, we will identify the five and the seven electrical parameters using the Levenberg-Marquardt combined with the Simulated Annealing method (LMSA). We have already proved the ability of this method to identify the five electrical parameters in previous work, published in the "solar energy" journal by Elsevier [1]. In this present work, we

S. Motahhir and B. Bossoufi (Eds.): ICDTA 2022, LNNS 454, pp. 684–692, 2022.
https://doi.org/10.1007/978-3-031-01942-5_68

have applied the LMSA algorithm on the two-diodes model for the first time, to test its identification ability facing the strong non-linearity that characterizes the model. The comparison was made between the error function values obtained by the single and the double diode models to decide which one is the most reliable. On other hand, the high accuracy results of parameter identification obtained by the proposed method are proved by comparing its values with those of other methods reported in the literature for each model. For single diode model we report as compared methods "Grouping-based Global Harmony Search" (GGHS) [8], "Artificial Bee Swarm Optimization" (ABSO) [9], "Chaos Particle Swarm Optimization" (CPSO) [10], Newton [7], "Particle Swarm" (PS) [12], "Simulated Annealing" (SA) [13]. While, the methods used to prove the high performance of LMSA technic for parametric identification of solar cell's double diode model are: "Innovative Global Harmony Search" (IGHS) [6], GGHS [8], "Harmony Search" (HS) [14] and PS [12] methods.

2 Single and Double Diode Models of Solar Cell

A solar cell can be modeled by an electrical equivalent circuit based on one diode (Fig. 1), the latter presents two electrical parameters: the saturation current Is and the ideality factor n. the diode is shunted with a current source that generates the photocurrent I_{ph}. Apart from this current feed a shunt resistance R_{sh} and a series resistance R_s. these two resistances respectively present the losses due to the various contacts and connections and the leaks caused by the imperfections of the material forming the cell.

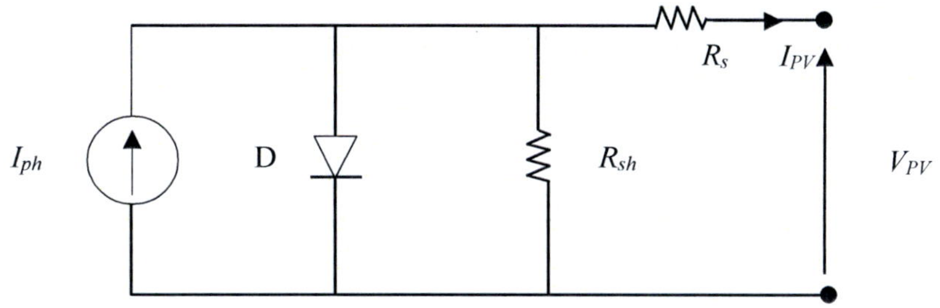

Fig. 1. Electrical circuit model of solar cell

To study from a close cell's behavior, we should use the mathematical equation derived from the electrical circuit presented in Fig. 1. The expression of this equation is given as follow:

$$I_{pv} = I_{ph} - I_s\left[\exp\left(\frac{V_{pv} + R_s I_{pv}}{nV_{th}}\right) - 1\right] - \frac{V_{pv} + R_s I_{pv}}{R_{sh}} \quad (1)$$

For the double diode model (Fig. 2), a solar cell is presented by two diodes and by Iph, Rs, and Rsh parameters. Each diode is modeled by its characteristic parameters. (Is1, n1) for the first one and (Is2, n2) for the second one.

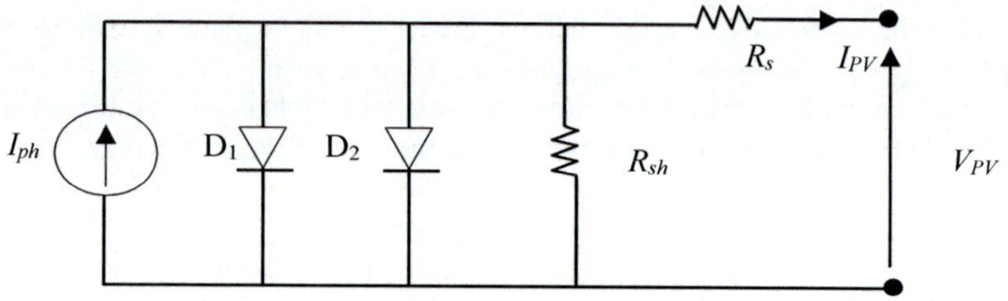

Fig. 2. Double diode model of solar cell

The mathematical equation for a double diode model linking between IPV, VPV, and the seven electrical parameters is expressed as follow [2]:

$$I_{PV} = I_{ph} - I_{s1}\left[\exp\left(\frac{V_{PV} + R_s I_{PV}}{n_1 V_{th}}\right) - 1\right] - I_{s2}\left[\exp\left(\frac{V_{PV} + R_s I_{PV}}{n_2 V_{th}}\right) - 1\right] - \frac{V_{PV} + R_s I_{PV}}{R_{sh}} \quad (2)$$

3 Parametric Identification Process

During the minimization of the Sum Squared Errors Function (SSEF) between the model and the experimental points of Ipv (Vpv), the values of the intrinsic electrical parameters of the model are adjusted until reaching their optimal values. The principle of adjusting these parameters is ensured according to the process represented by the diagram shown in Fig. 3.

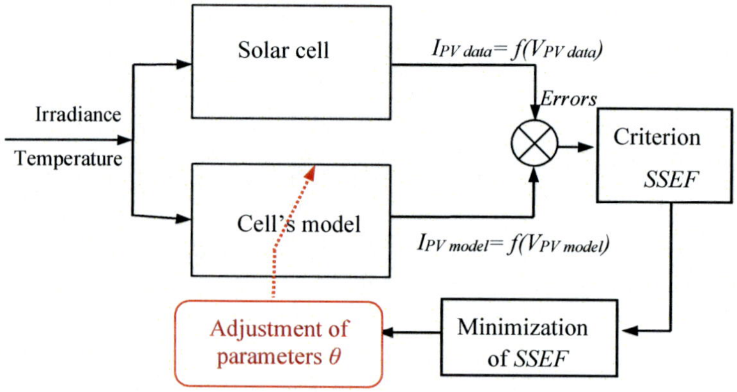

Fig. 3. Synoptic diagram of the non-linear optimization process of the errors for parametric identification

The SSEF function is defined by the following equation [3]:

$$g(\theta) = SSEF = \sum_{i=1}^{N} \left[(I_{PV\,data}(i) - I_{PVmodel}(i))^2 \right] \quad (3)$$

With:
IPV model: Equation of solar cell,
IPV data: Experimental values of the solar cell,
i: Index of data values,
N: Number of data values.

The deviations generated at each point between the model and the Ipv(Vpv) characteristic are presented graphically for the SEEF by squares, as illustrated in Fig. 4.

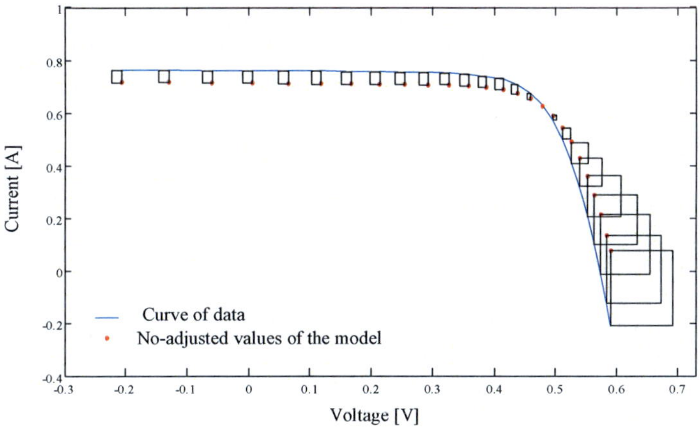

Fig. 4. The different squared errors generated between the model the $I_{PV}(V_{PV})$ data.

The optimization of SSEF consists of minimizing the sum of these squares for an optimal fit of the points calculated by the model to those of the experimental curve.

To illustrate schematically the optimization scenario of target function SSEF, we have taken the case of error function g(θ) with a single parameter θ (one-dimensional function).

The parabolic curve in Fig. 4 explains the square nature of the function g(θ).

Figure 5 presents normal convergence for initial values more or less close to the optimum, found by a well-defined initialization routine.

The vector θ represent the five or the seven intrinsic electrical parameters depending on the model, with one or two diodes (Eqs. (1) and (2)):

$\theta = [R_s, R_{sh}, I_{ph}, I_s, n]$
Or: $\theta = [R_s, R_{sh}, I_{ph}, I_{s1}, I_{s1}, n_1, n_2]$

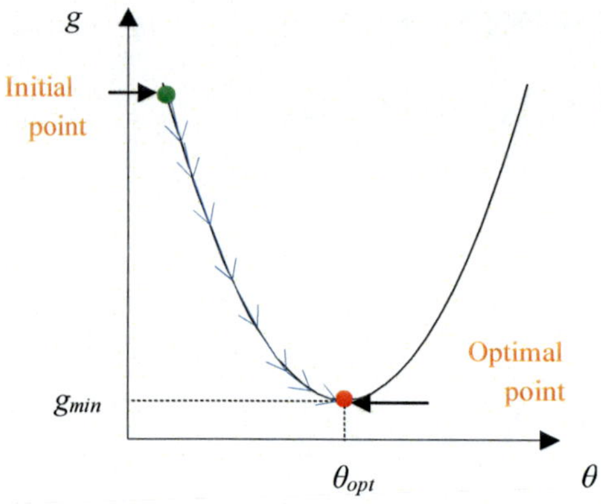

Fig. 5. Parabolic presentation of the objective function.

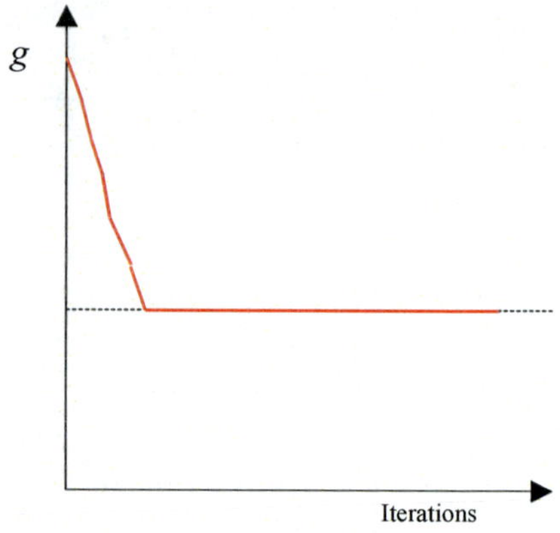

Fig. 6. Illustration of the g(θ) evolution according to iteration

4 Proposed Initialization Routine

Like any nonlinear optimization method, the initialization of the unknown parameters to be identified is essential to ensure the convergence of the corresponding algorithm. Known for their strong dependence on initial values, gradient descent algorithms only converge if they start with parameters close to the optimum. In this context, we used The "Linear Least Squares" (LLS) method, inspired by the literature [5], to initialize the five electrical parameters, represented by the vector θ.

5 Proposed Method LMSA

The Levenberg-Marquardt method consists in adjusting unknown parameters to find their optimal values. This is ensured by the characteristic Eq. (4) [6].

$$\theta_{k+1} = \theta_k - \left[\frac{-2.J^T(\theta_k).\varepsilon(\theta_k)}{2.J^T(\theta_k).J(\theta_k) + \lambda_k.I} \right]_{\theta=\theta_k} \tag{4}$$

With:
J: Jacobean of SSEF function,
ε: Error between data and model of the solar cell,
I: Matrix identity.
λ: Damping factor.

After a detailed study of the SSE function according to the variation of factor λ at every loop of the Levenberg-Marquardt method, we found that SSE includes several local minima. Aware of the importance of the Simulated Annealing (SA) method in the global convergence of functions with local minima, this method appears the most suitable for determining the target value of the factor λ at every loop of the LM algorithm. Where had the idea for a combination of the SA with LM method come from, which we called «LM associated with SA - LMSA».

5.1 Algorithm of Proposed LMSA

The algorithm of LMSA is shown in Fig. 7. g(θ) is the objective function to be minimized.

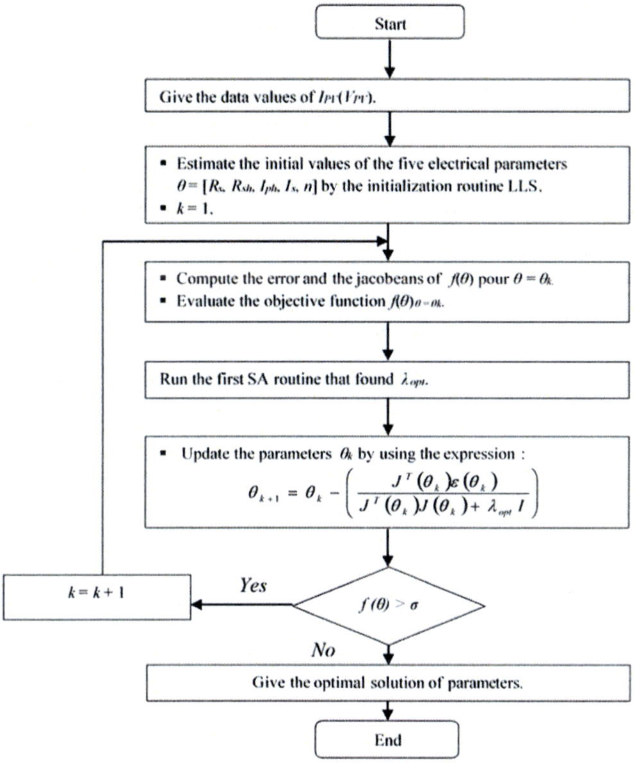

Fig. 7. Schema of the algorithm LMSA for the identification of the five and the seven parameters of the solar cells.

6 Results and Discussions

To prove the high accuracy of the proposed method LMSA, we identify the five and the seven parameters respectively of the single and the double diode models of solar cell by using the experimental measures of a commercial silicon solar cell (R.T.C France) with a diameter of 57 mm, measured at 33 °C and an irradiance (1000 W/m^2) [7]. The obtained results are compared with a set of methods reported in the literature.

6.1 Parametric Identification Results of Single Diode Model

The outcomes given by the proposed method are compared with others cited in the literature. These methods are GGHS, ABSO, CPSO, Newton, PS, and SA. These methods of different nature and principles of optimization were chosen for the comparison since they present the most precise and recent outcomes in parametric extraction of the solar cell's five parameters.

The parametric extraction methods are cited in Table 1 is placed from the highest to the lowest value of SSE.

Table 1. The outcomes obtained by our proposed method and those reported in the literature.

Parameters	SA	PS	Newton	CPSO	ABSO	GGHS	*LMSA*
$R_s(\Omega)$	0.0345	0.0313	0.0364	0.0354	0.0366	0.0363	***0.03643***
R_{sh} (Ω)	43.1034	64.1026	53.7634	59.012	52.2903	53.0647	***53.3264***
I_{ph} (A)	0.7620	0.7617	0.7608	0.7607	0.7608	0.7609	***0.76078***
I_s (μA)	0.4798	0.9980	0.3223	0.4000	0.3062	0.3262	***0.31849***
n	1.5172	1.6000	1.4837	1.5033	1.4758	1.4822	***1.47976***
Criterion							
SSEF	9.3839e−3	5.8005e−3	2.4445e−3	4.9952e−5	2.5547e−5	2.5528e−5	***2.53019e−5***
RMSE	1.8998e−2	1.4936e−2	9.6964e−3	1.3861e−3	9.9124e−4	9.9097e−4	***9.86485 e−4***

Table 2. The seven parameters obtained by the proposed method and those reported in the literature.

Parameters	PS	HS	GGHS	IGHS	*LMSA*
Rs (Ω)	0.0320	0.03545	0.03562	0.0369	***0.036872***
Rsh (Ω)	81.3008	46.82696	62.7899	56.8368	***56.911928***
Iph (A)	0.7602	0.76176	0.76056	0.76079	***0.760790***
Is1 (μA)	0.9889	0.12545	0.37014	0.97310	***1.139846***
Is2 (μA)	0.0001	0.25470	0.13504	0.16791	***0.185624***
n1	1.6000	1.49439	1.49638	1.92126	***1.994808***
n2	1.1920	1.49989	1.92998	1.42814	***1.435128***
Criterion					
SSEF	5.9886e−3	4.1254e−5	2.9676e−5	2.5306e−5	***2.526439e−5***
RMSE	0.01518	0.00126	0.00107	9.8635e−4	***9.857521e−4***

6.2 Parametric Identification Results of Double Diode Model

In this section, we discuss the use of LMSA for parametric identification for the double diode model. To shed light on the ability of this method to provide better precision in the parametric identification of the seven electrical parameters, we used the identified values obtained in the literature by the IGHS, GGHS, HS, and PS methods for comparison.

From the results reported in Table 2, the values obtained by LMSA which are marked in **bold** *italic* show that the proposed method is more accurate than IGHS, GGHS, HS, and PS methods. The minimal value of SSEF obtained for the double diode model

2.526439e–5 shows that this one is more reliable than the single diode model which generates an SSEF = 2.53019e–5. The accuracy in the parametric identification of the five and seven parameters obtained by the LMSA method is clearly shown in the two curves of $I_{PV}(V_{PV})$ and $P_{PV}(V_{PV})$ characteristics illustrated in Fig. 8 and Fig. 9.

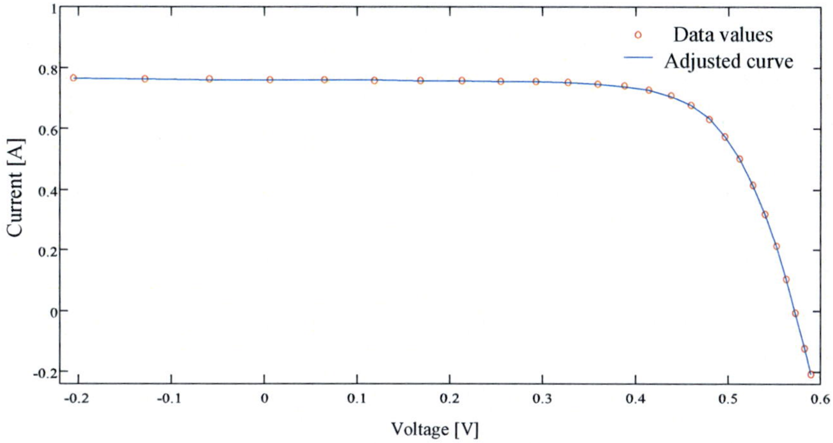

Fig. 8. The $I_{PV}(V_{PV})$ characteristic was found for single and double diode models of the solar cells.

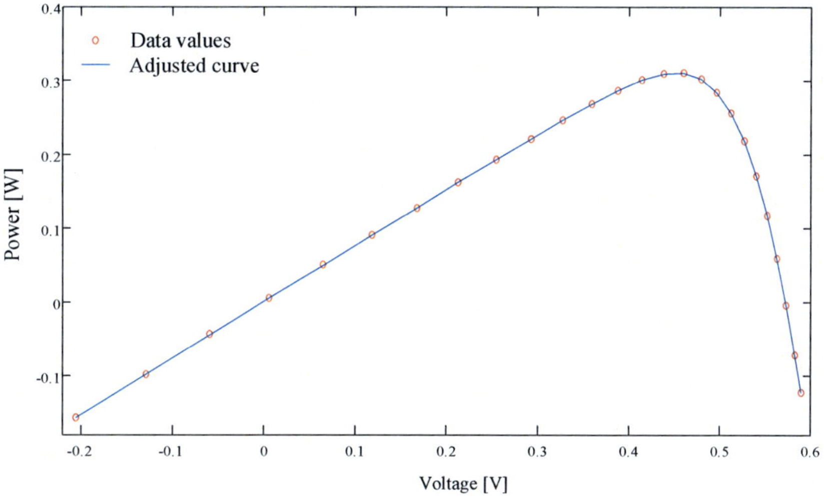

Fig. 9. The $P_{PV}(V_{PV})$ characteristic was found for single and double diode models of the solar cells.

7 Conclusion

The parametric identification ensured by LMSA method allows us to determine the five and the seven parameters with high accuracy. This ability is demonstrated by comparing our results with those of the latest optimization techniques reported in the literature. The purpose behind the identification of the double diode model is to evaluate the robustness of LMSA to face the double non-linearity in the model.

We conclude from this paper that the double diode model is more reliable in parametric identification than the single diode model. This is shown in the results of SSE of the double diode which is smaller than those found for the single diode. In a future paper, we will discuss the identification of the five and seven parameters of a photovoltaic module's model.

References

1. Dkhichi, F., Oukarfi, B., Fakkar, A., Belbounagiua, N.: Parameter identification of solar cell model using Levenberg–Marquardt algorithm combined with simulated annealing. Solar Energy **110**, 781–788 (2014)
2. Marios, T., Venizelos, V., George, M., George, G.E.: Energy yield in photovoltaic. systems. In: McEvoy's Handbook of Photovoltaics (Third Edition), Fundamentals and Applications, pp. 671–713 (2018)
3. El-Ahmar, M., Ahmed, A.H., Hemeida, A.M.: Mathematical modeling of photovoltaic module and evalute the effect of varoius paramenters on its performance. In: 2016 Eighteenth International Middle East Power Systems Conference (MEPCON), December 2016. https://doi.org/10.1109/MEPCON.2016.7836976
4. Johannes, F., Philip, C., Sammut, C., Craw, S.: Mean squared error. In Book: Encyclopedia of Machine Learning, January 2010. https://doi.org/10.1007/978-0-387-30164-8-528,
5. Bouzidi, K., Chegaar, M., Bouhemadou, A.: Solar cells parameters evaluation considering the series and shunt resistance. Sol. Energy Mater. Sol. Cells **91**(18), 1647–1651 (2007)
6. Askarzadeh, A., Rezazadeh, A.: Parameter identification for solar cell models using harmony search-based algorithms. Sol. Energy **86**(11), 3241–3249 (2012)
7. Jacobs, I.S., Bean, C.P.: Fine particles, thin films and exchange anisotropy. Magnetism, **III**, 271–350 (1963)
8. Easwarakhanthan, C., Bottin, T.J., Bouhouch, I., Boutrit, C.: Nonlinear minimization algorithm for determining the solar cell parameters with microcomputers. Int. J. Solar Energy **4**(1), 1–12 (2007)
9. Askarzadeh, A., Rezazadeh, A.: A grouping-based global harmony search algorithm for modeling of proton exchange membrane fuel cell. Int. J. Hydro. Energy **II**, 5047–5053 (2011)
10. Askarzadeh, A., Rezazadeh, A.: Artificial bee swarm optimization algorithm for parameters identification of solar cell models. Appl. Energy **109**, 943–949 (2013)
11. Wei, H., Cong, J., Lingyun, X., Deyun, S.: Extracting solar cell model parameters based on chaos particle swarm algorithm. In: IEEE XPLORE, 978–1–4244–8039–5 (2011)
12. AlHajri, M.F., El-Naggar, K.M., AlRashidi, M.R., Al-Othman, A.K.: Optimal extraction of solar cell parameters using pattern search. Renew. Energy **44**, 238–245 (2012)
13. El-Naggar, K.M., AlRashidi, M.R., AlHajri, M.F., Al-Othman, A.K.: Simulated annealing algorithm for photovoltaic parameters identification. Solar Energy **86**, (201), 266–274 (2012)
14. Askarzadeh, A., Rezazadeh, A.: An innovative global harmony search algorithm for parameter identification of a PEM fuel cell model. IEEE Trans. Ind. Electron. **9**(59), 3473–3480 (2012)

FOC-DPC Hybrid Structure of a Wind Energy Conversion System Based on PMSG

Hassna Salime[1(✉)], Badre Bossoufi[1], Nada Zine Laabidine[1], Ikrame Saady[1], Houda Elalami[1], Btissam Majout[1], and Youness EL Mourabit[2]

[1] Laboratoire LISTA, Faculty of Science Dhar El Mahraz- USMBA, Fez, Morocco
hassna1salime@gmail.com

[2] National School of Applied Sciences at Tetouan, Abdelmalek Essaadi University, Tetouan, Morocco

Abstract. Everywhere in the world, the improvement of the electrical energy production as well as the improvement of the electrical network quality has now become a crucial necessity to meet the expected requirements in terms of the energy produced quality. In addition, the shift towards renewable energies such as solar or Wind Power Systems (WPS) to generate electricity has become a current trend for many countries around the world. In this context, we are interested in the Permanent Magnet Synchronous Generator (PMSG) which remains a relevant solution in Wind Energy Conversion Systems (WECS). This machine has many advantages over other machines of the same power such as control of machine power and operation with less noise thanks to the absence of a gearbox in most cases of installation. The complete chain of WECS uses connection stages which serve to connect the generator with the network thanks to power converters mounted back to back. The power control is related to both the PMSG and the control algorithms applied to the Machine Side Converter (MSC) and the Grid Side Converter (GSC). In this present work, we will proceed to the analysis of Field Oriented Control (FOC) applied to MSC and Direct Power Control (DPC) applied to GSC. The simulation results of such controls are provided through the Matlab\Simulink framework.

Keywords: Wind Power Systems (WPS) · Permanent Magnet Synchronous Generator (PMSG) · Wind Energy Conversion Systems (WECS) · Machine Side Converter (MSC) · Grid Side Converter (GSC) · Field Oriented Control (FOC) · Direct Power Control (DPC)

1 Introduction

Currently, with the technological advancement of wind turbines, wind power has spread rapidly as a reliable solution to generate electrical power [1]. Several studies [2, 3] have attempted to prove that wind power is the fastest growing renewable energy technology in the world and continues to improve because it is readily available in nature and represents an immense deposit at the global level. Obviously, production based on renewable

S. Motahhir and B. Bossoufi (Eds.): ICDTA 2022, LNNS 454, pp. 693–702, 2022.
https://doi.org/10.1007/978-3-031-01942-5_69

energies and particularly that produced by wind energy plays a complementary role with conventional energy production.

Thanks to the new generation of wind turbines operating at variable speed [4], the cost of producing electrical energy has been continuously reduced compared to fixed speed wind turbines, which consequently makes it possible to improve energy efficiency and the quality of the power generated. The architecture corresponding to the variable speed wind turbine based on the PMSG is connected to the grid through a full-scale frequency converter [5]. Static converters make the connection to the grid smoother.

Obviously, the construction of the PMSG offers a rotor without winding and by default without excitation. This means the absence of the brushes and the sliding contacts necessary for the excitation, and consequently, a reduced maintenance and a greater robustness for the machine. The most important advantage of the PMSGs used in the wind energy sector lies in the absence of the gearbox for certain types of generators having a fairly large number of poles [6]. This topology of the machine allows a speed variation over a range of 100% of the nominal speed of rotation, and consequently, reduced maintenance and a limited volume of the wind turbine. These various advantages cover the high cost of magnets [7, 8]. However, the control of PMSG is not easy enough due to the non-linearity of the real model [9].

In this work, the command used to control the PMSG is the FOC control to independently control the machine electromagnetic torque and the rotor magnetic flux while decoupling the stator current into two distinct rotating components with the rotor flux vector. In this case, the control of the synchronous machine becomes similar to the control of a DC machine with independent excitation. This control strategy was founded by "Blaschke" in 1972 [10] and further developed by "Hasse" in 1979 [11]. To ensure this control, it is necessary to choose control laws guaranteeing the decoupling of the electromagnetic torque and the flux, as well as the d-q axis system as a rotary reference [12, 13].

The second control algorithm used for this study is the Direct Power Control (DPC) which applies to the Grid Side Converter (GSC). The DPC control was designed by "Noguchi" in 1996 [14] using the instantaneous power theory [15]. The principle of this control strategy does not require any coordinate transformation. The switching signals of the electronic converter switches is chosen directly on the basis of the information required from the hysteresis controllers and the voltage vector position of the electrical grid while taking into account the errors of instantaneous power. The combination of field oriented control and the direct power control is practically one of the new methods for controlling wind turbine systems [16]. With this combination, good steady state performance is generally achievable. Adjusting the controller parameters is also tedious and requires some expertise.

2 Wind Energy Conversion System Modeling

The general model of our system is schematized as shown in the following figure (Fig. 1).

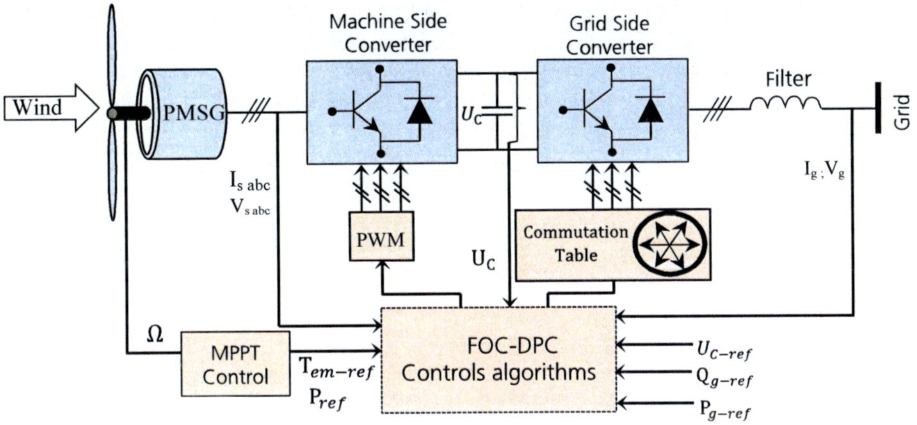

Fig. 1. General model of WECS

2.1 Wind Turbine Modeling

According to [17], the relation which connects the aerodynamic power of the wind with that of the wind speed V_v is as follows:

$$P_{wind} = \frac{1}{2}\rho S V_v^3 \tag{1}$$

The turbine only takes advantage of a fraction of what exists in the wind power, the portion is expressed by the coefficient of performance C_p:

$$P_t = C_P(\lambda, \beta) * P_{wind} = \frac{1}{2} C_P(\lambda, \beta)\rho S V_v^3 \tag{2}$$

The power extracted by a wind turbine is always only a fraction of what exists in the wind. The extracted portion is expressed by the coefficient of performance "C_p"

$$\lambda = \frac{R\Omega}{V_v} \tag{3}$$

The relationship between "C_p", "λ" and "β" is nonlinear and is obtained by curve fitting experimental data.

$$\begin{cases} \frac{1}{\lambda'} = \frac{1}{\lambda + 0.08\beta} - \frac{0.035}{\beta^3 + 1} \\ C_P(\lambda, \beta) = 0.5176\left(\frac{116}{\lambda'} - 0.4\beta - 5\right)e^{\frac{-21}{\lambda'}} + 0.0068\lambda \end{cases} \tag{4}$$

We can derive the formula of the torque which is proportional to the square of the rotor angular speed.

In ideal case, the power coefficient C_p reaches a maximum value that will be within the range 59.26% according to Betz's limit. This means that the extracted power from the wind is always less than this value [18].

$$T_{tur} = \frac{1}{2}\frac{C_P(\lambda, \beta)\rho S R^3}{\lambda^3}\Omega^2 \tag{5}$$

The objective is always will be the maximum extraction of the wind power, the coefficient "λ" to "λ_{opt}" must be maintained with an optimum rotor angular speed. The following equations show the values to be considered:

$$T_{tur}^{opt} = \frac{1}{2}\frac{C_{P-\max}(\lambda,\beta)\rho SR^3}{\lambda_{opt}^3}\Omega^2 \tag{6}$$

$$\Omega_{ref} = \frac{\lambda_{opt}}{R} * V_v \tag{7}$$

2.2 FOC Control for PMSG

The wind turbine is mechanically coupled to a generator in a WECS, in our case it is PMSG, then we will describe the model of this generator. The output voltages of PMSG can be represented in the synchronous rotating reference frame where the q-axis is 90° ahead of the d-axis with respect to the direction of rotation.

2.2.1 Mathematical Model of PMSG

After the necessary simplifications, we get:

$$\begin{cases} V_{sd} = -R_s.i_{sd} - L_d\frac{d(i_{sd})}{dt} - \omega_e(L_q.i_{sq}) \\ V_{sq} = -R_s.i_{sq} - L_q\frac{d(i_{sq})}{dt} + \omega_e(L_d.i_{sd}) + \omega_e.\Psi_f \end{cases} \tag{8}$$

The totalized fluxes in each phase are as follows

$$\begin{cases} \Psi_{sd} = L_d i_{sd} + \Psi_f \\ \Psi_{sq} = L_q i_{sq} \end{cases} \tag{9}$$

The mechanical equations are described:

$$\frac{T_t - T_{em}}{J} = \frac{d\Omega}{dt} \tag{10}$$

With:

$$T_{em} = \frac{-3}{2}p\left[\left(L_d - L_q\right)i_{sd}.i_{sq} + \Psi_f.i_{sq}\right] \tag{11}$$

The active and reactive powers of the machine can be expressed as follows:

$$\begin{cases} P_{gen} = T_{em}\Omega = \frac{3}{2}\left[V_{sd}.i_{sd} + V_{sq}.i_{sq}\right] \\ Q_{gen} = \frac{3}{2}\left[V_{sq}.i_{sd} - V_{sd}.i_{sq}\right] \end{cases} \tag{12}$$

2.2.2 Summary of the PI Regulator

The PI regulator used to control the PMSG has the expression $\frac{K_p s+K_i}{s}$. Where "K_p'' presents the proportional gain of the PI corrector, "K_i'' pr presents the integral gain and "s" the Laplace variable.

The first-order regulated system of stator currents defined by the following transfer function:

$$\begin{cases} \mathrm{V_{sd}} = -\mathrm{R_s.I_{sd}} - \mathrm{s.L_d.I_{sd}} - \mathrm{E_q} \\ \mathrm{V_{sq}} = -\mathrm{R_s.I_{sq}} - \mathrm{s.L_q.I_{sq}} + \mathrm{E_d} \end{cases} \tag{13}$$

The compensation terms E_d, E_q are considered as measurable disturbing elements. Hence, the transfer function of the machine:

$$FT(s) = \frac{1}{R_s + s.L_s} = \frac{1}{R_s}.\frac{1}{1+T_e s} \tag{14}$$

$$FT(s) = \frac{I_{d,q}(s)}{V_{d,q}(s) + E_{d,q}(s)} \tag{15}$$

With:
$T_e = \frac{L_s}{R_s}$: The time constant of the machine.
$L_s = L_d = L_q$: The stator inductance of the smooth pole machine.

2.3 DPC Control Principle

The DPC control algorithm set the reactive reference power $Q_{R-ref} = 0$ in order to achieve unity power factor operation. The active reference power P_{R-ref} is obtained by regulating the ($\mathrm{U_C}$) through a PI corrector. The set point power values are then compared with the estimated active and reactive power values $\left(\widehat{P}_R \; and \; \widehat{Q}_R\right)$.

2.3.1 Estimation of Active and Reactive Power

The implementation of direct power control requires a reliable estimation of active and reactive power. The instantaneous power values are defined by the product of the three-phase voltages and currents as shown by the following relationships:

$$P_R = Re\{\underline{u}.\underline{i}^*\} = V_{R\alpha} i_{R\alpha} + V_{R\beta} i_{R\beta} = V_{Ra} i_{Ra} + V_{Rb} i_{Rb} + V_{Rc} i_{Rc} \tag{16}$$

$$Q_R = Im\{\underline{u}.\underline{i}^*\} = V_{R\beta} i_{R\alpha} - V_{R\alpha} i_{R\beta} = \frac{1}{\sqrt{3}}(U_{Rbc} i_{Ra} + U_{Rca} i_{Rb} + U_{Rab} i_{Rc}) \tag{17}$$

The instantaneous values of the powers can be estimated according to Table 1 which groups together the different combinations of electronic switches [14]:

Table 1. Relations between the switching mode and the estimation of the instantaneous active and reactive power equations

S_{oa}, S_{ob}, S_{oc}	$\widehat{P}_R$	$\widehat{Q}_R$
1, 0, 0	$L\left(\frac{di_{Ra}}{dt}i_{Ra} + \frac{di_{Rb}}{dt}i_{Rb} + \frac{di_{Rc}}{dt}i_{Rc}\right) + U_C i_{Ra}$	$\frac{1}{\sqrt{3}}\left\{3L_F\left(\frac{di_{Rc}}{dt}i_{Rb} - \frac{di_{Rb}}{dt}i_{Rc}\right) - U_C((i_{Rb} - i_{Rc}))\right\}$
1, 1, 0	$L\left(\frac{di_{Ra}}{dt}i_{Ra} + \frac{di_{Rb}}{dt}i_{Rb} + \frac{di_{Rc}}{dt}i_{Rc}\right) - U_C i_{Rc}$	$\frac{1}{\sqrt{3}}\left\{3L_F\left(\frac{di_{Rb}}{dt}i_{Ra} - \frac{di_{Ra}}{dt}i_{Rb}\right) + U_C((i_{Ra} - i_{Rb}))\right\}$
0, 1, 0	$L\left(\frac{di_{Ra}}{dt}i_{Ra} + \frac{di_{Rb}}{dt}i_{Rb} + \frac{di_{Rc}}{dt}i_{Rc}\right) + U_C i_{Rb}$	$\frac{1}{\sqrt{3}}\left\{3L_F\left(\frac{di_{Ra}}{dt}i_{Rc} - \frac{di_{Rc}}{dt}i_{Ra}\right) - U_C((i_{Rc} - i_{Ra}))\right\}$
0, 1, 1	$L\left(\frac{di_{Ra}}{dt}i_{Ra} + \frac{di_{Rb}}{dt}i_{Rb} + \frac{di_{Rc}}{dt}i_{Rc}\right) - U_C i_{Ra}$	$\frac{1}{\sqrt{3}}\left\{3L_F\left(\frac{di_{Rc}}{dt}i_{Rb} - \frac{di_{Rb}}{dt}i_{Rc}\right) + U_C((i_{Rb} - i_{Rc}))\right\}$
0, 0, 1	$L\left(\frac{di_{Ra}}{dt}i_{Ra} + \frac{di_{Rb}}{dt}i_{Rb} + \frac{di_{Rc}}{dt}i_{Rc}\right) + U_C i_{Rc}$	$\frac{1}{\sqrt{3}}\left\{3L_F\left(\frac{di_{Rb}}{dt}i_{Ra} - \frac{di_{Ra}}{dt}i_{Rb}\right) - U_C((i_{Ra} - i_{Rb}))\right\}$
1, 0, 1	$L\left(\frac{di_{Ra}}{dt}i_{Ra} + \frac{di_{Rb}}{dt}i_{Rb} + \frac{di_{Rc}}{dt}i_{Rc}\right) - U_C i_{Rb}$	$\frac{1}{\sqrt{3}}\left\{3L_F\left(\frac{di_{Ra}}{dt}i_{Rc} - \frac{di_{Rc}}{dt}i_{Ra}\right) + U_C((i_{Rc} - i_{Ra}))\right\}$
0, 0, 0	$L\left(\frac{di_{Ra}}{dt}i_{Ra} + \frac{di_{Rb}}{dt}i_{Rb} + \frac{di_{Rc}}{dt}i_{Rc}\right)$	$\frac{1}{\sqrt{3}}\left\{3L_F\left(\frac{di_{Rb}}{dt}i_{Ra} - \frac{di_{Ra}}{dt}i_{Rb}\right)\right\}$
1, 1, 1	$L\left(\frac{di_{Ra}}{dt}i_{Ra} + \frac{di_{Rb}}{dt}i_{Rb} + \frac{di_{Rc}}{dt}i_{Rc}\right)$	$\frac{1}{\sqrt{3}}\left\{3L_F\left(\frac{di_{Rb}}{dt}i_{Ra} - \frac{di_{Ra}}{dt}i_{Rb}\right)\right\}$

2.3.2 Choice of Hysteresis Comparators

This type of regulator is widely used and offers good dynamic performance, in addition, it remains a fairly simple solution better suited to power regulation. The hysteresis regulator outputs digitized words $(H_{P_R}\ and\ H_{Q_R})$ according to the errors between the reference power values $(P_{R-ref}\ and Q_{R-ref})$ and those estimated $\left(\widehat{P}_R\ and\ \widehat{Q}_R\right)$.

Figures 2 and 3 illustrate the different types of hysteresis comparators used for DPC control [14].

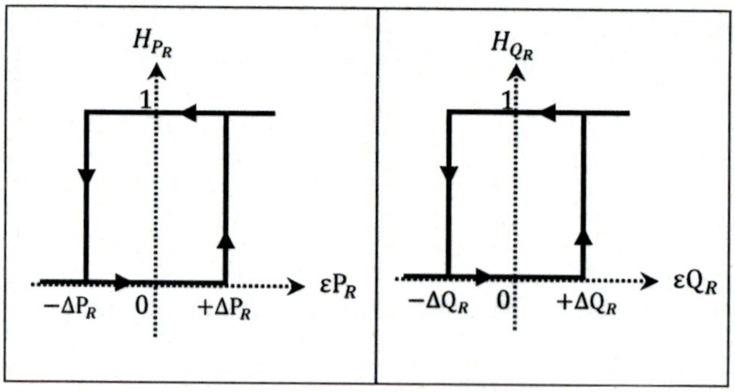

Fig. 2. Two-level hysteresis comparators

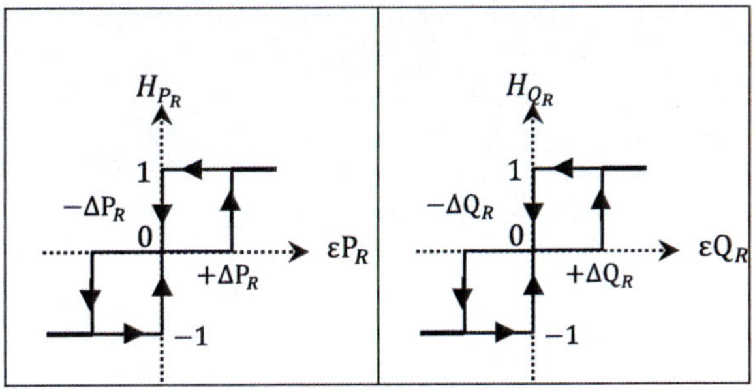

Fig. 3. Three-level hysteresis comparators

3 Results and Discussion

3.1 Simulation Results

The Matlab/Simulink framework was used to show the efficacy of the wind system controlled by an association of FOC-DPC control algorithms. The simulation results show the effects of the controllers during different values of wind speed.

3.2 Discussion of Results

In order to test the robustness of the controls applied to the WECS system, this test part is based on a step wind profile which shows the effect of the controls on the system during sudden changes in the wind.

The choice of the reference values for the stator currents are defined as: $i_{sd\text{-}ref} = 0$ to obtain a maximum torque at the minimum current, as shown in Fig. 4 (d), $i_{sq\text{-}ref}$ comes from the reference electromagnetic torque which is illustrated in Figs. 4 (e) and 4 (c). According to these two figures, the correct monitoring of the values of the quadrature stator current and of the electromagnetic torque is observed with their reference values. In addition, we see the similar formats of the current i_{sq} and Tem, which justifies the torque control via the control of the stator currents.

Vector control with MPPT provides better regulation of mechanical rotational speed, and the effect of regulation is clearly seen in Fig. 4 (b). This good regulation has a considerable effect on the regulation of other quantities in the complete conversion chain.

With regard to the DPC control, the figures of the active power injected into the network and of the DC bus voltage clearly show the effect of the control studied. The curves are quite adequate with their reference values. The minimum overshoots as well as the low ripples in the graphs justify the right choice of this robust control for the employee in a real site. The choice of sector for the use of the DPC is well presented in Fig. 4 (h) where one can see the good progress of the sectors in order in the interval [$-\pi$ $+\pi$].

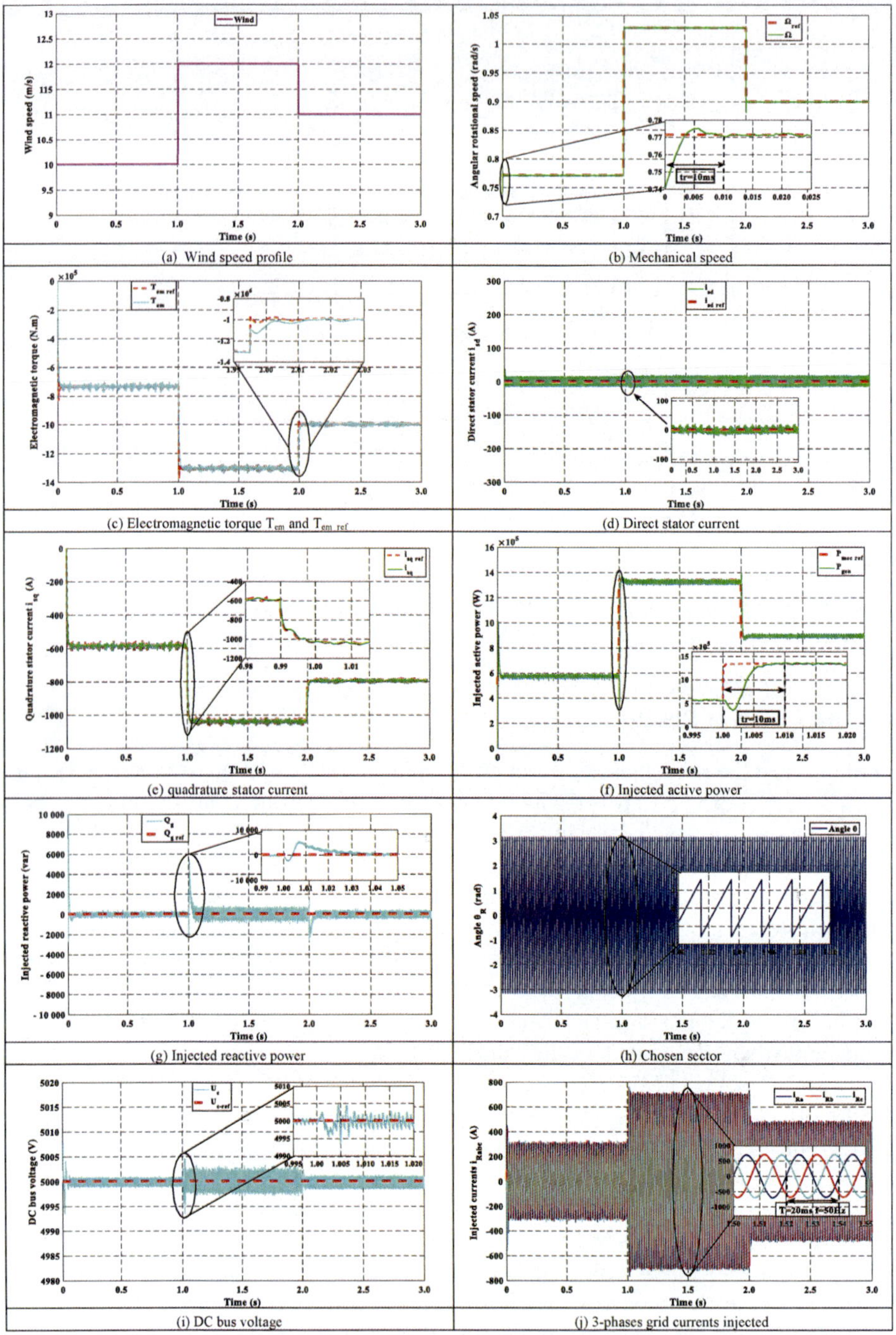

(a) Wind speed profile

(b) Mechanical speed

(c) Electromagnetic torque T_{em} and T_{em_ref}

(d) Direct stator current

(e) quadrature stator current

(f) Injected active power

(g) Injected reactive power

(h) Chosen sector

(i) DC bus voltage

(j) 3-phases grid currents injected

Fig. 4. Simulation results of FOC-DPC controls

4 Conclusion

This present work develops the dynamic model of the complete chain of a wind power system that employs a PMSG. To better control the wind chain and connect it to the grid, we are obliged to know beforehand the parameters of the turbine, the synchronous machine as well as the power converters. The two control algorithms used in this present work show certain robustness regarding the sudden changes in the wind speed profile. The performances of proportional-integral and hysteresis regulators are well-validated through to the results developed by the simulation platform. The low ripple rate and low overshoots shown in the simulation graphics attest to the high performance of the proposed control.

References

1. Bossoufi, B., Karim, M., Lagrioui, A., Taoussi, M., Derouich, A.: Observer backstepping control of DFIG-generators for wind turbines variablespeed: FPGA-based implementation. Renew. Energy. **81**, 903–917 (2015)
2. El Mourabit, Y., Derouich, A., ElGhzizal, A., El Ouanjli, N., Zamzoum, O.: Nonlinear backstepping control of variable speed wind turbine based on permanent magnet synchronous generator. In: Paper presented at: International Conference on Wireless Technologies, Embedded and Intelligent Systems (WITS), pp. 1–7. IEEE (2019)
3. Taoussi, M., Karim, M., Bossoufi, B., Hammoumi, D., Lagrioui, A., Derouich, A.: Speed variable adaptive backstepping control of the doubly fed induction machine drive. Int. J. Autom. Control **10**, 12–33 (2016)
4. Bossoufi, B., Karim, M., Lagrioui, A., Taoussi, M., Derouich, A.: Observer backstepping control of DFIG-generators for wind turbines variable-speed: FPGA-based implementation. Renew. Energy **81**, 903–917 (2015)
5. El Mourabit, Y., Derouich, A., ElGhzizal, A., et al.: Implementation and validation of backstepping control for PMSG wind turbine using dSPACE controller board. Energy Rep. **5**, 807–821 (2019)
6. El Mourabit, Y., Derouich, A., Allouhi, A., El Ghzizal, A., El Ouanjli, N., Zamzoum, O.: Sustainable production of wind energy in the main Morocco's sites using permanent magnet synchronous generators. Int. Trans. Electr. Energ. Syst. e12390 (2020). https://doi.org/10.1002/2050-7038.12390
7. Ackermann, T., (Ed.).: Wind Power in Power Systems. John Wiley & Sons, ISBN 0–470–85508–8 (2005)
8. Hau, E.: Wind turbines, Fundamentals, Technologies, Applications, Economics. Springer-Verlag, Berlin Heidelberg (2006)
9. Bianchi, F.D., De Battista, H., Mantz, R.J.: Wind Turbine Control Systems: Principles, Modelling And Gain Scheduling Design. Springer Science & Business Media (2006)
10. Blaschke, F.: The principle of field orientation as applied to the new transvector closed-loop system for rotating-field machines. Siemens Rev. **34**(3), 217–220 (1972)
11. Hasse, K.: On the dynamics of speed control of a static ac drive with a squirrel-cage induction machine. PhD,'TH Darmstadt (1979)
12. Pillay, P., Krishnan, R.: Modeling, simulation, and analysis of permanent-magnet motor drives. I. the permanent-magnet synchronous motor drive. IEEE Trans. Ind. Appl. **25**(2), 265–273 (1989)
13. De Fornel, B., Roye, D. : Principes généraux du contrôle vectoriel.Technical report, INP-ENSEEIHT (1993)

14. Noguchi, T., Tomiki, H., Kondo, S., Takahashi, I.: Direct power control of PWM converter without power-source voltage sensors. IEEE Trans. Ind. Appl. **34**(3), 473–479 (1998)
15. Akagi, H., Kanazawa, Y., Nabae, A.: Instantaneous reactive power compensators comprising switching devices without energy storage components. IEEE Trans. Ind. Appl. **3**, 625–630 (1984)
16. Li, S., Haskew, T.A., Xu, L.: Conventional and novel control designs for direct driven PMSG wind turbines. Electric Power Syst. Res. **80**(3), 328–338 (2010)
17. El Mourabit, Y., Derouich, A., ElGhzizal, A., Zamzoum, O.: Dynamic modeling and control of a wind turbine with MPPT control connected to the grid by using PMSG. In: Paper Presented at: 2017 International Conference on Advanced Technologies for Signal and Image Processing (ATSIP), pp. 1–6. IEEE (2017)
18. Taoussi, M., Karim, M., Bossoufi, B., et al.: Low-speed sensorless control for wind turbine system. WSEAS Trans. Syst. Control. **12**, 405–417 (2017)

TS-Fuzzy Model Based Predictive Controller for a Variable Speed Wind Turbine

Rime El Houti(✉), Soukaina Bougdour, Selma Sefriti, and Ismail Boumhidi

LISAC Laboratory, Faculty of Sciences Dhar El Mahraz, Sidi Mohamed Ben Abdellah University, Fez, Morocco
{Rime.elhouti,soukaina.bougdour,selma.sefriti, ismail.boumhidi}@usmba.ac.ma

Abstract. The purpose of this paper is to look into the design and implementation of a wind turbine system. Because the wind turbine system's linearized model is required for control synthesis, the Takagi-Suengo fuzzy technique is used to generate a robust linear model, which is then used to design the continuous-time model predictive controller (CMPC). The simulation results demonstrate the effectiveness of the suggested approach for variable speed wind turbine control.

Keywords: Wind turbine control · T-S fuzzy modelling · Continuous-time model · Predictive controller (CMPC)

1 Introduction

Wind turbines with variable speeds have become more popular in recent years than those with fixed speeds. Because of its variable speed operation, the wind turbine can effectively extract maximum power for the wind.

Model predictive control is an optimized control method based on optimal control in which a system efficiency is minimized in terms of a sequence of corresponding future states at each sampling time subjected to a mathematical model over a prescribed prediction horizon.

The fuzzy design of industrial processes is typically based on human knowledge and experience or prior knowledge. In the absence of this information, input/output data trying to describe the plant's behavior is required [3, 8]. However, we do not know what imputations are needed to identify a good fuzzy model based on predictive control. Some research has been conducted to address this issue, but the solutions offered are exhaustive model-based tests. Many authors use Takagi-Sugeno fuzzy models to design model-based controllers for nonlinear systems. We're curious about predictive control schemes. In this work, we are only interested in the region of full load, the wind speed is greater than the rated speed. This research focuses on the mechanical component, with the main objective of generating a controller that maximizes wind capture by controlling the predictive model (MPC). However, the high nonlinearity of the wind turbine system and the wind's fluctuation make the design of the MPC system a difficult task to achieve. The

S. Motahhir and B. Bossoufi (Eds.): ICDTA 2022, LNNS 454, pp. 703–711, 2022.
https://doi.org/10.1007/978-3-031-01942-5_70

model-based predictive control (MPC), has indeed been implemented to wind turbine control in several studies to reduce the computational load [9] relied on the Laguerre function. A continuous predictive control model (CMPC) is a systems work that predicts the future output of a precise estimated period using a control system. An inline solution is obtained in each case to compute the control action over the forecast horizon. The CMPC is built with a full-load linear model of a nearby wind power system. In many cases, valid lines are only used around a setpoint, that will not cover the entire variety of nonlinear operations [5]. In order to increase the validity domain, we consider the Takagi-Sugeno (T-S) fuzzy modeling approach to linearize the nonlinear wind system [2, 4]. And then obtain a linear model needed for the design of the CMPC.

This method enables is on the representation of dynamic systems [1] by using a suitable "blending" of linear subsystems, each of which corresponds to a different operating point. The idea is to create a local controller for each local linear model. By fuzzy aggregation of all premises the resulting global fuzzy controller, which is usually linear is acquired [6, 7].

The foregoing is how this paper is structured: Sect. 2 represents wind turbine modeling, Sect. 3 presents Takagi Sugeno fuzzy modeling, Sect. 4 discusses CMPC control design, and Sect. 5 discusses the outcomes of the simulation. Finally, Sect. 6 concludes this paper.

Nomenclature

R	Blade length (m)	J_t	The turbine's inertia (kg/m2)
β	Blade pitch angle (°)	N_g	Gear ratio
λ	Tip speed ratio	**υ**	Effective wind speed
ω_r	Turbine rotational speed (rad/s)	**λ**	Tip speed ratio
K_r	Shaft stiffness (N/m)	C_p	Power coefficient
β	Blade pitch angle	**ρ**	The air density

2 Wind Turbine Modeling

The wind turbine is an interaction of three subsystems, respectively known as the aerodynamics of the blades, which is composed of the drive train and in the generator unit. The generator unit that describes the conversion from the mechanical power into electrical power, As a result, the aerodynamic torque T_a can be calculated as follows:

$$T_a = \frac{1}{2\omega_r}\rho A V_\omega^3 C_P(\lambda, \beta) \quad (1)$$

As a result, each wind turbine has its power coefficient $C_p(\lambda, \beta)$, which should be a non - linear function affected by pitch angle and turbine efficiency. The power coefficient is represented by means:

$$\text{With}: C_p(\lambda, \beta) = d_1\left(\frac{d_2}{\lambda_i} - d_3\beta - d_4\right)e^{\frac{-d_5}{\lambda_i}} + d_6\lambda \tag{2}$$

$$\frac{1}{\lambda_i} = \frac{1}{\lambda + 0.08\beta} - \frac{0,035}{\beta^3 + 1}$$

where: $\lambda = \frac{\omega_r R}{V_\omega}$ and $d_1 = 0.5176;\ d_2 = 116;\ d_3 = 0.4;\ d_4 = 5;\ d_5 = 21;\ d_6 = 0.0068$.

To have a basic structure and minimize the system's model, the wind turbine is represented as a single mass drive-train system by accepting because of the low flow rate, the shaft is relatively rigid. The dynamic can be described using the first-order equation:

$$\dot{\omega}_r = \frac{T_a}{J_t} - \frac{k_r}{J_t}\omega_r - \frac{T_g}{J_t} \tag{3}$$

The differential equations that describe the entire dynamics of the transmission model are as follows:

$$\dot{\beta} = -\frac{1}{\tau_\beta}(\beta - \beta_{ref}) \tag{4}$$

The first degree delay model [10] characterizes the effectual wind speed, which is hypothesized to be a combination of a mean wind speed and a constantly changing turbulent wind speed:

$$\dot{v} = -\frac{1}{\tau_v}(v - \overline{v}) \tag{5}$$

where τ_v is the time constant and $\overline{v}$ is the turbulent wind.

By combining the subsystems, the global wind turbine model can be represented as shown in:

$$\begin{cases} \dot{x} = Ax + Bu + DT_a + E\overline{v} \\ y = Cx \end{cases} \tag{6}$$

Therefore $x = \begin{bmatrix}\omega_r & \beta & v^T\end{bmatrix}$ is state vector, $u = \begin{bmatrix}T_g & \beta_{ref}^T\end{bmatrix}$ is control imput vector obtained from the controller, y is measurable output.

$$A = \begin{bmatrix} -\frac{k_r}{J_t} & 0 & 0 \\ 0 & -\frac{1}{\tau_\beta} & 0 \\ 0 & 0 & -\frac{1}{\tau_v} \end{bmatrix} B = \begin{bmatrix} Jr & 0 \\ 0 & \frac{1}{\tau_\beta} \\ 0 & 0 \end{bmatrix} D = \begin{bmatrix} \frac{1}{Jr} \\ 0 \\ 0 \end{bmatrix} E = \begin{bmatrix} 0 \\ 0 \\ \frac{1}{\tau_v} \end{bmatrix}$$

$$C = [100]$$

As a result, T_a which is a nonlinear function that is affected by wind speed, rotor speed, and pitch angle, can be linearized using the given formula:

$$T_a = T_{av}v + T_{a\beta}\beta + T_{a\omega_r}\omega_r \tag{7}$$

where T_{av}, $T_{a\beta}$ and $T_{a\omega_r}$ are the rotor speed, wind speed, and pitch angle partial derivatives of the aerodynamic torque T_a at the operating point.

3 Takagi Sugeno Fuzzy Representation

The Takagi Sugeno fuzzy model is useful for expressing the nonlinear system (6). The following factors in the wind turbine are anticipated to change over time: $v_{min} \le v \le v_{max}$, $\beta_{min} \le \beta \le \beta_{max}$ as a result, we can start by taking these factors:

IF-THEN rules:

If v is N_2 and β is M_2 Then $\begin{cases} \dot{x} = A_1x + B_1u_1 + E_1v(t) \\ Y = C_1x \end{cases}$

If v is N_2 and β is M_1 Then $\begin{cases} \dot{x} = A_2x + B_2u_2 + E_2v(t) \\ Y = C_2x \end{cases}$

If v is N_1 and β is M_2 Then $\begin{cases} \dot{x} = A_3x + B_3u_3 + E_3v(t) \\ Y = C_3x \end{cases}$

If v is N_1 and β is M_1 Then $\begin{cases} \dot{x} = A_4x + B_4u_4 + E_4v(t) \\ Y = C_4x \end{cases}$

The corresponding T-S fuzzy model is then:

$$\begin{cases} \dot{x}(t) = \sum_{i=1}^{4} h_i(z(t)(A_ix(t) + B_iu_i(t)) + E_iv(t) \\ y(t) = \sum_{i=1}^{4} h_i(z(t))C_ix(t) \end{cases} \tag{8}$$

$$\text{With: } h_1(z) = M_1(\beta(t))N_1(v(t))$$

$$h_2(z) = M_1(\beta(t))N_2(v(t))$$

$$h_3(z) = M_2(\beta(t))N_1(v(t))$$

$$h_4(z) = M_2(\beta(t))N_2(v(t))$$

$$\text{With: } M_1(\beta(t)) = \frac{\beta - \beta_{min}}{\beta_{max} - \beta_{min}};\ M_2(\beta(t)) = \frac{\beta_{max} - \beta}{\beta_{max} - \beta_{min}};\ N_1(v(t)) = \frac{V - V_{min}}{V_{max} - V_{min}};$$

$$N_2(v(t)) = \frac{V_{max} - V}{V_{max} - V_{min}}$$

$$A_1 = \begin{bmatrix} -\frac{k_r + T_{aw_r}}{J_t} & \frac{T_{a\beta_{max}}}{J_t} & \frac{T_{aV_{max}}}{J_t} \\ 0 & -\frac{1}{\tau_\beta} & 0 \\ 0 & 0 & \frac{1}{\tau_V} \end{bmatrix} A_2 = \begin{bmatrix} -\frac{k_r + T_{aw_r}}{J_t} & \frac{T_{a\beta_{min}}}{J_t} & \frac{T_{aV_{max}}}{J_t} \\ 0 & -\frac{1}{\tau_\beta} & 0 \\ 0 & 0 & \frac{1}{\tau_V} \end{bmatrix}$$

$$A_3 = \begin{bmatrix} -\frac{k_r + T_{aw_r}}{J_t} & \frac{T_{a\beta_{max}}}{J_t} & \frac{T_{aV_{min}}}{J_t} \\ 0 & -\frac{1}{\tau_\beta} & 0 \\ 0 & 0 & \frac{1}{\tau_V} \end{bmatrix} \quad A_4 = \begin{bmatrix} -\frac{k_r + T_{aw_r}}{J_t} & \frac{T_{a\beta_{min}}}{J_t} & \frac{T_{aV_{min}}}{J_t} \\ 0 & -\frac{1}{\tau_\beta} & 0 \\ 0 & 0 & \frac{1}{\tau_V} \end{bmatrix}$$

$$B_i = \begin{bmatrix} -\frac{1}{J_r} & 0 \\ 0 & \frac{1}{\tau_\beta} \\ 0 & 0 \end{bmatrix} \text{ and } C_i = [100] \text{ with } i = 1,\ 2,\ 3,\ 4$$

Traditional state feedback control in the Takagi Sugeno model employs that a so distributed processing compensation state control method, whose it plant guideline is similar to that presented in (7) and is described in more detail:

$$u_i(t) = k_{mpci}x(t) \tag{9}$$

Figure 1 depicts the method and the fuzzy model in the MBPC scheme with an adapted version.

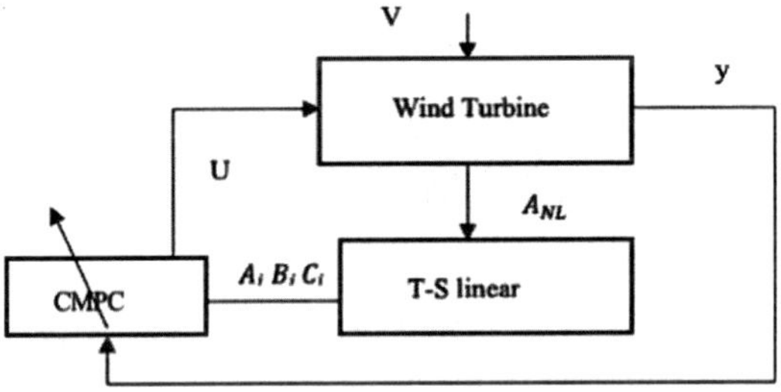

Fig. 1. The proposed control strategy

4 Continuous Time Model Predictive Control Design

The application of a continuous-time predictive model offers several advantages, this model is one of the most efficient and modern continuous-time technical methods. By considering the model:

$$x_m(t) = \dot{x}(t)$$
$$y(t) = C\,x(t)$$

With a new vector state $z(t) = \left[x_m(t)^T\ y(t)^T\right]^T$ in conjunction. An enhanced model can be acquired in the accompanying ways:

$$\dot{z}(t) = Az(t) + B\dot{u}(t) \tag{10}$$

where $A = \begin{bmatrix} A_{ml} & O_m^T \\ C_m & O_{q\times q} \end{bmatrix}$, $B = \begin{bmatrix} B_{ml} \\ O_{q\times m} \end{bmatrix}$ and $C = \left[O_{q\times n}\ I_{q\times q}\right]$

Laguerre roles are a set of orthonormal functions used to describe control projections or the development of continuous-time predictive controls. We can define the routes of future indicators using orthogonal base functions to detect all the con-strained coefficients to minimize them thus the optimization of the detected problems [11]. The control signal derivative $\dot{U}(\tau)$, $0 \le t \le T$ is described by:

$$\dot{U}(t) = L(t)\theta_{opt} \tag{11}$$

Specify the vector $L(t) = [l_1(t)\ l_2(t) \ldots\ldots l_N(t)]^T$.

The vector's initial summing condition L(t) as: $L(0) = \sqrt{2p}\,[11 \ldots. 1]^T$, the Laguerre functions achieve the formula below:

$$\begin{bmatrix} \dot{l}_1(t) \\ \dot{l}_2(t) \\ \vdots \\ \dot{l}_N(t) \end{bmatrix} = \begin{bmatrix} -p & \cdots & 0 \\ \vdots & \ddots & \vdots \\ -2p & -p & 0 \\ -2p & \cdots & -2p-p \end{bmatrix} \begin{bmatrix} l_1(t) \\ l_2(t) \\ \vdots \\ l_N(t) \end{bmatrix} \tag{12}$$

As a result of the differential equation's solution, we obtain a set of Laguerre functions for i = 1,…N as follows:

$L(t) = e^{Apt}\, L(0)$ where:

$$AP = \begin{bmatrix} -p & \cdots & 0 \\ \vdots & \ddots & \vdots \\ -2p & -p & 0 \\ -2p & \cdots & -2p-p \end{bmatrix} \tag{13}$$

Using the property of orthogonality of Laguerre function the cost function J is given by:

$$J = \int_0^T z(t_i+\tau)^T\, Q\, z(t_i+\tau)d\tau + \theta^T R\, \theta \tag{14}$$

By substituting the prediction equation:

$z(t_i+\tau) = e^{A\tau} z(t_i) + \varphi(\tau)^T\theta$ into the cost function (14) it becomes:

$$J = \theta^T\left[\int_0^T \varphi(\tau)\, Q\, \varphi(\tau)^T d\tau + R\right]\theta + 2\theta^T\left[\int_0^T \varphi(\tau)\, Q\, e^{A\tau} d\tau\right] z(t_i) + z(t_i)^T\left[\int_0^T e^{A^T\tau}\, Q\, e^{A\tau}\, d\tau\right] z(t_i) \tag{15}$$

where $\varphi_i(\tau)^T = \int_0^\tau e^{A(\tau-\gamma)} B_i L_i(\gamma)^T d\gamma$.

It is enough to use the derivative to reduce the value of J in the absence of constraints, of the cost function in terms of the Laguerre coefficients θ.

By assuming the existence of $\left(\int_0^T \varphi(\tau)\, Q\, \varphi(\tau)^T d\tau + R\right)^{-1}$, the optimal solution is:

$$\theta_{opt} = -\left(\int_0^T \varphi(\tau)\, Q\, \varphi(\tau)^T d\tau + R\right)^{-1}\left(\int_0^T \varphi(\tau) Q\, e^{A\tau} d\tau\right) z(t_i) \tag{16}$$

To make things easier, let's say:

$$\psi\_CMPC = \int_0^T \varphi(\tau)\, Q\, \varphi(\tau)^T d + R$$

$$\phi\_CMPC = \int_0^T \varphi(\tau) Q\, e^{A\tau} d\tau$$

Then:

$$\theta_{opt} = -\Omega^{-1}_{\_CMPC}\psi_{\_CMPC}\, z(t_i) \tag{17}$$

The primary goal of defining the Laguerre coefficient is to reduce the number of variables that must be optimized or, at the very slightest, to keep improving the conditioning of such an optimal solution. As part of the MPC design process in real-time. As part of the MPC design process in real time, the matrices $\psi_{\_CMPC}$ and $\phi_{\_CMPC}$ can be calculated offline.

We considered Model predictive controller based TS-fuzzy for a variable speed wind turbine:

$$\dot{\mathbf{u}}(t) \sum_{i=1}^{4} \mathbf{h_i}(\mathbf{z})\mathbf{k_{mpci}}\mathbf{x}(\mathbf{t}) \tag{18}$$

With: $\mathbf{k_{mpci}}(\mathbf{t}) = \mathbf{L_i}(t)\left(\frac{\psi_{\_CMPCi}}{\phi_{\_CMPCi}}\right)$, where:

If **v is** $\mathbf{V_{max}}$ and $\boldsymbol{\beta}$ is $\boldsymbol{\beta_{max}}$ then: $\mathbf{k_{mpc1}} = \mathbf{L}_1\left(\frac{\psi_{\_CMPC1}}{\phi_{\_CMPC1}}\right)$

If **v is** $\mathbf{V_{max}}$ and $\boldsymbol{\beta}$ is $\boldsymbol{\beta_{min}}$ then: $\mathbf{k_{mpc2}} = \mathbf{L}_2\left(\frac{\psi_{\_CMPC2}}{\phi_{\_CMPC2}}\right)$

If **v is** $\mathbf{V_{min}}$ and $\boldsymbol{\beta}$ is $\boldsymbol{\beta_{max}}$ then: $\mathbf{k_{mpc3}} = \mathbf{L}_3\left(\frac{\psi_{\_CMPC3}}{\phi_{\_CMPC3}}\right)$

If **v is** $\mathbf{V_{min}}$ and $\boldsymbol{\beta}$ is $\boldsymbol{\beta_{min}}$ then: $\mathbf{k_{mpc4}} = \mathbf{L}_4\left(\frac{\psi_{CMPC4}}{\phi_{\_CMPC4}}\right)$

5 Simulations Results

Simulations are performed in MATLAB to evaluate the proposed control. With the friction coefficient is $K_r = 3029{,}5$ N.m/rad, moment of inertia $J_t = 830000\ \text{kg/m}^2$, the air mass density $\rho = 1.225\ \text{kg/m}^3$, the length of the rotor blades R = 30 m, wind speed delay time $\tau_{ov} = 0,4$ s , pich angle delay time $\tau_{og} = 0$. The considered the range of wind speed is $10\ \text{m/s} \leq V \leq 25\ \text{m/s}$, the range of the pitch angle is $-2° \leq \beta \leq 24°$.

The following matrices are provided to check the system states of the wind turbine predictive model, which was solved using MATLAB:

$$K_{mpc1} = \begin{bmatrix} 0\,0\,0\,0 \\ 0\,0\,0\,0 \end{bmatrix}; K_{mpc2} = \begin{bmatrix} 0 & 0 & 0 & 0 \\ 1.7125 & 10.1274 & 0.0878 & 2.8941 \end{bmatrix}$$

$$K_{mpc3} = \begin{bmatrix} 0 & 0 & 0 & 0 \\ 2.6010 & 21.6466 & 0.1069 & 4.3982 \end{bmatrix}$$

$$K_{mpc4} = \begin{bmatrix} 0 & 0 & 0 & 0 \\ 0.9869 & 5.9824 & 0.0406 & 1.6677 \end{bmatrix}$$

Figure 2 depicts the rotational speed curve. The fuzzy controller is clearly attempting to maintain the rotor's optimum values based on the estimated wind speed.

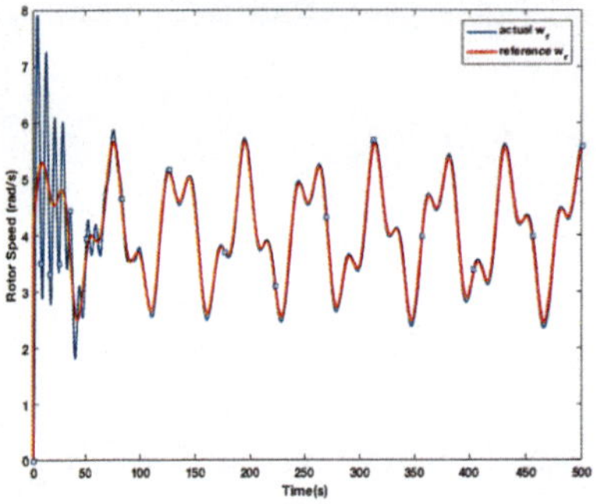

Fig. 2. Actual and reference rotational speed

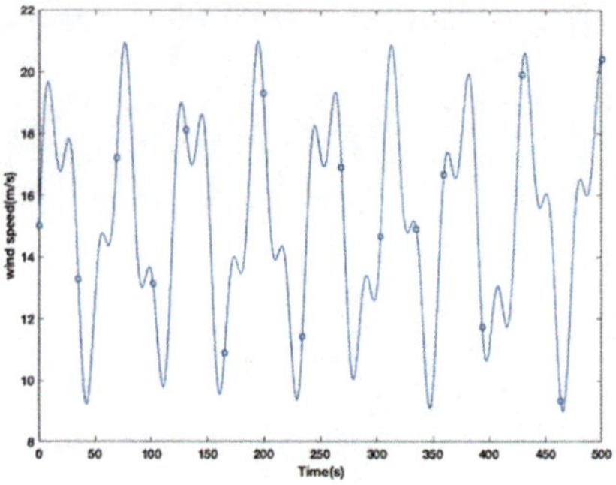

Fig. 3. The wind speed

Figure 3 depicts the approximated wind speed, provides a good traceability of the true wind speed without error in constant state based on the predictive control.

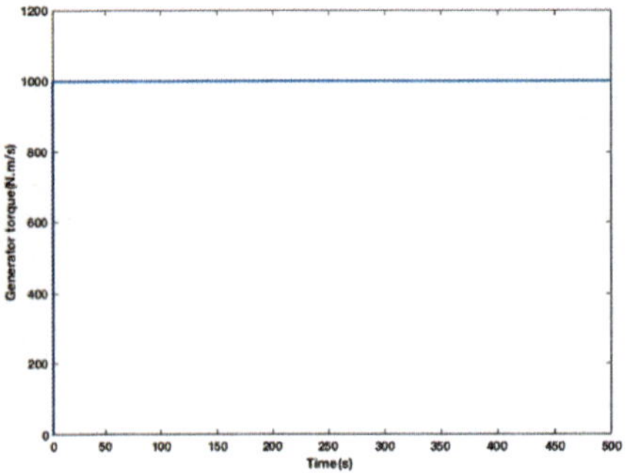

Fig. 4. The generator Torque

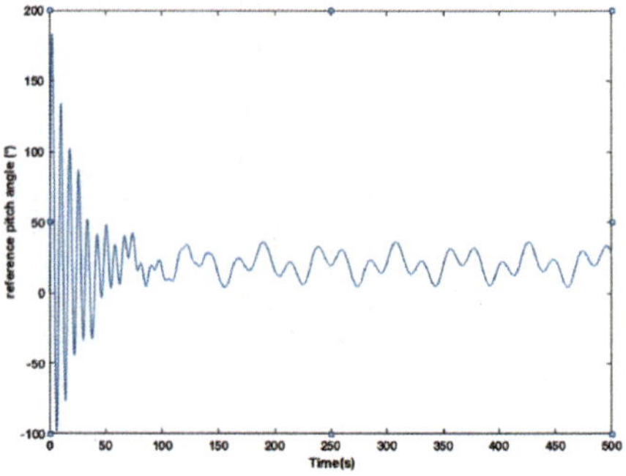

Fig. 5. Reference pitch angle

As shown in Fig. 4, the fuzzy-MPC controller response to generator torque based on predictive control. Finally, Fig. 5 shows the reference of pitch angle varies with wind speed.

However, as shown in the illustration, the Model predictive controller-based TS-generator fuzzy's torque response varies with wind speed variation. The proposed simulation demonstrates the effectiveness of obtaining the best tracking speed under wind speed fluctuations.

6 Conclusion

Within this study, we investigate the simulation of a nonlinear wind turbine using the Takagi-Suengo Fuzzy logic and then the design of predictive control in continuous time. The goal of this research is to get the most out of wind energy while maintaining the power's net value. To manage the control, we based on a predictive control to ensure the monitoring reference. It is widely assumed that good performance includes tracking the optimum rotor speed. The modeling experiments show that the proposed predictive control approach works well. However, the work to be followed is based on the application of the proposed command to the hybrid system [12].

References

1. Hadjili, M.L., Kara, K.: Modelling and control using Takagi-Sugeno fuzzy-model. In: Communications and Photonics Conference (SIECPC), vol. 5. Saudi International (2011)
2. Meda-Campana, J.A., JGomez-Mancilla, J.C., Castillo To-ledo, B.: Exact output regulation for nonlinear systems described by Takagi–Sugeno fuzzy models. J. IEEE Trans. Fuzzy Syst. **20**(2), 235-247 (2012)
3. Soliman, M., Malik, O.P., Westwick, D.T.: Multiple model multiple-input multiple-out-put predictive control for variable speed variable pitch wind energy conversion systems. IET Renew. Pow. Gener. **5**(2), 124-136 (2011)
4. Østergaard, K.Z., Brath, P.,Stoustrup, J.: Estimation of effective wind speed. J. Phys. **75**(1), 012082 (2007)
5. Jena, D., Rajendran, S.: A review of estimation of effective wind speed based control of wind turbines. Renew. Sustain. Energy Rev. **1**(43), 1046–1062 (2015)
6. EL Bakri, A., Boumhidi, I.: Finite-frequency observer-based fault estimation and fault-tolerant control for wind turbine. Int. J. Electr. Eng. Inform. **12** (2020)
7. Vall Hemeyine, A., Abbou, A., Tidjani, N., Mokhlis, M., Bakouri, A.: Robust Takagi sugeno fuzzy models control for a variable speed wind turbine based a DFI-generator. Int. J. Intell. Eng. Syst. **13**(3), 90–100 (2020)
8. Bougdour, S., Elhouti, R., Sefriti, S., Boumhidi, I.: Optimal predictive control model of wind turbine. In: 2021 Fifth International Conference on Intelligent Computing in Data Sciences, pp. 1–7 (2021)
9. Al-Hadithi, M., Jiménez, A., López, R.G.: Fuzzy optimal control using generalized Takagi–Sugeno model for multivariable nonlinear systems. Appl. Soft Comput. **30**, 205–213 (2015)
10. Gauterin, E., Kammerer, P., Kühn, M., Schulte, H.: Effective wind speed estimation: comparison between Kalman Filter and Takagi–Sugeno observer techniques. ISA Trans. **62**, 60–72 (2016)
11. Wang, L.: Model Predictive Control System Design and Implementation Using MATLAB. London: Springer, London (2009)
12. Camacho, E.F., Ramirez, D.R., Limon, D., Muñoz de la Peña, D., Alamo, T.: Model predictive control techniques for hybrid systems. J. IFAC Int. Fed. Autom. Control **34**(1), 21–31 (2010)

A FOSMC-DPC Control Strategy for Power Quality Enhancement of DFIG Wind Turbine by Using Three-Level Inverter

Mohamed Amine Beniss(✉), Hassan El Moussaoui, Ouadiâ Chekira, Tijani Lamhamdi, Hassane EL Markhi, and Ali Boharb

Electrical Department, Laboratory of Intelligent Systems, Georesources and Renewable Energies, Faculty of Sciences and Technologies, Sidi Mohamed Ben Abdellah University, Fez, Morocco

mohamedamine.beniss@usmba.ac.ma

Abstract. This paper proposes a control system for a DFIG power converter, applied by a Three-level NPC back-to-back converter, to maintain a unitary power factor injection to the network that to maximize wind power harvesting using a maximum power point tracking method (MPPT). The use of the FOSMC provides good robustness to the system while the insertion of the Three-level NPC inverter greatly reduces the total harmonic distortion (THD) of the currents injected to the power grid.

Keywords: Doubly Fed Induction Generator (DFIG) · Classical two-level inverter · Fractional order calculus · Sliding Mode Control (SMC) · NPC three-level Inverter · Control of renewable energy resources

1 Introduction

Nowadays, wind energy has become one of the fastest-growing renewable energy resources because of its clean nature, pollution-free, and gives a good opportunity to generate electricity [1]. From all the power generators used in gathering the wind energy, the application of a doubly fed induction generator (DFIG) in the global wind park its skyrocketing in past decades due to its changing speed operation, low price of inverter, and the possibility to control the reactive and active power independently [2]. The structure of a DFIG based wind energy conversion systems (WECS) presented in Fig. 1. In this topology, the stator circuit of DFIG windings are directly linked to the power network, and the rotor circuit windings are linked to the grid via a back-to-back power electronic converter PEC consisting of a rotor-side converter (RSC) and a grid-side converter (GSC) [3].

The idea of the controller design for the DFIG is to optimize the performances versus the wind speed disturbances and variations. The use of the DFIG can adopt many types that are well popular of control, for instance, a direct power control (DPC), direct torque control (DTC), and Vector control (VC) or strictly speaking, Field oriented control (FOC). The (VC) strategy is based on traditional controllers PI [4].

S. Motahhir and B. Bossoufi (Eds.): ICDTA 2022, LNNS 454, pp. 712–721, 2022.
https://doi.org/10.1007/978-3-031-01942-5_71

In this work, a DPC based on the Fractional-Order Sliding Mode control DPC-FOSMC technique is proposed to control the (DFIG) using a three-level neutral point clamped inverter (NPC). For this proposed strategy, the scheme proposed provides reduced active and reactive powers ripples, and a low current harmonics rate, compared to Tow-level DPC-FOSMC strategy.

The following is the work overall structure: Sect. 2 discusses the dynamics of the wind turbine. Then the DFIG modelling will be presented and describes the FOSMC technique in Sect. 3. models for back-to-back converters will be introduced in Sect. 4. Section 5 shows results and discussions of the DPC-FOSMC simulation using tow and 3-level NPC inverters. Finally, general conclusions are drawn in Sect. 6.

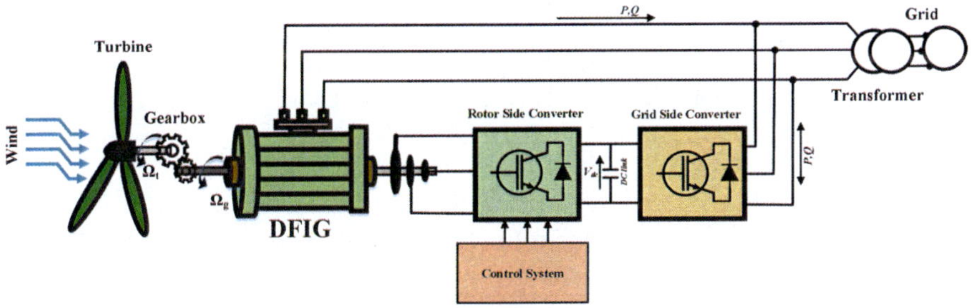

Fig. 1. Structure configuration of DFIG-based WECS

2 Wind Turbine Modeling

The wind power amount P_{av} available can be written as[5]:

$$P_{av} = \frac{1}{2}\rho_{air} A v^3 \tag{1}$$

The aerodynamic power P_{aer} can be expressed as follows [6]:

$$P_{aer} = P_{av} C_P(\lambda, \beta) = \frac{1}{2}\rho_{air} A v^3 C_P(\lambda, \beta) \tag{2}$$

where is A the swept area of the turbine blades; ρ represents the air density and $C_P(\lambda, \beta)$ is the turbine power coefficient.

$$C_P(\lambda, \beta) = 0.5176\left(\frac{116}{\lambda_i} - 0.4\beta - 5\right)e^{\left(\frac{21}{\lambda_i}\right)} + 0.0068 \tag{3}$$

With; $\lambda_i^{-1} = \frac{1}{\lambda+0.08\beta} - \frac{0.035}{\beta^3+1}$
Where is the tip speed ratio (TSR) that is written as:

$$\lambda = \frac{\omega_r R}{v_w} \tag{4}$$

where R is the turbine blade radius, V_w is the wind speed and, ω_r is the rotor rotational speed.

When the wind speed is less than the rated value and more than the cut-in value, the Maximum Power Point Tracker algorithm is being implemented in order to maximize the wind power harvesting procedure.

3 DFIG Modeling

The park representation of the DFIG model includes an induction generator, GSC, RSC, shafting, and DC bus link, which by the following dynamic equations[7]:

$$\begin{cases} V_{ds} = R_s i_{ds} + \frac{d\psi_{ds}}{dt} - \omega_s \psi_{qs} \\ V_{qs} = R_s i_{qs} + \frac{d\psi_{qs}}{dt} - \omega_s \psi_{ds} \\ V_{dr} = R_r i_{dr} + \frac{d\psi_{dr}}{dt} - (\omega_s - \omega_r)\psi_{qr} \\ V_{qr} = R_r i_{qr} + \frac{d\psi_{qr}}{dt} - (\omega_s - \omega_r)\psi_{dr} \end{cases} \tag{5}$$

where,V_{dr}, V_{qr}, V_{ds}, andV_{qs}, are the d and q axis rotor and stator voltages, i_{ds}, i_{qs}, i_{dr}, and i_{qr} are the d and q axis stator and rotor currents, ψ_{ds}, ψ_{qs}, ψ_{dr}, ψ_{qr}, are the d and q axis stator and rotor fluxes, respectively. ω_r and ω_s are the angular velocity and the synchronous angular velocity. R_s and R_r are the stator and rotor resistances, respectively.

The rotor and stator flux can be expressed as [8]:

$$\begin{cases} \psi_{ds} = L_s i_{ds} + M i_{dr} \\ \psi_{qs} = L_s i_{qs} + M i_{qr} \\ \psi_{dr} = L_r i_{dr} + M i_{ds} \\ \psi_{qr} = L_r i_{qr} + M i_{qs} \end{cases} \tag{6}$$

With the constant P refers to the number of pairs of poles.

The active and reactive power in the (d, q)-axis reference frame can be described by the following dynamic equations [9]:

$$\begin{cases} P_s = V_{sd} I_{sd} + V_{sq} I_{sq} \\ Q_s = V_{sq} I_{sd} - V_{sd} I_{sq} \end{cases} \tag{7}$$

The electric torque Tem and mechanical equations can be written as:

$$T_{em} = P\frac{M}{L_s}(\psi_{qs} I_{dr} - \psi_{ds} I_{qr}) \tag{8}$$

$$J\frac{d}{dt}\Omega_r = T_{em}\, T_r f_r\, \Omega_r \tag{9}$$

By aligning the d-axis to the direction of stator flux, we have:

$$\psi_{ds} = \psi_s, \psi_{qs} = 0 \tag{10}$$

Consequently, the Eqs. (11) can be expressed as:

$$T_{em} = -P\frac{MV_s}{\omega_s L_s} I_{qr} \tag{11}$$

if we suppose that the stator flux is constant and the resistance of the stator phase is small, subsequently voltage vectors on the stator axis as given in:

$$V_{qs} = V_s = \omega_s \psi_s,\ V_{ds} = 0 \tag{12}$$

With this supposition and using, the Eqs. (10) and (12) in (5) and (6) the quadrature and direct voltages on the rotor axis and stator reactive and active power can be written in terms with the rotor currents becomes.

$$\begin{cases} V_{dr} = R_r I_{dr} + \sigma L_r \frac{d}{dt} I_{dr} - \sigma L_r S \omega_s I_{qr} \\ V_{qr} = R_r I_{qr} + \sigma L_r \frac{d}{dt} I_{qr} - \sigma S \omega_s I_{dr} + S\frac{MV_s}{L_s} \end{cases} \tag{13}$$

$$\begin{cases} P_s = -\frac{MV_s}{L_s} I_{qr} \\ Q_s = \frac{V_s^2}{\omega_s L_s} - \frac{MV_s}{L_s} I_{dr} \end{cases} \tag{14}$$

With $S = \frac{\omega_s - \omega_r}{\omega_s}$, $\sigma = 1 - \frac{M}{L_r L_s}$
The nonlinear state-space model can be written form as given in Eq. (13).

$$\dot{X} = F(X) + GU \tag{15}$$

Where the nominal matrices can be represented as:

$$\begin{cases} U = \begin{pmatrix} V_{dr} \\ V_{qr} \end{pmatrix},\ X = \begin{pmatrix} I_{dr} \\ I_{qr} \end{pmatrix} \\ G = \begin{pmatrix} \frac{1}{\sigma L_r} & 0 \\ 0 & \frac{1}{\sigma L_r} \end{pmatrix} \text{and } F(X) = \begin{pmatrix} -\frac{R_r}{\sigma L_r} I_{dr} & + S\omega_s I_{qr} \\ -\frac{R_r}{\sigma L_r} I_{qr} - S\omega_s I_{dr} & + S\frac{MV_s}{\sigma L_s L_r} \end{pmatrix} \end{cases} \tag{16}$$

The control strategy's goal is to guarantee that the state variable X follows the reference trajectory X_{ref}:

$$X_{ref} = \begin{pmatrix} I_{dr_{ref}} \\ I_{qr_{ref}} \end{pmatrix} \tag{17}$$

According to the Eq. (12), to show the rotor currents in stator reactive and active powers, we set Qs = 0 to achieve a unity power factor.
The rotor current can be written in terms of active powers as follows:

$$\begin{cases} I_{qr_{ref}} = -\frac{L_s}{MV_s} P_{qr} \\ I_{dr_{ref}} = \frac{V_s}{\omega_s L_s} \end{cases} \tag{18}$$

The following are the definitions of current tracking errors:

$$e = X - X_{reff} \tag{19}$$

$$e = \begin{pmatrix} e_1 \\ e_2 \end{pmatrix} = \begin{pmatrix} X_1 - X_{1ref} \\ X_2 - X_{2ref} \end{pmatrix} = \begin{pmatrix} I_{dr} - I_{drref} \\ I_{qr} - I_{qrref} \end{pmatrix} \quad (20)$$

Fractional-Order Sliding Mode control (FOSMC) of the rotor side converter (RSC) is the chosen of a suitable sliding surface. The derivative of fractional sliding surface is chosen as follows:

$$S_i = \begin{pmatrix} S_1 \\ S_2 \end{pmatrix} = \begin{pmatrix} ke_1 + D^\alpha e_1 \\ ke_2 + D^\alpha e_2 \end{pmatrix} \quad (21)$$

With $k > 0$ is a positive real coefficient and S_i denotes the sliding surface.

4 Switching Principle of Back-To-Back Converters

Adjusting the frequency, amplitude, and phase of the rotor voltage create reactive and active power and generator torque, through the dc-link voltage, the rotor, and stator are controllable.

The power converter is composed of a back-to-back AC/DC/AC PWM converter attaching the machine to the network. The converters are generally composed of voltage source inverters (VSI) included with power semiconductor switches IGBTs supplied with flyback diodes, which allow for bidirectional power flow [4]. Further, the appropriate control techniques with the back-to-back converters can improve system reliability by averting the disturbances that influence power quality.

In this work, two topologies of the power converter are employed for further analysis of the system behavior.

4.1 Classical Two-Level Inverter

A two-level power converter is an AC/DC/AC converter, consisting of three switching commutation cells; each cell comprises two IGBTs (S1, S2) as an inverter, each cell includes two IGBTs and are commanded by the signal to terminals, $\overline{S_1}$, $\overline{S_2}$, $\overline{S_3}$, S_1, S_2, andS_3. It is expected that T1, T2, T3, and T4 also T5 and T6 can be switched similarly [8, 10].

The following mathematical model summarizes all of the previously mentioned information:

$$\begin{bmatrix} Va_n \\ Vb_n \\ Vc_n \end{bmatrix} = \begin{bmatrix} 2 & -1 & -1 \\ -1 & 2 & -1 \\ -1 & -1 & 2 \end{bmatrix} \begin{bmatrix} S_a \\ S_b \\ S_c \end{bmatrix} \quad (22)$$

It should be noted that the states of the switches could be represented by three Boolean variables, such as:

$$S_x = \begin{cases} Si & = 0 \text{ if } \overline{T_x} \text{ is open and } T_i \text{ is closed.} \\ Si & = 1 \text{ if } \overline{T_x} \text{ is closed and } T_i \text{ is open.} \end{cases}$$

where $(x = 1,\ 2,\ 3)$

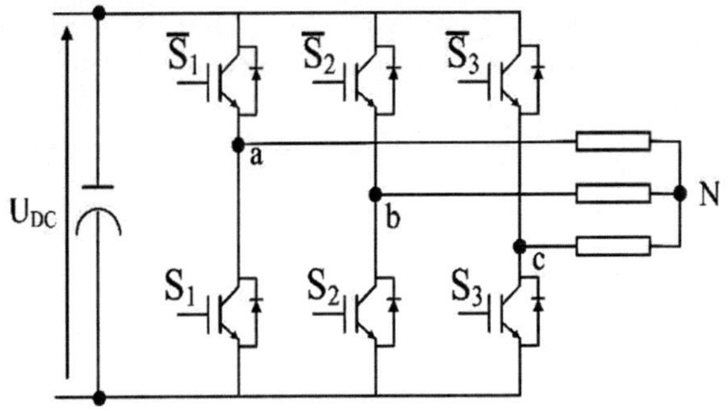

Fig. 2. Structure of the RSC of DFIG connected tow-level inverter

4.2 Three-Level NPC Inverter

An NPC three-level converter is also an AC-DC-AC power converter with 3-arms, and each arm includes four IGBTs Si1-Si4 and two neutral clamping diodes Di1-Di4. The output phase voltage of the converter is determined by controlling the turn-off and turn-on switching IGBTs (S_{x1}, S_{x2}, S_{x3}, and S_{x4}). The Switching instances (0; 1) that determine the off-turn and on-turn of the switching IGBTs are produced and regulated by the switching states that are generated by the sinusoidal PWM carrier based method [11]. Since these switches are in the same arm, they cannot turn off or turn on simultaneously. In this method, the triangular carrier signals, "lower carrier" and 'upper carrier' are compared with three-phase modulating reference signals. For each phase, the logical signals (+1, 0, −1) have been determined and produce voltage -VDC/2, 0, and, VDC/2 respectively. By a specific logical signals switch, the AC side creates a synthesis sine wave and phase voltage of three levels at the output. The configuration circuit diagram of an NPC 3-level inverter is presented in Fig. 3.

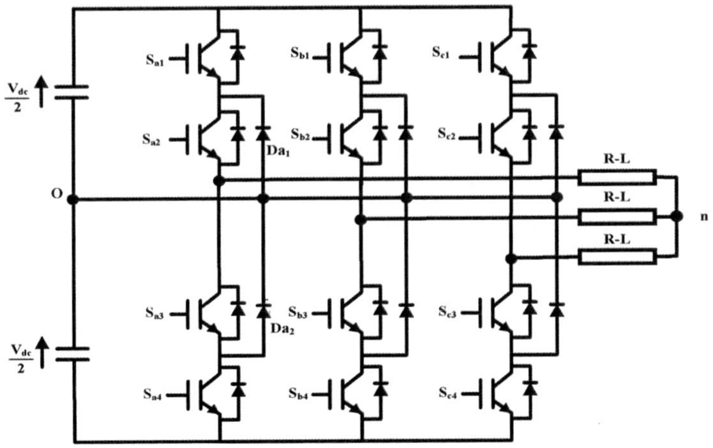

Fig. 3. Structure of the RSC of DFIG connected three-level inverter

The phase to phase output voltages (Va_b, Vb_c, Vc_a) and The phase to neutral (Va_o, Vb_o, Vc_o). It can be represented by the matrix form shown below:

$$\begin{cases} \begin{bmatrix} Va_b \\ Vb_c \\ Vc_a \end{bmatrix} = \frac{V_{dc}}{2} \begin{bmatrix} 1 & -1 & 0 \\ 0 & 1 & -1 \\ -1 & 0 & 1 \end{bmatrix} \begin{bmatrix} S_a \\ S_b \\ S_c \end{bmatrix} \\ \begin{bmatrix} Va_o \\ Vb_o \\ Vc_o \end{bmatrix} = \frac{V_{dc}}{2} \begin{bmatrix} 2 & -1 & -1 \\ -1 & 2 & -1 \\ -1 & -1 & 2 \end{bmatrix} \begin{bmatrix} S_a \\ S_b \\ S_c \end{bmatrix} \end{cases} \tag{23}$$

$$\text{With} : S_a = \begin{cases} 0\, if\ S_{a4} = on\, and\ S_{a3} = on \\ 1\, if\ S_{a3} = on\, and\ S_{a2} = on \\ 2\, if\ S_{a2} = on\, and\ S_{a1} = on \end{cases} \tag{24}$$

For instance, Standards IEEE-519 [68] impose limits for Total Harmonic Distortion (THD). The limit of THD for this standard is 5%, which is used as a guideline in this study to analyze the viability of the power quality injected into the network.

$$THD(\%) = 100\frac{\sqrt{X_H^2}}{X_F} \tag{25}$$

X_F is the root mean square (RMS) value of the fundamental component and X_H is the RMS value of the signal.

5 Results and Discussions

The DFIG-WECS model is built in MATLAB/Simulink software to validate the proposed DPC-FOSMC approach, and the results of the simulation has been carried out using An 7, 5 kW DFIG is used; its parameters involved are listed in Table 1.

Table 1. DFIG Parameters.

Parameters
P = 7.5KW; R =43,5 m; J = 0.3125Kg.m2; fr= 6.73.10e-3N.m.s-1; ρ=1.22 Kg.m^{-3}; p = 2; M = 0.078 H; Rr = 0.062Ω; Rs = 0.455Ω; Ls = 0.084 H; Lr = 0.081 H

5.1 Performance of WECS of the Proposed FOSMC Using 3-level

In order to evaluate the reliability, performance of WECS, reactive and many active power steps were carried out and using a variable wind profile. Figure 4 and Fig. 5 shows that the trajectories of the reactive and active power track each other perfectly when utilizing the 3-level inverter.

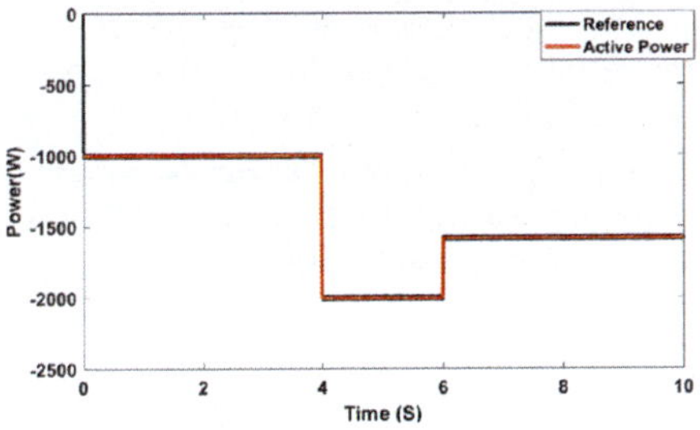

Fig. 4. Stator active power

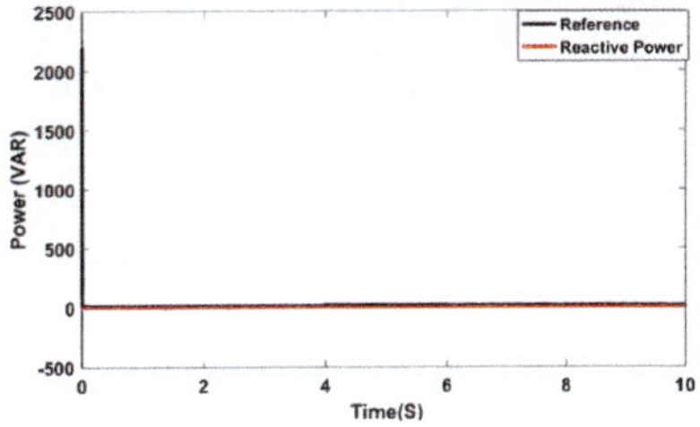

Fig. 5. Stator reactive power

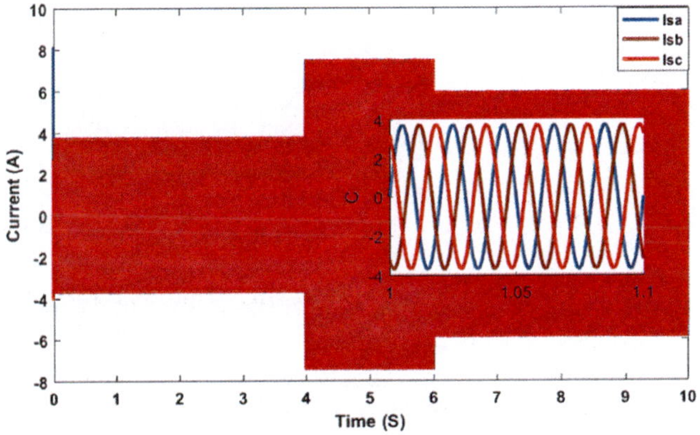

Fig. 6. Stator current versus time

Figure 6 represents that the stator currents injected into grid, this current has remains a sinusoidal waveform with a frequency of 50 Hz.

Figures 7 and 8 illustrate the stator active and reactive power simulation results under variable wind speed perfectly follow their reference.

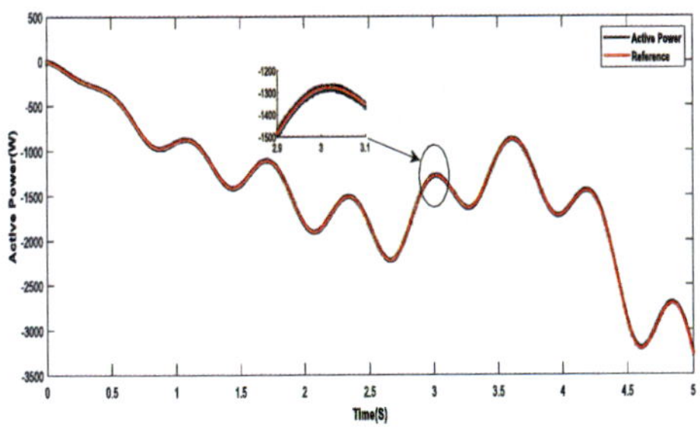

Fig. 7. Stator active Power using the variable wind profile

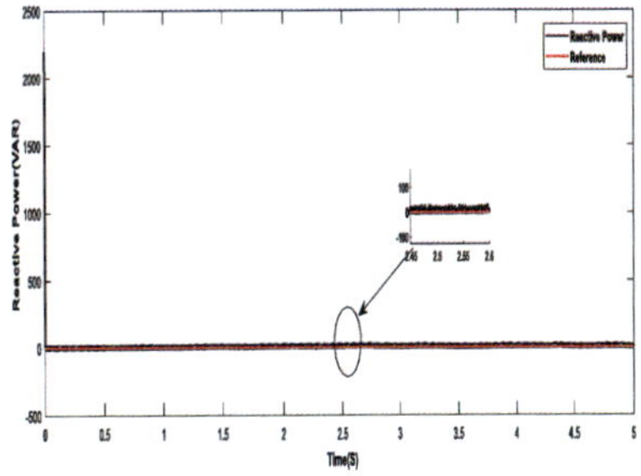

Fig. 8. Stator reactive Power

5.2 Comparative Study

In this subsection, Fig. 9 and Fig. 10 indicates that the harmonic distortion rate of THD has been ameliorated; it can be noted that THD has better quality in the two cases and remains under the limit of 5% According to IEEE-519 standard [18]. A THD rate of 0.51% for DFIG using 3-level and a THD rate of 1.13% for DFIG using 2-level (see Table2).

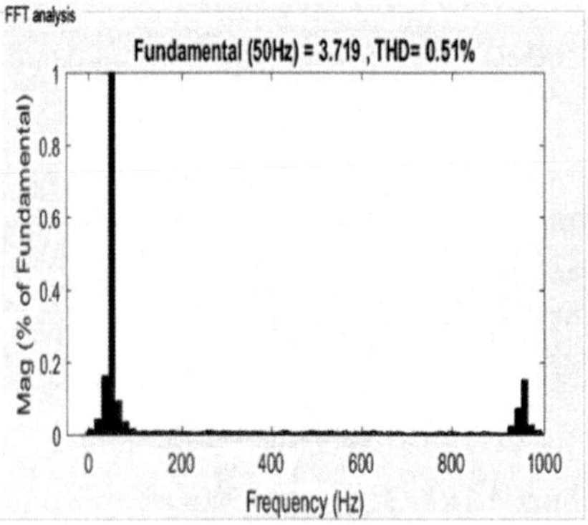

Fig. 9. FFT of the stator current for DPC-FOSMC using 3-level inverter

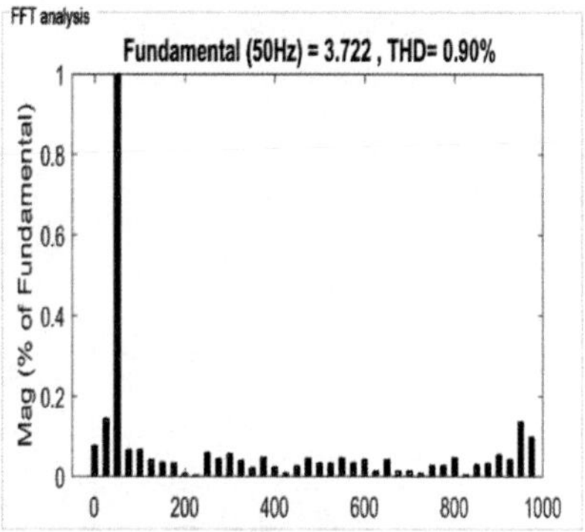

Fig. 10. FFT of the stator current for DPC-FOSMC using tow-level inverter

Table. 2. THD comparison of of 2-level and 3 level converter

	THD
3-level converter	0.51%
2-level converter	0.90%

6 Conclusion

In this work, the FOSMC control system is introduced. The control system's purpose is optimizing the performance of WECS-DFIG. In addition, it has been affirmed that the NPC three-level inverter technology enhances performance and reduces THD. Indeed, the use of NPC three-level inverter and DFIG gives a better reduction of THD in WECS when compared to the classical two-level inverter, and hence, the power quality produced by the DFIG is improved.

Future work can address the implementation of the proposed method to a three-level inverter based DFIG to validate how this can be clowned useful and practical.

References

1. Bossoufi, B., et al.: Rooted tree optimization for the backstepping power control of a doubly fed induction generator wind turbine: dSPACE implementation. IEEE Access **9**, 26512–26522 (2021)
2. Luo, J., Zhao, H., Gao, S., Han, M.: A low voltage ride through strategy of DFIG based on explicit model predictive control. Int. J. Electr. Power Energy Syst. **119**, 105783 (2020)
3. Yessef, M., Bossoufi, B., Taoussi, M., Lagrioui, A., El Mahfoud, M.: Evaluation of adaptive backstepping control applied to DFIG wind system used on the real wind profile of the Dakhla-Morocco city. In: Bennani, S., Lakhrissi, Y., Khaissidi, G., Mansouri, A., Khamlichi, Y. (eds.) WITS 2020. LNEE, vol. 745, pp. 661–671. Springer, Singapore (2022). https://doi.org/10.1007/978-981-33-6893-4_61
4. Beniss, M.A., El Moussaoui, H., Lamhamdi, T., El Markhi, H.: Improvement of power quality injected into the grid by using a FOSMC-DPC for doubly fed induction generator. Int. J. Intell. Eng. Syst. **14**(2), 556–567 (2021)
5. Taoussi, M., Karim, M., Bossoufi, B., et al.: Low-speed sensorless control for wind turbine system. WSEAS Trans. Syst. Control **12**, 405–417 (2017)
6. Yessef, M., Bossoufi, B., Taoussi, M., Lagrioui, A., Chojaa, H.: Improved hybrid control strategy of the doubly-fed induction generator under a real wind profile. In: Motahhir, S., Bossoufi, B. (eds.) ICDTA 2021. LNNS, vol. 211, pp. 1279–1290. Springer, Cham (2021). https://doi.org/10.1007/978-3-030-73882-2_117
7. Amrane, F., Francois, B., Chaiba, A.: Experimental investigation of efficient and simple wind-turbine based on DFIG-direct power control using LCL-filter for stand-alone mode. ISA Trans. (2021)
8. Beniss, M.A., El Moussaoui, H., Lamhamdi, T., El Markhi, H.: Performance analysis and enhancement of Direct power control Of DFIG based wind system. Int. J. Pow. Electron. Drive Syst. **12**(2), 1034–1044 (2021)
9. Xiong, L., Li, P., Wang, J.: High-order sliding mode control of DFIG under unbalanced grid voltage conditions. Int. J. Electr. Power Energy Syst. **117**, 105608 (2020)
10. El Ouanjli, N., Motahhir, S., Derouich, A., El Ghzizal, A., Chebabhi, A., Taoussi, M.: Improved DTC strategy of doubly fed induction motor using fuzzy logic controller. Energy Rep. **5**, 271–279 (2019)
11. Van, T.L., Lee, D.: Developing function models of back-to-back PWM converters for simplified simulation. J. Power Electron. **11**, 51–58 (2011)

Robust Control of a Wind Conversion System Based on a Doubly Fed Induction Generator: A Comparison Between Adaptive Backstepping and Integral Sliding Mode Controllers

Mourad Yessef[1] (✉), Badre Bossoufi[1], Mohammed Taoussi[2], Ahmed Lagrioui[3], Hamid Chojaa[2], Btissam Majout[1], and Houda El Alami[1]

[1] Laboratory of Engineering, Modeling and Systems Analysis, SMBA University Fez, Fez, Morocco
mourad.yessef@usmba.ac.ma

[2] Laboratory of Technologies and Industrial Services, Higher School of Technology, SMBA University Fez, Fez, Morocco

[3] Department of Electrical and Computer Engineering, Higher National School of Arts and Trades, Moulay Ismail University Meknes, Meknes, Morocco

Abstract. The voltage sag, caused by the short-circuit faults in transmission and subtransmission system of the wind energy conversion system (WECS), produces a transient stator flux (DC component) in the stator windings of the doubly fed induction generator (DFIG).This phenomenon damaging the fragile inverter and the system gearbox. In fact, the dynamic equations of the wind turbine are strongly nonlinear as are the ones of the DFIG. In addition, most of the WECS parameters are uncertain, and this system is always suffering from disturbance sources such the harmonics and the mechanical vibrations. To address these problems, robust power control strategies must be implemented. In this paper, two controllers are investigated. These strategies contribute to the elimination of the high peak currents and stabilize the active power at its desired value. First, an Adaptive Backstepping (AB) controller is adopted. Then, an integral sliding mode controller (ISMC) is designed. The performance of the two controllers is tested and compared under Matlab/Simulink environment. It has been proved that the Integral Sliding mode controller is more robust to minimizes the currents fluctuations and the parameters uncertainty.

Keywords: Adaptive Backstepping · DFIG · Integral sliding mode · WECS

1 Introduction

Nowadays, the consumption of electrical energy is reaching an unbearable level. However, the majority of electrical power production comes from petroleum, fossil fuels and natural gas. Recently, these natural resources are becoming scarce and their huge use creates economic and environmental damage all over the planet [1]. The main problems

S. Motahhir and B. Bossoufi (Eds.): ICDTA 2022, LNNS 454, pp. 722–732, 2022.
https://doi.org/10.1007/978-3-031-01942-5_72

caused by this exploitation are the global warming, the increase of the CO2 emission and other toxic gases, in addition to the increase of the gasoline and gas coasts which leads to skyrocketing electricity price. Hence, all the above elements negatively affect the activities in economic, industrial and agricultural fields [2].

In recent decades, the governments in most worldwide countries consider that renewable energies are the alternatives to polluting energy resources [3]. Some of these alternatives lay in photovoltaic energy, wind energy, geothermal energy, biomass, and hydroelectric power. In the last years, however, due to the profitability, durability, and cleanliness of wind turbines energy, most of these countries opt for the wind farms technology. Therefore, according to World Wind Energy Association, wind power production remarkably increases in many countries reaching over 600 GW with the potential level of around 1600 GW in 2030.

Thus, the improvement of the renewable energy production of the WECS becomes a primordial target for all researchers and inventors, due to its huge contribution in the global energy production.

Variable speed DFIG generator technology becomes the most important industrial solution in aero-generators design. The DFIG is characterized by its capacity to convert optimally the aerodynamic wind power with a low charge. It can also work suitably under a wide variable wind speed [4]. The turbine transforms the kinetic energy of the captured wind into mechanical energy delivering it to the electrical power machine through the transmission machine gearbox. The stator of the DFIG is directly connected to the electrical distribution grid, but the DFIG rotor is linked by a power converter. As the stator is linked through any electronic device to the grid, and when the stator currents high peaks arise at the windings of the stator, because the establishment of the high drop of the voltage, overcurrents are generated in the rotor side converter (RSC), due to the strong coupling between the stator and the rotor [5]. consequently, the value of the DC link voltage will increase and the fluctuations of the torque come by. Therefore, the high peaks transient currents damage the RSC and destroys the DFIG in the absence of protection measures [6, 7].

During the installation of wind farms based on WECS trained by the DFIG, several researchers have developed algorithms and control methods to ensure the stability and achieve the performance and robustness required for WECS wind turbines such as PIC (Proportional Integral Controllers) and the adaptive PIC [8, 9]. Other research papers have focused on other command strategies such as adaptive fuzzy logic, Direct Power Control (DPC) [10], integral sliding mode control (ISMC) and predictive control [11]. However, the operation of the DFIG, controlled by these control architectures, under a real wind profile, shows the inability of these controls to optimize the extraction of electrical energy and to improve the effectiveness of the wind turbine.

This research paper starts with the description of the mathematical model of the doubly fed induction generator. This model allowed us to develop two robust power controllers. Namely, the adaptive backstepping and the integral sliding mode controllers. These control strategies have been established by the equations of the rotor voltage. From the stator powers and their errors, the reference voltage values were calculated. However, the stator flux is considered to be variant. Therefore, the elimination of the high peak stator and rotor currents is reached by the control of the active power on the stator circuit

[12]. Besides, to push the voltage to return at its nominal value, the reactive power has to be controlled. Thus, the connectivity of the WECS to the electrical grid is ensured and the continuity of the electrical power production is guaranteed despite the voltage dips [10].

The two control methods in this article are validated with a doubly fed induction generator, which generates a nominal power of 1.5 KW. Thus, the implementation and the simulation were carried out with Matlab/Simulink on the wind profile of the Essaouira City of Morocco country.

This research work is organized as follows; after the introduction section, the wind turbine is presented and the DFIG general structure is modeled. Subsequently, the adaptive backstepping command and the sliding mode controller are approached. The last section of this research paper is reserved for obtained results of performance and THD. All results have been analyzed and commented to prove the robustness of the designed command on the real wind profile. In addition a comparison table is mentioned to highlight the proposed command strategy.

2 General Structure of the WECS Based on DFIG

The knowledge of the WECS dynamic equations is essential to design the appropriate power control laws. The general structure of the wind energy conversion system based on the DFIG is presented in the Fig. 1 [5].

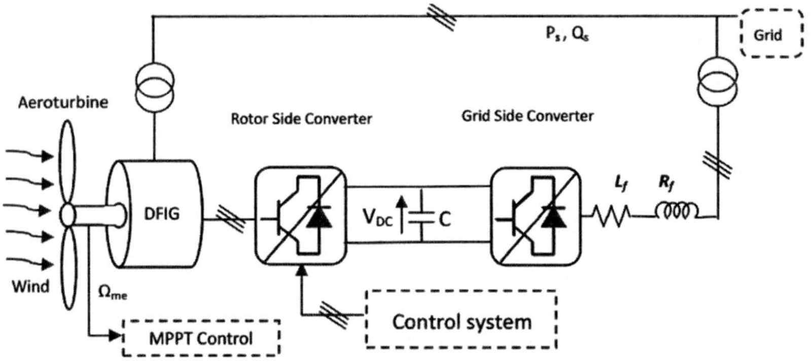

Fig. 1. General structure of the WECS.

The equations used to model the turbine are given by [5, 10]:

$$\begin{cases} P_{Rotor} = P_{wind}.C_p(\lambda,\beta) \\ P_{wind} = 0.5\rho A v^3 \\ T_{Rotor} = \frac{P_{Rotor}}{\Omega_t} \end{cases} \Rightarrow \begin{cases} P_{Rotor} = 0.5\rho A C_p.v^3 \\ T_{Rotor} = \frac{1}{2\Omega_t}\rho S.C_p v^3 \end{cases} \tag{1}$$

$$\text{where: } \begin{cases} C_p(\lambda,\beta) = k_1(\frac{k_2}{A} - k_3\beta - k_4)e^{(\frac{-k_5}{A})} + k_6\lambda \\ k_1 = 0.5872,\ \mathrm{k}_2 = 116,\ \mathrm{k}_3 = 0.4,\ \mathrm{k}_4 = 5,\ \mathrm{k}_5 = 21 \\ and\ \mathrm{k}_6 = 0.0085 \end{cases}$$

The equations of the mechanical torque, At the rotor, is [5]:

$$\begin{cases} T_{mec} = J\frac{d\Omega_{mec}}{dt} \\ T_{mec} = T_g - T_{em} - f\Omega_{mec} \end{cases} \Rightarrow J\frac{d\Omega_{mec}}{dt} = T_g - T_{em} - f\Omega_{mec} \quad (2)$$

The gearbox equations are modeled by:

$$\begin{cases} \Omega_g = G.\Omega_t \\ T_{Rotor} = G.T_g \end{cases} \quad (3)$$

The DFIG of the WECS converts the mechanical power to the electrical power. The modeling of DFIG is done through the following equations, in the (d, q) reference [7] (Table 1):

Table 1. The equations of the voltages and the currents.

The stator equations in the (d, q) reference	The rotor equations in the (d, q) reference
$V_{sd} = R_s.I_{sd} + \frac{d\varphi_{sd}}{dt} - \omega_s.\varphi_{sq}$	$V_{rd} = R_r.I_{rd} + \frac{d\varphi_{rd}}{dt} - \omega_r.\varphi_{rq}$
$V_{sq} = R_s.I_{sq} + \frac{d\varphi_{sq}}{dt} - \omega_s.\varphi_{sd}$	$V_{rq} = R_r.I_{rq} + \frac{d\varphi_{rq}}{dt} - \omega_r.\varphi_{rd}$
$I_{sd} = \frac{1}{\sigma.L_s}.\varphi_{sd} - \frac{M_{sr}}{\sigma.L_r}.\varphi_{rd}$	$I_{rd} = \frac{1}{\sigma.L_r}.\varphi_{rd} - \frac{M_{sr}}{\sigma.L_r.L_s}.\varphi_{sd}$
$I_{sd} = \frac{1}{\sigma.L_s}.\varphi_{sd} - \frac{M_{sr}}{\sigma.L_s.L_r}.\varphi_{rd}$	$I_{rq} = \frac{1}{\sigma.L_r}.\varphi_{rq} - \frac{M_{sr}}{\sigma.L_r.L_s}.\varphi_{sq}$

The rotor electromechanical torque is modeled by [7]:

$$T_{em} = p(\varphi_{rd}.\varphi_{sq} - \varphi_{rq}.\varphi_{sd}) \quad (4)$$

$$\text{With: } \begin{cases} \varphi_{rd} = L_r.I_{rd} + M_{sr}.I_{sd} \\ \varphi_{rq} = L_r.I_{rq} + M_{sr}.I_{sq} \\ \varphi_{sd} = L_s.I_{sd} + M_{sr}.I_{rd} \\ \varphi_{sq} = L_s.I_{sq} + M_{sr}.I_{rq} \end{cases}$$

The active and reactive powers are presented by the following equations [7]:

$$\begin{cases} P_s = V_{sd}I_{sd} + V_{sq}I_{sq} \\ Q_s = V_{sq}I_{sd} - V_{sd}I_{sq} \end{cases} \quad (5)$$

3 Power Control Strategy

3.1 Adaptive Backstepping (AB) Controller

The adaptive backstepping command doesn't require the system linearity. This strategy uses an adaptation law which estimates the unknown parameters and ensures the convergence to their respective desired values. The design of this command is based on three parts; the law of control, the Lyapunov function and the law of adaptation [5].

The expressions of the powers become [7]:

$$\begin{cases} \frac{dP_s}{dt} = -\left(\frac{3}{2.\sigma.L_s.L_r}\right).\begin{pmatrix} \frac{2}{3}(R_s.L_r + R_r.L_s).P_s + \\ (R_r + \omega_r.L_r).(v_{sd}.\psi_{sd} + v_{sq}.\psi_{sq}) \\ +M.(-v_{sd}.v_{rd} + v_{sq}.v_{rq}) + L_r.V_s^2 \end{pmatrix} - \omega_s.Q_s \\ \frac{dQ_s}{dt} = -\left(\frac{3}{2.\sigma.L_s.L_r}\right).\begin{pmatrix} \frac{2}{3}(R_s.L_r + R_r.L_s).Q_s + \\ (R_r + \omega_r.L_r).(v_{sd}.\psi_{sd} + v_{sq}.\psi_{sq}) + \\ M.(-v_{rd} + v_{rq}).v_{sd} \end{pmatrix} + \omega_s.P_s \end{cases} \tag{6}$$

The Formulation of the Control strategy for RSC converter is given by the following equation [4, 5, 7]:

$$e_{P_s} = P_{s\_ref} - P_s \tag{7}$$

The lyaponuv function [5]:

$$V_1 = \frac{1}{2}e_{P_s}^2 \tag{8}$$

$$\begin{cases} Q_{s\_ref} = Q_s \\ P_{s\_ref} = \frac{3}{2(R_s.L_r + R_r.L_s)}\left(\begin{array}{l}\left(\frac{2.\sigma.L_s.L_r}{3}\right).(-k_{P_s}.e_{P_s} + \omega_s.Q_{s\_ref}) \\ -\begin{pmatrix} (R_r + \omega_r.L_r).(v_{sd}.\psi_{sd} + v_{sq}.\psi_{sq}) \\ -M.(-v_{sd}.v_{rd} + v_{sq}.v_{rq}) + L_r.V_s^2 \end{pmatrix}\end{array}\right) \end{cases} \tag{9}$$

$$\dot{V}_1 = -k_{Ps}e_{Ps}^2 \le 0 \tag{10}$$

$$e_{Q_s} = Q_{s\_ref} - Q_s \tag{11}$$

$$V_2 = \frac{1}{2}\left(e_{P_s}^2 + e_{Q_s}^2\right) \tag{12}$$

$$\begin{cases} v_{rd} = \frac{1}{v_{sd}M}\begin{pmatrix} \left(\frac{2\sigma L_s L_r}{3}\right)\left(K_{Ps}e_{Ps} + \dot{P}_{s-ref} - \omega_s Q_s\right) \\ \frac{2}{3}(R_s L_r + R_r L_s)P_s + (R_r + \omega_r L_r)\left(v_{sd}\varphi_{sd} + v_{sq}\varphi_{sq}\right) + L_r V_s^2 \end{pmatrix} \\ v_{rq} = -\frac{1}{v_{sd}M}\begin{pmatrix} \left(\frac{2\sigma L_s L_r}{3}\right)\left(K_{Qs}e_{Qs} + \dot{Q}_{s-ref} - \omega_s P_s\right) \\ \frac{2}{3}(R_s L_r + R_r L_s)Qs + (R_r + \omega_r L_r)\left(v_{sd}\varphi_{sd} + v_{sq}\varphi_{sq}\right) \end{pmatrix} \end{cases} \tag{13}$$

$$\dot{V}_2 = -(K_{Ps}e_{Ps} + K_{Qs}e_{Qs}) \le 0 \tag{14}$$

3.2 Integral Sliding Mode Controller

The sliding mode control (SMC) is characterized by its robustness against system non-linearities and its capability to protect the controlled systems from uncertainties and active disturbances. This control strategy is a variable structure controller is known as a discontinuous control signal that pushes the state system trajectory to get to the sliding surface. whereas, the chattering phenomenon is the problem that limits the use of this control strategy. In fact, an integral sliding mode control (ISMC) is designed in this research work in order to rectify the phenomenon of chattering in the sliding mode controller [12].

This control is elaborated with the following equations [12]:

$$\overset{*}{X} = f(x,t) + g(x,t)u \tag{15}$$

$$g(x,t) = \begin{bmatrix} \frac{1}{\sigma L_r} & 0 \\ 0 & \frac{1}{\sigma L_r} \end{bmatrix} \tag{16}$$

$$f(x,t) = \begin{bmatrix} f_1 \\ f_2 \end{bmatrix} = \begin{bmatrix} -\frac{R_r}{(\sigma L_r)^2}(\sigma L_r i_{rd} + \frac{MVs}{\omega}) + \frac{R_r M \varphi_s}{L_s(\sigma L_r)^2} + \omega_r i_{rq} \\ -\frac{R_r}{(\sigma L_r)} i_{rq} - \frac{\omega_r}{\sigma L_r}(\sigma L_r i_{rd} + \frac{MVsq}{\omega_s L_s} \end{bmatrix} \tag{17}$$

where: $\varphi_{sd} = \varphi_s$; $\varphi_{sq} = 0$; $V_{sd} = V_s = \omega_s \varphi_s$; $V_{sq} = 0$.

The switching variables are given by the following equations [12]:

$$\begin{cases} S_d = e_d(t) + \alpha_d e_I(t) \\ S_q = e_q(t) + \alpha_q e_I(t) \end{cases} \tag{18}$$

where: $e_{dq}(t) = i_{rdq}$ - $i_{rdq\text{-}ref}$, α_d is a sliding constant coefficient, and $e^{\cdot}_I(t) = sign(e_{dq}(t))$

$$\begin{cases} \overset{*}{S}_d = \overset{*}{e}_d(t) + \alpha_d \overset{*}{e}_I(t) = 0 \\ \overset{*}{S}_q = \overset{*}{e}_q(t) + \alpha_q \overset{*}{e}_I(t) = 0 \end{cases} \tag{19}$$

$$\begin{bmatrix} V_{rd} \\ V_{rq} \end{bmatrix} = g^{-1} \begin{bmatrix} \overset{*}{i}_{rd-ref} - f_1 - \alpha_d sat(e_d(t)) \\ \overset{*}{i}_{rq-ref} - f_2 - \alpha_q sat(e_q(t)) \end{bmatrix} \tag{20}$$

We define the Lyapunov function as follows [12]:

$$V = \frac{1}{2} s_{dq}^T s_{dq} \tag{21}$$

To guarantee the convergence of V, its derivative must be negative [12].

$$\overset{*}{V} = \frac{s_d}{\sigma L_r}(\overset{*}{i}_{rd-ref} - f_1 \alpha_d sat(e_d(t))) + \frac{s_q}{\sigma L_r}(\overset{*}{i}_{rq-ref} - f_2 \alpha_q sat(e_q(t))) \tag{22}$$

$$\overset{*}{V} < 0 \tag{23}$$

4 Simulation Results and Discussions

To check out the performance of the two proposed controllers on the real wind speed profile and to choose the suitable controller, simulation tests have been realized with the 1.5 kW DFIG wind turbine under Matlab/Simulink environment. For this test, to guarantee a unity power factor and a good power quality, the reactive power will be kept zero. The chosen profile wind is that of the Essaouira city of Morocco.

The choice of this site was taken from the research works which made by the researcher Allouhi [4]. The machine parameters are described in the two Tables 3 and 4 in the appendix (Fig. 2).

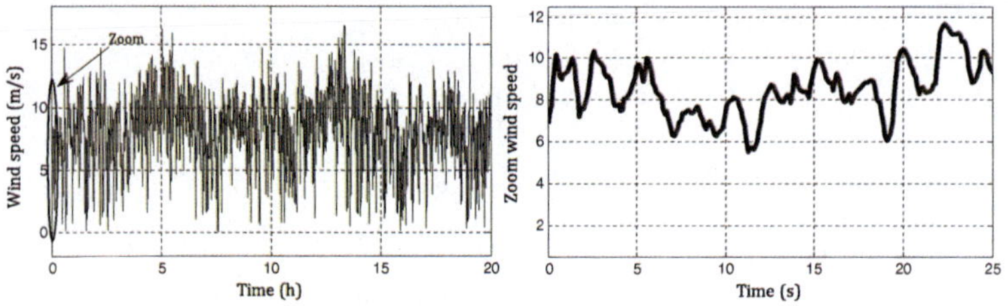

Fig. 2. Wind speed profile of Essaouira city of Morocco

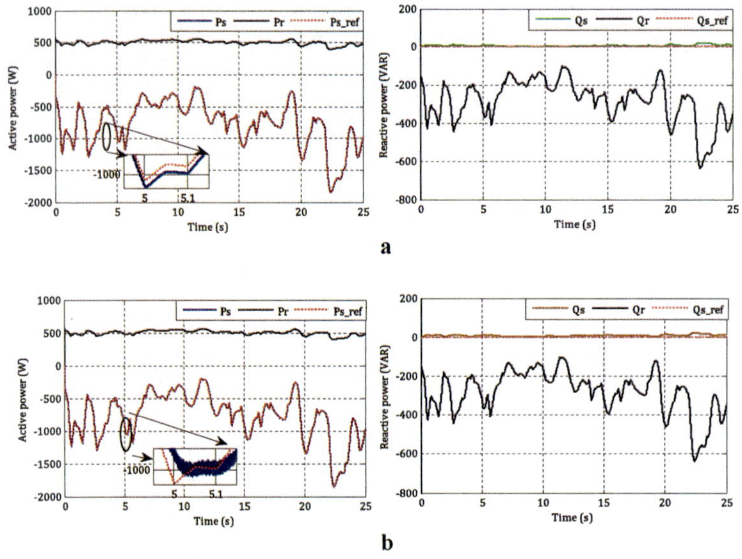

Fig. 3. a. Active and reactive power for ISMC, b. Active and reactive power for AB

The Fig. 3a confirms that the active power, by the ISMC, follows perfectly the desired reference power, calculated by the strategy of MPPT, with a dynamic error that does not exceed 0.51%.

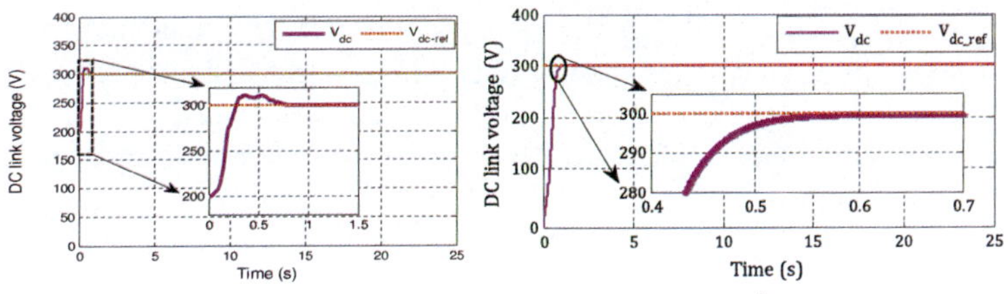

Fig. 4. a. DC link voltage for ISMC, b. DC link voltage for AB

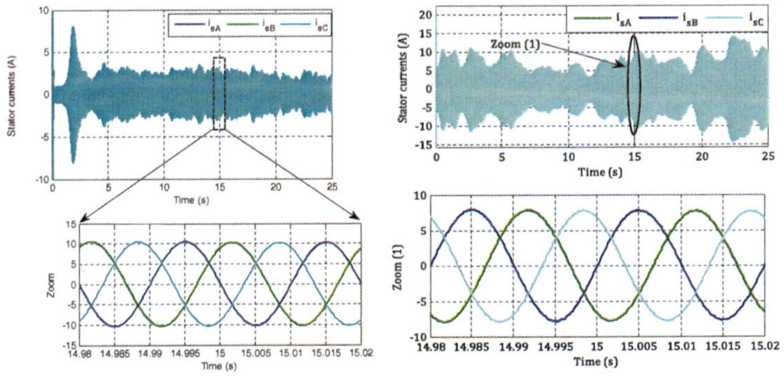

Fig. 5. a. Stator currents for ISMC, b. Stator currents for AB

The same Fig. 3a shows that the reactive power, of the ISMC, has always been kept at zero confirming that the DFIG operate with the unit power factor, that is reached after the operation transient regime of the machine.

Regarding the Fig. 4a, it can be seen that the DC link voltage, by the ISMC, reaches its reference value at 230 ms with a negligible static error a null overshoot. This voltage is characterized by a precise and fast dynamic.

It is clear from the Fig. 5a that the stator current remains sinusoidal with a 50 Hz frequency and it varies according to the captured power of the wind, which means that it is in relationship with the Essaouira-city real wind profile. These currents have less fluctuations despite the variable wind speed.

For the AB controller, the Fig. 3b shows that stator active power follows the reference power) with a 1.05% of dynamic error. This power always remains negative confirming that the DFIG is supplying the electrical power to the electrical grid. At startup, it can be noted that a slight fluctuations occur in high frequencies on the active and reactive powers. However, its amplitudes is decreasing when the reactive power become zero.

Moreover, the Fig. 4b shows that the DC link voltage, by the AB controller, reaches its reference value at 510 ms with a negligible static error a null overshoot. This voltage is characterized by a precise and fast dynamic and it reaches its reference value with an overshoot of D(%) = 1.01%.

The stator currents in the Fig. 5b, by the AB controller, are sinusoidal with a 50 Hz frequency, which allows the injection possibility into the electrical distribution grid. This currents varies in harmony with the wind turbine mechanical speed.

After the figures analyzing of the two controllers, a comparative study between the two controllers was realized in order to highlight the qualities of each command strategy and to select the right suitable command for the WECS giving the best performance (see Table 2).

Table 2. Comparison between AB and ISMC

		Adaptive Backstepping	Integral sliding mode	Improvement rate
Error	Static	0.4%	0.067%	83.25%
	Dynamic	1.05%	0.51%	51.83%
Response time	Active Power	170 ms	100 ms	34.61%
	Reactive Power	150 ms	50 ms	60.00%
	DC Link	510 ms	230 ms	36.25%
Overshoot	Active Power	0	0	0%
	Reactive Power	0	0	0%
	DC Link	1.01%	0.03%	99%
Variation band	Active Power	12 w	3 w	75%
	Reactive Power	8 VAR	2 VAR	75%

5 Conclusion

In this paper, at first, the WECS is modeled, then two controller are investigated. Namely, the adaptive backstepping and the integral sliding mode controllers. This controllers are used to control the rotor side converter in order to eliminate the high peak currents and stabilize the active power at its desired value. The simulation results have shown that the integral sliding mode controller gives the best dynamic performance in the reference tracking. Moreover, the ISMC can be considered as a concrete solution of the robustness problems encountered in the structure of the adaptive backstepping controller.

For the next research works, the hardware implementation on dSPACE 1104 and the FPGA co-simulation will be achieved to check out the proposed control strategy in the real time.

Appendix

Table 3. The DFIG parameters.

Name	Values
Nominal Rotor Voltages	12 V
Nominal Stator Voltages	220/380 V
Nominal Currents	3.64/6.31 A
Nominal Power	$P_n = 1{,}5$ kW
Nominal Frequency	$f_n = 50$ Hz
Nominal Torque	$C_n = 10$ N.m
Power Factor	0.89
Number of Pole Pairs	$P = 2$
Nominal Speed	1440 tr/min
Stator Resistance	$R_s = 0.435\ \Omega$
Rotor Resistance	$R_r = 0.816\ \Omega$
Stator Inductance Coefficient	$L_s = 7.1.10^{-2}$ H
Rotor Inductance Coefficient	$L_r = 7.10^{-2}$ H
Mutual Inductance Coefficient	$M = 6.9.10^{-2}$ H
Friction Coefficient	$f = 6.73.10^{-3}$ kg.m^2
Moment of Inertia	$J = 3.125.10^{-1}$ N.s/rad

Table 4. The wind turbine parameters.

Name	Value
Nominal Power	$P_n = 7.8$ kW
Density of Air	$\rho = 1.225$
Conversion Yield	$\eta = 1$
Number of Blades	3
Blade Radius	$R = 3$ m
Gearbox Ratio	$G = 10{,}38$
Blade Angle	$\beta = 2°$
Moment of Inertia	$J = 4.2.10^{-2}$ kg.m^2
Viscous Friction Coefficient	$f = 2.4.10^{-2}$ N.s/rad

References

1. Allouhi, A.: Energetic, exergetic, economic and environmental (4 E) assessment process of wind power generation. J. Clean. Prod. **235**, 123–137 (2019)
2. LacalArantegui, R.: Materials use in electricity generators in wind turbines – state-of-the-art and future specifications. J. Clean. Prod. **87**, 275–283 (2015)
3. Allouhi, A., et al.: Evaluation of wind energy potential in Morocco's coastal regions. Renew. Sustain. Energy Rev. **72**, 311–324 (2017)

4. Yessef, M., Bossoufi, B., Taoussi, M., Lagrioui, A., Chojaa, H.: improved hybrid control strategy of the doubly-fed induction generator under a real wind profile. In: Motahhir, S., Bossoufi, B. (eds.) ICDTA 2021. LNNS, vol. 211, pp. 1279–1290. Springer, Cham (2021). https://doi.org/10.1007/978-3-030-73882-2_117
5. Yessef, M., Bossoufi, B., Taoussi, M., Lagrioui, A., El Mahfoud, M.: Evaluation of adaptive backstepping control applied to DFIG wind system used on the real wind profile of the Dakhla-Morocco City. In: Bennani, S., Lakhrissi, Y., Khaissidi, G., Mansouri, A., Khamlichi, Y. (eds.) WITS 2020. LNEE, vol. 745, pp. 661–671. Springer, Singapore (2022). https://doi.org/10.1007/978-981-33-6893-4_61
6. Bossoufi, B., Karim, M., Lagrioui, A., Taoussi, M., Derouich, A.: Modeling and backstepping control of DFIG generators for wide-range variable-speed wind turbines. J. Electr. Syst. JES **10**(3), 317–330 (2014)
7. Bossoufi, B., et al.: DSPACE-based implementation for observer backstepping power control of DFIG wind turbine. IET Electric Power Appl. **14**(12), 2395–2403 (2020)
8. Bouderbala, M., et al.: Experimental validation of predictive current control for DFIG: FPGA implementation. Electronics **10**, 2670 (2021). https://doi.org/10.3390/electronics10212670
9. Taoussi, M., Karim, M., Lagrioui, A., Bossoufi, B., El Mahfoud, M.: The Fuzzy Control for Rotor Flux Orientation of the doubly-fed asynchronous generator Drive. Int. J. Comput. Technol. **13**(08), 4707–4722 (2014)
10. Bossoufi, B., Karim, M., Lagrioui, A., Taoussi, M., Derouich, A.: Observer backstepping control of DFIG-Generators for wind turbines variable-speed: FPGA-based implementation. Renew. Energy **81**, 903–917 (2015)
11. Taoussi, M., Karim, M., Bossoufi, B., Hammoumi, D., Elbakkali, C., El Ouanjli, N.: Comparative study between Backstepping adaptive and Field-oriented control of the DFIG applied to wind turbines. Processing IEEE Xplore, 23 October (2017)
12. Chojaa, H., Derouich, A., Chehaidia, S.E., Zamzoum, O., Taoussi, M., Elouatouat, H.: Integral sliding mode control for DFIG based WECS with MPPT based on artificial neural network under a real wind profile. Energy Rep. **7**, 4809–4824 (2021). https://doi.org/10.1016/j.egyr.2021.07.066

Fire Safety System Implementation for Controlling and Monitoring a Siren in Smart Farm Using Gas Sensor and Flame Sensor

Abdennabi Morchid[1](✉), Rachid El Alami[1], Hassan Qjidaa[1], Mustapha El Alaoui[1], Hassan Touati[1], Karim El Khadiri[2], and Mohammed Ouazzani-Jamil[3]

[1] Laboratory of Computer Science, Signals, Automation and Cognitivism (LISAC), Department of Physics, Faculty of Sciences Dhar El Mahraz, Sidi Mohamed Ben Abdellah University, B.P. 1796 Fes-Atlas, 30003 Fez, Morocco
abdennabi.morchid@usmba.ac.ma

[2] Laboratory of Computer Science and Interdisciplinary Physics, ENS, Sidi Mohamed Ben Abdellah University, B.P 5206 Bensouda Fez, Morocco

[3] Systems and Sustainable Environment Laboratory, Private University of Fez (UPF), Fez, Morocco

Abstract. This paper proposes designing and realising a Fire Safety System for Controlling and Monitoring a Siren using a new Embedded Board to solve fire control and extinguishing at the Smart Farm level. The new embedded fire safety system is developed by designing and modelling an embedded board called F3SF (Fire Safety System Smart Farm) based on the Atmega328p Microcontroller. It ensures the monitoring of the Smart Farm in the event of a fire by using specific sensors that detect the fire linked to the proposed board, which automatically controls and monitors (automatic maintenance of the siren and the sprinklers) the siren and sprinklers used to extinguish the fire. The proposed Fire Safety System implemented in the new embedded board F3SF-0X using a voltage sensor, an H-bridge, an ATMega328p microcontroller, a Flame Sensor, an MQ-135 gas sensor, a Water sprinkler, and an electronic relay. The new system is simulated in the EasyEDA web software tool.

Keywords: Fire Safety System · Smart Farm · Atmega328p · Embedded system · MQ-135 · H-bridge · Siren · PCB

1 Introduction

Agriculture is a crucial economic sector in Morocco. It generates around 14% of the gross domestic product (GDP), but with significant variations (11 to 18%) depending on the year and climatic conditions. Its performances even condition the economy: the country's growth rate is strongly correlated with agricultural production. Moreover, agriculture remains the country's primary source of employment, far ahead of other economic sectors, with 40% of the working population living in this sector. In 50 years, Moroccan agriculture has made significant progress in modernisation and diversification.

S. Motahhir and B. Bossoufi (Eds.): ICDTA 2022, LNNS 454, pp. 733–742, 2022.
https://doi.org/10.1007/978-3-031-01942-5_73

Yesterday as today, agricultural activity was one of the pillars of the Moroccan economy. Morocco has also improved its self-sufficiency in particular foodstuffs, in a context marked by the growing integration of the international market. The Kingdom ensures 100% of meat, fruit, and vegetables; 82% of milk, 50% sugar, 60% of cereals, and 20% of oil. Internationally, agricultural imports represent between 14 and 24% of global imports. As for agricultural exports, they represent between 15 and 21% of global exports.

Agriculture is one of the more important sources for Moroccan businesses and farmers. For this, its progress has been enhanced by the use of several new systems, technologies, and practices [1–3]. Agricultural practices are undergoing a significant change by integrating the image processing and analysis methods to provide an easy solution and overcome agricultural problems [4, 5]. Artificial intelligence and machine learning are among emerging tools and techniques used in the development and maintenance of the fields [6–8]. Also, the IoT (Internet of Things) is one of the modern techniques, massively used [9–12].

Today, fires have become a critical issue [13–15], which must be processed on time without unnecessary delay to avoid the loss of life and property. It is one of the unforeseen and unknown events that affect a farmer's productivity.

Fire represents a very significant risk to the lives of farmers, their agricultural products, and animals existing on the farm. One solution is to respond quickly to this emergency to avoid these losses. In such situations, early fire detection followed by a safety alert will protect the farm from loss of life and property by using a fire alarm system and the automatic activation of sprinklers to extinguish the fire.

Furthermore, the power to use the currently available technology is their benefit in the lives of farmers, which will be a considerable contribution to the community [16, 17]. Although in practical scenarios, several systems exist and are well developed and can be used [18–20]. However, these systems are unreliable in terms of their automatic response and are also not cost-effective, and they are not easy to implement. In the agricultural market, there are different types of fire alarm systems. Still, the complexity of their design requires maintenance to ensure proper operation, which increases the cost of using these systems [21–24]. Many researchers have worked on the fire safety system to decrease the risks becoming from the fire hazards in the farms [25] and [26].

To solve the fire control and extinguishing problem and make the conventional farm an intelligent farm, we proposed a new embedded Fire Safety System by designing and modelling an embedded board based on an At-mega328p Microcontroller and other electronic sensors. The proposed system is called F3SF (Fire Safety System Smart Farm). This Smart Farm is monitored in the case of a fire by using specific sensors that detect the fire linked to the proposed board. Automatically providing rapid control and monitoring (automatic maintenance of the siren) of the siren in all conditions, which is critical because in the event of a fire and the siren is out of order; also, sirens are the approved way to warn people so that they can call the firefighters. Here it would be a disaster on the farm. In our proposed system, it is sufficient for the farm manager to check the state of the siren on the LCD to be replaced as soon as possible to prevent human and agricultural losses. The new system was simulated in the EasyEDA Web Software Tool and programmed by the Arduino IDE.

We presented this article at the following: in Sect. 1 an Introduction, in Sect. 2 an Algorithm of the proposed Fire Safety System, in Sect. 3 a Simulation and PCB of the proposed Fire Safety System, in Sect. 4 a Realisation and Discussion of the proposed Fire Safety System, and Sect. 5 a Conclusion and Future Work are presented.

2 Algorithm of the Proposed Fire Safety System

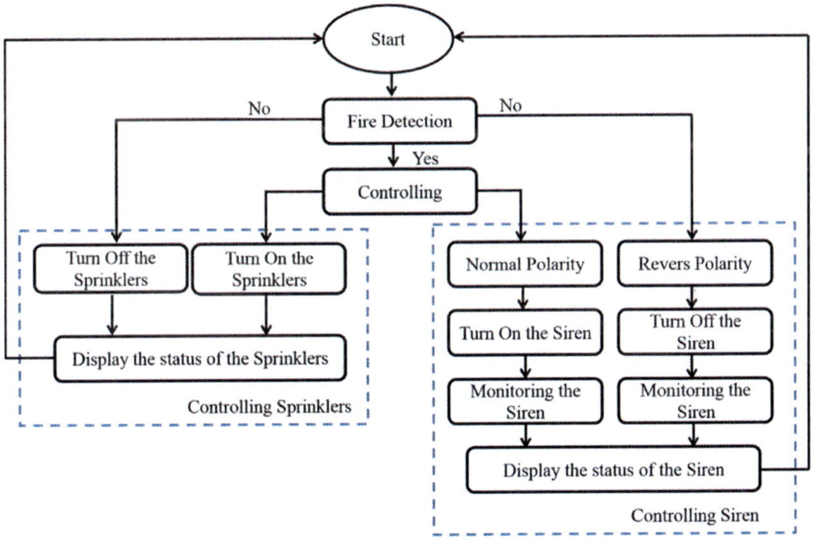

Fig. 1. The Algorithm of the proposed fire safety system.

The proposed Fire Safety System developed for the Smart Farm is described in Fig. 1 is based on fire detection; if the fire appears, the normal polarity will activate to turn on and monitor the siren and turn on the sprinklers simultaneously to extinguish the fire. Else if the fire does not appear, the reverse polarity will always act to turn off and monitor the siren, and also the performance of the sprinklers will also be disabled (turn off). All operations status of the siren and sprinklers will display in the new embedded card.

We measured voltage on the terminal of a resistor with the siren and their polarity resistance to monitor the siren. The H-bridge and the programmed microcontroller ATMega328p are used to reverse polarity direction and control the siren, respectively. As mentioned, the siren is controlled by two operations, normal polarity and reverse polarity; the processes of normally and reversely polarity are explained below.

A 24 V battery supplies the siren, and when the Fire Detector detects a fire, the normal polarity is applied. The positive terminal of the voltage source is connected to the positive terminal of the siren, while the negative terminal is connected to the negative terminal; this polarity allows for the measurement of a voltage that determines the monitoring state, as shown in Fig. 2.

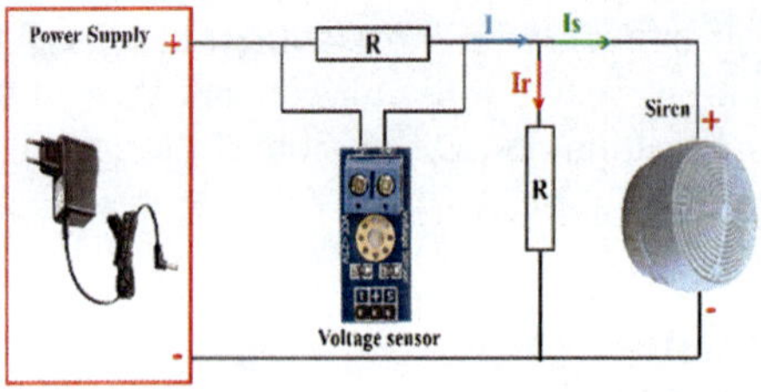

Fig. 2. Normal polarity

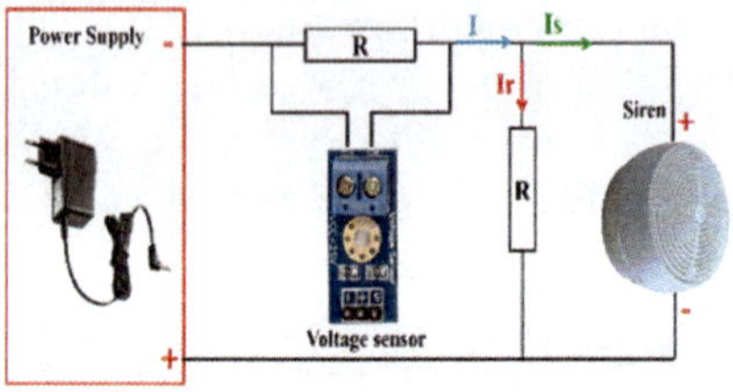

Fig. 3. Reverse polarity

Reverse polarity is used when the fire detector does not detect a fire. In that case, and as shown in Fig. 3, positive terminal is linked to the negative one of the siren.

3 Simulation and PCB of the Proposed Fire Safety System

3.1 Simulation

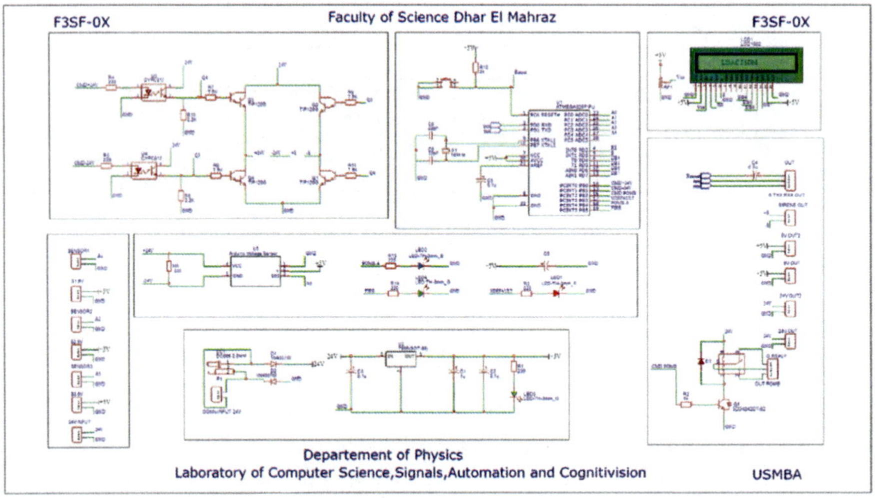

Fig. 4. Schematic circuit of the proposed fire safety system in EasyEDA.

Figure 4 shows the Schematic Circuit of the fire safety system for a Smart Farm. This card is based on two sensors to detect a fire in an agricultural area. A voltage sensor is used to monitor the siren in different cases, exist of a fire or absence of fire. Furthermore, an ATMega328p microcontroller processes the analogue values of the autonomous sensors according to the program.

When the detectors declare a fire, the microcontroller executes the analogue signals according to the program stored in that microcontroller. It activates the siren to inform

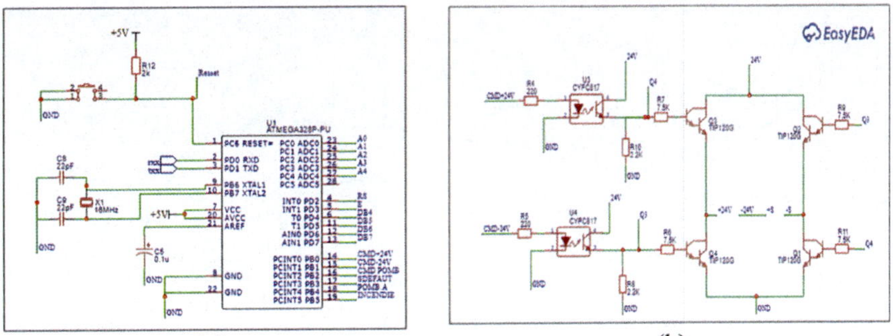

(a) (b)

Fig. 5. (a) Control by the microcontroller, (b) H-Bridge in EasyEDA.

people of the burning and starts the water sprinkler system to extinguish it. In the centre of the H-bridge is the siren in series with the polarity resistor.

Fig. 5(a) shows Atmega328p Microcontroller connection with all components, the exit door, and the part that declares a fire and disturbance in the siren.

The siren and polarity resistor will be put in the middle of the H-bridge to apply polarity inversion. Still, the voltage measurement at the resistor terminal will be situated outside the H-bridge, directly linked to the power supply. The values used to measure the two scenarios are different, but they are always positive, as illustrated in Fig. 5(b).

The voltage sensor for measuring the voltage at the resistor terminal is in series with the siren and the polarity resistor; this polarity reversal is used to monitor the siren.

A red LED to declare a malfunction in the siren, a green one to state that there is a fire, and a blue one to indicate that the water sprinkler system is on. The sensor and LEDs used and the relay block is illustrated in Fig. 4. When a fire in agriculture is detected by autonomous detectors such as flame sensors and gas sensors, the relay in the board is used to switch on the water sprinkler system to stop the fire. The operation of the water sprinkler system concerns as an average to make the fire safety of the intelligent farmer. The input and output connections, the output door, the section that informs the system that there is a fire, and the alarm and water sprinkler control are all included. This system's result is shown on an onboard display (LCD).

3.2 Print Circuit Board PCB

The PCB of the new board in 2D is illustrated in Fig. 6(a) and the PCB of the new board in 3D is shown in Fig. 6(b). After the PCB schematisation of the board in the EasyEDA software, we can click on the menu generation manufacturing file (Gerber) from the JLCPCB site linked automatically with EasyEDA to manufacture the board. We used two layers of 170.18 mm * 91.19 mm (17 cm * 9.1 cm) for our board.

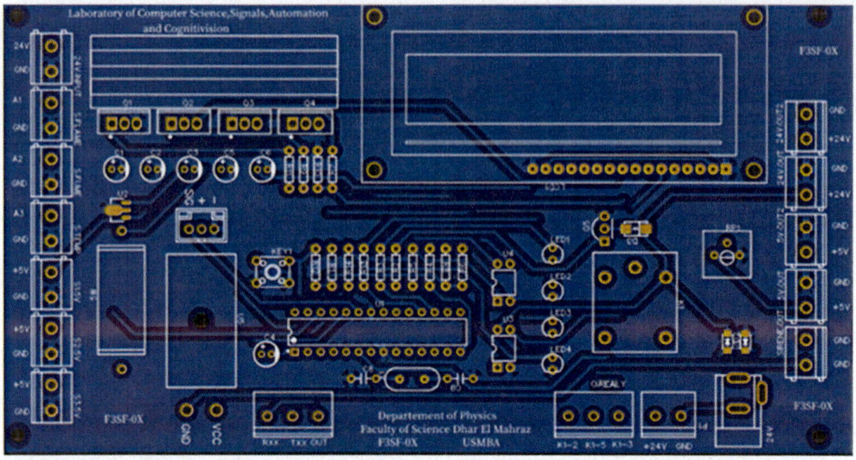

(a)

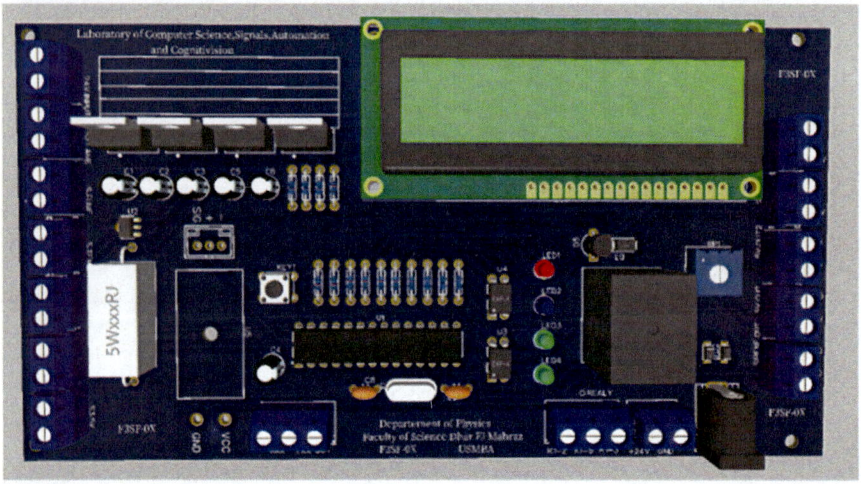

(b)

Fig. 6. (a) The PCB of the new board in 2D using EasyEDA, (b) The PCB of the new board in 3D using EasyEDA.

4 Realisation and Discussion

Figure 7 shows the realisation of the Fire Safety System using the different Hardware tools: Arduino Uno, MQ-135 gas sensor, Flame Sensor, Water sprinkler, LCD, and siren; we used the Arduino UNO board for the prototype. Still, for the PCB board, we used the atmega328 microcontroller.

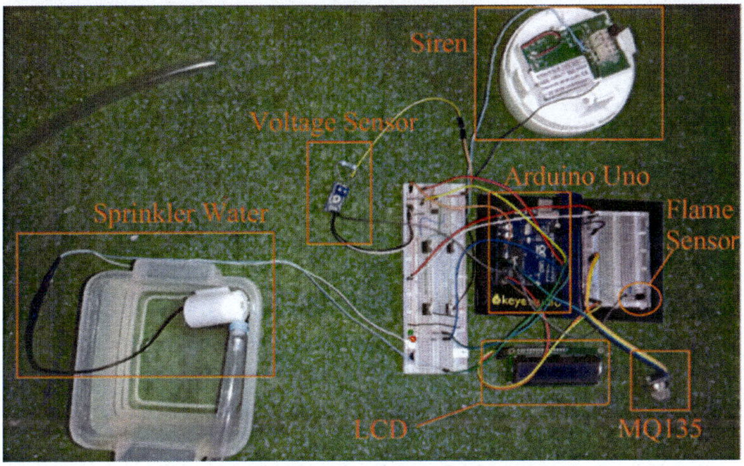

Fig. 7. The realisation of the fire safety system

Table 1. Experimental results of the system.

No	Flame sensor	Co2 value ppm	Voltage MQ-135	Status	Voltage siren	Green LED	Siren and Red LED	Sprinkler Water
1	13%	2.76	0.76 V	No Fire	7.20 V	ON	OFF	OFF
2	35%	9.46	1.20 V	Fire	8.64 V	OFF	ON	ON
3	40%	15.2	1.41 V	Fire	8.64 V	OFF	ON	ON
4	49%	25.4	1.68 V	Fire	8.64 V	OFF	ON	ON
5	15%	4.01	0.86 V	No Fire	7.20 V	ON	OFF	OFF
6	14%	2.95	0.78 V	No fire	7.20 V	ON	OFF	OFF

Table 1 demonstrates that six types of testing were conducted in a continuous attempt to observe the sensor changes. Continuous readings have been carried out to establish firefighting. In the beginning, the flame sensor was exposed to flame. However, no fire was detected in the first testing due to response time. All the testing is continuous and studied every ten seconds. In Tests 2, 3, and 4, a fire was detected in the siren, water sprinkler, and red LED were turned on. While in Tests 1, 5, and 6, they were turned off because no fire was detected.

If there is a fire and the siren is activated, the normal polarity of the H-bridge will be + 24 V. The recorded voltage for the siren is 8.64 V, the flame sensor value is more significant than 20%, the dioxygen concentration value is greater than seven ppm, and the voltage value of the MQ-135 gas sensor is greater than 1 V. If there is no fire and the siren is not activated, the application of reverse polarity on the H-bridge will be –24 V, and the measured voltage for the siren will be 7.20 V, as shown in the Table. 1.

Figure 8 shows the experiment results with the MQ-135 gas sensor shown in Table 1 with different cases, either in the event of a fire or in the absence of fire.

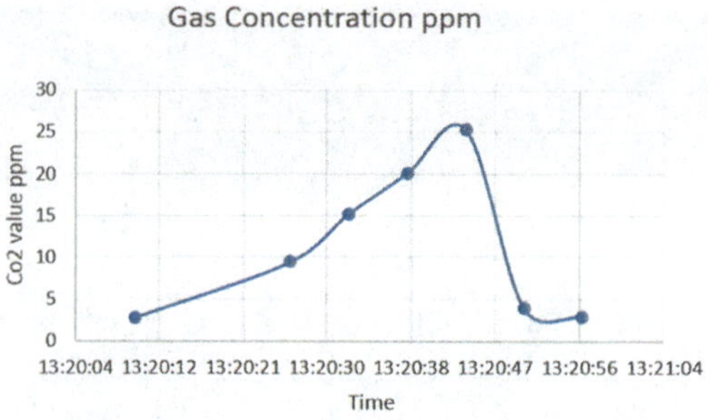

Fig. 8. Gas concentration ppm of MQ-135

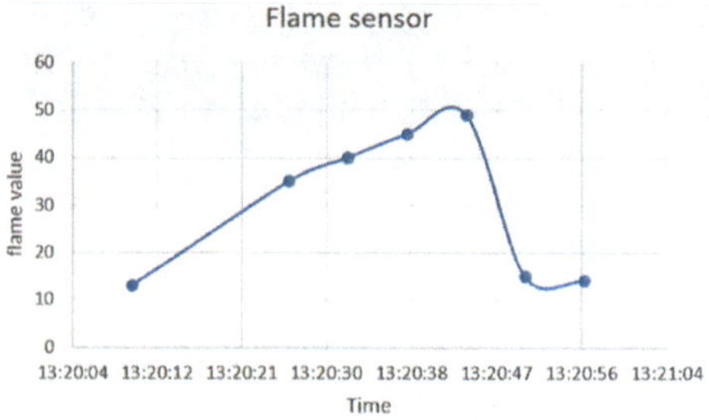

Fig. 9. Flame sensor.

Figure 9 shows the results of the flame sensor as presented in Table 1 in both cases (when there is a fire or when there is no fire). After studying and realising our proposed fire safety system on the farm, we achieved good results. Our system can detect the fire from its inception and extinguish it with sprinklers. It is also able to control and monitor the siren in all circumstances. It is essential because the siren is out of order in a fire is the approved method of informing people to call the firefighters. Here will be the disaster on the farm. In our fire safety system, it is enough for the person in charge of the system to go to the site of the system, and he will find the status of the siren on the LCD screen. Is it appropriate to change it in time or not to avoid loss of life and agricultural yield? That is why our research distinguishes from other research.

5 Conclusion and Future Work

Using a novel embedded board called F3SF, the suggested Fire Safety System was successfully built and implemented to control and monitor a Siren. This research includes simulation, analysis, and PCB (Printed Circuit Board) realisation; it considerably reduces the problem of fire control and extinguishment. The proposed Fire Safety System employs a voltage sensor, an H-bridge, an ATMega328p microcontroller, a Flame Sensor, an MQ-135 gas sensor, a water sprinkler system, and an electronic relay on the new embedded board F3SF-0X. The EasyEDA online software program was used to model the new system. The proposed approach is based on the principle of intelligent detection of farm fires based on the polarity of the reflection to monitor and control the siren. However, this reverse polarity still needs specific study to achieve better results and integrate the Internet of Things, and this is the following paper.

References

1. Blasch, J., Kroon, B., Beukering, P., Munster, R., Fabiani, S., Nino, P., et al.: Farmer preferences for adopting precision farming technologies: a case study from Italy. Euro. Rev. Agri. Econ. 1–49 (2020). https://doi.org/10.1093/erae/jbaa031
2. Meliani, M., El Barkany, A., El Abbassi, I., Moumen Darcherif, A., Mahmoudi, M.: Smart grid implementation in Morocco: case study. Mater. Today Proc. **45**, 7675–7679 (2021). https://doi.org/10.1016/j.matpr.2021.03.176
3. Idoje, G., Dagiuklas, T., Iqbal, M.: Survey for smart farming technologies: challenges and issues. Comput. Electr. Eng. **92**, 107104 (2021). https://doi.org/10.1016/j.compeleceng.2021.107104
4. Iftekharul, M., Abid-Ar-Rafi, M., Neamul, M.: An intelligent fire detection and mitigation system safe from fire (SFF). Int. J. Comput. Appl. **133**, 1–7 (2016). https://doi.org/10.5120/ijca2016907858
5. Pillai, A., David, V.J., Shibu, R., Fernandes, J., Shirke, A.: f-detect: early detection of forest fires using wireless sensor networks (2018)
6. Bao, J., Xie, Q.: Artificial intelligence in animal farming: a systematic literature review. J. Clean. Prod. **331**, 129956 (2022). https://doi.org/10.1016/j.jclepro.2021.129956
7. García, R., Aguilar, J., Toro, M., Pinto, A., Rodríguez, P.: A systematic literature review on the use of machine learning in precision livestock farming. Comput. Electron. Agric. **179**, 105826, (2020). https://doi.org/10.1016/j.compag.2020.105826
8. Dubois, A., Teytaud, F., Verel, S.: Short term soil moisture forecasts for potato crop farming: a machine learning approach. Comput. Electron. Agric. **180**, 105902 (2021). https://doi.org/10.1016/j.compag.2020.105902
9. Lal, A., Prabu, P.: Fire detection and prevention in agriculture field using IoT. J. Xi'an Univ. Archi. Technol. **XII**(IV), 3708–3719 (2020)
10. Boursianis, A.D., Papadopoulou, M.S., Diamantoulakis, P., Liopa-Tsakalidi, A., Barouchas, P., Salahas, G., et al.: Internet of Things (IoT) and agricultural unmanned aerial vehicles (UAVs) in smart farming: a comprehensive review. Internet of Things, 100187 (2020). https://doi.org/10.1016/j.iot.2020.100187
11. Haque, M.A., Haque, S., Sonal, D., Kumar, K., Shakeb, E.: Security enhancement for IoT enabled agriculture. Mater. Today Proc. (2021) https://doi.org/10.1016/j.matpr.2020.12.452
12. Mahbub, M.: A Smart Farming concept based on smart embedded electronics, internet of things and wireless sensor network. Internet of Things **9**, 100161 (2020). https://doi.org/10.1016/j.iot.2020.100161
13. Bu, F., Gharajeh, M.S.: Intelligent and vision-based fire detection systems: a survey. Image Vis. Comput. **91**, 103803 (2019)
14. Saeed, F., Paul, A., Rehman, A., Hong, W.H., Seo, H.: IoT-based intelligent modeling of smart home environment for fire prevention and safety. J. Sens. Actuator Netw. **7**(1), 11 (2018)
15. Saeed, F., Paul, A., Karthigaikumar, P., Nayyar, A.: Convolutional neural network based early fire detection. Multi. Tools Appl. **79**(13–14), 9083–9099 (2019). https://doi.org/10.1007/s11042-019-07785-w
16. Zeebaree, S.R., Yasin, H.M.: Arduino based remote controlling for home: power saving, security and protection. Int. J. Sci. Eng. Res. **5**(8), 266–272 (2014)
17. Piyare, R., Tazil, M.: Bluetooth based home automation system using cell phone. In: 15th International Symposium on Consumer Electronics (ISCE), IEEE, Singapore, pp. 192–195 (2011). https://doi.org/10.1109/ISCE.2011.5973811

18. Jinghong, L., Xiaohui, Z., Lu, W.: The design and implementation of fire smoke detection system based on FPGA. In: 24th Chinese Control and Decision Conference (CCDC), Taiyuan, pp. 3919–3922 (2012). https://doi.org/10.1109/CCDC.2012.6244626
19. Celik, T., Demirel, H., Ozkaramanli, H., Uyguroglu, M.: Fire detection in video sequences using statistical color model. In: IEEE International Conference on Acoustics Speech and Signal Processing Proceedings, Toulouse, pp. II–II (2006). https://doi.org/10.1109/ICASSP.2006.1660317
20. Angayarkkani, N., Radhakrishnan, N.: Efficient forest fire detection system: a spatial data mining and image processing based approach. Int. J. Comput. Sci. Netw. Secur. **9**(3), 100–107 (2009)
21. Suresh, S., Yuthika, S., Vardhini, G.A.: Home-based Fire Monitoring and warning system. In: 2016 International Conference on ICT in Business Industry & Government (ICTBIG), Indore, pp. 1–6 (2016). https://doi.org/10.1109/ICTBIG.2016.7892664
22. Tun, M.Z., Myint, H.: Arduino-based fire detection and alarm system using smoke sensor. IJASRE **06**, 89–94 (2020). https://doi.org/10.31695/IJASRE.2020.33792
23. Yadav, R., Rani, P.: Sensor-Based Smart Fire Detection and Fire Alarm System. Rochester, NY: Social Science Research Network (2020). https://doi.org/10.2139/ssrn.3724291
24. Morchid, A., El Alaoui, M., El Alami, R., Qjidaa, H., El Khadiri, K., Mehdaoui, Y.: Design and realization of fire safety system for controlling and monitoring a siren using arduino uno. In: Motahhir, S., Bossoufi, B. (eds.) ICDTA 2021. LNNS, vol. 211, pp. 433–445. Springer, Cham (2021). https://doi.org/10.1007/978-3-030-73882-2_40
25. Song, W.S., Hong, S.H.: A reference model of fire detection and monitoring system using BACnet. J. Build. Environ. **42**(2), 1000–1010 (2007). https://doi.org/10.1016/j.buildenv.2005.10.028
26. Sharma, S., Chand, K., Sharma, D., Guha, P.: Development of an early detection system for fire using wireless sensor networks and arduino. In: International Conference on Sustainable Energy, Electronics, and Computing Systems (SEEMS), Greater Noida, India, pp. 1–5 (2018). https://doi.org/10.1109/SEEMS.2018.8687333

Lightweight Hardware Architecture of EKF-SLAM and Its FPGA Implementation

Slama Hammia[1(✉)], Anas Hatim[2], Abella Bouaaddi[1], and Abdelilah Haijoub[3]

[1] LGEMS, ENSA, Ibn Zohr University, Agadir, Morocco
slama.hammia@edu.uiz.ac.ma, a.bouaaddi@uiz.ac.ma
[2] TIM Team, ENSA, Cadi Ayyad University, Marrakech, Morocco
a.hatim@uca.ma
[3] Mohammed VI Polytechnic University, Ben Guerir, Morocco

Abstract. Simultaneous localization and mapping (SLAM) estimate the position of a mobile robot and the locations of its landmarks from noisy measurements. EKF-SLAM is a very important class of algorithms that uses the extended Kalman filter (EKF) for SLAM. It is among the most widely used algorithms in the field of mobile robotics for localization and mapping missions for its capability to treat noise and its very high level of accuracy. In light of the computational complexity of SLAM algorithms, the major focus of research to meet real-time requirements is on reducing the computational complexity to develop embedded systems based on lower resource and computationally complex platforms. This paper presents the design of the hardware architecture of EKF-SLAM and its implementation on FPGA. The whole design is implemented through Cyclone 2 FPGA; it can achieve 114 MHz and uses 21512 LUTs, which leads to lightweight hardware architecture.

Keywords: Design · SLAM · EKF · EKF-SLAM · FPGA · Hardware architecture

1 Introduction

The problem of SLAM is one of the most important issues in mobile robotics. It is confronted when a robot is located in an unknown area and needs to discover the surrounding space and generate a map while keeping track of its position. This topic has been the focus of intensive study over the last decade, as its resolution is widely viewed as necessary for the creation of truly autonomous robots. The SLAM problem has traditionally been addressed using a Gaussian or nonlinear filter. The Gaussian approach has been preferred for more than a decade. Based on Gaussian filters, a variety of SLAM solutions have been developed to approximate the state of the vehicle in addition to the surrounding features in the map taking into account the uncertainty. Nowadays, the implementation of EKF-SLAM on modern embedded electronic devices is becoming a new trend that attracts many researchers. The SLAM community has not been paying a great deal of importance to the embedded systems and their impact on the potential implementations. We could mention the implementation of an EKF on FPGAs [1–3] as

S. Motahhir and B. Bossoufi (Eds.): ICDTA 2022, LNNS 454, pp. 743–752, 2022.
https://doi.org/10.1007/978-3-031-01942-5_74

well as several designs for the FPGA of EKF-SLAM [4–6]. Strasdat et al. discussed in [7] the advantages and disadvantages of SLAM methods; they concluded that filtering methods can be advantageous in cases of lower resource allocation implementations. In addition, a great deal of previous research has focused on the improvement of SLAM. Furthermore, the majority of these works have agreed on the important impact of reducing the hardware architecture resources of the applications to achieve the requirements of real-time [8–14].

SLAM algorithms become more and more powerful, but no full hardware or architectural implementation has taken place. Such an architecture implementation must take into consideration energy consumption and computing power. However, due to the computational complexity of SLAM algorithms, it is very difficult to realize a real-time implementation on low-resource platforms. Field-Programmable Gate Array (FPGA) devices are attractive real-time platforms for SLAM systems because they are high-performance and low power. This paper aims to design and implement the hardware architecture of the EKF -SLAM on the FPGA platform. The implementation realized in this work is very useful for platforms using mobile robots, with constraints of low power consumption and few hardware resources. It provides a lightweight and simple kernel to perform the computations of the prediction step and correction step of EKF-SLAM that represent heavy and expensive operations in SLAM. This implementation can be used as a coprocessor in a CPU/FPGA architecture or as a fully integrated hardware core in an FPGA architecture. The proposed computational core provides very accurate results and significantly reduces the error with minimal hardware resources. This approach is supported by our previously published work on this topic [15–17]. The contribution of this paper is the additional implementation on FPGAs of the following elements:

- The equation of the gain matrix of the Kalman filter.
- The equation of the update of the state vector estimate.
- The equation of the update of the covariance matrix estimate.

The remaining of the paper is organized as follows: Sect. 2 presents a general description of the EKF-SLAM algorithm. Section 3 presents the detailed description of the hardware architecture design of EKF-SLAM. Section 4 shows the implementation and the experiment result. Finally, Sect. 5 concludes this paper.

2 EKF-SLAM

This technique aims to represent the location of the robot and landmarks observed L_i using a state vector "(1)" with an associated variance-covariance matrix P "(2)" representing the Gaussian noise. EKF-SLAM aims to maintain the map $\{\widehat{X}_t, P\}$ continuously updated.

$$X = [R\ M]^T = [R\ L_1\ L_2\ \ldots\ L_n]^T \tag{1}$$

$$P = \begin{pmatrix} P_{RR} & P_{RL_1} & \dots P_{RL_n} \\ P_{L_1R} & P_{L_1L_1} & \dots P_{L_1L_n} \\ \vdots & \vdots & \vdots \quad \vdots \\ P_{L_nR} & P_{L_nL_1} & \dots \ P_{L_nL_n} \end{pmatrix} \quad (2)$$

R denotes the robot and M denotes the observed landmarks L_i. The state of the robot is formed by its position and orientation "(3)" with a control vector "(4)". When the robot moves its odometry provides the distance and direction of a landmark "(5)".

$$X_t = [x_t, y_t, \theta_t]^T \quad (3)$$

$$U_t = [x_u, y_u, \theta_u]^T \quad (4)$$

$$Y_t = \left[r_t^k, \varnothing_t^k\right]^T \quad (5)$$

There are two basic mathematical models considered in the EKF-SLAM namely the motion model "(6)" and the observation model "(7)".

$$X_t^* = f(X_t, U_t) = \begin{pmatrix} x_t + x_u \ Cos\ \theta_t - y_u \ Sin\ \theta_t \\ y_t + x_u \ Sin\ \theta_t + y_u \ Cos\ \theta_t \\ \theta_t + \theta_u \end{pmatrix} \quad (6)$$

$$Y_t^* = h^k(X_t^*) = \begin{pmatrix} \sqrt{(x_k - x_t)^2 + (y_k - y_t)^2} \\ Atan\left(\frac{y_k - y_t}{x_k - x_t}\right) - \theta_t \end{pmatrix} \quad (7)$$

X_t^* denotes the predicted state vector and Y_t^* denotes the predicted measurements of the observation. The SLAM algorithm solved by an EKF is stated by equations "(9)" to "(13)", all variables involved in the estimation process are considered as Gaussian random variables. EKF-SLAM is divided into two stages:

- **Prediction stage:**

$$X_t^* = f(\hat{X}_t, U_t) \quad (8)$$

$$P_t^* = A.\widehat{P}_t.A^T + B.Q.B^T \quad (9)$$

$$Y_t^* = h(X_t^*) \quad (10)$$

- **Correction stage:**

$$K = P_t^* H^T.(H.P_t^*.H^T + P_Y)^{-1} \quad (11)$$

$$\widehat{X}_{t+1} = X_t^* + K(Y - Y_t^*) \quad (12)$$

$$\widehat{P}_{t+1} = P_t^* - KHP_t^* \quad (13)$$

3 Hardware Architecture of EKF-SLAM

3.1 Design of Prediction Stage of EKF-SLAM

The hardware architectures of the EKF-SLAM's prediction stage are presented in this part. Our previous work on the subject in [15–17] are exploited for the implementation of the circuit blocks of trigonometric functions Sin(x), cos(x), Atan(x), and the circuit block of the Jacobian matrices A, B included in the prediction stage as well as the Jacobian matrix H that we will need in the correction stage. In Fig. 1 the hardware architecture of the evolution model "(6)" designed to calculate the prediction of the state vector X_t^* consisting mainly of functions sin(x) and cos(x). The observation model $h(X_t^*)$ "(7)" used to compute the prediction of the observation Y_t^* is composed principally of two functions: square root and Atan(x). The hardware architectures of these functions designed in [16] are exploited for the implementation of the circuit blocks of the hardware architecture of the observation model design $h(X_t^*)$. In the end, the prediction of the covariance matrix "(9)" is formed by two blocks of matrix multiplications composed by matrices A, $\widehat{P}_t$, A^T, B, Q and B^T (see Fig. 1).

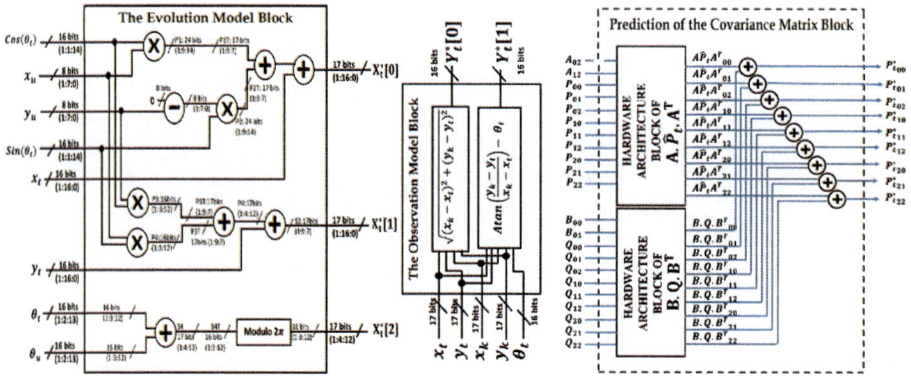

Fig. 1. Hardware architecture blocks of prediction stage.

3.2 Design of Correction Stage of EKF-SLAM

The following section describes the hardware architecture of EKF-SLAM's correction stage. It consists of computing the KALMAN gain K, the correction of $\widehat{X}_{t+1}$ and the correction of $\widehat{P}_{t+1}$ using three equations "(11)", "(12)" and "(13)" mainly formed by matrix multiplications. The key to successful design is to establish the hardware architectures of the blocks of the matrix multiplication figured in the correction stage such as $P_t^* H^T$, HP_t^*, $K(Y - Y_t^*)$ and KHP_t^*. Finally, the hardware architecture of the correction stage will be designed by encompassing all these circuit blocks.

Hardware Architecture Block of $\mathbf{HP_t^*H^T}$. The expression $P_t^* H^T$ is the multiplication of the two matrices P_t^* and H^T(the transpose of the Jacobian matrix H), the result is presented in "(14)". Matrix multiplication design $P_t^* H^T$ is made according to the calculation result obtained in "(14)", the 10 coefficients of this matrix compose the outputs of this circuit.

$$P_t^* H^T = \begin{pmatrix} P_{t00}^* & P_{t01}^* & P_{t02}^* & P_{t03}^* & P_{t04}^* \\ P_{t10}^* & P_{t11}^* & P_{t12}^* & P_{t13}^* & P_{t00}^* \\ P_{t20}^* & P_{t21}^* & P_{t22}^* & P_{t23}^* & P_{t00}^* \\ P_{t30}^* & P_{t31}^* & P_{t32}^* & P_{t33}^* & P_{t34}^* \\ P_{t40}^* & P_{t41}^* & P_{t42}^* & P_{t43}^* & P_{t44}^* \end{pmatrix} \begin{pmatrix} H_{00} & H_{10} \\ H_{01} & H_{11} \\ 0 & -1 \\ 0 & 0 \\ 0 & 0 \end{pmatrix}$$
$$= \begin{pmatrix} P_{t00}^* H_{00} + P_{t01}^* H_{01} & P_{t00}^* H_{10} + P_{t01}^* H_{11} - P_{t02}^* \\ P_{t10}^* H_{00} + P_{t11}^* H_{01} & P_{t00}^* H_{00} + P_{t11}^* H_{11} - P_{t12}^* \\ P_{t20}^* H_{00} + P_{t21}^* H_{01} & P_{t00}^* H_{00} + P_{t21}^* H_{11} - P_{t22}^* \\ P_{t30}^* H_{00} + P_{t31}^* H_{01} & P_{t00}^* H_{00} + P_{t31}^* H_{11} - P_{t32}^* \\ P_{t40}^* H_{00} + P_{t41}^* H_{01} & P_{t00}^* H_{00} + P_{t41}^* H_{11} - P_{t42}^* \end{pmatrix} \quad (14)$$

For the implementation of the circuit block $HP_t^* H^T$, the matrix multiplication of the two matrices H and $P_t^* H^T$ "(14)" is calculated in"(15)". The outputs of the Jacobian matrix H circuit block and the outputs of the $P_t^* H^T$ circuit blocks are exploited for the design of the architecture representing the circuit $HP_t^* H^T$ using multipliers, adders, and subtracters. The outputs of the hardware architecture $HP_t^* H^T$ will represent the four coefficients of the result matrix "(15)".

$$HP_t^* H^T = \begin{pmatrix} H_{00} & H_{01} & 0 & 0 & 0 \\ H_{10} & H_{11} & -1 & 0 & 0 \end{pmatrix} \cdot \begin{pmatrix} P_t^* H^T{}_{00} & P_t^* H^T{}_{01} \\ P_t^* H^T{}_{10} & P_t^* H^T{}_{11} \\ P_t^* H^T{}_{20} & P_t^* H^T{}_{21} \\ P_t^* H^T{}_{30} & P_t^* H^T{}_{31} \\ P_t^* H^T{}_{40} & P_t^* H^T{}_{41} \end{pmatrix}$$
$$= \begin{pmatrix} H_{00} P_t^* H^T{}_{00} + H_{01} P_t^* H^T{}_{10} & H_{00} P_t^* H^T{}_{01} + H_{01} P_t^* H^T{}_{11} \\ H_{10} P_t^* H^T{}_{00} + H_{11} P_t^* H^T{}_{10} - P_{t20}^* & H_{10} P_t^* H^T{}_{01} + H_{11} P_t^* H^T{}_{11} - P_{t21}^* \end{pmatrix} \quad (15)$$

Hardware Architecture Block of $\mathbf{HP_t^* H^T + P_Y}$. The expression $HP_t^* H^T + P_Y$ is obtained by applying the operation of the addition of the matrix $HP_t^* H^T$ "(15)" and the matrix representing the permanent measurement noise P_Y. The result is presented in "(16)".

$$HP_t^* H^T + P_Y = \begin{pmatrix} HP_t^* H^T{}_{00} & HP_t^* H^T{}_{01} \\ HP_t^* H^T{}_{10} & HP_t^* H^T{}_{11} \end{pmatrix} + \begin{pmatrix} P_{Y00} & P_{Y01} \\ P_{Y10} & P_{Y11} \end{pmatrix}$$
$$= \begin{pmatrix} HP_t^* H^T{}_{00} + P_{Y00} & HP_t^* H^T{}_{01} + P_{Y01} \\ HP_t^* H^T{}_{10} + P_{Y10} & HP_t^* H^T{}_{11} + P_{Y11} \end{pmatrix} \quad (16)$$

The circuit block $HP_t^* H^T + P_Y$ will have four outputs S_{00}, S_{01}, S_{10} and S_{11} representing the coefficients of the matrix "(16)".

Hardware Architecture Block of $(\mathbf{H.P_t^*.H^T + P_Y})^{-1}$. To calculate the inverse matrix, we assume that the matrix $HP_t^*H^T + P_Y$ "(16)" is invertible, and since its size is 2 × 2, then its inverse matrix is presented in "(17)". The outputs of the circuit block of $(H.P_t^*.H^T + P_Y)^{-1}$ will represent the four coefficients of the matrix "(17)".

$$(H.P_t^*.H^T + P_Y)^{-1} = \frac{1}{S_{00}S_{11} - S_{10}S_{01}}\begin{pmatrix} S_{11} & -S_{01} \\ -S_{10} & S_{00} \end{pmatrix} = \begin{pmatrix} \frac{S_{11}}{S_{00}S_{11}-S_{10}S_{01}} & \frac{-S_{01}}{S_{00}S_{11}-S_{10}S_{01}} \\ \frac{-S_{10}}{S_{00}S_{11}-S_{10}S_{01}} & \frac{S_{00}}{S_{00}S_{11}-S_{10}S_{01}} \end{pmatrix} \quad (17)$$

Hardware Architecture of KALMAN Gain K. The KALMAN gain "(11)" is presented by matrix multiplication of the expression $P_t^*H^T$"(17)" times the expression $(HP_t^*H^T + P_Y)^{-1}$ "(17)", the calculation result provides matrix "(18)".

$$K = P_t^*H^T.\left(HP_t^*H^T + P_Y\right)^{-1} = \begin{pmatrix} \mathrm{P_t^*H^T}_{00} & \mathrm{P_t^*H^T}_{01} \\ \mathrm{P_t^*H^T}_{10} & \mathrm{P_t^*H^T}_{11} \\ \mathrm{P_t^*H^T}_{20} & \mathrm{P_t^*H^T}_{21} \\ \mathrm{P_t^*H^T}_{30} & \mathrm{P_t^*H^T}_{31} \\ \mathrm{P_t^*H^T}_{40} & \mathrm{P_t^*H^T}_{41} \end{pmatrix}\begin{pmatrix} \frac{S_{11}}{S_{00}S_{11}-S_{10}S_{01}} & \frac{-S_{01}}{S_{00}S_{11}-S_{10}S_{01}} \\ \frac{-S_{10}}{S_{00}S_{11}-S_{10}S_{01}} & \frac{S_{00}}{S_{00}S_{11}-S_{10}S_{01}} \end{pmatrix} \quad (18)$$

To design the circuit block of the KALMAN gain, we exploit the circuit block$P_t^*H^T$, their outputs represent the coefficients $P_t^*H^T{}_{ij}$ of the matrix $P_t^*H^T$ "(14)". We also exploit the outputs I_{ij} of the hardware architecture block representing the coefficients of the matrix $(H.P_t^*.H^T + P_Y)^{-1}$ "(17)". The final hardware architecture represents the expression $P_t^*H^T(HP_t^*H^T + P_Y)^{-1}$ of KALMAN gain is presented in Fig. 2. The outputs K_{ij} represent the result of the matrix multiplication done in "(18)".

Correction of the State Design $\widehat{\mathbf{X}}_{\mathbf{t+1}}$. Following equation "(12)", the hardware architecture design of the update of the state vector $\widehat{X}_{t+1}$ must be started by the implementation of the circuit block $Y - Y_t^*$, such that Y_t^* is the prediction of the observation "(7)" designed in Fig. 1 and Y is the sensor measurement. Then the hardware block of $K\left(Y - Y_t^*\right)$ is designed through matrix multiplication of K and $\left(Y - Y_t^*\right)$ "(19)". Finally the final hardware architecture representing $\widehat{X}_{t+1}$ are, therefore, the outputs of the two hardware architectures: $K\left(Y - Y_t^*\right)$ "(19)" and X_t^* "(8)" summed. The hardware architecture block of $\widehat{X}_{t+1}$ is shown in Fig. 3.

$$K\left(Y - Y_t^*\right) = \begin{pmatrix} K_{00} & K_{01} \\ K_{10} & K_{11} \\ K_{20} & K_{21} \\ K_{30} & K_{31} \\ K_{40} & K_{41} \end{pmatrix}.\begin{pmatrix} Y_{00} - Y_{t00}^* \\ Y_{10} - Y_{t01}^* \end{pmatrix}$$

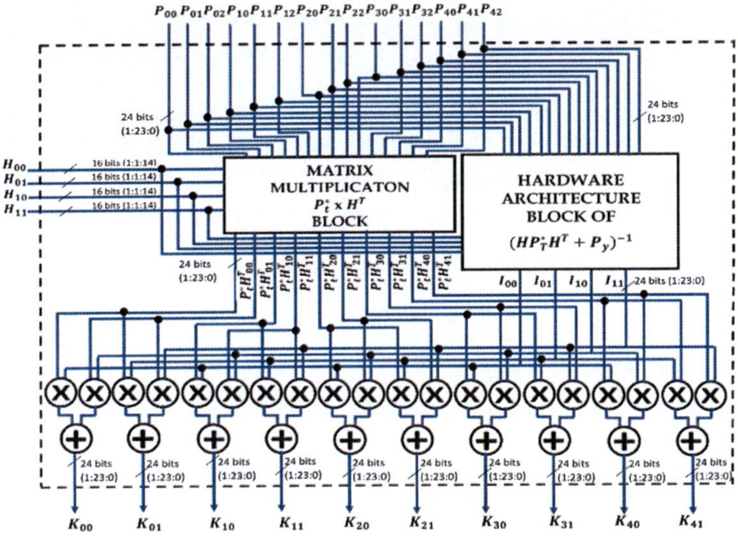

Fig. 2. Hardware architecture of KALMAN gain.

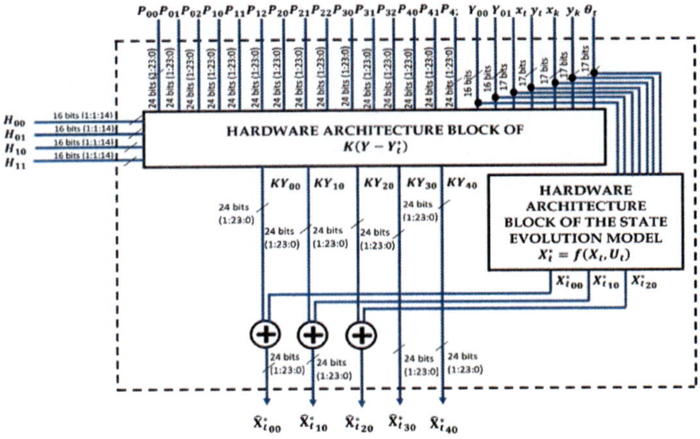

Fig. 3. Hardware architecture of correction of state vector $\widehat{\mathrm{X}}_{t+1}$.

$$
= \begin{pmatrix} K_{00}(Y_{00} - Y^*_{t00}) + K_{01}(Y_{10} - Y^*_{t01}) \\ K_{10}(Y_{00} - Y^*_{t00}) + K_{11}(Y_{10} - Y^*_{t01}) \\ K_{20}(Y_{00} - Y^*_{t00}) + K_{21}(Y_{10} - Y^*_{t01}) \\ K_{30}(Y_{00} - Y^*_{t00}) + K_{31}(Y_{10} - Y^*_{t01}) \\ K_{40}(Y_{00} - Y^*_{t00}) + K_{41}(Y_{10} - Y^*_{t01}) \end{pmatrix} \tag{19}
$$

Correction of the Covariance Design $\widehat{\mathbf{P}}_{t+1}$. Correction of the covariance matrix $\widehat{P}_{t+1}$ "(13)" is expressed by the subtraction of the prediction of the covariance matrix P^*_t "(9)" minus the expression KHP^*_t. The hardware architecture block of the covariance matrix prediction P_t^* is designed previously in Fig. 1, for this, we proceed to design the hardware architecture block KHP_t^*. At last, we will subtract the outputs of both architecture blocks to obtain the expression of the equation "(13)". HP_t^* represents the matrix multiplication of the Jacobian matrix H times P_t^* witch provides a matrix of size 2 × 5. Hardware block of HP_t^* is designed through the values of the 10 outputs $HP_{t\,00}^* \ldots HP_{t\,14}^*$ "(20)". On the other hand, the matrix multiplication KHP_t^* "(20)" is obtained by multiplying the KALMAN gain K times HP_t^* matrix. The hardware architecture of KHP_t^* is designed by implementing the result of this matrix multiplication "(20)". Finally, the subtraction of outputs of P_t^* minus the outputs of KHP_t^* is applied to get the expression of the update of the matrix $\widehat{P}_{t+1}$ "(13)".

$$KHP_t^* = \begin{pmatrix} K_{00} & K_{01} \\ K_{10} & K_{11} \\ K_{20} & K_{21} \\ K_{30} & K_{31} \\ K_{40} & K_{41} \end{pmatrix} \cdot \begin{pmatrix} HP_{t00}^* & HP_{t01}^* & HP_{t02}^* & HP_{t03}^* & HP_{t04}^* \\ HP_{t10}^* & HP_{t11}^* & HP_{t12}^* & HP_{t13}^* & HP_{t14}^* \end{pmatrix} \quad (20)$$

4 FPGA Implementation

The implementation of the designed hardware architecture of EKF-SLAM has been achieved on the Cyclone 2 FPGA, with the use of the Very High-Speed Integrated Circuit (VHSIC) Hardware Description Language (VHDL) and Intel's Quartus II synthesis and implementation tool. Quartus II is used for the design, synthesis, verification, and simulation. The implementation is entirely pipelined and takes 12 cycles. It utilizes 21512 logical elements LUT, 212 9-Bit embedded multipliers, and 1408 registers (Table 1). The overall frequency of the suggested architecture is 114 MHz. The FPGA implementation hardware utilization of the EKF-SLAM designed in this paper is presented in Table 1.

Table 1. FPGA area performance of designed EKF-SLAM

EP2C70F89618	Resources	Design usage
Logic elements	21512	68.416 (31.44%)
Registers	1408	68.416 (02.05%)
Embedded multipliers	212	300 (70.66%)

Table 2 shows a comparison of hardware resource consumption between our design and other recent methods used in [4–6, 11, 12]. Significant decreases in hardware resources have been achieved. The number of logical elements LUTs have been diminished to reach 94.85%, whereas the number of registers has also decreased to reach 96.44%. The comparison clearly shows that the implementation made in this paper consumes very few resources compared to the other techniques.

Table 2. Hardware resource consumption in different FPGA implementations.

Work	LUTs	Reduction in LUTs	DSP	Memory blocks	Embedded multipliers	Registers	Reduction in registers	Freq MHz
This paper	21512	–	0	0	212	1408	–	114
[11] 2021	22174	02,98%	43	3272051	–	–	–	717
[5] 2015	27889	22,86%	175	–	–	–	–	49,91
[6] 2020	28433	24,34%	87	3970	174	77650	95,05%	125,42
[4] 2018	39600	45,67%	–	–	252	39600	96,44%	–
[10] 2020	418080	94,85%	2520	–	–	–	–	200

5 Conclusion

This work presents a lightweight hardware architecture of EKF-SLAM and focuses on its implementation on an FPGA based on our previous work on this topic. The equations computing the gain matrix of the Kalman filter, the updates of the state vector, and the correction of the covariance matrix that constitute the correction step are designed as additional implementation on FPGA. Designed matrix multiplication calculation modules have been exploited several times in the three equations of the correction and update steps, this has resulted in a significant improvement in complexity and the achievement of highly efficient results. In this way, our design significantly minimized the resources of EKF-SLAM implementations compared to recent implementations (21512 of LUTs, 1408 registers, 212 embedded multipliers). Thus, this design turns out to be lightweight and has an excellent balance between performance and resource consumption.

References

1. Guo, H., Chen, H., Xu, F., Wang, F., Lu, G.: Implementation of EKF for vehicle velocities estimation on FPGA. IEEE Trans. Industr. Electron. **60**(9), 3823–3835 (2012)
2. Quang, N.K., Hieu, N.T., Ha, Q.P.: FPGA-based sensorless PMSM speed control using reduced-order extended Kalman filters. IEEE Trans. Industr. Electron. **61**(12), 6574–6582 (2014)
3. Ma, Z., Zhang, X.: FPGA-based sensorless control for PMSM drives using the stator/rotor frame extended Kalman filter. In: 2018 Chinese Control And Decision Conference (CCDC), pp. 102–107. IEEE (2018)
4. Contreras, L., Cruz, S., Mauricio, J., Motta, S.T., Llanos, C.H.: FPGA implementation of the EKF algorithm for localization in mobile robotics using a unified hardware module approach. In: 2015 International Conference on ReConFigurable Computing and FPGAs (ReConFig), pp. 1–6 (2015)

5. Bouhoun, S., Sadoun, R., Adnane, M.: OpenCL implementation of a SLAM system on an SoC-FPGA. J. Syst. Architect. **111**, 101825 (2020)
6. Wu, Y., et al.: An FPGA based energy efficient DS-SLAM accelerator for mobile robots in dynamic environment. Appl. Sci. **11**(4), 1828 (2021)
7. Strasdat, H., Montiel, J.M.M., Davison, A.J.: Real-time monocular SLAM: Why filter? In: 2010 IEEE International Conference on Robotics and Automation, pp. 2657–2664. IEEE (2010)
8. de Souza Rosa, L., Dasu, A., Diniz, P.C., Bonato, V.: A faddeev systolic array for EKF-SLAM and its arithmetic data representation impact on fpga. J. Sig. Process. Syst. **90**(3), 357–369 (2018)
9. Latif, R., Dahmane, K., Saddik, A.: SLAM algorithm: overview and evaluation in a heterogeneous system. Enabling Mach. Learn. Appl. Data Sci. 165–177 (2021)
10. Xu, Z., Yu, J., Yu, C., Shen, H., Wang, Y., Yang, H.: CNN-based feature-point extraction for real-time visual SLAM on embedded FPGA. In: 2020 IEEE 28th Annual International Symposium on Field-Programmable Custom Computing Machines (FCCM), pp. 33–37. IEEE (2020)
11. Gerlein, E.A., Díaz-Guevara, G., Carrillo, H., Parra, C., Gonzalez, E.: Embbedded system-on-chip 3D localization and mapping—eSoC-SLAM. Electronics **10**(12), 1378 (2021)
12. Ullah, I., Su, X., Zhang, X., Choi, D.: Simultaneous localization and mapping based on Kalman filter and extended Kalman filter. Wirel. Commun. Mob. Comput. **2020**, 12 (2020). Article ID 2138643. https://doi.org/10.1155/2020/2138643
13. Mamri, A., Abouzahir, M., Ramzi, M., Sbihi, M.: High-level synthesis implementation of monocular SLAM on low-cost parallel platforms. In: Motahhir, S., Bossoufi, B. (eds.) ICDTA 2021. LNNS, vol. 211, pp. 399–409. Springer, Cham (2021). https://doi.org/10.1007/978-3-030-73882-2_37
14. Saddik, A., Latif, R., El Ouardi, A.: Improving multispectral image processing for real-time agricultural indices tracking using an embedded system. In: Motahhir, S., Bossoufi, B. (eds.) ICDTA 2021. LNNS, vol. 211, pp. 411–420. Springer, Cham (2021). https://doi.org/10.1007/978-3-030-73882-2_38
15. Hammia, S., Hatim, A., Bouaaddi, A.: Novel hardware architecture of EKF-SLAM's Jacobian matrices and its FPGA Implementation. In: 2020 International Symposium on Advanced Electrical and Communication Technologies (ISAECT), pp. 1–5. IEEE (2020)
16. Hammia, S., Hatim, A., Bouaaddi, A.: Hardware architecture of EKF-SLAM's prediction stage and its FPGA implementation. In: 2021 International Conference on Digital Age and Technological Advances for Sustainable Development (ICDATA), pp. 140–146. IEEE (2021)
17. Hammia, S., Hatim, A., Bouaaddi, A., Najoui, M., Jakjoud, F., Ez-Ziymi, S.: Efficient EKF-SLAM's Jacobian matrices hardware architecture and its FPGA implementation. Int. Rev. Electr. Eng. (IREE) **16**(5), 484–496 (2021). https://doi.org/10.15866/iree.v16i5.20149

Scenarios for ADAS Testing: Modeling and Design

Abderrahmane Lyamani(✉), Tarik Hajji, Ibtissam Elhassani, and Tawfik Masrour

Laboratory of Mathematical Modeling, Simulations and Smart Systems (L2M3S), Artificial Intelligence Team for Engineering Sciences (IASI), ENSAM – Meknes, Moulay Ismail University, Meknes, Morocco
abd.lyamani@edu.umi.ac.ma

Abstract. The aim of this paper is to offer a novel way of creating a collection of automated scenario-based tests to evaluate the safety of autonomous cars, especially those that mix simulation and real-world testing. Our methodology is built on formal methods, which combine a mathematical model to describe a scenario, define a relevant metric, and generate a set of test scenarios for characterization of test scenarios with a formal situation description. However, any advanced driving system (ADS) may encounter an infinite number of situations and scenarios that influence its interaction. As a result, such scenarios can be characterized as an extensive collection of characteristics and parameters that define all potential scenarios and are investigated using a scenario-based method.

Keywords: Advanced Driver Assistance Systems (ADAS) · Autonomous vehicles · System verification · Test scenario · Automatic testing

1 Introduction

For decades, automated vehicles have been considered a key research area in the automotive industry [1]. Many constructors -such as Waymo self-driving cars [2]-already have examined their Driver Assistance Systems in real traffic, principally in the USA [3]. However, several industries, including aviation, manufacturing, and automotive, have seen a rise in automation in safety-critical systems. Furthermore, the deployment of automation in the automotive sector is motivated by a wide range of potential benefits, such as improved security and safety [4].

According to the World Health Organization, over 1.3 million people die each year as a result of road accidents [5]. In the future, autonomous driving vehicles (SAE Level 3 and higher) are expected to reduce this figure eventually. Because of the risks and flaws introduced by uninvolved individuals in the event of system failure, demonstrating a high level of safety is recommended for social acceptance. On the other hand, testing the safety of automated vehicles in an economically workable manner [6] is challenging due to the diversity of space parameters. However, a minimal function can be approved to avoid so-called test gaming and prove an adequate safety performance.

S. Motahhir and B. Bossoufi (Eds.): ICDTA 2022, LNNS 454, pp. 753–762, 2022.
https://doi.org/10.1007/978-3-031-01942-5_75

Furthermore, the ADAS test techniques are divided into three stages, each with its aims and priorities: road, site, and bench tests [7]. Nevertheless, standards and regulations for certifying automated vehicles are still needed. An economically viable method for performing automated vehicle safety assessments is necessary due to the large amount of plausible scenarios and critical traffic circumstances. Thus, a significant amount of effort must be invested in the development process for a considerable number of test cases [8].

On the other hand, scenario-based testing is confined to meaningful occurrences and is based on the idea that many traffic conditions are irrelevant and unimportant (scenarios). However, the lack of standardized methods for defining or classifying test scenarios, as well as international standards for determining ADAS and ADS safety requirements, has resulted in a subjective interpretation of test scenarios and desired safety levels [9].

Although ADAS validation is a complex matter, especially when verifying the functionality of a specific ADAS System [10], these functions may be exposed to an unlimited number of instances. Nonetheless, certain situations will have a more significant influence on the responsiveness of the ADAS function than others, and others will occur more frequently. However, a test scenario formulation is expected to mimic any event since many publications allude to a system test for the autonomous driving system without a formulation for the influencing parameters.

With this technique, the difficulty of choosing and selecting all relevant cases persists since the test scenario design of automated systems should refer to the influence of many kinds of traffic factors comprehensively and determine the critical characteristics of the traffic accident scenarios. As a result, this work proposes a novel technique for testing and verifying automated cars using system-specific modelization of relevant test scenarios.

Following that, the created approach is detailed in the METHODOLOGY section. The rest of the paper is laid out: Sect. 1 provides an introduction to the topic and goals for which this work was produced, followed by Sect. 2, which discusses some relevant works linked to autonomous cars and evaluating ADAS capabilities, and Sect. 3 explores the technique used to construct this study. Finally, the report wraps up with a summary and recommendations for further research.

2 Related Work

2.1 Motivating Case Study

Due to the infinite number of possible traffic situations, the safety assessment of automated vehicles can no longer be carried out economically in road tests. With the scenario-based approach, the required effort level can be decreased significantly if desirable traffic situations are too restricted to pertinent issues.

As a result, creating test scenarios is critical for assuring the safety and reliability of ADAS performance. Ulbrich [11] examines the term scenario in various contexts and proposes a uniform definition for the autonomous vehicle industry. However, in this work, the authors refer to the proposed definitions by Ulbrich [11], Fig. 1 illustrates a scene from a hypothetical situation in which a pedestrian crosses the street.

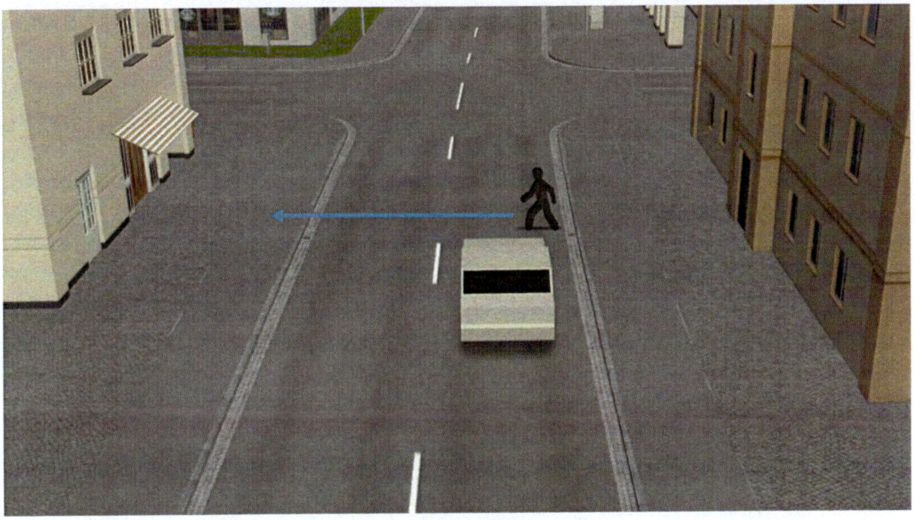

Fig. 1. Test scenario for pedestrian-detection

Several publications propose using test scenarios during the development phase of self-driving cars to create a set of scenarios to test and validate the performance of ADAS functionality. Bagschik [12] presents a methodology for generating potentially hazardous scenarios during the process phase of hazard analysis and risk assessment, as stipulated by the ISO 26262 standard [13]. This approach employs a natural-language abstract description of traffic participants and their surrounding environment. Additionally, Bach et al. [14] suggested a model-based scenario representation with spatial and temporal links as a general scenario notation.

Nevertheless, the challenging task of obtaining all relevant and appropriate scenarios persists with this approach. However, to generate more crucial scenarios, Zhao [15] and Huang [16] proposed AE, which applies the importance-sampling theory to advance the evaluation by finding the most critical scenarios. The given method considers controlled influence factors for some particular driving assistance function, restricting its application range. To directly extract the most critical and significant scenarios, a worst-case scenario evaluation method was proposed based on the traffic accidents database. Therefore, these extreme scenarios serve to identify ADAS errors and the attainable performance of some functions, but the coverage of all associated factors cannot be measured quantitatively.

SCHULDT [17] examined how these criteria might be applied to the complexity of traffic circumstances for automated cars and found that they could be employed in this application as well.

2.2 Background

Simulation-based development and novel concepts for virtual testing and validation are required instead of experimenting and testing new driving functions with the help of multiple thousand test kilometers. There are already many tools for simulation-based function improvement and validation, such as driving simulators with realistic, environmental-sensitive behavior of roadway users such as cyclists, pedestrians, trucks, and cars, which can be used to generate massive amounts of data. This data set contains new situations and scenarios suitable for testing various ADAS functions.

On the other hand, many tools are available to create and generate a test scenario for testing any functionality of ADAS, **CARLA** [18], an open platform for urban driving that simulates a static and dynamic world with a functional inter-face between the world and an agent that interacts with the world. In addition, automated Driving Toolbox on **MATLAB** provides algorithms and tools for simulating, designing, and testing ADAS functions and autonomous driving systems [19]. Also, there is **LGSVL Simulator** [20], built as an open platform. SVL Simulator provides a simulation engine and cloud simulation to test and validate automated vehicles at scale. **ASAM OpenX** [21] which is an open-file format that illustrates dynamic content in driving simulation applications, it refers to **OpenSCENARIO** which gives a description of dynamics and moving objects, their movements, and the parameters. For the static content is referenced as an **OpenDRIVE** road network, optionally combined with an **OpenCRG** road surface description [22]. This project is still in the beginning and will be available to the public shortly.

Accordingly, many tools and platforms define and offer some parameters to describe the test scenario that will evaluate any Advanced Driving System (ADS). Consequently, the scenario can be chosen as crucial based on these parameters. However, this paper aims to model a test scenario for autonomous driving systems (ADS). Before discussing the automatic generation of test scenarios for intelligent driving systems, a mathematical model must be introduced to describe the parameters that influence the perception of an intelligent driving system, which act as the input to the automatic generation process of test scenarios.

3 Methodology

To ensure that the System Under Test has reliable and robust ADAS functionality, it is critical to define test scenarios that can:

1. Build a real-world based on the sequence.
2. Insert the user input parameters.
3. Define and list "all" possible operational circumstances.

The elements used to build the test scenario are called scenarios-parameters, which are related to the functionality of the driving system under test or ADAS function. In general, the influencing factors of an intelligent driving system may be divided into three categories: environment, self-state, and other road users. To cover all possible scenarios and conditions that ADAS systems may encounter, a model of the environment, as well as the characteristics that influence the System Under Test's functioning, must be provided.

The goal is to build a test-scenario model that includes influential parameters that depict the variability of scenarios and situations that the system may encounter. Such a model necessitates considering numerous aspects and view-points, each of which has a distinct impact on the scene as viewed by the system.

This model must collect information about the ADAS's surroundings environment (landscape, road geometry, curvature, infrastructure, etc.), traffic scenarios (behavior of the equipped vehicle and adjacent cars), weather conditions (rain, fog, sun, etc.), and known troublemakers.

The environment should always be described in as much detail as is conceivable and feasible. Indeed, the test-scenario model is expected to replicate all the scenarios and conditions that the vehicle may face. However, we have identified numerous types of influential parameters and characteristics, which are listed below. These categories include parameters based on their nature and allow any function to be transposed to another [23]. The influencing parameters for a scenario are depicted in the diagram below (Fig. 2).

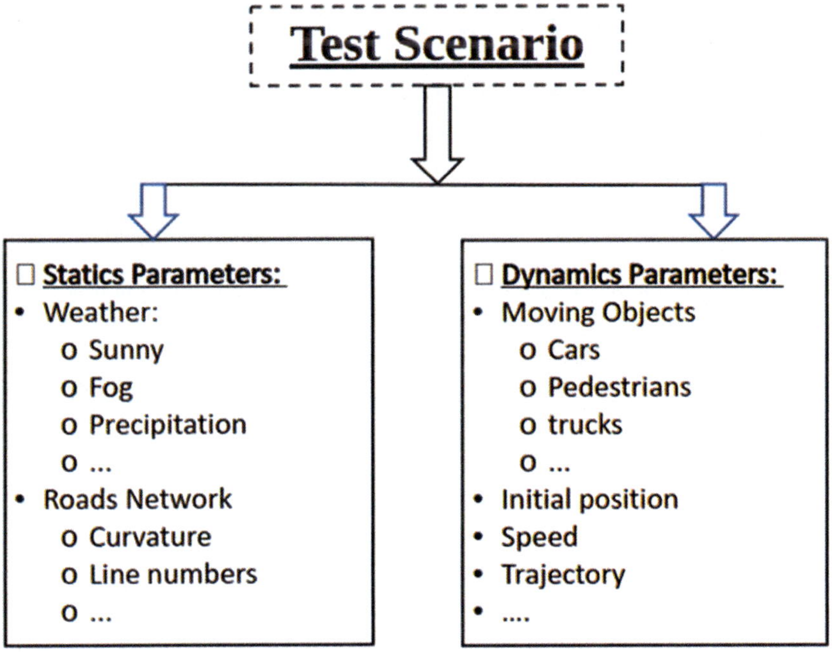

Fig. 2. Test-scenario categories

1. **Static Objects:** These are objects connected to weather and lighting conditions as well as the disruptions they generate because the weather affect show the ADAS views a scene. Additionally, the topology, curvature, the number of lanes, and the aesthetics and overall appearance of roads and their surroundings have an impact on this class (marking, surface, etc.)
2. **Dynamic Objects:** This category involves moving objects (vehicles, pedestrians, trucks, etc.) that influence how the ADAS or System Under Test (SUT) sees the scene. Other vehicles' behavior is defined by a set of characteristics that are identical to those supplied for equipped vehicles' behavior, with the addition of factors relating to their location in the scene, their numbers, and the trajectory variations they can perform.

In general, multiple characteristics that may be classified into various values can influence an intelligent driving system and impact its vision on the environment; nevertheless, in [24], Feng Gao gives a table (Table 3) of all influencing parameters that affect an ADAS system.

Mathematical Model. In this section, we formalize a Test-Scenario system and its surrounding environment. Our formalization is assigned to accurately represent all parameters that impact a test scenario for ADAS functionality testing.

The set of all Test-Scenarios will be contained in an Ωspace, since it is UNION of all the previous areas: Ω_{Static}which describes the parameters of the static elements present in the surrounding environment, $\Omega_{Dynamic}$contains all dynamic objects and their parameters, which is mathematically denoted by:

$$\Omega = \Omega_{Static} \cup \Omega_{Dynamic}$$

Although his work was established for LDW, Fen Gao [24] identify around 43 parameters that impact a specific scenario for static elements; nevertheless, the dimension of static space is dim Ω_{Static}= 43. While the parameters are cited in Table 5 [24]:

$$\mathbf{S} = (\mathbf{sunny}, \mathbf{fog}, \mathbf{precipitation}, \ldots, \mathbf{Road\ curvature}, \mathbf{lines\ number}, \ldots) \tag{1}$$

However, for dynamic sections, the quantity of moving things, such as vehicles, pedestrians, trucks, etc., plays a vital role since it affects the testing system's responsiveness to the environment. Furthermore, the parameters of the test scenarios are influenced by their starting position, speed, and trajectory. This maybe stated mathematically as:

$$D_i = (x_i, y_i, z_i, p_i, r_i, h_i, speed_i) \tag{2}$$

Each scenario can be modeled as vectors in the space of dynamic objects $\Omega_{Dynamic}$is:

$$\Omega_{Dynamic} = \{D_i \text{ such that } i \in [0, n]\} \tag{3}$$

where n denotes the total number of movable elements. However, the dimension of this space is dim $\Omega_{Dynamic} = 7$n while n present the moving elements. Since, the dynamic space can be presented as:

$$\Omega_{Dynamic} = \{(D_1\ D_2 \cdots D_n) \text{ while n defines movable elements}\} \tag{4}$$

Consequently, a test scenario may be represented as a matrix, with each parameter's value on each frame. Note that n represents the number of objects which are moving in the scenario:

$$\mathbf{\Omega} = \{(\mathbf{S\,Di})\ \mathbf{such\ that\ i} \in [\mathbf{0}, \mathbf{n}]\} \tag{5}$$

$$T = (\text{ sunny, fog, precipitation}, ..., \text{ n}, ..., x_i, y_i, z_i, p_i, r_i, h_i, speed_i, ...) \tag{6}$$

```
<ParameterDeclarations>
   <ParameterDeclaration name="HostVehicle" parameterType="string" value="car_white"/>
   <ParameterDeclaration name="Host_Speed" parameterType="double" value="9"/>
   <ParameterDeclaration name="laneId_initial" parameterType="double" value="3"/>
   <ParameterDeclaration name="offset_initial" parameterType="double" value="0.5"/>
   <ParameterDeclaration name="roadId_initial" parameterType="double" value="0"/>
   <ParameterDeclaration name="S_initial" parameterType="double" value="1.6"/>
   <ParameterDeclaration name="h_initial" parameterType="double" value="15"/>
   <ParameterDeclaration name="p_initial" parameterType="double" value="1.6"/>
   <ParameterDeclaration name="r_initial" parameterType="double" value="1.6"/>
   <ParameterDeclaration name="PedestrianSpeed_initial" parameterType="double" value="1.6"/>
</ParameterDeclarations>
```

Fig. 3. Parameters declaration for pedestrian crossing scenario

Study Cases. We opt to develop and test some ADAS functionality using simulation tools to improve the mathematical model. First, we use ASAM-OpenScenario tool [25] to run a pedestrian detection scenario. However, many parameters are missing since ASAM-OpenScenario is still in development, but they will be accessible in the future release [26]. To accomplish this, we choose to perform a primary function that involves detecting pedestrians as they cross the road. The figure below shows the parameter declaration, which allows us to generate the previous scenario (Fig. 3).

The element "ParameterDeclarations" presented in the XML file allows us to generate different Test-Scenarios by adjusting the value of the "child"; laneId1, offset1, roadId1,..., The resulting scenarios are illustrated in Fig. 4.

Fig. 4. ASAM-OpenSCENARIO implementation

However, the declarations of these parameters may have an impact on the scenario that the system under test will analyze.

On the other hand, we used Carla-Simulator to adjust the parameters in order to test the implementation; the results are shown in Fig. 5.

Figure 5 presents another scenario generation implementation using CARLA Simulator by parametrizing the statics elements.

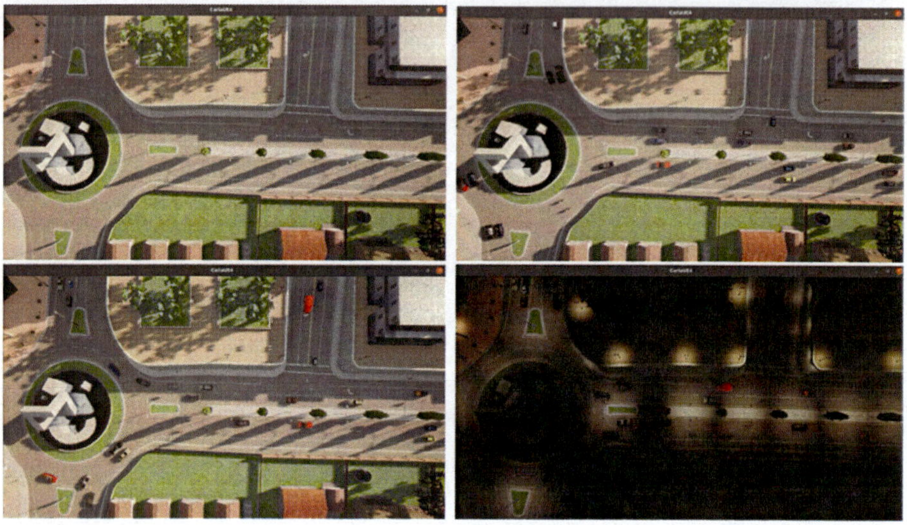

Fig. 5. Carla scenario implementation

4 Conclusion

This paper investigates the feasibility of using a Test-Scenario technique to create a collection of scenarios to test the given ADAS features.

A mathematical model has been developed to achieve this purpose, in which scenarios may be described as a set of parameters to construct a test scenario. However, a test scenario is divided into two categories: static components and dynamic objects, each described by a set of parameters.

Future work will include evaluating and judging our methodology for test-scenario meta-model design on more complex and higher-dimensional scenarios. Nevertheless, offering a more detailed automated analysis of failures in planning, perception, and prediction is still the most crucial aim of our formulation in order to develop more sophisticated test equipment that can better test the given ADAS functionality.

References

1. Bagloee, S.A., Tavana, M., Asadi, M., Oliver, T.: Autonomous vehicles: challenges, opportunities, and future implications for transportation policies. J. Mod. Transp. **24**(4), 284–303 (2016). https://doi.org/10.1007/s40534-016-0117-3
2. Hawkins, A.: Waymo is first to put fully self-driving cars on US roads without a safety driver. The Verge. 7 Nov 2017. www.theverge.com/2017/11/7/16615290/waymo-self-driving-safety-driver-chandler-autonomous
3. Voinea, G.-D., et al.: Driving performance and technology acceptance evaluation in real traffic of a smartphone-based driver assistance system. Int. J. Environ. Res. Public Health **17**(19), 7098 (2020)
4. Frigerio, A., Vermeulen, B., Goossens, K.G.: Automotive architecture topologies: analysis for safety-critical autonomous vehicle applications. IEEE Access **9**, 62837–62846 (2021)

5. Road Traffic Injuries: World Health Organization, World Health Organization. https://www.who.int/news-room/fact-sheets/detail/road-traffic-injuries
6. Go, K., Carroll, J.M.: The blind men and the elephant: views of scenario-based system design. Interactions **11**(6), 44–53 (2004)
7. Solmaz, S., Holzinger, F.: A novel testbench for development, calibration and functional testing of ADAS/AD functions. In: 2019 IEEE International Conference on Connected Vehicles and Expo (ICCVE). IEEE (2019)
8. ADAS Standards and Adas Safety Protocols: Dewesoft. https://dewesoft.com/daq/adas-standards-and-adas-safety-protocols
9. Khastgir, S., et al.: Towards increased reliability by objectification of Hazard Analysis and Risk Assessment (HARA) of automated automotive systems. Saf. Sci. **99**, 166–177 (2017)
10. Zhou, J., del Re, L.: Reduced complexity safety testing for ADAS & ADF. IFAC-PapersOnLine **50**(1), 5985–5990 (2017)
11. Ulbrich, S., Menzel, T., Reschka, A., Schuldt, F., Maurer, M.: Defining and substantiating the terms scene, situation, and scenario for automated driving. In: 2015 IEEE 18th International Conference on Intelligent Transportation Systems, pp. 982–988. IEEE (September 2015)
12. Graubohm, R., et al.: Functional safety concept generation within the process of preliminary design of automated driving functions at the example of an unmanned protective vehicle. In: Proceedings of the Design Society: International Conference on Engineering Design, vol. 1, no. 1. Cambridge University Press (2019)
13. Stolte, T., Bagschik, G., Maurer, M.: Safety goals and functional safety requirements for actuation systems of automated vehicles. In: 2016 IEEE 19th International Conference on Intelligent Transportation Systems (ITSC). IEEE (2016)
14. Bach, J., Otten, S., Sax, E.: Model based scenario specification for development and test of automated driving functions. In: 2016 IEEE Intelligent Vehicles Symposium (IV). IEEE (2016)
15. Zhao, D., Huang, X., Peng, H., Lam, H., LeBlanc, D.J.: Accelerated evaluation of automated vehicles in car-following maneuvers. IEEE Trans. Intell. Transp. Syst. **19**(3), 733–744 (2017)
16. Huang, Z., Lam, H., LeBlanc, D.J., Zhao, D.: Accelerated evaluation of automated vehicles using piecewise mixture models. IEEE Trans. Intell. Transp. Syst. **19**(9), 2845–2855 (2017)
17. Schuldt, F.: Ein Beitrag für den methodischen Test von automatisierten Fahrfunktionen mit Hilfe von virtuellen Umgebungen, Towards testing of automated driving functions in virtual driving environments. Diss. Ph.D. thesis, Technischen Universitaet Braunschweig (2017)
18. Dosovitskiy, A., Ros, G., Codevilla, F., Lopez, A., Koltun, V.: CARLA: an open urban driving simulator. In: Conference on Robot Learning, pp. 1–16. PMLR (October 2017)
19. Automated Driving Toolbox Documentation: Automated Driving Toolbox, MATLAB. www.mathworks.com/help/driving. Accessed 25 Nov 2021
20. Lgsvl: LGSVL/Simulator: A Ros/ROS2 Multi-Robot Simulator for Autonomous Vehicles. GitHub. https://github.com/lgsvl/simulator
21. Asam Launches Open Source Tooling Platform to Support Use of Asam Openx Standards for the Validation of Autonomous Driving Functions. ASAM OpenX. https://www.asam.net/news-media/news/detail/news/asam-launches-open-source-tooling-platform/
22. Beil, C., et al.: Detailed streetspace modelling for multiple applications: discussions on the proposed CityGML 3.0 transportation model. ISPRS Int. J. Geo-Inf. **9**(10), 603 (2020)
23. Zhang, X., Khastgir, S., Jennings, P.: Scenario description language for automated driving systems: a two level abstraction approach. In: 2020 IEEE International Conference on Systems, Man, and Cybernetics (SMC). IEEE (2020)
24. Gao, F., et al.: A test scenario automatic generation strategy for intelligent driving systems. Math. Probl. Eng. **2019**, 1–10 (2019)

25. ASAM e.V.: ASAM OpenSCENARIO. https://www.asam.net/standards/detail/openscenario/
26. ASAM OpenSCENARIO V2.0. https://www.asam.net/project-detail/asam-openscenario-v20-1/

Classification Framework and Performance Analysis of VANET Simulators

Brahim Driouch[1(✉)], Mohammed-Alamine El Houssaini[2], Souad El Houssaini[1], and Jamal El Kafi[1]

[1] Faculty of Sciences, Chouaib Doukkali University, El Jadida, Morocco
{Driouch.b,Elhoussaini.s,Elkafi.j}@ucd.ac.ma
[2] ESEF, Chouaib Doukkali University, El Jadida, Morocco
Elhoussaini.m@ucd.ac.ma

Abstract. Vehicular Ad Hoc Network (VANET) can be defined as a wireless communication Vehicle-to-Vehicle (V2V) or Vehicle to Infrastructure (V2I) insuring the exchanging of warnings and alerts to avoid car accidents as example.

VANET is a type of Mobile Ad Hoc Network (MANET); it can be used as an application in road safety domain to avoid traffic congestion as well as for driving comfort. The efficiency of wireless communication in VANET is related to the routing protocol performance in the network. In order to confirm the results and the efficiency of this protocol, there are many simulators that can be used before using the VANET routing protocols in the real word. VANET simulation is characterized by several challenges and completely different from MANET's simulation. Road topology, traffic flow, roadside vehicular speed, mobility, and driver's behaviors are some of these challenges.

In this paper, we give you an overview on the simulation tools and a study of the simulators performance. The aim of this study is to analyze strengths and weakness of simulators commonly used for VANET simulation, based on several criteria mainly considered in the reviewed contributions.

Keywords: VANET · Simulator · Performance · Mobility

1 Introduction

A VANET is a network where vehicles are considered as nodes to create a mobile network. The network performance evaluation is important in order to identify any problems that may occurred in real word, so the adequate way to achieve this evaluation is in effect to execute simulations that shows identical results to the real word conditions. We can found several simulation tools that are used to evaluate and simulate the performance of VANET routing protocols. The simulation of VANETs needs two different elements to achieve a practical simulation: a vehicular mobility simulator, which is capable to provide an accurate traffic simulation model for the nodes, and a network simulator for simulating the interaction of a wireless network. Vanet Mobility Simulation MobiSim [1] and Network Simulator NS-2 [2] are some examples of these tools, either C++ and simulators based on the Java programming languages.

S. Motahhir and B. Bossoufi (Eds.): ICDTA 2022, LNNS 454, pp. 763–771, 2022.
https://doi.org/10.1007/978-3-031-01942-5_76

The scoop of this paper is mainly to address an over-view of existent tools for VANET simulators and to analyze different programs so as to make decision based on performance analysis of the practical tools that we can use for next contribution in the VANET simulation specifically in the traffic safety.

To achieve this analysis we describe an analysis framework represented by several criteria that identify the strength, weakness, opportunities and challenges of each simulation tool escalated in the reviewed pa-pers.

The present study covers the following sections. In Sect. 2, we present a brief overview of VANET simulation tools, and then we give a description of the VANET mobility generators in Sect. 3. Section 4 will describe the networks simulators. Section 5 covers the VANET simulators. At the end, we finish with conclusion and next research orientations.

- **Related work**

  A set of contributions were reviewed during the preparation of this paper. The current simulation tools that are communally used are presented (Alba et al. n.d.). (Spaho et al. 2021). In order to better frame the scoop of this contribution, which is articulated around the classification of simulation tools according to their categories and their performance, based on several criteria that have been defined in the classification framework as part of the performance analysis of these tools.

2 Overview of Simulators for VANET

In recent years, several methodology highlighting the multiple challenges of VANET simulation trying to approach the reality context. First, a designer could use a mobility generator for traffic simulation by generating a close to reality vehicular mobility schemes used as an input for a MANET simulator. Second, the researcher could use a specially designed VANET simulator as shown in the Fig. 1.

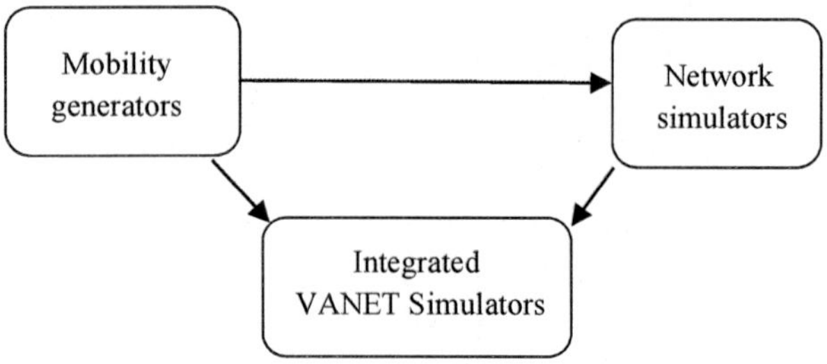

Fig. 1. Integrated VANET simulators

The first methodology used for VANETs simulation is based on a traffic simulator or a traffic model generator capable of generating schemes models. Then these schemes are evaluated and tested based on a specific MANET simulator. The free access and availability of these MANET simulators is the main strength for the use of this methodology.

However, the major weakness is that many of VANET applications specifically for traffic safety need interactions between vehicles and network events, alarms; and it is difficult to be framed with these simulation schemes. The majority of research community use ns-2 (network simulator) for MANET simulating for many reasons. First, the number of traffic simulators, which generates ns-2 format, is large: the most comprehensive is Vanet MobiSim; however, we can also find another. There are also traffic simulators that generate schemes for other MANET simulators as STRAW [3]. Finally, TraNS [4] and MOVE [5] combine the SUMO [6] mobility generator and ns-2 simulator combining in an integrated tool as shown in the Fig. 2.

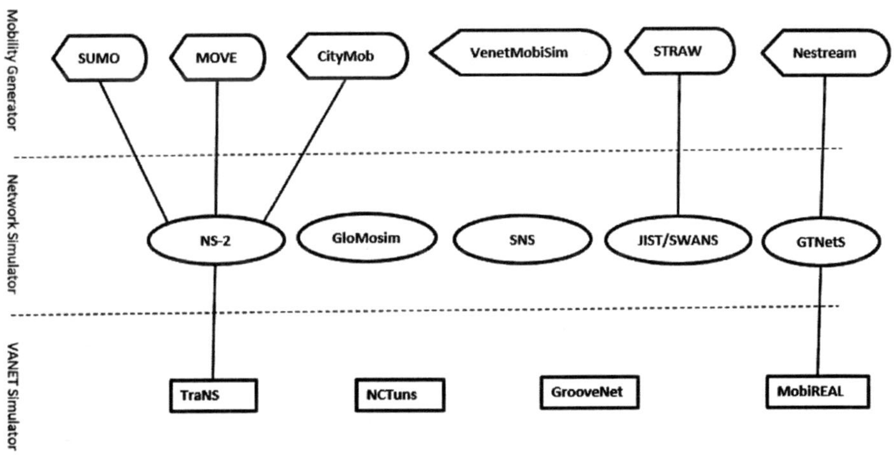

Fig. 2. Interaction of VANET simulation software's.

3 VANET Traffic Mobility Generators

The importance of a mobility generator is to increase the realistic aspect in the VANET simulations aiming to reinforce the results and conclusions for the use of VANET. In first part of our paper, we give you an overview of the existing mobility generators and a comparative study.

3.1 Overview of Mobility Generators

In recent years, several simulation software framework exists and enable us to generate trace files modelling the vehicles' movements and effects.

- *VanetMobiSim* focus on the vehicular mobility to provide a realistic vehicular interaction frames at the macroscopic level and microscopic level as well. At first view, VanetMobiSim provide maps from Toppologically Integrated Geographic Encoding and Referencing database (TIGER database). The TIGER is a database of geographical components in a macroscopic aspect for example roads, rivers, lakes and legal boundaries. Secondly, it is about mobility models as Intelligent Driving Model with Lane Changing (IDM/LM) and an overtaking model, which interacts with IDM/LM to manage lane changes and vehicle feedbacks, enabling a real vehicle-to-vehicle and vehicle-to-infrastructure communications needed for VANET simulation. VanetMobiSim can provide mobility traces in supported formats, supporting different simulation tools for mobility networks such as ns-2, GloMoSim [7], and QualNet [8].

- *SUMO* (Simulation of Urban MObility) is based on a program source that is open and portable, with a package of microscopic road traffic simulation supporting large road networks. SUMO a traffic provide simulation in several zones all over the world. Nevertheless, SUMO is limited on mobility generation; the existing network simulators cannot directly use its generated traces.
- *MOVE* (Modeling, Virtual Environments, and Simulation) is a mobility model generator for vehicular networks. Programmed under the Java language and based on SUMO (Simulation of urban mobility). MOVE has an advantage of being supported by the Graphical user interface (GUI) facilities. Mobility traces can be generated from the Google Earth or TIGER databases. Custom traces are supported; however, the car movement is limited to a grid in a random chart. MOVE is limited in networks simulation, but it can analyse traces files and then to be processed by a network simulator.
- *STRAW* (STreet RAndom Waypoint) It comes up with exact simulation results through a vehicular mobility model kept in place in original North American Countries. The disadvantage of this method is that the model is able to define the node movements to only the streets, which are predefined by the map knowledge as in the case of North American countries. This further makes sure that the mobility is limited for per vehicular congestion and simplified traffic management mechanisms

3.2 Performance Analysis of Mobility Generators

To achieve a performance analysis of the studied mobility generators, which we described above, we focus on several specifications and criteria's such as the software specifications, the uploaded maps quality, the interaction with the mobility in the network and the traces files usage. In the Table 1 below, we present the summary of the analysis results.

Table 1. Performance analysis of mobility generators software's.

Criteria	Sub criteria	*Vanet Mobisim*	*SUMO*	*MOVE*	*SRTRAW*
Software	Freeware	Yes	Yes	Yes	Yes
	Open source	Yes	Yes	Yes	Yes
	Available example	Yes	Yes	Yes	No

(*continued*)

Table 1. (*continued*)

Criteria	Sub criteria	*Vanet Mobisim*	*SUMO*	*MOVE*	*SRTRAW*
	Ease of set up	Moderate	Moderate	Yes	Moderate
	Ease of use	Moderate	No	Moderate	Moderate
	Continuous development	No	Yes	No	No
Maps	Real	Yes	Yes	Yes	Yes
	User defined	Yes	Yes	Yes	No
Mobility	Traffic models	Yes	Yes	Yes	Yes
	Multiline roads	Yes	Yes	Yes	Yes
	Lane changing	Yes	Yes	Yes	Yes
	Traffic signs	Yes	Yes	Yes	Yes
	Intersection management	Yes	Yes	Yes	No
	Overtaking criteria	Yes	No	No	No
	Large road networks	No	Yes	Yes	Yes
	Collision free movement	No	Yes	Yes	No
Traces	Ns-2 trace support	Yes	No	Yes	No
	GloMosim support	Yes	No	Yes	No
	QualNet support	Yes	No	Yes	No
	XML-based trace support	Yes	No	No	No
	Import different formats	Yes	Yes	Yes	No

4 Network Simulators

Designers and researchers use network simulators to test how the network can react under specific and various conditions. Then they can adapt simulator to achieve their customized needs regarding the time and effectiveness in setting up the multiple computers on the same network, routers and data links. Globally the network simulators are fasts and less expensive than using real hardware, they are particularly useful and adaptable to modifications to better simulate several conditions.

4.1 Overview of Network Simulators

Many network simulators are used to simulate an Inter-Vehicle (IV) communication systems or Vehicle-to-Vehicle communication (V2V). We present a sample of promising network tools for VANET network based on main characteristics and requirements.

- *NS-2* is an event simulator that include node mobility; it is a research-oriented open source in communication networks. Since its creation, NS2 has received considerable interest from industry, academia and government. Having been the subject of constant improvements. To study the performance of the network, researchers can simply use an easy-to-use script programming language for a network and observe the results provided by NS2. It has become the most used tool and one of the most popular open source network simulator.
- *NS-3* [9] is a different network simulator that is used mainly for research and academic sectors. NS-3 is a publicly available program and its license is giving under the GNU license, it is freely available for use, research and improvement. The main objective of the NS-3 is to create a favorable, open environment for networking research and simulation: it need to be adapted with the project simulation requirements of new research such as VANET and should enable researcher's contributions, content improvement, and development of the program as well.
- *GlomoSim* (Global mobile Simulator) is also the most used for simulating network, the second after NS2. GlomoSim was firstly introduced to achieve simulation of wireless network. GlomoSim can be used under shared-memory symmetric processor (SMP): a memory that is simultaneously accessible by several programs. It has an advantage to separate the network into different modules; each one can be run as an independent process. And as a result, it decreases the load on the processor by distributing its workload. GlomoSim can be run under several wireless technologies including IEEE802.11. GlomoSim was developed to be adapted to a large scale of Data (million of nodes as one simulation). Aggregation of nodes and layers are considered to be a limitation in many network simulators,
- *JiST/SWANS* [10] is a network simulator based on the JiST platform. It was developed initially regarding the need of network simulation tools for the research needs. SWANS is composed of distinct software elements that can be combined to frame the wireless network or sensor network. It has many similarities with ns-2 and GloMoSim, in addition SWANS can run a large simulation networks. With less memory needs, and run standard Java network applications over simulated networks. SWANS can run simulations that are one or two orders of magnitude larger than what we can do with GloMoSim and ns-2, respectively, using the same time and quantity of memory, and with the same level of detail.
- *SNS* [11] (Staged Network Simulator) has a faster rate of quick simulation. It performs quicker than regular NS-2. In this scenario, half of the hour is spent only on the staging process followed by rest of engineering process. SNS is enabled by the above described performance level to run a huge networks. However this implementation is predicated on NS-2 version 2.2, which doesn't mean to run simulation of VANET situations.

4.2 Performance Analysis of Network Simulators

NS-2 is a free and open program network simulator as shown in the Table 2. We can use NS-2 in wired or wireless networks, including several MANET routing protocols under the IEEE 802.11 MAC layer. It is the commonly used simulator for networking researches.

The source program of NS-2 is developed under C++ and users can interact with NS-2 using writing TCL scripts.

The NS-3 shows the better result and compatibility than the other simulator, which is clearly shown in Table 2.

GloMosim, JIST/SWANT and SNS are another open source network simulators used for simulations but still limited for continuous development and are not portables.

Table 2. Performance analysis of network simulators software's.

Criteria	NS-2	NS3	GloMosim	JIST/SWANT	SNS
Freeware	Yes	Yes	Yes	Yes	Yes
Continuous development	Yes	Yes	No	No	No
Portability	No	Yes	No	No	No
Including C++	No	Yes	No	No	No

5 VANET Simulators

Traffic simulator and network simulator are combined in an Integrated Simulators, usually is composed by two sub simulators: mobility simulator and network simulator, the communication between each other is described like shown in Fig. 1. These simulators offers a high level of simulation capability in the two areas: the mobility simulation and the network simulation.

5.1 Overview of VANET Simulators

In the following, we shows some of useful examples of VANET/integrated simulators:

- *TraNS* (Traffic and Network Simulator) is a simulation tool that combine network and mobility simulators to provide an efficient VANET simulation. TraNS provide the possibility that the information communicated in a VANET network can influence the nodes behavior in the traffic graph. Trying to frame the SUMO with ns-2. However, it is not easy that the SUMO language to be understood and unnecessary, because the communication designers needs only a parsimonious model, easy to develop and/or modify accordingly.
- *GrooveNet* [12] supports different models of communication, traffic and travel control providing a large scale of simulations. The principal weaknesses are the maps witch not include one-way and the altitude of the street. GrooveNet is developed under C++ and Qt graphics in Linux. GrooveNet is also based on the US TIGER database and is able to load counties at run-time. GrooveNet loads files from map database and converts the data into a binary file with a geographic structure.

- *NCTUns* 1.0 [13] initially used as a network simulator, but actually the new version integrates some travel and mobility simulation functions, such as configuring maps and managing cars mobility. A large amount of maps can be developed under different types of adapted road segments. The best feature for NCTUns is that its network protocol stacks includes the Linux kernel protocol stack, including TCP/IP and UDP/IP. In this simulator, the code for controlling the nodes movement is combined with the network simulation code, this fact makes it difficult to modify and/or to extend.
- *MobiREAL*[14] is mainly framed by two elements. Firstly a MobiREAL behavior simulator which simulates mobile nodes' behavior. Then MobiREAL network simulator, which simulates information communication among mobile nodes. The network and behavior simulators are two separated programs that periodically and independently communicate necessary data among a TCP channel. Being a network simulator which simulates the human as well as vehicular realistic mobility, it can allow the dynamic and variation on their behavior reckoning on a given environment and application context. Its mechanism is simple, that it simply describes the quality of nodes which were mistreated by C++. It actually visualizes the node movement in a dynamic mode, which further states its properties and packet transmission.

Through this, the simulation results can be understood intuitively. Mobility of nodes is simulated within the Behavior simulator. In addition to the above, there is a special rule which applies in the case of pedestrians to avoid collision. Using MobiREAL (Konishi et al. 2005), the vehicles getting congested can be further modeled. It has the capacity to simulate a combination of assorted quality models simultaneously.

5.2 Performance Analysis of VANET Simulators

The Table 3 shows a comparison between the network simulators that we presented in the last section. The analysis of the well-known VANET Simulators shows that the unavailability of freeware is one of the main common weaknesses.

Table 3. Performance analysis of network simulators software's.

Criteria	Trans	Groovenet	NCTuns	Mobireal
Freeware	NO	NO	NO	NO
Simulation type	Street speed	Street speed	Random	Street speed
Trafic model	Random	Dijkstras	Manually defined	Manually defined
Vanet in application	Detour and road warning	Vehicle warning and flexible rebroadcasting	NO	NO
Vanet communication	V2NV and manual V2V	V2V and V2NV	V2V and V2NV	Manually defined

6 Conclusion

The main scoop of the research achieved in this paper is enhancing the knowledge of the existing simulator for VANET environment. In this paper, we present the main categories of simulators used in VANET, which will be challenged when choosing the adequate simulation tool for the VANET application. In order to make this choice, the main challenge is the high cost for VANET simulators and the performance of each simulation tools. As a conclusion, we can use different network simulators and mobility generator in free cost by considering the interconnection and the adaptation between these tools categories. We wish that the actual classification would give reference as a ready comparison study for other researchers working in the same research area, specifically in the step of choosing the VANET simulator more adapted to demonstrate their studies results in a customized way.

References

1. VanetMobiSim Project. http://vanet.eurecom.fr
2. The Network Simulator - ns-2 (isi.edu)
3. STRAW - STreet RAndom Waypoint - vehiclar mobility model for network simulations 2021. http://www.aqualab.cs.northwestern.edu/projects/STRAW/index.php
4. Piorkowski, M., Raya, M., Lugo, A.L., Papadimitratos, P., Grossglauser, M., Hubaux, J.-P.: TraNS (Traffic and Network Simulation Environment). Ecole Polytechnique Fédérale de Lausanne, EPFL, Switzerland (2021). http://trans.epfl.ch/
5. MOVE (MObility model generator for VEhicular networks): Rapid Generation of Realistic Simulation for VANET (2020). http://lens1.csie.ncku.edu.tw/MOVE/index.htm
6. Krajzewicz, D., Rossel, C.: Simulation of Urban MObility (SUMO). German Aerospace Centre (2020). http://sumo.sourceforge.net/index.shtml
7. Martin, J.: GloMoSim, Global mobile information systems simulation library. UCLA Parallel Computing Laboratory (2021). http://pcl.cs.ucla.edu/projects/glomosim/
8. Network Technologies. Qualnet. Scalable Network Technologies, Inc. (2020). http://www.scalable-networks.com/products/download.php
9. ns-3 | a discrete-event network simulator for internet systems (nsnam.org)
10. SWANS: Java in Simulation Time/Scalable Wireless Ad hoc Network Simulator (2020). http://jist.ece.cornell.edu/
11. Walsh, K., Sirer, E.G.: A staged network simulator (SNS). Computer Science Department, Cornell University (2020). http://www.cs.cornell.edu/people/egs/sns/
12. Mangharam, R., Weller, D., Rajkumar, R., Mudalige, P., Bai, F.: GrooveNet: A Hybrid Simulator for Vehicle-to-Vehicle Networks. Carnegie Mellon University (2020). http://www.seas.upenn.edu/rahulm/Research/GrooveNet/
13. NCTUns 5.0 (2021). http://nsl10.csie.nctu.edu.tw/
14. MobiREAL (2021). http://www.mobireal.net/

Road Traffic Anomaly Detection: A Survey

Imane El Manaa(✉), Fadwa Benjelloun, My Abdelouahed Sabri, Ali Yahyaouy, and Abdellah Aarab

LISAC Laboratory, Faculty of Sciences, Dhar El Mahraz, Sidi Mohamed Ben Abdellah University, Fes, Morocco
imane.elmanaa@usmba.ac.ma

Abstract. Lately the number of vehicles circulating on roads had been widely increased especially in the big cities, leading instantly to congestion and consequently the accidents rate increasement. In order to reduce traffic issues and enhance road safety, so many researchers focus their vision on this era problem trying to look for solutions by developing algorithms and embedded applications based on image processing and artificial intelligence, therefore we aim to automate the traditional detecting systems used to spot traffic violations such as: speeding detection; crossing a red light; driving on a prohibited lane, etc. this offenses considered among the principal factors of road accidents, for this main reason the built of those embedded systems play a very important role for traffic accident reduction through monitoring the road safety and automating the penalty detection. In this article, we are trying to study the most recent techniques based on computer vision to monitor and detect traffic offenses, lightning the strengths and weaknesses of these techniques, additionally we provide research a comparison and study basis to facilitate their future work also improve robust approaches to detect road violations.

Keywords: Detection of traffic violations · Crossing a red light · Speed detection

1 Introduction

In these recent years, traffic management has become one of the main issues concerning public health, especially with the increasement of vehicles, good traffic management requires real-time embedded systems that are able to replace human roles, installed in most principal roads especially those with traffic jam. These monitoring systems allow us to detect all types of code infringement. To take precautions and make all the road users drive quietly and safely [1].

This article presents a study of different systems based on image processing and artificial intelligence, additionally a state of art concerning automation of code violation detecting systems, we have summarized some recent works, each one solves one of traffic problems, either red signal detection [2–5], speeding detection [6–8], sudden change of direction [9] or driving on a prohibited lane [10].

We aim to provide researcher with a preliminary data basis about various existing techniques, different pre-processing, and techniques to minimize noise in frames which

S. Motahhir and B. Bossoufi (Eds.): ICDTA 2022, LNNS 454, pp. 772–781, 2022.
https://doi.org/10.1007/978-3-031-01942-5_77

is a necessary step in most algorithms, the major step in all the proposed algorithms is the choice of technology used for detecting vehicles; additionally, we present the experimental results of each application to discuss the performance of each algorithm.

This work is organized as follow: Sect. 2 present a state of art about the different approaches based on image processing and artificial intelligence dedicated to detecting all types of traffic violations, Sect. 3 presents the results analysis of the different techniques discussed above, conclusion and future work are presented in Sect. 4.

2 Literature Review

In this section, we present an overview about the automated systems based on artificial intelligence and image processing dedicated for automatic detection of traffic violations. More precisely, speeding; crossing a red light; driving on a prohibited lane etc. These techniques require a high-resolution camera to facilitate the progress of different algorithms and get the best results, in addition to make the system apt in different lighting conditions (day and night). We have examined the powerful points of the techniques documented furthermore suggested different other research orientations in pursuance of improving the obtained results and the offered systems service. Most of these algorithms suggest adding new functionalities such as detect and recognize number of vehicles violating the Highway Code.

Akila et al. [9] presented an automated and real-time system for detecting traffic violations, more precisely detecting the vehicles that suddenly change their trajectory, to achieve this system, they have chosen the "Haar Cascade" detector because of its high precision and performances, additionally to employ it they have trained a classifier using images database which either contain vehicles or not. Subsequently this classifier checks all the image fractions to determine if there are one or more vehicles in it, furthermore, tracks every single vehicle using the previous detecting results (x and y coordinates, area, and width/height ratio), these values are inserted in a KalmanBoxTracker model while identifying each object separately, afterward the KalmanBoxTracker object is placed in list, this last contains all the information about tracked vehicle.

During the association phase each newly detected vehicle is added to the list while the others deemed obsolete are removed, and to detect vehicles road violations, they adopt for a new mechanism that calculate the trajectory angle of the vehicle, i.e., the angle between two lines connecting four points of the trajectory, considering that the trajectory makes a series of points pk, pk-1, pk-2 ……. With pk is the current position, then calculate the angle between the line pk and pk-2 with pk-2 and pk-4. The main raison for skipping a position is reducing the effects of noise and position errors, therefore while defining a threshold the violation is detected if the value found exceeds this threshold.

The system produces results with a large value of MOT = 0.90 (Multiple Object Tracking), towards result amelioration through increasing accuracy rate we should define the detection area to avoid tracking the discarded vehicles and therefore minimize the false negative rate.

Joel et al. [7] proposed a very fast traffic violation detecting system, to detect vehicles, they used the SSD model trained by the COCO dataset, and then define a line of interest regions i.e., the line where the vehicles are banned to cross when the light is red, afterwards inserting the image in the SSD model to generate a bounding box if there is a vehicle.

The coordinates of the latter used to determine if a vehicle goes by the line of interest, the condition of this line is defined by the condition of the red light i.e., by a timer, for example if a vehicle goes by the line while the red traffic light is on then this vehicle marked as a red-light offender.

Additionally, they proposed an algorithm to detect speeding vehicles based on OpenCV, which means calculating the Euclidean distance [11] between two successive frames. To find the current distance in meters we should multiply the resulted distance in pixels by the equation converting pixels to meters, therefore we divide the obtained result by the value of fps. The researchers conclude that the SSD model is excellent for detecting red light crossing with a rate of 100% and speeding with a rate of 92.1%.

Bordia et al. [5] presents a system dedicated to detects vehicles crossing the lines when the signal turns yellow or red, this system requires two cameras, one to detect the traffic light is placed in front of the light and the other to detect vehicles placed in front of traffic. Therefore, the development of this algorithm requires four main phases.

First phase, the detection of the traffic light contains two tasks: 1 detecting the position of the traffic light using TensorFlow object detection trained on (COCO dataset [12]), 2 color identification by Faster RCNN trained on Bosh database [13, 14]. The second phase is the detection of traffic line, to facilitate the calculation and eliminate all kind of noise, the algorithm starts with the identification of the interest region while taking just half of the image containing the stop line. Therefore, the image is cropped (zero, height/2) and (zero, width/2) and then converting each frame to grayscale to apply the Canny Edge algorithm.

Additionally, for vehicle detection the studies are based on the Mobile net SSD model and trained on the COCO dataset, to detect vehicles crossing the white security line they combined the results of the last two steps i.e., the intersection of the white line coordinates and the coordinates of the detected vehicles. Let's take the coordinates p (x1, y1), q (x2, y2) as the ends of the safety line and the coordinates d1, d2 as the car box diagonal, we decide a traffic violation if the triplet (p, q, d1) and (p, q, d2) have different orientations.

To summarize, concerning vehicle detection the SSD model according with the COCO database gives the best results, concerning the traffic lights detection, the faster RCNN gives an accuracy of 74.14% and it could be improved with more training, and for more system amelioration they have proposed adding the classification of various vehicles and reading car numbers.

In order to detect speeding vehicles, Neha et al. [6] have proposed an algorithm that goes through several stages, starting with the acquisition of the frames, then the pre-processing steps which have a role of minimizing the noise, This model consists of extracting the different ROIs, converting the image to grey level so as to reduce noise and computational complexity, therefore applying Gaussian Blurring to facilitate edge detection, and finally apply FastNIMeansDenoising, in order to simplify the analysis, we partitioned the frame by segmentation, next the classification is done using Haar Cascade while training a model to distinguish the positive images that contains the vehicles and the negative ones, afterwards the Haar Cascade becomes able to detect vehicles and draw a bounding box for each vehicle.

The last model of the proposed algorithm is responsible for speed calculating, therefore they used the VASCAR algorithm to estimate the speed of each vehicle, while knowing the displacement between two reference points on the trajectory, the principle of this algorithm is to start a timer when the vehicle passes the first point and doing the same thing for second point. We can conclude the time needed to move without breaking the road rules by subtracting the timer's values, therefore we can simply calculate the speed with the formula v = displacement/time, whilst comparing this result to a speed threshold we detect the vehicles that offences highway code. The authors concluded that their method compared to others is more efficient and eventually gives an accuracy of 0.928.

Pasit et al. [3] have presented a software technique to identifying traffic light colors, the proposed algorithm is very important because it represent the first step that launch the other traffic light violation step. First, define the region of interest to detect only the area containing the traffic light; afterwards, authors have proposed three methods to detect the red-light color. The first one consists in converting the images in grey level, doing a thresholding, then comparing the three colors with a predefined model to determine which color is on, the second is based on the conversion of RGB color images to HSV and compare the value of the hue to determine the color of the traffic light lit, and thirdly a step that makes a combination of the first two methods which means that the true color is when the results of the two methods are identical, and if it is red the violation detection routine automatically triggered, this technique detected the three color with an accuracy of 96%.

A one-way violation detection system by three-wheeled vehicles because they have a greater tendency to violate these rules is proposed by Helen et al. [10], to detect three-wheeled vehicles they used the pre-trained TensorFlow object detection model provided by the Zoo model, the next step is to follow this type of vehicle, by storing the centroid coordinates of each vehicle for the displacement calculation, and then if we notice that the direction of a vehicle is opposite to the direction of traffic, then we can conclude that this vehicle has violated the rules of one-way traffic. This method is cost-effective and easy to deploy and works well for one-way traffic areas.

In this study the authors Ruben et al. [8], presents a traffic violation detection system: traffic light violation and vehicle speed detection based on YOLOV3 as a detection algorithm [15], because of its detection speed and accuracy, This model allows us to classify moving objects, and therefore to classify vehicles, and then classified the other elements by the same model as the traffic lights, after detection of the vehicles we obtain the coordinates of the box and we calculate the intersection of these coordinates with the ROI which is the traffic signal line, the percentage of the intersection is the index that will show whether a car has violated the traffic law or not, and the speed is calculated using three principles: anchor calculation, pixel calculation and box area calculation. This system gives an accuracy of 89.24% for the detection of speed violations.

A complete application to detect vehicles in violation of the stop line is developed by Satadal et al. [2], to test the correct functioning of this application, the authors collected a real database by installing three cameras at an important intersection of the road, the cameras are synchronized with the traffic signals to take only the videos with the red light on. The videos are taken in different lighting conditions (day/night), to this end for the detection of the vehicles they used a method which adapts to the change of illumination, based on the adaptative background subtraction method. the algorithm is structured as follows: calculate the difference between the successive images, if the value is below than a predefined threshold, then the image is considered as background, and it is stored in the database as soon as a change is made to the database, then the average background image is generated by averaging the last five images. Then a new image is subtracted from the background image, in this image of subtraction the gray values of all pixels are added and compared to a threshold, to decide that there is a possibility of a vehicle intrusion. and to detect stop line violations, the authors have proposed a new technique: draw five lines hypothetically on the stop line, if all pixels along the lines are black, it is decided that there is no occlusion, and to decide that there is a violation of the stop line or red light, the vehicle must continuously obscure the stop line, which means that the longest average exceeds a given threshold, This technique successfully identifies vehicles in 92% of cases.

A region-based approach is proposed by Yanyue et al. [4], to detect that does not stop in front of the stop line, the authors define the detection zone by (the stop line until the front of the sidewalks) which is long enough to take a minimum number of frames to study the movement of the vehicles. and to detect moving vehicles, they proposed a method based on the background subtraction and also allows to distinguish the moving object from the shadow, This method relies on the difference between the hue frames based on the HSV color model to detect vehicles [16], by following the steps: transformation of the detection region from RGB to HSV, calculate the line-by-line average of the hue of the current image and the background, Do the subtraction by line for the average hue of the background frame to current frame. The authors defined two thresholds, The first one is used to judge the state of the line, if the result of the subtraction is greater than the threshold then we say that the detection region is incomplete, we count the incomplete and continuous lines, where SLmax1 is the maximum value, if the SLmax1 value is higher than the second predetermined threshold which is the maximum length of a continuous incomplete line, then the car is in the detection zone, This method is based on the initialization of the background during the red signal, and also to take image captures without moving objects quickly, They divided the detection area into three parts, and update each part separately. To update the background the authors, use Surendra's algorithm [17], in which we replace the background with the current frame for the region that does not contain a moving object, and to detect the movement of vehicles they proposed to set two thresholds, one is the Sstop threshold, and the other Srun. They assume that fw is the total pixel rows between the two frames of vehicle forward, If fw is less than Sstop then the vehicle is stopped, i.e. there is no violation, if fw is greater than Srun then the vehicle does not stop which means there is a violation, The algorithm proposed in this article detects vehicles efficiently and can eliminate the disturbances caused by shadow and illumination, and the weak points of these algorithms are, it does not address the problem of hidden vehicles, and also the quality of the video affects the veracity of the detection.

3 Results Analysis

The Table (1) evaluates the techniques described in the previous section:

Table 1. Representative work based on type of violation

Ref	Type of violation	Specific technique	Detected object	Method	Assessment
[2]	Stop line violation	Adaptative background subtraction method	Vehicle	Image processing	**Pros:** Identifies vehicles in 92% of cases **Cons**: a configurable solution, a threshold must be set
[3]	Red light violation	Compare the RGB and HSV color of traffic light to a threshold	Traffic light colors	Image processing	**Pros:** This technique is simple for implementation and give an accuracy of 96% **Cons**: sensitive to light changes
[4]	Stop line violation	Frame difference in hue based on HSV color of image	Vehicle	Image processing	**Pros:** detects vehicles efficiently **Cons**: it does not address the problem of hidden vehicles, and the quality of the video affects the veracity of the detection
[5]	Red light violation	TensorFlow object detection COCO dataset	Traffic light	Supervised learning	**Pros:** Good result of detection and classification **Cons:** Detect traffic light in all condition (angle of vision and lighting) need for data augmentation

(*continued*)

Table 1. (*continued*)

Ref	Type of violation	Specific technique	Detected object	Method	Assessment
[5]	Red light violation	Faster RCNN Bosh dataset	Traffic light colors	Supervised learning	**Pros:** gives an accuracy rate of 74.14% which is sufficient to validate the proper functioning of the violation detection algorithm **Cons**: requires more training of the model, and it is slower compared to the SSD algorithm
[5]	Red light violation	Mobile net SSD model	Vehicle	Supervised learning	**Pros**: SSD model suitable for detecting red light violation **Cons**: A large training dataset is required for high accuracy
[6]	Over speed detection	Haar Cascade	Vehicle	Supervised learning	**Pros:** image pre-processing + Haar cascade + VASCAR: the combination of the three treatments gives an accuracy of 0.928 for speed detection **Cons**: They tend to follow false positive vehicles

(*continued*)

Table 1. (*continued*)

Ref	Type of violation	Specific technique	Detected object	Method	Assessment
[7]	Red light violation	SSD model COCO Dataset	Vehicle	Supervised learning	**Pros**: SSD model is excellent for detecting red light crossing with a rate of 100%, very fast **Cons**: A large training dataset is required for high accuracy
[7]	Over speed detection	SSD model COCO dataset	Vehicle	Supervised learning	**Pros:** SSD model suitable for detecting over speed and give a rate of 92.1%, very fast **Cons**: A large training dataset is required for high accuracy
[8]	Traffic light violation	YOLOV3	Vehicle	Supervised learning	**Pros:** This algorithm based on YOLOV3 give an accuracy of 89.24% for the detection of speed violations, speed detection **Cons**: does not detect small objects correctly
[9]	Sudden change of trajectory	Haar Cascade	Vehicle	Supervised learning	**Pros**: provides high accuracy for the MOT which means that the algorithm will work very well with violation detection **Cons**: They tend to follow false positive vehicles, especially the far ones

(*continued*)

Table 1. (*continued*)

Ref	Type of violation	Specific technique	Detected object	Method	Assessment
[10]	A one-way violation detection	TensorFlow object detection model	Three-wheeled vehicles	Supervised learning	**Pros:** This method is cost-effective and easy to deploy **Cons**: this method works only for one-way traffic areas

All the methods we have studied give good results for the detection of violations in the roads, based on the disadvantages of the used methods we can conclude that most of the algorithms based on deep learning (supervised), require an increase of the training database, and for the other algorithms which are based on background subtraction it is necessary to always take into consideration the change of lighting, the shadow detection and the intersection of objects.

4 Conclusion and Future Work

In this research paper, we have presented an analysis study of different techniques dedicated to resolve traffic violation issues based on image processing and different artificial intelligence models. To automatically detect the road violations, we introduced the most recent systems apt to detect all types of traffic offences, through combining the best-proposed solutions that gives the most efficient results. This paper aims to provide researchers with the preliminary knowledge needed to develop robust approaches. And therefore, ameliorate different features or build new algorithms with innovative ideas, additionally improve both the accuracy and the computation time. The Future work consists of developing some ameliorated approaches based on artificial intelligence algorithms to detect vehicles that cross the stop signs further the duplication of vehicles in a banned place taking into consideration all real lighting conditions.

References

1. Mbarek, A., Jiber, M., Yahyaouy, A., Sabri, A.: Road traffic mortality in Morocco: analysis of statistical data. In: 2020 International Conference on Intelligent Systems and Computer Vision (ISCV), Fez, Morocco (2020)
2. Saha, S., Basu, S., Nasipuri, M., Basu, D.K.: Development of an automated Red Light Violation Detection System (RLVDS) for Indian vehicles. In: IEEE National Conference on Computing and Communication Systems, Burdwan (2009)
3. Wonghabut, P., Ung-arunyawee, R., Satiennam, T.: Traffic light color identification for automatic traffic light violation detection system. In: IEEE International Conference on Engineering, Applied Sciences, and Technology, Phuket, Thailand (2018)

4. Chen, Y., Yang, C.: Vehicle red-light violation detection base on region. In: 3rd International Conference on Computer Science and Information Technology, Chengdu, China (2010)
5. Bordia, B., Nishanth, N.: Automated traffic light signal violation detection system using convolutional neural network. In: Soft Computing: Theories and Applications, pp. 579–592 (2020)
6. Prakash, N., Kamath, C.U., Gururaja, H.S.: Detection of traffic violations using moving object and speed detection. Wutan Huatan Jisuan Jishu **XVI**(VI), 1001–1749 (2020)
7. de Goma, J.C., Bautista, R.J., Eviota, M.A.J.: Detecting Red-Light Runners (RLR) and speeding violation through video capture. In: IEEE 7th International Conference on Industrial Engineering and Applications, Bangkok, Thailand (2020)
8. Franklin, R.J., Mohana: Traffic signal violation detection using artificial intelligence and deep learning. In: Proceedings of the Fifth International Conference on Communication and Electronics Systems. IEEE, Coimbatore (2020)
9. Peiris, A., Edirisuriya, E.A.T.A., Athuraliya, C.D.: Computer vision based approach for traffic violation detection. In: 13th International Research Conference, Sri Lanka (2020)
10. Mampilayil, H.R., Rahamathullah, K.: Deep learning-based detection of one-way traffic rule violation of three-wheeler vehicles. In: Proceedings of the International Conference on Intelligent Computing and Control Systems. IEEE, Madurai (2019)
11. Aqel, S., Hmimid, A., Sabri, M.A., Aarab, A.: Road traffic: vehicle detection and classification. In: 2017 Intelligent Systems and Computer Vision (ISCV), Fez, Morocco (April 2017)
12. Coco Dataset Dataset. http://cocodataset.org
13. Behrendt, K., Novak, L., Botros, R.: A deep learning approach to traffic lights: detection, tracking, and classification. In: 2017 IEEE International Conference on Robotics and Automation (ICRA), pp. 1370–1377 (May 2017)
14. Heidelberg Collaboratory: Bosch Small Traffic Dataset (2017). https://hci.iwr.uni-heidelberg.de
15. Lee, Y., et al.: Fast detection of objects using a YOLOv3 network for a vending machine. In: IEEE International Conference on Artificial Intelligence Circuits and Systems (AICAS), Taiwan (2019)
16. Cucchiara, R., Grana, C., Piccardi, M.: Detecting objects, detecting objects, shadows, and ghosts in video streams by exploiting color and motion information. In: International Conference on Image Analysis and Processing (ICIAP) (2001)
17. Chengzhong, Y., Jun, Z., Yuan, X.: Video object detection based on background subtraction. J. Southeast Univ. (Nat. Sci. Ed.) **35**(Sup (II)) (2005)

Mechatronic, Robotic and Control System

Improvement of DTC Performance of Three Level Inverter Fed IM Drive with High Gain Flux Observer

Salma Jnayah(✉) and Adel Khedher

LATIS - Laboratory of Advanced Technology and Intelligent Systems, Université de Sousse, Ecole Nationale d'Ingénieurs de Sousse, 4023 Sousse, Tunisie
salmajnayah@gmail.com

Abstract. The present paper proposes an enhanced direct torque control (DTC) strategy for induction machine (IM) drive associated with a three-level neutral clamped (NPC) inverter. To ensure high-performance of the control scheme, accurate knowledge of flux, current, and speed values is crucial. However, some state variables, namely stator fluxes, are not accessible for measurement. To solve this problem, the high gain observer (HGO) has been integrated. This observer allows estimating the stator flux based on directly measured quantities (stator voltage and current) and also provides a robust control against machine parameter variation. The global control algorithm of the DTC with HGO has been designed and investigated via numerical simulations with Matlab/Simulink software.

Keywords: DTC · Three-level inverter · HGO · Parameter variation

1 Introduction

Direct Torque Control strategy was introduced and developed in the 1980s to provide a decoupled command of both flux and electromagnetic torque [1]. Thereafter, it has aroused the interest of many researchers because of its simple structure and fast dynamic response covering a large speed range [2]. It has also gained widespread use in academia as well as in industrial applications mainly in electric drive [3]. Although DTC shows good performances in IM drive applications, it suffers from significant torque and flux ripples caused by hysteresis controllers. Which can produce an annoying audible noise and degrade the efficiency of the controlled machine. Hence, many endeavors have been made to improve the dynamics of this technique. This paper is then dedicated to presenting an assay to improve the classical DTC strategy.

As reported in the literature, multiple attempts tried to handle the conventional DTC deficiencies such as the development of this approach with invariable switching frequency so that ripples can be tremendously diminished [4]. Similarly, authors in [5] treated the DTC_SVM strategy where hysteresis controllers are substituted by two classical PI regulators to generate the reference voltage vectors VSα and VSβ. Some other

S. Motahhir and B. Bossoufi (Eds.): ICDTA 2022, LNNS 454, pp. 785–794, 2022.
https://doi.org/10.1007/978-3-031-01942-5_78

methods for conventional DTC enhancement are based on intelligent techniques, principally neural networks and fuzzy logic. In this case, stator flux and torque hysteresis controllers are replaced by intelligent regulators [6]. Recently, some research work treating DTC of IM drive has been developed using multi-level inverters since they are considered as a promising approach in obtaining high power and voltage ratings. Furthermore, multilevel voltage source inverters (VSI) have the advantage of improving the harmonic profiles of output waveforms in terms of torque ripples reduction and flux smoothing [7]. Among the different commercially deployed inverter topologies, the three-level inverter with NPC structure seems to be the most commonly used in electric drive applications [8, 9]. Various work has been reported in the literature for DTC of the three-level VSI supplying IM. In [10], the authors presented a 7-level torque comparator to decrease undesirable undulations in the DTC of the association three-level five phases inverter/five-phase asynchronous machine. Simulation and practical results are exhibited to substantiate the proposed approach. Likewise, the association of a three-level VSI with a DTC technique based on a fuzzy PI controller is studied in this paper [11]. The regulator parameters are adjusted in a manner that the developed controller can adjust itself to diverse operating situations. In the DTC technique, the torque may be evaluated using flux and stator currents [12]. However, this latter cannot be directly measured. Consequently, in the last decade, numerous research works have treated the development of DTC based on observers to construct the stator flux. Another aim of inserting the closed-loop flux estimators is the sensitivity of the DTC strategy to the variance of the motor parameters mainly the stator resistance [13]. Many algorithms are available such as Extended Kalman Filter (EKF). In [14], load torque, motor speed, and stator resistance are esteemed based on EKF observer to enhance the motor performance based on DTC but empirical results show that EKF is unable to perform properly under a load [15]. Luenberger and sliding-mode observers are also utilized for estimating the states of IM [5]. They exhibit good robustness against the machine parameter variance, namely mutual inductance and stator resistance. These observers are efficient for low extraneous noise, but ripples in flux and torque responses tend to be greater as well as the noise increases [16].

In this work, simulation results are dedicated to investigating the DTC performance built on HGO, devoted to a three-level NPC converter supplying an IM drive. This observer is employed for stator flux and currents estimation based on measured quantities and to strengthen the control strategy against load torque and stator résistance variation.

The paper is structured in the following erection: after an introduction of the basic principles of DTC strategy of three-level inverter/IM drive in Sect. 2, the mathematical modeling of the high gain observer is exposed in Sect. 3. In Sect. 4, simulation results are exposed to evaluate the HGO performances. The conclusions are pondered in the last section.

2 DTC of IM Drive Dedicated to Three Level NPC Inverter

2.1 IM Mathematical Modeling

In the stationary frame (α, β), the state equivalent modeling of IM is expressed as:

$$\begin{cases} \dot{X} = (A_1)X + (B_1)V \\ Y_1 = C_1 X \end{cases} \quad (1)$$

X, V and Y are respectively the state, control and output vectors defined as:

$$X = \begin{bmatrix} i_{S\alpha} \\ i_{S\beta} \\ \varphi_{S\alpha} \\ \varphi_{S\beta} \end{bmatrix}, \; V = \begin{bmatrix} V_{S\alpha} \\ V_{S\beta} \end{bmatrix}, \; Y_1 = \begin{bmatrix} i_{S\alpha} \\ i_{S\beta} \end{bmatrix}, \; (B_1) = \begin{pmatrix} \frac{1}{\sigma L_S} & 0 \\ 0 & \frac{1}{\sigma L_S} \\ 1 & 0 \\ 0 & 1 \end{pmatrix}, \; C_1 = \begin{pmatrix} 1\,0 \\ 0\,1 \\ 0\,0 \\ 0\,0 \end{pmatrix}$$

$$[A_1] = \begin{pmatrix} -(\frac{R_s}{\sigma * R_s} + \frac{R_r}{\sigma * L_r}) & -\omega & \frac{R_r}{\sigma * L_s * L_r} & \frac{\omega}{\sigma * L_s} \\ \omega & -(\frac{R_s}{\sigma * R_s} + \frac{R_r}{\sigma * L_r}) & \frac{\omega}{\sigma * L_s} & \frac{R_r}{\sigma * L_s * L_r} \\ -R_s & 0 & 0 & 0 \\ 0 & -R_s & 0 & 0 \end{pmatrix}$$

Where σ denotes the Blondel coefficient determined as:

$$\sigma = 1 - \frac{M_{sr}^2}{L_s L_r} \quad (2)$$

$$\omega = p\Omega \quad (3)$$

where Ω is the mechanical motor speed.

The mechanical expression of the asynchronous machine is determined in the following formulation:

$$J\frac{d}{dt}\omega = T_{em} - fc\omega - T_r \quad (4)$$

with T_r is the load torque.

2.2 Direct Torque Control Principle

The DTC relied on the direct determination of, appropriately chosen, switching sequences of the voltage inverter feeding the asynchronous machine. This control strategy is built on hysteresis comparators: one for the stator flux and the other for the electromagnetic torque. These quantities had to be maintained within hysteresis limits. These controllers output on the one hand, and the stator flux vector stance, on the other hand, allow selecting the voltage vector that should be stratified to the machine, and consequently, the combination of control signals required for the power switches of the inverter. The elementary units needed for DTC algorithm implementation are illustrated in Fig. 1.

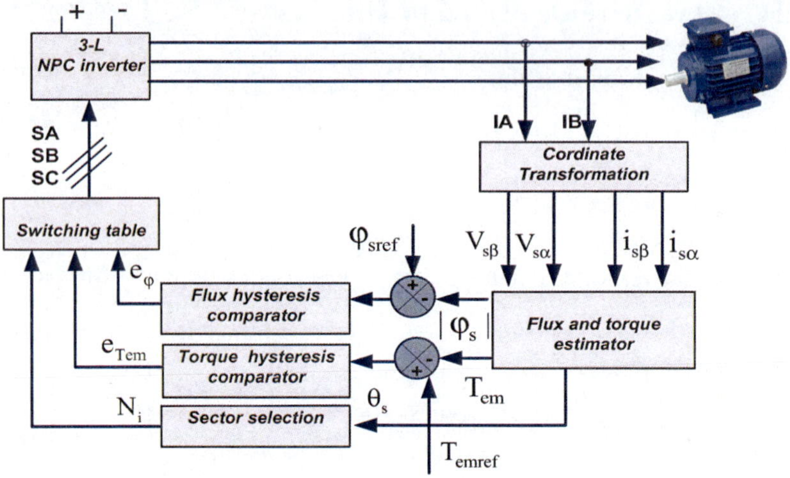

Fig. 1. DTC of IM controlled by 3-L inverter.

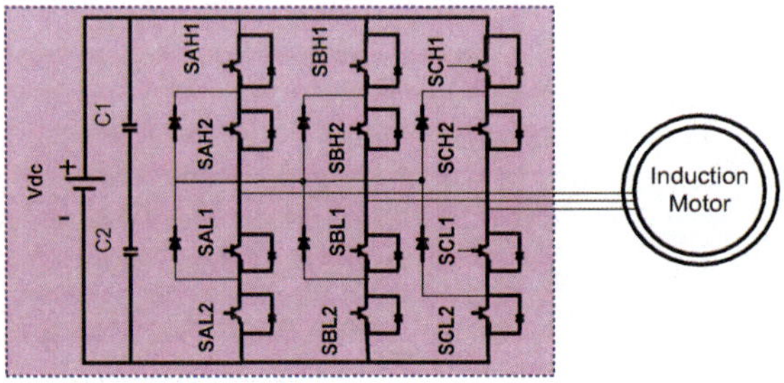

Fig. 2. IM/three-level VSI association.

Three-Level Inverter. Figure 2 shows the connections of the 3-L NPC inverter controlling the IM.

The NPC three-level inverter has the advantage of providing less distortions in the output waveforms due to the greater number of levels in comparison with the typical VSI [17]. Hence, the control system will be enhanced and its wave shapes will be better. There are three states in each phase. Consequently, a three level VSI generates 3^3 active vectors. The expression of the VSI voltage vector can be written as follows:

$$V_s = \sqrt{\frac{2}{3}} V_{dc}(S_a + S_b.e^{j\frac{2*\pi}{3}} + S_c.e^{j\frac{4*\pi}{3}}) \tag{5}$$

where $\overline{V}_s = V_{s\alpha} + jV_{s\beta}$ and S_A, S_B, S_C are control variables.

Flux and Torque Estimator. The stator flux is expressed as:

$$\varphi_{s\alpha} = \int (V_{s\alpha} - R_s * i_{s\alpha})\,dt \text{ and } \varphi_{s\beta} = \int (V_{s\beta} - R_s * i_{s\beta}) \tag{6}$$

Then the corresponding flux module and angle are easily calculated.

$$\begin{cases} |\varphi_s| = \sqrt{\varphi_{s\alpha^2} + \varphi_{s\beta^2}} \\ \theta_s = \arg(\varphi_s) \end{cases} \quad (7)$$

The estimation of the torque is formulated based on Eq. (8):

$$T_{em} = \frac{2}{3} p(\varphi_{s\alpha} * i_{s\beta} - \varphi_{s\beta} * i_{s\alpha}) \quad (8)$$

Switching Table. As illustrated in the DTC scheme, the adequate voltage vector choice is explained in the table hereunder. It leans on SECTORi, the error of stator flux eφ, and the error of torque eTem. These errors are the hysteresis comparators outputs (Fig. 3) (Table 1).

Table 1. The switching table

eφ	1		− 1	
eTem	1	−1	1	−1
SECTOR 1	V16	V26	V23	V19
SECTOR 2	V22	V15	V18	V25
SECTOR 3	V17	V21	V24	V20
SECTOR 4	V23	V16	V19	V26
SECTOR 5	V18	V22	V25	V15
SECTOR 6	V24	V17	V20	V21
SECTOR 7	V19	V23	V26	V16
SECTOR 8	V25	V18	V15	V22
SECTOR 9	V20	V24	V21	V17
SECTOR 10	V26	V19	V16	V23
SECTOR 11	V15	V25	V22	V18
SECTOR 12	V21	V20	V17	V24

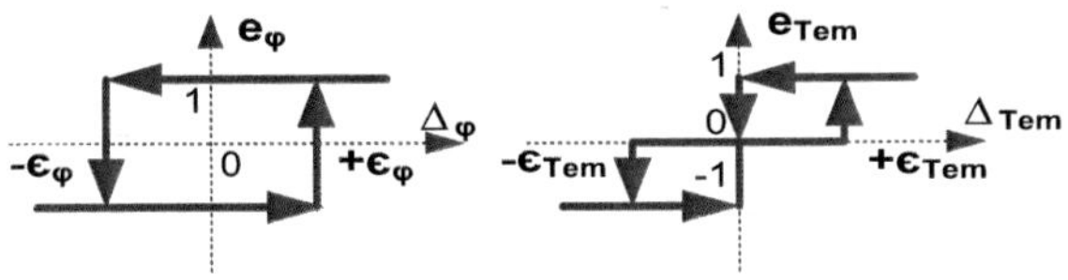

Fig. 3. Hysteresis flux and torque regulators.

3 High Gain Flux Observer

The high Gain Observer technique, which is considered one of the most suitable competitors in the state estimation problem, has attracted the interest of many researchers for sensorless IM drives. In the absence of external noise, the HGO can reconstruct quickly the system state variables and discard the incertitude of its modeling. The design of the HGO is inherently simple. Its adjustment is realized based on a single parameter. For more research papers related to HGOs, consult [18] and the references therein.

The developed HGO has the objective of constructing the stator flux components $\varphi_{s\alpha}$ and $\varphi_{s\beta}$. They are estimated from the directly measured quantities via sensors (stator currents and voltages). The rotor speed is also established based on observed variables applying the mechanical equation of the machine. The principle diagram of the HGO is given by the following figure (Fig. 4):

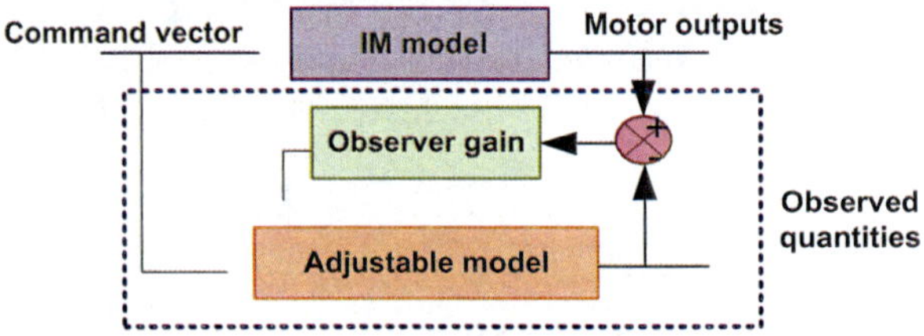

Fig. 4. HGO design.

The equation of the observed IM is written as follows:

$$\begin{cases} \frac{d}{dt}\hat{X} = [A]\hat{X} + [B]U + v(e) \\ v(e) = \theta\Lambda^{-1}(\hat{X})\Delta^{-1}S^{-1}C^{T}Ce \end{cases} \tag{9}$$

where e and $\hat{X}$ denote respectively the vector of stator current error and the observed state vector.

$$e = \left[i_{s\alpha} - \hat{i}_{s\alpha} \; i_{s\beta} - \hat{i}_{s\beta} \; 0 \; 0 \right]^{T} = \left[e_1 \; e_2 \; 0 \; 0 \right]^{T} \tag{10}$$

$$\Delta = \mathrm{diag}(I_2, \frac{1}{\theta} I_2) \text{ or } \theta > 0 \text{ so } \Delta^{-1} = \mathrm{diag}(I_2, \frac{1}{\theta} I_2) \tag{11}$$

$$\Lambda = \mathrm{diag}(I_2, \kappa F(\omega)) \text{ or } \theta > 0 \text{ so } \Lambda^{-1} = \mathrm{diag}(I_2, \frac{1}{\kappa} F^{-1}(\omega)) \tag{12}$$

Where $\kappa = \frac{1}{\sigma L_s}$ and $F(\omega) = \begin{bmatrix} \frac{R_r}{L_r} & \omega \\ -\omega & \frac{R_r}{L_r} \end{bmatrix}$.

S represents the peculiar solution of Lyapunov function:

$$\begin{cases} S_L + A^{T} * S_L + S * A = C^{T} * C \\ S^{-1}C^{T} = \left[2I_2 \; I_2 \right]^{T} \end{cases} \tag{13}$$

$$v(e) = \theta \begin{bmatrix} I_2 & 0 \\ 0 & \frac{F^{-1}(\omega)}{\kappa} \end{bmatrix} \begin{bmatrix} I_2 & 0 \\ 0 & \theta I_2 \end{bmatrix} \begin{bmatrix} 2I_2 \\ I_2 \end{bmatrix} \begin{bmatrix} I_2 & 0 \end{bmatrix} e \tag{14}$$

when replacing $v(e)$ with its expression in Eq. (9) we obtain the following system:

$$\begin{cases} \frac{d}{dt}\hat{i}_{s\alpha} = \left(\frac{R_s}{\sigma L_s} + \frac{R_r}{\sigma L_r}\right)\hat{i}_{s\alpha} - \hat{\omega}\hat{i}_{s\beta} + \frac{R_r}{\sigma L_r L_s}\hat{\varphi}_{s\alpha} + \frac{\hat{\omega}}{\sigma L_s}\hat{\varphi}_{s\beta} + \frac{1}{\sigma L_s}V_{s\alpha} + 2\theta e_1 \\ \frac{d}{dt}\hat{i}_{s\beta} = \hat{\omega}\hat{i}_{s\alpha} + \frac{1}{\sigma}\left(\frac{R_s}{L_s} + \frac{R_r}{L_r}\right)\hat{i}_{s\beta} - \frac{\hat{\omega}}{\sigma L_s}\hat{\varphi}_{s\alpha}\frac{R_r}{\sigma L_r L_s}\hat{\varphi}_{s\beta} + \frac{1}{\sigma L_s}V_{s\beta} + 2\theta e_2 \\ \frac{d}{dt}\hat{\varphi}_{s\alpha} = V_{s\alpha} - R_s\hat{i}_{s\alpha} + \theta^2\frac{a_1}{a_2^2+a_2^2}e_1 - \theta^2\frac{a_2}{a_1^2+a_2^2}e_2 \\ \frac{d}{dt}\hat{\varphi}_{s\beta} = V_{s\beta} - R_s\hat{i}_{s\beta} + \theta^2\frac{a_2}{a_1^2+a_2^2}e_1 - \theta^2\frac{a_1}{a_1^2+a_2^2}e_2 \end{cases} \tag{15}$$

Where $a_1 = \frac{R_r}{\sigma L_r L_s}$ and $a_1 = \frac{\omega}{\sigma L_s}$, $\theta > 1$ is the parameter of adjustment of the observer dynamics which is chosen arbitrarily.

4 Simulation Results and Discussion

The above-presented design is simulated with Matlab-Simulink Software. The simulation results were achieved for 1.5 kW IM with parameters addressed in the appendix (Fig. 5).

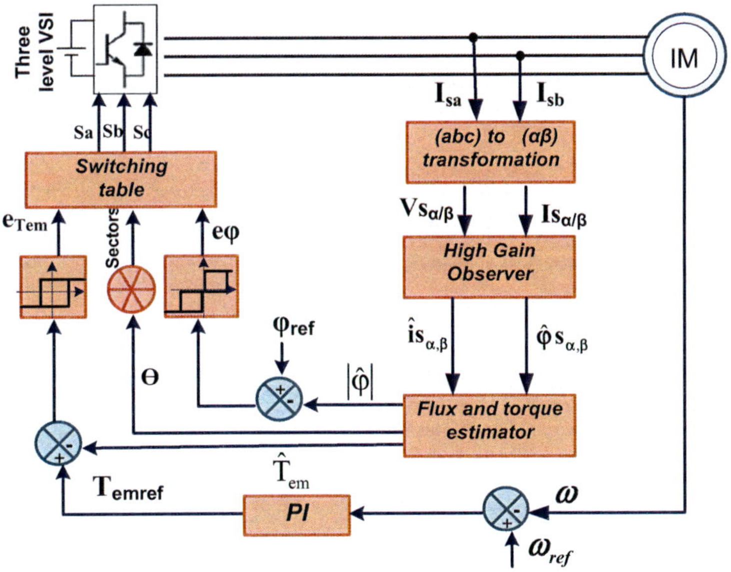

Fig. 5. Block diagram suggested DTC built on HGO.

To study the efficiency of the developed observer with the DTC strategy, simulation results were conducted using Matlab/Simulink software. The motor speed and flux references are respectively 157 rad/s and 0.91 wb. At t = 0.5 s, Rs is increased by 200% of its nominal value and at t = 1 s we inserted a sudden load change of 10 N.m. the motor speed is ensured by a PI regulator.

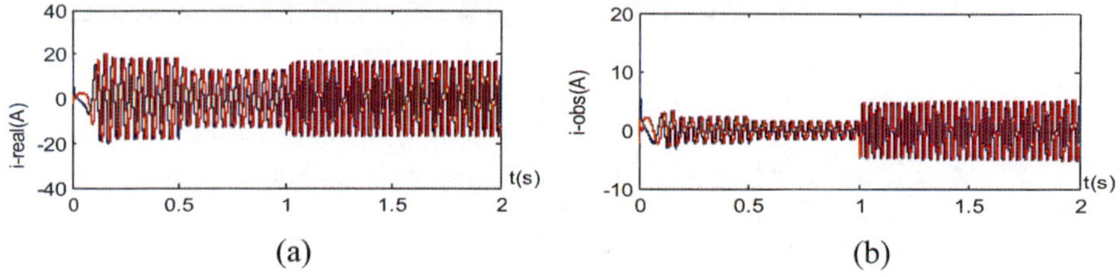

Fig. 6. Variation of stator current: (a) real stator current, (b) observed stator current

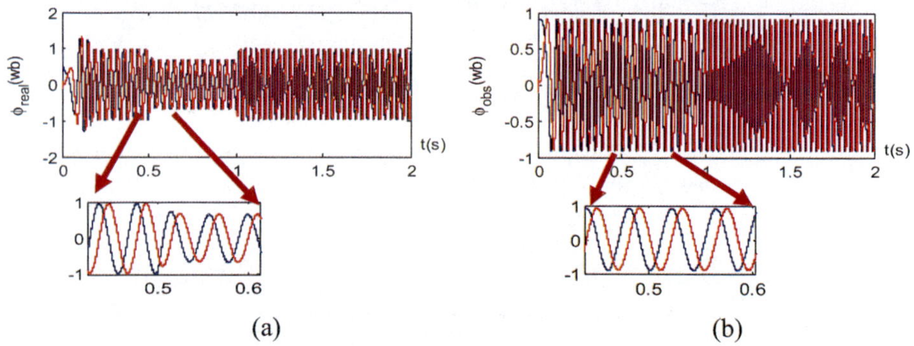

Fig. 7. Stator flux evolution: (a) real flux, (b) observed flux

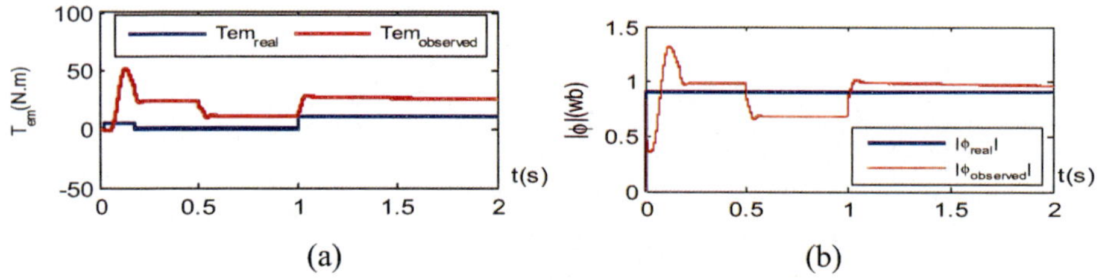

Fig. 8. Actual and estimated variables: (a) electromagnetic torque, (b) flux magnetude.

Figure 6 shows the waveforms of both real and observed stator current. The parameter variations led to undesirable effects on the control strategy as well as on the HGO observer. However, the deflection is less abrupt for observed currents. In Fig. 7(a), we can perceive at t = 0.5 s the real flux is influenced by stator resistance variation. It decreases sharply from 0.91 wb to 0.67 wb, albeit the observed flux kept its shape around its nominal value. Figures 8. a and 8. b illustrate respectively the electromagnetic torque and the flux module waveforms of both real and observed quantities (blue profile represents the observed quantities while the red line is dedicated to real ones). As it

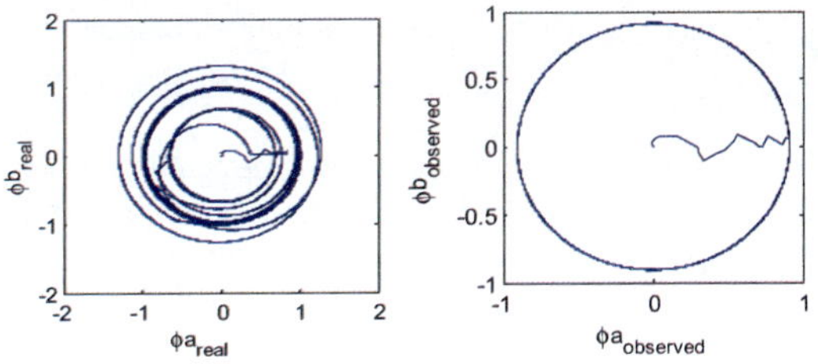

Fig. 9. Real and observed stator flux trajectories

can be noticed, waveforms of observed variables follow perfectly their references while the real ones are noticeably affected by the parameter variations. In Fig. 9, the real and observed flux shapes are conferred. It is remarkable that the observed flux trajectory has almost a circular form when compared to the real one.

5 Conclusion

This paper is devoted to developing the Direct Torque Control of IM controlled with a three-level inverter using High Gain Observer. The proposed closed-loop estimator has been suggested to reconstitute the stator flux since it is not accessible for measurement. The advocated observer is also utilized to enhance the control technique's robustness against parameter variations. The simulations results using Matlab/Simulink software have demonstrated that the suggested observation approach has better performances which are noticeable especially in the stator flux curves.

Appendix

The IM parameters, used for simulating the control strategy behavior are given below: Motor speed = 150 rad/s, Pole pair numbers = 2, Resistance of the stator Rs = 5.72 Ω, Resistance of the rotor Rr = 4.28 Ω, Stator and rotor inductance Ls = Lr = 0.464 H, Mutual inductance Msr = 0.44 H, Viscous friction coefficient fc = 0.002, Moment of inertia J = 0.0049 kg.m^2.

References

1. Takahashi, I., Noguchi, T.: A new quick-response and high-efficiency control strategy of an induction motor. IEEE Trans. Ind. Appl. **5**, 820–827 (1986)
2. Gallah, T., Bouzidi, B., Masmoudi, A.: Synthesis and performance investigation of a 30-degree bus-clamping based DTC strategy aimed to three-level inverter fed IM drives. In: 11th International Conference on Ecological Vehicles and Renewable Energies, Monte Carlo, Monaco, pp. 1–12 (2016)
3. Douib, L., Bencheikh, F., Metidji, B., Kheldoun, A.: FPGA-based DTC method for three level inverter-fed induction motor speed. In: 5th International Conference on Electrical Engineering-Boumerdes, Boumerdes, Algeria, pp. 1–5 (2017)

4. Krim, S., Gdaim, S., Mtibaa, A., Mimouni, M.F.: Implementation on the FPGA of DTC-SVM based proportional integral and sliding mode controllers of an induction motor: a comparative study. J. Circuits Syst. Comput. **26**(03), 1750049 (2017)
5. Ammar, A., Bourek, A., Benakcha, A.: Robust SVM-direct torque control of induction motor based on sliding mode controller and sliding mode observer. Front. Energy **14**(4), 836–849 (2017). https://doi.org/10.1007/s11708-017-0444-z
6. El Ouanjli, N., Motahhir, S., Derouich, A., El Ghzizal, A., Chebabhi, A., Taoussi, M.: Improved DTC strategy of doubly fed induction motor using fuzzy logic controller. Energy Rep. **5**, 271–279 (2019)
7. Jnayah, S., Khedher, A.: DTC of induction motor drives fed by two and three-level inverter: modeling and simulation. In: 19th International Conference on Sciences and Techniques of Automatic Control and Computer Engineering, Tunisia, pp. 376–381 (2019)
8. Tang, Q., Ge, X., Liu, Y.C., Hou, M.: Improved switching-table-based DTC strategy for the post-fault three-level NPC inverter-fed induction motor drives. IET Electr. Power Appl. **12**(1), 71–80 (2018)
9. Kiran, T.V., Amarnath, J.A.: Sliding mode controller based DTC of 3 level NPC inverter fed induction motor employing space vector modulation technique. In: International Conference on Advances in Engineering, Science and Management, pp. 372–377 (2012)
10. Tatte, Y.N., Aware, M.V.: Torque ripple and harmonic current reduction in a three-level inverter-fed direct-torque-controlled five-phase induction motor. IEEE Trans. Industr. Electron. **64**(7), 5265–5275 (2017)
11. Jnayah, S., Khedher, A.: Fuzzy-self-tuning PI speed regulator for DTC of three-level inverter fed IM. In: 17th International Multi-Conference on Systems, Signals & Devices, Tunisia, pp. 709–714 (2020)
12. Pan, Y., Chen, T.: A high-order sliding mode stator flux observer for asynchronous motor based on feedback linearization. In: 12th International Conference on Fuzzy Systems and Knowledge Discovery, China, pp. 1514–1518 (2015)
13. Krim, S., Gdaim, S., Mtibaa, A., Mimouni, M.F.: FPGA-based implementation direct torque control of induction motor. Int. J. Power Electron. Drive Syst. **5**(3), 293 (2015)
14. Usta, M.A., Okumus, H.I., Kahveci, H.A.: Simplified three-level SVM-DTC induction motor drive with speed and stator resistance estimation based on extended Kalman filter. Electr. Eng. **99**(2), 707–720 (2017)
15. Zhang, Y., Zhao, Z., Lu, T., Yuan, L., Xu, W., Zhu, J.: A comparative study of Luenberger observer, sliding mode observer and extended Kalman filter for sensorless vector control of induction motor drives. In: Energy Conversion Congress and Exposition, pp. 2466–2473 (2009)
16. Ouboubker, L., Khafallah, M., Lamterkati, J.: DTC two level and three level inverters of induction motors with speed control using fuzzy logic controller. Int. J. Enhanc. Res. Sci. Technol. Eng. **3**(3), 284–295 (2014)
17. Kadrine, A., Tir, Z., Malik, O.P., Hamida, M.A., Reatti, A., Houari, A.: Adaptive non-linear high gain observer based sensorless speed estimation of an induction motor. J. Franklin Inst. **357**(13), 8995–9024 (2020)
18. Astolfi, D., Marconi, L.: High-gain nonlinear observer with limited gain power. IEEE Trans. Autom. Control **60**(11), 3059–3064 (2015)

Model Reduction in Takagi Sugeno Systems: An LMI Method

Zineb Lahlou(✉), Abderrahim El-Amrani, and Ismail Boumhidi

LISAC Lab, SMBA University, Fez, Morocco
zineblalou@gmail.com

Abstract. The article is interested MR problem for continuous T-S systems. Sufficient conditions are then established for the error system (ES) are asymptotically stable (AS) has L2-gain performance. An example is provided to check effectiveness of the derived results.

Keyword: MR · LMI · Takagi Sugeno fuzzy scheme · H∞ performance

1 Introduction

Model reduction has received a lot of attention. The goal of such a problem is to design an inferior system to approximate that receives a high system based on specified criteria. Important results in the LMI method have already been reported in [1–12].

The practical systems are often disturbed by the noise signals, therefore, the MR issues for nonlinear systems have been studied extensively by the fuzzy model approach. In addition, in recent years, many results based on the fuzzy approach have been presented [8, 10–14].

This article has been proposed for MR design, with which the ES is AS and has a L2-gain performance. An explicit parameterization of the desired MR models are given. An example is presented to check effectiveness of the derived results.

2 System Description

Consider the nonlinear systems:

Rule r: if $\Phi_1(t)$ is $T_1^i \dots \Phi_p(t)$ is T_p^i then

$$\begin{cases} j(t) = A_i j(t) + B_i l(t) \\ \quad o(t) = C_i j(t) \end{cases} \tag{1}$$

With T_j^i, j = 1, 2, ..., p, i = 1, 2, ..., s; $x(.)$ is state vector; $y(.)$ is output vector.

S. Motahhir and B. Bossoufi (Eds.): ICDTA 2022, LNNS 454, pp. 795–801, 2022.
https://doi.org/10.1007/978-3-031-01942-5_79

A_i, B_i, C_i are known matrices of appropriate dimensions; l(t) is belong to $l_2\{[0, \infty), [0, \infty)\}$.

And

$$\zeta_i \geq 0,\ i = 1, \ldots ., s,\ \sum\nolimits_{v=1}^{s} \zeta_i = 1 \tag{2}$$

Then, we have:

$$\begin{cases} j(t) = A(\zeta)j(t) + B(\zeta)l(t) \\ o(t) = C(\zeta)j(t) \end{cases} \tag{3}$$

With

$$\begin{cases} A(\zeta) = \sum\nolimits_{v=1}^{s} \breve{\zeta}_i A_i \\ B(\zeta) = \sum\nolimits_{i=1}^{r} \breve{\zeta}_i B_i \\ C(\zeta) = \sum\nolimits_{v=1}^{s} \breve{\zeta}_i C_i \end{cases} \tag{4}$$

The work, a MR model as

$$\begin{cases} \dot{\breve{j}}(t) = \breve{D}(\breve{\zeta})\breve{j}(t) + \breve{M}(\breve{\zeta})\breve{l}(t) \\ \breve{o}(t) = \breve{N}(\breve{\zeta})\breve{j}(t) \end{cases} \tag{5}$$

With

$$\breve{D}\left(\breve{h}\right) = \sum\nolimits_{v=1}^{s} \breve{\zeta}_i \breve{D}_i,\ \breve{M}\left(\breve{h}\right) = \sum\nolimits_{i=1}^{r} \breve{\zeta}_i \breve{M}_i,\ \breve{N}\left(\breve{h}\right) = \sum\nolimits_{i=1}^{r} \breve{\zeta}_i \breve{N}_i \tag{6}$$

Augmenting the MR states with system (6), the following error system (ES) can then be obtained:

$$\begin{cases} e(t) = \overline{P}\left(\zeta, \breve{\zeta}\right)e(t) + \overline{Q}\left(\zeta, \breve{\zeta}\right)l(t) \\ \mu(t) = \overline{R}\left(\zeta, \breve{\zeta}\right)e(t) \end{cases} \tag{7}$$

With

$$\bar{P}(\zeta, \breve{\zeta}) = \begin{bmatrix} A(\zeta) & 0 \\ 0 & \breve{D}(\breve{h}) \end{bmatrix},\quad e(t) = \begin{bmatrix} j(t) \\ \breve{j}(t) \end{bmatrix},\quad \bar{Q}(\zeta, \breve{\zeta}) = \begin{bmatrix} B(\zeta) \\ \breve{M}(\zeta) \end{bmatrix}, \mu(t) = o(t) - \breve{o}(t),$$
$$\bar{R}(\zeta, \breve{\zeta}) = \left[C(\zeta) - \breve{N}(\zeta)\right] \tag{8}$$

The MR design for fuzzy models discussed in this article can be worded as follows:

- ES (5) is asymptotically stable (AS).
- Letting $\gamma > 0$, the equation

$$\sup_{0 \neq l(t)} \frac{\|\mu(t)\|_2}{\|l(t)\|_2} \leq \gamma^2 \tag{9}$$

Lemma 1 [15] **.** Let Δ, Φ and Λ with $\Lambda\Lambda\perp = 0$. Then:

(1) $\Delta^{T}\ \Phi\Delta < 0$: $\Lambda\Delta = 0$
(2) $\Lambda^{\perp T}\ \Phi\Lambda^{\perp} < 0$
(3) $\exists Z$: $\Phi + Z\Lambda + \Lambda^{T}Z^{T} < 0$

3 H∞ Model Reduction Analysis

Theorem 1. Assume that the ES (5) is AS has a L2-gain γ. If there exist matrices P, G, F, H satisfying.

$$\emptyset = \begin{bmatrix} \emptyset_3.\emptyset_2\ G\overline{Q}\left(\zeta,\breve{\zeta}\right) - H^T(\zeta) & 0 & \\ \emptyset_3 & \emptyset_4 & \overline{R}^T\left(\zeta,\breve{\zeta}\right) \\ * & \emptyset_5 & 0 \\ * & * & -I \end{bmatrix} < 0 \tag{10}$$

Where

$$\emptyset_1 = -G - G^T$$
$$\emptyset_2 = P - F^T + G\overline{P}\left(\zeta,\breve{\zeta}\right)$$
$$\emptyset_3 = -F\overline{P}\left(\zeta,\breve{\zeta}\right) + \overline{P}^T\left(\zeta,\breve{\zeta}\right)F^T$$
$$\emptyset_4 = F\overline{Q}\left(\zeta,\breve{\zeta}\right) + \overline{P}^T\left(\zeta,\breve{\zeta}\right)H^T(\zeta)$$
$$\emptyset_5 = -\gamma^2 I + H\overline{Q}\left(\zeta,\breve{\zeta}\right) + \overline{Q}^T\left(\zeta,\breve{\zeta}\right)H^T$$

Proof 1. Define

$$\emptyset = \begin{bmatrix} 0 & P & 0 \\ P\ \overline{R}^T\left(\zeta,\breve{\zeta}\right)\overline{R}\left(\zeta,\breve{\zeta}\right) & & 0 \\ 0 & 0 & -\gamma^2 I \end{bmatrix};\ Z = \begin{bmatrix} G \\ F \\ H \end{bmatrix};\ \Lambda = \left[-I\ \overline{P}\left(\zeta,\breve{\zeta}\right)\ \overline{Q}\left(\zeta,\breve{\zeta}\right)\right]$$

By schur complement, (10) is equivalent to

$$\emptyset + Z\Lambda + \Lambda^T Z^T < 0 \tag{11}$$

With

$$\Lambda = \begin{bmatrix} \overline{P}\left(\zeta,\breve{\zeta}\right) & \overline{Q}\left(\zeta,\breve{\zeta}\right) \\ I & 0 \\ 0 & I \end{bmatrix}$$

4 H∞ Model Reduction Design

Parameterize the slack matrices G, F and H as

$$G = \begin{bmatrix} G_1 & EW \\ G_2 & W \end{bmatrix};\ F = \begin{bmatrix} F_1 & EW \\ F_2 & W \end{bmatrix};\ H = \begin{bmatrix} H_1 & 0 \end{bmatrix};\ E = \begin{bmatrix} I & 0 \end{bmatrix}^T \tag{12}$$

Theorem 2. Assume that the ES (5) is as has a L_2-gain γ. If there exist matrices $P_1, P_2, P_3, \widehat{A}_i, \widehat{A}_i, \widehat{B}_i, \widehat{C}_i, G_1, G_2, F_1, F_2, H_1$, and V, satisfying.

$$\tilde{\emptyset}_{ij} = \begin{bmatrix} \tilde{\emptyset}_{11} & \tilde{\emptyset}_{12} & \tilde{\emptyset}_{13} & \tilde{\emptyset}_{14} & \tilde{\emptyset}_{15} & 0 \\ * & \tilde{\emptyset}_{22} & \tilde{\emptyset}_{23} & \tilde{\emptyset}_{24} & \tilde{\emptyset}_{25} & 0 \\ * & * & \tilde{\emptyset}_{33} & \tilde{\emptyset}_{34} & \tilde{\emptyset}_{35} & C_j^T \\ * & * & * & \tilde{\emptyset}_{44} & \tilde{\emptyset}_{45} & -\hat{C}_i^T \\ * & * & * & * & \tilde{\emptyset}_{55} & 0 \\ * & * & * & * & * & -I \end{bmatrix} < 0 \tag{13}$$

where

$$\begin{aligned}
\tilde{\emptyset}_{11} &= -G - G_1^T \\
\tilde{\emptyset}_{12} &= -EW - G_2^T \\
\tilde{\emptyset}_{13} &= P_1 - F_1^T - G_1 A_j \\
\tilde{\emptyset}_{14} &= P_2 - F_2^T - E\widehat{A}_i \\
\tilde{\emptyset}_{15} &= -H_1^T + G_1 B_j + \widehat{B}_i \\
\tilde{\emptyset}_{22} &= -W - W^T \\
\tilde{\emptyset}_{23} &= P_2 - W^T E^T + G_2 A_j \\
\tilde{\emptyset}_{24} &= P_3 - W + \widehat{A}_i \\
\tilde{\emptyset}_{25} &= A_j^T H_1^T + F_1 B_j + E\widehat{B}_i \\
\tilde{\emptyset}_{33} &= F_1 A_j + A_j^T F_1^T \\
\tilde{\emptyset}_{34} &= A_j^T F_2^T + E\widehat{A}_i \\
\tilde{\emptyset}_{35} &= F_1 B_j + E\widehat{B}_i \\
\tilde{\emptyset}_{44} &= \left(\widehat{A}_i + \widehat{A}_i^T\right) \\
\tilde{\emptyset}_{45} &= F_2 B_j + \widehat{B}_i \\
\tilde{\emptyset}_{55} &= \gamma^2 I + H_1 B_j + B_j^T H_1^T
\end{aligned}$$

We have a state space realization for MR (6) with

$$\breve{D}_i = W^{-1}\hat{A}_i,\ \breve{M}_i = W^{-1}\hat{B}_i, \breve{N}_i = W^{-1}\hat{C}_i \tag{14}$$

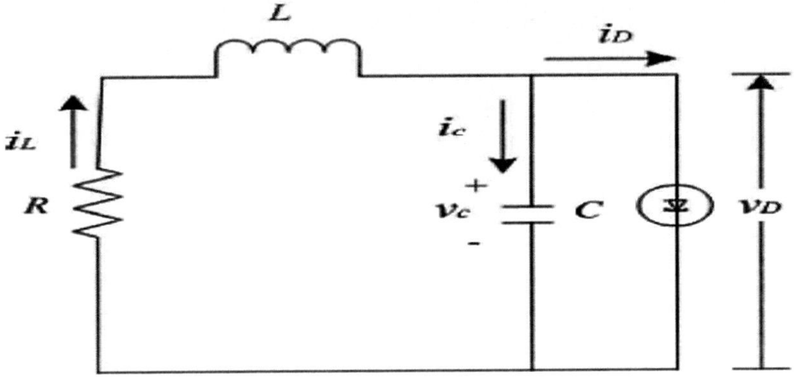

Fig. 1. Tunnel diode circuit.

5 Numerical Example

Suppose that a circuit illustrated in Fig. 1 are modeled as

$$\begin{cases} C\dot{x}_1(t) = -0.002x_1(t) - 0.01x_1^3(t) + x_2(t) \\ L\dot{x}_2(t) = -x_1(t) - Rx_2(t) + l(t) \\ o(t) = x_1(t) \end{cases} \tag{15}$$

With $R_1 = 10\Omega$, L $= 100$ mH, C $= 2$ mF.
The non-linear network model (17) can be approximated as:

$$A_1 = \begin{bmatrix} -0.2 & 48 \\ -1 & -10 \end{bmatrix};\ A_2 = \begin{bmatrix} -4.6 & 50 \\ -0.8 & -09 \end{bmatrix};\ B_1 = B_2 = \begin{bmatrix} 0.2 \\ 0.9 \end{bmatrix};\ C_1 = C_2 = \begin{bmatrix} 0.8\ 0.2 \end{bmatrix} \tag{16}$$

Figure 2 show the plot of the membership method. Via using the MR design (13), the L_2-gain γ are listed in the Table 1.

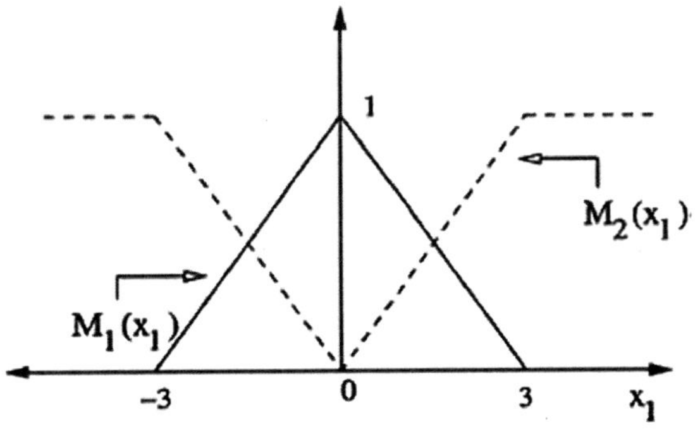

Fig. 2. Membership function.

Table 1. Comparison by different methods

Methods	γ
Corollary [8]	0.6845
Theorem 2	0.1287

From Table 1, we have the proposed method in this article leads to a more conservative filtering performance.

The MR parameters as:

$$\begin{bmatrix} \breve{D}_1 \ \breve{M}_1 \\ \breve{N}_1 \end{bmatrix} = \begin{bmatrix} -0.6125 \ -0.7625 \\ -1.0114 \end{bmatrix} \tag{17}$$

$$\begin{bmatrix} \breve{D}_2 \ \breve{M}_2 \\ \breve{N}_2 \end{bmatrix} = \begin{bmatrix} -0.5978 \ -0.7748 \\ -1.0215 \end{bmatrix} \tag{18}$$

can show the influence of the disturbance $l(t)$ on the ES $\mu(t)$, and the plot of the ratio is shown in Fig. 3 It can be seen that $m(t)$ tend to a value 0.1247, which is less than the prescribed value, i.e., 0.1287.

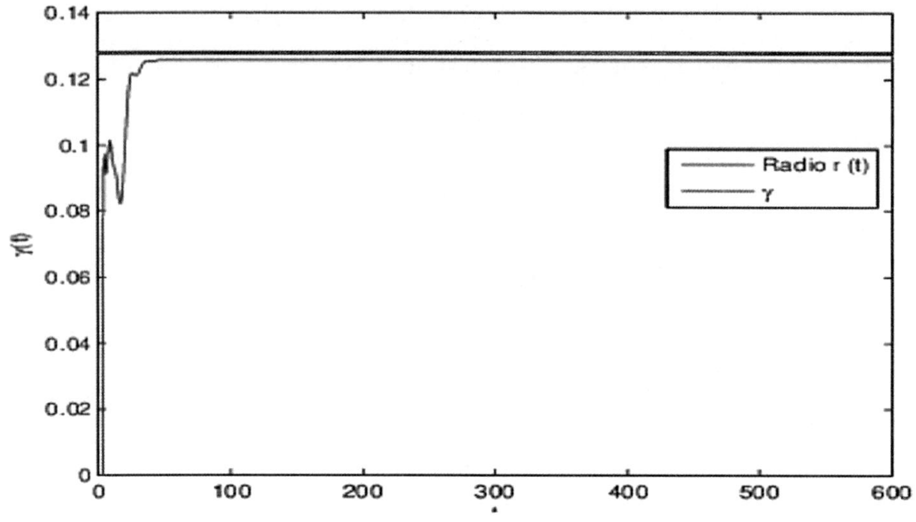

Fig. 3. Values of condition (21)

6 Conclusion

The MR design problem has been concerned for continuous fuzzy systems. Systematic method has been proposed for MR design, with which the ES is AS and has a L2-gain performance. A simulation case further shows validity of the results discussed above. Applying the techniques to MR problem and extending the proposed technique to the fuzzy systems with delays are important and challenging research topics, which deserve more future efforts.

References

1. Wang, Q., Lam, J.: Optimal H2 model reduction for mechanical systems. Int. J. Innov. Comput. Inf. Control **6**(5), 2045–2054 (2010)
2. Xu, S., Chen, T.: H∞ model reduction in the stochastic framework. SIAM J. Control Optim. **42**(4), 1293–1309 (2003)
3. Glover, K.: All optimal Hankel-norm approximations of linear multivariable systems and their L∞ error bounds. Int. J. Control **39**(6), 1115–1195 (1984)
4. Boyd, S., El Ghaoui, L., Feron, E., Balakrishnan, V.: Linear Matrix Inequalities in Systems and Control Theory. SIAM, Philadelphia (1994)
5. Wang, Q., Lam, J., Xu, S., Gao, H.: Delay dependent and delay independent energy-to-peak model approximation for systems with time-varying delay. Int. J. Syst. Sci. **36**(8), 445–460 (2005)
6. Zhang, L., Huang, B., Lam, J.: H∞ model reduction of Markovian jump linear systems. Int. J. Control **78**(15), 1198–1205 (2005)
7. Wu, L., Su, X., Shi, P., Qiu, J.: Model approximation for discrete-time statedelay systems in the T-S fuzzy framework. IEEE Trans. Fuzzy Syst. **19**(2), 366–378 (2011)
8. Su, X., Wu, L., Shi, P., Song, Y.-D.: H∞ model reduction of Takagi-Sugeno fuzzy stochastic systems. IEEE Trans. Systems Man Cybern.-Part B: Cybern. **42**(6), 1574–1585 (2012)
9. Yoo, S.-H., Choi, B.-J.: A balanced model reduction for T-S fuzzy systems with uncertain time varying parameters. In: Zhang, J., He, J.-H., Fu, Y. (eds.) CIS 2004. LNCS, vol. 3314, pp. 148–153. Springer, Heidelberg (2004). https://doi.org/10.1007/978-3-540-30497-5_24
10. Feng, G.: A survey on analysis and design of model-based fuzzy control systems. IEEE Trans. Fuzzy Syst. **14**(5), 676–697 (2006)
11. Kim, E., Lee, H.: New approaches to relaxed quadratic stability condition of fuzzy control systems. IEEE Trans. Fuzzy Syst. **8**(5), 523–534 (2000)
12. Takagi, T., Sugeno, M.: Fuzzy identification of systems and its applications to modeling and control. IEEE Trans. Syst. Man Cybern. **SMC-15**(1), 116–132 (1985)
13. Nguang, S.K., Assawinchaichote, W.: H∞ filtering for fuzzy dynamical systems with D-stability constraints. IEEE Trans. Circuits Syst. I Reg. Pap. **50**(11), 1503–1508 (2003)
14. El-Amrani, A., Boumhidi, I., Boukili, B., Hmamed, A.: A finite frequency range approach to H∞ filtering for TS fuzzy systems. Procedia Comput. Sci. **148**, 485–494 (2019)
15. Gahinet, P., Apkarian, P.: A linear matrix inequality approach to H∞ control. Int. J. Robust Nonlinear Control **4**(4), 421–448 (1994)

Fault Diagnosis of the Bearing Outer Ring of an Induction Motor Under DTC Control by Using Hilbert Filter

Abderrahman El Idrissi(✉), Aziz Derouich, and Said Mahfoud

Industrial Technologies and Services Laboratory, Higher School of Technology, Sidi Mohamed Ben Abdellah University, Fez, Morocco
abderrahman.elidrissi1@usmba.ac.ma

Abstract. Motor Current Signal Analysis (MCSA) has been the most requested method by researchers in recent years for the diagnosis of faults in three-phase Induction Motors (IM). While this method must always involve another field of research for it to be effective, it faces challenges imposed by the evolution of the controls. In this article, it is proposed to adopt this method of detecting Bearing Defects (BD) when the machine is controlled by Direct Torque Control (DTC). Spectral analysis of line current in the case of a direct start, then by the DTC using the Fast Fourier Transform (FFT) alone and with the Hilbert Transform (HT), which is an efficient and preferred tool for the extraction of the signal envelope, The extracted signals will clearly illustrate the interest of HT. This diagnostic system is simulated in MATLAB/ Simulink.

Keywords: Diagnosis · Induction machine · Bearing defects · Direct torque control · Hilbert transform

1 Introduction

The most popular and efficient industrial machines in the industrial sector are the three-phase squirrel-cage Induction Motors (IM) due to their well-known benefits (robustness, low cost, standardization, no sliding electrical contact between stator and rotor) [1–4]. Indeed, research work in recent years has become increasingly interested in this problem of IM fault diagnosis, considering the controls' evolution.

The most common IM faults are mechanical (40%–50%), followed by electrical faults in the stator windings (30%–40%), and finally electrical faults in the rotor cage (5% –10%) [4]. The Bearing Defect (BD) disturbs the machine operation by increasing the torque ripples [5], creating vibrations that can cause other fault types. The DTC is the most demanded control in the industry thanks to these advantages: fast response, direct action on torque, and simplicity. However, its major drawback is torque ripples [6]. Most researchers use vibration analysis to diagnose mechanical faults, but this method has limits in the case of torque variations. In fact, the popular method that makes it possible to extract useful information about this defect type is the MCSA [1, 5]. To

S. Motahhir and B. Bossoufi (Eds.): ICDTA 2022, LNNS 454, pp. 802–812, 2022.
https://doi.org/10.1007/978-3-031-01942-5_80

be effective, the MCSA must meet the requirement to adapt to the evolution of new closed-loop controls, in particular the DTC. This control causes harmonics which are added to those of the stator current, and then the FFT is not sufficient [3–5, 8]. In recent years, several studies who adopt the current signal analysis approach use advanced signal processing methods in the case of non-stationary signals, Abdelkarim et al. [5] complete the FFT with Park Vector Approach Orbital (PVAO) and the Short-Time Fourier Transform (STFT), Gangsar et al. chose the wavelet decomposition, and from 2010 to now, it is difficult to find a study that adopted the HT to diagnose the bearing defects [1], while it's been adopted and it's effective for the broken rotor bar in several research works [3, 4].

The Hilbert filter is the most advantageous when analyzing the Signal Envelope (SE), because it is less sensitive to the variation of the load torque. The envelope of the extracted signal gives the requested information on the low-frequency signal [3, 4]. The objective of this work is to detect BD by MCSA considering the DTC control and based on the HT filter.

This paper will be organized as follows: Sect. 2 introduces the model of IM equations, the DTC is presented in Sect. 3, the HT is addressed in Sect. 5, the specific frequency in the case of inner and outer races is given in Sect. 5, Sect. 6, presents and discusses the results, while Sect. 7 concludes the paper.

2 Modeling of the Three-Phase Squirrel Induction Motor

Given the non-linear behavior and the large number of equations modeling IM in a three-phase frame, all researchers use the famous three-phase-biased transformation, namely the PARK transformation, to model any kind of electric machine in a d, q frame, for different purposes. The IM equations related to the rotating field in the (α, β) frame are expressed as follows [7]:

- Electrical equations:

$$\begin{cases} v_{s\alpha} = R_s.i_{s\alpha} + \dfrac{d\psi_{s\alpha}}{dt} \\ v_{s\beta} = R_s.i_{s\beta} + \dfrac{d\psi_{s\beta}}{dt} \\ 0 = R_r.i_{r\alpha} + \dfrac{d\psi_{r\alpha}}{dt} + \omega_m.\psi_{r\beta} \\ 0 = R_r.i_{r\beta} + \dfrac{d\psi_{r\beta}}{dt} - \omega_m.\psi_{r\alpha} \end{cases} \tag{1}$$

- Magnetic equations:

$$\begin{cases} \psi_{s\alpha} = L_s i_{s\alpha} + L_m.i_{r\alpha} \\ \psi_{s\beta} = L_s i_{s\beta} + L_m.i_{r\beta} \\ \psi_{r\alpha} = L_r i_{r\alpha} + L_m.i_{s\alpha} \\ \psi_{r\beta} = L_r i_{r\beta} + L_m.i_{s\beta} \end{cases} \tag{2}$$

- Mechanical equations:

$$\begin{cases} T_{em} = p.(\psi_{s\alpha} i_{s\beta} - \psi_{s\beta} i_{s\alpha}) \\ J.\frac{d\Omega}{dt} + f.\Omega = T_{em} - T_r \end{cases} \quad (3)$$

3 Direct Torque Control

The Direct torque control (DTC) is proposed by Depenbrock and Takahashi. It is the most popular control method for asynchronous machines in the industry thanks to its main advantages, which are rapid response, direct action on torque, and non-sensitivity to parameter variations. However, its major disadvantage is the ripples of the flux and torque [4, 7, 8] (Fig. 1).

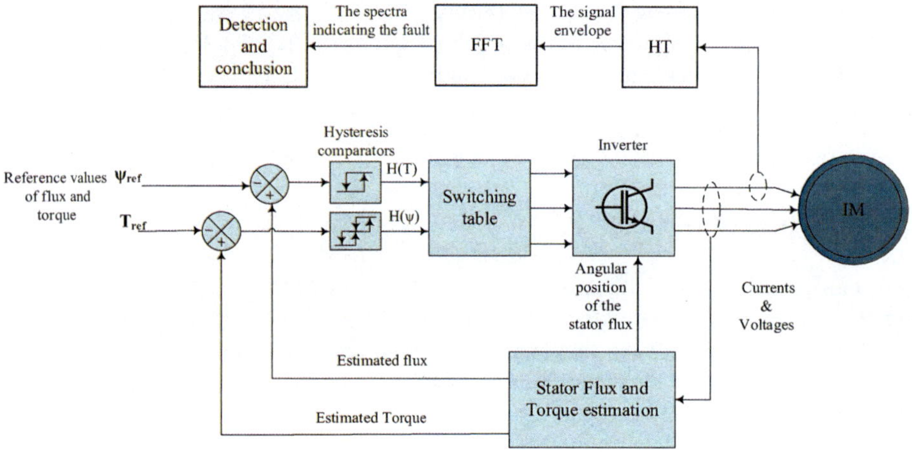

Fig. 1. Direct torque control strategy with detection method by HT.

3.1 Hysteresis Comparators

The two-level hysteresis comparator (Fig. 3b) has the role of maintaining the extremity of each flux vector in a circular ring as shown in Fig. 2. The other comparator, with three hysteresis levels, illustrated in Fig. 3a, makes it possible to control the motor's electromagnetic torque in both rotation directions by developing a positive torque or a negative one.

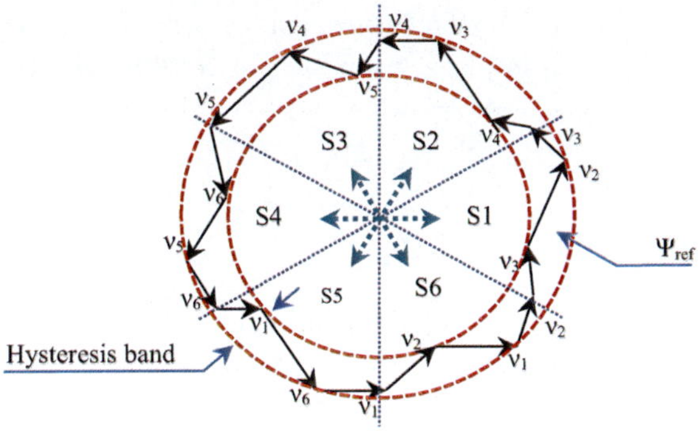

Fig. 2. Flux trajectory

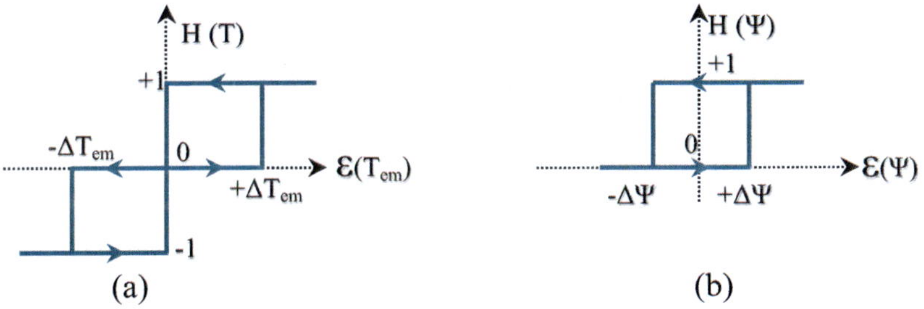

Fig. 3. Hysteresis comparators: (a) of the three-level torque, (b) the two-level flux

3.2 Elaboration of the Switching Table

The voltage vectors are determined as a function of the sector and of the flux and torque, to follow the reference values for flux and torque. Table 1 shows the possibilities for choosing the appropriate voltage vector.

Table 1. Inverter sequences

		Sectors					
H (Ψ)	H (T)	S1	S2	S3	S4	S5	S6
1	1	v_2	v_3	v_4	v_5	v_6	v_1
	0	v_7	v_0	v_7	v_0	v_7	v_0
	−1	v_6	v_1	v_2	v_3	v_4	v_5
0	1	v_3	v_4	v_5	v_6	v_1	v_2
	0	v_0	v_7	v_0	v_7	v_0	v_7
	−1	v_5	v_6	v_1	v_2	v_3	v_4

The ripples caused by the command have an immediate impact on the electric quantities to be processed, which makes the diagnosis more difficult. The analysis of the signal envelope is important in this case. This can be done thanks to the Hilbert transform. [3, 4].

4 Specific Frequency of Bearing Faults

The spectral content of signals is often obtained by the FFT. This transform allows passing from the time domain to the frequency domain by clearly discovering the main spectrum and the harmonic spectra. In the case of the MSCA for BD fault detection. This transforms the harmonic spectra obtained into the frequencies [9]:

4.1 External Ring Defect

It is characterized by the passage frequency of a rolling element on an external ring defect (the outer ring being assumed to be fixed), which is given by the following relation [10]:

$$f_{ex} = \frac{N_b}{2} f_r (1 - \frac{D_b}{D_c} cos\beta) \tag{4}$$

4.2 Internal Ring Defect

The passage frequency of a rolling element on a defect in an internal ring (the internal ring being supposed to be mounted on the rotating shaft) is given by the following relation [10]:

$$f_{int} = \frac{N_b}{2} f_r (1 + \frac{D_b}{D_c} cos\beta) \tag{5}$$

where:

- fr: is the frequency of rotor rotation.
- N_b: is the number of bearing balls.
- D_b: is the diameter of the balls.
- D_c: is the diameter of the cage.
- β: is the contact angle.

The frequency of the lines indicating the defect is:

$$f_d = f_s \pm k.f_c \tag{6}$$

- k: is the order of harmonic balls.
- f_s: is the power frequency.
- f_c: is the default characteristic frequency.

In the case of a bearing fault, the expression of the load torque can be written as the sum of a constant part and an oscillating part ΔT at the characteristic frequency fc:

$$Tr = T_{const} + \Delta T = \mathrm{T}_{\mathrm{const}} + \mathrm{T}_{\mathrm{c}}\cos(\omega_{\mathrm{c}}\mathrm{t}) \quad (7)$$

where:

$\mathrm{T}_{\mathrm{const}}$: Constant load torque.
$\omega_c = 2\pi f_c$: Pulsation of oscillation and bearing defect.
T_{c}: The amplitude of the oscillation related to the defect severity.
f_{c}: Characteristic frequency of the bearing defect (fc = fex, fin, fbi and fca depending on the defect type).

5 The Hilbert Filter

The processing of the electrical quantity signal is generally carried out by the FFT, but in the case of non-stationary signals, as in the case of closed-loop controls, in particular the DTC, one must use the methods of advanced signal processing. The extraction and analysis of the envelope of the signal to be processed is among the most preferable and efficient solutions [3, 4]. Indeed, in the literature and for the study of IM defects, several researchers use the HT [4, 11]. The HT is defined by Eq. (8) and its structure is illustrated by Fig. 4 [3]:

$$X(t) = \frac{1}{\pi t} * x(t) = \frac{1}{\pi}\int_{-\infty}^{+\infty} \frac{x(t)}{t-\tau} d\tau \quad (8)$$

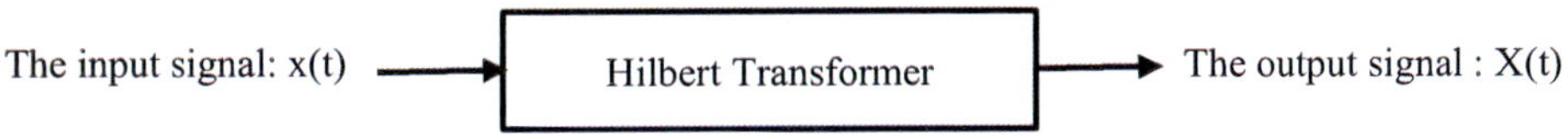

Fig. 4. Block diagram of the Hilbert filter

The resulting analytical signal contains only the positive frequency components while eliminating the negative components and doubling the DC component [3] (Fig. 5).

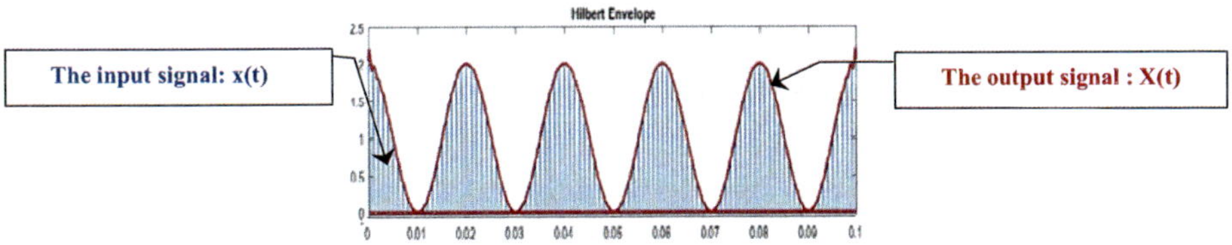

Fig. 5. A signal and its envelope extracted by the Hilbert filter

6 Simulation and Interpretation

To verify the effectiveness of the method adopted for the diagnosis of bearing faults in the induction motor controlled by DTC, which is configured by the parameters presented in Table 2 in the appendix, we tried to simulate the diagnostic system in MATLAB/Simulink in the event of a fault in the outer ring of a machine bearing whose characteristic frequency is 63 Hz.

The power supply frequency of the machine is 50 Hz, and the applied load torque is 5 Nm.

Initially, we simulated the machine without control in the healthy state and with failure of the outer ring of the bearing by taking the current spectrum for each case.

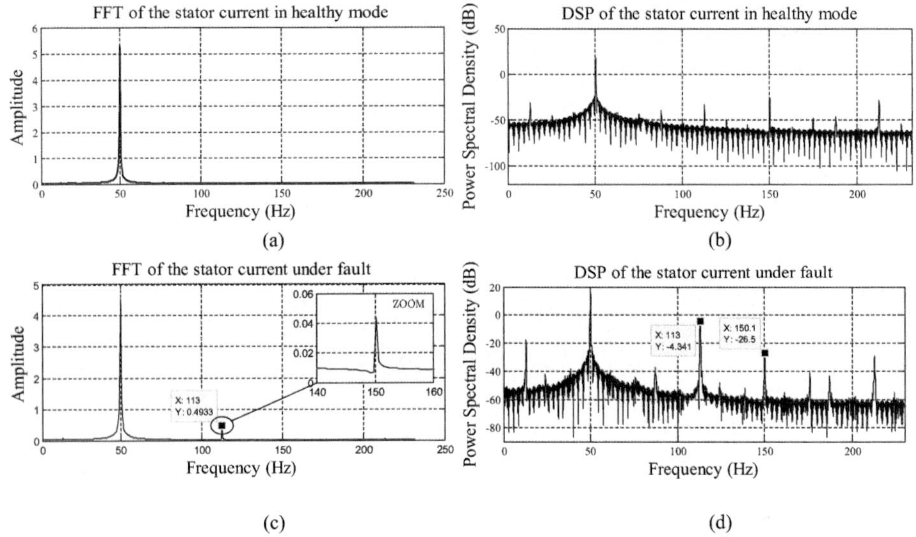

Fig. 6. (**a**) FFT of stator current in healthy mode. (**b**) Power Spectrum Density (PSD) of stator current in healthy mode (**c**) FFT of stator current under fault in the outer ring. (**d**) PSD of stator current under fault in the outer ring.

In Figs. 6c and 6d, in the event of a fault in the outer ring, we can clearly see the harmonic spectra linked to this characteristic frequency fault (113 Hz = 50 Hz + 63 Hz). However, in the absence of a fault in Fig. 6a, the spectrum of the 50 Hz power supply frequency is the only one that confirms the healthy state of the machine.

Secondly, and in the presence of the classic DTC control, this fast spectral analysis of the current is no longer effective because of the other harmonics and inter-harmonics that are added to the signal (see Fig. 7); in this case, the analysis of the signal envelope is very useful and efficient. This makes it possible to increase the amplitude resolution of the spectra around the fundamental, which often indicates rotor faults.

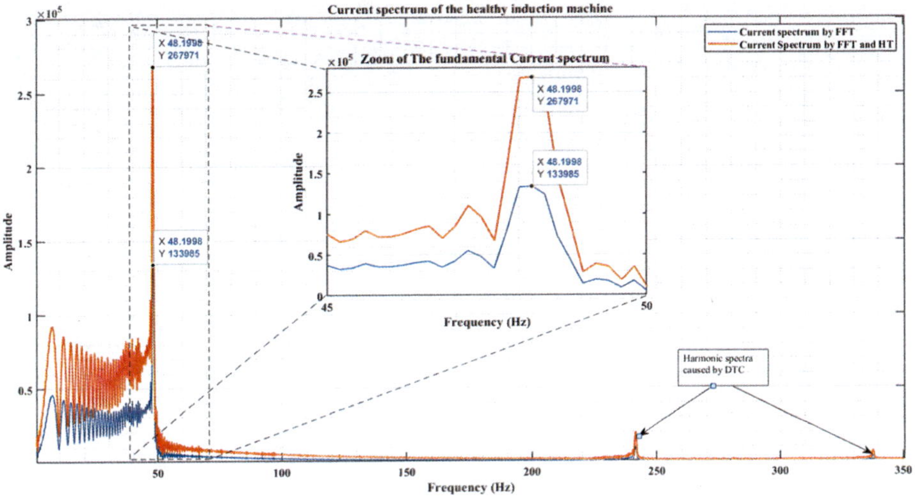

Fig. 7. Current spectrum of a healthy induction machine when it is controlled by the DTC.

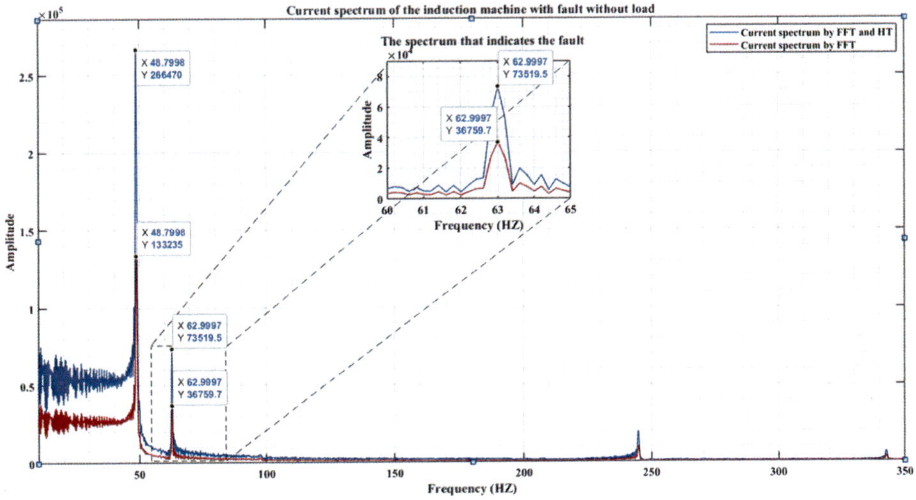

Fig. 8. Current spectrum where the faulty machine is controlled by DTC without load.

The three results obtained are compared and analyzed in order to detect the fault.

- Figure 8 shows the line current harmonics spectra when the faulty machine is controlled by the DTC and subjected to a load torque of 5 Nm in two situations:
- Use of the FFT alone: We notice the appearance of the fundamental at the point (X = 48.80; Y = 133235) and the harmonic, which indicates the fault is located at the point (X = 63; Y = 34648.1), but its appearance is just mediocre.

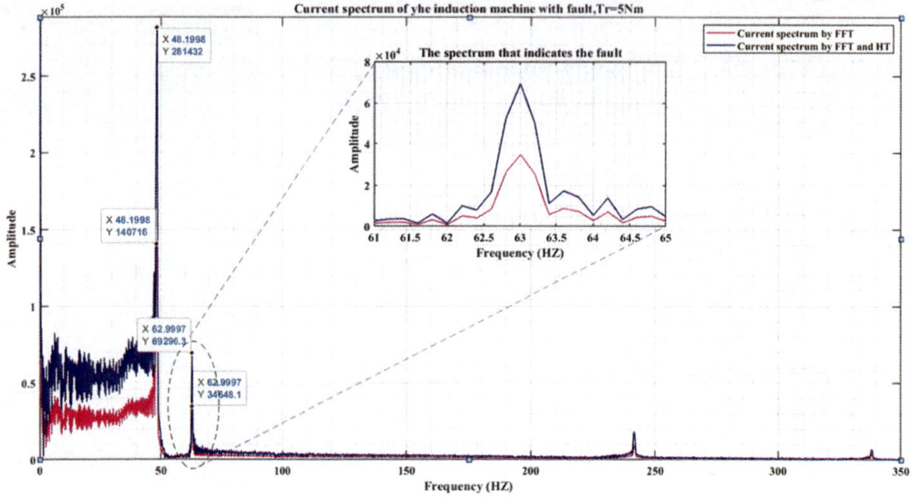

Fig. 9. Current spectrum where the faulty machine is controlled by DTC, with a load torque is 5 Nm.

- Use of HT and FFT: We notice the appearance of the fundamental at the point (X = 48.80; Y = 281432) and the harmonic, which indicates the fault is located at the point (X = 63; Y = 69296.3), so its appearance is clear.

where

- X: is the frequency of current signal.
- Y: is the amplitude.

In Fig. 9, the spectra are almost the same as in Fig. 8, with the difference that the load torque changes the frequency and amplitude of the fundamental harmonics and the others. The spectrum indicates the fault may change slightly in amplitude but not in its characteristic frequency. The increase in the spectra amplitude around the fundamental clearly proves the interest of the analysis of the signal envelope by the Hilbert filter.

These simulation results validate the important conclusion of the fault diagnosis of broken bars by this Hilbert filter. These conclusions mainly concern the non-sensitivity to the variation of load torque compared to other techniques [4]. It is clear from these figures that the DC component of the signal is doubled. For example, the spectrum amplitude that indicates the defect is increased by 100% using this Hilbert filter (see Fig. 9), this values is calculated by Eq. (9):

$$\left(\frac{69296.3 - 34648.1}{34648.1}\right) \times 100 = 100\% \tag{9}$$

7 Conclusion

The problem of diagnosis in the presence of DTC and of an induction machine is a challenge for researchers. The simulation results found in this work confirm the conclusions of other work interested in the diagnosis of faults and highlight the importance of the analysis of the signal envelope by HT.

It is very interesting to integrate artificial intelligence for the diagnosis of faults and to achieve a combination with this type of filter to properly estimate the severity of the fault. It is in this regard that our research team will propose future work such as follows:

- Detection of broken bar faults in the asynchronous cage machine using the Hilbert filter and neural networks.
- Detection of bearing defects in the asynchronous cage machine using the Hilbert filter and fuzzy logic .

Appendix

See (Table 3).

Table 2. Parameters of induction motor

Variable	Value
Nominal power	Pu = 2.2 kW
Stator and rotor resistances	Rs = 4.125 Ω; Rr = 2.486 Ω
Maximum of mutual inductance	Lm = 0.2848 H
Self-inductances	Ls = 0.01557 H; Lr = 0.01557 H
Viscous frictions	f = 0.00114 $kg.m^2/s$
Total inertia	J = 0.139 $kg.m^2$
Pair pole number	P = 4

Table 3. Nomenclature

Parameters	Description
$V_{s\alpha}$, $V_{s\beta}$, $V_{r\alpha}$, and $V_{r\beta}$	Stator and rotor voltages in (α, β) plan
$I_{s\alpha}$, $I_{s\beta}$, $I_{r\alpha}$, and $I_{r\beta}$	Stator and rotor currents in (α, β) plan
$\Psi_{s\alpha}$, $\Psi_{s\beta}$, $\Psi_{r\alpha}$, and $\Psi_{r\beta}$	Stator and rotor fluxes in (α, β) plan
R_s, R_r	Stator and rotor resistors
L_s, L_r	Stator and rotor inductors

(continued)

Table 3. *(continued)*

Parameters	Description
L_m	Mutual Inductance
p	Number of pairs of poles
ω_r	Rotor angular speed
ω_s	Stator angular speed
Ω	Rotation speed
T_{em}	Electromagnetic torque
T_r	Resistant torque
f	Viscous friction coefficient
J	Moment of inertia

References

1. Gundewar, S.K., Kane, P.V.: Condition monitoring and fault diagnosis of induction motor. J. Vibr. Eng. Technol. **9**(4), 643–674 (2021)
2. Jiageng, C., Niaqoning, H., Lun, Z., Ling, C., Bozheng, W., Yang, Z.: A method for broken rotor bars diagnosis based on sum-of-squares of current signals. Appl. Sci. **10**, 5980 (2020)
3. Pita, H., Zurita, G., Vllarroel, A.: Software development firmware system for broken rotor bar detection and diagnosis of induction motor through current signature analysis. JMES. **13**, 6920–6931 (2020)
4. Senthil, K.R., Gerald, C.R., Saravanan, S., Umashankar, S.: Broken rotor bar fault detection using Hilbert transform and neural networks applied to direct torque control of induction motor drive. IET Inst. Eng. Technol. **13**, 3328–3335 (2020)
5. Bouras, A., Bennedjai, S., Bouras, S.: Experimental detection of defects in variable speed fan bearing using stator current monitoring. SN Appl. Sci. **2**(5), 1–8 (2020). https://doi.org/10.1007/s42452-020-2687-2
6. Asad, B., Vaiman, T., Anouar, B., Ants, K., Anton, R., Iqbal, M.N.: Broken rotor bar fault detection of grid and inverter-fed induction motor by effective attenuation of the fundamental component. IET Elect. Power App. **13**, 2005–2007 (2019)
7. Mahfoud, S., Derouich, A., Ouanjli, E.L., Mahfound, E.L., Taoussi, M.: A new strategy-based PID controller optimized by genetic algorithm for DTC of the doubly fed induction motor. Systems **9**, 37 (2021)
8. Mahfoud, S., Derouich, A., Iqbal, A., El Ouanjli, N.: ANT-colony optimization-direct torque control for a doubly fed induction motor: an experimental validation. Energy Rep. **8**, 81–98 (2022)
9. Tarek, A., Arezki, M., Hicham, T., Youness, A.: Discrete wavelet transform and energy eigen value for rotor bars fault detection in variable speed field-oriented control of introduction motor drive. ISA Trans. **79**, 217–221 (2018)
10. Safin, N.R., Prakht, V.A., Dmitrievskii, V.A., Dmitrievskii, A.A.: Stator current fault diagnosis of induction motor bearings based on the fast Fourier transform. Russ. Electr. Eng. **87**(12), 661–665 (2016). https://doi.org/10.3103/S1068371216120063
11. Immovilli, F., Bellini, A., Rubini, R., Tassoni, C.: Diagnosis of bearing faults in induction machines by vibration or current signals: a critical comparison. IEEE Trans. Ind. Appl. **46**(4), 1350–1359 (2010)

New Printable Concrete Formulation Based on Fibers from Doum Palm Leaflets for 3D Printing of Energy-Efficient Buildings

Basma M'Lahfi[1](✉), Driss Amegouz[2], and Mostafa El Qandil[1]

[1] LERDD Laboratory, Faculty of Sciences Dhar El Mahraz, Sidi Mohamed Ben Abdellah University, Fez, Morocco
mlahfi.basma@gmail.com

[2] LPEDD Laboratory, EST, Sidi Mohamed Ben Abdellah University, Fez, Morocco

Abstract. Numerous countries throughout the world, including Morocco, are devoted to the process of sustainable construction in order to construct high-energy-efficiency structures capable of significantly reducing energy consumption during the various stages of construction. Among the innovative solutions for achieving sustainability in the field of construction is 3D printing technology, which allows for raw material savings, cost reductions on construction sites, and a significant reduction in CO_2 emissions.

The 3D printing of buildings, or additive manufacturing, is carried out by the injection of concrete, layer by layer, with a specific formulation and properties. This formulation could be developed by integrating ecological materials, which will contribute to the thermal inertia of buildings thanks to their low thermal conductivity.

Several insulating materials have been proposed in this regard, including minerals such as natural basalt fiber, pozzolana, and hemp, and biological fibers such as flax, rapeseed, and palm leaflets. The objective of this article is to propose a new formulation of injectable concrete for buildings made by the 3D technique, which must be fluid to be used in the printer and firm to hold the times printed. This is done by using aggregates characterized by their low thermal conductivity and reinforcing fibers based on natural materials, providing thermal comfort to users. These fibers are, in particular, the leaflets of the doum palm, which are very abundant in Morocco and in the Mediterranean region, and their characteristics will be discussed in this article.

Keywords: Sustainable construction · 3D printer · 3D printed concrete · Doum palm leaflets

1 Introduction

Construction has long been regarded as a resource-intensive industry that imposes severe environmental limits. Specifically, the construction industry uses more materials by weight than any other industry. Although energy is used in the project's life cycle at

S. Motahhir and B. Bossoufi (Eds.): ICDTA 2022, LNNS 454, pp. 813–824, 2022.
https://doi.org/10.1007/978-3-031-01942-5_81

different levels, it is during the construction operation (initial embodied energy) that most of the energy is consumed [1].

In aggregate, buildings utilize approximately 48% of global energy each year for embodied and operational energy. All products and processes used in the construction, maintenance, replacement, repair, and demolition of buildings use embodied energy. In addition, heating, air conditioning, lighting, and the functioning of building equipment all demand operating energy. Therefore, to significantly minimize energy consumption throughout the course of a product's life, both embodied and operating energy must be optimized [2].

The initial embodied energy is consumed directly during construction, installation, fabrication, transportation, management, and other related on-and off-site processes, as well as indirectly through the use of materials, assemblies, equipment, and vehicles, all of which consume energy during their manufacturing processes [3, 4].

This reflects the construction activity and a real track of energy consumption during these 3 main phases, namely: the supply phase of raw materials, during the construction phase, and during the phase of its operation. This problem could lead to real ecological nuisances in terms of the exhaustion of non-renewable construction materials and the non-rationalized exploitation of deposits, but above all, very severe environmental disturbances, illustrated by greenhouse gas emissions and global warming [5, 6].

In order to minimize these nuisances and to remedy them, the new strategies and policies of the state tend towards the notions of sustainability, especially in the field of construction, since a sustainable building is a building constructed with low criteria energy consumption, taking into account the significant reduction in embodied energy and operating energy while using ecological and renewable building materials, but also the construction processes and technologies that consume less energy [7, 8].

Among the innovative solutions that can bring sustainability to the construction industry, we find additive manufacturing, or 3D printing, which will become a key technology for the building of tomorrow. 3D printing ensures consumption efficiency as it does not waste raw materials and reduces the volume of waste on the job site. It also saves time.

Additive manufacturing is carried out by gantry printers, making it possible to print a building in 3 dimensions with concrete of a specific composition and properties.

Hence the objective of this article, which will focus more particularly on the composition of a new formulation of concrete intended to be printed in 3D and which will be composed of ecological aggregates, making it possible to achieve a thermal inertia to reduce air conditioning and heating requirements. This will be done by presenting these materials, as well as the tests that will be carried out.

In conclusion, avenues for reflection will be provided in order to open up other lines of research on the subject.

2 The 3D Printing System

2.1 Generality About Additive Manufacturing or 3D Printing

3D printing, often known as additive manufacturing, is not a new technique, having originated in the 1980s. It is the process of creating a three-dimensional model, or prototype, using layers of material from a computer-aided design. This is standard procedure in a wide variety of industries, including aerospace, architecture, and high-end medical manufacturing [9].

Among its most important advantages is the flexibility of this type of solution, which allows rapid adaptation to new designs and complex architectural forms, significantly reducing the number of intermediate actions on site. Along with saving time and money, 3D printing positively influences the environment by lowering the amount of materials used, trash generated, and the danger of accidents. Additionally, it may increase competition within the construction business, resulting in lower pricing and increased homeownership rates. Cities that are overcrowded and rapidly rising in population, in particular, stand to gain from this technology.3D printing can also help develop relatively inaccessible areas. In addition, in some cases, portable systems can be made, allowing on-site printing, reducing the transport needs of precast items and increasing response speed. It is therefore obvious that the development of this technology covers specific needs that are of great interest in the construction sector [10].

2.2 The Operating Principle of a 3D Printing System

Three-dimensional 3D printing (affectionately referred to as additive manufacturing or rapid prototyping) enables the fabrication of three-dimensional things by layering a material. A typical 3D printer employs an extrusion material in a print head that moves in three Cartesian axes (x, y, and z) relative to a print surface. A computer controls the print head. It goes through a sequence of positions on the print surface, depositing a tiny volume of material at each site to define a portion of the object created at that location. After printing a base layer directly onto the print surface, the printhead is sequentially elevated (z axis) to print further layers on top of it until the entire item is created [11].

In general, the 3D printing process can be summarized according to the steps mentioned in Fig. 1:

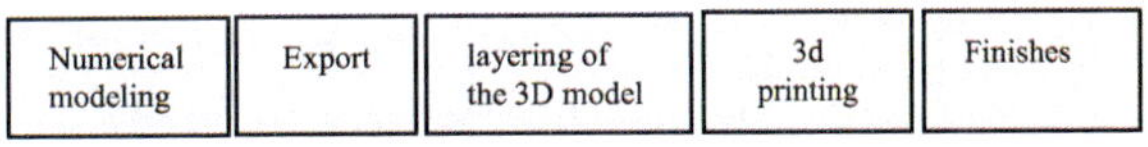

Fig. 1. The phases of the 3D printing process [6].

2.3 Materials used in 3D Printing of Buildings

In automated and robotic processes in construction, 3D printing additive manufacturing using various building materials has attracted great interest in recent years due to its possibilities in terms of flexibility and speed.

3D printing in construction is the combination of computer design with extrusion and infill processes. The main materials used for printing are concrete, plaster and hybrid materials, as well as various additives and fibers [12].

2.4 The Characteristics of the Moroccan Construction Materials to be used in the 3D Printing System

The construction materials available in Morocco must first undergo a general analysis of their characteristics in order to determine their advantages, their limits, as well as their specifications so that they can be adapted to the application when using construction technologies. 3D printing for custom prefabricated elements.

The following four main characteristics related to the extrusion process must be answered in depth [13]:

- *Pumpability* is defined as the ease and reliability with which material is moved through the distribution system.
- *Printability* (sometimes also called extrudability for materials used in extrusion processes) is defined as the ease and reliability of depositing material through a deposit device.
- *Constructibility* is defined as the resistance of the wet material deposited to deformation under loads. Alternatively, the number of layers of filament that can be added without significant deformation of the lower layers
- *Open time* is defined as the period during which the aforementioned properties remain constantly within acceptable tolerances.

Therefore, potential building materials must be defined physically and chemically, and their limits must also be analyzed.

3 Review of Innovative Eco-materials that Can Be Used in 3D Printing and that Provide Thermal Inertia in the Printed Building

As already mentioned in the previous chapters, the construction of buildings using the 3D printing technique can offer great energy savings in the various phases of construction. Furthermore, the development of a new formulation of extruded concrete based on local, ecological, and innovative materials and capable of providing thermal comfort to building users due to its thermal insulation capacity will be a great revolution in the Moroccan construction sector, particularly in terms of support for the national energy-saving strategy and compliance with the Moroccan Construction Thermal Regulations.

Therefore, the state of the art of eco-insulating materials available in Morocco and whose previous research has proven their effectiveness will be discussed.

3.1 Prefabricated Hemp Concrete Uses a Pozzolanic Binder

The author, DINH LE MANH [15], developed in his thesis a concrete based on a pozzolanic binder and hemp. The latter is characterized by its lightness (110.09 ± 0.7 kg/m^3) and its low conductivity thermal (0.0580 ± 0.0004 W/m/K). However, it has a high water absorption capacity, hence the need for upstream treatment. Therefore, two pozzolanic binders were prepared, the first consisting of 50% NHL 3.5 and 50% Metakaolin (MK) identified by HM and the second SM by 30% slaked lime and 70% MK. Two adjuvants were added separately to the two binders (glycerol carbonate and potassium sulphate) to improve mechanical resistance at a young age.

According to the tests carried out, the binders containing both the two admixtures could be applied for the manufacture of hemp concrete because they have better mechanical performance, a good shrinkage reducing effect, and low thermal conductivity.

Hence, it leads to the consideration of the use of pozzolan in the composition of 3D printed concrete, given its mechanical resistance and low thermal conductivity, which allows thermal inertia in the building (Fig. 2).

3.2 Insulating Panels Made from Natural Basalt Fiber

Fig. 2. Insulating panels based on basalt fibers [11]

Cinzia Buratti et al. [15] worked on the preparation of insulating panels based on basalt fibers. These are made from basaltic rock, which is a volcanic rock known for its mechanical resistance and high temperatures.

Three different panels were investigated for their density (145, 175 and 200 kg/m^3) and thickness (9 mm, 18 mm, and 27 mm). However, only panels with a total thickness of 9 mm were examined when three different densities were considered (BF-9-a, BF-9-b, and BF-9-c). The values found for samples BF-9-a, BF-9-b, and BF-9-C are 0.0320, 0.0335, and 0.0345 W/mK, respectively.

Innovative basalt fiber insulation panels may be a viable alternative to traditional rock wool or glass wool panels. The advantage is that they may be utilized at a low thickness (less than 3 cm), whereas rock wool and glass wool panels vary in thickness from 3 to 10 cm and in terms of tensile strength. It is approximately 350 kPa for basalt, but only 10 to 20 kPa for a dense rock wool panel (around 120 kg/m^3).

Hence, it leads to the consideration of the use of basalt as an integrated material in the composition of 3D printed concrete, given its high mechanical strength as well as its low thermal conductivity, which allows thermal inertia in the building.

3.3 The Fibers of the Date Palm

3.3.1 Structure and Morphology

According to studies by Agoudjil et al. [12], Fig. 3 presents SEM images of a typical petiole sample in the transverse direction of the fibers. These images show that the sample surface is not smooth, it is irregular with a lot of filaments, impurities, cells, and pores that allow it to adhere well to other material (the matrix) (Fig. 4).

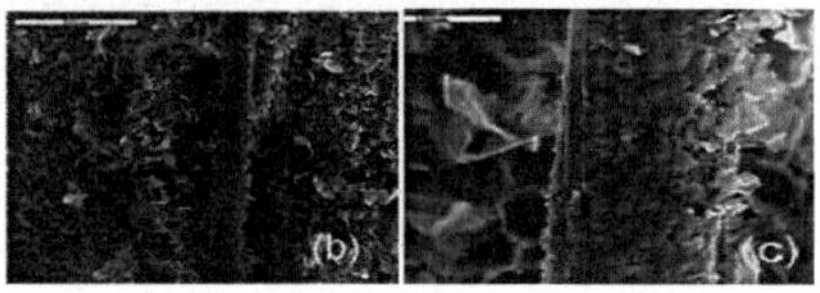

Fig. 3. SEM images of petiole samples in the transverse direction of the fibers [12].

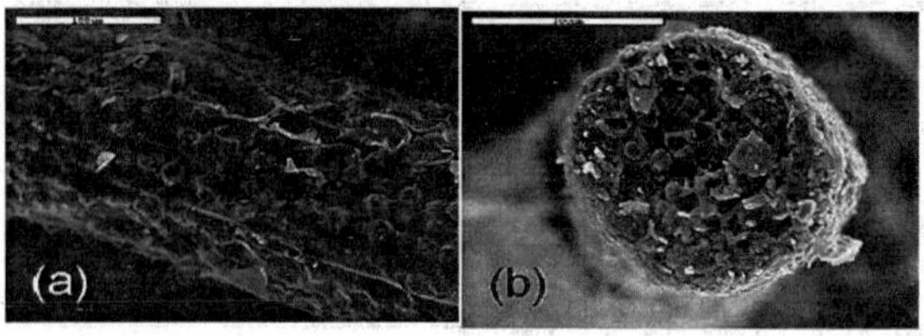

Fig. 4. SEM images for a single petiole fiber: (a) a longitudinal section, (b) a cross section [12].

3.3.2 Thermo-Physical Properties

The table shows the thermophysical and dielectric properties of various varieties of date palm [12]. According to Table 1, the thermal properties of the different varieties of date palm wood show low values. The thermal conductivity varies between 0.072 $W.m^{-1}.K^{-1}$ and 0.085 $W.m^{-1}.K^{-1}$.

Table 1. Thermal conductivity and relative permittivity properties of the date palm (petioles and clusters) [12].

Sample	K ($W.m^{-1}.K^{-1}$)	ε_r
1	0.046 ± 0.003	–
2	0.042 ± 0.003	–
3	0.084 ± 0.003	10.38 ± 0.52
4	0.083 ± 0.003	8.09 ± 0.26
5	0.073 ± 0.003	24.70 ± 1.61
6	0.072 ± 0.002	2.58 ± 0.02
7	0.085 ± 0.004	5.81 ± 0.08
8	0.084 ± 0.005	5.70 ± 0.10
9	0.074 ± 0.004	5.13 ± 0.09

With: K: thermal conductivity a: thermal diffusivity
ε_r: the relative dielectric permittivity.

4 Construction Materials Opted for 3D Printing and Allowed Thermal Inertia

The composition of injectable concrete for 3D printing is mainly composed of a specific formulation based on cement, mixing water, non-coarse aggregates, and additives, but also reinforcing fibers to improve the performance mechanics of the constructed structure.

Many mixture designs have been proposed by previous research [14, 15]. Therefore, to enhance the mechanical characteristics, namely the traction, while remedying the phenomena of shrinkage, drying, and cracking, it was recommended to add to the printable concrete mix polypropylene (PP) microfibers, given that the retarders and accelerators that modify the setting time can cause very high shrinkage.

Besides, natural fibers have received special attention in 3D construction technology due to their lightness and environmentally friendly nature, and they defy substitute artificial fibers that have been used as a reinforcing material for this concrete mix printable since its inception.

4.1 Characteristics to be Taken into Account for Aggregates

To select the most suitable materials to carry out the mission of concrete extrusion in buildings to be printed by the 3D technique. It will be necessary to take into account, in the first place, in the selection, the aggregates available in the different geographical areas of Morocco, which are derived from natural materials. As a result, they are tailored to the specific requirements of the 3D printing system, allowing them to have good mechanical resistance while also providing thermal inertia in the built environment (low thermal conductivity).

4.2 Characteristics to be Taken into Account for Natural Reinforcing Fibers

Natural plant, animal, and mineral fibers are the three primary classifications for natural fibers. Fibers with natural origins are classified in Table 2 and depicted in Fig. 3 as various plant fibers [12] (Fig. 5).

Table 2. Classification of natural fibers [13].

Natural Fibers	Natural Plant or Vegetable (Cellulose or Lignocellulose)	Seed	Cotton, Kapok, Milkweed
		Bast or Stem	Flax, Hemp, Jute, Ramie, Kenaf
		Leaf or Hard	Pineapple (PALF), Banana (Abaca/Manila-hemp), Henequen, Sisal
		Stalk	Wheat, Maize, Barley, Rye, Oat, Rice (Husk)
		Cane , Grass & Reed Fibers	Bamboo, Bagasse, Esparto, Sabei, Phragmites, Communis
		Fruit	Coir or Coconut Fibers
	Animal (Protein)	Wool/ Hair	Lamb'sWool, Goat Hair, AngoraWool, Cashmere, Yak, Horse Hair
		Silk	Tussah Silk , Mulberry Silk
	Mineral Fibers	Asbestos Fibrous brucite Wollastonite	

Fig. 5. Natural vegetable fibers (a) Banana, (b) sugar cane bagasse; (c) curaua; (d) flax; (e) hemp; (f) jute; (g) sisal; (h) kenaf; (i) Jute fabric; (j) ramie (k) jute [12].

Natural fibers that have been studied in the past are summarized in Table 2, where the thermal conductivity of each type of fiber is indicated, allowing for the selection of fibers that are acceptable for use in 3D printing and capable of providing good thermal inertia (low thermal conductivity) (Table 3).

Therefore, the main objective to be addressed will be the determination of the natural materials to be used in this construction technique, which must particularly meet the main characteristics [17]:

- Lightness;
- Sealing;
- Mechanical performance;
- Property of thermal insulation and, where appropriate, sound;
- Ability to damp vibrations;
- Renewable material origin, with a low CO_2 balance.

Table 3. Properties of natural plant fibers from past germane studies [14–17].

Fibre Type	Density (g/cm³)	Length (mm)	Diameter (micro)	Tensile Strength (MPa)	Elongation (%)	Cellulose (wt.%)	Hemi-Cellulose (wt.%)	Lignin (wt.%)	Pectin (wt.%)	Waxes (wt.%)	Thermal Conductivity (W/mK)	USD/Kg
Hemp	1.4–1.5	5–55	25–500	270–900	1–3.5	68–74.4	15–22.4	2.7–10	0.92	0.81	0.035–0.060	1.0–2.5
Flax	1.4–1.5	5–900	12–600	343–2000	1.2–3.3	62–72	18.6–20.6	2–6	2.4	1.45–1.8	0.035–0.080	2.0–4.4
Jute	1.3–1.49	1.5–120	20–200	320–800	1–1.8	59–71.5	13.6–20.4	10–15	0.1–0.5	0.45	0.035–0.060	0.25–2.0
Kenaf	1.4-1.5	–	–	223–930	1.5–2.7	31–72	20.3–21.5	7–20	2–6	–	0.033–0.053	0.15–0.66
Ramie	1.0–1.55	900–1200	20–80	400–1000	1.2–4.0	68.6–85	13–16.7	0.55–0.73	2.9	0.23	–	1.2–2.8
Oil palm	0.7–1.55	–	150–500	80–248	17–25	60–65	–	10–30	–	–	–	-
Coir	1.15–1.46	20–150	10–460	95–230	15–51.4	32–43.8	0.15–20	30–50	3.5–4.5	–	–	0.2–0.6
Banana	1.35	300–900	12–30	500	1.5–9	63–67.6	10–19	4	–	–	–	–
Sisal	1.33–1.5	900	8–200	363–700	2.0–7.0	60–78	10.0–14.2	–	–	–	–	0.5–0.8
Pineapple	0.8-1.5	900–1500	–	170–627	0.8–14.5	80–83	15–19	4-13	–	–	0.033–0.047	0.3–0.7
Bagasse	1.25	10–300	10–34	222–290	1.1	32-55.2	16.8	20-27.5	–	–	0.042-0.065	-
Bamboo	0.6–1.1	1.5–4	25–40	140–800	2.5–3.7	26–65	30	4–32	–	–	–	0.3–0.6
Rice (Husk)	–	–	–	–	–	–	–	–	–	–	0.042–0.568	-
Woods and Roots	–	1.2–3.6	5–50	–	–	–	–		–	–	0.035–0.055	-

4.3 Selected Materials

According to the results of previous research, we opt for this work for the formulation of concrete 3D printed basalt and pozzolan aggregates given their characteristics of rigidity as well as their low thermal conductivity, which allows them to participate in ensuring thermal inertia in the printed building.

In addition, and given the climate of Morocco, we note that the doum palm, or dwarf palm (Chamaerops humilis), is very abundant and available in the various Moroccan regions. Besides, and according to visual observation, the structure of its leaflets is characterized by rigidity and hydrophobia.

Therefore, we propose in this work the integration of the fibers of the leaflets of the doum palm into the formulation of the concrete printed in 3D as reinforcement, thanks to its structural characteristics which appear visually as well as thermal (the thermal conductivity of the fibers of jute is similar to that of the doum palm and in the interval (0.035–0.060 W/mK).

Moreover, these mechanical and thermal characteristics need to be scientifically proven by laboratory scale tests.

Natural fibers, on the other hand, face several restrictions. For example, Hom et al. [10] studied the water absorption properties of composites, including unsaturated polyesters with non-woven hemp fibers. They concluded that the increased cellulose content and amplified voids in the basic structure of composites lead to an increase in moisture retention and fiber volume fraction [11]. Other challenges include mechanical discrepancy, reduced elongation and nozzle flow issues, bubbles in the product, and poor weather resistance from natural fibers [12].

On the other hand, natural fibres can be treated to increase their adhesion to matrix materials. As a result, the next step is to investigate the behavior of its fibers in the formulation of injectable concrete, which will take into account the various exposure media (ambient, acidic, and marine).

5 Prospects for the Next Job

From a research point of view, we plan to experiment with natural fibers from the leaflets of the doum palm (saw palmetto), alone and incorporated into the formulation of 3D printed concrete, taking into account their main characteristics (Fig. 6).

Fig. 6. Doum palm (Chamaerops humilis)

As a result, we aim to conduct in the subsequent study the following primary experiments on fibers alone and incorporated with injectable concrete:

- Scanning Electron Microscopy (SEM);
- RX diffraction;
- Absorption capacity;
- Thermal conductivity;
- Mechanical tests (traction, bending, hardness, fatigue, stress at break, etc.);

To find out if this new material is capable of ensuring good thermal inertia for the building.

6 Conclusion

The technique of 3D printing on bio-based concrete constitutes an innovative avenue of research. Moreover, the application of this technology in the construction industry presents challenges to be overcome.

The main factor limiting 3D printing is the development of printable concrete capable of balancing the stability and required flow of concrete [12–14].

Among the components that can ensure the stability of concrete, there are reinforcements, including those that are natural and whose use is recommended in 3D printed concrete according to several studies, because they have thermal, mechanical, environmental, ecological, and economic advantages. Hence, the use of 3D printed concrete incorporated with natural fibers will be a sustainable solution for the new technology. In addition, among the advantages of using this type of reinforcement in the composition of printable concrete are the large dispersion of these fibers, which prevents plugging in the printing process, improving the mechanical and thermal resistance and providing good interfacial bonding.

Moreover, the disadvantage of the use of natural fibers in 3D printable concrete is its high water absorption capacity, which generates poor interfacial interaction between the matrix and the fiber. Hence the need to anticipate their use by chemical treatment to remedy this phenomenon.

As a conclusion, additional studies will be conducted within the framework of this research project by the use of doum palm fibers in order to study their mechanical and thermal behavior in the concrete, which will be used for 3D printing and which will be published in the next work.

Funding:. I would like to thank IRESEN for funding my participation in the 2022 edition of the International Conference on Digital Technologies and Applications (ICDTA'22) as part of the 3D Printer project, which integrates the research work being the subject of an oral presentation presented during this event.

References

1. Gardner, J.: Exploring the emerging design territory of construction 3D printing projected architectural research (2011)
2. Perrot, A., Rangeard, D., Pierre, A.: Structural built-up of cement-based materials used for 3D-printing extrusion techniques. Mater. Struct. **49**(4), 1213–1220 (2015). https://doi.org/10.1617/s11527-015-0571-0
3. M'Lahfi, B.: A new approach for the mandatory application of the thermal regulation of construction (RTCM) in the future moroccan buildings (2020)
4. THE MANH D: Contribution to the development of precast hemp crete using innovative pozzolanic binder, PhD thesis, University of Toulouse 3 Paul Sabatier (2014)
5. Cinzia, B., et al.: Thermal and acoustic performance evaluation of new basalt fiber insulation panels for buildings. In: 6th International Building Physics Conference, IBPC (2015)
6. Agoudjil, B., Benchabane, A., Boudenne, A., Ibos, L., Fois, M.: Renewable materials to reduce building heat loss: characterization of date palm wood. Energy Build. **43**, 491–497 (2011)
7. Bono, P.: Materials: New fields of research and development for the valorization of technical plant fibers (flax fiber and hemp) (2015)
8. Le, T.T., Austin, S.A., Lim, S., Buswell, R.A., Gibb, A.G.F., Thorpe, T.: Mix design and fresh properties for high-performance printing concrete. Mater. Struct. **45**, 1221–1232 (2012)
9. Kazemian, A., Yuan, X., Cochran, E., Khoshnevis, B.: Cementitious materials for construction-scale 3D printing: laboratory testing of fresh printing mixture. Constr. Build. Mater. **145**, 639–647 (2017)
10. Le, T.T., et al.: Hardened properties of high-performance printing concrete. Cem. Concr. Res. **42**, 558–566 (2012)
11. Salmabanu, L.: Sustainable and renewable bio-based natural fibres and its application for 3D printed concrete: a review (2020)
12. Júnior, C.P., De Carvalho, L., Fonseca, V., Monteiro, S., D'Almeida, J.: Analysis of the tensile strength of polyester/hybrid ramie–cotton fabric composites. Polym. Test. **23**, 131–135 (2004)
13. Akil, H., Omar, M., Mazuki, A., Safiee, S., Ishak, Z., Abu Bakar, A.: Kenaf fiber reinforced composites: A review. Mater. Des. **32**, 4107–4121 (2011)

14. Ku, H., Wang, H., Pattarachaiyakoop, N., Trada, M.: A review on the tensile properties of natural fiber reinforced polymer composites. Compos. Part B Eng. **42**, 856–873 (2011)
15. Bledzki, A.: Composites reinforced with cellulose based fibres. Prog. Polym. Sci. **24**, 221–274 (1999)
16. Dittenber, D.B., GangaRao, H.: Critical review of recent publications on use of natural composites in infrastructure. Compos. Part A Appl. Sci. Manuf. **43**, 1419–1429 (2012)
17. Saheb, D.N., Jog, J.P.: Natural fiber polymer composites: a review. Adv. Poly. Technol. **18**, 351–363 (1999)

Predictive Torque and Direct Torque Controls for Doubly Fed Induction Motor: A Comparative Study

El Mahfoud Mohammed[1(✉)], Bossoufi Badre[1], El Ouanjli Najib[2,4], Hilali Abdelilah[3], El Alami Houda[1], Majout Btissam[1], and Mahfoud Said[4]

[1] Laboratory of Engineering Modeling and Systems Analysis, SMBA University, Fez, Morocco
Mohfoudmohammed.el@usmba.ac.ma
[2] LMIET, Faculty of Sciences and Technology, Hassan First University, Settat, Morocco
[3] Electronics, Communication Systems and Energy Optimization Team, Faculty of Sciences, Moulay Ismail University, Meknes, Morocco
[4] Technologies and Industrial Services Laboratory, SMBA University, Fez, Morocco

Abstract. The aim of the study presented in this paper is to improve the performance of the direct torque control (DTC) applied to doubly fed induction motors (DFIM). The DTC is characterized by good speed tracking and high robustness against parametric variations of the motor and external disturbances. However, there are significant drawbacks to this method as well. On the one hand, the switching frequency is not controlled, with a high ripple in the stator flux and the torque. Therefore, a new approach is developed to improve the performance of DTC. Predictive torque control (PTC) is a technique that determines the appropriate voltage vector to apply in the next sampling period by constructing a unique cost function from a linear combination of objective functions. The torque and flux errors are accounted for utilising weight factors in a cost function. Finally, the proposed control is compared to DTC by simulation in the Matlab\Simulink environment.

Keywords: DFIM · DTC · MPC · PTC

1 Introduction

In the high power sector, one of the most extensively utilised pieces of equipment is the doubly fed induction machine (DFIM) [1, 2]. However, its modelling gives rise to a system of non-linear, strongly related and multivariate equations, and consequently the dynamic behaviour of the DFIM is often very complex [3, 4]. More powerful computing techniques are required to manage the torque and flux of this machine in order to address these challenges.

Nevertheless, due to technical advances in power electronics and microcomputers, the domain of variable speed electrical machines has developed considerably over the last decade. Indeed, direct torque control (DTC) has been shown to be a useful approach for achieving quick dynamic response [5]. This technique is theoretically simple, and it

S. Motahhir and B. Bossoufi (Eds.): ICDTA 2022, LNNS 454, pp. 825–835, 2022.
https://doi.org/10.1007/978-3-031-01942-5_82

does not need complicated circuits or computers to implement. Compared to FOC (field oriented control), DTC is simpler, as it does not use Pulse Width Modulation (PWM) or PI controllers. It uses hysteresis controllers and switching tables [6]. DTC also has several disadvantages: owing to the employment of hysteresis controllers, the occurrence of torque and flux ripples. In addition, it has problems during starting and low-speed operation. To solve these drawbacks, a lot of research has been done [7].

Among these proposals, Predictive Control aims to control complex industrial systems [8, 9]. The idea behind this method is to employ a dynamic model of the process within the controller in real time to predict how the process will behave in the future. This control strategy has proven to be effective, flexible and successful in industrial applications [10]. The use of predictive control in the realm of digital controls has shown favourable outcomes in terms of speed and accuracy [8, 11].

This research describes an improved DTC that achieves reduced torque ripples and steady switching frequency. In the proposed predictive torque control (PTC), hysteresis comparators and switching tables are replaced by prediction blocks and cost functions that optimise the voltages applied to the inverter. A comparison table contains the simulation findings.

This paper is organized as follows: Sect. 2 presents the modelling of the DFIM. The two control approaches are described in Sect. 3. Sections 4 present and discuss the results of the simulations. Finally, a conclusion and recommendations for future work are presented in Sect. 5.

2 Mathematical Model of DFIM

The DFIM, unlike the induction machine (IM), is powered by two voltage inverters, one linked to the stator windings and the other to the rotor windings, as illustrated in Fig. 1 [6, 7]. This configuration offers better efficiency of the static converters, the possibility to control the power distribution, and a wide speed range [5].

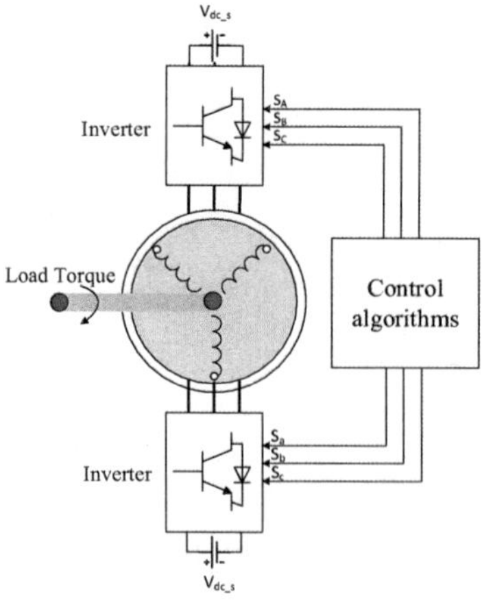

Fig. 1. Block diagram of DFIM

To make the suggested control scheme's implementation easier, the mathematical model of a DFIM presented in this section is developed in the stationary reference frame (α, β). The mathematical model of this machine is described as follows [2]:

$$v_s = R_s.i_s + \frac{d\psi_s}{dt} \tag{1}$$

$$v_r = R_r.i_r + \frac{d\psi_r}{dt} - j\omega_m\psi_r \tag{2}$$

$$\psi_s = L_s.i_s + L_m.i_r \tag{3}$$

$$\psi_r = L_m.i_s + L_r.i_r \tag{4}$$

$$T = p.Im(\psi_s.i_s) = T_L + J.\frac{d\Omega}{dt} + f.\Omega \tag{5}$$

$$\Omega = p.\omega_m \tag{6}$$

where $v_s = \left(v_{s\alpha}, v_{s\beta}\right)^T$, $\psi_s = \left(\psi_{s\alpha}, \psi_{s\beta}\right)^T$ and $i_s = \left(i_{s\alpha}, i_{s\beta}\right)^T$ are the stator voltage vector, stator flux vector, and stator current vector, respectively. $v_r = \left(v_{r\alpha}, v_{r\beta}\right)^T$, $\psi_r = \left(\psi_{r\alpha}, \psi_{r\beta}\right)^T$ and $i_r = \left(i_{r\alpha}, i_{r\beta}\right)^T$ are the rotor voltage vector, rotor flux vector, and rotor current vector, respectively. R_s, L_s, R_r, L_r And L_m are the stator resistance, stator inductance, rotor resistance, rotor inductance and mutual inductance, respectively. p presents the pair of poles, T denotes the electromagnetic torque, T_L denotes the load torque, Ω and ω_m corresponds to the rotor mechanical and electrical angular speed, respectively.

3 Algorithm Control

In this section, two torque control methods are presented, namely DTC and PTC. Both methods have advantages, but differ in structure as the PTC control loop does not include hysteresis controllers.

3.1 DTC

The DTC method allows the decoupling of flux and torque by regulating the magnitude and angle of the stator and rotor flux by applying an appropriate voltage vector to obtain the desired torque [5, 7]. For this purpose, the terms $R_s.I_s$ and $R_r.I_r$ are supposed to be negligible in front of the voltages V_s and V_r, so Eqs. 1 and 2 can be written in the (α, β) reference frame as follows:

$$\begin{cases} \Delta\psi_s = v_s.T_s \\ \Delta\psi_r = v_r.T_s \end{cases} \tag{7}$$

With $\Delta\psi$ is the flux variation and T_s the sample time. This indicates that a voltage vector may control the flux vector. [6]. The second hypothesis is to assume that the flux magnitude remains constant and follows these reference values, so the torque equation can be written as follows [5]:

$$T_{em} = \frac{p(1-\sigma)}{\sigma \cdot m} \psi_{s_{ref}} \cdot \psi_{r_{ref}} \cdot \sin(\gamma) \tag{8}$$

where γ is the angle formed by the stator and rotor flux vectors.

The torque of the DFIM is additionally controlled by three hysteresis regulators in DTC [6, 7]. Two 2-level controllers to control the flux. These flux regulators receive as input the error between the reference value ψ_{ref} and the estimated flux $\hat{\psi}$. The same thing applies to the torque regulator, which gets as input the reference torque $T_{em_{ref}}$ generated by a speed controller based on a PI regulator and the estimated electromagnetic torque $\hat{T}_{em}$ calculated using Eq. 10. H_{Tem}, H_{ψ_s} and H_{ψ_r} are computed by hysteresis controllers and used as inputs to switching tables (Table 1), which use the estimated flux vector argument $\hat{\theta}$ calculated by Eq. 11 to select the flux sector [5]. Figure 2 represents the overall DTC control structure.

$$\begin{cases} \hat{\psi}_s = \int (v_s - R_s.i_s).dt \\ \hat{\psi}_r = \int (v_r - R_r.i_r).dt \end{cases} \tag{9}$$

$$\hat{T}_{em} = p.Im(\psi_s.i_s) \tag{10}$$

$$\hat{\theta} = \arctan\left(\psi_\beta / \psi_\alpha\right) \tag{11}$$

Table 1. Switching table

$H_{T_{em}}$	$H_{\psi_{s,r}}$	Sector					
		S_1 [−π/6,π/6]	S_2 [π/6,3π/6]	S_3 [3π/6,5π/6]	S_4 [5π/6,7π/6]	S_5 [7π/6,9π/6]	S_6 [9π/6,11π/6]
1	1	v_2	v_3	v_4	v_5	v_6	v_1
0		v_7	v_0	v_7	v_0	v_7	v_0
−1		v_6	v_1	v_2	v_3	v_4	v_5
1	0	v_3	v_4	v_5	v_6	v_1	v_2
0		v_0	v_7	v_0	v_7	v_0	v_7
−1		v_5	v_6	v_1	v_2	v_3	v_4

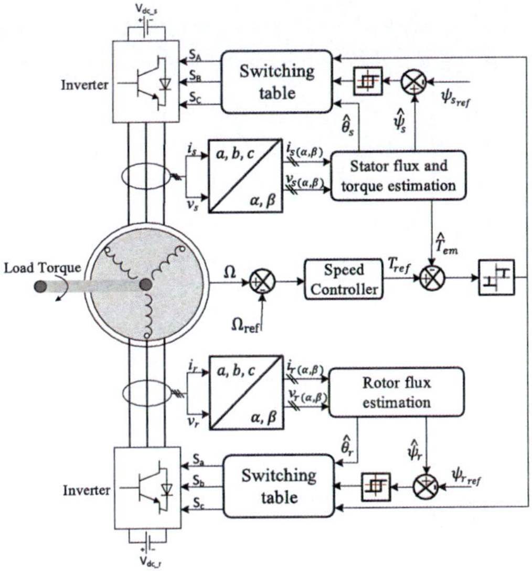

Fig. 2. . Structure global of DTC for DFIM

3.2 Predictive Torque Control

As in all predictive controls, the PTC control involves the following steps: measuring (or estimating) the voltage and speed, estimating the unmeasurable variables, predicting the variables to be controlled, and comparing these variables with the reference using a predefined cost function [10]. Figure 3 summarises the predictive control steps.

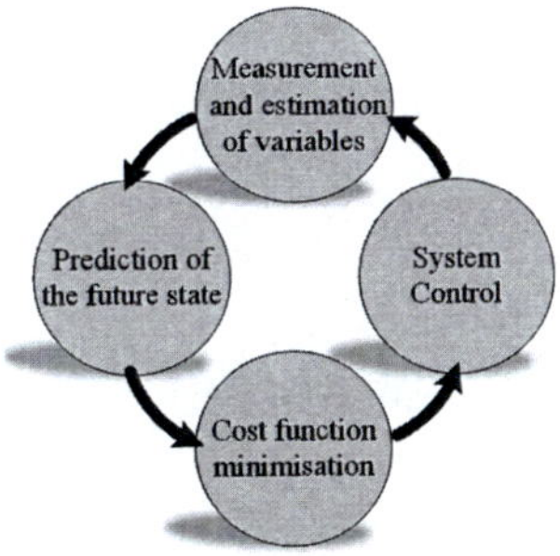

Fig. 3. MPC control steps

The principle of PTC proposed in this section is the same as that of DTC. But with this strategy, predictions of future values of flux and electromagnetic torque are taken into consideration [10, 12]. Therefore, the switching table, the hysteresis band controllers and the position information of the stator and rotor flux used in DTC are replaced by a cost function operating according to the future behaviour of the controlled variables [13]. Predictions are calculated for several possible causes of the voltage vectors V_s and

V_r. The cost function chooses the voltage vector with the optimal control of stator flux, rotor flux, and electromagnetic torque. The PTC control of a DFIM is shown in Fig. 4.

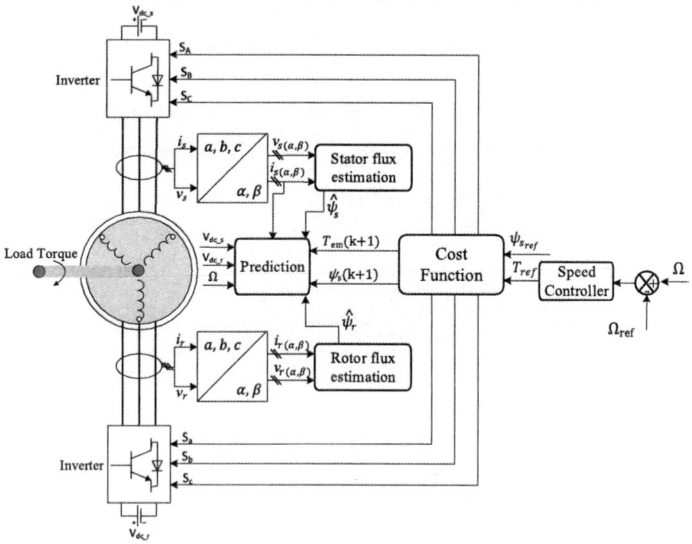

Fig. 4. Structure global of PTC for DFIM

The measurement and estimation blocks are used to measure and calculate the current values (at time k) of measurable and non-measurable variables, such as stator and rotor currents and stator and rotor flux. Then, the prediction model then determines the future values of the controllable variables at time k + 1. The stator flux _ $\psi_s(k+1)$ and the electromagnetic torque $T_{em}(k+1)$ are used in this application. These predictions are made for all potential switching states based on the configuration of the voltage inverters that feed the DFIM.

- Estimation and prediction

In the PTC, the flux are estimated by the DFIM model. Using Euler's formula to discretize Eqs. (1) and (2), the below expressions are used to estimate the stator and rotor flux:

$$\begin{cases} \psi_s(k) = \psi_s(k-1) + T_s.(v_s(k) - R_s.i_s(k)) \\ \psi_r(k) = \frac{1}{1+j.\omega.T_s}[\psi_r(k-1) + T_s.(v_r(k) - R_r.i_r(k))] \end{cases} \tag{12}$$

It is required to compute the prediction of the controlled variables after receiving the flux estimate. At the future sample period k + 1, the PTC predicts the electromagnetic torque T_{em} and the stator flux ψ_s. For the prediction of the stator flux $\psi_s(k+1)_i$. The following formula is used to determine the stator flux prediction:

$$\psi_s(k+1)_i = \psi_s(k) + T_s.\left(V_{dc_s}(k)_i - R_s.i_s(k)\right) \tag{13}$$

where $i = 0...7$, the switching states for the rotor inverter applied in this system.

The torque prediction is directly dependent on the flux and the stator current, according to the electromagnetic torque Eq. (5). As a result, predicting the stator current is necessary. From Eqs. 2, 3 and 4 and after discretization, At sampling time k + 1, the stator current forecast may be obtained

$$i_s(k+1)_{i,j} = \left(1 - \frac{T_s}{\tau_\sigma}\right).i_s(k) + \frac{1}{R_\sigma}.\frac{T_s}{\tau_\sigma}\left[\left(\frac{k_r}{\sigma_r} - j.k_r\omega(k)\right).\psi_r(k) + V_{dc_s}(k)_i - k_r.V_{dc_r}(k)_j\right] \quad (14)$$

where $j = 0 \ldots 7$, switching states for the rotor inverter, $k_r = L_m/L_r$, $\sigma_r = R_r/L_r$ and $R_\sigma = R_s + k_r^2 R_r$. The torque and stator flux predictions are expressed as functions of the inverter voltages V_{dc_r} and V_{dc_s}. This implies that 7 different predictions of the stator flux $\psi_s(K+1)_i$ are obtained as a function of the number of voltage vectors generated for a three-phase two-level inverter and 49 different predictions of the torque $T_{em}(K+1)_{i,j}$. Finally, the selection of the switching state of the voltage inverter is performed by means of a cost function that would contain the control law.

Hence, the electromagnetic torque predicted at time k + 1 is expressed by the following relation:

$$T_{em}(k+1)_{i,j} = p.Im\left(\psi_s(k+1)_i.i_s(k+1)_{i,j}\right) \quad (15)$$

- Cost function

Basically, the cost function is a comparison between the references for both variables under control and their forecast values. Using this cost function, each prediction will be assessed, and the one that produces the lowest value will be used to select which switching states of the voltage inverter to use. In direct control systems, several structures for cost functions are proposed in the literature. A variant of these cost function structures introduces the respective references of the two controlled variables with their different magnitudes. As it is the case of our study with the PTC control, we have therefore considered the following cost function:

$$g_{i,j} = \left|T_{em}(k+1)_{i,j} - T_{em_{ref}}\right| + \lambda.\left|\psi_s(k+1)_i - \psi_{ref}\right| \quad (16)$$

where λ is the weight factor that defines the stator flux's proportional contribution to torque. The ratio between the nominal values of electromagnetic torque and stator flux is represented by this coefficient, which is defined as:

$$\lambda = \left|\frac{T_{nom}}{\psi_{nom}}\right| \quad (17)$$

Also, to avoid overcurrent, a constraint is added to the cost function. The protection is activated when the absolute value of the predicted current is higher than the maximum current. So this constraint is defined as follows:

$$I_m = \begin{cases} \infty if : |i(k+1)| > i_{max} \\ 0 \;\; if : |i| < i_{max} \end{cases} \quad (18)$$

Therefore, the total cost function g is given by:

$$g_{i,j} = \left| T_{em}(k+1)_{i,j} - T_{em_{ref}} \right| + \lambda.\left| \psi_s(k+1)_i - \psi_{ref} \right| + I_m \tag{19}$$

Figure 5 shows the flowchart of PTC control.

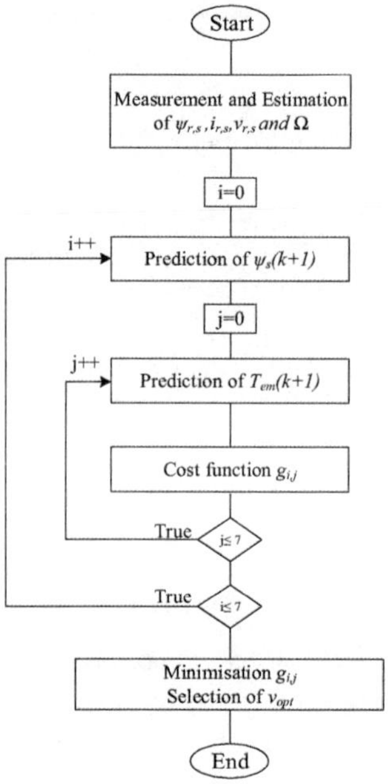

Fig. 5. Flowchart of the predictive torque control

4 Comparison Study

The dynamic and static properties of DTC and PTC control for the DFIM are compared in this section. This comparison is carried out in the Matlab/Simulink environment with the goal of determining which method best meets the requirements, which include:

- Improved tracking of control set point (speed);
- Torque and flux ripple have been reduced.

This simulation makes use of a 1.5 kw DFIM, the characteristics of which are shown in Table 2 below.

Table 2. Motor parameters

Variable	Value (unit)
Pair pole number	P = 2
Stator resistance	$R_s = 1.75\ \Omega$
Rotor resistance	$R_r = 1.68\ \Omega$
Maximum of mutual inductance	M = 0.165 H
Stator self-inductance	$L_s = 0.295$ H
Rotor self-inductance	$L_r = 0.104$ H
Total inertia	$J = 0.01\ kg.m^2$
Total viscous frictions	$f = 0.0027\ kg.m^2/s$

Figures 6, 7 and 8 show the simulation results of these two controls, which were tested in the same conduits.

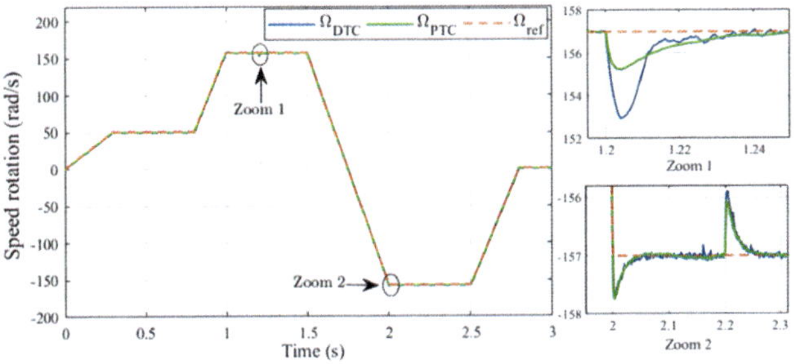

Fig. 6. Speed rotation

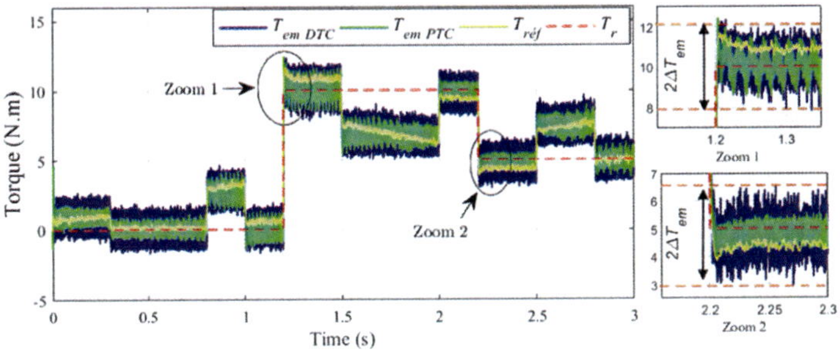

Fig. 7. Torque

When compared to the DTC, the PTC simulation findings promise higher performance. It is interesting to note that the PTC control differs from the DTC in that it

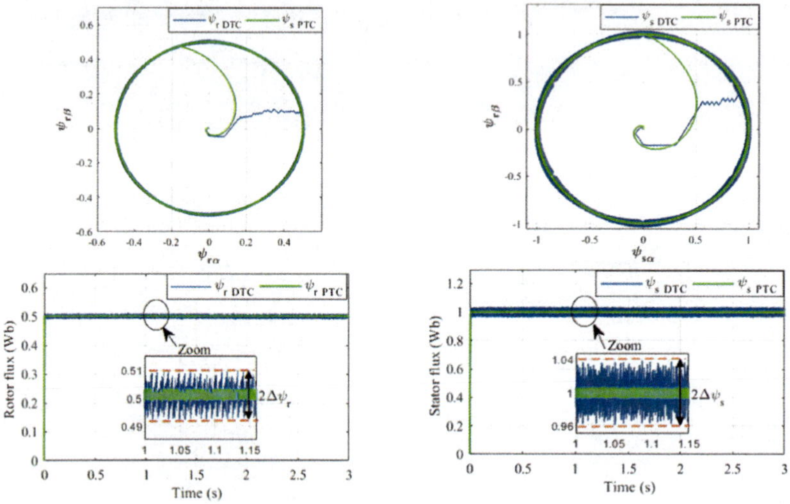

Fig. 8. Stator and rotor flux

significantly attenuates torque and flux ripples. In terms of rejection time, torque ripples, and flux ripples, Table 3 summarises the comparison analysis between the two forms of control.

Table 3. Comparison results between DTC and PTC

Performances		DTC	PTC	Improvement (%)
Speed perturbation (rad/s)		4	1.7	57.5%
Rejection time (s)		0.049	0.048	2.04%
Ripple	Torque (N.m)	1.785	0.58	67.5%
	Stator flux (Wb)	0.0383	0.0066	82.7%
	Rotor flux (Wb)	0.0092	0.0024	73.9%

The PTC has made significant progress, as seen in this table. These improvements include speed disturbance optimisation and a significant reduction in torque ripples and fluxes. As a result, predictive torque control offers better performance than DTC.

5 Conclusion

This study discusses the use of predictive control to enhance the DFIM control approach. This technique of improvement is based on future flux and electromagnetic torque forecasts, as well as a cost function that optimises the voltage vectors that drive the inverters. The capacity to cope with the imprecise, ambiguous, and hazy justifies the adoption of this strategy. The results of a comparison between PTC and DTC are provided. The examination of the acquired findings reveals that the suggested technique provides improved outcomes, such as better speed tracking and a decrease in torque and flux ripples.

References

1. Bossoufi, B., Karim, M., Lagrioui, A., Taoussi, M., Derouich, A.: Observer backstepping control of DFIG-Generators for wind turbines variable-speed: FPGA-based implementation. Renew. Energy **81**, 903–917 (2015)
2. El Mahfoud, M., Bossoufi, B., El Ouanjli, N., Mahfoud, S., Taoussi, M.: Three speed controllers of direct torque control for a doubly fed induction motor drive–a comparison. Electrica **21**(1), 129–141 (2021)
3. Bakouri, A., Mahmoudi, H., Abbou, A.: Intelligent control for doubly fed induction generator connected to the electrical network. Int. J. Power Electron. Drive Syst. **7**(3), 688–700 (2016)
4. Elmahfoud, M., Bossoufi, B., Taoussi, M., El Ouanjli, N., Derouich, A.: Comparative study between backstepping adaptive and field oriented controls for doubly fed induction motor. Eur. J. Electr. Eng. **22**(3), 209–221 (2020)
5. El Ouanjli, N., et al.: Modern improvement techniques of direct torque control for induction motor drives-a review. Prot. Control Mod. Power Syst. **4**(1), 1–12 (2019)
6. El Ouanjli, N., Derouich, A., El Ghzizal, A., Chebabhi, A., Taoussi, M.: A comparative study between FOC and DTC control of the Doubly Fed Induction Motor (DFIM). In: Proceedings of 2017 International Conference Electrionics and Information Technolpgy, ICEIT 2017, vol. 2018-Janua, pp. 1–6 (2018)
7. El Mahfoud, M., Bossoufi, B., El Ouanjli, N., Said, M., Taoussi, M.: Improved direct torque control of doubly fed induction motor using space vector modulation. Int. J. Intell. Eng. Syst. **14**(3), 177–188 (2021)
8. Davari, S.A., Khaburi, D.A., Kennel, R.: An improved FCS-MPC algorithm for an induction motor with an imposed optimized weighting factor. IEEE Trans. Power Electron. **27**(3), 1540–1551 (2012)
9. Kumar, R., Das, S., Bhaumik, A.: Speed sensorless model predictive current control of doubly-fed induction machine drive using model reference adaptive system. ISA Trans. **86**, 215–226 (2019)
10. Vazquez, S., Rodriguez, J., Rivera, M., Franquelo, L.G., Norambuena, M.: Model predictive control for power converters and drives: advances and trends. IEEE Trans. Ind. Electron. **64**(2), 935–947 (2017)
11. Rodríguez, J., Kennel, R.M., Espinoza, J.R., Trincado, M., Silva, C.A., Rojas, C.A.: High-performance control strategies for electrical drives: an experimental assessment. IEEE Trans. Ind. Electron. **59**(2), 812–820 (2012)
12. Wang, F., Xie, H., Chen, Q., Davari, S.A., Rodriguez, J., Kennel, R.: Parallel predictive torque control for induction machines without weighting factors. IEEE Trans. Power Electron. **35**, 1779–1788 (2019)
13. Zhang, Y., Yang, H., Xia, B.: Advanced control strategies of induction machine: field oriented control, direct torque control and model predictive control. IET Electr. Power Appl. (2015)

Control of a Bi-axial Solar Tracker by Observing the Power Supplied by the MPPT

Hicham Bouzakri(✉), Ahmed Abbou, Zakaria Abousserhane, and Rafika El Idrissi

Electrical Engineering Department, Mohammadia School of Engineers, Mohammed V University in Rabat, Rabat, Morocco
hichambouzakri@research.emi.ac.ma, abbou@emi.ac.ma

Abstract. The solar tracker is a mechanism that helps the photovoltaic panel maximize its efficiency during the day while keeping it oriented towards direct solar radiation. To manipulate the movement of the solar tracker, most use a solar collector that specifies the position of the sun before operating the motors. The idea of this paper is to replace the solar collector with an observation system of the PV output power (between the PV and the MPPT). The follower control card observes the variation of this power as long as it is increased. The system continues the correction of the tracking and it stops only if it detects a decrease in this power. In this way, we will have a perfect follow-up that looks for the peak of the power provided by the photovoltaic panel.

Keywords: Photovoltaic panel · Solar tracker · Solar radiation · Mppt

1 Introduction

Renewable energy is defined as the generation of energy from an infinite source, for example, the sun's beams or the wind. Solar photovoltaic energy is one of the most frequently utilized sources of energy. A photovoltaic panel's job is to convert solar energy into electricity [1]. This panel must be placed perpendicular to the sun to get the maximum amount of solar energy [2]. For this reason, we added a mechanical system called a solar tracker [3–6]. To specify the tracking, the majority of solar trackers use two motors, one for the horizontal axis and one for the vertical axis. [7–10].

Researchers in Turkey and Malaysia have proposed the use of artificial intelligence to improve the tracking systems of solar trackers; this method offers a prediction of the position of the sun in the sky with high-level accuracy [11].

Other researchers in Turkey deployed a two-axis horizontal and vertical tracking system capable of determining the azimuth and elevation angles using precalculated formulas for one year. A prototype generated 42.6% more energy than a fixed system [12]. This method of monitoring is known as "monitoring by programming." It does not use a solar sensor; It uses astronomical computations to predict where the sun will be. The problem with these tracking technologies is that they keep monitoring even when the sky is cloudy.

S. Motahhir and B. Bossoufi (Eds.): ICDTA 2022, LNNS 454, pp. 836–844, 2022.
https://doi.org/10.1007/978-3-031-01942-5_83

Researchers in Thailand made a prototype of a low-cost bi-axial solar tracker. The tracking system is a closed-loop system, which means they used a solar collector. The sensor is formed by four LDRs distributed on the sides of the PV. The advantage of using a sensor is that it stops the follower when the sky is covered. On the other hand, it is difficult to have a high level of precision with the LDRs. This will surely influence the performance of the PV [13].

The idea of this paper is to introduce to the solar tracker control board another tracking control system that consists of observing the power supplied by the PV between the PV and the MPPT (Maximum Power Point Tracking). The solar sensor offers the control system the orientation of the two motors to the East or the West, up or down. As the motor turns (horizontal or vertical motor), we will observe the variation of PV power via two sensors installed between the PV and the MPPT (current sensor and voltage sensor). As long as the power increases, the motor continues on its way, and it will not stop until the power reaches its maximum and begins to decrease.

A general introduction to the solar tracker and its control is given in the article's first paragraph. This is followed by a thorough explanation of the suggested approach and the flow chart that will be utilized on the biaxial solar tracker's control board.

2 Bi-axial Solar Follower

The bi-axial solar tracker is a device that is installed with a photovoltaic panel to keep it perpendicular to the sun during the day. The majority of solar trackers have two axes; one horizontal and one vertical (see Fig. 1).

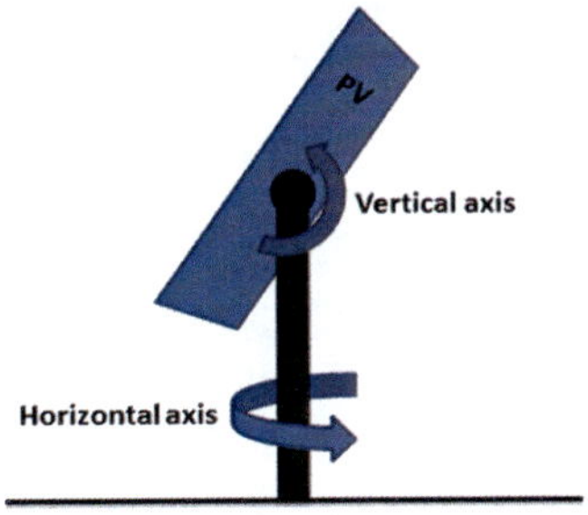

Fig. 1. Bi-axial solar tracker.Within a circular route, the sun seems to move from east to west during daylight hours in the Northern Hemisphere (see Fig. 2) and around the southern star for the southern hemisphere (see Fig. 3) [14].

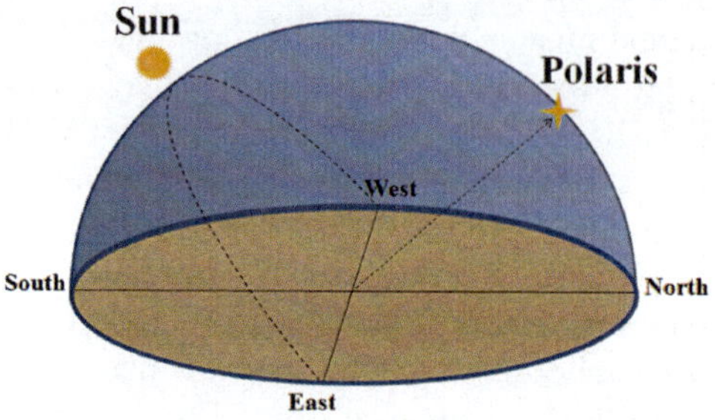

Fig. 2. In the northern hemisphere, the sun's apparent course during the day.

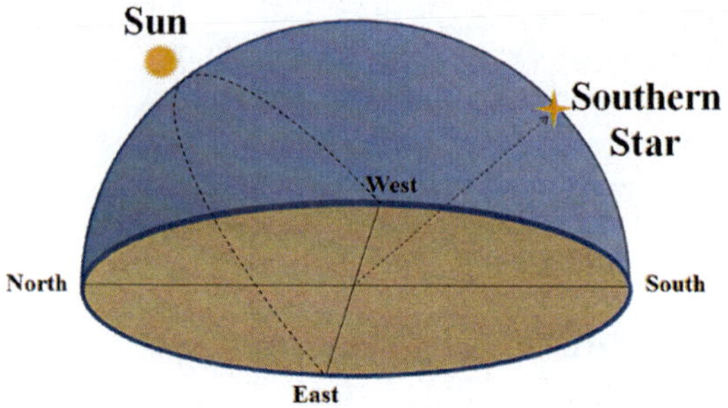

Fig. 3. The apparent path of the sun during the day in the southern hemisphere.

This apparent solar path changes its inclination from day to day throughout the year, it lies up to +24.45° concerning the celestial equator at the summer solstice and it descends to −24.54° at the solstice winter (see Fig. 4) according to the following "Eq. (1)": [13].

$$\delta = 23.45 \sin\left[\frac{360}{365}(d_n + 284)\right] \tag{1}$$

With δ in degree

d_n is a number the day of the year (1 = 1 January and 365 = 31 December). [13].

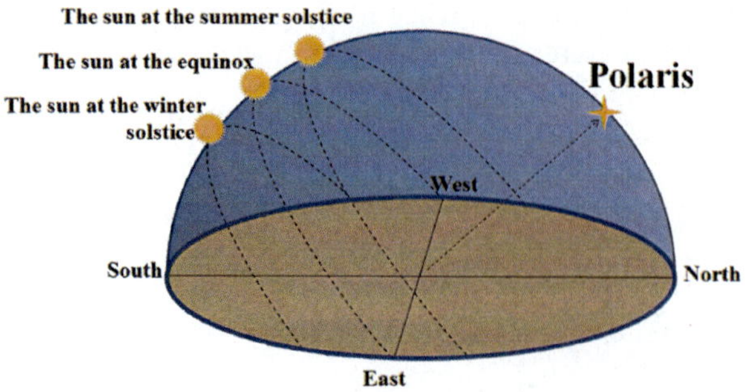

Fig. 4. The sun's apparent path throughout the year.

To perfectly follow this solar movement during the day, the bi-axial tracker operates two motors, One motor is mounted on the horizontal axis while the other is mounted on the vertical axis, as depicted in (see Fig. 5).

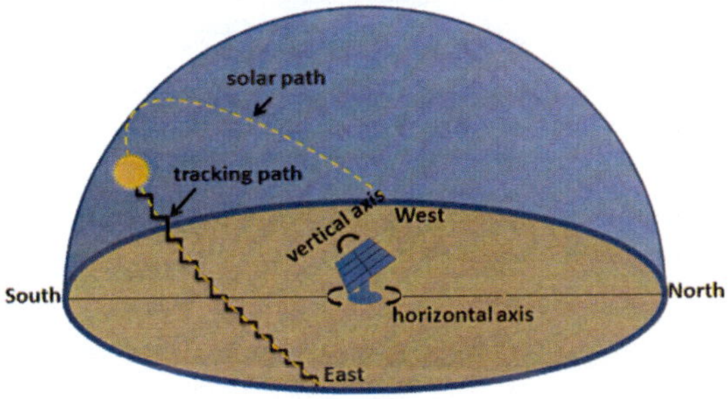

Fig. 5. The bi-axial solar tracker follows the sun.

Most systems employ two ways to determine the location of the sun during tracking: one by calculation and the other by the sensor.

The calculation technique involves determining the precise location of the sun's light at each instant of the day. To do this, the system uses an internal or external clock to determine the precise time and date before each tracking correction, and then it calculates how much each of the two motors should be rotated (see Fig. 6).

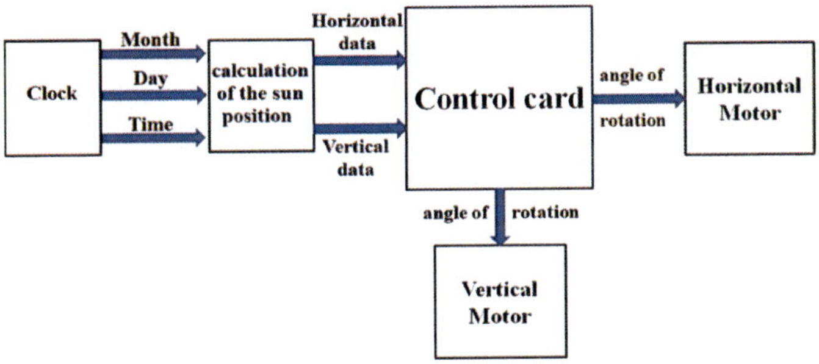

Fig. 6. Control logic by calculation.

The sensor method consists of using a solar collector that offers the instantaneous position of the sun during the day (see Fig. 7).

This sensor shown in Fig. 7 is a simple four-LDR sensor. If the LDR1 is illuminated while the LDR2 is under amber, it means that the sun is in the West (see Fig. 8 -a-), the control board operates the horizontal motor towards the West until the moment when the two LDRs will be illuminated (see Fig. 8 -b-), and vice versa. If the LDR3 is illuminated while the LDR2 is under amber, it means that the sun is at the top of the sensor (see Fig. 8 -c-), the control board operates the vertical motor towards the up until the moment when the two LDRs will be illuminated (see Fig. 8 -d-) and vice versa.

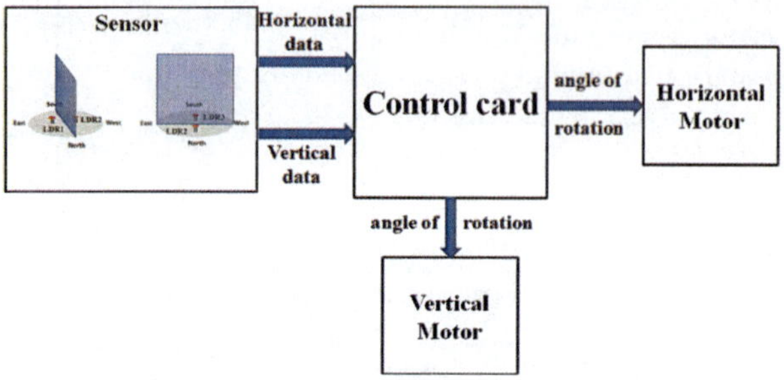

Fig. 7. Control logic by the sensor.

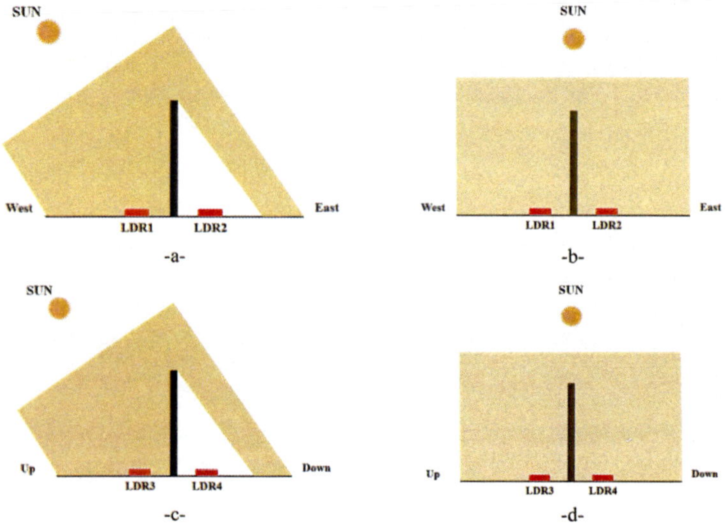

Fig. 8. The sun's position about the four LDR sensors.

3 The Proposed Bi-axial Solar Tracker Modification

As it is known, the photovoltaic panel should not be connected directly to the battery, the system will be damaged. As a result, between the PV and the battery, a voltage regulator was installed. Most systems use either a PWM (Pulse Width Modulation) or an MPPT (Multiple Power Point Tracking) to get the most power out of the PV module. The MPPT is a device introduced to the solar panel whose role is to control the suitable voltage for charging the battery while absorbing the maximum possible power from the PV module (see Fig. 9) [15, 16].

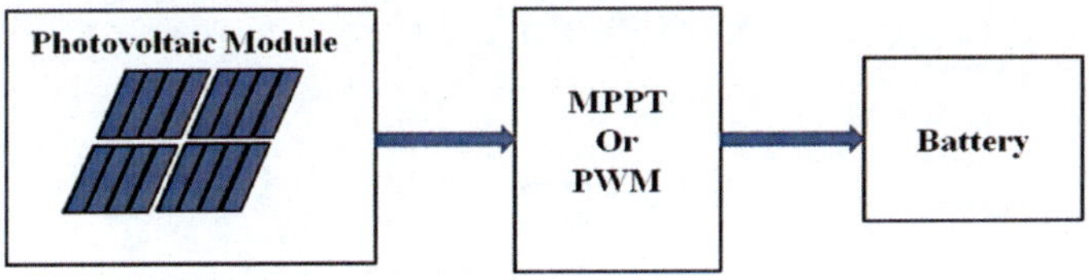

Fig. 9. Battery charging system through a PV module.

The idea proposed is to introduce between the PV and the MPPT a current sensor and a voltage sensor (see Fig. 10) to calculate at each moment the power supplied by the PV module.

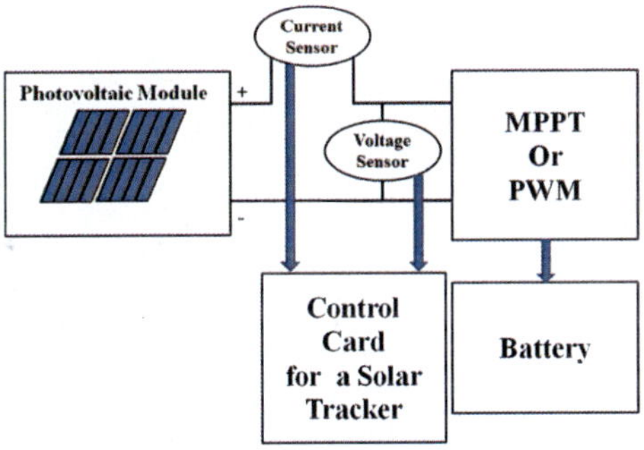

Fig. 10. The voltage and current sensors are installed between the PV and the MPPT.

To follow the progression of this power, this measurement will be sent to the control card of our tracking device. While adjusting the tracking, the control board uses the four LDR sensors to know the sun's position relative to the two axes (the vertical axis and the horizontal axis). Then the board activates the horizontal motor with the instantaneous observation of the evolution of the PV power. As long as the panel approaches the perfect orientation towards direct solar radiation, the measured power increases, but as it starts to drop, the horizontal motor tracking stops.

The same procedure applies to the tilt axis. After the correction of the horizontal axis, the system switches to the vertical axis according to the measurements of the sensor. The axis starts to rise or fall as the map indicates. The control panel observes the evolution of the PV power until it reaches its maximum, at which point the tilt motor stops. We are therefore in the perfect orientation towards direct solar radiation in which the PV module supplies the maximum power to the battery.

Note that the maximum power given by the PV module is impacted by the amount of sunlight and the temperature. Therefore, the exact orientation of the solar panel towards the sun does not necessarily indicate the maximum power supplied by the PV module. As you approach the exact orientation towards the sun, the amount of radiation increases, but the PV cells heat up.

4 The Flowchart of the Command Card

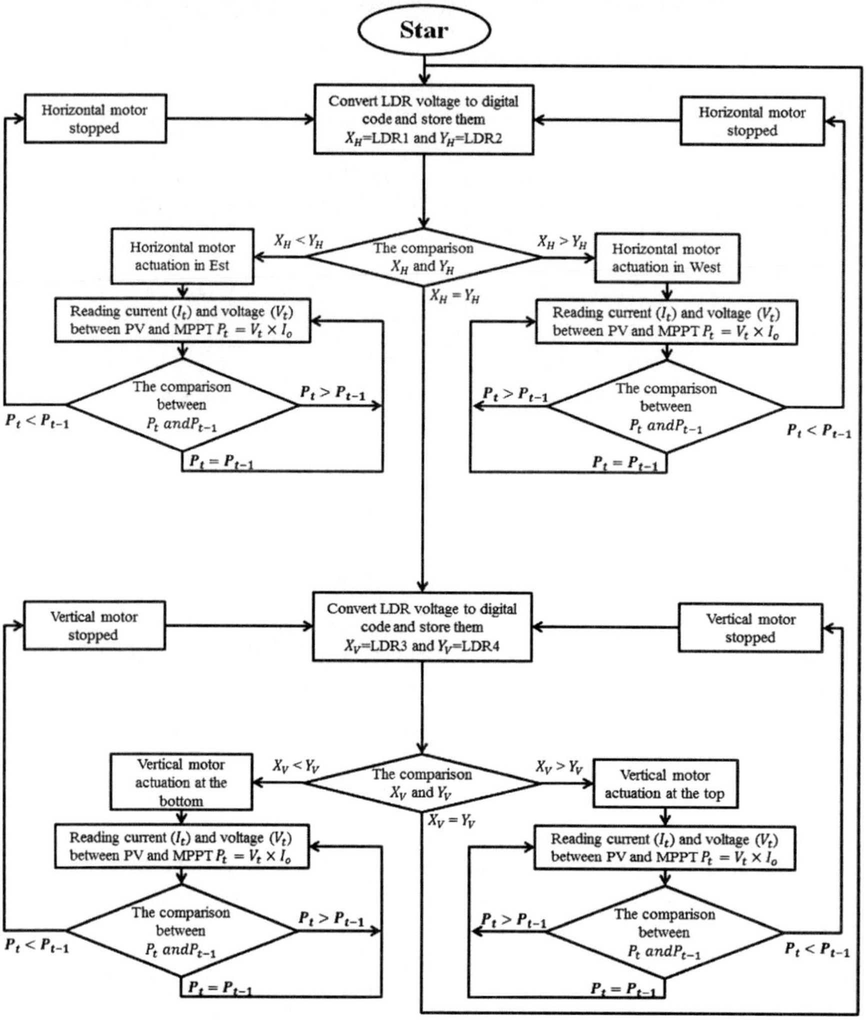

Fig. 11. Control board flowchart for a two-axis solar tracker.

The system starts with the conversion of the value of LDR1 and LDR2 in digital code via a digital-analog converter and stocks in two predefined variables X_H and Y_H, then it compares the two values, if $X_H > Y_H$ that is to say that the sun is west of the sensor the Horizontal motor will be operated in the West, on the other hand, if $X_H < Y_H$ that is to say that the sun is to the East of the sensor the Horizontal motor will be operated in the East, in both cases as the motor rotates left or right, the control board links the voltage and current between the PV and the MPPT to calculate and observe the power P_t, if the power is stable or it increases the system continues the actuation of the motor Horizontal, If the system detects that the power begins to decrease the Horizontal motor stops, The technique is the same for the vertical motor (see Fig. 11).

After completing the Horizontal tracking correction, the system proceeds in the same way to the Vertical tracking correction with the use of LDR3 and LDR4 (X_V and Y_V) and the Vertical motor actuation. The system starts with the conversion of the values of LDR3 and LDR4 in digital code via a digital-analog converter and stores them in two predefined variables, X_V and Y_V. Then it compares the two values. If $X_V > Y_V$ that is to say that the sun is the top of the sensor, the vertical motor will be operated at the top. On the other hand, if $X_V < Y_V$ that is to say that the sun is to the bottom of the sensor, the vertical motor will be operated at the bottom. In both cases, as the motor rotates top or bottom, the control board links the voltage and current between the PV module and the MPPT to calculate and observe the power. If the power is stable or it increases, the system continues the actuation of the horizontal motor. If the system detects that, the power is beginning to decrease; the vertical motor stops (see Fig. 11).

5 Conclusions

The use of the bi-axial solar tracker requires a solar sensor, which helps the control system observe the position of the sun at every moment of tracking during the day. This article proposes to change this solar observation procedure by another more precise one, which consists in observing during tracking the evolution of the power between the PV module and the MPPT (Maximum Power Point Tracking) to find the peak of this power during the correction of the regular. This manipulation was explained by a flowchart, which details the moment when the system starts operating the two motors until the moment when the motors stop.

6 Perspective

To practically value the system, we will move on to the realization of a follower with the command card according to the proposed flowchart. The findings of the tests will be compared to those of a stationary photovoltaic panel.

References

1. Abousserhane, Z., Abbou, A., Id-Khajine, L., Zakzouk, N.: Power flow control of PV system featuring on-grid and off-grid modes. In: 7th International Renewable and Sustainable Energy Conference (IRSEC) , Agadir, pp. 1–7 (2019)
2. Wenham, S.R., Green, M.A., Watt, M.E., Corkish, R., Sproul, A.: Applied Photovoltaics. Routledge, Abingdon (2013)
3. Bouzakri, H., Abbou, A.: Study and realization of a monoaxial solar tracker over an equatorial mount. In: 7th International Renewable and Sustainable Energy Conference (IRSEC), Agadir, pp. 1–6 (2019)
4. Bouzakri, H., Abbou, A.: Mono-axial solar tracker with equatorial mount, for an improved model of a photovoltaic panel. Int. J. Renew. Energy Res. (IJRER) **10**(2), 578–590 (2020)

5. Bouzakri, H., Abbou, A., Tijani, K., Abousserhane, Z.: Biaxial equatorial solar tracker with high precision and low consumption: modelling and realization. Int. J. Photoenergy (2021). https://doi.org/10.1155/2021/6679576
6. Bouzakri, H., Abbou, A., Chennoufi, K.: Efficiency enhancement of a fixed photovoltaic panel using simple motorized reflection model. Int. J. Intell. Eng. Syst. **14**(4) (2021). https://doi.org/10.22266/ijies2021.0831.08
7. Batayneh, W., Bataineh, A., Soliman, I., Abed Hafees, S.: Investigation of a single-axis discrete solar tracking system for reduced actuations and maximum energy collection. Autom. Constr. **98**, 102–109 (2019)
8. Yao, Y., Hu, Y., Gao, S., Yang, G., Du, J.: A multipurpose dual-axis solar tracker with two tracking strategies. Renew. Energy **72**, 88–98 (2014)
9. Oh, S.J., Burhan, M., Ng, K.C., Kim, Y., Chun, W.: Development and performance analysis of a two-axis solar tracker for concentrated photovoltaics. Int. J. Energy Res. **39**(7), 965–976 (2015)
10. Said, M.N.A.M., Jumaat, S.A., Jawa, C.R.A.: Dual axis solar tracker with IoT monitoring system using arduino. Int. J. Power Electron. Driver Syst. **11**(1), 451–458 (2020)
11. Al-Rousan, N., Isa, N.A.M., Desa, M.K.M.: Efficient single and dual axis solar tracking system controllers based on adaptive neural fuzzy inference system. J. King Saud Univ.-Eng. Sci. **32**(7), 459–469 (2020)
12. Sungur, C.: Multi-axes sun-tracking system with PLC control for photovoltaic panels in Turkey. Renew. Energy **34**(4), 1119–1125 (2009)
13. Jamroen, C., Komkum, P., Kohsri, S., Himananto, W., Panupintu, S., Unkat, S.: A low-cost dual-axis solar tracking system based on digital logic design: design and implementation. Sustain. Energy Technol. Assess. **37**, 100618 (2020)
14. Iqbal, M.: An Introduction to Solar Radiation. Elsevier, Amsterdam (1983)
15. El Idrissi, R., Abbou, A., Mokhlis, M., Salimi, M.: Adaptive backstepping controller design based MPPT of the single-phase grid connected PV system. Int. J. Intell. Eng. Syst. **14**(3), 282–293 (2021)
16. El Idrissi, R., Abbou, A., Mokhlis, M., Bouzakri, H., El Houm, Y.: Real- Time implementation of a PV system maximum power point tracking based on the ANN-Backstepping sliding mode control. Int. J. Renew. Energy Res. **11**(4), 1959–1967 (2021)

Control of Fluid Flow Coupled on Heat Transfer Around a Square Cylinder by Using Three Attached Partitions

Youssef Admi(✉), Mohammed Amine Moussaoui, and Ahmed Mezrhab

Mechanics and Energy Laboratory, Faculty of Sciences Oujda, Mohammed First University, Oujda, Morocco
admi_youssef1718@ump.ac.ma

Abstract. In this research, a numerical investigation of the fluid flow and thermal convection phenomena around a heated square cylinder controlled by three attached downstream partitions is conducted. The numerical simulations are performed at a fixed Reynolds number (Re = 150) by using the double lattice Boltzmann method with multiple relaxation times (MRT-DLBM). The insertion of the partitions is to control the vortex shedding (reduction of the drag coefficient) and improve the heat transfer around the squares cylinder.

The numerical results are presented as contours of the streamlines, the isotherms lines and the time and average drag coefficient. These results indicate that there is a critical length of the control partitions Lp = 4 where a maximum percentage reduction of the drag coefficient reached 17.33%. Also, regular and important heat exchange is observed for this crucial length.

The numerical code has been validated with existing numerical investigations in the literature for a flow around a square cylinder without and with a single detached partition.

Keywords: Double MRT thermal lattice Boltzmann method · Flow control · Drag coefficient reduction · Heat exchange · Partitions control

1 Introduction

Generally, fluid flow phenomena coupled with convective heat transfer around bluff bodies are frequently encountered in nature and/or also in various industrial applications, including heat exchangers, cooling of electronic components, risers in marine technology, road vehicles, aeroplanes, buildings and bridges etc.[1–4]. In these domains, the structures are considered as bluff bodies characterized by charges induced either by the strong fluid-structure interaction or by the presence of a temperature gradient. These interactions generate large unstable forces that can violently damage the structural integrity of the bluff body. These effects can be reduced by integrating passive and active methods proposed to control the wake vortex dynamics with the purpose of attenuating the vortex shedding, reducing the amplitude of fluctuating lift, drag, vibration and radiated sound, also, for improving the heat exchange between the incoming fluid and the square cylinder.

S. Motahhir and B. Bossoufi (Eds.): ICDTA 2022, LNNS 454, pp. 845–854, 2022.
https://doi.org/10.1007/978-3-031-01942-5_84

A multitude of pioneering numerical and experimental investigations are being conducted by researchers, scientists and engineers working in the thermofluids and digital technologies fields [5–11]. Moussaoui et al. [6] studied the flow around a square cylinder placed in a channel controlled by a bi-partition using the Boltzmann lattice method at a fixed Reynolds number (Re = 150). The results obtained by the authors show that the implementation of the bi-partition was beneficial in all cases studied. This was reflected in the reduction of the drag value and in the fluctuating lateral forces due to vortex shedding behind the square block. Rashidi et al. [7] conducted a review of existing numerical and experimental studies related to vortex shedding suppression and wake control methods. The authors classified these methods into two groups and presented the advantages, limitations, energy efficiency and specific applications of each method.

Wang et al. [8] present a paper on particle image velocimetry (PIV) of the two-dimensional flow past a bluff body confined in a channel. The authors study the effect of bluff body shape (three typical shapes) with the same height, but with varying blocking ratio (β = 0.25, 0.3, 0.4 and 0.5) for Reynolds numbers (Re = 3200, 4600 and 5400). They found that for each body shape, the vortex loss patterns are relatively insensitive to Reynolds number but strongly dependent on the blocking ratio. Also, they observed that the strength of the vortex effusion generally decreases with increasing blockage ratio except for circular cylinder at B = 0.25–0.3. On the other hand, they found that there is a proportional and monotonic increase between the Strouhal number (St) and the blocking ratio until the vortex loss is removed. Moussaoui et al. [9] carried out a numerical study of two-dimensional laminar flow and heat transfer characteristics around two different shaped tandem obstacles in a horizontal channel. This study is based on a coupling between the Lattice Boltzmann Method (LBM) and the Finite Difference Method (FDM). The authors studied the influence of the Reynolds number and the distance between the two obstacles on the fluid flow and heat transfer. From the results obtained, the authors concluded that the flow structure is strongly influenced by the Reynolds number and the spacing. Likewise, the heat transfer is clearly affected by the variation of the spacing between the two cylinders, especially at high Reynolds numbers. Admi et al. [11] perform a numerical simulation of the laminar flow and convective heat transfer characteristics around three heated square cylinders placed side by side and equally spaced in a 2D horizontal channel. A flat plate is inserted behind the obstacles in order to control the flow in the central obstacle wake. The fluid enters with a cold temperature in order to cool the three heated square obstacles. The authors found that the optimal position of the flat plate is g = 3 to obtain maximum vortex suppression. Also, the authors performed a study on the effect of plate length on fluid flow and heat transfer and found that to have a large and even heat exchange and maximum vortex shedding suppression, the length of the control plate should be equal to 4d.

The numerical approach used to perform the simulations is the Boltzmann lattice method. This method has been extensively developed and utilized by a large number of researchers in various disciplines such as physics, chemistry, health, biology, etc. Two models of LBM are widely used in the literature: the SRT model and the MRT model.

In the SRT model (also called the "BGK" model [12], all physical quantities relax at a single relaxation time. Whereas in the other model (MRT-LBM): the lattice Boltzmann method with multiple relaxation times [5, 9, 11, 13] where each physical quantity relaxes at its own relaxation time. Due to its convergence and stability, the LBM-MRT is the most widely used model in recent years and it is the one used for the discretization in this work.

2 Physical Problem Configuration and Condition Boundaries

2.1 Physical Problem Configuration

Figure 1 illustrates the configuration of the computational domain to study the flow around a heated square cylinder controlled by three attached plates. The assembly is placed in a two-dimensional horizontal channel of length L = 37d and height H = 11d where "d" represents the dimension of the square cylinder. The square cylinder is located at an upstream distance Lu = 6d (see Fig. 1). The control plates are attached downstream of the square cylinder. These plates have the same length "Lp" and height "h = 0.02d".

At the inlet of the channel, the incoming fluid with dimensionless temperature $\theta_c = -0,5$ is characterized by a parabolic velocity profile $\left(u = 1.5 * U_{max}\left[1 - \left({}^{y}/_{H}\right)^2; \text{v} = 0\right)\right.$, where U_{max} is the maximum inlet velocity, and y is the vertical dimension from the centerline, and u and v are the components of the velocity vectors. Also at the outlet, the parabolic velocity profile is imposed and the velocity and pressure gradients are considered zero.

Note that the block is considered heated with a dimensionless temperature $\theta_h = 0,5$. While the walls of the channel are taken as adiabatic.

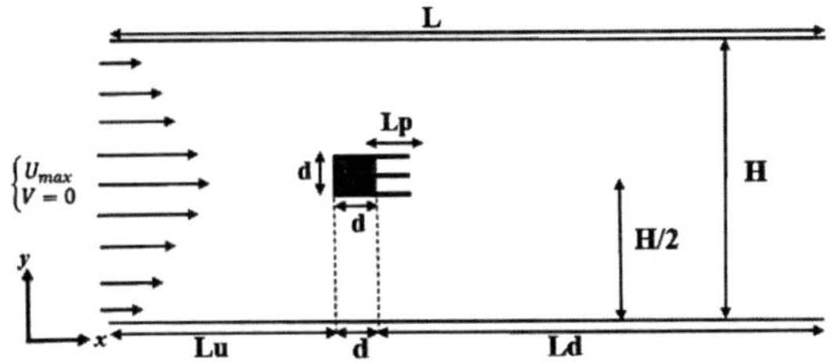

Fig. 1. Physical problem configuration

2.2 Condition Boundaries

The implementation of boundary conditions is a very important step in any numerical method, especially in the lattice Boltzmann method. These conditions are essential for the stability, accuracy and convergence of numerical calculations. For solid boundaries such as channel walls and square cylinder faces, the Bounce-Back boundary conditions [14] are employed. While at the inlet and outlet of the channel, the flow is completely developed with a parabolic velocity profile, so the implementation of Zou and He boundary conditions is most preferred [15].

The conditions proposed by Mezrhab et al. [16] are applied to treat the heat flow boundary conditions. Thus, adiabatic boundary conditions are imposed for the channel walls.

3 Results and Discussions and Code Validation

3.1 Code Validation

Since the present work consists of a square cylinder and control partitions, we present two validation studies. The first study treats the problem of flow through a square cylinder without a control plate with a blocking ratio of $\beta = 1/8$ and for a range of $20 \leq Re \leq 300$. The results obtained are compared with those found by Breuer et al. [17].

Figure 2-a. illustrates the graph presenting the variation of averaged drag coefficient values as a function of the Reynolds number. This graph shows that our LBM results are in good agreement with the reference results. Note that in this work, Breuer et al. employed two different methods (LBM and MVF).

A second validation study is performed to ensure the quality and precision of our numerical code. In this investigation, the simulation of the flow of a fluid around a square cylinder controlled by a flat plate at a blocking ratio $\beta = H/d = 11$ was performed and validated with the results found by Islam [18]. The graph in Fig. 2-b presents the average values of the drag coefficient for different gap spacing at a fixed Reynolds number (Re = 150). This figure shows that our results obtained by using our LBM-MRT code are almost identical to the reference results.

From these two validation studies, we can conclude that our LBM-MRT based code is reliable for studying our present problem.

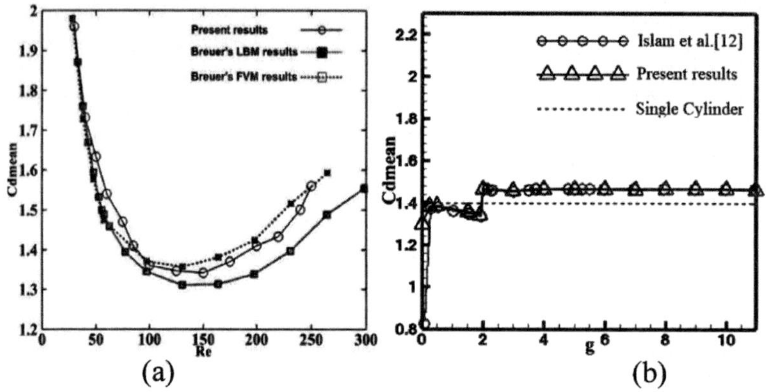

Fig. 2. Comparison of the average drag coefficient with previous works for a single obstacle: (a) without control; (b) with control partition

3.2 Results and Discussions

As previously mentioned, the physical problem considered in this work is to study the effect of the length of the three attached partitions on the velocity contours of the flow patterns, the isotherm contours and on the mean and time drag coefficient. The partition length Lp is varied from 0.25d to 6d.

3.2.1 Fluid Flow Results

Figure 3 shows the velocity contours around the square cylinder, heated and controlled by three attached parallel partitions placed horizontally upstream. In this figure (Fig. 3), an alternate Van Karman vortex street is seen in the upstream wake.

This wake can be considered as periodically undulating in the case where the control partitions are short lengths (cases tested Lp = 0.25; 0.5; 1). The vortex shedding number decreases with the increasing plate length until a critical length Lp = 4d. After this critical length, the wake becomes undulating again. Likewise, the size of detached vortexes behind the square cylinder decreases with the increasing length until Lpcr = 4d where the vortexes are smaller. Then, this size increases strongly with the increase of the length of the partitions. We also observe that the number of vortexes shedding in the upstream wake for the short lengths cases increase until the critical case. Then, these vortex increases with the increase of the length of the partitions.

In the case of short partition lengths, the shear layers produced by the extremities of the square cylinder are evacuated just behind the partitions, resulting in alternating positive and negative vortices. A decrease in the number and size of these vortices is observed with increasing partition length up to a critical length Lp = 4d. At this critical length, a minimization of the number and size of vortexes is observed. In this case, part of the chisel layers fall on the faces of the partitions, while the other part falls behind the partitions. Also, for long partitions, the shear layers falls totally on the top and bottom face of the partitions. This gives the possibility of a collision between two successive shear layers, which generates large vortexes.

3.2.2 Temporal Variation and Average Values of Drag Coefficient

The present section presents the effect of the length of the partitions on the temporal and averaged values of drag force acting on the square cylinder. Table 1 illustrates the mean values obtained from the drag coefficient Cdmean for each length of the control partitions while fixing the spacing between the partitions and the square cylinder. Thus,

Figure 4 illustrates the time variation of the drag coefficient for a certain length of the partition. From the results presented in the table and in this figure, it can be seen that the drag coefficient varies periodically for all lengths. Also, a diminution is observed in the average and temporal values of the drag coefficient. This diminution arrives at a critical value where a percentage reduction of the drag coefficient reaches 17.33%. This corresponds to the same critical length found previously where suppression or diminution of vortices is observed.

For short partition lengths, the incoming fluid applies a force almost on the full cylinder, while for the critical zone, the partition length increases and the fluid exerts a small force on the back face of the cylinder. If the length increases, the shear layers are discharged on the faces of the partitions and collide with each other. This increases the fluid forces on the top and bottom of the cylinder, which increases the time-averaged values of the drag coefficient.

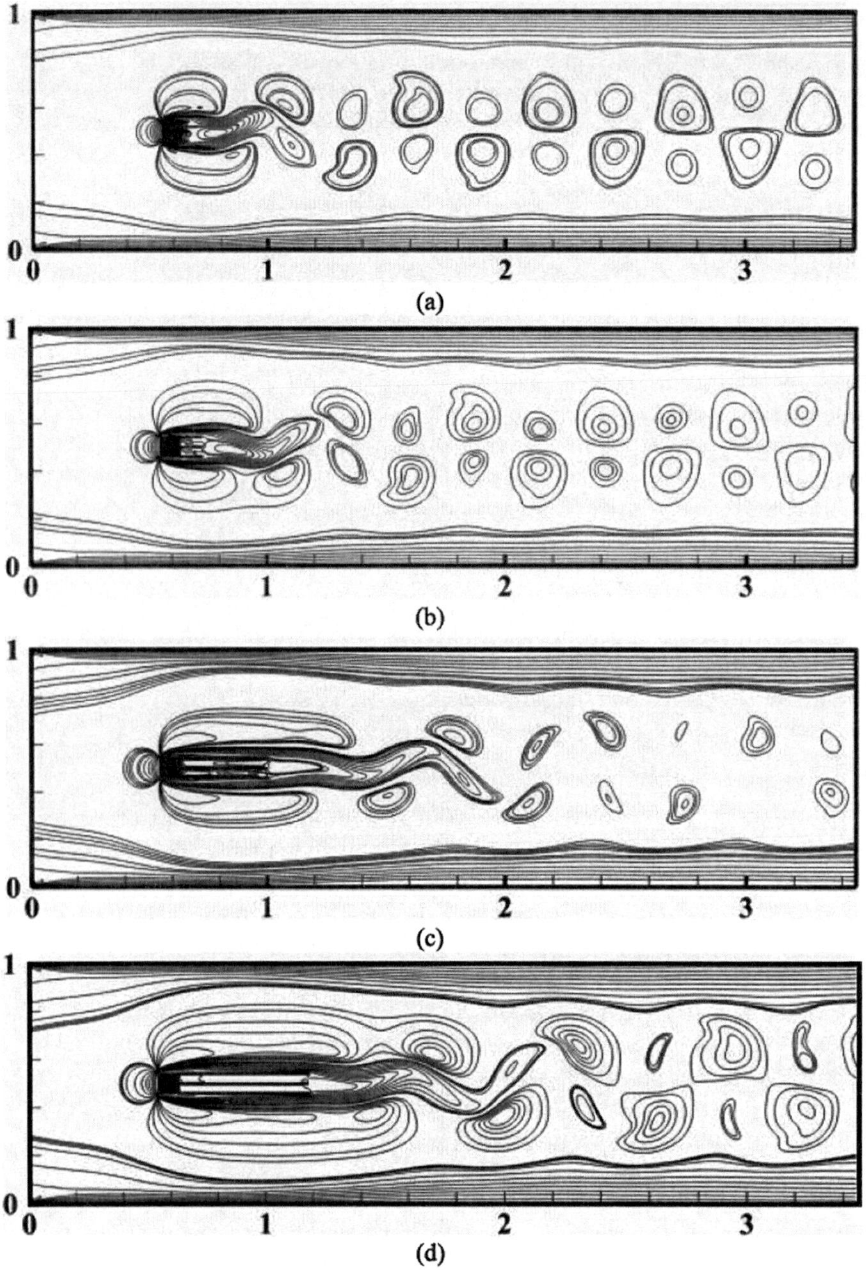

Fig. 3. Velocity contours visualization for: (a) Lp = 0.25d; (b) Lp = 1d; (c) Lp = 4d; (d) Lp = 6d.

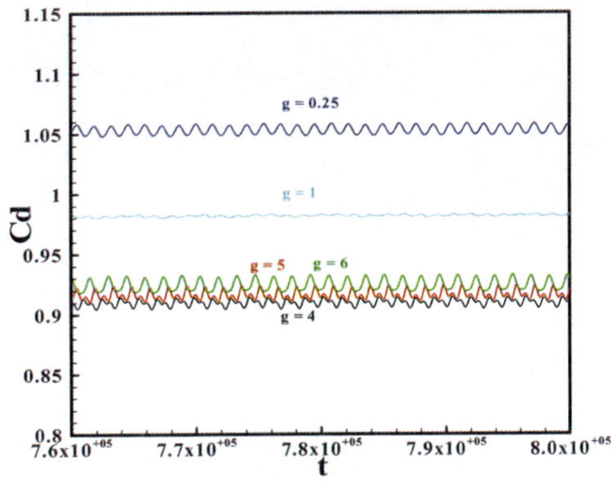

Fig. 4. Temporal variation of drag coefficient for different lengths.

Table 1. Average values of the drag coefficient for different lengths.

LP	Cd_{mean}	Reduction percentage
0.25	1.05325	04.25%
0.5	1.02315	06.98%
1	0.98171	10.75%
2	0.93717	14.80%
3	0.91436	16.87%
3.5	0.90966	17.30%
4	**0.90932**	**17.33%**
5	0.91680	16.65%
6	0.92555	15.85%

3.2.3 Heat Transfer Results

From this figure, it can be seen that the heat exchange between the incoming fluid with a cold temperature and the square cylinder with a hot temperature takes place in the same way for the front face and both the top and bottom faces in all the cases tested. While a small difference exists at the rear face such that the heat exchange decreases slightly with the increasing length of the control plates. The longer control plates impede the evacuation of part of the shear layers delivered by the top and bottom edge of the square cylinder. It is generally found that the heat exchange around the square cylinder is almost the same. However, the implementation of the control plates and the increase in their length has a positive influence on the suppression of the thermal vortex shed. Figure 5 shows that significant vortex generation is observed for small plates. When the length of these partitions reaches a critical length (Lp = 4D), the vortices observed previously are reduced. After this critical length, the form of the vortexes shed in the wake increases again.

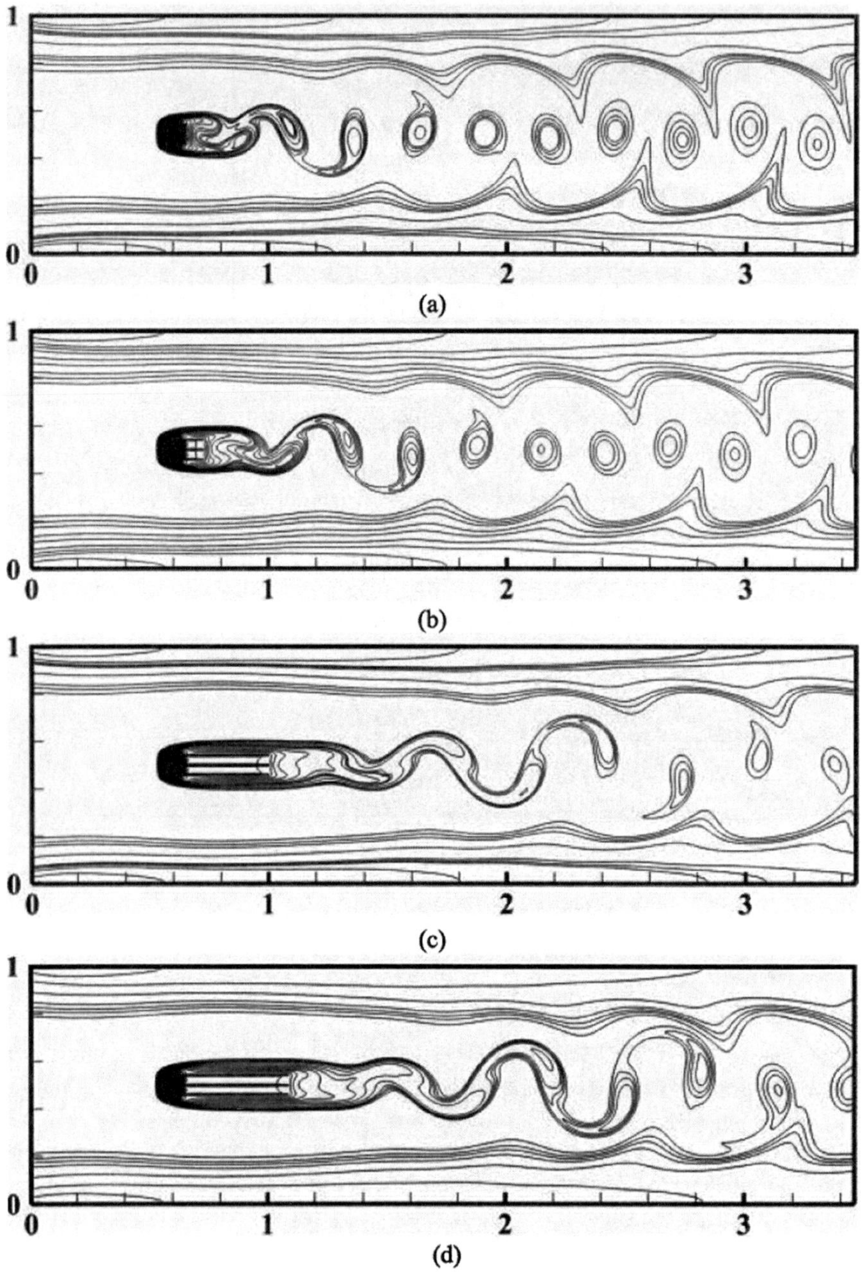

Fig. 5. Isotherms contours visualization for: (a) Lp = 0.25d; (b) Lp = 1d; (c) Lp = 4d; (d) Lp = 6d.

4 Conclusion

The present work presents a numerical study on drag force reduction, vortex suppression and studies the heat transfer characteristics on a heated square cylinder placed in a horizontal 2D channel and controlled by three attached partitions arranged in parallel downstream of the square cylinder.

The results obtained illustrate that there is a maximum reduction in drag force in a critical case where Lpcr = 4D. These results show that the drag force acting on the square cylinder decreases in all cases studied compared to the uncontrolled case. This explains the importance and necessity of flow control. Also, at this cruel length, large and regular heat exchange is observed. In addition, the number and size of the thermal vortexes and the width of the Von Karman Street decrease at this critical length. Therefore, for beneficial use of the three partitions attached to the back of the system under study, e.g. in the cooling of electronic components, heat exchangers, risers in marine technology, road vehicles, aircraft, it is preferable to use digital technologies. This technology allows several parameters to be varied digitally at the same time, such as length, inclination and optimum position. This leads to a significant energy gain.

References

1. Monat, J., Gannon, T.: Applying systems thinking to engineering and design. Systems **6**(3), 34 (2018)
2. Elshaer, A., Bitsuamlak, G., El Damatty, A.: Enhancing wind performance of tall buildings using corner aerodynamic optimization. Eng. Struct. **136**, 133–148 (2017)
3. Florides, G., Kalogirou, S.: Ground heat exchangers—a review of systems, models and applications. Renew. Energy **32**(15), 2461–2478 (2007)
4. Chapain, S., Aly, A.: Vibration attenuation in high-rise buildings to achieve system-level performance under multiple hazards. Eng. Struct. **197**, 109352 (2019). https://doi.org/10.1016/j.engstruct.2019.109352
5. Yao, G.Z., Yap, F.F., Chen, G., Li, W., Yeo, S.H.: MR damper and its application for semi-active control of vehicle suspension system. Mechatronics **12**(7), 963–973 (2002)
6. Moussaoui, M.A., Jami, M., Mezrhab, A., Naji, H., Bouzidi, M.: Multiple-relaxation-time lattice Boltzmann computation of channel flow past a square cylinder with an upstream control bi-partition. Int. J. Numer. Meth. Fluids **64**(6), 591–608 (2010)
7. Rashidi, S., Hayatdavoodi, M., Esfahani, J.: Vortex shedding suppression and wake control: a review. Ocean Eng. **126**, 57–80 (2016)
8. Wang, X., Chen, J., Zhou, B., Li, Y., Xiang, Q.: Experimental investigation of flow past a confined bluff body: effects of body shape, blockage ratio and Reynolds number. Ocean Eng. **220**, 108412 (2021). https://doi.org/10.1016/j.oceaneng.2020.108412
9. Moussaoui, M.A., Admi, Y., Lahmer, E.B., Mezrhab, A.: Numerical investigation of convective heat transfer in fluid flow past a tandem of triangular and square cylinders in channel. In: IOP Conference Series: Materials Science and Engineering, vol. 1091, no. 1, p. 012058. IOP Publishing (2021)
10. Kral, L.D.: Active flow control technology. ASME Fluids Engineering Technical Brief (2000)
11. Admi, Y., Moussaoui, M.A., Mezrhab, A.: Effect of a flat plate on heat transfer and flow past a three side-by-side square cylinders using double MRT-lattice Boltzmann method. In: 2020 IEEE 2nd International Conference on Electronics, Control, Optimization and Computer Science, ICECOCS 2020 (2020)

12. Bhatnagar, P.L., Gross, E.P., Krook, M.: A model for collision processes in gases. I. Small amplitude processes in charged and neutral one-component systems. Phys. Rev. **94**, 511 (1954)
13. Mohamad, A.A.: Lattice Boltzmann method: fundamentals and engineering applications with computer codes. AIAA J. (2011)
14. Bouzidi, M., Firdaouss, M., Lallemand, P.: Momentum transfer of a Boltzmann-lattice fluid with boundaries. Phys. Fluids **13**, 3452–3459 (2001)
15. Zou, Q., He, X.: On pressure and velocity boundary conditions for the lattice Boltzmann BGK model. Phys. Fluids **9**, 1591–1598 (1997)
16. Mezrhab, A., Moussaoui, M.A., Jami, M., Naji, H., Bouzidi, M.: Double MRT thermal lattice Boltzmann method for simulating convective flows. Phys. Lett. A **374**, 3499–3507 (2010)
17. Breuer, M., Bernsdorf, J., Zeiser, T., Durst, F.: Accurate computations of the laminar flow past a square cylinder based on two different methods: Lattice-Boltzmann and finite-volume. Int. J. Heat Fluid Flow **21**, 186–196 (2000)
18. Islam, S.U., Rahman, H., Abbasi, W.S., Shahina, T.: Lattice Boltzmann study of wake structure and force statistics for various gap spacings between a square cylinder with a detached flat plate. Arab. J. Sci. Eng. **40**(8), 2169–2182 (2015)

Author Index

S. Motahhir and B. Bossoufi (Eds.): ICDTA 2022, LNNS 454, pp. 855–858, 2022.
https://doi.org/10.1007/978-3-031-01942-5